INSECT PEST AND DISEASE MANAGEMENT

THE EDITOR

Dr. D. Prasad (M.Sc. Ag. Zool.-Ent.-1969-Allahabad University, Allahabad and Ph.D.-1978, Indian Agricultural Research Institute, New Delhi) is a distinguished Scientist having more than 35 years of research, teaching and extension of experience in the field of Nematology in particular and Plant Protection in general, presently working as Principal Scientist at Indian Agricultural Research Institute, New Delhi. Born on 20th January 1948 in Allahabad, U.P. having background of organic farming. He has published 120 research papers, 62 abstracts presented/published in several symposia of National and International repute and 50 chapters in various books. Dr. Prasad is known for his research contribution in nematode management through environmentally safe pesticides. A new ecofriendly molecule was developed and tested against plant parasitic nematodes and PATENT filed through ICAR. He won a number of prestigious awards and recognitions.

Dr. Prasad written books namely–*Environmental Pollution: Air, Environmental Pollution Water; Environmental Pollution: Land; Environmental Pollution: Solid Waste; Environmental Pollution: Radiation; Environmental Pollution: Thermal; Soil Environment and Pesticides; Potential IPM Tactics; Ecofriendly Pest Suppression; Crop Pest and Disease Management: Challenges for the Millennium and Advances in Plant Protection Sciences, Crop Protection: Management Strategies, Sustainable Pest Management*, during last few years.

Presently working in the project entitled Nemtaode Problems and their Management in peri urban areas. Dr. Prasad also founded a Society of Plant Protection Sciences in 1992 and publishing an International Journal "Annals of Plant Protection Sciences." He has been advisor to some of the apex bodies of GOI, referee of several scientific journals and magazines, examiner at different Universities, chaired scientific session in the symposia and offered trainings in plant protection subjects.

INSECT PEST AND DISEASE MANAGEMENT

Editor
Dr. D. Prasad
Division of Nematology, Indian Agricultural Research Institute, Pusa, New Delhi – 110 002

2016
Daya Publishing House®
A Division of
Astral International Pvt. Ltd.
New Delhi - 110 002

First Published, 2008
Reprinted, 2016

ISBN: 978-81-7035-552-6 (Hardbound)

Publisher's note:

Every possible effort has been made to ensure that the information contained in this book is accurate at the time of going to press, and the publisher and author cannot accept responsibility for any errors or omissions, however caused. No responsibility for loss or damage occasioned to any person acting, or refraining from action, as a result of the material in this publication can be accepted by the editor, the publisher or the author. The Publisher is not associated with any product or vendor mentioned in the book. The contents of this work are intended to further general scientific research, understanding and discussion only. Readers should consult with a specialist where appropriate.

Every effort has been made to trace the owners of copyright material used in this book, if any. The author and the publisher will be grateful for any omission brought to their notice for acknowledgement in the future editions of the book.

Published by : **Daya Publishing House®**
A Division of
Astral International Pvt. Ltd.
– ISO 9001:2008 Certified Company –
4760-61/23, Ansari Road, Darya Ganj
New Delhi-110 002
Ph. 011-43549197, 23278134
E-mail: info@astralint.com
Website: www.astralint.com

Laser Typesetting : **Classic Computer Services**
Delhi - 110 035

Digitally Printed at : **Replika Press Pvt. Ltd.**

PRINTED IN INDIA

डा. मंगला राय
सचिव एवं महानिदेशक
DR. MANGALA RAI
Secretary & Director-General

भारत सरकार
कृषि अनुसंधान और शिक्षा विभाग एवं
भारतीय कृषि अनुसंधान परिषद
कृषि मंत्रालय, कृषि भवन, नई दिल्ली - 110 001
GOVERNMENT OF INDIA
DEPARTMENT OF AGRICULTURAL RESEARCH & EDUCATION
AND
INDIAN COUNCIL OF AGRICULTURAL RESEARCH
MINISTRY OF AGRICULTURE,
KRISHI BHAVAN, NEW DELHI – 110 001
Tel.: 23382629, Fax: 91-11-23387293, E-mail: mrai.icar@nic.in

Foreword

Introduction of new crop varieties coupled with production technologies successfully ushered Green Revolution in the country and also placed our national food security on a sound platform. However, it is very disheartening to see that the crops grown with hard work and labour of the farmer are damaged by various pests and pathogens. The estimates are that of the reduction in crop yields due to various biotic stresses account for nearly 18 per cent.

A large number of institutions undertake research globally to combat the biotic stresses and the literature on efficient and effective management of a vast array of insect pests and diseases of agriculture crops is available in a number of journals. A handy compendium with updated information on plant protection can certainly prove to be useful for field workers and farmers to take appropriate plant protection measures against biotic stresses. The edited book entitled "Insect Pest and Disease Management" that brings together updated information on pests and disease Management is a very useful effort.

The editor of this book, Dr. D. Prasad, Division of Nematology, Indian Agricultural Research Institute, New Delhi has made a concerted effort to compile the work done at various institutions under 41 thought provoking chapters. In the days of growing concern for safe environment without sacrificing higher level of crop productivity, the compilation could serve as a reference book and also as a guide for students, development-officials and farmers associated with crop production.

MANGALA RAI

Date: 3rd March, 2008

New Delhi

Preface

With rapidly increasing human population that hardly matched by the rate of increase in agricultural production, the per capita availability of food and other products is bound to fall short. There is very little hope and scope for further increase in cultivated area. Therefore, increasing production per unit area is the most plausible alternative to sustain self sufficiency in food, fodder and fibre. Crop pests (insects, diseases, nematodes, weeds, rodents etc.) continue to cause losses of about 18 per cent of this crop yield worth more than Rs. 60,000 crores annually despite the facts that more than 48,000 metric tones of technical grade chemical pesticides are used to manage the pests in the country. Insecticides consume almost 60 per cent of the total pesticides used followed by fungicides and herbicides among the crops, cotton, rice and horticultural crops accounts for more than 80 per cent of the total pesticides used. Indiscriminate and injudicious use of pesticides is associated with a number of adverse effects on health and environments.

A number of cultural, biological and physical means, many of them are modifications of the agro-practices being followed by the farmers since ages, were also developed either for economic and production reasons. But very soon dawn realized that none of these techniques individually could serve as a panacea against target pests and pathogens. During the past few decades, the science of plant protection has developed rapidly resulting on many diverse techniques to manage a variety of pests, pathogens and nematodes. Therefore ,vigorous research efforts for exploring newer but pragmatic tools for the control of insect pests, diseases, nematodes and weeds have become imperative. National Agricultural Policy, GOI has laid special emphasis on IPM and use of biotic agents to minimize the indiscriminate use of pesticides in agriculture. IPM follows a system approach while combining a wide array of crop production and protection practices to reduce the economic losses caused by the pests.

The book "*Insect Pest and Disease Management*" is the efforts to compile a large number of control methods available in literature which are otherwise, scattered over a number of journals and periodicals.

Experts in the relevant fields have given a critical appraisal of the various aspects on management of crop pests, diseases nematodes, weeds etc. and extended useful suggestions. The book, not only provides a reference but also serves as guide and inspiration for future research into the realm of biological, chemical, physical and quarantine aspects.

The scientists, teachers, scholars, administrators and policy makers dealing with insect pest and disease control in particular and plant protection in general and expected to find this book very useful and informative. I am grateful to all the authors for their valuable articles and cooperation as and when required during the final preparation of the articles. I am also thankful to *Shri Anil Mittal* and *Shri Sunil Mittal*, *Daya Publishing House*, New Delhi for their technical guidance and help in bringing out this book.

D. Prasad

Contents

Insect Pest and Disease Management *Pages* **1–19**
Editor: **Dr. D. Prasad**
Published by: **DAYA PUBLISHING HOUSE (2016)**

Chapter 1

Adaptable IPM Technology for Vegetable Crops

H.R. Sardana and R.K. Tanwar
National Centre for Integrated Pest Management, New Delhi – 110 012

Introduction

Vegetables being a rich and cheap source of vitamins and minerals, occupy an important place in the food basket of Indian consumers, provide food and nutrition security, generate income and employment and also earn foreign exchange. India is bestowed with varied agro-climatic conditions, which allow growing of several types of vegetables in the country. India is the second largest producer of vegetables in the world, occupies 12 per cent of the total area and produces 10 per cent of the vegetable production in the world. But assuming a population growth rate of 1.7 per cent per annum and country's commitment of exports, atleast 220 million tonnes of vegetables would be required by 2020 AD. Therefore to meet this requirement of ever increasing burgeoning population, we need to produce about 100 millions tonnes of vegetables every year. Introduction of high yielding short duration varieties/hybrids has resulted in increased production of vegetables manifold, but has also brought changes in pest scenario as many minor pests have become major and occasional pests have become regular pests. Due to their tender and supple nature, vegetables are also more prone to pest attack and at a conservative estimate cause about 35-40 per cent losses. Cultivation of vegetables crops in an around cities in large areas and raising of short duration, off seasons/varieties/hybrids results in availability of continuous and abundant food supply for the build up, perpetuation and multiplication of insect pests, diseases and nematodes (Singh *et al.*, 2003).

To mitigate the losses due to these pests, a huge quantity of pesticides is used in vegetable crops by the farmers. On an average, 1.11 kilograms of pesticides per hectare of fruits and vegetables are applied. Vegetables alone consume 12–13 per cent of the pesticides used in the country though they occupy an area of about 2-3 per cent only. Thus the produce harvested at short intervals is prone to retain unavoidably high level of pesticide residues. Vegetables available in the market are laced with the pesticides. This affects export quality and hence puts non-tariff barrier on high value export of vegetables. Further, the excessive reliance on chemicals has created problem of pest resistance to pesticides and led to decline of useful fauna, flora and soil health. In addition, pesticide treadmill lures the farmers for ever increasing use of pesticide dosages, thereby encouraging a vicious cycle of environmental pollution. Because of lack of awareness among farmers about other alternatives of pest control, natural forces of pest suppression/bio-agents are not properly exploited in vegetables crops. Cultural practices such as healthy and hygienic cropping system, proper crop rotation, judicious use of fertilizers and pesticides are not being followed in their right perspective so as to check the severity of the pest problems. In the past there have been numerous studies on the pest management aspects of vegetables, which have mostly been dealt in isolation and individually. Our country with rich agricultural base, traditional and poor farm resource is badly in need of integration of all the traditional and scientific knowledge to develop IPM strategies and promotion of their large scale utilization in a participatory approach so that the problems caused due to harmful pesticides could be reduced to a great extent and IPM could prove to be a noble technology for improving the socio-economic status of small and marginal farmers. Thus the lopsided aim of just controlling the pest has to be replaced with a comprehensive and holistic view of pest control approach and that is integrated pest management (Singh *et al.*, 2005).

Production of Major Vegetable Crops in India

The vegetable production has touched new height in recent years with total production of 105.00 million tonnes. Among the various vegetables, okra occupies maximum area (0.43 m ha) followed by onion (0.39 m ha), while potato has the maximum production and cabbage has the highest productivity of 17.7 tonnes/ha followed by potato, being 16.99 tonnes/ha. Though okra has the maximum area but it has minimum productivity (Table 1.1). UP is the leading state in most of the vegetables. Chillies occupy the maximum area in AP State, Orissa and Bihar are the leading brinjal produciong states. Punjab occupies a leading place in the cabbage production.

Table 1.1: Area, Production and Productivity of Major Vegetable Crops

Crops	*Area (m ha)*	*Production (m t)*	*Productivity (t/ha)*
Brinjal	0.39	7.96	14.84
Cabbage	0.22	6.62	17.71
Cauliflower	0.22	5.07	11.24
Okra	0.43	4.43	9.37
Onion	0.39	4.78	10.32
Peas	0.22	2.75	10.49
Potato	0.11	24.80	16.99
Tomato	0.36	08.02	15.32
Others	0.19	40.08	12.35
Total	5.34	104.51	13.41

Integrated Pest Management Tools

IPM is basically a philosophy for the optimal utilization of techniques in an integrated way so as to economise in terms of inputs, environment and efforts. In that endeavour it becomes a decision support system wherein the clientele has to opt a line of decision from among various options, depending upon the immediate problem or future strategy and the limitations in terms of resources and knowledge. In that sense, it is decision of when to apply, why to apply and how much to apply. That is why IPM is also called as Intelligent Pesticide Management or Intelligent Pest Management Broadly speaking following are the major IPM tactics which can be followed:

1. Host plant resistance
2. Crop husbandry/Cultural and Mechanical control
3. Biological control
4. Chemical control.

Before any control measures are taken, habitat management is the most important activity in the IPM. This is to know the field situations or the ground realities with regard to pests, plant health, soil conditions, climatic factors etc. These can be employed in making decisions on pest management.

Host Plant Resistance (Resistant Varieties)

It is the most effective, economical, practical and easiest way of tackling pest problems and is free from all problems of environmental pollution. By growing resistant varieties we are not only reducing the use of pesticides, but they are often compatible with other methods of pest management. As such a lot of work has been done on vegetable resistance breeding in the country and number of varieties have been developed in major vegetable crops (*See* Tables 1.2–1.4) (Das Gupta and Ganguly, 1986). One most important achievement in vegetables research, is that, many off-season and short-duration varieties/hybrids have been developed over the years which can be grown through out the year albeit the fact that they may be resistant during one season while susceptible during the other.

Table 1.2: Vegetable Varieties Tolerant/Less Susceptible to Insect Pests

Crops	*Insects*	*Varieties*
Brinjal	Shoot and fruit borer	SM17-4, PBr 129-5, Punjab Barsati, ARV 2-C, Pusa Purple Round, Punjab Neelam
	Aphid, jassids, thrips	Kalyanpur, Punjab chamkila, GB-I
	White fly	GB-6
Onion	Thrips	Pusa Ratnar, Arka Niketan, PBR-2, PBR-5, PBR-6
Cabbage	Aphid	All Season, Red drum Head, Sure Head
Cauliflower	Stem borer	Early Patna, Kathmandu local
Okra	Jassid	New selection, Punjab Padmini
Bitter gourd	Fruit fly	Hissar-II
Round gourd	Fruit fly	Arka tinda

Crop Husbandry (Cultural Practices)

Cultural practices are virtually the Cinderella of pest managmenent programmes as their potential in pest management is grossly underestimated. They constitute the cheap, effective and eco-safe components of IPM and require no additional budget for the farmers. But they do need careful consideration with sufficient planning in advance. Routine agronomic conditions are modified to create uncongenial conditions for pests and favourableness for natural enemies thus giving opportunity to the naturally existing environmental resistance factors (ERFs).

Table 1.3: Vegetable Varieties Tolerant/Resistant to Diseases

Crops	*Varieties*	*Diseases*
Tomato	Pant Bahar, BT-1, BWR-5, LE 79-5	Bacterial wilt (*Pseudomonas solanacarium*), Fusarium wilt, leaf curl
Potato	Kufri Jyoti	Late blight
Pea	PRS-4, FC-1, JP-83	Powdery mildew
Musk melon	Arka Rajhans	Powdery mildew
Onion	Pusa Red, Pb Sel-385, Pusa Ratnar, Arka Kalyan	Purple blotch
Cowpea	Pusa Komal	Bacterial blight
Okra	Pusa A-4, Parbhani Kranti, Arka Abhay, Varsha Uphar	YVM Virus
Cabbage	Pusa Mukta (Sel-8)	Black rot
Cauliflower	Pusa Shubhra, Pusa Snowball	Black rot

Table 1.4: Important Vegetable Varieties/Lines Showing Resistance/ Tolerance/Less Susceptibility to Nematode Pests

Crops	*Varieties*	*Nematode Pests*
Tomato	SL- 120	*Root knot nematode*
Pota to	Kufri Swarna	*Globodera rostochiensis*, G. *pallida*
Pea	C-50, A-7O, B-58	M. *incognita*
Musk melon	France D, Goldstar, Perlita	M. *incognita*
French bean	Banat, Blue lake, Stringless, Sea ferer, Brown beauty	*M. incognita*
Cowpea	line M57-13N, C-I 52, V-16	*M. incognita, Rotylenchulus reniformis*
Watermelon	Dixie Queen	*M. incognita*
Pepper	Kalluvally California Wonder	M. *incognita*

Cultural practices may include:

1. Ploughing of fields during summer to expose the resting stages of soil inhabiting insect pests, diseases and weeds. Deep ploughing of the field after harvest reduces the activity of fruit fly, red pumpkin beetle and cut worms as they remain in the soil in earthen cocoons to complete the diapause stage of their life-cycle.
2. Soil solarization, *i.e.*, covering the soil surface with thin transparent polythene sheet (60 to 100 guage) for 2-3 weeks to retain soil moisture and trap the heat during hot summer plays a crucial role in management of soil pests in nurseries in crops such as cabbage, cauliflower, brinjal and tomato.
3. Soil amendments like organic manure cakes, decomposed plant residue, city waste and vermicompost are used especially for nematodes suppression.

4. Many-a-times, pruning is done to provide sufficient breathing space to the plants so that micro climate dose not permit development of harmful organisms and insects.
5. Optimum fertilizer and spacing, especially in cucurbits as overlapping of these creepers encourages the incidence of pests like red pumpkin beetle
6. Sowing/planting times are also altered to avoid the critical stage of crop from pest attack. Early planting of cucurbits in November escapes the attack of red pumpkin beetles while delayed sowing of bitter gourd suffers less from fruit fly. Sowing of okra during peak summer suffers less from borer and YMV. July planted brinjal suffers more from shoot and fruit borer so planting date should be suitably altered.
7. Inter cropping and trap crop also limit the infestation from insect pests. Inter crop acts as a barrier for egg laying and also release volatile allelochemicals which deter the insect from damaging the other crop. It also creates uncongenial micro-climate for the pest. Inter-crop has a diverse geometry (Table 1.5).

Table 1.5: Inter Crop Combination Effective in Vegetable Pest Management

Crop Combinations	*Target Pests*
Cabbage + Carrot	Diamond back moth
Brocolli + French bean	Flea beetle
Okra + Cowpea	YVM
Okra + Onion	Borer
Cabbage + Tomato	Diamond back moth

Trap crop should be more attractive to the major pests than the main crop. Trap crop protects by preventing the pest from reaching the main crop or by attracting them for egg laying etc. which can be destroyed later. Mustard as trap crop has been successfully utilized for the control of diamond back moth and leaf webber. Marigold in tomato traps the adults of fruit borer. It also attracts the leaf miner for egg laying. Maize plants applied with bait spray when sown in bitter gourd trap fruit fly adults (Table 1.6).

Table 1.6: Effective Trap Crop Combinations for Vegetable Pest Management

Main Crops	*Trap Crops*	*Target Pests*
Cabbage	Mustard	Diamond back moth
Tomato	Marigold	Fruit borer
Bitter gourd	Maize	Fruit fly

Biological Control

India is the home of several vegetable crops and is therefore endowed with many natural enemies of the respective pests associated with them. Several natural enemies recorded on important insect pests of vegetable crops from different parts of the country are presented in Table 1.7. There is an endless scope for the use of natural parasites for their exploitation in the management programmes in vegetables. Among the egg parasites, *Trichogramma* spp. have been utilized to some extent for the control of fruit borers. Inundative releases of *Trichogramma* have been made to control the fruit borers convincingly. Predator *Chrysoperla carnea* needs attention. This is very effective for the control of aphids, jassids, and eggs of some borers. Its mass rearing has to be economized. The larval parasites of diamond back moth like *Apanteles plutellae* and *Diadegma semiclausum* can be incorporated into biological pest management because of their potential. The feasibility of biological control in vegetables is comparatively less because of the short growing period. However, considering the wide spread use of chemicals and residue problem, biocontrol has become now more viable and relevant in vegetables. Among the microbial insecticides, *Bt* has been extensively used against *Plutella* and *Helicoverpa* to the extent that resistance has been reported to have been developed in *Plutella xylostella.* Application of NPV reduces the population of target pests to some extent. But the lack of commercial production and its high specificity has suppressed their popularity.

Table 1.7: Natural Enemies of Insect Pests of Some Important Vegetable Crops

Pests	*Parasites/Predators*	*Parasitisation (per cent)*
Plutella xylostella	*Apanteles plutellae, Diadegma semiclausum*	3.5–20
Crocidolomia binotalis	*Apanteles crocilodomia*	–
Hellula undalis	*Bracon* spp.	16.7
Helicoverpa armigera	*Trichogramma* spp.	80
Earias sp.	*Trichogramma* spp.	1–56
Aphis spp.	*Coccinella septumpunctata, Chrysoperla carnea*	–

Table 1.8: Important Bioagents Recommended in Vegetables

Target Pests	*Bioagents*	*Dose*
Okra shoot and fruit borer	*Trichogramma chilonis*	1.0–1.5 lakh *eggs/ha*
Tomato fruit borer	*Trichogramma pretiosum*	1.0–1.5 lakh *eggs/ha*
Diamond back moth	*Trichogrammatoidea bactrae*	1.0–1.5 lakh *eggs/ha*
Brinjal shoot and fruit borer	*Trichogramma chilonis*	1.0–1.5 lakh eggs/ha
Tomato fruit borer	*Ha*NPV	250 LE/ha
Tomato fruit borer, diamond back moth	*Bacillus thuringiensis*	500 g a.i./ha

Some of the methods to conserve natural enemies

1. Avoiding pesticide use at the time of abundance of natural enemies.
2. Avoiding broad spectrum insecticide and use only selective insecticides. Application of insecticide as seed treatment or seed soaking.
3. Bare-root dip treatment before planting.

For disease management, the research efforts in biocontrol are scanty except that seed treatment by *Trichoderma* is given in the early stages to control soilborne/seed-borne diseases. A few antagonists are effective for some of the diseases. However, the efficacy of these organisms is limited especially under field conditions.

Chemical Control and Pesticide Residue

Chemical pesticides are still the most reliable methods of pest control as they give quick results, are easily available and are easy to apply. To avoid residues, choice of chemical plays a very important role. The chemical should be chosen considering the waiting period. Vegetables with less picking interval should be sprayed with insecticides having less waiting period (Table 1.9). Chemicals of plant origin can fit well in IPM of vegetables. Neem products have shown efficacy against insects like jassids, aphids and leaf miners. Neem products are non-persistent and safe to animals. Diamond back moth and okra pests are controlled very well. Okra pests can be controlled effectively by using a mixture of neem and half the dose of endosulfan. Instead of going for spray schedule, need based spray based on ET levels be applied. ETL for few important vegetable pests are given in the Table 1.10.

Table 1.9: Waiting Period of Important Insecticides in Vegetable Crops

Pesticides	*Okra*	*Cabbage*	*Brinjal*	*Cauliflower*	*Tomato*	*Pea*
Malathion	3	7	1	7	–	3
Fenvalerate	5	10	5	3	3	–
Endosulfan	8	10	3	10	5	–
Cypermethrin	3	10	1		1	–
Monocrotophos	18	–	–	12	–	9
Phosphamidon	10	–	–		–	6
Carbaryl	–	25	25	30	–	–
Dimethoate	–	10	–	7	–	–

Table 1.10: Economic Threshold Levels for Some Vegetable Pests

Crops	*Pests*	*Economic Threshold Level*
Tomato	Fruit borer	1 egg/one larva per plant
	White fly	4 adults/leaf
	Root-knot nematodes	2 second stage juveniles/g soil
Cabbage	Diamond back moth	0.5 larva/plant
	Leaf webber	0.3 egg mass/plant
Okra	Jassid	4.46 nymphs/plant
	Root-knot nematodes	2 second stage juveniles/g soil
Potato	Aphids	20 aphids/100 leaves

Farmers' Participatory Validation of IPM Technology: A Case Study

A farmer-centered and farmer-driven participatory programme was put in operation for the validation of IPM technology in four major vegetable crops *viz.*; tomato, cabbage, okra and brinjal, in a total area of 250 hectares covering 1350 farming families in the states of Uttar Pradesh, Karnataka and Jharkhand over a period of four years (2001–2006) with the support of Indian Institute of Horticultural Research, Bangalore; Indian Institute of Vegetable Research, Varanasi and Rama Krishan Mission, Krishi Vigyan Kendra, Ranchi.

Baseline Information of IPM Adopted Villages

Initially, the base line information on adopted farming families in ten IPM villages in Bangalore (Karnataka); Mirzapur (Uttar Pradesh); Ranchi (Jharkhand) and Ghaziabad (Uttar Pradesh) was collected in participatory rapid appraisal techniques for prioritization of the issues (Table 1.11).

Table 1.11: Baseline Information on Selected IPM Villages

Indicator		*Bangalore (Kesthur, Neraleghatta, Somasettahalli, Maralenahalli and Mandi Byaderahalli)*	*Varanasi (Dhauraghud and Jayapur)*	*Ranchi (Hesalpidi, Barandi, Khakhra and Kharkutali)*	*Ghaziabad (Raispur, Harsawan, Sadarpur and Mainapur)*	
Number of villages		5	2	4	4	
Area under crops (ha)		85.58	40.80	180	80	
Population of village(s)		3317	7020	1064	8500	
Number of farming families under IPM programme		64	57	479	80	
Literacy rate (per cent)		60	45	47	65	
Key pests	Cabbage	Diamond back moth (DBM), aphids, stem borer, downy mildew	DBM, aphids, Early and late blight	DBM, aphids, Downy mildew	Okra	Shoot and fruit borer, leaf hopper, Yellow Vein Mosaic Virus
	Tomato	Fruit borer, mites, leaf curl, bacterial spot	Fruit borer, powdery mildew, early and late blight	Leaf eating caterpillars, early and late blight	Brinjal	Shoot and fruit borer, little leaf, *Phomopsis* fruit rot
Major source of plant protection information		Fellow farmers and pesticides dealers	Fellow farmers and pesticides dealers	Fellow farmers and pesticides dealers	Fellow farmers and pesticides dealers	
Commonly used pesticides in different locations		Endosulfan, carbendazim, dicofol, cypermethrin	Bavistin, dithane, dimethoate, chlorpyriphos, endosulfan	Endosulfan, bavistin, chlorpyriphos	Monocrotophos, imidacloprid, acetamiprid	
Number of pesticide sprays	Cabbage	9-12	5	3	Okra	10-12
	Tomato	14	5	3	Brinjal	5-7
Average yield (t/ha)	Cabbage	63.43	50	22.9	Okra	9.75
	Tomato	46.61	32.35	30	Brinjal	32.0

Key pests were fruit borer and whitefly in tomato in Varanasi and Bangalore and *Spodoptera litura* in Ranchi but Diamond back moth (DBM) and downey mildew in cabbage are the common problems at all the three locations. Shoot and fruit borer, leaf hoppers and yellow vein mosaic virus (YVMV) in okra; and shoot and fruit borer, little leaf and Phomopsis fruit rot are the key problems in brinjal crop in Western Uttar Pradesh. On an average, farmers applied 3-5 chemical sprays in cabbage and tomato in Varanasi and Ranchi; where as in Bangalore farmers applied 9-14 sprays of pesticides in tomato and cabbage. Ten to fourteen sprays in okra and brinjal were applied in western Uttar Pradesh in crop duration. Monocrotophos, endosulfan, cypermethrin, imidacloprid and thiamethoxam were the commonly used insecticides in these crops. Farmers were ignorant of IPM approach and alternatives to chemicals.

Development and Validation of Integrated Pest Management

Appropriate IPM modules were synthesized for each of the four locations and validated. The classical individual components reported effective by the researchers were put together in order of crop growing stages for all the locations and crops. The critical inputs like seeds of disease resistant hybrids, bio-pesticides, pheromone traps and quality chemical pesticides were provided to IPM farmers. The initiatives were also taken to strengthen the bioagents production at the project site for timely availability to the farmers. The regular monitoring of insect pests and diseases and finally need-based application of pesticides were some of the unique features of the programme (Tables 1.12–1.14).

Table 1.12: Validated IPM Modules in Cabbage Crop

Banaglore, Karnataka	*Varanasi, U. P.*	*Ranchi, Jharkhand*
Nursery	**Nursery**	**Nursery**
• Preparation of a raised seed bed upto 10 cm height	• Preparation of a raised seed bed of 15 cm height	• Preparation of a raised seed bed of 10 cm height
• Soil drenching with copper oxychloride (COC)	• Soil solarization for three weeks with 45 μm polythene sheet	• Soil solarization for three weeks with 45 μm polythene sheet
• Seed treatment with *Trichoderma viride* @ 4g/kg	• Seed treatment with *Trichoderma harzianum* @ 4g/kg	• Soil treatment with *Trichoderma harzianum* mixed with FYM @ 1 kg/t.
• Use of nylon net to avoid entry of sucking pests	• Use of nylon net for preventing entry of sucking insects	
• Spray of *Bt* @ 0.5 ml/L one day before transplanting		
• Spray of Dithane M-45/Ridomyl @ 0.2% against downy mildew		
Main field	**Main field**	**Main field**
• Growing of two rows of mustard after every 25 rows of cabbage as a trap crop at the time of planting.	• Spray of Neemarin @ 5% when DBM appeared early	• Spray of Neemarin @ 5% when DBM appeared early
• Wider spacing of 60 × 45 cm	• Spray *B. t* (1 g/L) if DBM 1.0/plant is noticed	• Spray *B. t* (1 g/L) if DBM 1.0/plant is noticed
• Use light traps @ 5/ha for trapping adult DBM	• Spray of pongamia soap @ 5% at 15-20 days after planting	• Spray of pongamia soap @ 5% at 15-20 days after planting
• Spray of *B. t* (1 g/litre) if DBM 1/plant is noticed early or spray of NSKE 5% or Pongamia soap 1% at primordial stage. Repeat 3-4 sprays.	• Installation of pheromone traps @ 5/ha for *Spodoptera litura* monitoring	• Installation of pheromone traps @ 5/ha for *Spodoptera litura* monitoring
• Rogue out disease affected bottom leaves from time to time.	• Need based application of fungicide like chlorothalonil for *Alternaria*	• Need based application of fungicides like chlorothalonil for *Alternaria*
• Need based sprays of chlorothalonil/ mancozeb @ 0.2% for *Alternaria* and Blitox + streptomycin @ 0.2% for black rot.	• Spray of insecticides like endosulfan @ 0.07% or Triazophos @ 0.05%.	• Spray of insecticides like endosulfan @ 0.07% or Triazophos @ 0.05%.

Pest Monitoring

Regular field monitoring and surveillance of pests is one of the most important activities of IPM programme for timely and accurate intervention commensurate to the pest density and damage levels using standard sampling techniques. Control measures were applied when felt loss was observed. Pest population monitoring was done by observation boys/technicians/scientists and farmers in a participatory mode.

Table 1.13: Validated IPM Modules in Tomato Crop

Banaglore, Karnataka	*Varanasi, U.P.*	*Ranchi, Jharkhand*
Nursery	**Nursery**	**Nursery**
• Preparation of raised seed bed of 15 cm height	• Preparation of raised seed bed of 10 cm height to avoid flooding of bed.	• Preparation of raised seed bed of 15 cm height to avoid flooding of bed.
• Soil solarization for 3 weeks using transparent polythene sheet of 45 µm thickness.	• Soil solarization using transparent polythene sheets.	• Soil solarization using transparent polythene sheets.
• Grow leaf curl resistant hybrid/ varieties like Avinash-2.	• Soil treatment with *Trichoderma harzianum* @ 4g/ kg of seed.	• Grow leaf curl resistant hybrid/ varieties like- Avinash-2.
• Seed treatment with *Trichoderma viride* @ 4g/ kg of seed.	• Spray of copper oxy chloride.	• Soil treatment with *Trichoderma harzianum* @ 100 g/ kg of FYM 1 kg/t.
• Use of nylon nets to prevent entry of white fly etc.		
Main field	**Main field**	**Main field**
• Apply neem cake @ 250 kg/ha while planting or at 20 DAP	• Spray imidacloprid at 15 DAP	• Spray imidacloprid @ 0.5 ml/L at 20 DAP
• Wide spacing @ 90 × 60 cm for hybrids	• Installation of pheromones traps @ 5/ha for monitoring of *H. armigera.*	• Installation of pheromone traps @ 5/ha for monitoring of *H. armigera* population
• Dip seedling roots for 15 minutes in imidacloprid @ 0.5 ml/L.	• Spray of *Ha*NPV 250 LE/ha	• Monitor top three leaves for *Helicoverpa* eggs.
• Transplant 1 row of marigold as trap crop/14 rows of tomato.	• Release of *T. bactrae* @ 1.0 lakh/ ha six times.	• Release of *T. chilonis* @ 1.0 lakh/ ha after appearance of adults
• Spray imidacloprid/thiomethoxam @ 0.5 ml/L at 15 DAP for whitefly or 5% NSKE at 15 DAP against leaf-miner.	• Regular collection and destruction of damaged fruits.	• Spray of *Sl*NPV @ 250 LE/ha twice at 15 day interval or *Ha*NPV for *H. armigera*
• Installation of pheromone traps @ 5/ha at 20 DAP for monitoring of *H. armigera*	• Need based application of pesticides like thiomethoxam and mancozeb @ 0.02%	• Regular collection & destruction of damaged fruits from time to time.
• Monitor top three leaves for *Helicoverpa* eggs at flowering.	• Need based application of pesticide like endosulfan @ 0.07%.	
• Release of *T. pretiosum* @ 1.0 lakh/ ha six times after appearance of adults.		
• Spray of NSKE 5% at 15 DAP against leaf miner		
• Spray of *Ha*NPV 250 LE/ha or neem soap 3 times at 28, 35 and 42 DAP.		
• Regular collection and destruction of damaged fruits.		
• Spray of endosulfan @ 650 g a.i./ha against *H. armigera.*		
• Destruction of leaf-curl disease and wilt affected plants.		
• Spray 0.02% mancozeb/captan for the control of early and late blight.		

Singh *et al.*, 2005.

Table 1.14: Validated IPM Modules for Okra and Brinjal

Okra	*Brinjal*
• Growing of popular yellow vein mosaic resistant hybrid *viz;* Sun-40 and Makhmali	**Nursery**
• Sowing of sorghum/ maize as border crop (Plate 1.3)	• Raised seed bed of 10-15 cm height to avoid flooding of bed
• Yellow sticky polythene traps smeared with castor oil and Delta traps set up for white fly and other small sucking pests	• Soil solarisation for three weeks during June using polythene sheets of 45 mm width
• Erection of bird perches @ 25/ha for facilitating predation of borer larvae	• Soil treatment with *T. viride* @ 100gm/ kg FYM. Enrichment of *T. viride* for three weeks before mixing in soil (Plate 1.1).
• Pheromone traps for monitoring of *Earias vittella* installed @ 5/ ha	**Main Field**
• Three sprays of NSKE @ 5% for hopper, white fly and mites starting at 28 days after sowing (Plate 1.2)	• Erection of bird perches @ 25/ha for facilitation of predation of insects
• Five releases of *Trichogramma chilonis* starting from 42 days after sowing at weekly interval were made	• Installation of Delta traps and yellow sticky traps @ 5/ha for hopper and white fly
• YVM affected plants, if any, and borer affected shoots rouged out from time to time	• Pheromone traps installed @ 12/ ha for *Leucinodes orbonalis* mass trapping as well as monitoring
• 3-4 sprays of Imidacloprid, cypermethrin, chlorpyriphos and fenvalerate insecticides	• Soil application of neem cake @ 250 kg/ha along the plant rows at planting and 30 days after planting
	• Three sprays of NSKE @ 5% against leaf hoppers, aphids, mites depending upon the appearance of the pest
	• Six releases of egg parasite, *T. brasiliensis* @1.0 lakh/ha at weekly interval for shoot and fruit borer
	• Collection and destruction of egg masses, larvae and adults of hadda beetle
	• Clipping of borer damaged shoots and collection and destruction of damaged fruits
	• Rouged out little leaf affected plants at monthly interval
	• One spray each of imidacloprid @ 0.5 ml/L and cypermethrin @ 1 ml/L in the season

Sardana *et al.*, 2004.

Damage by lepidopteran feeders or defoliators was also monitored by quick visual damage ratings. As in cabbage, it was seen that if damage was confined to older leaves, cabbage could tolerate damage as outer leaves were removed at harvest. Emphasis was given on the efficient utilization of scouting programme in determining the larval presence to make decision not only when to apply a material but also what material in term of control to be used, based on utilizing action threshold. Phenological stages *i.e.* time of flowering in tomato, primordial stage in cabbage, shoot formation in okra and brinjal to time the application of pesticides etc. along with population count resulted in reduction in time and monitoring. In tomato, the scout took 'V' shaped route through the field and examined 6 stops in a 1–5 acre field. Monitoring by watching `hot spots' for diseases was also followed in tomato and cabbage. It was also noticed that wherever the economic threshold level was higher, the range of control options was much greater than when it was low.

IPM Components

Plate 1.1: *Trichoderma* Mixing with FYM in Soil

Plate 1.2: NSKE Preparation

Plate 1.3: Sorghum as a Border Crop in Okra

IPM Validation Results

Cabbage

The major pests included diamond back moth, *Plutella xylostella*; stem borer, *Hellula undalis*; tobacco caterpillar, *Spodoptera litura*; downy mildew, *Perenospora parasitica*; bacterial black rot, *Xanthomonas campestris* pv. *Campestris* and root knot nematode, *Meloidogyne* sp. etc.

The IPM technology was field validated in a total area of more than 45 ha. The modules with pragmatic alternatives, as per required needs, gave very successful results. Both insect pest and diseases were recorded to be comparatively lower in IPM fields as compare to non-IPM field.

Tomato

In Bangalore, amongst various pests, tomato fruit borer was the major problem and its incidence was 9.33 per cent in non-IPM fields as compared to only 1.41 per cent in IPM. Among diseases, leaf curl incidence was as low as 0.9 per cent in IPM and 12.2 per cent in non-IPM fields (Baranwal *et al.*, 1996). The lower incidence of leaf curl in IPM fields was mainly due to growing of resistant hybrids and raising of healthy nursery in a protected environment. At IIVR, (Varanasi), leaf miner was also more abundant as compared to the above mentioned borer as 19.3 per cent of leaves per plant were infested by this pest in non-IPM fields as compared to 4.1 per cent infestation in IPM fields. This was mainly brought down by NSKE (5 per cent) sprays. About 7 per cent of plants were infested with bacterial wilt in non-IPM fields while only 3 per cent were infested in IPM fields. *Spodoptera litura* was the important pest of tomato at Ranchi and was observed at 49 larvae/100 plants in IPM as compared to 95 larvae/100 plants in non-IPM fields (Table 1.15).

Table 1.15: Pest Scenario in Cabbage and Tomato in IPM and Non-IPM Fields at Different Locations

Crop/Pest Incidence	*IIHR Bangalore*		*IIVR, Varanasi*		*RKM, KVK, Ranchi*	
	IPM	*Non-IPM*	*IPM*	*Non-IPM*	*IPM*	*Non-IPM*
			Cabbage			
DBM/plant	0.49	4.0	–	–	–	–
Black rot (%)	–	–	28.2	49.0	–	–
Alternaria leaf spot (%)	–	–	24.0	47.5	8	14
			Tomato			
Fruit borer (%)	1.41	9.33	2.3	7.3	–	–
Leaf miner (% infested leaves)	–	–	4.1	19.3	0.67	1.35
Spodoptera litura/100 plants	–	–	2.3	7.3	49	95
Soft root rot (%)	–	–	–	–	7	9
Bacterial wilt (%)	–	–	–	–	4	12
Phytopthora rot (%)	–	–	–	–	9	19
Leaf curl (%)	0.88	12.23	–	–	–	–

Among diseases soft root rot, bacterial wilt and *Phytophthora* were observed in both IPM and non-IPM fields at Ranchi. Incidence of these diseases was, however, less than 10 per cent in IPM fields while it was more than 10 per cent in non-IPM fields.

Okra

The traditional varieties like Super Anamika were replaced with disease resistant hybrids like Sun-40 and Makhmali. The incidence of majority of the insect pests and diseases was observed to be higher in non-IPM fields as compared to IPM fields. It was observed that YVMV alone was one of the major biotic components that affected yield directly. YVMV was significantly high in non-IPM fields while it was almost negligible in IPM fields (Table 1.16). Among natural enemies, spider population was abundant and was higher in IPM fields.

The studies further revealed that adoption of IPM technology which comprised of spray of NSKE 5 per cent, release of *Trichogramma* based on monitoring of pheromone traps and need based application of chemical pesticides was highly effective in keeping the pest population below ETL while increasing the yield substantially and resulting into higher C: B ratio. The IPM studies lowered expenditure on chemical pesticides to 5.8 per cent in IPM and 23.9 per cent in non-IPM fields, resulting in reduction of pesticide hazards to soil. The increase in number and population of natural enemies especially predatory spiders diversity indicated the overall increase in the fauna in IPM fields.

Table 1.16: Pest Incidence and Population of Natural Enemies in IPM and Non-IPM fields of *kharif* Okra at Raispur, Ghaziabad

Pest/Natural Enemy	2004-05		2005-06	
	IPM	Non- IPM	IPM	Non-IPM
Leaf hopper/top 3 leaves/plant	2.20	3.06	1.57	2.76
Whitefly/top 3 leaves/plant	4.24	5.36	3.48	6.07
Red spider mite/top 3 leaves/plant	74.70	83.50	4.82	11.86
Borer incidence (per cent)	9.09	12.85	12.88	15.01
YVMV (per cent)	0.54	23.16	0.15	21.62
Predatory spider/plant	0.40	0.008	0.24	0.07

Brinjal

The mean population of leaf folder and hadda beetle remained below ETL in IPM as well as non-IPM fields. After spray of NSKE (5 per cent), required for leaf hopper and hadda beetle, their population gradually vanished from the fields. In addition to NSKE spray, adverse environmental condition might have played some role in reducing its population. However, in some of the fields population rose above ETL, which was brought down by NSKE and imidacloprid insecticide spray. Shoot and fruit borer, *Leucinodes orbonalis*, one of the key pests of brinjal was present throughout the crop season and affected the fruit yield and quality directly (Table 1.17).

Table 1.17: Pest Scenario in IPM and Non-IPM Fields of Brinjal at Raispur, Ghaziabad

Pests	2004-05		2005-06	
	IPM	Non- IPM	IPM	Non-IPM
Leaf hopper incidence (ten leaves/plant)	5.42	7.95	6.13	8.01
Leaf folder incidence/plant	0.21	0.99	0.25	0.44
Hadda beetle/plant	0.20	2.98	0.26	3.23
Aphid (ten leaves/plant)	5.48	8.79	5.81	9.84
Red spider mite (ten leaves/plant)	1.79	3.84	4.06	7.42
Borer (per cent)	15.25	24.14	10.97	19.50
Little leaf incidence (per cent)	0.25	0.32	0.48	0.74
Number of predatory spiders/plant	1.50	0.18	1.72	0.36
Number of Coccinellid/plant	5.95	0.89	4.93	0.68
Number of Chrysoperla eggs/plant	1.20	0.10	1.83	0.08

However, its proper monitoring through pheromone traps, five releases of egg parasite, three sprays of NSKE and clipping of affected shoots and disposal of damaged fruits and intermittent sprays of cypermethrin significantly recorded lower mean incidence of borer in IPM fields than non-IPM fields on number and weight basis respectively. This resulted in lowering the number of sprays in IPM fields to only one as compared to 3-4 sprays in non-IPM fields (Sardana *et al.*, 2004).

Only one disease *viz.,* little leaf was the major disease that appeared only in two or three fields about a month after transplanting and continued till the end of December. Early transplanted fields had relatively higher incidence than late transplanted fields. The practice of rouging out of infected plants at monthly interval and three spraying of NSKE (5 per cent) in IPM fields and pesticide sprays in non-IPM fields might have helped in reducing the vector population and check the further infection of the disease in IPM fields as well as in non-IPM fields. Application of neem cake in IPM fields and three sprays of NSKE might have resulted in lower incidence of minor disease in fruit rot compared to higher incidence in non-IPM fields. As the above mentioned diseases appeared very late in the crop season, no management practices were advocated to farmers for further spread of disease and moreover it was not affordable to undertake any control measure when the crop season was almost over. Neem cake application resulted in lowering the rabbit foraging (Sardana, 2005).

A large build up of natural enemies *viz;* predatory spiders; coccinellids and *Chrysoperla* eggs was observed in IPM fields of brinjal where only 1-2 sprays of chemical pesticides were given. Lower number of sprays also helps in conservation of natural enemies whereas about 4-7 sprays in non-IPM fields resulted in decimation of natural enemies (Table 1.17).

Therefore, the natural enemy population build ups in different crops was a clear indicator of positive impact of the IPM technology.

Human Resource Development (HRD)

Validation of IPM technology on a large scale was taken up in mission mode and Adhoc scheme of ICAR and hence the same was coordinated with massive efforts on HRD and extension fronts. Large number of farmers' meetings, field days and exposure visits of farmers were organized before the crop season, during the crop season and even after the harvest. Several farmers and pesticide dealers were exposed to IPM programmes during 2001-2006. Resource manual on IPM, technical bulletins in regional languages (Kannad and Hindi), hand bills and crop based success stories were published in large number and distributed to the technology users. A single window approach was followed to provide the message to avoid any confusion in the farmers mind. 'Self-help groups' were formed for sustaining the programme and also to take new initiatives for dissemination of knowledge through farmers' field school and other village functions. The major strength of human resource development was the hundreds of farmers trained at Bangalore, Ranchi, Varanasi and Ghaziabad especially on handling of the biological control inputs and ill effect of injudicious pesticide usage.

Economic Viability and Impact of IPM Programmes

The major objective of the IPM validation programmes was to create a visibility of IPM at fields level keeping in mind the environmental and social impact factors. The data on number of chemical sprays, total cost of cultivation, expenditure on plant protection, yield and cost benefit ratio was worked out for IPM and non-IPM farmers to find out the economic viability and sustainability of the programme at all the locations for all the vegetables. The structured and unstructured impact analysis as increase in yield, cost of production, nature of plant protection measures, profitability level, and increase of bio-diversity has been studied.

The major success of the IPM technology was realized due to reduction in number of chemical pesticide sprays in all the vegetable crops. In cabbage, the number of chemical sprays was reduced to 3.5 and 6 in IPM from 11.7 and 16.0 in non-IPM, respectively, in Varanasi and Bangalore. Surprisingly, in Ranchi cabbage was grown without application of single insecticide in one of the seasons. Crop yield in IPM fields was significantly greater than non-IPM at 620.13 q/ha in IPM and 580.13 q/ha in non-IPM at IIHR, Bangalore. Similarly the yield performance was much better in favour of IPM than in non-IPM of Varanasi and Ranchi where it was recorded to be 240.8 and 347.50 q/ha in IPM and 211.7 and 287.60 q/ha in non-IPM fields, respectively (Table 1.18). Therefore, an account of both the increase in yield as well as by virtue of reduced number of pesticide sprays, IPM technology yielded better cost benefit ratio of 1: 3.04, 1:3.70 and 1:2.8 as against lower CBR of 1:0.61, 1:1.89 and 1:1.84 in non-IPM fields at three locations of Bangalore, Varanasi and Ranchi, respectively.

Table 1.18: Yield and Economics of IPM in Cabbage at Different Locations

Parameters	*IIHR Bangalore*		*IIVR, Varanasi*		*RKM, KVK, Ranchi*	
	IPM	*Non-IPM*	*IPM*	*Non-IPM*	*IPM*	*Non-IPM*
Number of chemical sprays	6	16	3.50	11.17	0	2
Cost of plant protection (Rs/ha)	5259	21806	3000	8000	2094	5090
Total cost of production (Rs/ha)	26414	66196	9000	16667	81942	85140
Yield (q/ha)	620.13	580.13	240.8	211.7	347.50	287.60
Gross returns (Rs/ha)	107348	106857	42333	48167	312750	263340
Net returns (Rs/ha)	80934	40661	33333	37500	231386	156900
C:B ratio	1: 3.04	1: 0.61	1:3.70	1: 1.89	1: 2.8	1:1.84

In tomato, at all the three locations *i.e.* Bangalore, Varanasi and Ranchi, the number of chemical sprays were significantly reduced to 5, 3.83 and 1 in IPM fields, whereas the numbers of chemical sprays were 28, 14.17 and 5 in non-IPM fields, respectively. The IPM fields also had lower cost of plant protection and recorded higher yields at all the three locations. The IPM fields also had better yield return of 740.38, 142.5 and 222.93 q/ha compared to 450.56, 130.0 and 187.72 q/ha in non IPM fields, respectively, in Bangalore, Varanasi and Ranchi (Table 1.19). Net returns and CBR obtained in IPM fields were also higher as compared to non-IPM fields.

Table 1.19: Yield and Economics of IPM in Tomato at Different Locations

Parameters	*IIHR Bangalore*		*IIVR, Varanasi*		*RKM, KVK, Ranchi*	
	IPM	*Non-IPM*	*IPM*	*Non-IPM*	*IPM*	*Non-IPM*
Number of chemical sprays	5	28	3.83	14.17	1	5
Cost of plant protection (Rs/ha)	3048	27668	3667	8917	3664.75	36000.00
Total cost of production (Rs/ha)	51844	114678	12083	18833	30237.58	31454.83
Yield (q/ha)	740.38	450.56	142.5	130.0	222.93	187.72
Gross returns (Rs/ha)	301565	184382	52000	57000	86942.7	73210.8
Net returns (Rs/ha)	249721	69704	39917	38167	56705.12	41755.97
C:B ratio	1: 4.82	1: 0.61	1: 3.30	1: 2.02	1: 1.87	1: 1.32

In okra crop, the average number of sprays was reduced to 4-5 in IPM as compared to 10-11 in non-IPM fields. Thus the residue content of the pesticides was reduced in fruits at harvest (Sardana, 2005). Though the cost of cultivation was higher in IPM fields as compared to non-IPM fields during

2004-05, but on account of higher yields obtained in IPM trial (103.05 q/ha) as against 72.46 q/ha in non- IPM fields; a better cost: benefit ratio of 1: 1.28 was recorded in IPM fields as against a lower cost: benefit ratio of 1: 0.72 obtained in non- IPM fields (Table 1.20). Similarly, there was a high reduction in number of sprays from 6-7 in non-IPM trials to 2-3 in IPM trial in brinjal crop resulting in lowering of pestices residues in the brinjal fruits at harvest. Just like okra, a slightly higher cost of cultivation *i.e.* Rs. 27310/ha was observed in IPM fields as against Rs. 26781/ha in non-IPM fields during 2004-05; but this higher cost was compensated by the higher yields obtained in IPM fields (613 q/ha) than non-IPM fields (509 q/ha). This resulted in higher cost: benefit ratio of 1: 2.27 in IPM than non-IPM fields (1: 1.02). Same trend was observed during 2005-06. IPM farmers earned a net income of Rs. 62700 and 66488/ha as compared to farmers who did not adopt IPM (Rs. 28043 and 45061/ha) (Table 1.20).

Table 1.20: Yield and Economics of IPM in Okra and Brinjal

Parameters	*Okra*				*Brinjal*			
	2004-05		*2005-06*		*2004-05*		*2005-06*	
	IPM	*Non-IPM*	*IPM*	*Non-IPM*	*IPM*	*Non-IPM*	*IPM*	*Non-IPM*
Number of sprays (Pestici)	4.1	11.25	4.40	10.60	2	6.6	2.70	7.20
Cost of plant protection (Rs/ha)	6609.5	8560.62	-	-	5140	4407	6820	5200
Total cost of production (Rs/ha)	49242.5	45037.5	44780.0	44851	27310	26781	24340.38	25863.0
Yield (q/ha)	103.05	72.46	110.71	76.19	613	509	394.53	302.61
Gross returns (Rs/ha)	114039.8	79715.63	121772.0	83809	122601	101794	93670.63	75652.5
Net returns (Rs/ha)	64797.25	34678.12	76993.0	38958	62700	28043	66487.56	45061
C:B ratio	1:1.28	1:0.72	1:1.71	1:0.87	1:2.27	1:1.02	1:2.69	1:1.73

The increased income from adoption of IPM was used for various social activities and general welfare of the family like education, housing, health and purchase of additional farm machinery and other facilities.

Constraints in Adoption of IPM

The validation trials have proved IPM as an effective method of controlling pests. Yet, it is not being adopted by farmers to the desired extent because of a number of constraints faced by them. Some important constraints are as follows:

1. Higher cost of hybrid seeds of vegetable crops than the traditional varieties.
2. Farmers' lack of awareness about IPM technology
3. High labour input
4. Non-availability of inputs like bio-agents and pheromone traps and true pest resistant varieties is a major problem in adoption of IPM in vegetable crops. Besides, there is a lack of trained personnel for mass multiplication of bioagents, their maintenance and distribution at proper time
5. Not convinced with the practices and their effectiveness

6. Complex nature of IPM technology and farmers lack understanding of biological interactions between the pests and their predators
7. Time consuming nature of IPM
8. Difficulty in implementing IPM practices
9. Inadequate credit facilities
10. Non-adoption of neighbouring fields
11. Problem of survival of bio-agent in the field
12. Difficulty in identifying threshold level
13. Weak extension service and inappropriate technology transfer mechanism are the critical impediments
14. Lack of forewarning systems and inadequate pesticide application technology.

Conclusions

The participatory approach of interdisciplinary team with multi-institutional support was the primary strength of the programme. The large area IPM validation has impacted on the visibility of its effectiveness for further diffusion of the technology in different regions of implementation. However, it has been experienced that the promotion of IPM in vegetable crops is a challenge because of small holdings under the crops, their raw consumption, harvest at short intervals and short time available for reaching from the harvest to table. The availability of short duration hybrids which, in addition to 2 to 3 crops, can even be grown in off season which in turn provide regular niche to various pests for the whole year. The confidence of farmers with the bioagents and bio-pesticides has not been very encouraging, therefore, the indiscriminate use of chemical pesticides and mixtures are not only well established but still garner sufficient favours with the farmers due to easy availability. However, the scenario has been changing with the introduction of disease resistant hybrids developed by private firms for several vegetable crops and initial awareness about biocontrol agents. The number of chemical pesticide sprays generally ranging from 8-10 were considerably reduced to 4-5 by providing viable alternatives like NSKE, Pongamia soap, *Ha*NPV, *Trichogramma*, pheromone traps, neem cake and crop geometry.

The challenges ahead are formidable but equally gratifying would be to sustain vegetable production partly through IPM adoption leading to comprehensive food and nutritional security to the people of India.

References

Baranwal, V.K., Kumar, Surender and Puri, S.N., 1996. Leaf curl disease of tomato and its managemant. *Technical Bulletin*-1, National Centre for Integrated Pest Management, New Delhi.

Faleiro, J.R., 1982. Investigations on the economic aspects of leaf hoppers (*Amrasca bigutulla bigutulla*) control on bhindi (*Abelmoschus esculentus* L.) Moench. *M.Sc. Thesis*, PG School, IARI, New Delhi.

Jayarathnam, K., 1977. Studies on the population dynamics of the diamond back moth, *Plutella xylostella* (Linnaeus) (Lepidoptera: Yponomeutidae) and crop loss due to the pest in cabbage. *Ph.D. Thesis*, Univ. of Agril. Sci., Banaglore, 220 pp.

Krishnaiah K, Ramachander, P.R., Jagan Mohan, N. and Wahi, S.D., 1979. Sampling technique for estimation of jassid population on okra. *Indian J. Ent.*, 41(2): 200–202.

Sardana, H.R., Arora, Sumitra, Singh, D.K. and Kadu, L.N., 2004. Development and validation of adaptable IPM in egg plant, *Solanum melongena* L. in a farmer's participatory approach. *Indian J. Plant Prot.*, 32(1): 123–128.

Sardana, H.R., Bambawale, O.M., Singh, D.K. and Kadu, L.N., 2005. A novel repellent action of neem cake against rabbit foraging on brinjal (*Solanum melongena* L.) crop. *Pestic. Res. J.*, 17(1): 23–24.

Sardana, H.R. Bambawale, O.M., Singh, D.K. and Kadu, L.N., 2005. Monitoring of insecticide residues in IPM and non-IPM fields of okra and brinjal. *Indian J. Pl. Prot.*, 33(2): 197–201.

Sardana, H.R., Trivedi, T.P. Bambawale, O.M., Sabir, N., Singh, R.V., Arora, S. and Krishnamoorty, P.N., 2004. Vegetable crops. In: *Validated IPM Technologies in Selected Crops,* (Eds.) Singh, Amerika, Sardana, H.R. and Naved Sabir. NCIPM, New Delhi, pp. 119–147.

Singh, Amerika, Sardana, H.R. and Sabir, Naved, 2004. *Validated IPM Technologies for Selected Crops.* National Centre for Integrated Pest Management, New Delhi, p. 253.

Singh, Amerika, Trivedi, T.P., Sardana, H.R., Dhandapani, A., Sabir, N., Singh, R.V., Prasad, R.N., Singh, Joginder, Sharma, P.D., Bhosle, B.B., Mathur, Y.S., Benagi, V.I., Srivastava, C.P., Porwal, N.K., Krishnamoorthy, P.N., Pandey, P.K., Rao, N.S., Sashankanand, Swami, Thakur, V.S. and Shukla, R.P., 2003. Validation and promotion of IPM technology in selected crops in different agro-ecological regions. In: *Recent Advances in Integrated Pest Management,* (Eds.) Singh, A., Trivedi, T.P., Sardana, H.R., Sharma, O.P. and N. Sabir. NCIPM, New Delhi, pp. 167–190.

Trivedi, T.P. and Misra, J.P., 2001. *IPM in Vegetable Crops*. Directorate of extension, Ministry of Agriculture, Govt. of India, Krishi Vistar Bhavan, Pusa, New Delhi, pp. 46.

Insect Pest and Disease Management *Pages 20–35*
Editor: **Dr. D. Prasad**
Published by: **DAYA PUBLISHING HOUSE (2016)**

Chapter 2

Insect Pheromones in IPM: Problems and Prospects

H.P. Misra
Department of Entomology, College of Agriculture, Orissa University of Agriculture and Technology, Bhubaneswar – 751 003, Orissa

Introduction

At present there is hue and cry all over the world about sustainability in agricultural production that always demand safe, effective and novel crop protection technologies to satisfy changing human needs of food, feed, fodder and fibre while protecting the environmental quality and safeguarding the economics of production. Production of high quality and globally competitive agro products with effective pest management and less environmental ill effects are the main objectives of modern agriculture. Over dependence on single (chemical) control tactic has created several problems in pest management. Therefore, researchers and scientists evolved Integrated Pest Management (IPM) as the only valuable option to be adopted worldwide in order to reach the goals. The IPM which is a multidisciplinary approach utilizes all available and appropriate farming techniques and methods in as compatible a manner as possible and maintains the pest population at levels below those causing economically acceptable damage (Meerman *et al.*, 1996). Integrated Pest Management is often referred as a biorational pest control that relies heavily on cultural and biological tactics that are supplemented carefully with timed application of highly selective chemicals. Of the several approaches being followed for IPM, application of semiochemical based products especially insect sex pheromones offer considerable promise and solutions to modern issues in pest management.

Insect pheromones are chemicals that are released by one individual which affect the physiology or behaviour of another individual of the same species (intraspecific). Since these are excreted by insects to outside of the body they were earlier called ectohormones. Later two German chemist Karlson and Butenandt (1959) isolated the sex pheromone of the silk worm, *Bombyx mori* L., characterized it and gave the present name pheromone based on the Greek words *Pherein*–to carry and *horman*–to excite. The first attempt at using pheromones for insect control was made in 1940s when crude extracts of abdominal tips of female Gypsy moth were used in monitoring traps in USA. There was rapid growth in the identification of insect pheromones during 1970s and by the end of 1980s pheromone and pheromone mimics were known for over thousand species of insects (Dhaliwal and Arora, 2006). Insect pheromones fall into several categories which are usually related to functions such as sex pheromones, aggregation pheromones, alarm pheromones, trail pheromones etc. (Srivastava, 1996). Since pheromones cause behavioural changes in the receiving individuals they are also termed as Behaviour Modifying Chemicals (BMC). Attempts for exploitation of BMC as pest management agents proved them as invaluable tools in modern IPM programmes (Beroza, 1972). Unlike conventional pesticides these chemicals are ecologically safe, do not damage non-target animals, nor do they cause health risks to people. Resistance to these chemicals has not been reported to date and it is also unlikely to occur, possibly due to low-level field concentrations that do not exert high selection pressure. Moreover, handling and application of these chemicals is very simple and the entire programme of work from research to application can be carefully framed and taken up under one roof.

It is the sex pheromones of insects that are of particular interest to agricultural IPM practitioners. Sex pheromones are usually produced by one sex, either male or female to attract their conspecifics for mating. In most instances it is the males that respond to the chemical elicited by the females. Scientists have isolated, identified these natural compounds and optimized the synthesized chemicals by blending them to attain maximum attraction of the opposite sex. Considerable work was carried out in developed countries to isolate and identify the pheromone systems during the past 30 years. Over 1500 moth sex pheromones have been identified to date (Arn *et al.*, 2000) and a valuable web based list is updated frequently (http/www-nysaes.cornell.edu/pheronet). Because of their natural occurrence, lack of toxicity, high biological activity, species specificity, long potency and compatibility with other pest control measures, currently pheromones are being commonly used as a component of many IPM programmes. Practical application of pheromones in IPM can be categorized as follows.

Indirect Control (Monitoring)

1. Trapping to collect species that are otherwise difficult to obtain
2. Trapping for monitoring and survey of key pests to target timing of pesticide application, release of parasitoids and early warning
3. Determination of timing of control measures can be done through finding action thresholds/ decision support
4. Knowing the population dynamics
5. Mapping pest distribution
6. Investigating dispersal and migrating behaviour of key pests of national and international importance
7. Quarantine work
8. Risk assessment
9. Knowing the effects of control measures
10. Enhancing the effectiveness of natural enemies
11. Detecting the occurrence of insecticide resistance in insects.

Direct Control

1. Luring the insects to traps that are treated with pesticides or chemosterilants with the purpose of male annihilation or male sterilization
2. Luring to areas that are treated with pathogens which can then be spread by the infected individuals to rest of the populations
3. Mass trapping for population reduction
4. Disruption of communication by permeation of areas of insect population, thereby resulting mating disruption and ultimately population suppression.

An initial survey (till 1990) indicated that world wide 1.3-2.0 million hectares of a variety of crops are treated with pheromones through monitoring, mass trapping and killing and mating confusants as a part of IPM strategy (Table 2.1).

Table 2.1: World Wide Pheromone Use

Pest	*Area (ha x 10^3)*	*Crop*	*Area (ha x 10^3)*
Moths	1250	Cotton	550
Beetles	45	Delicious fruits	322
		Forests	295
		Grapevine	108
Flies	17	Olives	17
		Cocoa	10
Scales, aphids, mites	01	Vegetables	NA
		Citrus	1.6
Types of Pheromone		Rice	0.1
Sex	1280	Kiwi	0.1
Aggregation	32	Stored products	NA
Alarm	<1	Roses	NA
Technique		*Geographical Zone*	
Monitoring	421	Europe	690
Mass trapping	306	North America	300
Confusion	556	Oceania	250
Attract and kill	30	Central and South America	53
		Africa and middle East	20

NA: Not available; *Source:* Yadav *et al.*, 2003.

Prospects

Fortunately the pheromone components of many of the problem insect pests of field, store and horticultural crops have been commercially manufactured in India and abroad and are available in India (Table 2.2). Interested persons can collaborate with laboratories like Tropical Products Institute and Chemical Entomology Unit, Southampton University (U.K.), Max-Plank Institute (Germany), Biologically Active Natural Products Laboratory, Maryland (USA), Laboratory for Research on

Insecticides, Wageningen (Netherlands), National Institute of Agricultural Sciences, Tokyo, Japan and Indian Institute of Chemical Technology, Hyderabad to identify, characterize and isolate pheromone compounds of problem insect pests.

Table 2.2: Pheromones of Important Agricultural Pests of India

Sl.No.	*Pest*	*Host*	*Pheromone Component*	*Trade Name*
1.	Tobacco caterpillar *Spodoptera litura*	Tobacco, Cotton, Groundnut, Castor	(Z,E)-9,11-14:AC and (Z,E)-9,12-14:AC (9:1)	Spodolure Litlure (Sex Pheromone)
2.	Indigo caterpillar *Spodoptera exigua*	Jute, Sugar beet, Tobacco, Lucerne	(Z,E)-9,12-14:AC	NA (Sex pheromone)
3.	Gram pod borer *Helicoverpa armigera*	Gram, Arhar, Tomato, Maize, Cotton	(Z) 11-16:AL and (Z) 9-16:AL (97:3)	Helilure (Sex pheromone)
4.	Maize stem borer *Chilo partellus*	Sorghum, Maize	(Z) 11-16AL and (Z) 11-16 OH (88:12)	NA (Sex pheromone)
5.	Pink boll worm *Pectinophora gossypiella*	Cotton	(Z,Z) and (Z,E) 7,11-hexa deca dienyl acetate(1:1)	Gossyplure (Sex pheromone)
6.	Spotted boll worm *Earias vitella*	Cotton, Okra	(E,E)10-12-hexadecadienal, Z,11-hexadecenal and Z, 11-Octadecenal (10:2:2)	Ervit (Sex pheromone)
7.	Spiny boll worm *Earias insulana*	Cotton, Okra	(E,E)10-12-hexadecadienal	Erin (Sex pheromone)
8.	White stripped borer *Chilo suppressalis*	Rice, Sugarcane	Z 9-16 AL, Z 11-16 AL and Z 13-18AL (1:10:1)	NA (Sex pheromone)
9.	Yellow stem borer *Scirpophaga incertulas*	Rice	Z 11-16 AL and Z 9-16 AL (3:1)	NA (Sex pheromone)
10.	Pink borer *Sesamia inferens*	Rice, Sugarcane	Z 11-16 AC, Z 11-16 OH and Z 11-16 AL (40:10:1)	NA (Sex pheromone)
11.	Leaf-folder *Cnaphalocrocis medinalis*	Rice	Z 13-18 AC and Z 11-16 OH (10:1)	NA (Sex pheromone)
12.	Climbing cut worm *Mythimna separata*	Rice	Z 11-16 AC and Z 11-16 OH (8:1)	NA (Sex pheromone)
13.	Greasy cut worm *Agrotis ipsilon*	Sugarcane, Gram, Potato	Z 7-10 AC, Z 9-14 AC and Z 11-16 AC (3:1:15)	NA (Sex pheromone)
14.	Cabbage looper *Trichoplusia ni*	Cabbage	(Z) 7-dodecenylacetate	Looplure (Sex pheromone)
15.	Diamondback moth *Plutella xylostella*	Cabbage,Cauliflower, Mustard, Radish	Z 11-16 Al, Z 11-16 AC and Z 11-16OH (50:50:1.5μg/ septum) or 1st two in 1:1 or 1st and 3rd in 7:3	Kongacon (Sex pheromone)
16.	Codling moth *Laspeyresia pomonella*	Apple	(E,E) 8,10 dodecadien-1-ol	NA (Sex pheromone)
17.	Oriental fruit moth *Grapholitha molesta*	Citrus	(E,Z) 8-dodecenylacetate (7:43)	NA (Sex pheromone)
18.	Melon fruitfly *Bactrocera cucurbitae*	Cucurbits	4(p-hydroxy phenyl) 2-butanone acetate	Cuelure (Food lure), Aggregation pheromone

Contd...

Table 2.2–Contd...

Sl.No.	Pest	Host	Pheromone Component	Trade Name
19.	Oriental fruitfly *Bactrocera dorsalis*	Mango	4,allyl-1,2-dimethoxybenzene Methyl Eugenol (Food lure)	Aggregation pheromone
20.	Mediterraanean fruitfly, *Ceratits capitata*	Stone fruits	t-butyl-2-methyl-4-chloro-cyclohexane carboxylate	Trimedlure (Food lure)
21.	Red palm weevil *Rhynchophorus ferrugineus*	Palms	4-methyl-5-nonanol (Ferrugineol)	Ferrolure, Rhynchophorol (Aggregation pheromone)
22.	Banana weevil *Cosmopolites sordidus*	Banana	NA	Cosmolure (Aggregation pheromone)
23.	White grub, *Holotrichia consanguinea*	Groundnut, pearlmillet	Anisole, Indole, Phenol	NA (Sex pheromone)
24.	Red hairy caterpillar, *Amsacta albistriga*	Groundnut	Octadecadienal, Z 9-12 Octadecadienal, Z 9,12,15 Octadecatrienal and Z 3,6,9 heneicosatriene (1:1:11.5:24)	NA (Sex pheromone)
25.	Groundnut leafminer, *Aproaerema modicella*	Groundnut	(Z) 7,9 decadienyl acetate, (Z) 7,9,10-decadienyl acetate and (Z) 7-10 deca-dienyl acetate (100:20:14)	NA (Sex pheromone)
26.	Green peach aphid, *Myzus persicae*	Tobacco, Chillies	(E) β-farnesene	Alarm pheromone
27.	Gypsy moth *Porthetria dispar*	Forest trees	NA	Gyplure, Disparlure
28.	Rhinoceros beetle *Oryctes rhinoceros*	Coconut	NA	Sime RB pheromone
29.	Coconut black headed caterpillar, *Opisina arenosella*	Coconut	1,2-dimethyl cyclopentane Methyl cyclohexane	NA
30.	Coffee white stem borer, *Xylotrechus quadripes*	Coffee	(S)-2-hydroxy-3-decanone, (R)-3-hydroxy-2-decanone and 2,3-dihydroxy octanone	NA
31.	Mango leaf webber, *Orthaga exvinacea*	Mango	Pentadecadienyl acetate Nonadecenyl acetate Heptadecenyl acetate Hepta decadienyl acetate	NA
32.	Cardamom fruit borer, *Conogethes punctiferalis*	Cardamom	(E)-10-hexadecenal and (Z)-10-hexadecenal (9:1)	NA
33.	Potato tuber moth, *Gnorimoschema operculella*	Stored potatoes	(E,Z) 4,7-tridecadien-1-Ol and (E,Z,Z)4,7,tridecatrien-1-Ol (1:4)	NA
34.	Brinjal shoot and fruit borer, *Leucinodes orbonalis*	Brinjal	(E)-11-hexadecenyl acetate and (E)-11-hexadecenol (10:1)	Leucinlure (Sex pheromone)
35.	Bihar hairy caterpillar, *Spilosoma obliqua*	Polyphagous	(Z_3Z_6)-cis-9,10-epoxy-heneicosadiene(1.0),(Z,Z)-1,3,6-cis-9,10-epoxy-heneico-satriene(0.5),(Z_9Z_{12})-octadecadienal (1.0),($Z_3Z_6Z_9$)-heneicosatriene(2.6)	NA

Contd...

Table 2.2–Contd...

Sl.No.	*Pest*	*Host*	*Pheromone Component*	*Trade Name*
36.	Caster semi-looper, *Achaea janata*	Caster	Heneicosane(1.0),(Z_9Z_{12})-octadecadienal(1.0), (Z_6Z_9)-heneicosadiene(1.0), ($Z_3Z_6Z_9$)-heneicosatriene(60/70)	NA
37.	Sorghum shoot fly, *Atherigona soccata*	Sorghum	Dihydro-paracymene, Limonene, 2,4-Diphenyl-4,methyl-2(E)-pentene	NA
38.	Teak defoliator, *Hyblaea puera*	Teak	Decadienyl acetate, Pentadeca diene acetate, heptadecenyl acetate	NA
39.	Teak skeletonizer, *Hapalia machaeralis*	Teak	Decadienyl acetate, Pentadeca diene acetate, heptadecenyl acetate	NA
40.	Khapra beetle, *Trogoderma granarium*	Stored wheat	(Z) and (E) 14-methyl 8-hexadecenal (92:8)	NA
41.	Cigarette beetle, *Lasioderma sericorne*	Stored tobacco	4,6-dimethyl 7-hydroxy nonan-3-one	NA
42.	Pulse beetle, *Callosobruchus chinensis*	Stored pulses	A mixture of several hydrocarbons and a dicarboxylic acid	Erectin
43.	Almond moth, *Cadra cautella*	Stored nuts	(Z,E) 9,12-tetredecadienyl acetate	NA
44.	Indian meal moth, *Plodia interpunctella*	Stored grain	(Z,E) 9,12-tetredecadienyl acetate	NA
45.	Angoumois grain moth, *Sitotroga cerealella*	Stored rice	(Z,E) 7, 11-hexadecadienyl acetate	NA
46.	House fly, *Musca domestica*	House hold pest	Z_9-Tricosene	NA

The data on pheromone usage (Table 2.1) indicated that the principal target pests are mainly moths. This is probably because of the fact that bulk of research work has been largely focused on this group and the components of moth pheromones belong to very limited class of organic compounds *i.e.* C 10–C 18 carbon chain molecules containing alcohol, acetate and aldehyde functional groups. This feature has not only facilitated the rapid characterization of the components, but led to similar methods to produce them. Pheromone chemistry ranges from the blends of aliphatic alcohols, aldehydes, esters and epoxides exploited by Lepidopterous insects to alkenoic acids and aldehydes, branched alkanones, esters, monoterpene alcohols and aldehydes, to a furanone employed by beetles (Justum and Gordon, 1989; Dhaliwal *et al.*, 1994). At present nearly 200 commercial pheromone formulations for the indirect control of pest by monitoring technique and about 20 formulations for direct control by mass trapping or mating disruption technique are being marketed by several reputed international companies. Notables among them are Agri Sense-BCS (UK), International Pheromone Systems (UK), Shin-Etsu (Japan), Chemtech BC (The Netherlands), Trace and Mycotech (USA), Bioprax (France), Agrimont (Italy), Flora (Estonia) and Pheronet (Sweden). The pheromones are and will contribute to a small part of agrochemical market compared to the insecticide trade world wide. It is estimated that at present pheromones account for 10–12 per cent of the insecticide market. However, it is expected to touch 20–30 per cent within a decade due to change in consumer attitude towards organic agricultural products, fruits and vegetables rather than insecticide treated and residue borne edible products. At the same time restructuring government policies towards hard pesticides vis-à-vis pheromones and regulatory guidelines for pheromone registrations by Central Insecticide Board, Government of India would lead to greater emphasis on pheromone use in pest management in a variety of crops.

The biological activity of many pheromone compounds have already been tested in the field and stores and standardization of dose and number of lures and traps required for unit area has been done. This will facilitate the farmers to resort to pest management by using pheromones thereby avoiding toxic chemical sprays on crops which have adverse effects on natural enemies of crops and the environment. Earlier, there was problem for finding out a suitable dispenser for pheromone carrying and releasing. The dispensers should sustain the local agro-environmental factors such as temperature, humidity, wind speed and direction etc. Until now, however, there is no suitable dispenser adaptable to Indian climatic conditions is available. However, for few insects dispensers and dosages have been standardized. For example two types of dispenser's *viz.*, shoelace type dispenser and laminated dispenser prepared at IICT with the help of polymer chemistry group were used with the commercially available rubber septa for rice yellow stem borer. All the dispensers were loaded with pheromone blend of rice yellow stem borer and exposed for field evaluations. Rubber septa loaded with 10mg of pheromone component and an equal amount of antioxidant-butylated hydroxytolune (BHT) proved to have an effective field life time for several weeks. For gram pod borer (*H. armigera*) yellow rubber septa loaded with 8mg of synthetic pheromone proved to be better than laminated blue, laminated coffee and white rubber septa for monitoring experiments. Similarly, locally manufactured plastic vial dispenser with a loading of 3 mg of synthetic pheromone and an equal amount of antioxidant (BHT) was standardized for groundnut leafminer, *A. modicella* than those vial dispensers from Aldrich Co (Yadav *et al.*, 2003).

Pheromones work on bad weather like rain, frost, dew, high temperature, wind etc. without having been washed out unlike chemical pesticide spray. It can also be used in difficult areas like forests avoiding costly air craft spray for forest pest control, orchards avoiding regular surveillance and spraying difficult plant parts like tall trees of coconut, areca nut and other palms.

Available pheromones may be used to monitor the population fluctuations of key pests/pests of national and international importance during different times of the year or for several years to understand the causes of epidemics of a given pest and develop a scientific forecasting system. This will be a cheap, efficient and easy to handle method. Important studies made in India with sex pheromones of key crop pests have been presented in Table 2.3.

Monitoring

Efficient monitoring is the foundation of sound IPM programmes, without which proper pest management decisions can not be taken. Pheromone baited traps for monitoring pest populations provide a highly sensitive means of detecting the insect pests with many advantages over conventional methods such as light traps or scouting (Dhaliwal and Arora, 2001). Pheromone traps can be used to detect both the presence and density of pest species. Insect populations can, thus, be estimated and new areas of infestation detected at a very early stage. The trap catches can be used to forewarn the outbreak of important pests. The economic threshold levels can be determined in terms of pheromone trap catches to decide about the timing of insecticide application (Justum and Gordon, 1989). Establishing correlations between trap catches and subsequent infestations is often difficult and requires close understanding of the agroecosystem. It took 10 years of study in case of Olive moth, *Prays oleae* (Bernard) to establish a working correlation between male trap catch and subsequent infestation of fruit. A high degree of asynchrony between moth emergence and tree phenology meant that late emerging moths could attack fruits (Jones, 1994).

Table 2.3: Important Studies Made in India with Sex Pheromones of Insects

Sl.No.	*Pests*	*Crop*	*References*
		MASS TRAPPING	
1.	*Spodoptera litura, Helicoverpa armigera*	Groundnut	Dhandapani (1985), Patel *et al.* (1985), Krishnaiah (1986), Kumari and Reddy(1992)
2.	Sweet potato weevil, *Cylas formicarius*	Sweet potato	Pillai *et al.* (1993), Pillai *et al.* (1996)
3.	Diamondback moth, *Plutella xylostella*	Cole crops	Reddy and Urs,1997
4.	Brinjal shoot and fruit borer, *Leucinodes orbonalis*	Brinjal	Cork *et al.,* 2003
5.	Yellow stem borer, *Scirpophaga incertulas*	Rice	Varma (1999), Krishnaiah *et al.* (2000), Gururaj Katti *et al.* (2001)
6.	Sugarcane borers	Sugarcane	Jayanth and Bhanu, 2004
		MATING DISRUPTION	
1.	*Spodoptera litura*	Chillies	Rao *et al.,*1989
2.	*Lymantria obfuscata*	Apple	Masoodi *et al.,*1990
3.	*Spodoptera litura*	Groundnut	Srinivas and Rao.1999
4.	*Scirpophaga incertulas*	Rice	Ganeswara Rao *et al.* (1994), Cork and Basu (1996), Cork *et al.* (1998)
5.	*Pectinophora gossypiella*	Cotton	Sohi *et al.,*1999
		MONITORING	
1.	Potato tuber moth, *Gnorimoschema opercullela*	Potato	Chandla *et al.* (1987), Trivedi (1989), Chandramohan(1995a)
2.	Tobacco caterpillar, *Spodoptera litura*	Groundnut, Vegetables, Soybean	Prasad *et al.* (1985), Gupta and Pawar (1989), Singh and Sachan (1993)
3.	Sweet potato weevil, *Cylas formicarius*	Sweet potato	Teli and Salunke,1993
4.	Pink bollworm *Pectinophora gossypiella*	Cotton	Sangappa *et al.* (1985),Thangaraj and Uthamasamy (1990), Dhanorkar and Puri (1993), Dhawan and Sidhu (1993), Korat and Lingappa (1996b)
5.	Spotted bollworm, *Earias vitella*	Cotton	Korat and Lingappa (1996a), Naik *et al.* (1997)
6.	American bollworm, *Helicoverpa armigera*	Cotton	Dhanorkar and Puri (1993), Verma and Sakhyan (1993), Prasad (1996)
		TIMING OF INSECTCIDAL SPRAYS USING PHEROMONES	
1.	Diamondback moth, *Plutella xylostella*	Cole crops	Prasad Reddy and Guerrero (2001) (Spain)
2.	Pink bollworm, *Pectinophora gossypiella*	Cotton	Dhawan and Sidhu,1984
3.	Cotton bollworms	Cotton	Prasad *et al.,*1993
		INTEGRATION OF PHEROMONES WITH OTHER STRATEGIES	
1.	*Scirpophaga incertulas, Cnaphalocrocis medinalis*	Rice (Pheromone + biocontrol)	Gururaj Katti *et al.,* 2001
2.	Cotton bollworms	Cotton (Insecticides + Pheromone)	Gupta and Srivastava (1998)
3.	*Helicoverpa armigera*	Chickpea	Krishnakant *et al.,* 1998

Contd...

Table 2.3–Contd...

Sl.No.	Pests	Crop	References
		IPM STUDIES	
1.	Cotton bollworms	Pheromones + pesticides + egg parasitoid	Butter and Kular, 2002
2.	Sweet potato weevil	IPM	Sethi *et al.*, 2003
3.	Groundnut leafminer	Pheromone + *B.t.* + insecticide	Muthia, 2003
		KAIROMONAL RESPONSE	
1.	Yellow stem borer	Wolf spider	Sinha *et al.*,1998
2.	*Helicoverpa armigera*	*Trichogramma chilonis*	Padmavathi and Paul, 1996
		GEOGRAPHICAL/SEASONAL VARIATION	
1.	Diamondback moth	–	Chandramohan,1995b
2.	*Helicoverpa armigera*	–	Bhadauria and Bhadauria (1998), Thamankar *et al.*, 2003
3.	Potato tuber moth	–	Kumar and Sivakumar, (2003), Raj (1998)
		SYNERGIZING ATTRACTION THROUGH PLANT EXTRACTS	
1	Yellow stem borer, *Scirpophaga incertulas*	Rice	Krishnaiah *et al.* (1998b), Ram Gopala Varma *et al.* (2002)

Source: Pasalu *et al.*, 2005.

In cereal crops several parameters influence efficient pheromone based monitoring for insect pests (Krishnaiah, 1995). Krishnaiah *et al.* (1998) reported that three traps per acre as trap density for monitoring of rice yellow stem borer. Various parameters of an efficient pest monitoring system for rice pests are presented in Table 2.4. Varma *et al.* (2000) conducted monitoring studies on rice yellow stem borer using natural blend (3:1) of (Z)-11-hexadecenal and (Z)-9-hexadecenal and noticed three peaks of moth activity during July, August and September in *Kharif* and two during January and March or April during Rabi. It is evident from the temporal relationship between peak periods of pheromone trap catches and field infestation that peak infestation occur 2-3 weeks after peak/sudden increase in moth capture. It is found that 30 male moths/week/trap seem to result in 10 per cent dead hearts which is taken as capture threshold for taking up control measures. Thus, pheromone traps can be utilized as an aid in judging the need for chemical intervention.

In India the trap catches of *H. armigera* were found to be positively correlated with larval stages and egg population which provide a basis for initiating control strategies. These studies prompted the establishment of trapping network in the areas where correlations between trap catches and larval population were high for several years. Continuous deployment of trap revealed three periods of moth activity appearing between 29 and 38, 47 and 48 and 22 standard weeks. It was possible to workout temporal and spatial variation in the pest abundance using pheromone traps in different agro climatic zones. Population trends of *H. armigera* were studied in Karnataka on cotton crop, in Haryana on chickpea crop, in Himachal Pradesh on chickpea and tomato crop, of *H. armigera* and *Pectinophora gossypiella* on cotton crop in Maharastra and on chickpea, pigeonpea, maize and sunflower in Uttar Pradesh. Extensive studies on monitoring the population levels of *H. armigera* have been conducted in Nepal. The monitoring of *H. armigera* population in Pakistan showed the maximum catches during first fortnight of April with no obvious peak in Dokri region on chickpea crop. The relationship between male moth catches and larval infestation was established in tomato and groundnut crops in Taiwan (Kanaujia, 2000).

Table 2.4: Parameters of an Efficient Rice Pest Monitoring System through Pheromones

Insect	*Scirpophaga incertulas*	*Chilo suppressalis*	*Sesamia inferens*	*Cnaphalocrocis medinalis*
Pheromone components	(Z) 11-16Al and (Z) 9-16AL	(Z) 11-16Al, (Z) 9-16AL and (Z) 13-18AL	(Z) 11-16AC, (Z) 11-16 OH and (Z) 11-16AL	(Z) 13-18 AC and (Z) 11-16 AC
Best blend	3:1	10:1:1	40:10:1	10:1
Dispenser	Rubber	Rubber	Rubber	Polyethylene vial
Dose	2 to 5 mg	0.3 to 0.6mg	2mg	1mg
Replacement period	3-4 weeks	4 weeks	3-4 weeks	4 weeks
Trap design	Sleeve	Plastic funnel	Sleeve	Delta sticky
Sleeve colour	White	Yellow	–	–
Trap elevation	0.5–1.0m	0.5m	1.0m	Canopy level

Source: Pasalu *et al.*, 2005.

Pheromone traps were found effective in monitoring the population of *Spodoptera litura* at Ludhiana, India. The trap data indicated about 10 generations annually in Southern India (Hyderabad) and 7 generations at Pantnagar (Kanaujia, 2000). Significant correlations between trap catches and oviposition on soyabean and sugar beet crops have been observed with no correlation with larval population. Traps provide reliable information about population levels of *S. litura* in Taiwan. A system based on moth catches for monitoring and forecasting of *S. exempta* was developed in Yemen Arab Republic.

Monitoring of pink bollworm, *P. gossypiella* population using 1:1 mixture of (Z,Z) and (Z,E)-7-11 hexadecadienyl acetate (the sex pheromone named gossyplure) was done throughout India and Spain. The correlation between moth catches and larval population in cotton crop have been established. Pesticide application at a threshold of 4-8 moths/trap/night proved economical and saved 1-2 insecticide sprays. Sex pheromone traps are reported to be potentially useful for monitoring *Earias vitella* and *E. insulana* populations in Pakistan; *E. insulana* and *P. gossypiella* in Egypt; *Cylas formicarius* in India; *Cydia molesta* in Bulgaria and stored grain pests in India. Deployment of 20 pheromone traps per hectare not only improved the quality of fruits but also significantly reduced chemical applications against *C. pomonella* and *C. molesta* in apple orchards in Ukraine. The peak flight period of cutworm, *Agrotis ipsilon* is closely related to feeding and consequently higher release of female pheromone (Kanaujia, 2000).

Forecasting Population Trends and Timing of Control

Pheromone traps provide a good estimate of population fluctuations including migratory trend and early arrival of the pest which help in initiating control measures. Field observations on *H. armigera* indicated that higher larval density in chickpea crop is preceded by peak moth catches in pheromone traps. In case of *S. litura*, however, peak moth activity synchronized with oviposition showing significant correlations. Application of pheromone traps against *P. gossypiella, Plutella xylostella* and stored grain pests have yielded useful results for predicting the pest population. The population estimates based on trap catches and other parameters provide useful information regarding population density and fluctuations. This information has been used for deciding the timing of control operations in *H. armigera* and *S. litura*. Constant monitoring minimized use of pesticides against *S. litura* and *H. armigera*. In Pakistan, pheromone formulations significantly reduced the abundance of *P. gossypiella*, *E. vitella* and *E. insulana* and provided protection equal to insecticide application against *P. gossypiella* on cotton in Egypt (Kanaujia, 2000).

Mating Disruption

By widespread application of synthetic pheromones over the treated crop, insect pests can be controlled by mating disruption technique. Various slow release pheromone formulations have been developed either to permeate the air with relatively high levels of pheromone compounds so as to achieve sensory adaptation and habituation or provide numerous discrete point sources for masking trail following or create false trail. Various synthetic chemicals with molecular structures similar to the naturally occurring pheromones have been examined and tested for their ability to elicit similar behavioural responses. Successful compounds are termed *parapheromones* or *antipheromones* depending on whether they attract the insects or interfere with the way in which the insects perceive the natural pheromones. Sprayable formulations introduced in 1980s suffered from short persistence in the field. Hand applied dispensers have season-long effectiveness and have opened up the technology for many cropping systems (Jones, 1994). Another recent approach involves the development of haloacetate analogues of natural pheromones of a number of insects including *Plutella xylostella* (L.), *Heliothis virescens* (Fab.), *Ostrinia nubilalis* (Hubner). These analogues act as competitive inhibitors of the male response in laboratory and field studies (Riba *et al*., 1994)..

A number of insects utilize multi-pheromone component system. They are polyphagous and migratory in nature. Hence, mating disruption technique in such cases is complex and usually successful due to several factors such as low release rate of pheromone in *H. virescens, H. zea* and *H. armigera* in Egypt and Taiwan. Limited field studies conducted in India revealed effective mating disruption of *S. litura* resulting in significantly lesser number of egg masses in treated area and the control of the pest. Similarly, mating disruption by using microencapsulated formulations (40-80g a.i./ha) in Egypt was found to be economically non-viable. Therefore, the lure and kill approach using 2g a.i./ha of pheromone per application with insecticides was proposed. Higher efficacy of mating disruption against *S. littoralis* was demonstrated. A large scale experiment conducted against *P. gossypiella* in cotton fields of Punjab, using aerial application of gossyplure @ 50-75g a.i./ha at interval of 8-20 days, showed significant suppression of pest population with an average increase of 33.7 per cent in yield over untreated area. Similar result was reported in Pakistan and Israel against *P. gossypiella* and *H. armigera*. A higher level of communication disruption (89-90 per cent) has been achieved by using 40-50g a.i./ha of pheromone against *S. incertulas* infesting rice. The results on *Cydia pomonella* in Chile and *Lymantria obfuscata* in Kashmir, India have shown significant suppression of mating. In China, the mating disruption rates of *Ostrinia furnocalis* is reported to be 79.3-85.5 per cent (Kanaujia, 2000).

Various attempts have been made in rice on farmer's fields to exploit mating disruption technique for direct control of the yellow stem borer utilizing a controlled release formulation with polyvinyl chloride (PVC) matrix as base through the Directorate of Rice Research, Hyderabad and Natural Resources Institute collaborative programme. Initial small scale on-farm trials with PVC based formulation of Agriscens BCS with pheromone blend of (Z) 9-hexadecenal and (Z) 11-hexadecenal in

1:10 ratio yielded promising results. The pheromone applied @ 40 g a.i./ha through 625 point sources disrupted 98 per cent communication for mating of Yellow stem borer during Rabi, 1992 (Ganeswara Rao *et al.*, 1994).

Mass Trapping/Male Annihilation

The rationale behind this technique is to place enough traps to concentrate insect pests to a restricted space (catch enough males) and leave the females of the species without mates. Mass trapping is possible in case of adult insects with highly developed ability to respond to attractants. The technique is mostly restricted to crops where the insecticides are not effective or environmentally acceptable. Mass trapping technique offers great promise against monophagous pests like yellow stem borer on rice (Karthikeyan and Purushothaman, 2003) with crop often cultivated extensively and contiguously. A trapping efficiency of more than 90 per cent is required in case of Lepidoptera males to achieve the desired results (Justum and Gordon, 1989). Studies done at Directorate of Rice Research to optimize the trap density, time of application for yellow stem borer mass trapping taking a 3:1 blend of (Z) 11-hexadecenal and (Z) 9-hexadecenal @ 5mg/dispenser revealed that a trap density of 20 traps/ha is optimum (Pasalu *et al.*, 2005). A modification of mass trapping is the male annihilation (lure and kill) technique where instead of being trapped; the responding insects come in contact with a conventional insecticide. This technique overcome the problem of low trapping efficiencies, since the responding insects need only to alight in the vicinity of the attractant source to pick up a lethal amount of the toxicant (Jones, 1994). Recent studies also showed that trap catches of YSB can be increased by addition of co-attractants like Pusa Basmati plant extracts and Vitamin E to YSB pheromone (Pasalu *et al.*, 2005).

Further studies conducted at Directorate of Rice Research revealed that pheromone mass trapping can be effectively integrated with chemical control (Table 2.5) and biological control (Table 2.6) for effective management of YSB and leaf-folder of rice (Katti *et al.*, 2001).

Table 2.5: Effect of Pheromone Mass Trapping and Chemical Control on YSB Infestation and Yield

Treatment	*Kharif, 1997 (Karjat)*			*Kharif, 1998 (Karjat)*		
	DH (%)	*WEH (%)*	*Yield (Kg/ha)*	*DH (%)*	*WEH (%)*	*Yield (Kg/ha)*
PT	11.91	12.42	3176	7.88	5.84	3302
PT + I	7.64	9.49	4208	6.38	4.79	4149
FP	17.21	20.54	2320	13.60	20.79	2193

DH: Dead heart; WEH: White ear head; PT: Pheromone treatment alone;
PT+I: Pheromone treatment + need based insecticide application, FP: Farmers practice.

Table 2.6: Influence of Mass Trapping and Bio-control on YSB and Leaf-folder Management

Treatment	*Nellore, Kharif, 1995*		
	WEH (%)	*LFDL (%)*	*Yield (Kg/ha)*
Pheromone + Bio-control	8.9	8.5	4747
Farmers Practice	28.0	22.0	3421

Pheromone traps @ 20/ha; *Trichogramma chilonis* @ 1 lakh/ha (5 releases).

WEH: White Ear Head; LDFL: Leaf-folder damaged leaves.

Male annihilation studies on *H. armigera* and *S. litura* revealed masking effect due to immigration of males and mated females from outside area. Optimum density of traps for effective trapping is in the range of 16-25 traps/ha for *H.armigera*. Deployment of 1000 traps over 175ha area against *S. litura* and *H. armigera* in cotton reduced insect population significantly and the number of insecticidal application is reduced by 3-6 sprays. In Israel and Egypt, the studies indicated the possibility of reduction of the number of egg masses through deployment of traps. Similar results were reported on *Sesamia nonagriodis* infesting early sown maize and on *Bactrocera oleae* in Greece. The feasibility of mass trapping was demonstrated in stored grain insects. The use of sex pheromone traps against *Rondotia meniciana*, an important pest of mulberry in China has been reported to be a valuable tool in the management (Kanaujia, 2000).

However, pheromones, when employed alone, may sometimes provide ineffective or insufficient pest control. This is because both pheromone releases by females and peak male responsiveness are dependant on several other parameters like age of the insect, host selection, temperature, humidity, wind speed etc. Of all these exogenous factors female habitat (where it oviposits) is found to be significant and volatiles released in the habitat are responsible for this phenomenon. These volatiles are inter-specific and if released to the advantage of the receiver are called kairomones. Kairomones interfere in a great way in sex pheromone production and release in insects as well as on the bio-active responses to the synthetic pheromones.

It is generally believed that the insect pheromones are too expensive compared to synthetic chemicals. This not true in the sense that, even the cost of the traps and lures is initially high but subsequently the farmer will only change the lure and the same trap is used. Traps of various kinds are used to catch insects such as delta trap, funnel trap, bucket trap, gallon trap with ramp, pit fall trap, bottle trap, Mc Phail trap, open pan trap, Morocco trap, Jackson strainer trap and polyethylene sleeve trap etc. The polyethylene sleeve trap is cheaper among all and widely used for various field pests. Bucket trap and pit fall trap are used in perennial orchards. Secondly, the safety to environment, non target organisms, wild lives and human community with the use of pheromones is never calculated and compared with respect to chemical insecticides.

Problems

1. Identification of pheromone components of few agricultural pests of national and international importance is done. There is a long way to go for identifying pheromone components of pests of regional importance.
2. Field testing of available pheromones in various land situations spread over India should be carried out to determine the role of various factors which can affect the performance of pheromones like release rates, dispensers, pheromone blends, amount of antioxidants and synergists, number of traps in an area, trap design, size, colour and height of trap placement to obtain maximum benefit out of their use which has not been done baring few instances.
3. The biological activity of the synthetic pheromones, especially their competitiveness with live virgin females in the fields is of great interest. The preliminary evaluation found promising for a pest in small area be further examined in large areas and under optimum population (not low) conditions to get confirmatory results, which has not been done for many pests of agricultural crops.
4. Many field problems arise while using pheromone traps such as the layman or children steal the trap and destroy since, they are not properly educated. Similarly, the polyethylene

sleeve traps placed at low heights in certain crops are destroyed by carnivorous jackals with the aim of feeding live insects trapped in them.

5. Many of the lures obtained through reputed firms in India for specific pests do not work in the field. Instead, other insects are caught in the trap except for the one it is employed. This creates lack of confidence among the leading farmers on the efficiency of control of insect pests using pheromones.
6. The methods of transport, storage of pheromone lures are not known to the distributor or farmers for which reduced efficiency is seen in the field.
7. Moreover, farmers do not get traps at their door steps unlike synthetic chemical pesticides and unless large scale orders are placed with the manufacturer, these are not available on retail, baring some cities.

Future Thrust Areas of Pheromones as a Component of IPM

1. Isolation, identification and synthesis of pheromones of insect pests for which pheromone system is not yet developed (such as rice gall midge, rice hispa and gundhi bug, mango stone weevil etc.).
2. Evaluation and optimization of monitoring and direct control methods with identified chemicals.
3. Indigenous synthesis of all pheromone components against insect pests of relevance in India to make them readily available to the farmers.
4. Development of new controlled release formulations of pheromones for the purpose of mass trapping and mating disruption.
5. Studies to utilize pheromones in enhancing natural enemies' performance.
6. Investigating the role of pheromones in tri-trophic interactions of food web consisting of host, pests and natural enemies.
7. Development of technologies to increase the persistence of dispensers.
8. Finding out the synergists for different pheromones utilizing plant extracts or other materials as co-attractants.
9. Investigating variations in populations in composition, response to pheromones and the long-term effects of extensive use of pheromones *e.g.* habituation, ecological effects etc.
10. Integrating pheromones as important component of IPM in future IPM programmes.

Conclusion

Though pheromones are extremely potent biological agents, they are not panaceas, any more than was DDT, but their use as direct pest control agents will probably be limited to certain key pests that have great economic importance. In general, sex pheromones would continue to serve a promising role in the management of monophagous and oligophagous pests through monitoring, mass trapping and mating disruption techniques. In polyphagous insects, however, the pheromones are valuable as supplementary measures or as a component of IPM being integrated with other methods of pest management. For successful exploitation of pheromones in integrated pest management, extensive work on chemical aspects involving identification and chemical synthesis followed by their field testing has to be undertaken. An emphasis has to be laid down for large scale adoption of pheromone use through generating awareness among farmers.

References

Arn, H., Toth, M. and Priesner, E., 2000. List of sex pheromones of female Lepidoptera and related male attractants. *Internet Edition http://phero.net/pherolist*.

Beroza, M., 1972. Insect sex attractant pheromones: A tool for reducing insecticide contamination in the environment. *Toxicology and Environmental Chemical Reviews*, 1: 109–134.

Dhaliwal, G.S. and Arora, R., 2001. *Integrated Pest Management: Concepts and Approaches*. Kalyani Publisher, pp. 335–337.

Dhaliwal, G.S. and Arora, R., 2006. *Integrated Pest Management*. Kalyani Publishers, pp. 285–288.

Dhaliwal, G.S., Bhathal, J.S. and Bhathal, S.S., 1994. Novel approaches in insect pest management: An overview. In: *Trends in Agricultural Insect Pest Management*, (Eds.) G.S. Dhaliwal and Ramesh Arora. Commonwealth Publishers, New Delhi, India, pp. 469–511.

Ganeswara Rao, A., Krishnaiah, K., Cork, A.and Hall, D.R., 1994. Disruption of sex pheromone mediated communication in yellow stem borer, *Scirpophaga incertulas* Walker with a controlled release pheromone formulation. *J. Ent. Res.*, 18: 287–288.

Jones, O.T., 1994. The current and future of pesticides for semiochemicals in integrated management of insect pests. In: *Brighton Crop Protection Conference, Pests and Diseases*, Vol. 3. British Crop Protection Council, Farnham, UK, pp. 1213–1222.

Justum, A.R. and Gordon, R.F.S., 1989. *Insect Pheromones in Plant Protection*. John Wiley and Sons, New York, USA.

Kanaujia,K.R., 2000. Current status and prospects of pheromones in IPM. In: *Sustainable Insect Pest Management*, Summer School held on June 20–July 10 at Department of Entomology, G.B.P.U.A.T., Pantnagar, pp. 82–86.

Karlson, P. and Butenandt, A., 1959. Pheromones (ectohormones) in insects. *Ann. Rev. Entomol.*, 4: 39.

Karthikeyan, K. and Purushothaman, S.M., 2003. Pheromone traps: An effective tool in rice stem borer management. *Indian Farming*, 53: 14 and 32.

Katti, G., Pasalu, I.C., Varma, N.R.G. and Krishnaiah, K., 2001. Integration of pheromone mass trapping and biological control as an alternate strategy for management of yellow stem borer and leaf folder in rice. In: *Proceedings of the 88th session of the Indian Science Congress*, New Delhi. Indian Agricultural Research Institute, New Delhi, Abstract No. 38, p. 38.

Krishnaiah, K., 1995. Insect pheromones in integrated pest management of cereals. In: *Pheromones in Integrated Pest Management*, (Eds.) Krishnaiah, K. and B.J. Divakar. Plant Protection Association of India, pp. 33–39.

Krishnaiah, K., Zainulabeuddin, S., Ganeswara Rao, A., Kumar, D.V.S.S.R. and Rama Gopala Varma, N., 1998. Pheromone monitoring system of rice Yellow stem borer, *Scirpophaga incertulas* Walker. *Indian J. Plant Protec.*, 26(2): 99–106.

Meerman, F. Van De Ven, G.W.J., Van Keulen,H. and Breman, H., 1996. Integrated crop management: An approach to sustainable agriculture and development. *International Pest Management*, 42(1): 13–14.

Pasalu, I.C., Varma, N.R.G. and Katti, Gururaj, 2005. Insect sex pheromones in integrated pest management with special reference to rice yellow stem borer. In: *Gleanings in Entomology*, (Eds.) V.V. Ramamurthy, V.S. Singh, G.P. Gupta and A.V.N. Paul. Division of Entomology, IARI, New Delhi, pp. 20–40.

Riba, M., Eizaguirre, M., Sans, A., Quero, C. and Guerrero, A., 1994. Inhibition of pheromone action in *Sesamia nonagriodis* by haloacetate analogues. *Pestic. Sci.*, 41: 97–103.

Srivastava, K.P., 1996. *Text Book of Applied Entomology*. Kalyani Publishers, pp. 238–255.

Varma, N.R.G., Krishnaiah, K, Pasalu, I.C. and Reddy, D.D.R., 2000. Monitoring of rice Yellow stem borer, *Scirpophaga incertulas* Walker using pheromone traps and thermal summation. *Indian J. Plant Protec.*, 28(1): 84–93.

Yadav, J.S., Prasad, A.R., Prasuna, A.L. and Jyothi, K.N., 2003. Semiochemicals as potential tools in pest management. In: *Frontier Areas of Entomol. Res.*, (Eds.) B. Subrahmanyam, V.V. Ramamurthy and V.S. Singh. Division of Entomology, IARI, New Delhi, pp. 411–422.

Insect Pest and Disease Management *Pages 36–48*
Editor: **Dr. D. Prasad**
Published by: **DAYA PUBLISHING HOUSE (2016)**

Chapter 3

Role of Sex Pheromones in Management of *Helicoverpa armigera* (Hubner)

Krishna Kant
KVK/ETC, (SKUAST-K) Malangpora, Pulwama – 192 308, J&K

Introduction

Helicoverpa armigera (Hubner) is a polytypic species comprising of three distinct sub species viz; *H. armigera armigera* attacking pea, gram and tomato; *H. armigera sorghi* attacking wheat, jowar and maize and *H. armigera hibisci* attacking cotton and Bindi. In *H. armigera armigera*, valva of male genetelia is broad, width being nearly one forth of the length clubbed, aedeageal cornuti with 13 stout spines (Jadhav *et al.*, 1985). In India it is a dominant pest of chickpea, pigeonpea, cotton, and also causes serious losses to sorghum, maize, tomato and many other crops. (Bhatnagar *et al.*, 1985)

The realization that the chemical insecticides, as a quick and easy solution to insect problem, are not with out considering cost of several fronts. Insecticides more than any other pollutant have been the subject of violent controversy because of their wide spread effect on environmental quality in the last 30-40 years. The use of insecticides has resulted in the development of resistance in 504 species of insect and mites. (Georgion, 1990) and has almost inevitably been followed by pest resurgence and an outbreak of secondary pests. (Georgion and Taylor, 1977; Metcalf, 1986). In India *H. armigera* is also reported to have developed resistance against many of common used insecticides of various groups. Among wide array of potential tools available for pest control, semiochemicals, particularly sex pheromones, have figured quite prominently and have proven most useful in the management of different pest species (Bakke and Lie, 1989). Sex pheromone being highly specific and active at molecular level, elicit behavioural responses over a long distances resulting in to mating and reproduction of the species. Pheromones have been used to monitor the infestation, assess population level and devise management strategies for mass trapping and mating disruption of different pest species. The use of pheromone trap has proved to be an excellent tool to monitor *H. armigera* population that help prediction of population for management including insecticidal uses.

The application of pheromone in pest control started in the 1940s in the U.S.A. when the crude extracts of abdominal tips of female Gypsy moth were used in monitoring traps (Holbrook *et al.*, 1960). However, it was not until thirty years ago that the possibility of using pheromone in crop protection was explored and the first meaningful field trial on mating disruption of a lepidopterous pest Cabbage looper, *Trichoplusia ni* was reported (Shorey *et al.*, 1967).

Chemical identification and characterization of first pheromone of silk moth, *Bombyx mori* by Butenandt *et al.* (1959) was followed by identification of pheromone in hundreds of species of insect and thus opened a new era of behavioural control of insect population.

Chemical Components of Sex Pheromone of *H. armigera*

Chemical analysis of the pheromone of *H. armigera* (Table 3.1) suggests that the pheromone is a multiple component and comprises 1 to 6 different components in various ratios though (Z)–ll Hexadecenal is the major component and ranged from 90-99 per cent with minor components in the lower amounts.

Pheromone Production in *H. armigera* Female Moth

A comparison made over 10 days for trapping of male by a virgin female showed that 2 days old female captured maximum male moths (Annon, 1987). Similarly laboratory tests indicated maximum level of pheromone production in 2 to 3 days old virgin females at 4.5 hr. after the start of scotophase (Rafaeli and Soroker, 1989). Sex pheromone production in virgin female of *H. virescens* peaked in 3 day old females (180 mg) after enclosion. (Mbato and Ramaswamy, 1990; Mbato and Ramaswamy 1991).

Application of Pheromone for Monitoring Population Dynamics

The most useful characteristic of a pheromone trap is that it catches selectively adults of a species even at lower population densities earlier than other available sampling methods (Cranham, 1980). Pheromone traps proved reliable for early detection and estimation of *H. armigera* population and establishing economic threshold (Rothschild, 1978; Kehat *et al.*, 1982; Mamedov and Velieu, 1984), resulted in the moth capture before light trap (Cranham, 1980; Porsaev, 1986), improved the timing of insecticidal sprays (Nymbo, 1989) and monitored the population during the remainder of crop season to provide an additional information on the potential oviposition period (Campion and Nesbitt, 1981; Witz *et al.*, 1985).

Trapping of *H. armigera* adult population in chickpea, lucern and cotton crop showed 3 peaks of activity (Tahhan and Hariri, 1982; Sherman and Lokhonya, 1982) and present of a single peak in December-January in tobacco crop (Chari *et al.*, 1985). In chickpea and pigeonpea crop, maximum activity of moth occurred during March-April in U.P. The larval density preceded the peak in moth catches. Accordingly, control measures were suggested at population of 1-2 larvae/m row in chickpea and 1-2 larvae/plant in pigeonpea (Lal *et al.*, 1985). Moth catches in pigeonpea in Tamil Nadu showed peaks in February and April and October and December. (Balasubrimaniam *et al.*, 1985) while population of *H. armigera* in the pheromone traps reached its peak in July to August in tomato field in U.S.S.R. (Eletskii, 1988).

Appearance of moths of *H. armigera* in the pheromone traps started from the first week of January till last week of April in chickpea field with peak number of >100 moth/trap/week caught from the 4th week of February to first week of March (Sah *et al.*, 1988). During the peak period of the pest, the chickpea crop was either at full blooming stage or at the initiation of podding. Yadav and Lal (1988) also reported two peaks in the population during the 11th to 15th and 47th to 50th standard week. Pattern of trap catches of *H. armigera* in India is characterized by almost similar pattern in any given agro climatic zone though catches were generally higher and had more sharply defined peak at northern locations than at southern locations. The catches at eastern locations were lower than those reported elsewhere (Srivastava *et al.*, 1990). In Pakistan maximum catches were recorded in the first fortnight of April. However, in the southern region, the pest was trapped throughout the growing season with no obvious peak (Srivastava *et al.*, 1991). Kononenko *et al.* (1986) reported that more than 30 adult males captured in pheromone traps over 3 days is to be followed with control measures against larvae 3-6 days after trapping.

Table 3.1: Components of Sex Pheromone of *H. armigera*

Sl.No.	Components of Sex Pheromone of H. armigera	References
1.	(Z)–11 hexadecenal	Piccardi, 1977; Soroker, 1989
2.	(Z)–9 tetradecenyl formate (Z)–11 hexadecenal	Gothilf, 1978
3.	(Z)–11 hexadecenal (Z)–9 tetradecenyl (in 20:1 ratio)	Rothschild, 1978
4.	(Z)–11 hexadecenal (Z)–I I tetradecenal	Gothilf, 1978b; Gothilf *et al.*, 1979
5.	(Z)–11 hexadecenal (Z)–11 hexadecen-1-o1	Nesbitt, *et al.*, 1979
6.	(Z)–11 hexadecenal 90-99 per cent (Z)–9 hexadecenal 1-10 per cent	Kehat *et al.*, 1980
7.	CiS–9 tetradecenal CiS–11 hexadecenal	Konyukhov *et al.*, 1980 Minyailo and Kovalev, 1980
8.	(Z)–9 hexadecenal 3 per cent (Z)–11 hexadecenal 87 per cent Hexadecenal 4 per cent Hexadecenal 6 per cent	Dunkelblum *et al.*, 1980
9.	(Z)–11 hexadecenal (Z)–11 hexadecen-1-o1 (Z)–9 hexadecenal Hexadecenal 1–hexadecenal	Nesbitt *et al.*, 1980
10.	3, 7 dimethyl 6 Octenal (Z)–11 hexadecenal (8E, 10E) 8, 10 dodecadienal with Girard reagent	Vrkoch *et al.*, 1988
11.	(Z)–11 hexadecenal (Z 11-16: Ald) (Z)–9 hexadecenal (Z9-16: Ald) Hexadecanal (Z)–11 hexadecenal (Z 11-16:OH) (Z)–7 hexadecenal (Z7-16: Ald) (Z)–9 tetradecenal (Z9-14: Ald)	Kehat and Dulkelblum, 1990

Pheromone Traps for Studies on the Distribution of *Helicoverpa armigera* Moth

The distribution of *H. armigera* and *H. punctigera* through trap catches was found to be highly clumped, both temporarily and spatially. Pattern of dispersion was species specific and reflected density dependent pattern of movement towards and away from center of abundance (Fitt *et al.*, 1989). In India abundance of *H. armigera* was found to be similar in any given agro climatic zone but there was an obvious change with latitude in pattern of trap catches. Different locations in any given agro climatic zone were characterized by almost similar pattern of trap catches (Srivastava *et al.*, 1990).

Factors Affecting Trapping Efficiency

In Lepidoptera, when adult moths are highly mobile, mass trapping in any limited area would probably be masked by the immigrants and mated females from outside the trapped area. Therefore, use of sex pheromones would be best utilized for monitoring of pest population (Pawar *et al.*, 1983). The catches of male moths of *H. armigera* were reduced when the trap was baited with the lure of two noctuids *Spodoptera litura* and *H. armigera* together than that of a trap containing lure of only one species (Pawar and Srivastava, 1986). Catches of males of *H. armigera* in pheromone trap were reduced when the field saturated with slow release formulation of pheromone at the rate of 100 mg/ha (Bulyginskaya *et al.*, 1989) while crop protection system in filed showed no effect on trapping efficiency although an increase in number of male catches was observed following treatment of cotton with "Phasalone". (Parsaev, 1986, Krishna Kant *et al.*, 1998).

Effect of Septa and Pheromone Load on Trapping

The use of pheromone formulations and their dose are critical. Lures used in monitoring trap should release attractant at a constant rate throughout the trapping period whether or not it is necessary to change lure regularly during the trapping period depends upon the formulation and response characteristic of an species (Wall and Perry, 1981; Senders, 1984; Wall, 1989). The loading of pheromone components of *H. armigera* on two different septa showed that dental roll needed only 0.5 mg/trap for sufficient trapping of *H. armigera* whereas rubber septa required 5mg for sufficient catches (Gothilf *et al.*, 1979). A rubber septa at a load of 2 mg effectively attracted males of *H. armigera* for 31 days (Kehat *et al.*, 1980). Small rubber septa with a load of 2 mg of sex pheromone were most attractive and were adopted as standard lure for 4 weeks of trapping (Pawar *et al.*, 1988). At Pant nagar loading of 1 mg/septum catches maximum number of male moths (71.16 moths/trap/night) in chickpea crop, also rubber septa (2.0 × 1.5 cm) were most efficient than cork, cigarette filter and filter paper (Krishna Kant *et al.*, 1998).

Yen *et al.* (1989) reported that attractant of *H. armigera* at 4 mg/trap was most effective and lasted for 730 days and little difference was observed between plastic and rubber dispensers. For mating disruption in tomato field, 5mg pheromone per trap was sufficient. Sinha and Mehrotra (1992) reported that one rubber septum with 5 mg lure was as effective as six rubber septa (30 mg) or one septa with 30 mg lure. Izquierdo *et al.* (1992) tested six pheromone dispensers for capturing males of *H. armigera* and found significant differences between dispensers of Agrisense and Bioprox of which the latter was most effective.

Mc Kibben *et al.* (1980) found that commercially available polyester coated filter was most effective formulation tested against boll weevil eradication trial in North Carolina. Cellulose acetate filter in glass vials was used in the development of a monitoring system for *H. virescens* incorporating the two components (Z)-11 hexadecenal and (Z)-9 tetradecenal virelure attractant (Hendricks *et al.*, 1977; Shaver *et al.*, 1981). A mixture of cis-11 hexadecenoic and cis-11 tetradecenoic acid (3:1), one plastic tube septum (15mm in length and 8mm in diameter) at a loading of 2000 ug, trap proved most effective for trapping of male moth (Sattar-Zade *et al.*, 1981). A laminated plastic lure loaded with appropriate pheromone blend provided higher and prolonged attraction of *H. armigera* and *H. punctgera* (Wilson, 1984).

Effect of Trap Design, Height and Distance on Trapping

Among the tunstall water trap and montedison sticky trap tested against *H. armigera*, water trap proved most effective and pheromone remained active for 3 months while it lasted only 1 month in sticky trap (Bourdouxhe, 1982). Trapping performance of two different traps (Delta trap and Italian trap) and two dispensers (Hollow cone harcone dispenser and rubber septum) showed little difference but more males were caught in delta trap than in Italian trap (Pawar *et al.*, 1984). A hollow cane (Texas) trap gave higher catches than the standard dry funnel trap for *H. armigera*, and *H. punctigera* trapping. Dry funnel was, however, more robust and less expensive and more effective than sticky trap (Wilson, 1984). A while funnel trap was more efficient and was adopted as the ICRISAT standard trap. It was also reported that trap placement just above the crop canopy (at 3 meter height) in sorghum, millet, pigeonpea, chickpea and groundnut caught more moths than other heights (1 or 2 meter) (Pawar *et al.*, 1988). Sinha and Mehrotra (1992) reported that double cone trap (Texas model) caught 3 times higher *H. armigera*, moths than sleeve trap. Improved sleeve traps were also as effective as double cone. Trap mounted at a height of 1.0-2.2 meter above the ground captured significantly more moths than the trap fixed at 0.4m. Nandihalli *et al.* (1991) reported that sleeve type of trap yielded highest number of catches (105 adults/week) than funnel trap (73.25 adults/week) whereas sticky traps were considered to be ineffective (2.63 adults/week). In chickpea pheromone trap (funnel trap, texas trap and sticky trap) placed at 1.8 meter height cached maximum number of moths of *H. armigera* than trap placed at 1.2 and 0.6 meter height. At initial build up of male moth population distance between two trap at 50 meter cached more moths than 100 meter distance. However at peak of moths activities there was no significant difference was observed (Krishna Kant *et al.*, 1999).

Interaction of Pheromone Vs Light Trap

Pheromone traps were found to have several advantages over light trap as the former detected the presence of low density nocturnal species prior to letter (Cranham, 1980; Srivastava and Srivastava, 1989). A significant positive correlation existed between pheromone and light trap catches. Pheromone traps caught 1.9 times as many males as light trap (Balasubriamainan *et al.*, 1985) while Pawar and Srivastava (1986) reported light traps to be more attractive than pheromone trap and catches of male and female of *H. armi*gera in light trap increased when the pheromone traps were operated within 20 meter of light traps. Moth catches in pheromone trap were higher in northern India compared to southern India whereas intermediate correlations were observed for traps located in the central part of India (Srivastava *et al.*, 1992).

Virgin Female Trap

Traps containing virgin female of *H. zea* were used to monitor population on the St Croix, US Virgin islands (Snow *et al.*, 1968). Hendricks *et al.* (1972) showed utility of *H. viriscens* virgin female as bait in trap. Comparison of population index of *Heliothis* spp. was made in the light Vs virgin female trap (Snow *et al.*, 1972; Hendrick *et al.*, 1973). Roach (1975) showed that the males of *H. virescens* were not very responsive to light trap while *H. zea* responded well to light trap and reflected seasonal fluctuation of population of *H. zea* more accurately than virgin female trap. However, the virgin female traps were more sensitive at low population early in the season and declined in efficiency with higher populations late in the season.

Virgin female of both *H. virescens* and *H. zea* placed together in trap resulted in catch reduction of >4.2 per cent for *H. virescens* and > 7.5 per cent for *H. zea* over catches made with single species lure (Haile *et al.*, 1973). Pheromone trap baited with synthetic compound captured 2-5 times as many moths over light traps and about twice as many as a virgin female trap (Rotundo and Tremblay, 1985).

A trapping performance of *H. armigera* through virgin female trap in tomato crop showed that water trough trap captured more moths than sticky paper trap and pet bottle trap. Trap height at 1 meter above the ground caught higher moths than other heights. Female of 3-7 days old baited trap captured highest number of male moths to which 4-5 days old male moths responded more than others. It was suggested that the traps baited with 5 and 15 virgin female/trap could be used for trapping. (Annon, 1987).

In Flight Orientation to Odour Source

Males of *S. litura* oriented to pheromone source in fluctuating wind condition showed that males approached intermittently the trap. The largest period when none of the moths attracted was 5-6 min (Nakamura and Kawasaki, 1984). Orientation of *Cydia modesta* and *H. virescens* in laboratory wind tunnel and in field revealed zigzag orientation flight of males (Keyserlingk and Keserlingk 1984).

Orientation of male potato tuber moth, *Phthorimaea operculella*, toward pheromone source consisted of a chain of short flights. The pattern of flight changed when male came within 1 meter of lure, with a decrease of flight speed and an increase in tortuousity (Ono and Ito, 1989). Under field conditions, the male of *H. armigera* responded to pheromone anemotactically. In the vicinity (~ 1m) of pheromone, the male oriented by zigzag flight pattern, showing 8-10 intermittent flight with wing fluttering each lasting to 5 to 10 seconds fallowed by resumption of short flights. On approaching to pheromone source, the male encircled the septum repeatedly (10 to 15 times) and the sequence culminated in copoulatory attempts on pheromone source that led to falling in the trap. The orientation period ranged from 7 to 10 minutes/moth (Krishna Kant *et al.*, 1999).

Relationship Between Pheromone Trap Catches of *H.armigera* Males and Larval Population

Significant correlation was found between male catches of *H. virescens* in the trap and larval infestation and damage level in tobacco one and two weeks after deployment of traps later (Tingle and Mitchell, 1981). Similarly positive correlation was found between total number of *H. zea* and *H. virescens* male moths catches in pheromone trap and oviposition in the field (Roger *et al.*, 1988; Lateef *et al.*, 1991).

A significant positive correlation was found between male moth catches and larval population of *H. armigera* increase of male moths in the trap was usually followed by increase in larval densities in cotton, groundnut and chickpea (Kehat *et al.*, 1982; Yen *et al.*, 1989; Srivastava *et al.*, 1990; Srivastava *et al.*, 1991) the peak moth catches between 8 and 12 standard week (53 to 137moths/trap/week) coincided with the peak larval density from 9 to 14 standard week (10-42 larvae/10m. row length) in chickpea in Kanpur (AICRIP, 1991-92).

In Citrus crop male population of *H. armigera* in trap catches showed significant relationship between larval populations with the size of egg population 3 to 4 week earliar (Newton, 1987). Significant correlation between peak moth catches and oviposition of *S. litura* was showed in soyabean (rainy) and sugarbeet (post rainy) field in Pantnagar (Singh and Sachan, 1991).

Nymbo (1989) reported that there was a one week drift between the timing of peak moth catches of *H. armigera* and the increase in egg density in 2 to 3 season of cotton crop. However, the relationship between moths catches larval population was not clear. The relationship between egg and larval population varied between season and sites. Wilson (1989) observed a poor relationship between trap catches of *H. armigera* and egg count on cotton crop in North West New South Wales USA.

Effect of Weather Factors on the Pheromone Trap Catches

Temperature

Maximum, minimum and average temperature affected directly the moth catches in pheromone trap. The temperature affected both the flight activity of male and release rate of pheromone from the trap. However, there was no evidence of any relationship of lower temperature threshold and catches in pheromone trap. (Rothschild *et al.*, 1981; Dent and Pawar, 1988; Patil *et al.*, 1992; Sinha and Jain 1992) Pheromone trapping of wild species of *S. litura*, *H. armigera* and *Chilo partellus* showed no correlationship with any of the weather factors although, trap catches give some indication of seasonal and annual variation in the population. The catches almost varied with change in temperature, wind velocity and others agro climatic factors (Pawar *et al.*, 1984). A negative correlation was found in trap catches of *Pectinophora gossipiella* and minimum temperature (Jayaraj *et al.*, 1988) Krishna Kant *et al.*, 2004 observed none of any environmental factors affects significantly on the development of adult moths of *H. armigera* in chickpea crop.

Relative Humidity

Relative humidity both morning and evening showed significant effect on trap catches of *H. armigera*, *P. gossypiella* and *Earias vitella*. Moth of *H. armigera* were not captured in the trap below 80.3 per cent relative humidity. (Afral *et al.*, 1985; Patil *et al.*, 1992; Sinha and Jain, 1992). An increase in morning humidity by 1 per cent reduce the male moth catches between 0.66 per cent (Jayaraj, *et al.*, 1988).

Rainfall: Rothschild *et al.* (1991) showed significant effect of trap catches of *H. armigera* in cotton crop. A positive correlation existed between rainfall deficit and out break of *H. armigera* in South India. (Srivastava *et al.*, 1990.

Wind Velocity

Wind speed enhanced the trapping significantly of *H. armigera* when trap were downward of the crop (Rothschild *et al.*, 1981; Umapathy *et al.*, 1988). Morten *et al.* (1981) found an exponential decline in catches at wind speed above 1.7m/s.

References

Afzal, M., Ahmed, K., Azhar Hussain, S., Malik, B.A. and Kalique, F., 1985. Effect of environmental factors on pheromone trap Catches of *Heliothis armigera* (Hubn.). *Chemosphere*, 14: 1921–1924.

AICPIP (All India Coordinated Project on Improvement of Pulses), 1991–92. *Annual Report on 'Rabi' Pulses (Entomology)*. Directorate of Pulses Research, ICAR, Kanpur, India.

Anonymous, 1987. Asian Vegetable Research and Development Center. Mass Rearing and Preliminary sex pheromone studies on tomato fruit worm. In: *Progress Report* 1985, Shanhua, Taiwan, pp. 110–113.

Bakke, A. and Lie, R., 1989. Mass tapping. In: *Insect Pheromone in Plant Protection*, (Eds.) Justum, A.R. and R.F. Gorden. John Wiley and Sons, New York, pp. 67–83.

Balasubramanian, G., Chelliah, S. and Balsubramanian, M., 1985. Comparison of *Heliothis armigera* moth catches in pheromone and light trap at Coimbator, India. *International Pigeon pea News Letter*, 4: 48–50.

Bourdouxhe, L., 1982. Comparison of two type of trap for sex trapping of *Heliothis armigera* in Senegal. *Plant Protection Bulletin FAO*, 3(3/4): 131–136.

Butenandt, A., Beckmann, R., Stamm, D. and Hecker, E., 1959. Uber den sexual-Lockstaff des Seidenspinness Bombyx mori. Reidanstellung und Konstitution Z. *Naturforsch B.*, 14: 283–284.

Bulyginskaya, M.A., Grichanov, I.Ya. and Shamshev, I.V., 1989.Use of synthetic sex attractant for disturbance of the olfactory communication of adult of the cotton moth *Heliothis armigera* Hb. (Lepidoptera: Noctuidae) in field conditions. *Entomologicheskoe Obozrenie*, 68(2): 273–275.

Campion, D.G., 1989. Mating disruption. In: *Insect Pheromone in Plant Protection*, (Eds.) Justum, A.R. and R.F. Gorden. John Wiley and Sons, New York, pp. 89–116.

Campion, D.G. and Nesbitt, B.F., 1981. Lepidopteran sex pheromone and pest management in developing countries. *Tropical Pest Management*, 27: 53–61.

Chari, M.S., Patel, A.R., Rao, B.S., Bharpoda, T.M. and Patel, N.M., 1985. Population studies on tobacco capsule borer *Heliothis armigera* (Hubner). *Tobacco Research*, 11(2): 98–104.

Cranham, J.E., 1980. Monitoring codling moth (*Cydia pomonella* L.) with pheromone trap. *Bulletin OEPP.*, 10: 105–108.

Dent, D.R. and Pawar, C.S., 1988. The influence of moonlight and weather on catches of *Helicoverpa armigera* (Hubner) (Lepidoptera: Noctuidae) in light and pheromone trap. *Bull. Entomol. Res.*, 78: 365–377.

Dunkelblum, E., Gothilf, S. and Kehat, M., 1987. Identification of sex pheromone of the cotton boll worm, *Heliothis armigera* in Israel. *Phytoparasitica*, 8: 209–211.

Elatskii, S.V., 1988. The use of sex pheromone to observe some noctuid causing damage to tomatoes. In: *Feromoney nasekomykh irazrabotka putei ikh prakticheskogo ispolzovanoya*, (Ed.) A.P. Sazonov. Leningrad, USSR, pp. 50–57.

Fitt, G.P., Zaluki, M.P., Twine, P., Taylor, L.R. and Woiwood, I.P., 1989. Temporal and Spatial pattern in pheromone trap catches of Helicoverpa spp. (Lepidoptera : Noctuidae) in cotton growing areas of Australia. *Bull. Entomol. Res.*, 79(1): 145–161.

Georghion, G.P., 1990. Resistance potential to biopesticides and consideration of countermeasures. In: *Insecticides and Alternatives: Innovative Chemicals and Biological Approaches to Pest Control*, (Ed.) John E. Cesida. Elsevier, New York, pp. 409–420.

Georghion, G.P. and Taylor, C.E., 1977. Operational influence in the evaluation of insecticides resistance. *J. Eco. Entomol.*, 70: 653–658.

Gothilf, S.M., Kehat, M., Dunkelblum, E. and Jacobson, S.M., 1979. Efficacy of (Z)-11-hexadecenal and (Z)-11-tetradecenal as sex attractant for *Heliothis armigera* on two different dispenser. *J. Eco. Entomol.*, 72: 718–720.

Gothilf, S.M., Kehat, M., Jacobson, M. and Gulan, R., 1978a. Sex attractant for male *Heliothis armigera* (Hbn.). *Expermentia*, 34: 863.

Gothilf, S.M., Kehat, M., Jacobson, M. and Gulan, R., 1978b. Screening pheromone analogue by EAG technique for biological activity on male of *Earias insulana, Heliothis armigera* and *Spodoptera littoralis. Environmental Entomol.*, 7: 31–35.

Haile, D.G., Snow, J.W. and Goodenough, J.L., 1973. Reduced capture of tobacco bud worm and corn earthworm male in electrical grid traps baited simultaneously with virgin female of both species. *J. Eco. Entomol.*, 68: 739–740.

Hendrick, D.E., Graham, H.M., Guerra, R.J. and Hartstack, A.W. Jr., 1972. Catches of tobacco budworm moths influenced by colour of sex lure trap. *Environmental Entomology*, 1: 48–51.

Hendrick, D.E., Graham, H.M., Guerra, R.J. and Perez, C.T., 1973. Comparison of number of tobacco budworm and bollworm caught in sex pheromone traps vs. black light trap in the lower Rio Grande valley, Texas. *J. Eco. Entomol.*, 64: 361–364.

Hendrick, D.E., Hartstack, A.W. and Shaver, T.N. 1977. Effect of formulations and dispensers on the attractiveness of virelure to the tobacco budworm. *J. Chemical Eco.*, 3: 497–506.

Holbrook, R.F., Beroza, M. and Burgess, E.D., 1960. Gypsy moth *Porthetria dispar* (L.) detection with natural female sex lure. *J. Eco. Entomol.*, 53: 751.

Izquierdo, J., Sampol, K. and Sorribas, X., 1992. Comparison of capsule of sex pheromone of *Heliothis (Helicoverpa) armigera* (Hubner) (Lepidoptera: Noctuidae). *Boletin de Sanidad Vegetal Plagas,* 18(2): 427–434.

Jadhav, D.R., Bhatnagar, V.S. and Pawar, C.S., 1985. The species status of *Heliothis armigera* (Hub.) (Lepidoptera: Noctuidae) in Andhra Pradesh, India on the basis of aedegal cornutal spines. *Current Science,* India, 54(4): 239–240.

Jayaraj, S., Uthamaswamy, S., Gopalan, M. and Rabindra, R.J., 1988. *Heliothis* management. In: *Proceeding of the National Workshop*, Center of Plant Protection Studies, Tamil Nadu Agricultural University, Coimbator, India.

Kehat, M. and Dunkelblum, E., 1990. Behavioural response of male *Heliothis armigera* (Lepidopera: Noctuidae) moth in flight tunnel to combination of the compound identified from female sex pheromone gland. *J. Insect Behaviour*, 3(1): 75–83.

Kehat, M., Gothilf, S., Dunkelblum, E. and Greenburg, S., 1980. Field evaluation of female sex pheromone component of cotton bollworm, *Heliothis armigera. Entomologia Experimentalis de Applicata*, 27(2): 188–193.

Kehat, M., Gothilf, S., Dunkelblum, E. and Greenburg, S., 1982. Sex pheromone trap as a means of improving control programme for the cotton bollworm *Heliothis armigera* (Lepidoptera: Noctuidae). *Environmental Entomol.*, 11(3): 727–729.

Keyserlingk, H.C. Von. and Van Keserlingk, H.C., 1984. Close range orientation of flying Lepidoptera to pheromone source in a laboratory wind tunnel and in the field. *Mededelingen Van de Faculteit Landbouwvetenschappen Rijksuniversiteit*, 49(3a): 683–689.

Kononenko, A.P., Grichanov, I.Ya, Karov, E.I. and Maiorov, V.I., 1986.Correlation between trap catches of adult of cotton moth and abundance of pre adult stage. *Izestiya Akademii Nauk Tadzhikshoi USSR Biologicheskikh Nauk,* 3: 51–55.

Konyukhov, V.P., Kavalev, B.G., Minyilo, V.A., Stan, V.V. and Oprunenko, Yu.E., 1980. The synthesis of cis-9 tetradecenol and cis-11 hexadecenal and determination of their effect on cotton bollworm using electroantennogram method. *Khemoretseptsiya Nasekomykh*, 3: 37–40.

Krishna Kant, Kanaujia, K.R. and Kanaujia, S., 1998. Influence of pheromone dispensers, loading and insecticides on the trapping of the *Helicoverpa armigera* (Hubner). *Shashpa*, 5(1): 77–82.

Krishna Kant, Kanaujia, K.R. and Kanaujia, S., 1999. Rhythmicity and orientation of *Helicoverpa armigera* (Hubner) to pheromone and influence of trap design and distance of moth trapping. *J. Insect Sci.*, 12(1): 6–8.

Krishna Kant, Kanaujia, K.R. and Kanaujia, S., 2004. Population assessment of *Helicoverpa armigera* moth through sex pheromone trap. *Ann. Pl. Protec. Sci.*, 12(2): 431–432.

Lal, S.S., Sachan, J.N. and Yadava, C.P., 1985. Sex pheromone trap a novel tool for monitoring gram pod borer population. *Plant Protect. Bull., India*, 37(3–4): 3–5.

Lateef, M.A., Witz, J.A. and Lopez, J.D., 1991. Relationship among pheromone trap catches of male corn earworm moths, egg numbers and phenology in corn. *Canadian Entomologist.*, 123(2): 271–281.

Mamadov, M.S. and Velieu, A.Yu., 1984. A complex system in action. *Zashchita rastenii*, 8: 10–12.

Mbato, G.M. and Ramaswamy, S.B., 1990. Rthymicity of sex pheromone content in female *Heliothis virescens* impact of mating. *Physiological Entomology*, 15(4): 423–432.

Mbato, G.M. and Ramaswamy, S.B., 1991. Relationship between pheromone titer and mating in *Heliothis virescens* F. (Lepidiptera: Noctuidae). *J. Applied Entomol.*, 111(3): 237–242.

McKibben, G.H., Johnson, W.L., Edwards, R., Kearny, J.F., Davich, T.B., Lloyd, E.P. and Gayard, M.C., 1980. A polyester wrapped cigarette filter for dispensing grandlure. *J. Economic Entomol.*, 73: 250–251.

Metcalf, R.L., 1986. The ecology of insecticides and the chemical control of the insects. In: *Ecology Theory and Integrated Pest Management Practices*, (Ed.) Marcos Kogen. John Wiley and Sons, New York, pp. 251–292.

Minyailo, V.A. and Kovalev, B.G., 1980. Electroantennogram of adult of the cotton bollworm in response to a number of aliphatic acetate, aldehyde and alchocals. *Khemeretseptsiya Nasekomykh*, 3: 103–106.

Morton, R., Iuart, L.D. and Wardhaugh, K.C., 1981. The analysis of standardization of light trap catches of *Heliothis armigera* (Hubner) and *H. puntigera* Wallengren (Lepidoptera: Noctuidae). *Bull. Ent. Res.*, 71: 207–295.

Nakamura, K., and Kawasaki, 1984. Male catches of *Spodoptera litura* (F) in pheromone trap under fluctuating wind velocity: Validation of active space model. *Applied Ent. Zool.*, 19(2): 192–201.

Nandihalli, B.S., Hugar, P. and Patil, B.V., 1991. Efficacy of three type of pheromone trap in trapping *Heliothis armigera* Hubner. *Madras Agric. J.*, 78(1–4): 91–92.

Nesbitt, B.F., Beevor, P.S., Hall, D.R. and Lester, R., 1979. Female sex pheromone components of the cotton boll worm, *Heliothis armigera* (Hubner). *J. Insect Physiol.*, 95: 531–541.

Nesbitt, B.F., Beevor, P.S., Hall, D.R. and Lester, R., 1980. (Z)-9 Hexadecenal: A minor component of the female sex pheromone of *Heliothis armigera* (Hubner) (Lepidoptera: Noctuidae). *Entomologia Experimentalis de Applicata*, 27(3): 306–308.

Newton, P.J., 1987. The potential of synthetic sex pheromone trap for monitoring out break of *Heliothis armigera* (Hubn.) (Lepidoptera: Noctuidae) on Citrus. *Citrus and Sub tropical Fruit J.*, 634: 7–10.

Nymbo, B.T., 1989. Assessment of pheromone trap for monitoring and early warning of *Heliothis armigera* (Hubner) in western cotton growing areas of Tanzania. *Crop Protect.*, 3: 188–192.

Ono, T. and Ito, M., 1989. Pattern of pheromone oriented flight in male potato tuber moths. *J. Chemical Ecology,* 15(9): 2357–2368.

Patil, B.V., Nandihalli, B.S., Hugar, P. and Somasekar, 1992. Influence of weather parameter on pheromone trap catches of cotton bollworm. *Karnataka J. Agric. Sci.*, 5(4): 346–350.

Pawar, A.D., Prasad, J. and Singh, Raj, 1984. Field evaluation of traps and sex pheromone of *Heliothis armigera* Hb. (Lepidoptera: Noctuidae) and effect of release of parasite on male moth catch in pheromone trap. *Plant Proctect. Bull., India*, 36(4): 29–30.

Pawar, C.S., Sithanantham, S., Bhatnagar, V.S., Srivastava, C.P. and Reed, W., 1988.The development of sex pheromone trapping of *Heliothis armigera* at ICRISAT, India. *Tropical Pest Mang.*, 34: 39–43.

Pawar, C.S., Srivastava, C.P. and Reed, W., 1983. Pheromone trap network for *Heliothis armigera* in India. *Int. Chickpea Newst.*, 2: 60–61.

Piccardi, P., Capizzi, A., Cassani, G., Spnelli, P., Arsura, E. and Massordo, P., 1977. A sex pheromone component of the old world bollworm *Heliothis armigera*. *J. Insect Physiol.*, 23(11/12): 1443–1445.

Porsaev, M.M., 1986. The use of sex attractant to monitor the state of population of the cotton moth *Helicoverpa armigera* (Hubner) Trudy. *Vsesoyuznogo Entomologicheskogo Obshchestva*, 68: 148–152.

Rafaeli, A. and Sorokar, V., 1989. Influence of dial rhythm and brain hormone on pheromone production of two lepidopteron species. *J. Chemical Ecol.*, 15(2): 447–455.

Roach, S.H., 1975. *Heliothis zea* and *H. virescens* moth activity as measured by black light and pheromone traps. *J. Economic Entomol.*, 68: 17–21.

Roger, B., Jerry, B., Graves, E. and Burris, 1989. *Heliothis* spp. (Lepidoptera: Noctuidae) captured in pheromone traps: Species composition and relationship to oviposition in cotton. *J. Economic Entomol.*, 82: 574–579.

Rothschild, G.H.S., 1978. Attractant for *Heliothis armigera* and *H. punctigera*. *J. Australian Entomol. Soc.*, 102: 389–390.

Rothschild, G.H.S., Wilson, A.G. and Malafant, K., 1981. Preliminary studies on the female sex pheromone of *Heliothis* species and their possible use in control programme in Australia. In: *Proceeding of the International Workshop on Heliothis Management*, (Eds.) W. Reed and V. Kunble, ICRISAT Centre, Patencheru, A.P., India, 15–20 November, pp. 319–328.

Rotundo, G. and Tremblay, E., 1985. Field evaluation of the female sex pheromone of the artichoke moth, *Gortyna xanthenes*. *Entomologia Experimentalis de Applicata*, 38(3): 201–204.

Sah, L.N., Sahu, R. and Neupane, F.P., 1988. Monitoring of chickpea pod borer, *Heliothis armigera* (Hubner) by a pheromone trap. *J. Inst. Agric Ani. Sci., Nepal*, 9: 107–109.

Sanders, C.J., 1984. Sex pheromone trap and lures for monitoring spruce bud worm population: The Ontario experience. *USDA Forest Service General Tech. Rep.*, 88: 17–22.

Sattar-zade, N.R., Konyukhov, V.P. and Kovalev, B.G., 1981. Attractant for cotton noctuid. *Zashehita Rastenii*, 5: 31.

Shaver, T.N., Hendricks, D.E. and Hartstack, Jr.A.W., 1981. Dissipation of virelure: A synthetic pheromone of the tobacco budworm from laminated plastic and cigarette filter dispensers. *South Western Entomologist*, 6: 205–210.

Sherman, L.V. and Lokhonya, R.N., 1982. The pheromone method for the Lucerne surveiliance. *Zashehita Rastenii*, 1: 47.

Shorey, H.H., Gasten, L.K. and Kaae, R.S., 1967. Air permission with gossyplure for control of the pink bollworm. In: *Pest Management with Insect Sex Attractant*, (Ed.) M. Beroza. Ser. 23, American Chemical Society, Washington, DC, pp. 67–74.

Singh, K.N. and Sachan, G.C., 1991. Relationship between pheromone trap catches of male *Spodoptera litura* and oviposition in soybean and sugarbeet at Pantnagar, Uttar Pradesh, India. *J. Plant Protect. Tropics*, 8(2): 121–124.

Sinha, S.K. and Jain, D.K., 1992. Field studies in pheromone trap catches of *Helicoverpa armigera* Hub. Monitoring seasonal abundance and larval infestation of chickpea and the effect of climate on trap catches. *Indian J. Plant Protect.*, 20(2): 149–157.

Sinha, S.N. and Mehrotra, K.N., 1992. Factors influencing pheromone trap catches of Helicoverpa armigera (Hubner). *Insect Science and its Application*, 1: 1–6.

Snow, J.W., Cantelo, W.W., Burton, R.L. and Hensley, S.D., 1968. Population of fall armyworm, corn earworm and sugarcane borer on St. Croix, US Virgin Island. *J. Eco. Entomol.*, 61: 1757–1760.

Snow, J.W., Sparks, A.N. and Lewis, W.J., 1972. Seasonal capture of corn earworm adult in light traps near Tiflon, Georgia: Compared to in trap baited with virgin females. *J. Georgia Entomol. Soc.*, 7: 85–89.

Sorokar, V. and Rafaeli, A., 1989. *In vitro* hormonal stimulation of (14 C) acetate incorporated by *Heliothis armigera* pheromone glands. *Insect Biochemistry*, 9(1): 1–5.

Srivastava, C.P., Pimbet, M.P. and Ahmed, K., 1991. Monitoring of *Helicoverpa armigera* (Hubner) in Pakistan with pheromone traps. *Int. Chickpea News.*, 25: 20–22.

Srivastava, C.P., Pimbert, M.P. and Reed, W., 1990. Spatial and temporal changes in the abundance of *Helicoverpa (= Heliothis) armigera* (Hubner) in India. *Insect Sci Application*, 11(6): 869–876.

Srivastava, C.P., Pimbert, M.P. and Reed, W., 1992. Monitoring of *Helicoverpa (= Heliothis) armigera* (Hubner) moths with light and pheromone trap in India. *Insect Sci Application*, 13(2): 205–210.

Srivastava, C.P. and Srivastava, R.P., 1989. Comparison of *Heliothis armigera* (Hubner) male moth catches in light and pheromone traps at Udaipur, Rajasthan, India. *Insect Sci Application*, 10(5): 565–568.

Tahhan, O. and Hariri, G., 1982. Preliminary studies of trapping *Heliothis armigera* (Hb.) with pheromone at ICARDA, Syria. *Int. Chickpea Newst.*, 6: 19–20.

Tringle, F.C. and Mitchell, E.R., 1981. Relationship between pheromone trap catches of male tobacco budworm larval infestation and damage level in tobacco. *J. Eco. Entomol.*, 74(4): 437–440.

Umapathy, G., Balasubramanian and Gopalan, M., 1988. Influence of abiotic factors on the incidence of *Heliothis armigera* and *Spodoptera litura* (Noctuidae) and their natural enemies. In: *National Workshop on Diseases Surveillance for IPM*, 16–17 September, TNAU, Coimbatore.

Vrkoch, Y.A., Shrants, L., Romanyuk, M., Konyukhov, V.P. and Kavalev, V.G., 1988. Pheromone derivatives with an aldehyde group, Propheromone. *Khemoretseptsiya Nasekomykh,* 10: 21–24.

Wall, C., 1989. Monitoring and spray timing. In: *Insect Pheromone in Plant Protection,* (Eds.) Jutsum, A.R. and R.F.S. Gordon. John Wiley and Sons, New York, pp. 39–40.

Wall, C. and Perry, J.N., 1981. Effect of dose and attractant on interaction between pheromone trap for pea moth, *Cydia nigrisigna* (F.). *Entomologia Experimentalis de Applicata,* 30: 26–30.

Wilson, A.G.L., 1984. Evaluation of pheromone trap design and dispenser for monitoring *Heliothis punctigera* and *H. armigera* In: *Proceeding of Fourth Australian Applied Entomological Research Conference,* Adelaide, 24–28 September, pp. 74–81.

Wilson, A.G.L., 1989. Some factor affecting the reliability of pheromone trap for measurement of relative abundance of *Heliothis punctigera* (Wallengren) and *H. armigera* (Hubner) (Lepidoptera: Noctuidae). *Bull. Entomol. Res.,* 79: 265–273.

Witz, J.A., Hartstack, A.W., King, E.G., Dickerman, W.A. and Phillips, J.R., 1985. Monitoring and prediction of *Heliothis* spp. *Southwest Entomologist,* 8: 56–70.

Yadava, C.P. and Lal, S.S., 1988. Relationship between certain abiotic and biotic factors and the occurrence of gram pod borer *Heliothis armigera* (Hub.) on chickpea. *Entomon.,* 13(3–4): 259–263.

Yen, F.C., Chang, S.H. and Huang, T.F., 1989. Application of sex pheromone of tomato fruit worm (*Helicoverpa armigera*) in the field. *Research Bulletin,* Taiwan District Agriculture Improvement Station, 23: 13–15.

Insect Pest and Disease Management *Pages* ***49–70***
Editor: **Dr. D. Prasad**
Published by: **DAYA PUBLISHING HOUSE (2016)**

Chapter 4

Integrated Approach for Management of Major Insect-pests of Sugarcane

M.K. Gupta, A.K. Sarma and K.M. Singh
Department of Entomology, College of Horticulture and Forestry, Central Agricultural University, Pasighat – 791 102, Arunachal Pradesh

Introduction

Sugarcane, *Saccharum officinarum* L., an industrial crop, is an integral component of agricultural economy of India. It is cultivated in over 4 mill ha spread over a wide range of agro-ecological situations, both in tropical and sub-tropical regions and contributes nearly 1.9 per cent of the national GDP (Shahi, 2002). India is the largest producer of sugar and second largest producer of sugarcane in the world. At present, the total production of sugarcane in the country is around 300 mill tones with an average productivity of around 72 tonnes ha^{-1}. By the year 2020, the country will have to produce 415 mill tonnes of sugarcane from 4.2 million hectare with an increased productivity of 100 tonnes ha^{-1} (Shahi, 2002). Although a considerable rise in productivity has been made during past, a jump of 30 million tonnes ha^{-1} within a period of less than two decades is a very big task before us. In this context, high level productivity without impairing the soil environment can be achieved only through well-tested improved production systems, which are less input intensive and environment friendly. The insect pest problem is one of the important limiting factors in the production of sugarcane. In India, insect alone is responsible for the loss of 8.20 to 12.60 per cent in cane yield and 10.76 to 48.55 per cent in sucrose content in endemic area (Butani, 1961). The insect pest associated with sugarcane in India has been listed by many workers. Box (1953) recorded a total of 125 insect species as pest of sugarcane in India. Gupta (1957) classified 18 insects as major and 21 as minor on sugarcane. Butani (1961) reported 32 insect species as major pest out of 50 species recorded on sugarcane crop in Bihar. In subsequent years, additional information has been added from the states of Assam (Avasthy, 1984),

Madhya Pradesh, Orissa, Rajasthan and Tamil Nadu (Kalra, 1967) and Karnataka (Balikal *et al.*, 1998) giving a comprehensive idea on the status, distribution and economic importance of the pests. The various control tactics for managing the major insect-pests of sugarcane are briefly presented in this article.

Changing Scenario of Insect-pests Problems in Sugarcane

Before the advent of improved hybrid sugarcane varieties during the early part of the 20th century, the clones of *Saccharum officinarum* were being grown for hundreds of years in tropical India and the indigenous varieties of *S. barberi* and *S. sinensis* in the north Indian sugarcane belt. The life span of these varieties was long with lower incidence of pest and disease. These indigenous varieties possessed the inherited mechanism of resistance to many pests. By the beginning of 20th century, the hybrid varieties with tremendous yield potential replaced the indigenous ones. But simultaneously there was increase in incidence of insect-pests leading to remarkable crop loss. There was increase in the number of new insect pests on this crop every year due to rapid migration of pests from one region to another.

Fletcher (1920) listed 28 species of insect of sugarcane. Of these, *Scirpophaga xanthogastrella* Walker and *Chilo simplex* were sometimes serious. 79 species were recorded on this crop in 1939 in India and by the year 1953 it increased to 125. In 1964, 43 major insect-pests including 10 species of borer and 19 species of sucking pests have been recorded in different parts of India. At present, more than 200 species have now been recorded feeding on sugarcane in India. There are as many as 25 species of borer (14 lepidopterans and 11 coleopterans), 82 species of sucking pests, 45 subterranean pests and 63 species of defoliators. The new pests recorded during the past decade include *Trochorrhopalus sacchari* Marshal from Nagaland, *Ariboridia* sp. and *Leucopholis lepidophora* Blanchard from Maharastra and *Panchnephorus* sp. from Gujarat (Verma *et al.*, 1993).

Table 4.1: Original Habitat and Spread of Important insect pests of sugarcane in India

Sl.No.	Insect Pest	Original Habitat (1st Report)	Present Distribution
1.	Stem borer, *Chilo infuscatellus* Snellen	Bengal (1957)	All over India
2.	Internode borer, *Chilo sacchariphagus indicus* (Kapur)	South India, (1914)	South India, Bihar, U.P., Maharashtra, Haryana
3.	Stalk borer, *Chilo auricilium* Dudgeon	Bengal (1905)	Bihar, Nagaland, Haryana, Bengal, Himachal Pradesh, Rajasthan
4.	Gurdaspur borer, *Acigona steniella* (Hampson)	Pakistan (1923) Punjab (1925)	Assam, Himachal Pradesh, Haryana, U. P., Punjab, Maharashtra, Rajasthan
5.	Scale insect, *Melanaspis glomerata* (Green)	Bihar (1942)	Andhra Pradesh, Himachal Pradesh, Orissa, Haryana, U. P., Maharashtra, Punjab, Tamil Nadu, Karnataka
6.	Wooly aphid, *Ceratovacuna lanigera* Zehnter	North Eastern India (1974)	North Eastern India, West Bengal, U. P., Sikkim
7.	Plassey borer, *Chilo tumidicostalis* Hampson	Assam (1919)	Assam, Nagaland, Bengal, Bihar, U. P.

Many of the borers and the scale insects have spread to new areas by indiscriminate transport of infested cane and setts (Table 4.1). Wooly aphid, *Ceratovacuna lanigera* Zehnter reported for the first time in 1974 as minor pest of sugarcane in Assam, Nagaland, Sikkim, Tripura, Uttar Pradesh and West Bengal and has spread to Maharashtra and Karnataka posing a serious threat to sugarcane crop in recent years. Similarly, Plassey borer, *Chilo tumidicostalis* Hampson which was confined to Assam, Nagaland, Bengal, Bihar and Orissa has also started spreading to eastern U.P. (Pandey *et al.*, 2005).

Management of Major Insect-pests of Sugarcane

Top Borer, *Scirpophaga excerptalis* Walker

This is considered to be the major pest of sugarcane in Punjab, Haryana, U.P., Bihar, West Bengal, Assam, M.P., Gujarat and some parts of Maharashtra. In Southern states, it was considered as a major pest till 1970, but by the introduction of *Isotima javensis* in late sixties and early seventies from north India, it has now became a minor pest. Incidence of this pest varies from 6–87 per cent, besides complete destruction of about 10–20 per cent shoots attacked during pre-monsoon period (April-June). Infestation by top borer results in a loss of 6.8–40 per cent or 10.0–37.4 tonnes/ha in cane yield and 0.25–3.0 units in sugar recovery (Chaudhary, 1983). The stalk length is reduced by 68.0 and 30.0 per cent, internodes by 67.00 and 29.00 per cent, cane weight by 86.0 and 34.5 per cent and CCS by 25.90 and 13.48 per cent, respectively (Singh *et al.*, 1997). The presence of nitrogen in the mid-ribs, growing points and leaf blades has positive correlation with the incidence of top borer (Chaudhury and Yadava, 1995a). There is decrease in brix (2.10 per cent), pol (4.28 per cent), sucrose (16.26 per cent), glucose (19.55 per cent) and CCS (4.76 per cent) in top borer damaged canes (Gupta and Singh, 1997).

The first two broods of this pest attack young plants before the formation of canes. These plants are killed and are a total loss. In subsequent broods, the pest attacks the terminal portions of the canes causing bunchy tops. Damage by 3rd and 4th broods may result in more than 25 per cent reduction in cane weight and decrease in quality of juice. The loss in cane weight in different varieties vary from 21–37 per cent and loss in sugar recovery is 0.2–4.1 units.

Management

Cultural Control

1. Autumn planting should be preferred to spring one since the former one can stand attack of top borer better because of its more vigorous growth (Kalra, 1990).
2. Harvesting of crop by the middle of February (before the moth emergence from over wintering broods) in sub-tropical India causes considerable reduction in the borer incidence due to reduction of initial incidence of the 1st brood during March–April.
3. Avoidance of frequent irrigation during peak period of moth emergence during summer protects the crop from heavy borer incidence as the high humidity during summer attracts large number of emerging moths in the irrigated fields (Gill, 1963; Kalra, 1980).
4. Suitable drainage to avoid water logging (Gupta, 1959a) and application of higher dosage of nitrogenous fertilizers (Kalra and Sidhu, 1965; Gupta and Roy, 1982).
5. Intercropping of sugarcane with radish, sunflower, cauliflower and turnip causes lower incidence of this pest (Singla *et al.*, 1994). Lower incidence of this pest is also reported under intercropping of sugarcane with various spices crop (Varun *et al.*, 1990; Singh *et al.*, 2003).

Mechanical Control

1. Collection and destruction of egg masses.

2. Cutting of shoots about 5cm below the ground level so as to bring borer larvae along with shoots and their regular destruction from March to June by organizing campaign have been reported to give very effective control of this pest (Gupta, 1959a).
3. Removal of first leaf sheath, dead-heart and killing of larvae.
4. Use of light-trap to collect and kill moths.

Biological Control

Introduction of *Isotima javensis* Rohwer from north India to Southern states have given very effective control of this borer in Andhra Pradesh, Karnataka and Tamil Nadu. This parasitoid has good scope of augmentation even in other sugarcane tracts of rest of the country. The egg parasitoid, *Telenomus dignus* and *Isotima javensis, Rhaconotus scirpophagae* and *Stenobracon nicevillei* are effective to check the population of top borer (Tanwar *et al.*, 1997). Moreover, conservation of coccinellid predator like *Micraspis univittata* (Hope) is also a promising step towards the suppression of top borer (Easwaramoorthy *et al.*, 2001).

Resistant Variety

1. Growing of resistant varieties with higher spindle length (Yadava, 1985) provide considerable protection of the crop.
2. Varieties like CoJ 67, Co 1007, Co 1158, Co 745, Co 1111, Co 1330, Co 7724, E 50, E128 and E 276 (Mukunthan, 1986), WL 22 (Gupta, 1997), CoJ 73, CoH 3, Co 6307, CoH I4, Co S 767 and CoH 20 (Chaudhury and Yadav, 1995b) and CoH 94201 (Bhoopathi and Karnataka. 2003) have been identified as highly resistant to this borer and have good scope of their cultivation in top borer prone area.

Chemical Control

The pest has been effectively controlled by the application of granules of Carbofuran @1.0 kg a.i./ ha or Phorate @ 2-3 kg a.i./ha by applying in the last week of June and first week of July (Sandhu *et al.*, 1979; Chaudhury *et al.*, 1986). However in states like Bihar, soil application of above mentioned granule does not provide effective control due to early rain, hence, spraying of Endosulfan 35 EC @ 1.75 litre/ha twice, first in the beginning of July and then in August has been recommended (Hameed and Singh, 1984). Foliar application of synthetic pyrethroids like Cypermethrin, Fenvalerate, Permethrin @ 100g a.i./ha and Deltamethrin @ 25g a.i./ha also give effective control of top borer (Dhura *et al.*, 1986).

Shoot Borer, *Chilo infuscatellus* Snellen

Incidence of this borer is found throughout India. Its attack becomes more serious from April to late June or mid July, killing 6 to 85 per cent shoots in different states, causing a loss of 0.6 to 38.78 tonnes/ha in cane yield and 0.4 to 4.2 units in sugar recovery (Choudhary, 1983). The attack of this pest produces dead heart from April to June leading to compete drying of the attacked shoots. A loss of 10–20 per cent of young shoots is very common during this period and in the years of serious infestation, it may be as high as 70 per cent. Attack after the formation of cane does not produce dead-heart and the damage is confined to few internodes only. Even then, there is considerable reduction in cane yield and sugar content. At harvest, loss of 22–33 per cent in yield, 12 per cent in sugar recovery, 2 per cent in commercial cane sugar (CCS) and 27 per cent in jaggery have been estimated.

Management

Cultural Methods

1. Frequent irrigation of the crop and trash mulching during summer provide effective control of this borer during summer. Irrigation at 10 days interval and application of recommended dose of urea are fruitful (Jaipal and Jaipal, 2000).
2. Earthing up gives considerable reduction of this pest (Malik and Chaudhary, 1988).
3. Intercropping with spices like coriander is recommended, since it reduces the population of shoot borer by increased activity of *Trichogramma chilonis* (Rachappa and Naik, 2004).

Mechanical Method

Spiking in the shoots after removal of dead hearts by organizing campaign has been suggested for the control of this pest (Gupta, 1959b). In those areas, where it acts as stem borer, removal of dried leaves at monthly interval after cane formation provides effective control of this pest (David, 1981).

Biological Control

1. Inundative release of *Trichogramma chilonis* Ishii against egg and *Sturmiopsis inferens* Tns. against larval stages have been recommended in southern states (Avasthy and Tiwari, 1986). *Heterorbabditis indica, H bacteriophora* and *Steinernema glasseri* caused a very high mortality of the pest (Sankaranarayanan *et al.*, 2003).
2. *Bacillus thuringiensis* formulation like Delfin, Biobit and Dipel and granulosis virus significantly controls shoot borer (Kesavan *et al.*, 2003; Mala and Solayappan, 2001)
3. Conservation and augmentation of predatory spider like *Lycosa poonaensis* appears to be promising (Patil *et al.*, 2001).

Resistant Variety

Sugarcane clone G 99721, G 99731 and G 99771 are least susceptible to shoot borer.

Chemical Control

1. Soil application of endosulfan @ 1.0-1.25 kg a.i/ha by diluting the insecticides in 2500-3000 liters of water at the time of planting in furrows over the cane sett, before covering them with soil have been recommended in different states (Chaudhaury *et al.*, 1987).
2. At the time of planting, soil application of phorate granules @ 1.0kg a.i./ha in Bihar (Singh *et al.*, 1981), phorate granules @ 0.7kg a.i./ha or sevidal, or carbofuran granules @ 1.0kg a.i./ha in Orissa (Balasubramanian, 1988) have been recommended.
3. Application of pressmud + dhaincha intercropping + chlorpyriphos 10G caused 60.25 per cent reduction in pest damage (Subramaniam and Kannappan, 2003).
4. Neem cake @ 250 kg/ha considerably reduces the shoot borer incidence (Thirumurugan and Koodalingam, 2005).
5. Trash mulching and spraying of Endosulfan 0.07 per cent at 4, 6 and 9 weeks after planting @ 450, 675 and 900 litres of spray fluid per hectare is very effective against shoot borer (Jhansi and Rao, 2004).
6. Endosulfan (0.7 per cent) applied at 4, 6, and 9 weeks after planting reduces the shoot borer incidence and increases the percent juice sucrose and yield (Jhansi and Babu, 2000).

7. Combination of NSKE and leaves of *Mentha spicata* and tobacco at 5 per cent reduces infestation of shoot borer and promotes germination (Solayappan *et al.*, 2000).
8. Regent (Fipronnil) 0.3G and 5 SC at 75g a.i/ha significantly reduces early shoot borer (Sardana, 2001).
9. Sett dipping in chloropyrifos 20 EC or fenvalerate 20 EC (0.2 per cent a.i/ha) combined with soil treatment with cartap 4G (1.5kg a. i/ha) are more effective (Maan Singh *et al.*, 2002).

Stalk Borer, *Chilo auricilius* Dudgeon

The stalk borer is a serious pest of sugarcane in Bihar, Uttar Pradesh, and some parts of Punjab and Himachal Pradesh. In the early stage of the crop growth (April to June) before internode formation, it acts as a shoot borer forming dead heart in cane sprouts mostly in ratoon crops. But from July onwards after internode formation it acts as stalk borer damaging both plant and ratoon crops causing heavy damage to the crop specially those which are luxuriant in growth. Choudhary (1987) estimated 12.5-98 per cent infestation of cane stalks which resulted in 2-29 per cent internode damage, causing 4 to 33 per cent or 12.9 to 97 quintal/ha loss in cane yield and 0.3 to 3.7 units in sugar recovery. In some of the lodged and water logged fields, almost all the cane stalk may be found attacked by this pest. There is negative correlation between percentage brix, pol, purity and commercial cane sugar and percentage intensity *C. auricilius* (Pal *et al.*, 2001).

Management

Cultural and Mechanical Control

1. Avoid taking seed canes from the heavily infested cane field.
2. Over dosing of nitrogenous fertilizers should be avoided to prevent pest build up.
3. Autumn planting should be avoided in stalk borer prone areas.
4. Water logging and lodging of canes should be avoided.
5. Traps with virgin females/pheromone-lure may be set up.
6. Remove dry leaves and late shoots from the standing crop from October onwards.
7. All water and late shoots should be harvested along with the harvesting of crop.
8. All trash including left out canes should be burnt immediately after harvest.
9. Crop residues after harvest should be burnt immediately.
10. Removal and destruction of alternative host plants from the vicinity of sugarcane field.

Chemical Control

1. Kalra and Gupta (1963) suggested the soaking of cane setts for 2 hours in diazinon emulsion.
2. The pest can also be suppressed by high volume spray of monocrotophos @ 0.75 kg a.i/ha following de-trashing of dry leaves during September-October (Varma and Mitra, 1981).
3. Application of monocrotophos granules @ 3.0kg a.i/ha in July was found more effective than foliar spray in reducing the borer incidence (Kalra, 1981; Varma and Mitra, 1981).
4. Granular application is more effective than foliar application.

Biological Control

1. Release of *Trichogramma chilonis* @ 50,000 adults/ha from July to October reduces the incidence of stalk borer by 55 to 66 per cent (Shenhmar *et al.*, 2003).
2. *Cotesia flavipes* @ 2000 mated females ha^{-1} reduces its incidence by 69 per cent (Brar *et al.*, 1996; Tanwar *et al.*, 2001).

Internode Borer, *Chilo sacchariphagus indicus* (Kapur)

This borer is mainly the pest of sugarcane in southern states including Andhra Pradesh, Karnataka, Tamil Nadu and some parts of Maharashtra, although its presence has also been reported from Orissa, Bihar and Uttar Pradesh. The neonate larvae feed on the leaf spindle or the leaf sheath by scrapping the tissues and characteristics white streaks are formed on the leaf lamina when it opens. Subsequently the larvae bore in to the tender cane top. About 85 percent of fresh borer attack is found in the top five immature internodes. In young and weak shoots, especially in the ratoons, the formative internodes are damaged badly resulting in the formation of dead-hearts. The larvae tunnel upwards in a characteristics spiral fashion. It rarely feeds downwards. The damaged internodes become hard and crushing of these becomes difficult. Incidence as reported by different workers varied from 15.8 to 39.2 per cent on stalk basis with percent loss in cane yield being 3.8 to 15.8 per cent or 3.1 to 8.8 tonnes ha^{-1} with loss in sugar recovery being 0.9 to 9.8 per cent (David, 1986).

Management

Cultural and Mechanical Control

1. Collection and destruction of egg masses.
2. De-trashing of cane leaves at 5th, 7th and 9th months after planting.
3. Removal of water shoots at 8th and 9th month after planting and burning of trash with a view to destroy larvae and pupae are suggested to control this pest (David, 1986).

Varietal Resistance

Cultivation of varieties having hard rind and loose and self trashing leaves has been suggested to reduce borer infestation particularly in those areas where this pest is serious problem (Jayanthi and David, 1986). Sugarcane cultivars *viz.*, C98732, C 98741, C 98751, G 99704, G 99706, G 99730, G 99750, G 99771 and G 99780 are relatively less susceptible to internode borer (Thirumurugan *et al.*, 2004).

Biological Control

1. Releasing *Trichogramma chilonis* Ishi, *Telenomus* sp. against eggs; *Phaconotus* sp. against larvae and *Goryphus* (*Mecha*) *ornatipanis* Cam. and *Tetrastichus israeli* M and K have been reported to provide effective control of this pest (Agarwal and Kumar, 1973; Solyappan, 1980).
2. *Cotesia flavipes* at 1600 mated female/ha has been reported to reduce internode borer by 28.4 per cent (Tanwar *et al.*, 2002.)

Chemical Control

Foliar as well as soil application of insecticides usually does not prove effective in reducing the incidence of this borer, however, according to David (1981) spraying of 0.1 per cent monocrotophos gives some reduction of the initial borer population when the crop is 4–6 months old.

Gurdaspur Borer, *Acigona steniella* (Hampson)

This borer is a serious pest of sugarcane in the Punjab, Haryana, Western parts of Uttar Pradesh and some parts of Rajasthan and Himachal Pradesh. Its damage to the crop is initiated from end June or early July along with the initiation of internode formation, which continues up to end September. In the attacked canes, top become dry and field gives blasted appearance. Accordingly to Bains and Dev Roy (1981) normally the pest destroys about 20-25 per cent cane but in severe infestation the loss had gone up to 70-75 per cent. Singh *et al.* (1957) estimated a loss of 5-15 per cent in cane yield and these losses were still higher when the attack occurred in the early stage of the crop growth. In recent years, however, with the improvement of parasites status, regular mechanical control campaign, introduction of less susceptible varieties, early harvesting of crops, proper drainage, control of flood, reduction in areas under ratoon crop, increase in the irrigational facilities (causing early emergence of moths) are some of the reasons which caused significant reduction in pest incidence and nowadays the incidence of this borer both in Punjab and Haryana has gone below 5 per cent or even less (Chaudhury *et al.*, 1987).

In the initial stages, larvae feed gregariously on the top portion of canes by making spiral galleries and then kill the plant. At about the same time the larvae leave the canes and move on to new ones. Later on, large patches of dry canes appear.

Infestation by various generations has been reported to a loss of 10.40–29.74 per cent in polarity percentage, 10.34–50.51 per cent in total solid, 10.0–40.42 per cent in CCS, 0.11–21.70 per cent in purity and 10.0–30.32 per cent in sucrose content (Mahla *et al.*, 2001).

Management

Cultural and Mechanical Control

1. Heavily infested fields should not be kept for ratooning purpose.
2. The stumps of the harvested sugarcane fields, not meant for ratooning should be ploughed up and burnt to kill the diapausing larvae.
3. Regular removal and destruction of infested canes by cutting at weekly interval from July to September by organizing campaign on whole area basis should be followed (Chaudhury *et al.*, 1987). If all the attacked canes are removed carefully in gregarious phase of attack, then chances of further attack are much less.
4. Heavily infested cane-fields should be identified and harvesting of such field should be done on priority basis.
5. Spiking in the stump containing diapausing larvae should be done with ice breakers soon after harvesting.
6. Setts free from borer infestation should be used for plantings.

Biological Control

1. Inundative release of *Trichogramma chilonis* @ 1.25 lac adult/ha against egg stage in each brood provide effective control of this borer in Punjab (Varma and Sindhu, 1979). Releasing of other species of *Trichogramma* has also been proved effective.
2. Conservation of larval parasite like *Apanteles flavipes* (Cam.), *Isotima javensis* (Rohw.), *Rhaconotus scirpophagae* Wlkn., *R. signipennis* (Wlk.) and *Stenobracon nicevillei* (Bingham); pupal parasites like *Tetrastichus israeli* (Mani and Kurian), *T. ayyari* Rohw. and *Xanthopimpla*

stemmator (Thnb.) may be helpful (Varma *et al.*, 1981). *Spathius elaboratus* Wilkinson has also been recorded as potential larval parasite of gurdaspur borer (Saxena, 1992).

Chemical Control

1. Soaking of cane setts in 0.2 per cent Trichlorphon for 2 hours before planting kills the larvae within (Kalra, 1963).
2. Hibernating larvae can be killed by injecting 0.5 per cent endosulfan in the stubbles (Agarwal *et al.*, 1966).

Root Borer, *Emmalocera depresella* Swinhoe

This pest is reported to be distributed in almost all the cane growing tracts of North and Central India and has now been taking the status of major pest in Punjab, Haryana, Uttar Pradesh and Madhya Pradesh. Besides causing direct damage by boring the under ground portion of the cane stalks, it also helps in the infestation by several fungi, causing wilting of canes. In recent years, more than 80 per cent canes particularly in the draught years are reported to be attacked by this borer in Haryana and in some parts of U. P. and Rajasthan. In some cases, whole of the field become complete dried by the end of October or November in Haryana state.

Management

Cultural and Mechanical Control

1. Deep harvesting of cane meant for ratooning helps in the reduction of this borer (Cheema, 1952).
2. Plough up the sugarcane field and dig the stubbles soon after harvesting if not required to keep for ratooning. Burn the trash and all the uprooted stubbles to kill the pest population. This will help in checking the carry over of population from one season to another (Thakur *et al.*, 1957).
3. Give profuse and timely irrigation to ratoon and plant crops and remove borer infested shoots 3-5 cm below the ground level and kill them with pointed spokes at weekly interval. These help in the reduction of pest incidence (Avasthy, 1981).
4. Non-insecticidal management is more effective against root borer (Sardana, 1996).

Varietal Management

Growing of relatively resistant or less susceptible variety like Co 6808 (Patel, *et al.*, 1996).

Chemical Control

1. At planting time: All those chemicals which have been suggested to control *C. infuscatellus* provide effective control of this borer from April to June or July.
2. At post harvesting time: Application of aldrin @ 2.25 kg a.i./ha (7.5 lit of 3 E.C) with irrigation water in standing crop provides considerable reduction in the infestation of this borer. Mrig (1989) suggested one application of 5 per cent quinalphos or helptachlor granules each 1 kg a.i./ha or chlordane 20 EC @ 1.25 kg a.i/ha during August for the very effective control of this borer. These insecticides protect the crop during monsoon and post monsoon month from root borer.

Integrated approach by adopting various techniques like flooding of field, use of light traps, application of Immidacloprid insecticide and release of *Trichogramma chilonis* together or in sequential manner is effective against *E. depressella* (Sardana, 2000).

Plassey Borer, *Chilo tumidicostalis* Hmpsn

In India, its occurrence has been reported from Purnea and adjoining parts of Bhagalpur, Monghyr and Darbhanga district of Bihar (Khanna *et al.*, 1957), Plassey in Nadia district of West Bengal, Jorhat and Cachar district of Assam (Fletcher and Ghosh, 1919) and Dimapur district of Nagaland (Avasthy, 1982). It was referred to as plassey borer by Gupta (1957) because of its occurrence in an epidemic form at Plassey (West Bengal) during 1956. Recently it has also been reported from Eastern Uttar Pradesh (Pandey *et al.*, 2005) and also from other countries like Bangladesh (Karim and Islam, 1977), Thailand (Suasa-ard *et al.*, 2000), Nepal and Burma (David *et al.*, 1986).

Losses caused by the insect in yield and sucrose are more due to primary than secondary infestation. Khanna *et al.* (1957) reported the loss in tonnage to vary from 8.2-12.6 per cent and loss in sugar from 10.7-48.6 per cent in Purnea district of Bihar. In West Bengal, the loss in yield varies from 35-100 tonnes per hectare in different varieties, while sugar recovery is reduced by 3-26 per cent (Gupta and Avasthy, 1959). In Assam, the pest causes damage to an extent of 60 to 70 per cent of sugarcane every year throughout the cropping period (Anonymous, 1993). Brix content in cane damaged by this borer is reduced by 4.21 per cent, pol by 10.0 per cent, sucrose by 9.36 per cent, glucose by 5.20 per cent and CCS by 1.28 per cent (Gupta and Singh; 1997).

Primary infestation is caused by the newly hatched larvae aggregating in the top 3-5 internodes of the cane. As many as 156 caterpillars could be collected from one internode, although the average number of borer larvae per cane is generally 50 (Gupta and Avasthy, 1957). As the larvae feed inside the internodes, fresh wet frass shiny red in colour is pushed to the exterior through the bore holes in the top internodes. Top leaves of infested canes dry completely. Tunneling of internodes is so severe that the dried top portion of the cane easily breaks off at the slightest jerk. Sett-roots are formed in the infested internodes and envelop the stalk in a network of root hair. The lateral buds also sprout.

In the phase of secondary infestation, grown up borer larvae migrate to adjacent canes or to the lower healthy portion of the canes showing primary infestation. One caterpillar may bore one to five internodes in a cane. However, cane top do not dry up in this case.

Management

1. Collection and destruction of egg masses and removal and destruction of cane tops, infested by primary infestation from mid February to mid July is recommended for the control of this pest (Gupta and Avasthy, 1959).
2. Using light traps to collect moths gives effective control.
3. Co 356 and 513 are the varieties reported to be resistant while Co 243, 453 and 617 are less susceptible (Agarwal, 1959). Co 740 has been identified as less susceptible variety in the agro-ecosystem of Assam and the variety WL-22 has shown resistance to insect due to ovipositional non-preference on it (Kalita and Gupta, 1998). Local cultivar of Assam *viz.*, "Ekora Kuhiyar" is almost completely free from Plassey borer infestation (Gupta, 1997).
4. Conservation of natural enemies like *Cotesia flavipes.*
5. Malathion 10 per cent dust @20kg a.i/ha and fenvalerate 0.4 per cent dust give effective control of this pest (Deka *et al.*, 1999).

Sugarcane Leaf Hopper, *Pyrilla perpusilla* (Walker)

Pyrilla perpusilla is the most destructive sucking pest of sugarcane in India. It is widely distributed in the oriental region. Its occurrence has been reported from India, Pakistan, Nepal, Bangladesh, Burma, Sri Lanka, Cambodia, Laos, Vietnam and Indonesia. Of the different species, *Pyrilla perpusilla* Walker is widespread in India while *P. aberrons* Kby. occurs in the peninsular India and Sri Lanka.

The insect was first recorded from eastern India as *Fulgora pallida* by Donovan in 1800. Since then it has been reported from different sugarcane growing areas of India. Records show that first epidemic of pyrilla occurred in Western districts of Uttar Pradesh during 1930-31, during 1934 in Punjab and in Deccan Canal area in 1939. Since then its epidemics have been observed in sub-tropical sugarcane growing tracts at interval of 5 to 8 years.

Adults and nymphs suck sap from the under side of the lower leaves. Continuous de-sapping results yellowing and withering of leaves, development of sooty moulds over the honey dews which falls on upper surface of the leaves below the affected ones. In severe cases of infestation the top leaves dry up and lateral buds on cane stalks germinate.

Pyrilla attack resulted in reduction in sucrose contents in juice and purity, ranging from 2 to 34.2 per cent and 5 to 12 units respectively, and reduction in sugar recovery in mills to an extent of 0.2 to 5 units. Infestation of pyrilla also affected the well setting of jaggery as well as difficulties in milling processes.

Management

Cultural and Mechanical Control Measures

1. *Burning of cane trash*: All the cane trash should be removed from the field after harvest and utilized either for composting or burnt. This operation should be carried out up to the middle of March in order to destroy the unhatched eggs and over-wintering nymphs (Gupta, 1948).
2. *Removal of sprouts*: Removal of sprouts from the stubbles atleast once by the end of April where ratooning is desired and facilities for irrigation exist (Gupta, 1948).
3. *Stripping of dry leaves*: Removal of dry leaves from August onwards effectively reduces pyrilla population in Uttar Pradesh (Gupta, 1948).
4. Bagging of pyrilla adult and nymphs with hand nets (made of thick cloth) provides effective control during the pre-monsoon period provided it is carried out simultaneously by all the cultivators of an infested locality (Gupta, 1948, Wagle, 1951).
5. Collection and destruction of adult and nymphs in light traps also appears promising (Francis, 1933; Dhiman, 2001).
6. *Destruction of pyrilla egg masses*: During pre-monsoon period, pyrilla eggs are laid on the two lower most leaves of the shoots. The egg masses can be easily detected as they are covered with white waxy threads. Leaves bearing egg may be destroyed after the collection. In cases where there are few egg masses, the egg may be destroyed with fingers wrapped with cloth or covered with metallic cap (Gupta, 1948).

Biological Control

1. Conservation, colonization and redistribution of *Epipyrops* cocoons/egg masses have given encouraging results in minimizing the pest problem in peninsular India.

2. Supplementing the population of *Epiricania melanoleuca* (Fletcher) is effective against this pest (Shenhmar *et al.*, 2002).
3. Augmentative release of egg parasite *Parachrysocaris javensis* @ 1 lakh/ha is effective in suppression of *Pyrilla* (Hugar *et al.*, 2003).
4. Conservation and augmentation of predatory spider like *Lycosa poonaensis, Palystes sp., Argiope sp.* appears to be promising (Patil *et al.*, 2001).

Chemical Control

1. Application of methyl parathion (2 per cent) at the rate of 20-30 kg/ha during pre-monsoon period is effective when the pest attack is restricted to certain pockets only. Power duster can be used for dusting the tall crop in monsoon and post-monsoon periods.
2. Spraying with one of the chemical insecticides *viz.*, malathion (1.25 lit/ha), quinalphos (0.2kg/ha), endosulfan (0.3kg/ha), carbaryl (0.1 to 0.25 per cent), monocrotophos (0.03 to 0.04 per cent) and dimethoate (0.04kg/ha) was found effective for the control of pyrilla in monsoon and post monsoon periods. Application of malathion and dimethoate was reported to be least harmful to the parasite *Epipyrops melanoleuca.*
3. Foliar application of Endosulfan has been reported to be most effective in respect to toxicity to the pest as well as safety to ectoparasitoid (Tripathy and Katiyar, 1998).

Mealy Bugs

These are ubiquitous in all the sugarcane growing tracts of India but seldom assume a pest status. Seven different species of mealy bugs have been reported from the country. These are *Antonina graminis* (Maskel), *Chorizococcus rostellum* (Lobdell), *Dysmicoccus carens* (Williams), *Kiritshenkella sacchari* (Green), *Phenacoccus saccharifoli* (Green), *Pseudococcus saccharicola* (Takasahi) and *Saccharicoccus sacchari* (Cockrell).

S. sacchari and *K. sacchari* are widely distributed while *A. graminis* reported from Delhi and Coimbatore from nodal portion below leaf sheath, underground stem portion and root portions. Mealy bug are gregarious in habit and are occasionally covered with white mealy or filamentous waxy secretion. Young mealy bugs crawls into the soil around the stubbles and start its development and multiplication and attack the emerging sprouts in the ratoon fields. The crawlers are active and have a tendency to move towards the upper portions of the canes.

Feeding is done with the help of feeding apparatus known as rostralis which consist of four chitinous stylets forming a tube like structure for sucking the sap. Stylets pierce the epidermis and penetrate in the parenchyma cells to suck its food material avoiding vascular bundles. Severe infestation by pink mealy bug creates stunting of cane and yellowing of leaves. Young shoots are killed due to loss of sap in the case of *P. saccharicola* (Rao, 1942), while in case of *P. saccharifolii* there was total drying of the crop (Mahmood, 1962). Its main attack started after internode formation of the cane. In case of severely infested canes by mealy bugs the sucrose content decreases by 24.1 per cent, while the reduction in brix is 16.2 per cent, besides this, the severely infested canes poorly germinate.

Management

Cultural Method

1. *Planting of un-infested setts*: Before planting the adhering leaf sheaths from the setts may be removed.

2. For keeping ratoon crop harvest the canes close to the ground, remove and destroy the crop residues from heavily infested fields. Burning of the trash has also been suggested to reduce the population considerably.
3. Repeated ratooning practice may add in the rapid multiplication of the pest; hence practice of regular ratooning may be avoided specially in areas prone to this pest.
4. In case of heavy infestation in grown up crops, de-trashing should be advocated followed by spraying of insecticides, if necessary.

Varietal Management

Varieties Co 671 and CoA 7601 has been found to be relatively less effected by mealy bugs as reported from south India. Variety Co 7704 is prone to severe infestation of mealy bug especially in drought hit areas.

Biological Control

Conservation and augmentation of natural enemies like *Anagyrus punctulatus* (Kapadia *et al.*, 1995) and *Dicrodiplosis* sp. (Reddy *et al.*, 2000) are promising.

Chemical Method

1. Spraying with 0.05-0.1 per cent malathion and foliar application of phosphomidon @ 3kg a.i/ha has been found effective against the pest.
2. Swabbing of cane pieces with 0.1 per cent malathion has been quite helpful in reducing the infestation.
3. Carbofuran 3G at 1kg a.i./ha applied three times is effective against *S. sacchari* (Pandya, 1997).

White Flies

These are major pests of un-manured ratoon crop, the nymphs of which suck the sap from the undersurface of the leaves resulting in yellowing of the same, ultimately the leaves dry up. Heavily infested leaves are covered by the sooty moulds (*Capnodium* sp.) which adversely affect photosynthetic activity of the plant. There are three species of whiteflies: (i) *Aleurolobus barodensis* (Mark) reported from Andhra Pradesh, Bihar, Tamil Nadu, Uttar Pradesh (Gupta, 1954) and Karnataka (Balikai *et al.*, 1998); (ii) *Neomaskollia bergii* Sign in Bihar, Karnataka, Maharastra, Madhya Pradesh and Tamil Nadu; (iii) *N. andropogonis* Corbet, in Uttar Pradesh (David and Banerji, 1981). Among the three species of white flies *A. barodensis* is widely distributed and was minor pest of crop. Recently it has assumed serious proportions in Bihar, Gujarat, Haryana, Maharastra, Punjab and Tamil Nadu and in mild form in Andhra Pradesh, while in case of other two flies, *i.e. N. bergii* and *N. andropogonis*, the attack is mild and sporadic.

Average loss to sucrose per cent in juice ranges from 1.0 to 3.0 units in plant and ratoon crop. Yield of the crop is reduced by 15 to 20 per cent while there is loss of 30 to 40 per cent in sucrose.

Management

Cultural Method

1. Discouraging ratooning in low lying areas.
2. Early harvesting of effected ratoon crop.

3. Adequate manuring of ratoon crop not less than 100kgN/ha
4. Proper drainage to avoid water logging in low lying areas.

Biological Control

Conservation of predatory coccinelid like *Serangium parcesetosum* (Patel *et al.*, 1996).

Chemical Method

Spraying the effected crop either with malathion 50 EC @ 0.1 per cent or with endosulfan 35EC @ 1.5 litres per hectares after stripping the leaves bearing nymphs and puparia. Besides these insecticides, triazophos @ 0.08 per cent and fluvalinate @ 0.05 per cent should be preferred (Patil and Hapse, 1987). Hasbe *et al.* (1985) reported that single spray of dichlorovos @ 0.08 per cent had been found to control the pest effectively. Moreover, carbosulfan at 0.1 per cent has been reported to be effective against *A. barodensis* (Patel *et al.*, 2003).

Aphids

These are soft bodied insects, apterous or alate, antennae 3 to 6 segmented. These are largest group of insect vectors transmitting diseases particularly mosaic. In northern Indian sugarcane belt, aphids like, *Aphis gossypi* (Glover), *Lipaphis pseudobrassicae* (Davis), *Melanaphis sacchari* (Zehnt), *Myzus persicae, Rhopalosiphum maidis* (Fitish), *R. rufiabdominalis, Sachizaphis graminum* (Rondani) have been reported from and under ground sugarcane field through out the year. Ghosh (1974) reported wooly aphid, *Ceratovacuna lanigera* Zehnter as a minor pest of sugarcane in Assam, Nagaland, Sikkim, Tripura, Uttar Pradesh and West Bengal. However, this pest attained already the status of major pest in recent years in north eastern India (Tripathy, 1995). In recent years, wooly aphid has been found to spread to Maharastra and Karnataka causing serious infestation on sugarcane (Galande *et al.*, 2002; Patil *et al.*, 2005). Yield loss due to wooly aphid is about 17.50–34.50 per cent in south Karnataka (Chakrabarty, 2004).

R. maidis are greenish or yellow in colour and generally found within the spindle of young shoots during January to April. *M. sacchari* is observed on senescent leaves from May to December. It is yellow in colour and turns purple when crowded. Besides this *M. indosacchari* prefers third to fifth leaf in young crop during May to October. Colour of adult and nymphs is brown. High build up of alate aphid population of *Melanaphis sp.* occurs during May to July. The incidence of *C. lanigera* in Assam is low in June, maximum in September and negligible in December-January (Gupta and Goswami, 1995).

Management

1. Growing of sugarcane clones *viz.*, SNK 44, SNK 61, and SNK 754 are promising against wooly aphid (Patil *et al.*, 2005).
2. Conservation of natural enemies like coccinellids and brown lace wing is fruitful.

Thrips

These are active insects taking flight with great agility having four feather like similar wings without vein, mouth part are asymmetrical and adapted for lacerating and sucking sap from plant tissues. Reproduction is sexual and also parthenogenetic. Hot weather favours multiplication of thrips while with the onset of monsoon their population declines (Agarwala, 1953). The rasping of upper leaf surface causing oozing and sucking up of the sap by nymphs result in the rolling up of the leaf tips. Damage by thrips, *Stenochaetothrip saccharicidus* (Ramk) on a large scale is reported in Bihar on two varieties *viz.*, Co 453 and Bo 11 (Agarwala, 1953) and in Assam (Gupta, 1996). Many valuable seedlings may be lost under heavy infestation by species *viz. Stenochaetothrips minutus* (Van Deveter), *Bragmatothrips ramakrishnae* (Bagri), *Anaphothrips sacchari* (Shumsher Singh), *A. flavicinctus* (Karny), *Baliothrips serratus* (Kobus), *Exothrips sacchari* (Shumster), *F. serrata* (Kobus) and *Haplothrips tolerabilis* (Priesner).

Management

Application of Fenvalerate 20 EC @ 0.2lit/ha, Imidacloprid 200SL @ 0.51lit/ha and dimethoate 30Ec @ 0.75lit/ha were found effective against thrips (Singh *et al.*, 1996).

References

Agarwal, R.A., 1959. Insect resistance in Coimbatore canes: A review. *Indian Sugar*, 8: 785–796.

Agarwal, R.A. and Kumar, Sunil, 1973. Integrated control of Internode borer (*Proceras indicus*) in Sugarcane. *Indian Sugar*, 23: 453–454.

Agarwal, R.A., Singh, O.P. and Bhatia, N.L., 1966. Control of Gurdaspur borer in Punjab (Pt–II) during active period. In: *Proc. All India Symp. Sug. Cane Dev.*, Delhi, pp. 31–38.

Agarwal, S.B.D., 1953. A note on the control of sugarcane thrips, *Fulmekiola saccharicida* (Ramk and Marg.) in Bihar. *Indian J. Ent.*, 15: 379–381.

Anonymous, 1993. *Annual Report*, Sugarcane Research Station, Assam Agricultural University, Buralikson, Golaghat, Assam, India.

Avasthy, P.N., 1981. Pest management in sugarcane ratoons. In: *Natl. Sem. on Ratoon Management*, 1: 75–81.

Avasthy, P.N., 1982. A Report, Scheme on Bionomics and Control plassey borer (*Chilo tumidicostalis* Hmpsn) in North Eastern state of India, pp. 9.

Avasthy, P.N., 1984. Survey and collection of insect pests of sugarcane and pathogenic micro-organism. *A Rep. Res. Scheme*, IISR, Lucknow, pp. 5.

Avasthy, P.N. and Tiwari, N.K., 1986. The shoot borer *Chilo infuscatellus* Snellen. In: *Sugarcane Entomology in India*, (Eds.) H. David, S. Easwaramoorthy and R. Jayanthi. Sugarcane Breeding Institute (ICAR), Coimbatore, pp. 69–92.

Bains, S.S. and Dev Roy, T.C., 1981. Integrated management of sugarcane pests in Punjab. *Proc. Natl. Symp. on Stalk Borer*, Karnal, pp. 147–155.

Balasubramaniam, M., 1988. Strategy for insecticidal efficiency in sugarcane pest management. In: *Biological Control Technology for Sugarcane Pest Management*, (Eds.) H. David, S. Easwaramoorthy and R. Jayanthi. Sugarcane Breeding Institute (ICAR), Coimbatore, pp. 328–344.

Balikal, R.A., Lingappa, S., Biradar, A. and Teggelli, R.G., 1998. Incidence of sugarcane whitefly, *Aleurolobus barodensis* (Mark) in Karnataka. *Karnataka.J. Agril. Sci.*, 11(3): 820–821.

Bhoopathi, R. and Karnataka, A.K., 2003. Screening of sugarcane germplasm to *Scirpophaga excerptalis* Walker *Hieroglyphus banian* Fabricius in Tarai region of Uttaranchal. *Annals of Agril. Res.*, 24(3): 679–680.

Box, H.E., 1953. List of Sugarcane insects. *Commonwealth Instt. Ent.*, London, pp. 101.

Brar, K. S., Maninder Shenhmar, Bakhetia, D.R.C., Doomra, S., Sharma, D.K., Duhra, M.S., Singla, M.L. and Shenhmar, M., 1996. Bio-efficacy of *Trichogramma chilonis* Ishii (Hymenoptera: Trichogrammatidae) for the control of *Chilo auricilius* Dudgeon on sugarcane in Punjab. *Pl. Prot. Bull., Faridabad*, 48(1–4): 9–10.

Butani, D.K., 1961. Insect pest of sugarcane in Bihar–II. *Indian Sugar*, 10(12): 765–781.

Chakravarthy, A.K., 2004. Impact of woolly aphid, *Ceratovacuna lanigera* Zehntner (Hemiptera: Aphididae) on sugarcane yield and other attributes in South Karnataka. *Insect Environment*, 10(3): 142–144.

Chaudhary, J.P., 1983. Losses due to borer in sugarcane. In: *Paper presented in All India Workshop on Crop Losses Due to Insect Pests*, Andhra Pradesh Agricultural University, Hyderabad, January, 7–9.

Chaudhhury, J.P., 1987. Key pests of sugarcane and their management. In: *Recent Advances in Entomology*, (Eds.) Y.K. Mathur, A.K. Bhattacharyya, N.D. Pandey, K.D. Upadhay and J.P. Srivastava. Gopal Prakashan, Parade, Kanpur, pp. 260–285.

Chaudhury, J.P., Singh, S.P., Mrig, K.K. and Bhardwaj, S.C., 1986. Evaluation for different control schedules for the suppression of major insect pest of sugarcane. *Indian J. Agric. Res.*, 20: 197–203.

Chaudhury, J.P., Yadav, S.R. and Singh, R.P., 1987. Declining trend in the incidence of Gurdaspur borer, *Acigona steniella* (Hmpsn). due to mechanical control. *Indian J. Ent.*, 49(4): 460–470.

Chaudhary, J.P. and Yadav, S.R., 1995a. Effect of nitrogen, phosphorus, potash and sulphur present in mid-ribs, growing points and leaf blades of sugarcane genotypes on the incidence of top borer, *Scirpophaga excerptalis* Walker (Lepidoptera: Pyralidae). *Indian-Sugar*, 45(9): 671–682.

Chaudhary, J.P. and Yadav, S.R., 1995b. Screening of sugarcane genotypes for the incidence of sugarcane top borer, *Scirpophaga excerptalis* Walker (Lepidoptera: Pyralidae) under natural infestation conditions. *Indian Sugar*, 45(2): 113–121.

Cheema, P. S., 1952. Deep harvesting of sugarcane as a measure for extermination of root borer, *Emalocera depresella* (Swon hoe). *Indian J. Ent.*, 14: 77–79.

David, H., 1981. Current Research on sugarcane internode borer, *Chilo sacchariphagus indicus* (Kapur) and its control. In: *Proc. Natl. Symp. on Stalk Borer*, Karnal, pp. 117–121.

David, H., 1986. Internode borer, *Chilo sacchariphagus indicus* (Kapur). In: *Sugarcane Entomology in India*, (Eds.) H. David, S. Easwaramoorthy and R. Jayanthi. Sugarcane Breeding Institute (ICAR), Coimbotore, pp. 121–134.

David, H. and Banerji, D.K., 1981. Some observations on the whitefly, *Neomaskelia andropogonis* Corbet infesting sugarcane. *Pestology*, 5: 7–8.

David, H., Easwaramoorthy, S. and Jayanthi, R., 1986. In: *Sugarcane Entomology in India*. Sugarcane Breeding Institute (ICAR), Coimbatore, p. 564.

Deka, M.K., Gupta, M.K. and Singh, S.N., 1999. Effect of different dust formulation of insecticides on the incidence of sugarcane insect pests. *Indian Sugar*, 49(5): 357–361.

Dhiman, S.C., 2001. Incidence of mass attraction of *Pyrilla perpusilla* Walker on flourescent light at Saharanpur, Uttar Pradesh. *J. Appl. Zool. Res.*, 12(2–3): 142–143.

Dhura, M.S., Singla, M.L. and Singh, D.P., 1986. Chemical control of top borer sugarcane. *Sug. Cane.*, 5: 3–5.

Easwaramoorthy, S, R. Nirmala and G. Shantilakhmi. 2001. Biology and predatory potential of *Micraspis univittata* (Hope) for shoot borer, internode borer. *J. Biol. Control.* 15(1): 97 –100.

Fletcher, T. B., 1920. The annotated list of Indian crop pests. *In: Proc. 3rd Entomological Meet, Pusa, India.* pp. 33–313.

Fletcher, T. B. and Ghosh, C. C., 1919. Borers in sugarcane, rice etc. *Proc. 3rd Entomological Meet., Pusa.*1: 380–389.

Francis, C.B., 1933. A note on Sugarcane leaf hopper (*Pyrilla*) in South Arcot district. *Madras Agril. J.* 21 (10): 510–514.

Galande, S. M, U. N. Mote, S. S. Shelke and P. G. Bhoi. 2002. Role of bio–agents in the management of sugarcane wooly aphid, *Ceratovacuna lanigera* Zehnter. Paper presented in the conference on *"Bio–technology in Agro–Industry and Environment"* organized by Shivaji University and Arts, Science and Commerce College, Loanand on 27 December, 2002.

Gill, P. S., 1963. Influence of hot weather irrigation and trash mulch on the incidence of certain sugarcane borers and weeds. *Indian J. Sug. Cane Res. Dev.* 7: 161–163.

Ghosh, A. K., 1974. Aphids (Homoptera: Insecta) of economic importance in India. *Indian Agric.* 18 (2): 135.

Gupta, B. D., 1948. Control of sugarcane leaf hopper (*Pyrilla sp*) in the UP. *Indian Farming.* 9 (10): 401–406.

Gupta, B. D., 1954. Aerial spraying with Endrin against pyrilla in sugarcane. *I.I.S.R. Newsletter.* 1 (i): pp.1.

Gupta, B. D., 1957. A note on the scientific and common names of sugarcane pests in India. *Indian J. Sug. Cane. Res. Dev.* 2: 9 –14.

Gupta, B. D., 1959a. Insect pest of sugarcane in India III, the top borer, *Scirpophaga nivella* Fabr. *Indian Sugar.* 9: 127–149.

Gupta, B. D., 1959b. Insect pest of sugarcane in India V, the shoot borer, *Chilo infuscatellus* Snell. *Indian Sugar.* 9: 445–465.

Gupta, B. D. and P. N. Avasthy., 1957. An epidemic of stem borer, *Chilo tumidicostalis* Hampson. *I.I S.R. Newsletter.* 3: 4 pp.

Gupta, B.D. and P. N. Avasthy., 1959. Biology and control of stem borer, *Chilo tumidicostalis* Hmpsn. *Proc: Int. Soc. Sug. Cane Technol.* 10: 886–901.

Gupta, M. K., 1996. Incidence of sugarcane thrips, *Stenocheatothrips saccharicida* (R&M) in Assam. *J: Agric. Sci. Soc. of NE India* 9(2): 226–227.

Gupta, M. K., 1997. Varietal reaction of sugarcane to top borer (*Scirpophaga excerptalis* Walker) and plassey borer (*Chilo tumidicostalis* Hampsn) infestation. *J. Agril Sc. Soc. of N E India.*, 1997, 10 (1): 40–142.

Gupta, M. K and P. K. Goswami., 1995. Incidence of sugarcane woolly aphid and its effect on yield attributes and juice quality. *Indian Sugar.* 44(12):883–885.

Gupta, M. K. and S. N. Singh., 1997. Qualitative losses in sugarcane by plassey borer and top borer damage. *Indian Sugar*. 47(4): 275–277.

Gupta, S.C. and Roy, C.S., 1982. An integrated approach to chemical control of the sugarcane top borer, *Tryporyza nivella* Fab. in Bihar. *Maharashtra Sugar, 7*: 35–41.

Hameed, S.F. and Singh, B., 1984. *Pest Control Recommendations*. Department of Entomology, Rajendra Prasad Agricultural University, Pusa, Bihar, pp. 53

Hasbe, H., Gajare, B.P. and Yadav, S.B., 1985. Control of whitefly (*Aleurobus barodensis* Mask) on sugarcane in Maharashtra. In: *Proc. 35th D.S.T.A. Conv.*, A93–A102.

Hugar, P. S., Rachappa, V., Hanumanth, V.N.Y., Rajendra Hegde, S. Lingappa and Hegde, R., 2003. Bio-efficacy of *Parachrysocaris javensis* (Girault) in the suppression of *Pyrilla perpusilla* (Walker) in sugarcane. *Indian J. Pl. Protection*, 31(1): 112–114.

Jaipal, Saroj and Jaipal, S., 2000. An IPM module for the management of major insect pests of sugarcane in Indian subtropics. *Sugar Tech.*, 2(3): 1–8.

Jayanthi, R. and David, H., 1986. Varietal resistance. In: *Sugarcane Entomology in India*, (Eds.) H. David, S. Easwaramoorthy and R. Jayanthi. Sugarcane Breeding Institute ICAR, Coimbatore, pp. 363–381.

Jhansi, K. and Rao, B.R., 2004. Management of early shoot borer Chilo infuscatellus Snellen in sugarcane. *Cooperative Sugar*, 35(11): 849–851.

Jhansi, K. and Babu, P.R., 2000. Chemical control of early shoot borer, *Chilo infuscatellus* Snellen in sugarcane under late planted conditions. *Indian J. Pl. Prot.*, 28(2): 204–205.

Kalita, J. and Gupta, M.K., 1998. Pubescence in sugarcane as a plant resistance character affecting oviposition by plassey borer *Chilo tumidicostalis* Hmpsn. (Lepidoptera: Crambidae). *Cooperative Sugar*, 33(1): 35–37.

Kalra, A.N., 1963. Control of Gurdaspur borer of sugarcane. *Sug. Cane Herald.*, 5: 1–3.

Kalra, A.N., 1967. Consolidated report of the entomological out-posts scheme. *I.I.S.R. Lucknow*, pp. 30.

Kalra, A.N., 1980. Factors leading to increased incidence of top borer in recent years in some areas of India. *Indian Sugar Crops J.*, 7: 42–43.

Kalra, A.N., 1981. Strategy for fighting a menace of sugarcane stalk borer, *Chilo auricilius* Ddgn. In: *Proc. Natl. Symp Stalk Borer*, Karnal, pp. 57–62.

Kalra, A.N., 1990. Top borer of sugarcane and its management. In: *Recent Advances in Management of Insect, Mite and Rodent Pests of Sugarcane in India*. Proc. of the Summer Instt., IISR, Lucknow, 4–23 June, pp. 15–30.

Kalra, A.N. and Sidhu, A.S., 1965. Biology of sugarcane top borer, *Scirpophaga nivella* in Punjab. *Indian Sugar*, 15: 37–43.

Kalra, A.N. and Gupta, M.C., 1963. Treatment of seed cane against *Chilotraea auricilia* Ddgn, a borer of sugarcane. *Indian J. Sug. Cane Res. Dev.*, 7: 158–160.

Kapadia, M.N., Prasana G.J. and Butani, P.G., 1995. Field recovery of *Anagyrus punctulatus* Agarwal a parasitoid of sugarcane mealy bug. *Indian Sugar*, 45(6): 361–362.

Karim, M.A. and M.N. Islam, 1977. Stem borers of sugarcane in Bangladesh. In: *Proc. 22nd Annual Scientific Conference*. Bangladesh Association for Advancement of Science, pp. 29–30.

Kesavan, R., Easwaramoorthy, S. and Santhalakshmi, G., 2003. Evaluation of different formulations of *Bacillus thuringiensis* against sugarcane early shoot borer, *Chilo infuscatellus* Snellen. *Sugar. Tech.*, 5(1–2): 51–55.

Khanna, K.L., Nigam, L.N. and Puri, V.D., 1957. *Chilo tumidicostalis* Hampson: A serious stem borer pest of sugarcane in Bihar. *Proc. Indian Acad. Sci.*, 46(B): 75–95.

Maan Singh, Madan, Y.P. and Singh, M., 2002. Effect of sett dipping and soil application of various insecticides on sugarcane shoot borer, *Chilo infuscatellus* Snell. in Haryana. *Indian Sugar*, 52(2): 105–107.

Mahmood, Ali S., 1962. Coccids affecting sugarcane in Bihar. *Indian J. Sug. Res. Dev.*, 6: 72–75.

Mala, S.R. and Solayappan, A.R., 2001. Screening of certain effective microbial insecticides for the control of sugarcane early shoot borer larvae *Chilo infuscatellus* Snell. *Co-operative Sugar*, 32(8): 631–633.

Malik, V. S. and Chaudhury, J.P., 1988. Integration of Heptachlor with light earthing for reduction of shoot borer (*Chilo infuscatellus* Snell) incidence in sugarcane. *Indian Sugar*, 38: 189–192.

Mrig, K.K., 1989. Evaluation of some important management practices in sugarcane ratoon. *Ph.D. Thesis*, Haryana Agril. University, Hisar, pp. 151.

Mukesh, Mahla, Singh, Dharm, Jaipal, Saroj, Mahla, M., Singh, D. and Jaipal, S., 2001. Assessment of damage-field and factory losses in sugar due to Gurdaspur borer, *Acigona steniella* (Hampson) infestation. *Indian J. Ent.*, 63(3): 227–232.

Mukunthan, N., 1986. The top borer, *Scirpophaga excerptalis* Walker. In: *Sugarcane Entomology in India*, (Eds.) H. David, S. Easwaramoorthy and R. Jayanthi. Sugarcane Breeding Institute, ICAR, Coimbatore, pp. 93–119.

Pandey, K. P, M. N. Pandey, Vinay Kumar and D. N. Singh. 2005. Occurrence of plassey borer (*Chilo tumidicostalis*) Hampsn in eastern U.P. *Indian Sugar*. 54(10):857– 858.

Pandya, H. V., 1997. Control of sugarcane borers, mealy bugs and scale insects in Gujrat. *Cooprative Sugar. 28* (10): 745–746.

Patel, C.B., Pandya, H.V., Patel, M.B., Patel, J.R., Radadia, G.C., Bhatt, T.A., Vyas, S.T., Mehta, V.R., Patel, K.K. and Patel, J.M., 1996. Varietal screening against sugarcane root borer. *Bharatiya Sugar*, 22(5): 19–20.

Patel, M.B., Patel, C.L., Kalaria, G.B., Makawana, K.V., and Purohit, M.S., 2003. Evaluation of some insecticides against sugarcane whitefly, *Aleurolobus barodensis* Mark. *Indian J. Sug. Cane Tech.*, 18(1–2): 72–75.

Patil, A.S. and Hapse, D.G., 1987. Losses caused by sugarcane whitefly, *Aleurolobus barodensis* Mark. in Maharashtra. *Cooperative Sugar*, 18: 731–736.

Patil, A.S., Magar, S.B., Shinde, V.D. and Hograth, D.M., 2001. Feeding potential of predatory spider on sugarcane pest. In: *Proc of the XXIV Congress of Intl. Soc. of Sug. Cane Technologists, Brisbane, Australia.* 17 –21 September, 2: 408–409.

Patil, S.B., Tippannavar, P.S., Patil, S.A., Balasundaram, N., Lingappa, S., Gowda, M.V.C., Khadi, B.M., Salimath, P.M., Nayakar, N.Y., Bhat, B.N. and Kambar, N.S., 2005. Identification of sugarcane clones resistant to the sugarcane woolly aphid (*Ceratovacuna lanigera* Zehntner). *Current Science*, 88(2): 284–288.

Rachappa, V. and Naik, L.K., 2004. Integrated management of early shoot borer, *Chilo infuscatellus* (Snellen) in sugarcane. *Ann. Pl. Protec. Sci.*, 12(2): 248–253.

Rao, Ram and Mohan S., 1942. Life history and bionomics of *Pseudococcus saccharicola* Takh.: A new pest of sugarcane in India. *Proc. Indian Acad. Sci.*, 16(B): 79–85.

Reddy, D.J. and Aziz, S.A., 2000. Record of *Dicrodiplosis sp.* on sugarcane mealy bug. *Insect Environment*, 6(3): 104.

Sandhu, J.S., Dhura, M.S. and Tripathi, G.M., 1979. Further Studies on the control of top borer (*Tryporyza nivella* F.) with systemic insecticides. *Proc. Intl. Soc. Sug. J.*, 79: 277–279.

Sankaranarayanan, C., Easwaramoorthy, S. and Nethi Somasekhar, 2003. Infectivity of entomopathogenic nematodes *Heterorhabditis* and *Steinernema* spp. to sugarcane shoot borer (*Chilo infuscatellus* Snellen) at different temperatures. *Sugar Tech.*, 5(3): 167–171.

Sardana, H.R., 1996. Evaluation of insecticides against sugarcane root borer, *Emmalocera depressella* Swinhoe. *Indian J. Ent.*, 58(2): 165–167.

Sardana, H.R., 2000. Integrated management of sugarcane root borer, *Emmalocera depressella* Swinhoe. *Cooperative Sugar*, 32(4): 271–274.

Sardana, H.R., 2001. Evaluation of a new insecticide Regent (Fipronil) against sugarcane shoot borer, *Chilo infuscatellus* Snellen and root borer, *Emmalocera depressella* Swinhoe. *Pesticide Res. J.*, 13(1): 74–78.

Saxena, H., 1992. *Spathius elaboratus* Wilkinson (Hymenoptera: Brachycera: Doryctinae): A new larval parasite of *Acigona steniella* (Hampsn) from Uttar Pradesh. *Indian J. Ent.*, 54 (4): 482.

Shahi, H.N., 2002. Crashing prices cause concern. In: *The Hindu Survey of Indian Agriculture*, Chennai, pp. 119–124.

Shashi, Pal, Singh, Maan Madan, Y.P., Pal, S. and Singh, M., 2001. Assessment of quality losses caused by sugarcane stalk borer (*Chilo auricilius* Ddgn) in Haryana. *Indian Sugar*, 51(4): 229–234.

Shenhmar, M., Singh, J., Singh, S.P., Brar, K.S. and Singh, D., 2003. Effectiveness of *Trichogramma chilonis* Ishii for the management of *Chilo auricilius* Dudgeon on sugarcane in different sugar mill areas of the Punjab. In: *Biological Control of Lepidopteran Pests. Proc. Symp. Biol. Control of Lepidopteran Pest*, July 17–18, 2002, Bangalore, pp. 333–335.

Shenhmar, Maninder, Brar, K.S. and Shenhmar, M., 2002. Effect of redistribution of *Epiricania melanoleuca* (Fletcher) cocoons against *Pyrilla perpusilla* (Walker) on sugarcane in Punjab. *Insect Environment*, 8(1): 33–34.

Singh, A.K., Prasad, A.R. and Gupta, S.C., 1981. Brief resume of work done on stem borer pests of sugarcane in Bihar (1935–80). *Proc. Natl. Symp. Stalk Borer, Karnal*, pp. 161–164.

Singh, Avtar, Solanki, R.S., Singh, Manager, Agarwal, M.C., Singh, A. and Singh, M., 1996. Sugarcane thrips and their control. *Cooperative Sugar*, 28(1): 41–45.

Singh, H., Kalra, A.N. and Sandhu, J.S., 1957. The new pyralid borer of sugarcane. *Indian Sugar*, 6: 737–742.

Singh, Manager, Singh, R.A. and Singh, M., 1997. Losses due to top borer, *Scirpophaga excerptalis* Wlk. in sub tropical India. *Indian Sugar*, 46(10): 799–801.

Singh, Suchita, Dixit, Anil, Singh, S.B., Yadav, S.S., Singh, S. and Dixit, A., 2003. Effects of spices intercropping on the incidence of top borer (*Scirpophaga excerptalis* Wlk.) and its biological control agents in sugarcane. In: *Proc. of the 65th Annual Convention of the Sugar Technologists' Association of India, Bhubaneshwar, Orissa, India, 22–24 August,* A178–A185.

Singla, M.L., Duhra, M.S., Dhaliwal, Z.S. and Sandhu, S.S., 1994. Effect of intercropping in autumn planted sugarcane on the incidence of *Scirpophaga excerptalis* Wlk. and termites. *J. Insect Sci.,* 7(2): 199–201.

Solayappan, A.R., 1980. Mass production of *Trichogramma* for release in factory area. In: *Biological Control of Sugarcane Pests in India,* Tamil Nadu Cooperative Sugar Federation Ltd, Madras, pp. 29–35.

Solayappan, A.R., Radha, M., Jesudasan, R.W.A. and Sujatha, T., 2000. NEMENTO: A green alternative to control early shoot borer (*Chilo infuscatellus* Snell.) in sugarcane. *Cooperative Sugar,* 32(1): 35–40.

Suasaard, W., Wiwat, Suasaard and Allsopp, P.G., 2000. *Chilo tumidicostalis* (Hampson) (Lepidoptera: Pyrallidae) and its natural enemies in Thailand. In: *Sugarcane Pest Management in the New Millennium. Proc. 4th Sugarcane Entomology Workshop,* International Society of sugarcane Technologists, Khon Kaen, Thailand, 7–10 February, pp. 10–16.

Subramanian, S. and Kannappan, K., 2003. Evaluation of IPM components in the management of shoot borer of sugarcane in Cauvery delta region, Tamil Nadu. *Cooperative Sugar,* 34(11): 879–882.

Tanwar, R.K., Varma, Ashok and Varma, A., 1997. *Scirpophaga excerptalis* Walker infestation in relation to its natural parasitoids and sugarcane cultivars in eastern Uttar Pradesh. *Uttar Pradesh J. Zool.,* 17(1): 33–37.

Tanwar, R.K., Varma, Ashok and Varma, A., 2001. Field evaluation of *Cotesia flavipes* Cameron (Hymenoptera: Braconidae) (Indonesian strain) against sugarcane stalk and internode borers. *J. Biol. Control,* 15(2): 127–131.

Tanwar, R.K., Varma, Ashok and Varma, A., 2002. Field trials with *Cotesia flavipes* Cameron against sugarcane borers in sub tropical India. *Sugar Tech.,* 4(3–4): 153–156.

Thakur, C., Butani, D.K. and Prasad, A.R., 1957. Pests of sugarcane (special supplement). *Eakh Samachar,* 1: 6–7.

Thirumurugan, A. and Koodalingam, K., 2005. Ecofriendly approaches for the management of sugarcane shoot borer. *Cooperative Sugar,* 36(7): 571–575.

Thirumurugan, A., Koodalingam, K. and Baskaran, T.L., 2004. Relative behaviour of selected sugarcane clones for mid-late season and their level of reaction to borers. *Cooperative Sugar,* 36(4): 335–338.

Tripathy, G.M., 1995. Record of parasite and predator complex of sugarcane woolly aphid, *Ceratovacuna lanigera* Zehnt. in Nagaland. *Indian Sugar,* 44(11): 839–841.

Tripathy, G.M. and Katiyar, R.P., 1998. Evaluation of some insecticides against sugarcane leaf hopper, *Pyrilla perpusilla* and its ectoparasitoid, *Epiricania melanoleuca. Indian J. Ent.,* 60(4): 391–395.

Varma, A. and Mitra, Krishna, 1981. Recent Researches leading to evolve a suitable control programme for stalk borer, *Chilo auricilius* Dudgn. In: *Proc. Natl. Symp. Stalk Borer, Karnal,* pp. 63–68.

Varma, A., Tiwari, N.K. and Tanwar, R.K., 1993. Changing scenario of insect pests in sugarcane and their management. In: *Pests and Pest Management in India: The Changing Scenario*. Plant Protection Association of India, Hyderabad, India, pp. 88–95.

Varma, G.C. and Sindhu, A.S., 1979. *Annual Report* (1978–79) of the Scheme *"All India Co-ordinated Research Project on the Biological Control of Crop Pests"*, Department of Entomology, Punjab Agricultural University, Ludhiana, pp. 26.

Varma, G.C., Mohanty, S.N. and Singh, Maninder, 1981. Natural enemies of *Acigona steniella* (Hmps.) (Crambidae: Lepidoptera) in Punjab. *J. Res. Punjab Agril. Univ*, 18(2): 163–169.

Varun, C.L., Singh, Suchita and Singh, H.N., 1990. Effect of intercropping of spices on the incidence of top borer (*Scirpophaga excerptalis* Wlk.) in sugarcane under bhat soil condition. *Indian Sugar*, 39(10): 751–756.

Yadava, S.R., 1985. Screening of sugarcane genotypes for infestation by top borer, *Tryporyza nivella* (F). and characterization for susceptibility. *Ph.D. Dissertation*, Haryana Agricultural University, Hisar, India, pp. 153.

Wagle, P.V., 1951. Pyrilla problem in Bombay Deccan. In: *Proc. 8th Ann. Con. Deccan Sug. Tech. Assoc.*, 1: 127–137.

Insect Pest and Disease Management *Pages 71–80*
Editor: **Dr. D. Prasad**
Published by: **DAYA PUBLISHING HOUSE (2016)**

Chapter 5

Integrated Ecofriendly Management of Jute Pests

U.S. Yadav[1] and S.S. Prasad[2]
[1]N.D. University of Agriculture and Technology, Crop Research Station, Bahraich – 271 801, U.P.
[2]C.R.S., Ghaghraghat, Bahraich – 271 901, U.P.

Introduction

The integrated pest management programmes depend on multicomponent suppression of pest population and reduce the frequent use of broad-spectrum pesticide. The main objective of IPM is to minimize the use of pesticides to control pests. Jute is an important fibre crop extensively cultivated in eastern India and Bangladesh. Its production and productivity is severely hampered mainly by insect-pest and diseases at different stages of its development. Some of important pests of jute in eastern India are semilooper, *Anomis sabulifera* Guen; stem weevil, *Apion corchori* Marshall; ash weevil, *Myllocerus discolor* Boh; yellow mite, *Polyphagotarsonemus latus* Banks, and stem rot and root rot diseases caused by *Rhizoctonia bataticola*. The biocides like *Beauveria bassiana* caused 68 and 14 per cent mortality in semilooper and stem weevil, respectively (Pandit and Som, 1988). Das and Singh (1998) found that though endosulphan was most effective but the bio-pesticides like neem oil and *Bacillus thuringiensis* were also equally effective against semilooper. Das and Singh (1998), Bhattacharya and Parmanik (2002) also reported that the bio-agents such as neem products and entomopathogenic fungus, *Beauveria bassiana* are effective against jute insect-pests and antagonistic fungus, *Trichoderma* spp. are effective against *Macrophomina* diseases (Srivastava *et al.*, 2004; and Roy and Pac, 2005). Keeping in view the above, various integrated approach for pests management of jute in which integration of all cultural, chemical insecticides and fungicide, botanicals and bio-pesticides to control pests were tried to bring the damage of pests below economic injury level.

First experiment with five treatments during 1999-2001, and second experiment with 10 treatments during 2003-2006 were conducted at the Crop Research Station, Bahraich N. D. University of Agriculture and Technology to find out the most effective and economically viable control measures against jute pests and diseases complex. The experiments were laid-out in randomized block design with four replications.

The plot size for each treatments in both experiments was 18 m^2 with 1m left as border all around each plot with var. JRO- 524. The first experiment was sown on 4th week of April (T_2 and T_3) and 2nd week of May (T_1, T_4 and T_5) and second experiment was sown in second week of May during all years. The jute crop was raised adopting standard package of practices and following integrated management schedule for different treatments. The plant to plant distance was maintained at 5-7 cm by thinning after germination of crop and line to line 30 cm. The Five treatments comprised of, IPM for disease only (T_1): Soil application of *Trichoderma viirde* @ 2.5 kg/ha mixed in FYM@ 5t/ha+ seed treatment with carbendazim @ 2 g/kg seed + Foliar application of carbendazim @ 1.5 g/lit of water at 2 per cent disease infestation, *i.e.* 11 weeks after sowing, IPM for insect and mite pests only (T_2): Sowing of yellow mite resistant variety JRO 524 + early sowing in the month of April+ foliar application of neem oil @ 4 ml/lit of water at 11 WAS + foliar application of endosulfan @ 400 ml/ha mixed with fungal culture *Beauveria bassiana* @ 10^{-10} spore suspension/ml at 14 WAS, IPM for insect and mite pests and disease both(T_3): Treatments schedule of T_1 and T_2 have been followed, Recommended practices (T_4): seed treatment with carbendazim @ 2 g/kg seed+ foliar application of endosulfan @ 750 ml/ha and carbindazim @ 1.5 g/lit of water at 11 WAS.Untreated Control (T_5)

Second experiment with 10 treatments including 8 IPM treatments, which are combinations of 2 weed managements, 2 bio-agents against diseases and 2 bio-agents against insect pests along with farmers practice and untreated control were tried. The details of treatments are comprised of; T_1-. Pre-emergence application of quizalofop-p-ethyl 5 EC @ 2.0 ml/lit followed by one hand weeding + soil application of *Trichoderma viride* @ 2.5 kg/ha mixed in compost + spraying of neem oil @ 2.0 per cent, T_2-. Pre-emergence application of quizalofop-p-ethyl 5 EC @ 2.0 ml/lit followed by one hand weeding + soil application of *Trichoderma viirde* @ 2.5 kg/ha mixed in compost + spraying of *Beauveria bassiana* @ 10^{6-8} spore suspension/ml, T_3-. Pre-emergence application of quizalofop-p-ethyl 5 EC @ 2.0 ml/lit followed by one hand weeding + seed treatment with *Trichoderma viride* @ 4 g/kg seed + spraying of neem oil @ 2.0 per cent, T_4-. Pre-emergence application of quizalofop-p-ethyl 5 EC @ 2.0 ml/lit followed by one hand weeding + seed treatment with *Trichoderma viride* @ 4 g kg^{-1} seed + spraying of *Beauveria bassiana* @ 10^{6-8} spore suspension/ml, T_5-. Two hand weeding + application of *Trichoderma viride* @ 2.5 kg/ha mixed in compost + spraying of neem oil @ 2.0 per cent, T_6-. Two hand weeding + application of *Trichoderma viride* @ 2.5 kg ha^{-1} mixed in compost + spraying of *Beauveria bassiana* @ 10^{6-8} spore suspension/ml, T_7-. Two hand weeding + seed treatment with *Trichoderma viride* @ 4 g/kg seed + spraying of neem oil @ 2.0 per cent, T_8-. Two hand weeding + seed treatment with *Trichoderma virdae* @ 4 g/kg seed + spraying of *Beauveria bassiana* @ 10^{6-8} spore suspension/ml, T_9-. Farmer's practice (seed treatment with carbendazim @ 2 g/kg seed followed by spraying of endosulfan 36 EC @ 2 ml/lit at 10 per cent insects' infestation, and T_{10}-. Untreated control

The data on different pests and diseases were recorded one week before and after foliar application of pesticides/bio-agents, and plant infestations were counted in 4 randomly selected spot of (0.5 × 0.5) m^2 in each plot. For plant infestation, over three puncture per plant was considered as stem weevil infested plant, whereas over three damage leaves/plant has been considered as semilooper, ash weevil and yellow mite infested plant. The incidence of stem weevil during 2005-06, yellow mite during 2003-04 and 2006-07 and ash weevil during 2004-05 was very low, hence, not reported. The fibre yield data were recorded for each plot separately by excluding 2 border rows from all sides. The data have been statistically analyzed after transforming it by square root transformation formula, a = ✓ p + 0.5, where 'a' is transformed value and 'p' is per cent infestation. The marginal benefit cost ratio (MBCR) for each treatment was computed by dividing added return from a treatment by its added cost against untreated control and farmer's practice separately.

The infestation of yellow mite, stem weevil, semilooper and ash weevil was significantly low in the treatments IPM for insects and mite pests only (T_2), and IPM for both insects and mite pest and diseases (T_3) in comparison to other treatments. However, the incidence of these pests were lowest in T_2 but was on par with T_3 (Table 5.1). The occurrence of ash weevil during 2001–02 was negligible, hence, not reported. It was also noted that the incidence of yellow mite and stem weevil was comparatively low during 2001-02 than 1999-2000 and 2000-01. The incidence of yellow mite and stem weevil varied from 0.5 to 1.83 per cent and 9.74 to 29.84 per cent during 2001–02 in comparison to 5.65–4.94 per cent and 17.57–29.22 per cent during 1999–2000, and 6.07–12.97 and 15.19–37.43 per cent during 2000-01, respectively. While the occurrence of semilooper was more during 2001–02 than 1999–2000 and 2000–01. The incidence of semilooper varied from 15.82 to 59.44 per cent during 2001–02 in comparison to 9.24–22.25 per cent during 1999-2000, and 6.07–12.97 per cent during 2000–01 respectively.

Table 5.1: Incidence of Semilooper, Stemweevil, Ashweevil and Yellow Mite During 1999–2000, 2000–01 and 2001–02 on Olitorius Jute

Treatments	*Plant Infestation (%)*										
	Semilooper			*Stemweevil*			*Yellow Mite*			*Ashweevil*	
	1999-02	*2000-01*	*2001-02*	*1999-02*	*2000-01*	*2001-02*	*1999-02*	*2000-01*	*2001-02*	*1999-02*	*2000-01*
T_1	5.21	12.54	35.13	27.78	30.35	15.43	5.61	12.54	1.19	7.57	10.83
	(2.36)*	(3.60)	(5.96)	(5.31)	(5.53)	(3.99)	(2.47)	(3.58)	(1.25)	(2.81)	(3.25)
T_2	12.04	6.07	15.82	17.57	15.19	9.74	7.41	6.07	0.5	2.89	7.18
	(3.30)	(2.53)	(4.05)	(4.23)	(3.96)	(3.18)	(2.63)	(2.54)	(0.98)	(1.82)	(2.75)
T_3	9.24	9.64	22.79	18.47	17.2	11.29	7.04	9.64	0.51	3.11	7.92
	(3.11)	(3.18)	(4.80)	(4.35)	(4.19)	(3.41)	(2.81)	(3.17)	(0.99)	(1.89)	(2.88)
T_4	19.31	10.65	35.78	27.46	32.66	17.12	11.91	10.65	1.14	8.97	8.18
	(4.42)	(3.32)	(6.01)	(5.28)	(5.75)	(4.19)	(3.48)	(3.33)	(1.19)	(3.07)	(2.95)
T_5	22.25	12.97	59.44	29.22	37.43	29.89	14.94	12.97	1.83	15.62	15.2
	(4.75)	(3.65)	(7.74)	(5.43)	(6.16)	(5.49)	(3.93)	(3.64)	(1.49)	(4.01)	(3.96)
CD (P = 0.50)	0.92	0.86	0.87	0.85	0.83	0.68	0.77	0.78	NS	0.69	0.59
CV (%)	11.49	12.04	6.92	7.85	7.35	7.67	11.45	10.87	-	11.6	7.32

Figures in parentheses are transformed values.

The occurrence of stem rot and root rot diseases was low in the treatments IPM for diseases only (T_1), and IPM for both insect and mite pest and diseases (T_3) followed by the treatment IPM for insect and mite pests only (T_2) in comparison to untreated control (Table 5.2).

Table 5.2: Effect of IPM Treatments on the Incidence of Stem Rot and Root Rot and Fibre Yield During 1999–2000, 2000–01 and 2001–02 on Olitorius Jute

Treatments	Plant Infestation (per cent)						Fibre Yield (Q/ha)		
	Stem Rot			Root Rot					
	1999-02	2000-01	2001-02	1999-02	2000-01	2001-02	1999-02	2000-01	2001-02
T_1	0.00 (0.71)	1.33 (1.29)	2.15 (1.60)	0.72 (1.06)	0.97 (1.13)	1.63 (1.45)	9.50	7.07	16.50
T_2	9.60 (3.16)	5.17 (2.37)	6.27 (2.58)	8.00 (2.91)	7.27 (2.76)	5.37 (2.41)	10.33	12.63	26.33
T_3	1.64 (1.45)	2.67 (1.72)	3.01 (1.86)	1.25 (1.31)	3.79 (2.06)	2.89 (1.82)	9.83	11.87	24.17
T_4	2.00 (1.50)	3.85 (2.07)	5.71 (2.49)	1.20 (2.49)	2.75 (1.79)	3.58 (2.01)	8.67	6.33	13.50
T_5	8.16 (2.94)	10.75 (3.34)	12.33 (3.57)	14.88 (3.92)	17.54 (4.24)	18.79 (4.38)	7.67	5.69	9.94
CD (P = 0.50)	0.80	0.93	0.72	0.70	0.81	0.69	NS	3.58	6.04
CV (%)	18.69	19.57	13.93	15.19	15.32	12.93	–	18.69	10.15

In general, the fibre yield was more during kharif 2001-02 in comparison to the yield obtained during 1999-2000. This may be attributed to low incidence of yellow mite, stem weevil and ash weevil during kharif 2001-2002. Mohapatra (1977) reported over 20 per cent avoidable loss due to stem weevil in jute.

The maximum fibre yield was obtained in treatment IPM for insect and mite pest only (T_2) followed by the treatment IPM for both insect and mite pests and diseases (T_3) as compared to untreated control and other treatments.

The average fibre yield recorded were 10.33, 12.63 and 26.33 q/ha in T_2 and 9.83, 11.87 and 24.17 q/ha in T_3 and 7.67, 5.69 and 9.94 q/ha in untreated control during kharif season 1999-2000, 2000-01 and 2001-02, respectively.

The results regarding second experiment, infestation of semilooper, stem weevil, ash weevil and yellow mite summarized in Tables 5.3 and 5.4 revealed that all the IPM treatments were significantly superior to farmer's practice and untreated control during the *kharif* seasons of 2003-04, 2004-05, 2005-06 and 2006-07. The average infestation indicated that T_5 (3.37 per cent) followed by T_8 (4.10 per cent) and T_6 (4.39 per cent) against semilooper, T_5 (9.43 per cent) followed by T_6 (9.77 per cent) and T_7 (12.04 per cent) against stem weevil, T_8 (6.32 per cent) followed by T_5 (7.34 per cent) and T_6 (7.98 per cent) against ash weevil and T_8 (2.31 per cent) followed by T_1 (3.79 per cent) and T_2 (3.98 per cent) against yellow mite were most promising IPM treatments. The average infestations of semilooper, stem weevil, ash weevil and yellow mite in farmer's practice and untreated control were 10.39, 19.31, 14.00, 6.05, and 12.29, 22.44, 16.20 and 7.83, respectively. Thus, over all T_5 comprising of two hand weeding, application of *Trichoderma viride* @ 2.5 kg/ha mixed in compost and spraying of neem oil @ 2.0 per cent and T_8 comprising of two hand weeding, seed treatment with *Trichoderma viride* @ 4 g/kg seed and spraying of *Beauveria bassiana* @ 10^{6-8} spore suspension ml/lt were most promising IPM practices against insect and mite pests. Earlier, Pandit and Som (1988) have observed that *Beauveria bassiana* caused 68 and 14 per cent mortality in semilooper and stem weevil, respectively. Das and Singh (1998) reported that neem oil was as effective as endosulfan against semilooper. Bhattacharya and Pramanik (2002) also observed larval mortality of *Diacrisia obliqua* (Walker) to the tune of 91.11 per cent due to azadirachtin (5000 ppm) @ 0.03 per cent.

All the IPM treatments were significantly superior over farmer's practice and untreated control in all years. The treatments T_5 followed by T_3 and T_4 with average infestation of 0.92, 2.24 and 4.91 per cent against stem rot, and T_4 followed by T_5 and T_3 with average infestation of 2.01, 2.89 and 5.42 per cent against root rot, respectively were most promising IPM practices. The average infestations of stem rot and root rot diseases in farmer's practice and untreated control were 15.79 and 21.96 and 10.53 and 15.31 per cent, respectively. Thus, over all IPM practices, viz; T_5, T_3 and T_4 treatments were most effective in suppressing stem rot and root rot diseases. Bhattacharya *et al.* (1985) reported up to 56 per cent suppression of *Macrophomina* (*Rhizoctonia*) *phaseolina* causing root rot by growing antagonistic fungus, *Aspergillus versicolor* in soil compost medium for 10 days under diffused light. The management of *Macrophomina* diseases of *capsularis* jute, viz; stem rot and root rot has been done by seed and soil inoculation of *Trichoderma viride* in potted plant as well as in field experiments (Srivastava *et al.*, 2004). Roy and Pac (2005) have evaluated antagonistic potential of wild and mutant isolates of *Trichoderma harzianum* against *Macrophomina phaseolina* (Tassi) Goid causing stem rot disease of jute (Table 5.5).

Treatment T_8 resulted maximum fibre yield of 28.46 30.42, 27.39 and 35.22 q/ha followed by T_5 with 27.43, 29.24, 26.06 and 32.95 q/ha in respective years. These treatments were significantly superior to farmer's practice and untreated control which produced 20.04, 22.62, 19.23; and 23.10 q/ha and 16.67, 21.62, 15.38 and14.39 q/ha fiber yield in respective years. Thus, T_8, T_5 and T_6 with average fibre production of 30.37, 28.92 and 26.27 q/ha were most effective IPM practices. The economic analysis of different IPM treatments against untreated control also indicated that T_8 was most economic and fetched average net profit of Rs. 9484/per ha with marginal benefit cost ratio (MBCR) of 2.45 (Tables 5.6 and 5.7) which was followed by T_5 and T_3 with average net profit of Rs 6714/- and 4555/ha and MBCR of 1.89 and 1.86, respectively. However, the economic analysis of IPM treatments over farmer's practice revealed that T_8 followed by T_5 were the economic IPM treatments with net profit of Rs. 7979/- and 5208/ha and MBCR of 3.68 followed by 3.34 with T_4 treatment.

On the basis of above studies, it is concluded that in first experiment, IPM practice for insect and mite pest only consisting of early sowing of yellow mite resistant variety JRO- 524 followed by foliar spraying of neem oil @ 4 ml/lit at 11 WAS, and endosulfan @ 400 ml/ha mixed with fungal culture of *Beauveria bassiana* at 14 WAS was most effective management strategy and may be recommended for field use on Olitorious jute, and in second experiment the IPM treatments comprising of seed treatment with *Trichoderma viride* @ 4 g/kg seed + soil application of *Trichoderma viride* @ 2.5 kg/ha mixed in compost followed by two hand weeding + spraying of neem oil @ 2.0 per cent or spraying of *Beauveria bassiana* @ 10^{6-8} spore suspension per ml was most effective and economic against diseases, insect and mite pests of *olitorius* jute and may be recommended for farmer's field.

Table 5.3: Effect of IPM Treatments on the Incidence of Semilooper and Stem Weevil on *Clitorius* Jute Variety, JRO-524 During *Kharif* 2003–04, 2004–05, 2005–06 and 2006–07

Treatments	*Plant Infestation (%)*								
	Semilooper					*Stem Weevil*			
	2003-04	*2004-05*	*2005-06*	*2006-07*	*Average*	*2003-04*	*2004-05*	*2006-07*	*Average*
T_1	5.34 (2.39)	3.06 (1.68)	9.44 (3.15)	3.33 (1.96)	5.29	5.09 (2.36)	25.45 (5.03)	15.36 (3.98)	15.3
T_2	8.71 (3.01)	8.21 (2.94)	10.00 (3.24)	3.88 (2.09)	7.7	6.82 (2.70)	27.44 (5.24)	17.30 (4.25)	17.18
T_3	5.14 (2.37)	9.78 (3.20)	7.22 (2.77)	3.33 (1.92)	6.36	4.10 (2.14)	23.53 (4.86)	19.07 (4.42)	15.56
T_4	7.50 (2.83)	11.68 (3.48)	11.67 (3.48)	4.99 (2.33)	8.96	5.56 (2.46)	19.14 (4.39)	14.67 (3.89)	13.12
T_5	3.24 (1.92)	2.51 (1.61)	4.99 (2.33)	2.77 (1.79)	3.37	2.67 (1.77)	12.79 (3.61)	12.84 (3.65)	9.43
T_6	3.31 (1.93)	2.60 (1.63)	6.67 (2.68)	5.00 (2.34)	4.39	2.68 (1.75)	12.51 (3.59)	14.12 (3.82)	9.77
T_7	4.84 (2.30)	6.77 (2.66)	7.22 (2.77)	4.44 (2.22)	5.81	2.79 (1.81)	13.59 (3.71)	19.74 (4.50)	12.04
T_8	3.92 (2.09)	4.17 (2.10)	6.11 (2.57)	2.22 (1.63)	4.1	2.87 (1.83)	28.25 (5.08)	10.10 (3.20)	13.74
T_9	9.67 (3.18)	13.56 (3.72)	12.78 (3.63)	5.55 (2.46)	10.39	7.79 (2.86)	31.04 (5.46)	19.11 (4.43)	19.31
T_{10}	13.01 (3.67)	15.60 (3.97)	13.89 (3.78)	6.66 (2.77)	12.29	8.07 (2.90)	31.08 (5.57)	28.19 (5.35)	22.44
CD (P=0.05)	0.61	0.72	0.42	0.42		0.48	1.26	0.31	
CV (%)	11.73	15.53	7.99	11.56		10.61	15.78	4.31	

Figures in parentheses are transformed values.

Table 5.4: Effect of IPM Treatments on the Incidence of Ash Weevil and Yellow Mite on *Olitorius* Jute Variety, JRO-524 During *Kharif* 2003–04, 2004–05, 2005–06 and 2006–07

Treatments	Plant Infestation (%)						
	Ash Weevil				Yellow Mite		
	2003-04	2005-06	2006-07	Average	2004-05	2005-06	Average
T_1	11.59 (3.46)	9.44 (3.15)	10.00 (3.24)	10.34	2.58 (1.72)	4.99 (2.33)	3.79
T_2	12.03 (3.53)	11.11 (3.40)	10.00 (3.24)	11.04	2.39 (1.69)	5.56 (2.42)	3.98
T_3	13.89 (3.79)	12.78 (3.64)	11.11 (3.39)	12.59	2.82 (1.81)	5.96 (2.46)	4.19
T_4	13.10 (3.68)	11.11 (3.40)	10.55 (3.32)	11.58	5.73 (2.48)	7.22 (2.75)	6.48
T_5	11.48 (3.46)	2.78 (1.80)	7.77 (2.85)	7.34	5.08 (2.35)	6.11 (2.57)	5.6
T_6	5.61 (2.46)	8.89 (3.06)	9.44 (3.15)	7.98	5.52 (2.42)	3.89 (2.09)	4.71
T_7	10.56 (3.32)	7.22 (2.77)	8.88 (3.05)	8.88	8.60 (2.69)	4.44 (2.20)	6.52
T_8	7.31 (2.78)	4.99 (2.33)	6.66 (2.66)	6.32	1.29 (1.27)	3.33 (1.92)	2.31
T_9	16.46 (4.11)	13.33 (3.72)	12.22 (3.56)	14	2.65 (1.63)	9.44 (3.15)	6.05
T_{10}	17.52 (4.24)	14.99 (3.93)	16.11 (4.07)	16.2	4.54 (2.23)	11.11 (3.40)	7.83
CD (P=0.05)	0.51	0.35	0.42		0.77	0.55	
CV (%)	7.2	6.51	9.09		9.96	12.77	

Figures in parentheses are transformed values.

Table 5.5: Effect of IPM Treatments on the Incidence of Stem Rot and Root Rot Diseases on *Olitorius* Jute Variety, JRO-524 During *Kharif* 2003–04, 2004–05, 2005–06 and 2006–07

Treatments	Plant Infestation (%)									
	Stem Rot					Root Rot				
	2003-04	2004-05	2005-06	2006-07	Average	2003-04	2004-05	2005-06	2006-07	Average
T_1	6.37 (2.62)	9.86 (3.22)	8.83 (3.04)	3.47 (1.96)	7.13	5.23 (2.38)	6.07 (2.56)	8.57 (3.01)	2.53 (1.71)	5.6
T_2	4.47 (2.22)	7.36 (2.79)	6.70 (2.68)	4.26 (2.16)	5.69	7.03 (2.73)	6.76 (2.69)	8.13 (2.93)	1.45 (1.33)	5.84
T_3	0.77 (1.03)	2.56 (1.73)	3.02 (1.87)	2.61 (1.75)	2.24	7.27 (2.78)	6.58 (2.64)	5.57 (2.69)	2.28 (1.50)	5.42
T_4	5.97 (2.51)	6.70 (2.68)	5.17 (2.37)	1.81 (1.58)	4.91	0.00 (0.71)	2.41 (1.70)	3.33 (1.95)	2.31 (1.66)	2.01
T_5	0.00 (0.71)	1.08 (1.18)	1.88 (1.54)	0.72 (1.07)	0.92	4.93 (2.31)	2.36 (1.68)	2.57 (1.74)	1.71 (1.48)	2.89
T_6	6.51 (2.64)	10.99 (3.38)	9.20 (3.10)	6.97 (2.72)	8.41	9.11 (2.08)	9.53 (3.16)	10.85 (3.36)	2.50 (1.72)	7.99
T_7	5.51 (2.39)	8.83 (3.04)	8.33 (2.96)	2.03 (1.59)	6.17	7.71 (2.84)	7.41 (2.81)	10.38 (3.30)	0.55 (0.90)	6.51
T_8	6.49 (2.57)	7.53 (2.83)	9.70 (3.19)	0.36 (0.89)	6.02	9.03 (3.08)	9.55 (3.15)	12.27 (3.56)	0.00 (0.71)	7.71
T_9	20.50 (4.56)	19.95 (4.52)	19.02 (4.41)	3.72 (2.03)	15.79	11.97 (3.39)	16.18 (4.06)	13.29 (3.70)	0.69 (1.01)	10.53
T_{10}	21.93 (4.73)	28.43 (5.37)	30.53 (5.57)	6.97 (2.73)	21.96	19.18 (4.43)	18.81 (4.39)	20.28 (4.56)	2.97 (1.82)	15.31
CD (P=0.05)	0.63	0.4	0.26	0.5		0.33	0.62	0.39	0.45	
CV (%)	14.04	7.7	6.95	18.78		9.72	12.58	10.32	18.9	

Figures in parentheses are transformed values.

Table 5.6: Effect of IPM Treatments on Fibre Yield Due to Incidence of Insect Pests and Diseases on *Olitorius* Jute Variety, JRO-524 During *Kharif* 2003–04, 2004–05, 2005–06 and 2006–07

Treatments	*Fibre Yield (q/ha)*				
	2003-04	*2004-05*	*2005-06*	*2006-07*	*Average*
T_1	20.52	22.98	22.22	29.16	23.72
T_2	20.28	22.77	21.15	27.65	22.96
T_3	22.55	26.11	23.5	28.78	25.23
T_4	20.81	24.85	20.72	28.4	23.69
T_5	27.43	29.24	26.06	32.95	28.92
T_6	26.47	27.78	24.35	26.51	26.27
T_7	24.19	26.55	23.82	25.94	25.12
T_8	28.46	30.42	27.39	35.22	30.37
T_9	20.04	22.62	19.23	23.1	21.24
T_{10}	16.67	21.62	15.38	14.39	17.01
CD (P=0.05)	2.1	5.17	1.45	4.54	
CV (%)	4.16	11.85	3.78	9.73	

Table 5.7: Economic Analysis of IPM Treatments Over Untreated Control and Farmer's Practice on *Olitorius* Jute Variety, JRO-524 (mean of 4 years)

Treatments	*Untreated Control*					*Farmer's Practice*			
	Fibre yield q/ha	*CPP Rs./ha*	*EP q./ha*	*NP Rs./ha*	*MBCR*	*CPP Rs./ha*	*EP q./ha*	*NP Rs./ha*	*MBCR*
T_1	23.72	5793.00	6.71	2251.00	1.38	2222.00	2.48	754.00	1.34
T_2	22.96	5247.00	5.95	1893.00	1.36	1676.00	1.72	388.00	1.23
T_3	25.23	5309.00	8.22	3555.00	1.86	1738.00	3.99	3050.00	2.75
T_4	23.69	4548.00	6.68	3468.00	1.76	977.00	2.72	2287.00	3.34
T_5	28.92	7578.00	11.91	6714.00	1.89	4007.00	7.68	5208.00	2.30
T_6	26.27	7032.00	9.26	4080.00	1.58	3461.00	5.03	2575.00	1.74
T_7	25.12	7094.00	8.11	26.38	1.37	3523.00	3.88	1133.00	1.32
T_8	30.37	6548.00	13.36	9484.00	2.45	2977.00	9.13	7979.00	3.68
T_9	21.24								
T_{10}	17.01	3571.00							

CPP: Cost of plant protection; EP: Extra produce over untreated control; NP: Net profit; MBCR: Marginal benefit cost ratio

Cost of jute fibre: Rs. 1200/- ha^{-1}.

Conclusion

Two field experiments were conducted on *olitorius jute* var. JRO 524 for seven kharif seasons to evaluate the effect of different IPM practices on jute pests and fibre yield. The first experiment was conducted during 1999-2001 with five treatments including three IPM modules revealed that occurrence of stem weevil, ash weevil, semilooper and yellow mite were significantly low in IPM module for insect and mite pests only but did not significantly differ with the IPM module for both insect and mite pests and diseases in comparison to untreated control. However, the incidence of stem rot and root rot diseases were low in IPM module for diseases only and did not significantly differ with the IPM module for both insect and mite pests and diseases in comparison to untreated control. The maximum fibre yield was obtained in the IPM module for insect and mite pests only, which did not significantly differ IPM for insect and mite pests and diseases both as compared to untreated control. Second experiment was conducted to find out the most effective IPM practices against diseases and insect-pests during 2003-2006. The pooled data of the four seasons indicated that the IPM treatments viz; two hand weeding + seed treatment with *T. virdae* @ 4 gm/kg seed + spraying of *Beauveria bassiana* @ 10^{6-8} spore suspension/ml, and two hand weeding + soil application of *T. virdae* @ 2.5 kg/ha + spraying of neem oil @ 2.0 per cent with average fibre yield of 30.37 and 28.92 q/ha, respectively were significantly superior to farmer's practice (21.24 q/ha) and untreated control (17.01 q/ha).The average incidence of semilooper, stem weevil, ash weevil and yellow mite as well as stem rot and root rot diseases in these treatments were significantly low in comparison to farmer's practice and untreated control. The marginal benefit cost ratio for these IPM treatments were 2.45 and 1.89 over untreated control; and 3.68 and 2.30 over farmer's practice, respectively.

Acknowledgements

The authors are thankful to AICRP on Jute and Allied Fibers, Barrackpore for financial assistance and supply of materials.

References

Bhattacharya, D., Basu, S., Chattapadhya, J.P. and Bose, S.K., 1985. Biocontrol of *Macrophomina* root rot disease of jute by an antagonistic organism, *Aspergillus versicolor*. *Plant and Soil*, 7: 435–438.

Bhattacharya, S. and Pramanik, A., 2002. Comparative efficacy of botanicals alone and in combination with synthetic insecticides against *Diacrisia obliqua* (Walker) under laboratory condition. *Environment and Ecology*, 20: 727–731.

Das, L.K. and Singh, B., 1998. Integrated management of jute pests. *Environment and Ecology*, 16: 218–219.

Mohapatra, L.N., 1997. Avoidable loses in fibre yield due to stem weevil in jute. *Madras Agric. J.*, 84: 390.

Pandit, N.C. and Som, D., 1988. Culture of *Beauveria bassiana* and its pathogenicity to insect pest of jute (*Corchorus capsularis* and *C. olitorius*) and mesta (*Hibiscus cannnabinus* and *H. sabdariffa*). *Indian Journal of Agricultural Sciences*, 58: 75–76.

Prasad, S.S., Yadav, U.S. and Srivastava, R.K., 2002. Integrated management studies against *olitorius* jute pests. *Annals of Plant Protection Sciences*, 10: 248–251.

Roy, A. and Pac, S., 2005. Evaluation of some antagonistic potential of some wild and mutant isolates of *Trichoderma harzianum* against *Macrophomina phaseolina* (Tassi) Goid causing stem rot of jute. In: *Proceedings of Global Conference II, Maharana Pratap University of Agriculture and Technology, Udaipur*, 171 pp.

Srivastava, R.K., Singh, S., Pandey, S.M., Prasad, S.S., Yadav, U.S. and Pandey, B.P., 2004. Biological management of *Macrophomina* diseases of *capsularis* jute. In: *Proceedings of 74th Annual Session of National Academy of Sciences India*, University of Rajasthan, Jaipur, 35 pp.

Insect Pest and Disease Management *Pages* **81–117**
Editor: **Dr. D. Prasad**
Published by: **DAYA PUBLISHING HOUSE (2016)**

Chapter 6

Insect Pest of Mungbean and Urdbean and their Integrated Management

S.K. Singh[1] and D.K. Yadav[2]
[1]Division of Crop Protection, Indian Institute of Pulses Research, Kanpur – 208 024
E-mail: singhskkanpur@indiatimes.com
[2]Project Co-ordinating Unit (Linseed), C.S.A.U.A.&T., Kanpur – 208 002, U.P.

Introduction

Mungbean is spread all over the country in about 3.4 million hectares. India is a leading country in the production of mungbean with about 65 per cent of the world hectarage and about 54 per cent of the world production (Pohelman 1991). Urdbean is a highly priced pulse and very rich in phosphoric acid. The important states producing these pulse crops are Madhya Pradesh, Uttar Pradesh, Maharastra, Punjab, Andhra Pradesh, Karnataka and Tamil Nadu. Under total pulses, both mungbean and urdbean, occupy more than 30 per cent of the area and about 69 per cent of total dry bean grown in the country. As regard the productivity levels, the former crop has 0.38 t ha^{-1} while the later has 0.42 t ha^{-1} (Anonymous, 1993) are extremely low in view of the realistic on farm yield potential of >1 to 2 t ha^{-1} for most of the cultivated varieties of these crops. Low crop yield is largely due to crop losses by major insect-pests. These crops can be grown on a wide range of soil types in semi arid tropical environment and several soil moisture regimes but a warm, deep well drained loam or sandy soil is most suitable for the higher yield. Successful cultivation of mungbean and urdbean crops must, therefore, include their efficient management; and this alone has the potential to more than double the present levels of on farm productivity.

Pest Complex of Mungbean and Urdbeans

A thorough scan of literature revealed that among various insect pest of mungbean and urdbean two and half dozen (Table 6.1) are assumed as major pests of economic importance (Arora and Pajni,

1957; Babu *et al.*, 1978, Sinha *et al.*, 1975; Srivastava *et al.*, 1977, Singh and Singh 1977; Saxena, 1978; Chhabra and Kooner, 1980, 1985a, Chhabra *et al.*, 1981 a&b, 1993a; Dhuri and Singh, 1983; Hameed and Dubey, 1983; Sehgal and Ujagir, 1988; Sontakke and Muduli 1990; Thapa and Timsina 1999; Natrajan and Subbarao, 1992; Hussain and Saharia, 1994; Borah 1995; Ahmad *et al.*, 1998; Lion-Xuming *et al.*, 1998; Ujagir *et al.*, 1999; Reddy *et al.*, 1999; Chandrakar and Srivastava, 2000; Lambrides and Imrie, 2000; Sethuram *et al.*, 2001; Dhakshinamoorthy and Selvanarayanan, 2002 and Jagruti *et al.*, 2002).

Table 6.1: Insect Pests of Mungbean and Urdbean

Common Name	*Scientific Name*	*Family*	*Reference*
Order: Orthptera			
Deccan Grasshopper	*Colemania sphenariodies Bol.*	Acridiidas	Nair, 1975
Grasshopper	*Oxya velox Fb(=O chinensis Thunb)*	Acridiidas	Singh and Singh, 1977
Acridaid	*Chortogonus trachypterus Blanch*	Acridiidas	Lal, 1983
Order: Hemiptera			
Mealy bug	*Ceococcuscoffeaie Green*	Coccidae	Subbao Roa *et al.*, 1976
Cow bug	*Tricentrus bicolor Dist.*	Mambracidae	Mayar *et al.*, 1976
Lygaeid bug	*Chauliops tallax Scott.*	Lygaeidae	Rawat and Sahu, 1968
Bean aphid	*Aphis craccivora Koch*	Aphididate	Nayar *et al.*, 1976
Aleyrodid	*Zaphanera publica (Singh)*	Aleyrodidate	Nayar *et al.*, 1976
Cotton whitely	*Bemisia tabaci Genn.*	Aleyrodidate	Nena, 1972
Coreid bug	*Riptortus pedestric Fb.*	Coreidae	Nayar *et al.*, 1976
	R. Linearis Fb.	Coreidae	Nayar *et al.*, 1976
	R. Fuscus Fb.	Coreidae	Nayar *et al.*, 1976
Brown pod bug	*Anoplocnemis phasiana F*	Coreidae	Nayar *et al.*, 1976
Lab bug	*Coptosoma cribaria Fb.*	Pentatomidae	Nayar *et al.*, 1976
Green bug	*Nezara viridula L.*	Pentatomidae	Nayar *et al.*, 1976
Leal hopper	*Empoasca kerri Pruthi*	Jassidae	Pruthi *et al.*, 1981
	E. Moti Pruthi	Jassidae	Chhabra *et al.*, 1981
	E. Terminalis Distant	Jassidae	Chhabra *et al.*, 1981
Order: Thysanoptera			
Thrip	*Calliothrips indicus (Bagnall)*	Thripidae	Lal *et al.*, 1981
	Megalurothrips distalis (Karny)	Thripidae	Lal *et al.*, 1981
Order: Lepidoptera			
Shoot borer	*Anarsia ephippias Meyr.*	Gelechiidae	Fletcher 1914
Leaf webber	*Eucosma melanaula Meyr.*	Eucosmidae	Nayar *et al.*, 1976
	Lamprosema indicate F.	Pyralidae	Nair, 1975
		Pyralidae	Nair, 1975
Spotted caterpillar	*Maruca (testulalis) Vitrate*	Pyralidae	Nair, 1975
Pirk borer	*Cydia ptychora Meyr.*	Pyralidae	Lal *et al.*, 1980
Pyralid	*Nacoleia vulgalis Gn.*	Pyralidae	Nair, 1975

Contd...

Table 6.1–Contd...

Common Name	Scientific Name	Family	Reference
Blue butterfly	*Euchrysops cnejus F.*	Lycaenidae	Nayar *et al.*, 1976
	Lampides boeticus L.	Lycaenidae	Nayar *et al.*, 1976
Sphingid caterpillar	*Herse convolvuli L.*	Sphingidae	Nayar *et al.*, 1976
Bihar hairy caterpillar	*Diacrisia obliqua Wlk.*	Arctiidae	Nair, 1975
Red hairy caterpillar	*Amsacta moorei Butl.*	Arctiidae	Bindra and kittur, 1961
	A. albistriga Wlk.	Arctiidae	Anon, 1970
	A. lactinea (Cam)	Arctiidae	Subba Rao *et al.*, 1976
	(=Estigmene lactinea)		
Tussock caterpillar	*Euproctis scintillans (Wlk.)*	Lymantridae	Subba Rao *et al.*
Leaf miner	*Gracillaria soyella V.D.*	Gracillariidae	1976 Nair, 1975
	Acrocercops phaseospora Myer.	Gracillariidae	Yyas, 1978
	Acrocercops sp.	Gracillariidae	Gupta and Singh, 1980
Green leaf caterpillar	*Antricarsia irrotata* B.	Noctuidae	Nayar *et al.*, 1976
Leaf caterpillar	*Eublema hemirhoda Wlk.*	Noctuidae	Nayar *et al.*, 1976
Pod borer	*heliothis armigera Hb.*	Noctuidae	Nayar *et al.*, 1976
Semilooper	*Diachrysia orichalcea (F.)*	Noctuidae	Babu *et al.*, 1978
	Plusia peponis F.	Noctuidae	Nair, 1975
	P. chalcytis F.	Noctuidae	Nair, 1975
	P.daubei F.	Noctuidae	Nair, 1975
	Autographa nigrisigna Wlk.	Noctuidae	Nair, 1975
Tobacco caterpillar	*Spodoptera litura (F.)*	Noctuidae	Nair, 1975
			Babu *et al.* 1978
	S. exigua (Hb.)	Noctuidae	Singh and Singh, 1978
Order: Diptera			
Stemfly	*Ophiomyia phaseoli Coq.*	Agromyzidae	Saxena, 1973b
Pea leaf miner	*Phytomyza horticola Gour*	Agromyzidae	Singh and Singh, 1978
Order: Hymenoptera			
Ant	*Ceratina binghami Ckll.*	Apidae	Nayar *et al.*, 1976
	Ceratina binghami Ckll.		
Order: Coleoptera			
Blister beetle	*Mylabris pustulata Th.*	Meloidae	Nayar *et al.*, 1976
Weevil	*Apion ampulum Fst.*	Apionidae	Subramaniam, 1959
Galerucid beetle	*Luperodes sp.*	Chrysomelidae	Lal *et al.*, 1980
	Madurasia obscurella Jacoby	Chrysomelidae	Menon and Saxena, 1970
Girdle beetle	*Oberea brevis* S.	Cerambycidae	Nair, 1975; Odak and Thakur, 1978; Singh *et al.*, 1978

Insect Pest Problem of Mungbean and Urdbean Crops vis-à-vis Crop Phenology

By and large mungbean and urdbean are grown during rainy season, however, in India these are also sown during spring/summer seasons, mid February to mid April in irrigated areas and harvested by mid-June before the onsect of monsoon. During this period, the insect pest activity is restricted to merely those which can with-stand hot and dry weather. Seedling mortality is caused by stem flies at the initial stage of crop growth. Further, defoliators like flea bettle, hairy caterpillar, semiloopers, hawk moth and till caterpillar are spectacular at vegetative stage of the crops. The thrips initiates at the bud stage which pose a serious threat at the peak flowering stage, eventually it resulted flower drop. Subsequently, pod borer damage initiate at flowering and become alarming at the green pod stage. Flea beetle damage commenced from 2nd std. week and continued till 7th std. week with maximum incidence (12.7 per cent) in 5th std. week at Vamban where lepiderpteran borer initiated from 5th std. week having maximum population during 8th std. week and Blister beetle damage initiated from 6th std. week which persisted upto 9th week when senescence set in (Annonymous, 2002). Singh and Rai (2005) reported the *C. chinensis* as serious pest and found its reproductive potential success as high as 77.87 per cent in young parents matings. Bruchid ravages the standing crops in the field as primary infestation and subsequent damage in the storage.

After the initiation of monsoon showers, sowing of rainy season crop starts. So, major area with these legumes sown during July and harvested by the end of October. Since in rainy season profuse vegetative growth is attained during September, hence it coincides with the insect pests *viz.* defoliators, flower and pod feeders activity during the period.

Estimation of Losses Caused by Insect Pest Complex

The assessment of crop losses is very essential to infer and determine the relative importance of insect pests so as to supervise the resource allocation for research and pest management of any crop. Since the reliable and accurate estimation of crop losses due to pest is very difficult as the pest attacks vary not merely across the space but in time as well, hence the pest inflicted losses differ from one agro-ecological zone to another and even in a particular zone between the localities in different years.

In India, serious attempts have not so far been made towards comprehensive assessment of crop losses in pulse except in chickpea and pigeonpea. However, individual attempts have been made to assess crop losses which are enumerated as here under.

General Assessment

Sinha (1981) reported form Dholi (Bihar) that urd (variety Navin) yielded 1056 kg/ha under protected condition as compared to only 689 kg/ha in check. The avoidable loss was calculated to be 35 per cent. At Raipur, grain yield of mungbean resulted 577 kg/ha under full protection, 433 kg/ha need based protection and 267 kg/ha in untreated control (Anon., 2002). Assessment of yield losses in mungbean due to insect pest reflected 465 kg/ha by seed treatment with carbofuran @ 8 ml/kg of seed + 2 spray of monocrotophos at 30 and 50 DAS and control plot registered merely 315 kg/ha (Anon., 2002). Saxena (1983) found that urdbean variety S_1 yielded 874 kg/ha when protected against pest complex in comparison to only 259 kg/ha without protection. Increase in yield was manifold and the avoidable loss was calculated as 71 per cent. Avoidable losses in Haryana have been estimated to be 48 per cent in mungbean, and 27 per cent in urdbean (Yadav and Yadav, 1979). In Bihar, avoidable losses were found to vary from 7 to 35 per cent in urdbean against pest complex (Hameed and Durbey, 1983). In Karnataka nearly 15 per cent crop was lost to pest complex in mungbean and urdbean (Thondadarya and Hiremath, 1983). Patel and Patel (1983) reported that in Maharashtra one crore rupees worth pulses (including mungbean and urdbean) are lost annually due to pest complex. Acephate application (0.07 per cent) registered 625 kg/ha yield against 416 kg yield in control in urdbean (Anon., 2002) by suppressing pod borers at Berhampore (Orissa).

Pest-wise Assssment

Stemfly, *Ophiomyia centrosemalis* de Mejre

Stemfly infestation begins as soon as the plants appear above ground. The attacked plant becomes stunted and yellow, which finally dies. The plants that survive an attack produce few pods, many of which are empty or contain small seeds. Stemfly damage is more severe during the rainy season than in summer season crop (Sehgal and Ujagir, 1988). Talekar (1990) reported 20 per cent yield loss in mungbean in Taiwan. The pest under population pressure can cause withing of plants to the extent of 99.3 per cent (Brar *et al.*, 1993b). Stemfly in mungbean was to the extent of 39 per cent infestation and in urdbean was as high as 47.7 per cent during *Kharif* 2002 at Coimbatore (Anonymous, 2002). At Pantnagar, infestation due to this pest was as high as 80 per cent (Anonymous, 2001). Stemfly damage was 26.6 per cent during Rabi at Vamban (Anonymous, 2002).

Leaf Hopper, *Empoasca kerri* Pruthi

The infested plant becomes stunted and a definite reduction in yield occurs. Gupta and Saxena (1971) reported that every unit increase in number of Jassids per plant reduces the yield by 443 kg/ha in mungbean.

Whitefly, *Bemisia tabaci* Gennadius

The pest sucks the sap of leaves and secrets honey-dew which causes shooty moulds and hampers the photosynthetic activity of leaves. The whitefly is a potent vector of yellow mosaic virus (YMV) both in mungbean and urdbean as most of the crop loss is due to YMV. The YMV can cause losses to the tune of 30-70 per cent in mungbean (Marimuthu *et al.*, 1981) In case of severe infection of YMV a very few pods are produced. The pods are reduced in size and grains are shriveled and reduced in size as well. About 57.7 per cent MYMV damage was reported during *Kharif* when the pest intensity was 41.3 whiteflies/cage, in BM 4 (AVT) genotype of mungbean (Anonymous, 2001) in Hisar while 36.5 per cent infestation was recorded in the same season in following year (Anonymous, 2002). The maximum whitefly population/cage was recorded 24.7 in Pusa 105 (Anonymous, 2002). Basheer *et al.*, 2002 found virus disease incidence ranging from 0 to 66.6 per cent.

Black Aplid, *Aphis craccivora* Koch

The feeding by aphids results in depletion of plant sap by sucking tender leaves, and yield may be upto 30 to 70 per cent. Besides, aphid can transmit banding cucumber mosaic virus in mungbean. Patel (1976) reported 40 to 80 per cent damage in pea by *A. craccivora*. The greenish black aphid, *Aphis craccivora* Koch was reported from Kanpur in Uttar Pradesh (Srivastava and Singh, 1976), from Punjab (Chhabra *et al.*, 1986a) and from Orissa (Raju and Panda, 1983) in India, causing damage to mungbean and blackgam. While recording reaction of mungbean entries against major insect pest during *Kharif* 2002 at Badnapur, aphid infestation was found with an average of 2.63 aphid (Anonymous, 2002). The damage varied form 4.26 to 34.78 per cent at Shillongani (Anonymous, 2002).

Galerucid Beetle, *Madhurasia obscurella* Jacoby

This is one of the major pest of mungbean and urdbean, found in Uttar Pradesh, Haryana, Orissa and Delhi in India and cause 20 to 60 per cent damage to the crop from germination till harvest

(Srivastava and Singh, 1976). The larvae also damage root nodules of the crop at rates between 25 to 60 per cent (Sahoo and Patnaik, 1994). While screening of mungbean entries (AVT_1 and AVT_2) for resistance to *Madurasia obscurella* during Kharif 2002 at Lam, the maximum mean infestation was recorded 66.02 per cent against Pusa 0172 (Anonymous, 2002) and the number of holes/leaf by the beetle was 3.76 at Badnapur in TARM-18 entry of mungbean. At Khargone per cent damaged by this pest was found varying significantly at all the three intervals of 15, 30 and 45 DAS. The infestation varied from 30.83 and 39.27 per cent in JV_2 and IU 31-13, respectively (Anonymous, 2002) and from 37.96 to 70.86 per cent infestation by the pest was reported from Lam (Anonymous, 2002).

Spotted Borer, *Maruca (testulalis) vitrata* Fab.

The legume pod borer or spotted pod borer *M. (testulalis) vitrata* is a serious pest of grain legumes in tropics and subtropics. This pest is able to establish on the crop early. Its caterpillar feed on buds, flower, bore the pods make webs and damage. According to an estimates the extent of damage to pods alone varies widely from 5.7 to 89.4 per cent in Bihar (Sinha *et al.*, 1979). Mungbean cultivars registered 7.52 to 8.93 per cent damage by the pest in Vamban (Anonymous, 2002).

Gram Pod Borer, *Helicoverpa armegera* Hubner

The young larva feeds on tender leaves, buds flowers and subsequently it bores into the pods and feeds on the seeds with its head and part of the body only thrust inside, the nest remains outside. A single larva may destroy 30-40 pods before its reaches maturity and 44.96 per cent damage was reported from Ranchi (Anonymous, 2002) against genotype MH 96-1.

Thrips, *Scirtothrips dorsalis and Megalurothrips distalis* Karney

This is one of the major insect pests of mungbean and urdbean in India, especially in northern states of the country. Both adult and nymph suck the buds, the flower shed before opening. Infested inflorescence becomes abnormal showing symptoms of flower drop. Excessive dry and drought weather resulted in severe infestation. Losses in summer mungbean in the Punjab state, India were dominated by thrips, *Megalurothrips distalis* (Karny) and have been reported to be 54.3 per cent losses by this pest (Chhabra and Kooner 1985a). Chhabra (1992) reported avoidable losses in mungban from 20 to 64.7 per cent and in urdbean from 18.1 to 61.5 per cent from reports of different research stations of the All Inida Coordinated Pulses Improvement Project AICPIP), India.

Bihar Hairy Caterpillar *Spilaractia obliqua* Walker

Damage is caused by caterpillars. When young, they feed gregariously but on few plants in scattered manner and eat the green matter of leaves, leaving behind only the mid rib. The grown up caterpillars feed voraciously on leaves, soft stems, branches, and may totally denude the crop within few days.

Bruchid *Callosobruchus* spp.

Among the insect pests that attack in pulses, bruchids are most destructive pests which inflict damage not merely in the standing crops in the field but also do substantial loss in storage also. Grubs damage the pods in the field and seeds under storage condition. Preharvest infestation, usually of low level, is very serious as it multiplies in severe proportion in storage Sparse information is available to quantify field infestation in different crops. Prevett (1961) reported that in cowpea bruchid complex caused 3.1 and 11.0 per cent infestation of pods collected as they dried or as single harvest, respectively. The extent of pod damage by *C. maculatus* varied from 0.3 per cent to 10 per cent (Lal *et al.*, 1980; Latif and Reed 1983; Bindra and Jakhmola (1967) reported damage due to pulse beetles varied from 1.2 to

10.2 per cent. Dias (1987) reported on the basis of samples collected from 8 different ecological zones that seeds were infected to the extent of 19.3 per cent by Bruchids. Damage due to stored grains pest, *Callosobruchus chinensis* in mungbean starts from the field. The secondary infestation is more damaging (Talekar, 1990). The infestation has been reported upto 100 per cent of the stored seeds of mungnbean and cowpea with 3 to 5 months and under ordinary storage conditions. (Caswell and Akibus, 1980). The losses are almost 100 per cent, if pulses are stored beyond rainy season (Anonymous, 2004). Gujar and Yadav (1978) reported a weight loss in mungbean of 55.6 to 73.5 per cent in individual seed damaged by single *C. maculatus* and 30 to 70 per cent by *C. chinensis* in one generation. The losses in seed by bruchid due to improper storage have been reported to be 50 per cent after only three months of storage (Hussain and Abdel, 1982). Rathore and Sharma (2002) reported in estimated loss of nearly 0.21 million tonnes of pulses bruchids, costing 315 million rupees. *Vigna* pulses are ravaged by bruchid genera of *Callosobruchus, Bruchidius, Speularius* and *Zabrotes.*

Damaging Stages

Insect Pests of Root and Seedling Stage

Stemflies (Agromyzid) cause serious damage at seedling stage to the crop in India, Malaysia and Taiwan (Ho 1967; Sehgal and Ujagir 1988; Talekar, 1990). Stemfly damage in mungbean is mainly due to *Ophiomyia phaseoli* (Tryon). (Srivastava and Singh, 1976; Saxena 1978; Islam *et al.*, 1984, Gupta and Singh 1984b; Lal 1985; Rajapakse and Jayasena 1989). Five species of agromizid stemflies, *Melanagromyza sojae* (Zehntner), *M. dolichostigma* de Meijere and *M. obtusa* Malloch *Ophiomyia phaseoli* (Tryon) and *O. spencerelia* and *O. centrosemalis* (Sehgal and Ujagir, 1988; Talekar, 1990). This pest under population pressure can cause wilting of plants to the extent of 99.3 per cent (Brar *et al.*, 1993b). The flies are active in summer and mate 2-6 days after emergence (Panwar, 1995). The pest is also recorded to remain active during winter as larvae or pupae in peas (Ram Ujagir, 1997). Singh (2003) reported summer crop in northern irrigated plains to be more vulnerable to stemfly and it is one of the major pest in India.

As regards the grubs of galerucid beetle, *Madurasia obscurella*, it feeds on the root hair nodules of the plants. (Saxena 1973; Gupta and Singh 1984a). Manjunath *et al.* (1988) reported a wireworm, *Agriotes* sp. feeds on the roots of mungbean plants. Nath and Singh *et al.* (1998) noticed its peak during the flowering stage. Singh (2003) reportd the nature of damage of adult beetles to feed root hairs and root nodules and also on the foliage and make small circular holes in the leaves at seedling stage. The caterpillars of cutworms cut stems and branches of seedling at night (Anonymous, 1976). *Amsacta moorei* also cause extensive damage.

Insect Pests at Vegetative Stages

Lepidoptera, Coleoptera, Orthoptera, Foliage Feeders

Insect pest species of the order Lepidoptera, Coleoptera and Orthoptera cause maximum damage at the vegetative stage. Insect pests defoliating these crops, feed primarily on leaves, flower and even pods as well. It is difficult to estimate damage inflicted by the defoliating insects as it depends upon the extent and duration of damage to the production of photosynthate. Normally, it will be rather little to that of the proportion etc. loss in foliage as new leaves, flowers and pods may develop, compensating partially for the destroyed parts. Removal of 25 to 50 per cent of the area of each leaf at the flowering stage of the mungbean plant to stimulate loss for insect feeding did not reduce the total dry weight or seed yield, but seed yields were reduced with 75 per cent reduction in leaf area (Robertson, 1981).

As regards the foliage damage, numerous lepidopteran have been reported damaging mungbean and urdbean. Ooi (1973) reported from Malaysia about the damage of leaf rollers and leaf folders. Besides, *Hedylepta (Lamprosema) indicala* (F) also damage mungbean crop in India (Sehgal and Ujagir, 1988) and Philippines (Catipon, 1986). *Homona coffearia* Nietner, *Sylepta sabinersalis*, Wlk., *Agrius convolvuli* Linn) and *Helicoverpa armigera* (Hub.) are the other foliage feeders in India and Philippines (Chhabra and Kooner 1980; Nair 1986; Litsinger *et al.*, 1988; Sehgal and Ujagir 1988). Pupation of *A. convoluli* takes place in soil, sometimes 8-10 cm. down and pupal period ranges from 17-26 days to 4-6 months depending upon the climate (Hill, 1993). Damage by *Cydia ptychora* (Meyrick) has been reported from Uttar Pradesh (Lal *et al.*, 1980) and from Southern India (Nair, 1986). Subba Rao *et al.* (1974) reported caterpillars of *Amsacta lactinea* (Cramier) as a major defoliator in South India. A severe outbreak of hairy caterpillars *A. moorei* and *Spilosoma obliqua* (Wlk.) has been reported from Rajasthan (Parihar, 1979) and northern Bihar (Sinha *et al.*, 1985). The seriousness of *Spilosoma obliqua* has been reported from Bihar, M.P., U.P. and Punjab (Atwal and Dhaliwal, 1986) and (Raman Goud, 1993), Delhi (Singh *et al.*, 1990), from Myanmar, China, Japan and Bangladesh (Hazarika, 1952). Singh and Pandey (1995) studied the biology of *A. moorei* on different varieties of groundnut in Delhi and common incidence was recorded from A.P., T.N., Karnataka and Maharashtra.

Adult galarucide beetle, *Madurasia obscurella*, feed on the leaves (Yein and Singh, 1981; Gupta and Singh, 1984a). Galerucid beetle in green gram infestation was reported during Rabi 2001-02 at Manipale (Anonymous, 2002). Various species of beetles are leaf feeders or they may serve as virus vectors. Lal *et al.* (1980) reported damage by *Epilachna* sp. from Kanpur, Uttar Pradesh. Urdbean mottled virus reported from India and Thailand, was transmitted by bean leaf beetle, *Cerotoma trifurcata* (Forsti) and Mexican bean beetle, *Epilachna varivestis* (Mulsi), *Monolepta signatata* Oliver, was found to transmit these disease, while *Aphis craccivora* Koch did not transmit this disease (Honda *et al.*, 1982).

Lal *et al.* (1980) reported that the weevil *Apion* sp. to feed on foliage, stem, buds, flowers and pod of mungbean and urdbean (Dhuri and Singh, 1983) in India. *A. ampulum* Fst and *A. collaris* Mshill have been reported by Sakamoto *et al.* (1983), feeding on buds, flowers and pods of mungbean in Japan. Three chrysomelid *viz. Altica coerulea* Oliver, *Luperodes* sp. and *Madurasia obscurella* have been reported to feed on foliage of mungbean and urdban (Lal *et al.*, 1980; Dhuri and Singh, 1983).

In the crops, grasshopper damage is mostly confirmed to the rainy season in India. So, *Chrotogonus trachypterus* Blanch, and *Oxya velox* (Feb.) usually cause damage as reported by Singh and Singh 1977; Dhuri and Singh 1983; Sehgal and Ujagir (1988). Patnaik, Mohanti and Chhabra (2000) reported *Gastrimagus marmoratus* (Thunb); *Colemania sphenariodes* Boliv; *Chrotogonus trachypterus* Blanch as minor pests of mungbean and urdbean. Foliage feeding Coccinelids like *Epilachna dodecastigma* Wied; *E. Varivestis* Muls; *Leis dimidiata* Fabr; are also of minor importance prevailing on these legumes.

Hermipteran, Thysanopteran and Acarina Floage, Flower and Pod Feeders

Insect pests of order Hamiptera, Thysanoptera and mite species (order: Acarina) damage the tender foliage, flowers and pod by sucking the cell sap. Their damage by and large lead to flower drop and poor seed quality. Attack of sucking pests on summer/spring season crops in India usually occur with the emergence of reproductive stage of the plant. This stage coincides with the dry periods from May to mid-June and later with rainy season and it causes major yield loss in mungbean and urdbean. However, rainy season crop can withstand the feeding stress, caused by the sucking pests due to increased vigour and relatively greater compensating ability than the summer season crops.

Thrips species attacking urdbean crop in Delhi (Dhuri and Singh, 1983). Two species of thrips *Caliothrips indicus* (Bognall) and *M. distalis* have been reported to attack tender buds and flowers of mungbean in Punjab, Uttar Pradesh and Maharashtra in India (Kooner *et al.*, 1983; Lal 1985; Dahaya *et al.*, 1986; Sehgal and Ujagir 1988). Chhabra and Kooner (1985b) reported extensive thrips damage to summer mungbean crop in Punjab and observed 19 thrips/25 flower, resulting 82.4 per cent flower dorp. Three species of thrips *viz.*, *Caliothrips* sp., *Scirtothrips* sp. and *Thrips palami* Karney have been found to cause damage to mungbean crop in Philippines (Litsinger *et al.*, 1988). *M. usitatus* Zehntrer has been reported by Sepswasdi *et al.* (1990) causing leaves and pod damage to mungbean in Thailand. As regards biology and bionomics of *M. distalis,* it has not been studied in detail and its one generation is completed within three weeks (Reed *et al.*, 1989). Sehgal *et al.* (2000) reported thrips *Caliothrips indicus* and *Megalurothrips distalis* are the dominant. SML 112 and SML 103 were rated resistant and had the least inflorescence deformity due to *M. distalis* (Chhabra and Kooner, 1994).

Whitefly, *Bemisia* tabaci Gennadius–a polyphogous pest has been reported to cause direct and indirect damage by transmitting mungbean yellow mosaic virus (MYMV) (Srivastava and Singh, 1976; Chhabra and Kooner 1980; Singh and Sehgal 1983). The host range, biology and field incidence has been widely studied in India (Nene, 1972; Murugesan and Chelliah 1978). Kooner and Chhabra (1980) observed 12 overlapping generations year^{-1} and it was most active during May to September in Punjab. Chhabra *et al.* (1984 and 1988) reported that the mechanism of resistance in the crops against whitefly, *B. tabaci;* Jassids, *Empoasca moti* and *E. terminalis* and MYMV is due to antibiosis. Incidence of MYMV caused by *B. tabaci* was less in summer than in the rainy season crop due to higher temperature (40+5°C) (Sharma and Verma, 1982). Chhabra and Kooner (1993) tested thirty nine accession of mungbean *Vigna radiata,* for their reaction to *Bemisia tabaci* and found ML 370, ML 428, ML 459, ML 506, ML 508 AND ML 537) as being resistant against the insect pests and MYMBV.

The greenish black aphid, *Aphis craccivora* Koch was reported from Kanpur in Uttar Pradesh (Srivastava and Singh, 1976), from Punjab (Chhabra *et al.*, 1986a) and from Orissa state (Raju and Panda, 1983) in India causing damage to mungbean and blackgram. *Aphis gossipi* Glover, *A. craccivora* and *Myzus persicae* Sulz were found to transmit the mosaic mottling virus in mungbean (Rama Krishna *et al.*, 1973). *A. craccivora, Acyrthosiphon pisum* Harris and *A. sesbaniae* David were reported to transmit Iranian strain of YMV from Iran (Kaiser, 1979). Detailed biology of *A. craccivora* was studied by Sri Kanth and Lakkundi (1988) on mungeban and urdbean at Bangalore and revealed that the life cycle of aphid on urdbean was completed earlier (10.34+4.11 days) than on mungbean (12.47+3.09). Li Dux *et al.* (1994) studied bionomics of *A. craccivora* in mungbean and found it was closely related with the crop growth period.

Three species of leaf hoppers *Empoasca kerri* Pruthi and *E. terminalis* Distant are known to damage mungbean and urdbean in Haryana, Delhi, India (Chaudhary *et al.*, 1980; Dhuri and Singh, 1983), while *E. moti* and *E. terminalis* have been recorded as serious pest of these crops at Ludhiana, India (Chhabra *et al.*, 1981a). Nath and Singh (1998) reported that Jassids (*Cicadellids*) first appeared at the seedling stage of the crop and maintained their population up to pod formation. Prakash and Rao (2000) reported three Jassid species *viz. Empoasea kerri* Pruthi, *E. moti* Pruthi and *E. terminalis* Dist. which suck the leaves, the former species is prevalent throughout India while the later are of regular occurrence in north India.

In mungbean and urdbean, four species of *Riptortus viz. R. fuscus* (Fab.), *R. linearis* (Fab.), *R. padestris* (Fab.) and *R. Pilosus* (Thunb) have been reported to damage in India and Philippines (Dhuri and Singh 1983; Sehgal and Ujagir 1988; Litsinger *et al.* (1988); Sharma and Dutta 1997). *Clavigralla gibbosa* (Spinola) has been reported to cause extensive damage in India to mungbean (Sehgal and Ujagir, 1988). Damage by *Chauliops fallax* Scott. to leaves, flowers and pods have been reported from Madhya Pradesh (Rawat and Sahu, 1973) and Uttar Pradesh (Lal, 1985) in India. *Nezara virudula* (Linn.) has been reported to cause damage to developing pods of mungbean in Madhya Pradesh, India (Singh and Rawat, 1982); in Philippines (Litsinger *et al.*, 1988). This pest has also been reported as destructive pest of mungbean in India, Philippines and Taiwan that damage pods and seeds. Highly toxic saliva is injectd into the plant tissue as it feeds. Pods may fail to develop or seed may be shrivelled from the punctures. Necrotic areas often develop at the feeding sites (Talekar and Chen, 1983). Damage by *N. virudula* is a major cause of flower drop and drying of pods during summer in India (Sehgal and Ujagir, 1988). Sharma and Dutta (1996) studied population dynamics of *Nezara viridula* (L.) on summer munbgean and found it to be appearing during 3rd and 4th week of April at the pod formation stage. Peak numbers of *N. viridula* coincided with the period of pod maturity and as the crop reached full maturity, pest numbers declined. There was a positive correlation between the minimum temperature and number of *N. viridula*. Natural enemies of *N. viridula* included the predatory thomisid *Thomisus* sp. and the *Scelionid Trissolcus* sp. Comparison of the population trends of pod sucking bugs on two summer green gram varieties was done in which the mean *R. linearis* population on AAU 34 during 1994 was 9.83 bugs/10 sweeps while the level was 11.87 bugs/10 sweeps during 1995. The population levels of both the bugs were higher on AAU 34 than on SG-1 variety, AAU 34 having stouter and taller plants with more branchs and bigger leaves and thus forming a larger canopy than SG-1, provided a more favourable micro climate to the bugs for survival and multiplication (Sharma and Dutta, 1997). *Riptortus pedestris* L. and *Nezara* sp. was reported from Vamban (Anonymous, 2002).

Piezodorus hybneri Gmelin and *P. rubrofusciatus* Fab. pentatomid bugs have been reported to damage leaves and pods of mungbean in Thailand and India, respectively whereas, *P. hybneri* is an economically important pest of mungbean and soyabean (Poonsavasde and Sukonthabhirom, 1990; Singh and Sehgal, 1993). The bugs normally suck the sap from the developing pods causing shrivelled, deformed and undersized pods and seeds.

In mungbean, mites belonging to family Tetranychidae are common and damage the crop by sucking the cell saps of leaves. Mites are of minor importance, but a severe infestation of *Tarsonemus* sp. has been reported by Singh and Chauhan (1984) from Hisar, India. *Tetranychus tilarius* Linn. *T. urticae* Koch, *T. cinnabarinus* (Boisd.) and *T. neocaldonicus* Andre have been reported to cause leaf and pod damage in mungbean in Tamil Nadu, Andhra Pradesh, India and South East Asia (Babu and Rajasekaran, 1988; COPR, 1981). *Polyphagotarsonemus latus* Banus has been reported to damage to mungbean at Ludhiana, India (Chhabra *et al.*, 1993a) and in SE Asia (Sepswadi *et al.*, 1990) by sucking sap of leaves of the crop.

Insect Pests at Bud, Flower and Pod Stages

The adult of blister beetle *Mylabris pustulata* Thunb feed on the flower on mungbean crop at Pantnagar (Sehgal and Ujagir, 1988), whereas Chhabra *et al.*, 1993a, also found this beetle to feed on the anthers of mungbean and urdbean crops in Punjab, India. *M. phalerata* Pallos has been reported to damage flowers of urdbean in Delhi (Dhuri and Singh 1983) and mungbean at Pantnagar (Singh and Sehgal, 1993).

Prakash and Rao (2000) reported blister beetle distribution throughout India and both the spp. *viz. M. pustulata* and *M. phalerata* are prevalent and they are flowers feeder. *Mylabris* sp. reported damaging mungbean crop during *Rabi* 2001-02 (Anonymous, 2002) which damage initiated from 6th week and persisted upto 9th week (7.3) at the time of senescence set in at Vamban; interestingly the beetles population was in increasing trend from its initiation (1.7 beetles/5 plants).

While considering bruchid in mungbean and urdbean, the major three species *viz. Callosobruchus chinensis* (L.) *C. analis* Fab., and *C. maculatus* are prevalent and have been reported to damage the pods and grain in field in India (Pruthi and Singh 1950; Gujar and Yadav 1978; Chhabra *et al.*, 1993a). These species also damage mungbean pods in Taiwan (Talekar 1988) and in Philippines (Litsiger *et al.*, 1988).

Singh and Singh (1990) tested the response of various cultivars of urdbean to field infestation of pulses beetle and found JU 78-3, RU-2, T-9 and JU 78-27 were free of *C. chinensis* infestation. Likewise, Dharmasena and Subasinghe (1986) studied seed characteristics contributing resistance to *Callosobruchus* spp. and found varieties with small seeds and glossy seed coats to be associated with higher degree of resistance than large seeds with a dull surface. Sarkar *et al.* (1991) screened sixteen genotypes of mungbean to test the resistance against *Callosobruchus maculatus* on the basis of the number of healthy seeds (resistance), number of eggs deposited (oviposition preference) and total emergence holes for 50 seeds. All three characters were highly heritable, but ovipositional preference was slightly lower in heritability than the other two. Resistances was negative function of adult survival, showing the importance of this character for selecting bruchid–resistant varieties. Haque *et al.* (1992) screened 150 genotyeps for succeptibility to *Callosobruchus maculatus* at 30°C and 70 per cent RH. The genotypes SBG 3 and UH 83-2 were the least susceptible (3.0–3.5 per cent weight loss) whereas PU-26 and UHB 2-26 were the most susceptible. Talekar and Lin (1992) also characterized mungbean resistance line against *C. chinensis.* Raghunathan and Ram (1992) screened mungbean varieties and found lowest feeding damage and adult emergence with cv. PDM-14. The work of Singh *et al.*, 1994; Singh *et al.*, 1995; Negi *et al.*, 1997; Verma 1998; Singh and Rine, 2000; Yadav *et al.*, 2000; Singh and Pandeey 2001; Srivastava and Kaur (2002) deserve special mention who screened a number of genotypes of different pulses against this dreaded stored pest.

Larvae of Lycaenids, *Euchrysops* spp. have been reported to damage pods of mungbean in Tamil Nadu (Jayaraj 1988) and Uttar Pradesh, India (Lal *et al.* (1980) and in North America (COPR 1981). Pods of mungbean and urdbean crops have been reported to be damaged by *Lampides boeticus* (L.) in India (Chhabra *et al.*, 1980; Dhuri and Singh, 1983). Damage to flowers and pods of mungbean by *Exelastis atomosa* (Wlsm) has been reported from Uttar Pradesh (Lal *et al.*, 1980). Pod borer *Maruca (testulalis) vitrata* is widely distributed in tropics. Eggs of this pest are deposited on petals and sepals, the larvae feed on the flowers and bore into the pods of *vigna* spp. crops. Young larvae of *M.* (testulalis) *vitrata* may cause flower bud shedding or destroy reproductive parts of flowers in mungbean in India (Chhabra *et al.*, 1980; Sehgal and Ujagir, 1988), in Philippines (Litsinger *et al.*, 1988) and in urdbean in India (Dhuri and Singh, 1983).

The other pests of bud flower and pods like *Ostrinia furnacalis* Guenae, *Etiella behrii Zeller* and *E. zinckennela* have been reported to web flowers and feed on pods of mungbean in Philippines. (Litsinger *et al.*, 1988). *Helicoverpa amigera* has been reported to feed on buds, flowers, pollen or immature pods of mungbean in India and is a serious pest at post flowering stage in Philippines (Chhabra *et al.*, 1980; Chaudhary and Sharma 1982; Litsinger *et al.*, 1988; Sehgal and Ujagir, 1988).

Pest Management

The mungbean and urdbean pulses of *vigna* group are by and large cultivated in the subsistence farming systems of hot, humid and semi-arid environments of torpical and subtropical parts of the globe. Besides, management of the scarce available resource for crop production, the farmers has to fight the on slought of a large and diverse pest fauna and comaplex of plant diseae pathogens. Despite availability of improved cultivars, which promise yield potentials of >1–2 t ha^{-1}, the farmers' average on – farm yields are often <0.4 t ha^{-1}.

For the successful cultivation of mungbean and urdbean crops, it is imperative to have some efficient management system with the farmers against the damaging effects of the large and diverse insect pest fauna. Under the new concept of IPM of harmful insect pest population, we must exploit the cultural practices, host plant resistance lines against different insect pests and disease pathogens, the endemic natural enemy fauna and options of biological control, and above all need based insecticides used to keep pest population below the economic threshold levels (ETL). Efforts should also be made to develop suitable and precise forecasting modules against the major pests of mungbean (Mathur *et al.*, 1997; Yadav and Singh, 2006). Almost similar objectives were put forth in an earlier review of mungbean pests in India by Sehgal and Ujagir (1988), but a lot more new information that has become available on different component strategies of IPM is documented here.

Cultural Control

The cultural practices evolved by farmers have sustained cultivation of *vigna* spp. crops, even under minimum input conditions and often in semi-arid environments. These practices which help destroy different stages of major insect pests or prevent them from causing economic injury, include tilling and cultivating relatively higher seed rate to compensate for seedling mortality, regularized irrigation and fertilizer application, low risk cropping schemes, manipulation of crop phonology by adjustments in sowing/harvesting dates and destruction of crop residues. (Asthana and Kim, 1996).

Manipulation in date of sowing in urdbean helped reducing the population of galerucid beetle and of jassid *Empoasca kerri* (Saxena *et al.*, 1973). Inter-cropping of mungbean with non-host plants like sorghum and pearl millet decreased incidence of jassids and other major pests populations which were high in mungbean and urdbean intercropping at Delhi. Similarly, inter-cropping of urdbean with pigeonpea delayed the appearance of green jassid, green bug and membracid (Singh and Singh, 1977). Relatively higher seed rate and synchronous planting of mungbean by the farmers in Philippines, together with crop rotation of grain legumes with rice and/or corn adopted by the farmers, have reduced damage due to flea beetles, major pests and of soil pests (Litsinger *et al.*, 1978b). In India too, mungebean sown at optimum time had reduced damage due to jassids, whitefly, galerucid beetle, pod borers and MYMV (Lal, 1983). Dasgupta and Chaudhary (1985) observed that intercropping of mungbean with urdbean had reduced MYMV incidence. Similarly, growing of trap crop, clean cultivation and crop rotation and destroying altrernate host, *Solanum nigrum*, had minimized beanfly, *Ophiomyia phaseoli* damage in India (Ramakrishnan, 1986).

Application of potash at 50 kg ha^{-1} reduced the pod borer infestation and increased yield by 36 per cent (Ramakrishnan, 1986). Dhuri *et al.* (1986) observed that urdbean intercropped with sesame, pigeonpea, mungbean and sorghum had influenced the incidence of galerucid beetle, jassid, whitefly, semi-looper, Bihar hairy caterpillar and pod borders. It was reported that there was reduction in jassid population in urdbean- sorghum and urdbean-sesame intecropping, however, intercropping with mungbean and pigeonpea further enhanced the population of this pest. Similarly, intercropping urdbean with sorghum and pigeonpea delayed the appearance of semi-looper, *Plusia orichalcea.* Intercropping with sorghum brought about significant reduction in the incidence of pod borers in the urdbean. Mungbean variety CO_4 intercroped with cotton variety MCU 10 registered a minimum incidence of leaf hopper (*Empoasca kerri*) and thrips (Janagarajan, 1987). Maize, sorghum and pearl millet when grown as barrier crops for mungbean, significantly reduced the population of whitefly, *B. tabaci* and incidence of MYMV and resulted in higher yield (Ravindrababu, 1987).

Different combinations of carbofuran 3G and phorate 10G along with nitrogen and phosphorus fertilizer, significantly reduced the leaf damage due to *M. obscurella* and *Anomis flava* (Gatoria and Singh 1988). Hand picking of egg masses and young larvae play an important role in controlling the Bihar hairy caterpillar (*D. obliqua*), the banded beetle (*M. phalerata*), the brown bug (*C. gibbosa*) and the tobacco caterpiller (*S. litura*) (Lal, 1987).

Thus, manipulation of dates of sowing and harvest to escape peak pest population; intercropping preferably with non-legume hosts; hand picking of egg masses of hairy caterpillars; proper soil preparation techniques and application of high dose of fertilizers together with carbofuran treatment had helped achieve optimum control of harmful insect pest population.

As regards cultural control measures of bruchid, time and frequency of harvesting, removal of alternate host plants, multiple cropping line, intercropping beans with maize reduced the abundance of the bruchid probably because of the shade given by the maize canopy (Labeyrie and Maison, 1957). Besides, field infestation may result from bruchids migrating from infested seeds in adjacent grainaries to the ripening crop (Prevett, 1961 and Caswell, 1973). Therefore, infested seeds in the warehouses or grainaries should be treated or disposed of before the planting season.

High insect proliferation rates in the tropics make good storage hygiene particularly important (De Lima, 1979a) when farmers in Nigeria after harvest, leave the pods in heaps or harvest the whole plant and remove the pods later pods are open to attack by bruchids. Storage in the grainaries immediately harvest gave physical protection and considerably reduced infestation (Caswell, 1968).

Host Plan Resistance (HPR)

The success of HPR to insect pests, as IPM stragegy depends on the efficacy of other available control options. Litsinger *et al.* (1978a) had suggested that in Philippines, priority in breeding for insect resistance in mungbean be given to flea bettles, leaf hopper and stemflies, as these are difficult to control by chemicals. HPR studies in India have largely focused on mungbean and urdbean resistance to key pests like whitefly, jassids, thrips and lepidopteran pod borers (Chhabra and Kooner 1988; Chhabra *et al.*, 1988). Genotypes resistant or least susceptible to major insect pests of mungbean and urdbean at glance are given in Table 6.2.

Screening of mungbean resistant cultivars against stemflies at Asian Vegetable Research and Development Centre (AVRDC), Taiwan, has led to the identification of 3 AVRDC, accessions, V 2396, V 3495 and 4281 as resistant to *M. Sojae, O. Centrosematis* and *O. phaseoli* (Talekar 1990). Balboa (1972) reported Mungong Tagalog as most resistant cultivars of mungbean in Philippines against *O. Phaseoli,* and two AVRDC accessions V 1360 and V 2816 were reported as least susceptible in Indonesia (Tengkano and Soekarna 1976), whereas V 2535 and V 2533 were identified as moderately resistant in China (Lin, 1981). Gunasekaran *et al.* (1977) screened 160 urdbean cultivars against *O. phaseoli* and identified No. 55 highly resistant to the fly and also to pod borer, *Catechrysops cnejus.* High trichome density on stems and leaves, purplish small diameter stem and small unifoliate leaves contributed to resistance against *Ophiomyia* spp. in munbgean (Talekar and Lee 1988), whereas a positive correlation was found between the thickness of mungbean leaf and stemfly infection. This implied that stemflies prefer to eat or lay eggs in thick leaf varieties (Lin and Rose, 1976).

Extensive research programme on HPR studies at Ludhiana, India showed that out of about 600 mungbean genotypes screened against whitefly–MYMV complex, jassids and pod borers (*L. boeticus, M. testulalis* and *H. armigera*), ML 1, ML 3, ML 5, ML 170, ML 194, ML 195, ML 235 and 15127 were identified as resistant. (Chhabra and Kooner 1980; Chhabra *et al.*, 1980). While studying the mechanism of resistance, Chhabra *et al.* (1981b) found the resistant cultivars having higher contents of phenols, free amino acids and non-reducing sugars at the green leaves stage than the 'infestor'. Three genotypes ML 337, ML 423 and ML 428 were resistant not only to whitefly, jassids and MYMV, but to three pod borer complex also, and the basis of resistance was antibiosis (Chabbra *et al.*, 1988).

Table 6.2: Genotypes Resistant or Least Susceptible to Major Insect-Pests in Mungbean and Urdbean at Glance

Pests	*Mungbean*	*Urdbean*
Stemfly	V 2396, V 3495, V 4281, V 2335, V 2533, COGG 407, -408, -424, KM 2, PUSA 107, -16, -8731, 8733, MH 81-7, MG 212, MUG 154, -249, G 65, NARP I, SML 82, -32, -69, -100, -134, -154, -168, -169, -178, -26, 269.	UG 218, PDU 1, -5, Sel 37, COGB 305, ACC 305, Trichy, M 10, NP 7, CO 2-100, TU 94-2, UPU 95-1, AKU 7, LBG 707, -689, -174, -694, UG 737, AKU 7.
Thrips	MG 815, -331, -423, -218, -368, G 65, MUG 212, UPM 83-6, -82-4, -84-4, M 81, MU 9161, PUSA 119, K 851, NARP 280, SML 7, -69, -77, -99, 100, -97, -103, -110, -112, -168, -130, -191, -258, -264, 84047, PDM 30, -36.	PDU 5, KB 63, PDU 88-23, -2, -5, UG 567, DU 4, T9, UH 80-4, UH 90-9,UH 80-7.
Jassids	Phule M 9037, -9505 PDM 40, PDM 84, -143.	–
Pod Borers	ML 131, OUM 6, PUSA 104, -105, -119, PDM 84-139, PDM 85-199, PDM 11, ML 5, -370, PDM 84-115, PDM 54, AB 19, PS 933, PDM 91-249, LGG 458, MGG 538, MG 316, GM 83-13, PDM 84, PDM 84-143, PMB 27, PUSA 104, -117, ML 131, -332, P 102, -103, JRVM 133, NPRG 3, B 24-88, MGG 341, ML 5, -131, PUSA 9531, PB 11, PMB 34, RMG 353, K 851, PUSA 9531, RMG 353, ML 5, PMB 27, PMB 14, LGG 460.	V 2135, 3274, -3276, -3229, -3277, -3812, -3949, -3281, -4411 Sarla, UG 737, PLU 557, TAU 1, -2
Madhurasia obscurella Galerucid Beetle (Flea Beetle)	LGG 470, -471, -475, M 2, PDM 54, ML 131, OVM 7, B 24-8-4, B 3-8-8, B 12-4-4, HY 4-3, TT 9-E, PDM 11-80	LAB 697, -611, -702, Pant U 35, UPU 95-1, TU 94-2, KU 92-1, BG 12/15, Sarla, PDU 2.
Bruchids	V 2709, V 2802	VM 2011, 2164

Field screening of mungbean breeding material during different years, Chhabra and Kooner (1992a, 1992b) let to the identification of ML 484, MH 309 and PDM 84-146 out of 67 and ML 370, ML 459, ML 506, ML 508 and ML 536 out of 48 test entries as resistant to whitefly MYMV complex and jassids. In further screening of mungbean material, Chhabra and Kooner (1993) contributed four more genotypes ML 267, ML 326, ML 372 and Pusa 326 as resistant to insect pests and MYMV.

In summer mungbean cultivars SML 77, SML 99, SML 100, SML 103, SML 112, Pusa 107, UPM 82-4 and UPM 83-6 have been identified as resistant to thrips at Ludhiana, India (Chhabra and Kooner 1988; Chhabra *et al.*, 1993b).

Out of > 475 urdbean accessions, Chhabra and Kooner (1981) and Chhabra *et al.* (1984), identified 4-5-2, LU 15, LU 178, LU 190, LU 196, LU 330, LU 397, LU 426, LU 436, UL 29,UL 257, UG 119, UG 187, UG 302, UG 407 and Pant U 30 as resistant to whitefly–MYMV and jassids. Chhabra *et al.* (1986a), in the field testing of 30 urdbean genotypes, identified LU 335, LU 274, LU 332, LU 470 as moderately resistant to *A. craccivora* and M 1-1 as highly resistant to aphids.

Ecological Approach

Every effort is being made to raise yield of pulse by adopting modern agricultural practices *viz.*, use of high yielding varieties, heavy manuring and assured irrigation in order to meet the growing demands of pulses. The ecological approach as a component to the pest management is also an imperative and indispensable component like developing forecasting modules against the major pests on different hot spots in varying agro-eco-system by studying pests population fluctuation in relation to agro-eco-factors (Kalaiyarasan and Palanisamy, 2004; Khan and Ramamurthy, 2004; Reddy and Kumar, 2005 and Yadav and Singh, 2006).

Chaudhary and Senapati (2004); Badjena and Mandal (2005) found positive association of thrips with temperature during summer. Yadav and Singh (2006) reported a unit increase in maximum daily temperature caused an increased of 1.97 thrips/plant at Kanpur and 2.7 thrips/plant at Etawah. A unit increase in minimum temperature caused an increase of 1.30 jassids/plant at Kanpur. The peak incidence of YMD (64.5-78.7 per cent) average temperature raised from 29.0-30.5°C and RH 52.52-72.0 per cent during rainy season was recorded at both the location while during summer average temperature ranged from 31.0-38.5°C alongwith average relative humidity ranging between 20.50-59.2 per cent was more conductive for the spread of YMD transmitted by whitefly (*B. tabaci*).

Active bruchid beetles days eggs on soft and smooth pods of mungbean (Doria and Raros, 1977, Dias and Yadav, 1988). Talekar and Lin (1981) reported that resistance in mungbean accession 2011 was due to dense hairs on pods and *C. chinensis* adults avoided alighting on these pods. They found hardiness of seed coat in mungbean was behind the resistance against *C. chinensis.* Rathore and Chaturvedi (1997) postulated that combination of several factors could be responsible other than texture alone for preference for oviposition by bruchids.

Seed protein of *Dolichos lablab, Lablab purpureum* and *Rhyncocea cana* revealed the presence high level of non-nutritional chemicals such as viclins, α-amylase inhibitors and lectins in quantities that may impart resistance to *C. maculatus* (Fory *et al.*, 1996; Ignacimuthu *et al.*, 2000). Dongre *et al.* (1996) evaluated cultivated and wild accessions of mungbean and urdbean and resistance to *C. maculatus* was observed in wild progenitor of urdbean, *vigna mungo* Var. *silvestris.* Cross of wild type with urdbean accession TU1 revealed presence of two dominant duplicate genes and designated as Cmr1 and Cmr2. Resistance to *C. chinensis* in an accession of TC 1966 of *Vigna radiata* var. *Sublobata* was reported by Fujii and Miyazaki (1987) and resistance found to be controlled by single dominant gene-R (Kitamura *et al.*, 1988).

Varieties/genotypes reported to possess low to moderate level of resistance bruchids were reported *viz.*, IGG2 and YKR 415/2 against *C. maculatus* in mungbean (Eduard and Gynathilagaraj, 1994); Ganapathy and Janarthanan (1995); and urdbean genotypes *viz.*, Vamban 1 against this pest was reported by Edward and Gunathilagaraj (1994).

As regards resistance accessions of wild progenitors, V 2709, V 2802, *Vigna sublobata* Acc. TC 1966 in mungbean and VM 2011 (pod resistance) and VM 2164 in urdbean have been reported by Talekar, 1996.

Biological Control

Studies on natual enemies have been limited to the identification of natural control agents. Extensive research will, therefore, be required before these could be exploited for the control of insect pests of mungbean and urdbean.

Out of the 123 natual control agents, 88 are parasitoids, 28 predators and 7 pathogens recorded on whitefly, jassid, aphid, stink bug, green bug, beanfly, red hairy caterpillar, Bihar hairy caterpillar, tobacco caterpillar, pod borers and bruchids.

Bemisia tabaci (Genn.) predation of immature stages by *Brumus* sp., *Chrysopa* spp., *C. cymbela* Banks, *Cybocephalus* sp. and parasitization of nymph and pupa by *Eretomocerus masii* Silvi *Prospaltella flava* Shaffee *and Geocoris tricolor* Fab. have been reported from India and Philippines (Nair 1986; Lal 1987; Litsinger *et al.*, 1988). Adult mortality due to NPV and fungus, *Poecilomyces farinosus* was reported by Nene (1973a, 1973b).

While considering *Empoasca kerri* Pruthi, five nymphal predators *viz. Chrysopa cymbela, Camponotus* sp., *Acantholepis simplex* and *Coccinella septumpunctata* and a nymphal and adult predator, *Distina albida* were reported from India (Nayar *et al.*, 1976; Dhuri and Singh 1983; Faleiro and Singh 1990) and two egg parasitoids, *Gonatocerus* spp. and *Oligosita* sp. have been recorded from Philippines (Litsinger *et al.*, 1988).

As regards *Aphis craccivora* Koch, predation of nymph and adult by 15 predators *viz. Brumus suturalis, Chilocorus nigritus, Coccinella repanda transveralis, Harmonia arcuata, Ischiodan javana, I. scutellaris, Menochiles sexmaculatus, Vernia* sp., *Cyclosa insulana, Lencauge celebesiana* and *Theridion* sp., *Chrysopa carnea, C. septumpunctata, M. sexmaculatus, Xantogramma scutellare,* was reported by Khan and Hussain (1965), Patel *et al.* (1976) Raychaudhari *et al.* (1979). Parasitization of symphs and adults by *Diaenetiella rapae, Praon* spp. and *Thrioxys indicus* have been reported by Stary and Ghosh (1978) and Sinha and Singh (1980). On *Piezodorus hybneri* Genelin, two egg parasitoids *Telenomus Telenomus* sp. and *Trissolucus* sp. have been reported in Thailand (Poonsavasde and Sukonthabhirom, 1990). As regards *Nezaa viridula* (Linn.) it is parasitized by a nymphal parasitoid, *Trissolucus basalis,* which has been reported from Austalia (Turner, 1983).

So far as the *Melanagromyza sojae* (Zehntner), is concerned, its two larval/pupal parasitoids *Afrostilba* sp. and *Opius oleracei,* have been recorded on mungbean in India by Singh (1982). While on *Ophiomyia phaseoli,* parasitization of larva and pupa by 14 parasitoids *viz. Plutarchia* sp., *Chrysomotomyia douglasi, Hemiptarsensus* sp. *H. agromyzae, H. biclavus, Opius oleracei, Poycystomyia benefica, Sphegigaster agromyzae, Crytoprymna* sp. *Diaentiella rapae, Eurytoma* spp. *E. poloni, Halticoptera* sp. and *Parattrigonogastra rugosa* from Australia, Egypt, Malaysia and Philippines has been reported by Ooi (1973), Kleinschmidt (1970), Abdul Nasr and Assem (1968), Otanes (1918).

NPV has been reported to cause larval mortality in *A. moorei* (Nene 1973b) and four parasitoids, *Apanteles* spp. *Carcelia evolens, Sturania inconspicuella* and *Tachina* (*Exorisia*) fallax and a predator *Mermis indica* have been recorded as larval biocontrol agents (Lal, 1987; Dhuri and Singh 1983; Mathur and Singh, 1998).

On *Helicoverpa armigera* (Hubner) Lazarevic (1971) recorded a larval parasitoid, *Disophrys nigricornis* in Sudan. Bhatnagar *et al.* (1982) reported 27 species of Bethylids, Braconids, Ichneumonides, Tachinids and Trichogrammatids parasitoids from the states of south central India. These workers recorded 5 egg parasitoids, 2 egg/larval parasitoids, 14 larval parasitoids and 4 larval/pupal parasitoids. These workers further reported 19 species of insect predators of eggs and larvae of *H. armigera.* A bacterial biocontrol agent *Bacillus thuringiensis* has been reported to cause larval mortality (COPR 1981). Singh and Mathur (1998-99) while working on life table studies found a high degree of death by parasitoid are pathogen on *H. armigero.* A predator *Ovomermis albicans* has been reported by Usha and Singh (1978) in India on *Maruca* (*testulalis*) *vitrata.*

On *Spilosoma* (=*Diacrisia*) *obliqua* (Walker), Krishnamurthi and Usman (1954) reported *Apanteles creatonoti*, parasitizing larvae. These workers also recorded two egg parasitoids *Trichogramma chilonis* and *T. minutum*, and 9 species of larval parasitoids *A. flavips*, *A. jayanagorensis*, *A. obliqua*, *A. ruidus*, *A. ater*, *A. vitripennis*, *Charops* sp. and *Drino* sp. *S. obliqua* (Kalra *et al.*, 1962; Shetgar *et al.*, 1990).

While accounting *Spodoptera litura* (Fab.), a larval parasitoid *Bracon hebetor*, an egg and larval parasitoid *Chelonus blackburni* and pupal parasitoid *Tetrastichus israeli* have been reported from Tamil Nadu by Udhyakumar and Jayaraj (1987). Krishnaiah *et al.* (1984) reported NPV causing larval mortality. Mohan *et al.* (2000) reported *Gryon* sp. a potential parasitoid on *Spodoptera litura* in Kanpur.

On storage legume pest, *Callosobruchus* spp., an egg parasitoid *Chaetostricha mukerji* and larval parasitoid *Bruchobius laticeps of C. chinensis* have been reported by Pruthi and Singh (1950). Biological control of bruchids has yet received very little attention. It is recommended that possibilities of classical biological control are assed. Among the natural enemies of bruchids, braconids, Trichogrammalid, Eulophid, Eupelmid Torymid, Pteromalid, Eurytomalid and mite are and the major which kills various stages of the pest, needs their inoculation, augmentation, inundation and conservation (Huis, 1991).

Botanical Control

Use of botanical under field conditions are not so promising as under storage condition. The pulse beetle *Callosobruchus* spp. is the major insect pest of mungbean and urdbean and other pulses causing substantial damages in the storage. In recent past, the use of plant materials and vegetable oils has acquired an important approach for the control of pests. Few significant contributions made so far using plant materials (Pandey and Singh, 1995; Thawsatine, 1987; Rajpakse *et al.*, 1998; Ram Singh *et al.*, 2001 and Singh *et al.*, 2004. The vegetative oils have been reported to inhibit oviposition, which results reduction in multiplication of bruchids (Invigaro, 1990 and Reddy *et al.*, 1994). Ginger rhizome powder and neem leaf powder has been reported to be effective in reducing the egg laying of beetle (Echendu *et al.*, 1998 and Rajpakse *et al.*, 1998; Singh *et al.*, 2004 found Bidang leaf powder (*Embelia ribes*) was reported to be significantly more effective on mungbean than neem leaf powder (*Azadiracta indica*).

The treatment with easter and neem oil prevents the mungbean seeds upto 75 days after treatments Ketker *et al.*, 1987; Reddy *et al.*, 1994 and Yadav *et al.*, 2004.

Chemical Control

In general, this option of control is favoured by farmers due to its quick response. Evidence of substantial increase in yield by the use of chemical insecticides in controlling insect pests of the crops, are on records with several research publications and reports globally.

Chemical Control of Insect Pests at Root and Seedling Stages

In mungbean and urdbean, *Ophiomyia phaseoli* cause extensive damage at seedling stage in India, Philippines, Taiwan, Indonesia, Malaysia, Thailand and African countries. Soil application of granular insecticides *viz.* aldicarb, carbofuran, disulfoton and phorate at 1.0 kg, 1.5 kg and 2.0 kg alongwith seed treatment with carbofuran @ 2 g and 4 g per 100 g of seed gave effective control of fly and substantial increase in yield (Roongsook *et al.*, 1973; Su *et al.*, 1976; Vyas and Saxena 1980; Gupta and Singh 1984b; Lal 1985; Catipon 1986; Sehgal and Ujagir 1988). In the spray formulation, carbaryl, chlorpyriphos, dimethoate, endosulfan, malathion, mavinphos, monocrotophos and thiometon were identified to be very effective for the control of beanfly (Harnoto 1976; Ibrahim 1977; Akinbohungbe, 1982; Mehrotra *et al.*, 1989). A comprehensive review on the chemical control of agromyzid flies giving details of presowing application of insecticides, seed treatment, soil treatment and post-sowing application of insecticides have been dealt by Talekar (1990).

Chemical Control of Insect Pests at Vegetative Stage

Foliage Feeder

For the control of galerucid beetle, whitefly and jassid on mungbean and urdbean, soil application of aldicarb, carbofuran, disulfoton, mephospholon at the time of sowing gave significant control of these pests (Chaudhary *et al.*, 1981; Yein and Singh 1981; Vyas and Saxena 1983; Singh and Gupta 1986). Similarly, seed treatment of mungbean with carbofuran 4 per cent followed by disulfotan and phorate granules at 1.5 kg a.i. ha^{-1} proved consistently effective against *M. obscurella* (Vyas and Saxena 1981). Sprays of chlorphyriphos, dimethoate, monocrotophos reduced the damage caused by *M. obscurella* (Yadav and Yadav, 1979; Chander and Singh 1989a). Endosulfan sprays 6 and 8 weeks after sowing gave the successful control of the beetle and semilooper, *Anomis flava* (Yein and Singh, 1981). Dust and emulsion formulations of fenvalerate rendered effective control of *S. obliqua* on mungbean and urdbean (Chhabra and Kooner, 1987).

Vekaria and Vyas (1985) evaluated the bioefficacy of certain insecticides for the control of *H. armigera* on mungbean and reported that fenvalerate, monocrotophos, endosulfan, quinalphos and phosalone alone or in combination with miraculon were effective in controlling the pest. Gupta and Singh (1988) reported that phorate 10G, mephoslon 5G, disulfoton 5G, carbofuran 3G and aldicarb 10G at 2 kg a.i. ha^{-1} at sowing, reduced population of leaf feeding caterpillar. Dust formulation of quinalphos 1.5 per cent, sevithion 4 per cent, carbaryl 5 per cent and 10 per cent, BHC 10 per cent and parathion 2 per cent were effective in controlling foliage flower-feeding weevil, *Apion collaris* (Kapoor *et al.*, 1984).

Sucking Pests

Effective control of whitefly, *B. tabaci* was obtained by the application of granular insecticides aldicarb 10G, disulfoton 5G, phorate 10G at 2 kg a.i. ha^{-1} and two sprays of monocrotophos 0.04 per cent also resulted in good control of *B. tabaci* and reduced the incidence of YMV (Murugesan and Chelliah 1978; Chaudhary *et al.*, 1981; Sharma and Verma, 1982). Mixing 2 per cent orchid oil with monocrotophos enhanced the efficacy of whitefly control (Vadodaria and Vyas, 1987), whereas Singh *et al.* (1981) reported the adverse effective of oil application alone. Mathur and Singh (1998) reported the granular effect of phorat or Carbofuran as basal application with two foliar spray of either with Mythyl-o-demeton or Monocrotophos or Cypermethrin suprss YMV intensity significantly.

Sprays of monocrotophos, endosulfan, fenitrothion, chlorfen-vinphos, malathion, phasalone, vamidothion, dimethoate at the concentrations of 0.075 per cent, 0.07 per cent, 0.05 per cent, 0.04 per cent and 0.03 per cent were found effective by Sidhu and Sandhu (1971); Vyas and Saxena (1983); Yein and Singh (1981); Gatoria and Singh (1984, 1988) for the control of jassids, *E. kerri, E. moli, A. biguttulu, A. devastans.* Galceron and dimethoate were most effective and gave maximum kill of mite *Polyphagotarsonemus latus* (Banks) (Gupa and Dhooria 1973; Chhabra *et al.*, 1993a).

In summer mungbean effective control of thrips, *M. distalis* has been obtained by the use of systemic insecticides dimethoate, monocrotophos, oxydemeton methyl and phosphamidon, malathion and endosulfan as per spray by Chhabra and Kooner (1985a, 1985b, 1986), Chhabra *et al.* (1986b). Damage of thrips has been reported to be reduced with a combination of aldicarb at 1 kg a.i. ha^{-1} and spray of monocrotophos 0.04 per cent (Chander and Singh, 1989b). Mathur and Singh (1998) reported the combination of phorate granules as basal application with two foliar sprays of cypermethrin or methyl-o-demeton at 35 and 50 days after sowing was most effective in reducing the population of thrips and jassids.

Chemical Control of Insect Pests at Flowering and Podding Stages

Bioefficacy of fenvalerate, monocrotophos, endosulfan, quinalphos and phosalone alone or in combination with miraculon were effective in controlling the pod borers *C. ptychora, M. vitrata* and *H. armigera* on mungbean. The highest yield was obtained in the treament with fenvalerate 0.01 per cent + miraculon (Vekaria and Vyas, 1985). Whereas, spray application at flower initiation with endosulfan 0.07 per cent, dimethoate 0.03 per cent, methyldemeton 0.025 per cent, monocrotophos 0.04 per cent, deltamethrin 0.01 per cent, fenvalerate 0.004 per cent also gave significant control of pod borer complex in mungbean (Bhalani and Parsana, 1986; Bhat *et al.*, 1988; Chander and Singh, 1989b). Synthetic pyrethroids, permethrin, cypermethrin and deltamethrin were found equally effective for the control of pod borers by Srivastava *et al.* (1984) and Sukul *et al.* (1987). Combination of disulfoton of 1 kg a.i. ha^{-1} as basal application followed by endosulfan 0.07 per cent spray too reduced the pod borer complex damage in mungbean and urdbean (Sundararaju and Rangarajan 1987a). Dusting with endosulfan 4 per cent, folidol 2 per cent and phosalone 4 per cent was also effective against pod borers and burchids on mungbean and urdbean crops (Rajendran *et al.*, 1981). Chlorpyriphos and cynofenphos were found superior in reducing *Etiella zinckenalla* infestation on mungbean in Indonesia (CRIA, 1971), whereas Litsinger *et al.* (1980) found Decis at 0.05 kg a.i. ha^{-1} applied either twice at 30 and 42 days after emergence (DAE) or three timss at 30, 40 and 50 DAE gave excellent control of post flowering insects on mungban in Philippines.

Combination of Insecticides and Fertilizers for the Control of Insect pest Complex of Mungbean and Urdbean

Application of phorate 10G, carbofuron 3G and aldicarb 10G at 1.5 kg a.i. ha^{-1} alone or in combination with fertilizers in mungbean, reduced jassid and whitefly population significantly (Yein and Singh 1981 Gatoria and Singh, 1984). Similarly, application of carbofuron 3G alone or in combination with fertilizers, reduced the incidence of galerucid beetle, *M. obscurella* (Gatoria and Singh, 1988) and disulfoton 5G at 1 kg a.i. ha^{-1} or endosulfan 0.07 per cent in combination with fertilizers or alone were effective in controlling pod borer in mungbean (Sundararaju and Rangarajan, 1987b).

Phytotonic effect of carbofuran soil application and seed treatment of greengram significantly improved plant growth by controlling the pest complex in India and gave maximum green fodder and dry matter yields (Patel and Srivastava 1990).

Effect of Soil Application and Seed Treatment of Insecticides on Germination

For the control of beanflies, sucking pests like whitefly, jassids, aphids, thrips and mites, presowing application or seed treatment is quite common. Work done on this aspects has revealed that persistence of carbaryl, a soil insecticide for the control of insect pests in mungbean, at a concentration of > 30 ppm, markedly reduced the germination and decreased survival of plants (Gupta and Mishra 1970). Similarly, seed treatment of mungbean with insecticides carbofuran, monocrotophos, dimethoate, phosphamidon and formothion at 1, 3 5 a.i. on 100 g of seed, considerably reduced germination percentage in all the treatments of dimethoate and with 5 per cent a.i. of monocrotophos (Pawar and Chowan 1979). Venkataramaiah *et al.* (1980) found that the blackgram seedlings treated with DDT showed retarded growth of the seedlings. Presowing soil application of HCH and Aldrin at 50 kg ha^{-1} also reduced the plant stand of mungbean (Verma *et al.*, 1983).

As regards bruchid its established infestation can only be controlled in bulk and bag storage in moderate to large quantities by fumigation with safe chemicals to eliminate the infestation. Treatments kill all stages of brunchids. EDB (ethylene dibromide) @ 3 ml/q with an exposure period of 7 days proved very effective. Ampules are available and farmers can use them easily. The only prerequisite is that the space treated should be made air-tigh by placing mud on the lid or cover gunny bags with thick polythene sheet. When quantities more than 2 tonnes to be furmigated, use AlP (Aluminium phosphide). AlP, available in lablets, is highly effective in controlling bruchids in longer space, should be used under expert guidance. To reduce the risk of above furmigants and to utilize rural based technology, the biogas produced at the village level can be used for the control of bruchids. Ensure air light store before furmigation. Exposure period of 7 days provide 100 per cent kill of eggs, grubs and adults (Mohan *et al.*, 1989).

Since no single method of control is effective for pests control, it is imperative to adopt holistic approach by integrating individual components to evolve an effective, efficient, cost effective and safe IPM strategy.

Future Agenda for Thrust Areas

Concept of Integrated Pest Management (IPM) implies effective use of as many control measures as are compatible to suppress pest populations below damaging levels and optimize yields with minimum disruption of or damage to the environment. Much debate exists on the status of IPM for the *Vigna* spp. corps in India, Africa and elsewhere in the tropics, but little efforts have been made, to implement the viewpoints that have often emerged. The possible reasons often putforth are that not sufficient information is available on various components necessary for the formulation of IPM programme. However, this review clearly shows that considerable knowledge already exists concerning pest management systems in mungbean and urdbean to begin to formulate IPM strategis. Modifications to IPM approaches can be made later on as our knowledge of HPR, ETL and EIL for different pests in the agro-ecosystem of the locality increases. It is to be underlined here that IPM strategies need to be initiated urgently to counteract the disastrous effects of possible over dependence on chemical pesticide to obtain quick control of different pests. Thus, results and experience gained so far in the *Vigna* spp. crops, the followinhg further agenda for IPM is suggested.

Information on the extent of damage by major pests of *Vigna* spp. crops in different agro-ecological zones need to be fully documented to focus our research on pest management and HPR studies on the key pests in each area. For this, an ideal survey can be planned coupled with methodology combining systematic sampling with statistics.

Manipulation of date of sowing, seed rate and crop rotation have given significant reduction in the lepidopteran and coleopteran foliage feeders and soil pests. Promoting the right sort of cropping system's diversity by raising companion and/or barrier crops such as sorghum, pearl-millet, maize, cassava to minimize the population of jassid, flower thrips, spotted pod borer has given a lead to identity many other compatible crops and other methods of cultural control, not only to reduce the incidence and damage by the pests and diseases but to improve the economy of farmers.

Studies on HPR against key endemic pests and the biochemical/biophysical basis, there of can make rapid progress in developing insect-resistant cultivars. Resistant markers in mungbean and urdbean in India for beanfly, whitefly–YMV complex, thrips and jassids are known. Plant breeders may choose the best parents for their crosses to consolidate insect resistance in high yielding genotypes. Development of more genetic materials with multiple resistance or atleast moderate resistance for key pests of the area should be kept on research priorities.

Although certain biocontrol agents have been identified for use in the control of some major pests like whitefly, jassid, aphids, *M. vitrata,* gram pod borer, hairy caterpillars which attack *Vigna* spp. crops. Among them *Chrysops* spp., *Apanteles* spp., *Trichogramma* spp., *Bacillus thuringiensis* and NPV are suitable candidates for improving the biological control of some of these pests (Table 6.2). Its potential should be assessed for other major pests in the areas/zones of the country where they are virulent.

Research on botanicals and on plant resources with pest suppressing qualities that are within the reach of farmers in rural areas, should be intensified. This will require identification of more such plants, evaluation of extracts for insecticidal properties and application methodology.

Studies on proper sampling methodology, spatial distribution pattern ETL, EIL, life tables and forecasting modules of key endemic pests of *Vigna* spp. crops are needed to generate data base for judicious need base application of eco-friendly chemical insecticides and to minimize risks of insecticide resistance and pest resurgence.

Conclusion

Pulses are an important component of the predominantly vegetarian diet of the Indian Society. Besides being rich in protein, they are also important for sustainable agricultural production system. Mungbean, *Vigna radiata* (L.) Wilczek and urdbean *V. mungo* (L.) Hepper are widely grown food legumes, in the entire Indian subcontinent, during the spring/summer season and the rainy seasons. In India, mungbean and urdbean are grown in about 60 lac ha land (Singh *et al.,* 2006). About 271 species of insect pests are known to damage these crops during the entire growing season and also as store grain pests (Bhatnagar and Dahia, 2005). The number of insect species found infesting mungbean is more than that in urdbean, which varies with the stages of crop growth (Patnaik *et al.,* 2000). Severe losses are caused by a complex of stem flies during the seedling stage, a large defoliator fauna consisting of grasshopper, flea beetles, hairy caterpillars, semi-loopers and other lepidopteran caterpillars during the vegetative stage and by a varied flower and pod feeding fauna consisting of sucking bugs, thrips, flower and pod webber and lepidopteran pod borers. Singh and Yadav (2002-03) reported bruchid infesting mungbean with 1-4 eggs/grain and egg hatchability of 85.6 per cent on it. This large insect pest fauna is normally regulated by an equally large and varied parasitoid and predator fauna and also by disease pathogens, which if conserved, provide an effective natural control of crop losses. Insect pest problem is relatively much less in summer crops due to hot and dry weather conditions as compared to the rainy season crop. In rainy season crops, profuse vegetative growth is attained during September which coincides with the peak activities of a variety of insects. The article encompasses a review of pest array on mungbean and urdbean and their management.

References

Abul-Nasr, S. and Assem, M.A.H., 1968. Studies on biological process of the beanfly, *Melanagromyza phaseoli* (Tryon). *Bull. Soc. Entomol., Egypt,* 52: 283–295.

Agyen-Sampong, M., 1978. Pests of cowpea and their control in Ghana. In: *Pests of Grain Legumes: Ecology and Control,* (Eds.) S.R. Singh, H.F. van Emden and T.A. Taylor, p. 85–92.

Ahmad, R., Yadava, C.P. and Lal, S.S., 1998. Evaluation of spray schedule for the control of insect pest of mungbean. *Ind. J. of Pul. Res.,* 11(2): 146–148.

Akingbohungbe, A.E., 1982. Seasonal variation in cowpea crop performance at Ile-Ife, Nigeria and the relationship to insect damage. *Insect Sci. Appl.,* 3: 287–296.

Allen, D.J, Anno-Nyako, F.O., Ochieng, R.S. and Ratinam, M., 1981. Beetle transmission of cowpea mottle and southern bean mosaic viruses in West Africa. *Trop. Agric.* (Trinidad), 58: 171–175.

Amoako-Atta, B., Omolo E.O. and Kidega, E.K., 1983. Influence of maize, cowpea and sorghum intercropping systems on stem/pod borer infestations. *Insect Sci. Appl.*, 4: 47–57.

Anonymous, 1976. Outbreak of pests and diseases. *Quarterly Newsletter*, FAO Plant Protection Committee for the South-East Asia and Pacific Region, 19(3): 5–6.

Anonymous, 1993. Project Director's Report, *Kharif* Pulses Group Meeting, AICPIP (ICAR), QUAT. Bhubaneswar, 12–14 May, Annexure III and IV.

Anonymous, 2001. *Annual Reports Kharif*, AICRP on MULLARP (ICAR), IIPR, Kanpur, pp. 96.

Anonymous, 2002. *Annual Report (Kharif)* All India Coordinated Research Project on MULLARP (India Council of Agricultural Research) IIPR, Kanpur, pp. 92–109.

Anonymous, 2004. Low cost technology for safe storage of pulses. *National Agricultural Technology Project, Research Achievements* (1999–2003). Indian Institute Pusles Research, Kanpur, pp. 49–52.

Asthana, A.N. and Kim, Doo Hwan, 1996. *Recent advance in Mungbean Research*. Indian Society of Pulses Research, IIPR, Kanpur, p. 219.

Atwal, A.S. and Dhaliwal, G.S., 1986. Polyphagous pests. In: *Agricultural Pests of South Asia and their Management*. Kalyani Publishers, Ludhiana, p. 359–376.

Babu, S.C. and Rajasekaran, B., 1988. Economics of integrated pest management in mungbean production: Research of a linear programming model. In: *Mungbean Proceedings of Second International Symposium*, Bangkok, (Ed.) S. Shanmugas-undaram, pp. 343–349.

Balboa, F.C., 1972. Comparative resistance of five varieties of mung, *Phaseolus aureus* Roxb., to the attacks of the beanfly, *Agromyza phaseoli Coq. Araneta Research Journal*, 19: 66–79.

Bashir, M., Iqbal, M.S., Ghafoor, A., Ahmad, Z. and Qureshi, A.S., 2002. Variability in cowpea germplasm for reaction to virus infection under field condition. *Pakistan J. of Botany*, 34(1): 47–48.

Bhalani, P.A. and Parsana, G.J., 1986. Chemical control of gram pod borer, *Heliothis armigera* Hubner infesting greengram (*Vigna radiata* L.). *Pesticides*, 20(1): 24–25.

Bhat, N.S., Manjunatha, M., Raju, G.T.T., Nagbhushan, G.G. and Deshpande, V.P., 1988. Chemical control pod borer complex of greengram (*Vigna radiata*). *Indian J. Agric. Sci.*, 58(11): 864–866.

Bhatnagar, P. and Dahia, B., 2005. Reaction of mubgean and urdbean genotypes against insect-pests at yellow mosaic virus. *Indian J. Pulses Res.*, 18(1): 90–91.

Bhatnagar, V.S., Lateef, S.S., Sithanantham, S., Pawar, C.S. and Reed, W., 1982. Research on *Heliothis* at ICRISAT. In: *Proceedings of the International Workshop on Heliothis Management*, 15–20 November 1981, ICRISAT Centre, Patancheru, Andhra Pradesh, India, pp. 385–396.

Bindra, O.S. and Jakhmola, 1967. Incidence and losses caused by some pod-infesting insects in different varieties of pigeonpea. *Indian J. Agric. Sci.*, 37(3): 177–186.

Borah, R.K., 1995. Effect of synthetic pyrethroids and organo phosphorus insecticides on the incidence of whitefly, *Bemisia tabaci* (Genn.) and yellow mosaic virus in greengram, *Vigna radiata* (L.) Wilczek. *Indian Journal of Virology*, 11(1): 75–76.

Caswell, G.H., 1968. The storage of cowpea in the Northern States of Nigeria. *Proc. Agric. Soc., Nigeria,* 5: 46.

Caswell, G.H., 1973. The storage of cowpea. *Samaru Agric. Newsl.,* 15: 73–75.

Caswell, G.H. and Akibu, S., 1980. The use of primiphos methyl to control bruchids attacking selected varieties of stored cowpea. *Trop. Grain Legume Bull.,* 17/18: 9–11.

Catipon, E.M., 1986. *Mungbean Plant Industry Production Guide* (41), Ministry of Agriculture and Food, Manila, Philippines.

Chalfant, R.B. and Gaines, T.P., 1973. Cowpea curuculio: Correlation between chemical composition of southern pea and varietal resistance. *J. Econ. Entomol.,* 66: 1011–1013.

Chalfant, R.B., Leigh, T.F., Renwick, J.A., Messina, F.J. and Schalk, J.M., 1985. Entomological research on cowpea pests in the USA. In: *Cowpea Research, Production and Utilization,* (Eds.) S.R. Singh, K.O. Rachiee, London, pp. 267–273.

Chander, S. and Singh, Yesbir, 1989a. Control of galerucid beetle *Madurasia obscurella* Jacoby infesting greengram, *Vigna radiata* (L.) Wilczek. *Indian J. Entmol.,* 51(2): 185–191.

Chander, S. and Singh, Yesbir, 1989b. Control of thrips *Megalurothrips distalis* and pod borer *viz. Euchryrops cnejus* Fab. and *Maruca testulalis Geyer* infesting greengram, *Vigna radiata* (L.) Wilczek. *Indian J. Entomol.,* 51(2): 131–135.

Chandrakar, H.K. and Srivastava, S.K., 2000. Efficacy and economics of some insecticides to pod borer complex of urdbean *Vigna mungo* (L.) Hepper. *Environment and Ecology,* 18(4): 1014–1016.

Chaudhary, J.P., Singh, A.K. and Yadav, L.S., 1981. Effect of disulfoton granules on pest complex, crop growth and yield of green-gram. *Indian J. Entom.,* 43(4): 369–372.

Chaudhary, J.P. and Sharma, S.K., 1982. Biology of gram pod borer, *Heliothis armigera* (Hubnr) in the Haryana State. *Bull. Entomol.,* 22(1/2): 101–112.

Chaudhary, J.P., Yadav, L.S., Poonia, R.S. and Rastogi, K.B., 1980. Some observation on field population of *Empoasca kerri* Pruthi, a jassid, pest on mungbean crop in Haryana. *Haryana Agric. Univ. J. Res.,* 10(2): 250–251.

Chhabra, K.S., 1992. Advances in management of insect pests of *Vigna* species. In: *Proceedings National Symposium on New Frontiers in Pulses Research and Development,* (Ed.) J.N. Sachan. Directorate of Pulses Research, Kanpur, India, pp. 178–186.

Chhabra, K.S., Kooner, B.S. and Brar, J.S., 1980. Resistance behaviour of mungbean, *Vigna radiata* (L.) W. cultivars against insect-pests complex and yellow mosaic virus. *Indian J. Ecology,* 7 (2): 276–280.

Chhabra, K.S., Kooner, B.S., Sharma, A.K. and Saxena, A.K., 1988. Sources of resistance in mungbean (*Vigna radiata*) to insect pests and mungbean yellow mosaic virus. In: *Mungbean Proceedings of the Second International Symposium,* Bangkok, (Ed.) S. Shanmugasundaram, pp. 308–314.

Chhabra, K.S., Kooner, B.S., Sharma, A.K. and Saxena, A.K., 1993b. Role of allomones in varietal resistance in summer mungbean against thrips, *Megalurothrips distalis* (Karny). In: *Proceedings of the International Symposium on Pulses Research,* 2–6 April 1994, New Delhi, India (In Press).

Chhabra, K.S., Brar, J.S. and Kooner, B.S., 1981a. Jassid species recorded on green gram, black gram and red gram in Punjab. *Pulse Crops Newsletter,* 1(2): 65.

Chhabra, K.S. and Kooner, B.S., 1980. Sources of whitefly, *Bemisia tabaci* G. and yellow mosaic virus resistance in mungbean, *Vigna radiata* W. *Tropical Grain Legume Bulletin*, 19: 26–29.

Chhabra, K.S. and Kooner, B.S., 1985a. Losses in summer mungbean due to insect pest in Punjab. *Indian J. Ent.*, 47(1): 103–105.

Chhabra, K.S. and Kooner, B.S., 1985b. Problem of flower shedding caused by thrips, *Megalurothrips distalis* (Karny) on summer mungbean, *Vigna radiata* (L.) Wilczek and its control. *Tropical Pest Management*, London, 31(3): 186–188.

Chhabra, K.S. and Kooner, B.S., 1986. Comparative efficacy of different insecticides in controlling thrips, *Megalurothrips distalis* (Karny) on summer mungbean, *Vigna radiata* (L.) Wilczek. *Legume Research*, 9(1): 39–42.

Chhabra, K.S. and Kooner, B.S., 1986–92. *Research on Kharif Pulses*, Consolidated Reports on Entomology, All India Coordinated Pulses Improvement Project, Punjab Agricultural University, Ludhiana.

Chhabra, K.S. and Kooner, B.S., 1987. Susceptibility of different stages of Hairy caterpillar *Diacrisia obliqua* (Walker) to some insecticides. *Pesticides*, 21(6): 22–24.

Chhabra, K.S. and Kooner, B.S., 1988. Varietal resistance in mungbean [*Vigna radiata* (L.) Wilczek] to Thrips (*Megalurothrips distalis* Karny. *TVIS NEWS (The Tropical vegetables Information Service)*, 3 (2): 10–11.

Chhabra, K.S. and Kooner, B.S., 1992a. Sources of resistance in mungbean against major insect-pests and yellow mosaic virus. *Legume Research*, 14(4): 175–184.

Chhabra, K.S. and Kooner, B.S., 1992b. Response of some promising mungbean genotypes towards whitefly, jassids and mungbean yellow mosaic virus. In: *Proceedings of National Symposium on 'Recent Advances in Integrated Pest Management'*, PAU, Ludhiana, 12–15 October, IV, 13: 136.

Chhabra, K.S. and Kooner, B.S., 1993. Screening of mungbean germplasm against insect pests and yellow mosaic virus. *Indian J. Pulses Res.*, 6(1): 69–75.

Chhabra, K.S. and Kooner, B.S., Saxena, A.K. and Sharma, A.K., 1981b. Effect of biochemical components on the incidence of insect pest complex and yellow mosaic virus in mungbean. *Crop Improv.*, 8(1): 56–59.

Chhabra, K.S., Kooner, B.S. and Sharma, A.K., 1986b. Check flower drop in summer mungbean by controlling thrip, *Megalurothrips distalis* (Karny). *Indian Farming*, 36(9): 25–26.

Chhabra, K.S., Kooner, B.S. and Cheema, H.S., 1980a. Incidence of *Aphis craccivora* Koch on grain legumes in the Punjab. In: *Aphidology in India, Proceeedings of the National Symposium*, (Ed.) B.K. Agarwala, 2–4 November 1985, Agartala, pp. 23–28.

Chhabra, K.S., Kooner, B.S. and Verma, M.M., 1993a. *Insect Pests of Pulses: Identification and Control Manual.* Co–published by Punjab Agricultural University, Ludhiana and Directorate of Pulses Research, ICAR, Kanpur, pp. 88.

Chhabra, K.S., Kooner, B.S., Saxena, A.K. and Sharma, A.K., 1984. Influence of biochemical components on the incidence of insect pest complex and yellow mosaic virus in blackgram. *Indian J. Entomol.*, 46(2): 148–156.

Chhabra, K.S., Kooner, B.S. and Sharma, A.K., 1986b. Check flower drop in summer mungbean by controlling thrip, *Megalurothrips distalis* (Karny). *Indian Farming*, 36(9): 25–26.

COPR, 1981. *Pest Control in Tropical Grain Legumes.* Center for Overseas Pest Control, Overseas Development Administration, London, pp. 127–206.

CRIA, 1971. Mungbean research: Insecticidal research. In: *Annual Report for 1971.* Central Research Institute for Agriculture, Bogor, Indonesia.

Dalaya, V.P., Rajput, S.G. and Awate, B.G., 1986. Occurrence of thrips (*Thrips* spp.) on greengram in Northern Maharashtra and its control. *Pesticides,* 20(8): 18–19.

Daoust, R.A., Roberts, D.W. and Das Neves, B.P., 1985. Distribution, biology and control of cowpea pests in Latin America. In: *Cowpea Research: Production and Utilization,* (Eds.) S.R. Singh and K.O. Rachie, London, pp. 251–266.

Dasgupta, B. and Chauhdury, A.K., 1985. Use of inter-varietal intercropping to minimize the yellow mosaic virus of urd and mungbeans. *Indian J. Plant Proct.,* 3(1): 100–101.

De Lima, C.P.F., 1979b. A review of the use of physical storage procedures in East Africa: Aspects for improvement and extension. In: *Proceedings of the Second International Working Conference on Stored Product Entomol.,* September 10–16, 1978, Ibadan, Nigeria, pp. 10–16.

Dhakshinamoorthy, G. and Selvanarayanan, V., 2002. Evaluation of certain natural products against pulse beetle, *Callosobruchus maculatus* (Fab.) infesting stored green gram. *Insect Environment,* 8(1): 29–30.

Dharmasena, C.M.D. and Subasinghe, S.M.C., 1986. Resistance of mung (*Vigna radiata*) to callosobruchus spp. (Col: Bruchidae). *Trop. Agric.,* 142: 1–6.

Dhuri, A.V. and Singh, K.M., 1983. Pest complex and succession of insect pests in blackgram, *Vigna mungo* (L.) Hepper. *Indian J. Ent.,* 45(4): 396–401.

Dhuri, A.V., Singh, K.M. and Singh, R.N., 1986. Effect of intercropping on population dynamics of insect pests of blackgram, *Vigna mungo* (L.) Hepper. *Indian J. Ent.,* 48(3): 329–338.

Dias, C.A.R., 1987. Ecological studies on *Callosobruchus maculatus* Lab., *Callosobruchus chinensis* Linn. and *Callosobruchus analis* Fab. Infesting legumes in India. *Ph.D. Thesis,* I.A.R.I., New Delhi, India.

Dias, C.A.R. and Yadav, T.D., 1988. Incidence of pulse beetles in different legume seeds. *Indian Journal of Entomology,* 50: 457–461.

Dongre, T.K., Pawar, S.E., Thakare, R.G. and Harwalkar, M.R., 1996. Identification of resistant sources to cowpea Weevil (*Callosobruchus maculatus* (F.) in black gram (*Vigna nungo* Var. mungo). *Journal of Stored Products Research,* 32(3): 201–204.

Doria, R.C. and Roros, R.S., 1975. Varietal resistant of mung to the bean weevil, *Callosobruchus chinensis* (Linn.) and some characteristics of field infestation. *Philippines Entomology,* 2: 399–409.

Echendu, T.N.C., Njoku, B.O., Oli, E., Odurukwe, S.O. and Ene, L.S.O., 1988. Preliminary investigations into the use of ginger, Neem and Gaswewnut shell liquid (CNSL) to reduce damage caused to stored cowpea by *Callosobruchus maculatus.* In: *Proceeding of the First National Ginger Workshop Umuadike, Nigeria,* p. 124–128.

Edward, Y.S.J.T. and Gunathilagraj, K., 1994. Relative susceptibility of greengram, black and cowpea cultivars to *Callosobruchus maculatus* (F.) in the field. *Madras Agricultural Journal,* 81(10): 575–577.

Faleiro, J.R. and Singh, K.M., 1990. Pest relationship on summer cowpea crop in Delhi. *Indian J. Ent.,* 52(4): 711–712.

Fory, L.F., Finhardi-Filho, F., Quintero, C.M., Osborn, T.C., Cardona, C., Christopeels, M.J. and Mayer, J., 1996. Alpha amylase inhibitors in resistance to common beans to the Mexican bean Weevil and the bean weevil (Coleoptera: Bruchidae). *Journal of Economic Entomology*, 89: 204–210.

Fujii, K. and Miyazaki, S., 1987. Infestation resistance of wild legumes (*Vigna sublobata*) to azukibean weevil, *Callosobruchus chinensis* (L.) (Coleoptera: Bruchidae) and its relationship with cytogenetic classification. *Applied Entomology* and *Zoology*, 22: 229–230.

Ganapathy, M. and Janarthanan, R., 1995. Phenetic resistance in blackgram (*Vigna mungo* L. Hepper) against *Callosobruchus maculatus* Fabricius. *Journal of Entomological Research*, 19(2): 119–122.

Gaswell, G.H. and Akibu, S., 1980. The use of pirimiphos methyl to control bruchids attacking selected varieties of stored cowpea. *Trop. Grain Legume Bull.*, 17/18: 9–11.

Gatoria, G.S. and Singh, H., 1984. Effect of the insecticidal and fertilizer application on jassid complex in greengram, *Vigna radiata* (L.) Wilczek. *J. Entomol. Res.*, 8(2): 154–158.

Gatoria, G.S. and Singh, H. 1988. Effect of the insecticidal and fertilizer application on chewing insect pests of greengram. *J. Res. Pb. Agril. Univ.*, 25(2): 225–232.

Gujar, G.T. and Yadav, T.D., 1978. Feeding of *Callosobruchus maculatus* (Fab.) and *Callosobruchus chinensis* (Linn.) in green gram. *Indian J. Ent.*, 40: 108–112.

Gunasekaran, G.R., Kumaresan, D., Balasubramanian, G., Jayaraj, S. and Marappan, P.V., 1977. Relative susceptibility of blackgram varieties to stemfly and pod borer. *Madras Agric. J.*, 64: 56–57.

Gupta, B.N. and Saxena, H.P., 1971. Chemical control of pests of moong (*Phaseolus aureus* Roxb.) leads to spectacular rich plant growth and high yield. In: *Proc. of the 5th Workshop* (Kharif Pulses) H.A.U., Hissar, held on 18–19 March, p. 18–19.

Gupta, D.S. and Mishra, R.C., 1970. Carbaryl affects germination of seeds. *Indian J. Entomol.*, 32(4): 396–397.

Gupta, P.K. and Singh, J., 1984b. control of *Ophiomyia phaseoli* Tyron in greengram with granular insecticides. *Indian J. Agric. Sci.*, 54: 321–324.

Gupta, S.K. and Dhooria, M.S., 1973. Occrrence of *Hemitarsonemus latus* (Banks) on blackgram and its control in Punjab. *Curr. Sci.*, 39(9): 401–402.

H.O., T.H., 1967. The beanfly, *Melanagromyza phaseoli* Coq. and experiments on its control. *Malaysian Agricultural Journal*, 46: 149–157.

Hameed, S.E. and Dubey, S.L., 1993. Losses due to insect pests in North Bihar. In: Special issue Vol. I. *Indian J. Ent.*, 1983, pp. 136–146.

Hill, D.S., 1993. *Agricultural Insect Pests of Tropics and their Management*. Cambridge University Press, pp. 746.

Honda, Y.M. Iwaki, Thongmeearkam, P., Deema, N. and Srithongchai, W., 1982. Blackgram mottled virus occurring on mungbean and soybean in Thailand. *JARQ*, 16(1): 72–77.

Huis, A. van, 1991. Biological methods of bruchid control in the tropics: A review. *Insect Sci. Applic.*, 12(1/2/3): 87–102.

Hurje, R.V., Kulkarni, G.N., Shashidhara, S.D. and Vyakarnahal, B.S., 1990. Effect of insecticide and fungicide treatment on cowpea seed quality. *Seed Research*, 18(1): 90–92.

Hussain, M.H. and Abdel-Ali, V.A.I., 1982. Toxicology of some compound against the cowpea bettle, *Callosobruchus maculatus* (Fab.) (Coleoptera: Bruchidae). *International Pest Control*, 24: 12–17.

Ibrahim, A.G.B., 1977. Increased production of leguminous crops through control of beanfly. In: *Food and Agriculture Malaysia* 2000, Faculty of Agriculture, Univedrsity Pertanian, Malaysia, pp. 301–309.

Ignacimuthu, S., Janarthanan, S. and Balachandran, B., 2000. Chemical basis of resistance in pulses to *Callosobruchus maculatus* (F.) (Coleoptera: Bruchidae). *Journal of Stored Products Research*, 36: 89–99.

Islam, W., Ahmad, E.N. and Joarder, O.I., 1984. Timing and extent of damage caused by insect pests of greengram in Bangladesh. *Crop Prot.*, 31(3): 343–348.

Jackai, L.E.N. and Daoust, R.A., 1986. Insect pests of cowpeas. *Ann. Rev. Entomol.*, 95: 119.

Jackai, L.E.N., Singh, S.R., Raheja, A.K. and Wiedijk, F., 1985. Recent trends in the control of cowpea pests in Africa. In: *Cowpea Research: Production and Utilization*, (Eds.) S.R. Singh and K.O. Rachie, London, pp. 233–234.

Jagginavar, S.B., Kululkarni and Lingappa, S., 1990. A note on seasonal abundance of cowpea pod borer complex. *Karnataka J. Agric. Sci.*, 3(3–4): 272–275.

Jagruti, Pattnaik, Kole, C. and Pattnaik, J., 2002. Detection of protein marker for screening of MYMV resistant mungbean genotypes. *Indian J. Genetics Plant Breeding*, 62(1): 77–78.

Janagarajan, A., 1987. Effect of intercropping on the incidence of in cotton and sorghum. In: *Proc. Reg. Res. Council*, Southern Zone, Tamil Nadu Agric. Univ., Aruppukkottai, p. 13.

Jayaraj, S., 1988. Non-chemical management methods of some key mungbean pests. In: *Mungbean Proceedings of the Second International Symposium*, Bangkok, (Ed.) S. Shanmug-asundaram, pp. 366–371.

Kaiser, W.J., 1979. Natural infection of cowpea and mungbean by alphalpha mosaic virus in Iran. *Plant Dis. Rep.*, 63(5): 414–418.

Kalra, A.N., David, H. and Srivastava, T.N., 1966. *Drino* (Prosturmia sp.) (Trachinidae) as a parasite of *Diacrisia obliqua W. Indian J. Entomol.*, 28: 556–557.

Kalra, V.K., 1985. Record of two new species of *Apanteles* parasitizing of *Spilosoma obliqua* (Wks.). *Indian J. Plant Protec.*, 12(2): 155.

Kapoor, K.N., Dhamdhere, S.V., Singh, O.P., Mishra, U.S. and Deole, J.Y., 1984. Evaluation of some insecticidal dust against Alcidodus collaris infesting greengram, *Vigna radiata* (L.) Wilczek, with records of its alternate host. *J. Entomol. Res.*, 8(1): 17–19.

Karel, A.K. and Maligna, Y., 1980. Leaf hopper and aphid resistance in cowpea varieties. *Trop. Grain Legume Bull.*, 20: 10–11.

Ketkar, C.M., Schmuttere, R. and Ascher, K.R.S., 1987. Use of tree derived non-edible oils as surface protectants for stored legumes against *Callosobruchus maculatus* and *C. chinensis.* Natural pesticides from the neem tree (*Azadirachta indica, A. juss*) and other tropical plants. In: *Proceeding of the 3rd International Neem Conference*, Nairobi Kenya, 10–15 July, 1986, p. 532–542.

Khamala, C.P.M., 1978. Pests of grain legumes and their control in Kenya. In: *Pests of grain Legumes: Ecology and Control*. Academic, London/New York, pp. 454.

Khan, M.Q. and Hussain, M., 1965. Role of coccinellid and syrphid predators in biological control of groundnut aphid. *Indian Oilseeds Journal*, 9: 67–70.

Kitamura, K.M., Ishimoto, M. and Sawa, M., 1988. Inheritance of resistance to infection with Azuki Bean weevil in *Vigna sublobata* and successful incorporation to *V. radiata. Japanese Journal of Breeding* 38: 459–464.

Kleinschmidt, R.P., 1970. Studies of some Agromyzidae in Queensland. Queensl. *J. Agric. Anim. Sci.*, 27: 341–384.

Kooner, B.S. and Chhabra, K.S., 1980. Pests of pulse crop and their control. In: *Breeding Methods for the Improvement of Oulse Crops*, (Ed.) K.S. Gill. Punjab Agricultural University, Ludhiana, India, pp. 132–141.

Kooner, B.S., Chhabra, K.S., Sekhon, H.S., Dhingra, K.K. and Cheema, H.S., 1983. A new deformity in summer mungbean, *Vigna radiata* (L.) Wilczek. *Pulses Newsletter*, 3(1): 40–42.

Krishnaiah, K., Ramakrishnan, N. and Reddy, P.C., 1984. Further trials on control of *Spodoptera litura* Fab. By Nuclear Polyhedrosis Virus on blackgram and groundnut. *Indian J. Plant. Prot.*, 12(1): 81–83.

Krishnamurthi, B. and Usman, S., 1954. Some insect parasites of economic importance notes in the Mysore State. *Indian J. Entomol.*, 16: 327–344.

Labeyrie, V. and Maison, P., 1957. Techniques culturals concernant la preservation du haricot contre la bruche *Acanthoscelides obtectus* Say. *Rev. Zool. Agric.*, 10: 1–18.

Lal, S.S., 1983. Studies on the feasibility of integrated control of major insect pests attacking *Vigna radiata. Ph.D. Thesis*, Kanpur University, Kanpur, pp. 1–10.

Lal, S.S., 1985. A review of insect pests of mungbean and their control in India. *Trop. Pest Management*, 31(2): 105–114.

Lal, S.S., 1987. Insect pests of mung, urd, cowpea and pea and their management. In: *Plant Protection in field crops*, (Eds.) Veerabhadra Rao and S. Sithanantham, pp. 185–201.

Lal, S.S., Yadav, C.P. and Dias, CAR, 1980. *Important Pests of Pulse Crops in India and their Management*. AICPIP, Kanpur, Publication No. 2.

Lambrides, C.J. and Imrie, B.C., 2000. Susceptibility of mungbean varieties to the bruchid species *Callosobruchus maculatus* (F. *C. phaseoli* (Gyll.) *C. chinensis* (L.) and *Acanthoscellides obtectus* (Say.) (Caleoptera: Chrysomelidae. *Australian Journal of Agricultural Research*, 51(1): 85–89.

Lateef, S.S. and Reed, W., 1983. Review of crop losses caused by insect pests in pigeonpea internationally and in India. In: *Crop Losses Due to Insect Pests, Vol. II. Special* Issue, *Indian J. of Entomology*. The Entomological Society of India, Hyderabad Branch, India, pp. 284–291.

Lawn, R.J. and Ahn, C.S., 1985. Mungbean *Vigna radiata* (L.) Wilczek, *Vigna mungo* (L.) Hepper. In: *Grain Legume Crops*, (Eds.) R.J. Summer Field and E.H. Roberts. Collins London, pp. 854–623.

Li, D.X., Tiang, J. and Zhang, Z.Y., 1994. A study of the bionomics and control techniques of *Aphis craccivora* (Koch) on mungbeans. *Plant Protection*, 20(6): 4–5.

Lin, C.S., 1981. Resistance of beans to insects. *Chinese Journal of Entomology*, 1(2): 1–6.

Lin, C.S. and Rose, R.I., 1976. Screening of mungbean varieties for resistance to beanfly and study for resistance mechanism. *Plant Protection Bulletin (ROC)*, 17(4): 14.

Litsinger, J.A., Barrion, A.T., Badong J.P., Gadela Cruz, C., Canapi, B.L. and Apostol, R.F., 1988. Food web, yield loss and chemical control of insect pests of wetland rice based mungbean in the Philippines. In: *Mungbean Proceedings of the Second International Symposium,* (Eds.) C.A. MacFoy and Z.T. Dabrowski. 1984. Preliminary studies on cowpea resistance to *Aphis craccivora* Kock (Hom: Aphididae). *Z. Angew. Entomol.,* 97: 202–209.

Litsinger, J.A., Quirino, C.B., Lumbanm, M.D. and Bandong, J.P., 1978b. The grain legume pest complex of rice based cropping system at three locations in the Philippines. In: *Pests of Grain Legumes Ecology and Control,* (Eds.) S.R. Singh, H.F. Van Emden and A.A. Taylor. Academic Press, London, pp. 309–320.

Litsinger, J.A., Price, E.C., Herrera, R.T., Bandong, J.P., Lumban, M.D., Quirino, C.B. and Castillo, M.D., 1978a. Introducting pest control technology to a low management crop: The mungbean example in the Philippines. In: *Proc. the First International Mungbean Symposium,* 16–19 August 1977, Los Banos, pp. 183–191.

Litsinger, J.A., Bandong, J.P. and Paragna, F., 1980. Field evaluation of foliar insecticides for post flowering mungbean insects. Pangasinan, Philippines, 1977–78. *Insecticide and Acaricide Tests,* 5: 137–138.

Liu, Xuming, Dasheng, Jin,, Xuzhen, Cheng, Xiaofel, Wu, Suhua, Wang, Liu, X.M., Jin, D.S., Cheng, X.Z., Wu, X.F. and Wang, S.H., 1998. Preliminary evaluation on their resistance to *Callosobruchus chinensis* (L.) in mungbean germplasm. *Crop Genetic Resources,* 2: 35–37.

Manjunath, M., Bhat, N.S., Raju, G.T.T. and Raju, M. and Viraktamoth, M., 1988. A new predatory ant *Aenactus pachycerus* Smith (Hymenoptera: formicidae) on wire worm, *Agriotes* sp. (Coleoptera: Elateridae. *Curr. Sci.,* 57(B): 7449.

Marimuthu, T.S., Subramanian, C.L. and Miohan, R., 1981. Assessment of yield losses due to yellow mosaic infection in mungbean. *Pulse Crops Newsletter,* 1: 104.

Mathur, Y.K. and Singh, S.K., 1998. Field evaluation of various insecticides against major pest of mungbean. In: *Abst. National Symposium on Management of Biotic and Abiotic Stresses in Pulse Crops* I.I.P.R. Kanupr, 26–28 June, pp. 86.

Mathur, Y.K. and Singh, S.K., 1998. Some important of natural enemies of *Amsacta moorei* Butler in Rajasthan. *Indian J. Ecol.,* 25(1): 77–78.

Mathur, Y.K., Singh, S.K. and Kaushal, P.K., 1997. Frecasting model of the key pests of mungbean in Uttar Pradesh, India. In: *Abst. International Conference on Ecological Agriculture* organized by Indian Ecological Society and CRRID, held at Chandigarh from November 15–17, pp. 89–90.

Mehrotra, P., Singh, S.D. and Kaul, C.K., 1989. A note on the chemical control of *Melanagromyza obtusa* Malloch on cowpea. *Indian J. Entomol.,* 51(3): 355–356.

Mohan, Krishna, Bisen, R.S. and Singh, S.K., 2000. New record of *Gryon sp. on Spodoptera litura* Fab. From Kanpur, Uttar Pradesh. *Insect Environment,* 6(3): 100–101.

Murugesan, S. and Chelliah, S., 1978. Effect of yellow mosaic infestation of the host greengram on the biology of *Bemisia tabaci. Gen. Entomol.,* 3(1): 41–43.

Nair, M.R.G.K., 1986. *Insects and Mites of Crops in India.* Indian Council of Agricultural Research, New Delhi, pp. 408.

Natarajan, N. and Subbarao, P.V., 1992. Suitable time of sowing and insecticide for the management of stemfly (*Ophiomyia phaseoli*) in blackgram (*Vigna mungo*). *Madras Agricultural Journal*, 79(2): 119–122.

Nath, P. and Singh, A.K., 1998. Effect of intercropping of groundnut with millets and pigeon pea on the relative incidence of insect pests. *Ann. Pl. Prot. Sci.*, 6(2): 151–154.

Nayar, K.K., Ananthakrishnan, T.N. and David, B.V., 1976. *General and Applied Entomology*. Mc Graw Hill, New Delhi, pp. 581.

Ndoye, M., 1978. Pests of cowpea and their control in Senegal. In: *Pests of Grain Legumes: Ecology and Control*. Academic, London/New York, pp. 454.

Negi, R.S., Srivastava, Meera and Saxena, M.M., 1997. Egg-Laying and adult emergence of *Callosobruchus chinensis* on green gram (*Vigna radiata*) treated with pongam oil. *Indian J. Ent.*, 59(2): 170–172.

Nene, Y.L., 1972. A survey of viral diseases of pulse crops in Uttar Pradesh. Final Technical Report of PL 480 Adhoc Project, p. 88–108.

Nene, Y.L., 1973a. Viral diseases of some warm weather pulse crops in India. *Plant Disease Reporter*, 57: 463–467.

Nene, Y.L., 1973b. Note on a fungal parasite of *Bemisia tabaci*. Genn.: A vector of several plant viruses. *Indian J. Agric. Sci.*, 43(5): 514–516.

Nilakhe, S.S. and Chalfant, R.B., 1982. Cowpea cultivars screened for resistance to insect pests. *J. Econ. Entomol.*, 75: 223–227.

Ooi, P., 1973. Some insect pests of greengram. *Phaseolus aureus: Malaysian Agricultural Journal*, 49: 131–142.

Otanes, F.Q., 1918. The beanfly. *Philipp. Agric.*, 7: 2–31.

Pandey, N.K. and Singh, S.C., 1995. Effect of neem leaf powder on survival and mortality of pulse beetle *Callosobruchus Chinensis* (L.) infesting blackgram Uttar Pradesh. *Journal of Zoology*, 15(3): 162–164.

Panwar, V.P.S., 1995. *Agricultural Insect Pests of Crops and their Management*. Kalyani Pubs., Ludhiana, Punjab, pp. 286.

Parihar, D.R., 1979. Outbrak of Kutra, *Amsacta moorei* pest in Rajasthan desert. *Ann. Arid Zone*, 18: 140–141.

Patel, M.B. and Srivastava, K.P., 1990. Phytotonic effect of insecticides on cowpea *V. unguiculata* (Linn.) and green gram *V. radiata* (Linn.). *Indian J. Entomol.*, 52(4): 583–588.

Patel, R.C. and Patel, J.R., 1983. Crop losses due to insect pest in Gujarat State. in Special issue Vol.I, *Indian J. Ent.*, pp. 27–32.

Patel, R.C., Yadava, D.N. and Patel, J.R., 1976. Natural control of groundnut aphid, *Aphis craccivora* Koch. in Central Gujarat. *Current Science*, 45: 34–35.

Patnaik, N.C., Mohanty, J.N. and Chhabra, K.S., 2000. In: *Insect Pests of Pulses and Oilseeds and their Management*, (Eds.) Anand Prakash and Jagadiswari Rao. Applied Zoologists Research Association, CRRI, Cuttak, India, pp. 11–27.

Pawar, C.S., 1983. *Cropping Entomology*, Report of Work 1982–83, ICRISAT, Patancheru PO, Andhra Pradesh, India, pp. 1–21.

Pawar, V.M. and Chowan, K.M., 1979. Effect of systemic insecticides on germination and seedling development in greengram (*Vigna aureus* Hepper). *Indian J. Entomol.*, 41: 405–408.

Perrin, R.M., 1978. Varietal differences in the susceptibility of cowpea to larvae of the seed moth; *Cydia ptychora* (Meyrick) (Lepidoptera: Tortricidae). *Bull. Entomol. Res.*, 68: 47–56.

Pohelman, J.M., 1991. *The Mungbean.* Oxford and IBH Publishing Co. Pvt. Ltd., New Delhi, pp. 1–375.

Poonsavasde. P. and Sukonthabhirom, S., 1990. Evaluation of the scelionid egg parasitoid for controlling *Piezodorus hybneri* Gmelin. In: *Proceedings of the Mungbean Meeting* 90, Chiang Mai, Thailand: p. 277–282.

Prakash, A. and Rao, 2000. *Insect Pests of Pulses and Oilseed and their Management.* C.R.R.I. Cuttack, pp. 1–259.

Prevett, P.F., 1961. Field infestation of cowpea (*Vigna unquiculata*) pods by beetles of the families Bruchidae and curculionidae in Northern Nigeria. *Bull. Entomol. Res.*, 52: 635–645.

Pruthi, H.S. and Singh, Mohan, 1950. Pests of stored grains and their control. Special No. *Indian J. Agric. Sci.*, 18(4): 18–23.

Rachie, K.O., 1985. Introduction: Cowpea research, production utilization. In: *Cowpea Research, Production and Utilization*, (Eds.) S.R. Singh and K.O. Rachie eds.. John Wiley and Sons, New York, xi–xii, p. 460.

Raheja, A.K., 1985. Present status of research in cowpea entromology at IAR (Institute of Agricultural Research), Ahmadu, Bello University, Zaria, Research Report: 1–23 (Personal Communication).

Rajapaksa, R.H.S. and Jayasena, K.W., 1989. Field pest problem of mungbean, *Vigna radiata* in Southern region of Sri Lanka. *Entomon.*, 14(12): 159–164.

Rajendran, R., Gopalan, M. and Janarthanan, Jr. 1981. A note on the control of pod borers in blackgram and green gram with dust formulations. *Pesticides*, 15(10): 35–36.

Rajpakse, R., Senanayake, S.G.J.N. and Ratnasekra, D., 1998. Effect of fice botani on oviposition adult emergence and mortality of *Callosobruchus maculatus* (Fab.) (Coleoptera: Bruchidae) infesting cowpea (*Vigna unguiculata* L. Walp.). *Journal of Entomologial Research*, 22(2): 117–122.

Raju, A.K. and Panda, N., 1983. Observations on the biology of *Aphis croecivora* Koch. on five varieties of green gram. *Indian J. Agric. Sci.*, 53: 868–869.

Ram Ujagir, 1997. Insect pests of peas and their management. In: *Integrated Pest and Disease Management*, (Eds.) R.K. Upadhyay *et al.* APH Publishing Corp., New Delhi, pp. 210–233.

Ramakrishnan, C., 1986. Management of early season pests of pulses. In: *Pest and Diseases Management: Oilseedsd, Pulses, Millets and Cotton*, (Ed.) Jayaraj. Tamil Nadu, Agril. Univ., Coimbatore, pp. 120–124.

Raman Goud, T., 1993. Sesamum and sunflower pests and their management. In: *Summer Institute on Advances in Entomology: Oilseeds and Pulses*, May 17 to June 5, Compd. of Lectures, p. 255–267.

Raman, K.V., Singh, S.R.and van Emden, H.F., 1982. Mechanism of resistance to leaf hopper damage in cowpea. *J. Econ. Entomol.*, 73: 484–488.

Ramkrishnan, G., Konadaswamy, T.K., Damodaran, A.P.S. and Ayjaroo, R., 1973. Studies on new mosaic viruses occurring on *Phaseolus aureus* Roxb. *Madras Agric. J.*, 60(7): 465–468.

Rao, S.S., Rao, S.N. and Azam, K.M., 1990. Occurrence of *Tetranychus urticae* Koch on greengram (*Vigna radiata. Indian J. Entomo.*, 52(1): 158–159.

Rathore, Y.S. and Chaturvedi, S.K., 1997. Developmental pattern of *Callosobruchus chinensis* (L.) on chickpea seeds of advanced breeding lines. *Indian Journal of Pulses Research*, 10(2): 180–184.

Rathore, Y.S. and Sharma, V., 2002. Management of bruchid infestation in pulses. ISPRD, 2002. Pulses for Sustainable Agriculture and Nutritional Security. In: *Proceeding of National Symposium*, (Eds.) Masood Ali, S.K. Chaturvedi and S.N. Gurha. Indian Institute of Pulses Research, Kanpur, pp. 111–124.

Ravindrababu, R., 1987. Studies on mungbean yellow mosaic virus disease. *Ph.D. Thesis*, Tamil Nadu Agril. Univ., Coimbatore, pp. 191.

Rawat, R.R. and Sahu, H.R., 1973. A study on the Bionomics of *Chauliops fallax* Scott (Heteroptera, Lygacidae) at Sehore (Madhya Pradesh). *J. Bombay N. History Soc.*, 70(3): 475–479.

Rawat, R.R. and Kausik, V.K., 1983. Crop losses due to insect pests in the north east Madhya Pradesh. In: *Crop Losses Due to Insect Pests.* Special Issue, Vol. I. *Indian J. Ent.*, 1983: 73–83.

Ray Chandhari D.N., Dutta, S., Agrawal, B.K., Raha, S.K. and Ray Chaudhury, D., 1979. Some parasites and predators of aphids in north east India and Bhutan II. *Entomol.*, 4(2): 163–166.

Reddy, M.U., Bharti, S.R. and Reddy, D.D.R., 1999. Efficacy of some vegetable oils as protectants against the pulse beetle (*Callosobruchus chinensis*) in green gram (*Phaseolus aureus*) during storage, *Indian J. Nutrition Dietetics*, 36(10): 436–442.

Reddy, V.S., Babu, T.R., Hussaini, S.H. and Reddy, B.M., 1994. Effect of edible and non-edible oils on the development of pulse beetle, *Callosobruchus chinensis* L. and on viability of mungbean seeds. *Pest Management and Economic Zoology*, 2(1): 15–17.

Reed, W., Lateef, S.S., Sithanathan, S. and Panwar, C.S., 1989. Pigeonpea and chickpea insect identification handbook. *ICRISAT Information Bulletin*, 26: 51.

Robertson, G.A., 1981. Response of mung to mechanical reduction of leaf area at flowering. *Pulse Crops Newsletter*, 1(4): 37–38.

Roongsook, D.T. Sitchawat, Knapp, F.W. and Tiemtaisong, A., 1973. The effect of phorate and malathion on beanfly control and on yield of mungbean. *Thai. J. Agric. Sci.*, 6: 283–286.

Sahoo, B.K. and Patnaik, N.C., 1994. Insect pests of greengram and blackgram in the south central region of Orissa with notes on their seasonal activity. *Orissa J. Agric. Res.*, 7 (Suppl.): 74–76.

Sakamoto, Y., Sasaki, H. and Kikuta, H., 1983. Ecological studies on the subfamily Apioninae injurious to leguminous crop in Japan. *I.J. Coll. Dairying*, 10(1): 153–160.

Sarkar, T.K., Gayen, P. and Somchoudhury, A.K., 1991. Some genetic parameters of resistance to *Callosobruchus maculatus* (F.) (Coleoptera: Bruchidae) in mungban, *Vigna radiata* (L.) Wilczek. *Journal of Entomological Research*, 15(1): 36–39.

Saxena, H.P., 1983. Losses in blackgram due to insect-pests. In: *Crop Losses Due to Insect Pests*, Special Issue Vol. II. *Indian J. Ent.*, pp. 294–297.

Saxena, H.P., 1973. Major pest problems in pulses. In: *Pest Management and Plant Protection Training*, New Delhi, Deptt. of Agriculture, Ministry of Agriculture and Irrigation, pp. 191–194.

Saxena, H.P., 1978. Pests of grain legumes and their control in India. In: *Pests of Grain Legumes; Ecology and Control*, (Eds.) S.R. Singh, H.F. van Emden and T.A. Taylor. Academic Press, London, pp. 15–23.

Saxena, H.P., Singh, Y. and Phokela, A., 1973. Effect of date of sowing on pest incidence. In: *Proceedings IX Workshop (Kharif Pulses)*, New Delhi, pp. 115.

Sehgal, V.K. and Ram Ujagir, M., 1988. Insect pest management of mungbean in India. In: *Mungbean Proceedings of the Second International Symposium*, Bangkok, (Ed.) S. Shanmugasundaram, pp. 315–328.

Sehgal, V.K., Rathore, Y.S. and Ram Ujagir, 2000. Insect pest management in mungbean and urdbean in India. In: *Proceeding of National Symposium on "Management of Biotic andAbiotic Stresses in Pulse Crops"* Kanpur, June 26–28, 1998, pp. 87–92.

Sethuraman, K., Manivannan, N. and Natrajan, S., 2001. Management of yellow mosaic disease of urdbean using neem products. *Legume Research*, 24(3): 197–199.

Sharma, S.R. and Verma, A., 1982. Control of yellow mosaic of mungbean through insecticides and oils. *J. Entomol. Res.*, 6(2): 130–136.

Shetgar, S.S., Bilapate, G.G., Patil, V.V. and London, G.M., 1990. A note on the natural enemies of Bihar hairy caterpillar, *Spilosoma obliqua (Walker)*. *Indian J. Ent.*, 52 (1): 158.

Sidhu, A.S. and Sandhu, A.S., 1971. Control of jassid infesting moong crop with new insecticides. In: *Proceedings, Fifth Workshop on Pulse Crops*, New Delhi, ICAR, p. 138–142.

Singh, S., 1982. Ecology of Agromyzidae (Diptera) associated with leguminous crops in India. *Memoirs of the School of Entomol.* No. 8, St. John's College, Agra, India, pp. 126.

Singh, P., Bhadauria, N.S. and Jakhmola, S.S., 2004. Efficacy of different plant materials as powder form against pulse beetle *Callosobruchus maculatus* in green gram. *Indian Journal of Entomology*, 66(4): 364–365.

Singh, B.B., Gupta, S., Singh, R.A., Singh, S.K. and Singh, K.K., 2006. *Mung aur Urd Ki Unnat Kheti.* Technical Bulletin. Indian Institute of Pulses Research, Kanpur, India, pp. 1–20.

Singh, B.K and Sehgal, V.K., 1993. Insect fauna of mungban and effects of crop protection of field incidence of major pests and grain yield at Pantnagar in Northern India. *Crop Protection*, UK (personal communication).

Singh, C.P. and Pandey, M.C., 1995. Studies on the biology of *Amsacta moorei* Butler on groundnut. *J. Insect Sci.*, 8(2): 207–208.

Singh, D.P. and Singh, B.B., 1994. A new variety of mungbean Narendra mung^{-1}. *Indian Farming*, 43(11): 3–4.

Singh, J. and Gupta, P.K., 1986. Relative efficacy of some granular insecticides against galerucid beetle (*Madurasia obscurella* Jacoby) and important pest of greengram. *Indian J. Entomol.*, 46(2): 221–231.

Singh, K.M. and Singh, R.N., 1977. Succession of insect pests in mungbean and blackgram under dryland conditions at Delhi. *Indian J. Entomol.*, 39: 365–370.

Singh, O.P. and Rawat, M.M., 1982. Some observations on the greengram on the green stink bug *Nazara viridula* (L.) on summer greengram crop at Jabalpur, Madhya Pradesh. *Sci. Cul.*, 48(3): 108–109.

Singh, R.A., Gurha, S.N., Mishra, D.P. and Gangal, L.K., 1981. Role of systemic insecticides in augmenting yield with reduced yellow mosaic virus incidence in mungbean. *Indian J. Plant Prot.*, 8(2): 167–169.

Singh, Ram and Chauhan, R., 1984. Screening greengram genotypes against *Tarsonemus mite. Indian J. Agric. Sci.*, 54(8): 669–670.

Singh, Ram, Singh, Basant and Verma, R.A., 2001. Effect of different indigenous plant products as grain protectant against, *Callosobruchus Chinensis* Linn. on pea. *Indian Journal of Entomology*, 63(2): 179–181.

Singh, S., 1982. Ecology of Agronomyzidae (Diptera) associated with leguminous crops in India. *Memoirs of the School of Entomol.* No. 8, St. John's College, Agra, India, pp. 126.

Singh, S.K. and Rai, A.B., 2005. Is the reproductive physiology of Adzuki been Weevil *Callosobruchus chinensis* (L.) Governed by aging? *Indian J. of Ent.*, 67(1): 1–6.

Singh, S.K. and Mathur, Y.K., 1998–99. Life table studies on *Helicoverpa armigera* Hubner on gram. Rabi programme report Deptt. of Entomology, CSAUA&T, Kanpur pp.

Singh, S.K. and Yadav, C.P., 2002–03. Bio-ecology of *Callosobruchus* spp. and identification of sources of resistance in stored chickpea. *Annual Report*, IIPR, Kanpur, pp. 36–38.

Singh, S.R., 1977. Cowpea cultivars resistant to insect pests in world germplasm collection. *Trop Grain Legume Bull.*, 9: 3–7.

Singh, S.R., 1978. Resistance to pests of cowpea in Nigeria. In: *Pest of Grain Legumes: Ecology and Control*, Academic, London/New York, pp. 454.

Singh, S.R. and Jackal, L.E.N., 1985. Insect pests of cowpeas in Africa: Their life cycle, economic importance, and potental for control. In: *Cowpea Research, Production and Utilization*, (Eds.) S.R. Singh, K.O. Rachie. London, p. 217–231.

Singh, S.R. and Taylor, T.A., 1978. Pests of grain legumes and their control in Nigeria. In: *Pests of Grain Legumes: Ecology and Control*. Academic, London/New York, pp. 99–111.

Singh, T.B. and Singh, R., 1980. Studies on the bionomics of *Trioxys indicus* S&S. a parasitoid of *Aphis craccivora* Koch: The area of discovery of the prasitoid. *Zeithchrift fur Ang. Ent.*, 89: 173–178.

Singh, T.V.K., Singh, K.M. and Singh, R.N., 1990. Groundnut pest complex: III. Incidence of insect pests in relation to agroclimatic conditions as determined by graphical super imposition technique *Indian J. Ent.*, 52(4): 686–692.

Singh, T.V.K., Singh, K.M. and Singh, R.N., 1990. Groundnut pest complex IV: Regression studies to determine the associationship between jassid and thrips and weathr parameters. *Indian J. of Ent.*, 52(4): 693–701.

Singh, V.N. and Pandey, N.D., 2001. Growth and development of *Callosobruchus chinensis* (Linn.) on different gram varieties. *Indian J. Ent.*, 63(2): 182–185.

Singh, V.N., Pandey, N.D. and Singh, Y.P., 1994. Effectiveness of vegetable oils on the *Callosobruchus chinensis* (Linn.) infesting stored grains (Chickpea). *Indian J. Ent.*, 56(3): 216–219.

Sinha, M.M., Yadav, R.P. and Kumar, A., 1975. Outbreak of the Bihar hairy caterpillar, *Diacrisia obliqua* Walker in North Bihar. *Entomologists Newsl.*, 5: 97.

Sinha, M.M., Yadav, R.P. and Kumar, A., 1979. Multidirectional approach for pest management in arhar, *Cajanus* cajan in Bihar. *Pesticides*, 13(11): 14–16.

Sinha, M.M., Yazdani, S.S., Kumar, A. and Singh, R., 1985. Observations on the seasonal occurrence of insect pests of greengram in Northern Bihar. *Bull. Entomol.*, 28(2): 210–211.

Sinha, T.B. and Singh, R., 1980. Studies on the bionomics of *Trioxys indicus* S&S a parasitoid of *Aphis craccivora* Koch: The area of discovery of the parasitoid. *Zeitchrift fur Ang. Ent.*, 89: 173–178.

Sivaprakasam, N., Balasubramanian, G., Venkatesan, S. and Chandrasekaran, J., 1988. Control of *Spodoptera litura* F. with nuclear polyhedrosis virus and chloropyriphos aon cowpea. *Madras Agric. J.*, 75(9–10): 366–368.

Speswasdi, P., Pitaksa, S., Chareonrak, T., Phapoom, V. and Heuel-Rof, B., 1990. Crop loss assessment for major mungbean pests in rice based cropping system. In: *Proceedings of the Mungbean Meeting*, Chaing Mai, Thailand, pp. 259–269.

Srikanth, J. and Lakkundi, N.H., 1988. Instar period, fecundity and longevity of cowpea aphid. *Aphid craccivora* Koch. on seven leguminous hosts. *J. Aphidology*, 2(1/2): 18–19.

Srivastava, A.S., Awasthi, B.K. and NIGAM, P.M., 1977. Incidence of *Myllocerus maculous* Desb. on leguminous crops (Coleoptera: Curceulionidae). *Labdev J. Sci. Technol.*, 15B: 215–216.

Srivastava, K.M. and Singh, L.N., 1976. A review of the pest complex of *Kharif* pulses in UP. *PANS*, 22: 333–335.

Srivastava, K.P., Gajblije, V.T., Jain, H.K. and Agnihotri, N.P., 1984. Efficacy of synthetic pyrethroids against the pod borer, *Euchrysops cnejus* Fab. and their residue in greengram, *Vigna radiata* (L.) and cowpea, *Vigna unguiculata* (L.) *Walp. J. Entomol. Res.*, 8(1): 36–41.

Srivastava, Meera and Kaur, A., 2002. An evaluation of efficacy of extracs of plant *Peganum harmala* against pulse beetle *Callosobruchus chinensis. Indian J. Ent.*, 64(2): 138–147.

Stary, P. and Ghosh, A.K., 1978. Further records of aphid parasitoid (Hym. Aphididae) from Meghalaya, India. *Oriental Insects*, 12(1): 77–80.

Su, G.Y., Kung, S.P. and Rose, R.I., 1976. Control of beanflies in mungbeans and asparagus beans, 1975. *Insecticide and Acaricide Tests*, 1: 53.

Subha Rao, P.V., Rangarajan, A.V. and Azeer, A.B., 1974. Record of new host plants for some important crops pests in Tamil Nadu. *Indian J. Entomol.*, 36: 227–228.

Sukul, P. S.K. Handa and Singh, K.M., 1987. Comparative efficacy of synthetic pyrethroids against pod borer complex, crop growth and yield of greengram. *Indian J. Entomol.*, 49(3): 408–413.

Sundaraju, D. and Rangarajan, A.V., 1987a. Insecticidal control of pests of greengram and blackgram. *Pesticides*, 21(12): 18–19.

Sundaraju, D. and Rangarajan, A.V., 1987b. Effects of insecticides in combination with fertilizer in controlling yellow mosaic disease and pod borer of greengram. *Pesticides*, 21(12): 20–21.

Talekar, N.S., 1988. Biology, damage and control of bruchid pests of mungbean. In: *Mungbean Proceedings, Second International Symposium*, Bangkok, (Ed.) S. Shanmugasundaram, pp. 329–342.

Talekar, N.S., 1990. *Agronomyzid Flies of Food Legumes in the Tropics*. Wiley Eastern Limited, New Delhi, India, pp. 297.

Talekar, N.S., 1996. Sources of resistance to major insect pests of mungbean in Asia, In: *Recent Advances in Mungbean Research*, (Eds.) Asthana, A.N. and D.H. Kim. Indian Society of Pulses Research, IIPR, Kanpur, India, p. 40–50.

Talekar, N.S. and Chen, B.S., 1983. Seasonability of insect pests of soybean and mungbean in Taiwan. *J. Eco. Entomology*, 76: 34–37.

Talekar, N.S. and Lin, C.L., 1992. Characterization of *Callosobruchus chinensis* (Coleoptera: Bruchidae) resistance in mungbean. *Journal of Economic Entomology*, 85(4): 1150–1153.

Talekar, N.S. and Lin, C.P., 1992. Mechanism of *Callosobruchus chinensis resistance in mungbean. Journal of Economic Entomology*, 85: 1150–1153.

Talekar, N.S. and Lin, Y.H., 1981. Two sources with differing modes of resistance to *Callosobruchus chinensis* in mungbean. *Journal of Economic Entomology*, 74(5): 639–642.

Tangkano, W. and Soekarna, D., 1976. Screening of mungbean (*Phaseolus radiata*) varieties for resistance to the beanfly *Agromyza phaseoli* Coq. In: *Annual Report Central Research Institute for Agriculture*, Ministry of Agriculture, Bogor, Indonesia, pp. 39–47.

Taylor, T.A., 1978. *Maruca testulalis:* An important pest of tropical grain legumes. In: *Pests of Grain Legumes: Ecology and Control*, (Eds.) S.R. Singh, H.F. van Emden and T.A. Taylor. Academic Press, London, pp. 193–200.

Telekar, N.S., 1990. *Agromyzid Flies of Food Legumes in the Tropics*. Wiley Eastern Limited, New Delhi, India, pp. 297.

Telekar, N.S. and Lee, Y.H., 1988. Biology of *Ophiomyia centrosematis* (Diptera: Agromyzidae): A pest of soybean. *Ann. Entomol. Soc. Am.*, 81(16): 938–942.

Thapa, R.B. and Timsina, J., 1990. Incidence of the beanfly, *Ophiomyia phaseoli* (Tryon) in mungbean and soybean varieties after late rice. *Journal of the institute of Agriculture and Animal Science*, 11: 83–88.

Thawstine, S., 1987. Effect of neem on the development of cowpea weevil *Callosobruchus maculatus* (Fab.) on some leguminous seeds. *Munoz. Nueva Ecija (Philiphines)*, 7: 9–11.

Thontadarya, T.S. and Hiremath, I.G., 1983. Losses due to insect pests in Karnataka. In: *Crop Losses Due to Insect Pests*. Special Issue, *Indian J. Entomol.*, 1: 43–58.

Turner, J.W., 1983. Influence of plant species on the movement of *Trissolucus basalis woolaston* (Hymenoptra: Scelionidae) a parasite of *Nezara viridua. J. Aust. Ent. Soc.*, 22(3): 271–272.

Udhaya Kumar, A. and Jayaraj, S., 1987. Influence of host plants on certain parasites of *Spodoptera litura* (F.), *J. Biol. Control.*

Ujagir, Ram, Singh, R.K. and Ujagir, R., 1999. Management of insect pests of mungbean by insecticides using seed treatments, soil and foliar application. *Indian Journal of Pulses Research*, 12(1): 82–91.

Usua, E.J. and Singh, S.R., 1978. Parasites and predators of the cowpea pod borer, *Maruca testulalis* Gey. (Lep. Pyralidae). *Nigerian Journal of Entomology*, 3: 100–102.

Vadodaria, M.P. and Vyas, H.N., 1987. Control of whitefly, *Bemisia tabaci* (Gennadius) and its impact on yellow mosaic virus (YMV) in greengram, [*Vigna radiata* (L.) Wilczek] and the grain yield. *Indian J. Agril. Res.*, 21(1): 21–26.

Venkataramaiah, C., Reddy, A.J. and Chandrashekharan, V., 1980. On DDT-induced Radicular proliferation in *Phaseolus mungo* Linn. *Current Science*, 49(6): 242–243.

Verma, A.N., Bhanot, J.P. and Khurana, A.D., 1983. Effect of pesowing soil applicationof BHC and aldrin on plant stand in wheat and *moong, Haryana Agricultural Univeristy Journal of Research*, 13(1): 150–152.

Vyas, H.N. and Saxena, H.D., 1980. Bio-efficacy of some synthetic insecticides on seed treatment and soil application to control stemfly, *Ophiomyia phaseoli* Cogi (Diptera: Agromygzidae) in pea, *Pisum sativum*. *Pesticides,* 14(1): 12–13.

Vyas, H.N. and Saxena, H.P., 1981. Toxicological and persistence studies of some insecticides in relation to pest complex and plant growth of greengram, *Vigna radiata* (L.) Wilczek and fieldpea, *Pisum sativum* L. *Pulse Crops Newsletter,* 1: 98.

Vyas, H.N. and Saxena, H.P., 1983. Bioefficacy of some insecticides against *Empoasca kerri* Pruthi in greengram, *Vigna radiata* (L.) Wilczek. *Pesticides,* 17(5): 15–17.

Yadav, A.S., Bhaduria, N.S. and Jakhmola, S.S., 2004. Efficacy of edible non-edible vegetable oils against pulse beetle, *Callosobruchus maculatus* (Fab.) in green gram. *Indian Journal of Entomology,* 66(4): 365–366.

Yadav, D.K. and Singh, S.K., 2006. Forecast model of major insect pests of mungbean. *Ann. Pl. Protec. Sci.,* 14(2): 323–328.

Yadav, L.S. and Yadav, P.R., 1979. Avoidable losses due to insect pests in pulses. *Military Farm J.,* 11: 17–19.

Yadav, T.D., Singh, S., Jha, A.N. and Yadav, S., 2000. Host suitability of chickpea cultivars to *Callosobruchus chinensis*. *Indian J. Ent.,* 62(2): 203–204.

Yein, B.R. and Singh, H., 1981. Effects of some agricultural chemicals on the control of chewing insect pests of greengram. *J. Res. Punjab Agril. Univ.,* 18(1): 30–36.

Yein, B.R. and Singh, H., 1982. Effect of pesticides and fertilizers on the population of whitefly and incidence of yellow mosaic virus in greengram. *Indian J. Agric. Sci.,* 52(12): 852–855.

Insect Pest and Disease Management *Pages* **118–129**
Editor: **Dr. D. Prasad**
Published by: **DAYA PUBLISHING HOUSE (2016)**

Chapter 7

Status and Strategies on Management of Coconut Eriophyid Mite

C. Muthiah
Agricultural College and Research Institute, T.N.A.U., Madurai – 625 104, T.N.

Introduction

Coconut is an important horticultural crop, grown across the tropics, covering an area of 12 million hectare, with total production of about 56 billion nuts, of which Asia and Pacific region contributes 90 per cent in area and production. Over the years the coconut palm has been affected by various common as well as less known pests and disease causing serious concern to those whose livelihood depends on it. In India, the coconut palm is affected by serious pests like red palm weevil, leaf eating caterpillars and rhinoceros beetle. Eriophyid mite of coconut, a microscopic insect observed first in Mexico way back in 1965 has become a serious pest in Asia and Pacific region, which was first observed in Sri Lanka in 1997 and then in India in 1998. The pest continues to be a cause of concern in both these countries, due to high economic losses affecting large number of small and marginal farmers. Although nine species of eriophyid mites have been reported to infest coconut leaves and nuts, only *Aceria* (Eriophyes) *guerreronis* (Keifer) (Eriophyidae: Acari) feeding on tender nuts has been found to cause heavy damage in different parts of India in recent years, resulting in loss of production of nuts.

Taxonomy and Distribution

Nine species of eriophyid mites have been recorded on coconut from different parts of the world (Amrine, Jr. and Stasny, 1994) and of these only two occur on the nuts. Among them, *A. guerreronis* is considered to be serious. Though symptoms of damage on nuts were first described by Martyn (1949) from Columbia they were attributed to eriophyid mite by Ortega *et al.*, 1965 in Guerrero on the west coast of Mexico. The mite was first described in 1965 from Guerrero State, Mexico by Keifer (1965) and since then it has been reported from many coconut growing areas of the America, West Africa and the Caribbean islands from Cuba to Trinidad and St. Lucia in the West Indies (Moore, 1986).

Asia and Pacific Region

Eriophyid mite has become a serious pest in Asia and Pacific region, which was first observed in Sri Lanka in 1997 and then in India in 1988. This pest continues to be a cause of concern in both these countries, due to high economic losses affecting large number of small and marginal farmers. In India, it was first observed in Ernakulam district of Kerala State and then spread to North and East, in Tamil Nadu and Karnataka from where it moved to southern parts of Andhra Pradesh, in the district of Chittoor and then to Northern Andhra Pradesh in East and West Codawari district. Now the occurrence of this pest is also noticed in the state of Orissa, West Bengal in Eastern India.

Biology

The eriophyid mites are vermiform with elongate body, distinctly divisible into a cephalothorax and a long, ringed tapering body. The mite possesses two long sinous setae arising from the posterior part of the body. The adult mite is very minute measuring 200–250 µm in length and 36–52 µm in width with two pairs of legs found in the anterior region of the body. It is pale in colour with elongate body and worm like in appearance. The female mite lays approximately 30–50 eggs. The eggs are globular, shiny and round in shape under the microscrope, which hatch into larva (protonymph) in two days. The second instar larva is called nymph which subsequently moult into adults. The life cycle of this mite, which consists of egg, two larval instars and an adult stage, is completed in 8–10 days. The mites are found under the bracts of fertilized female flowers and are not usually found in unfertilized flowers. The distribution of eriophyid mite colony is not uniform inside the perianth. Normally in two or three places the mite colonies are congregated under the tepals. In most of the cases old and fresh colonies are found beneath the fourth tepal. Each colony contains hundreds of eggs, larvae, nymphs and adults. The colonies vary in size and shape. After causing damage in a particular spot/area the mites tend to move to fresh areas in meristematic tissue and start colonization.

Symptoms of Damage

The mites pierce the tender tissues and develop on the meristematic tissues of the growing nuts under the perianth by desapping the soft tissues of the buttons. In the damaged nuts, initially triangular pale white or yellow patches close to perianth can be seen. Different stages of mites live in the white, tender portion covered by the inner bracts of the perianth and suck the sap continuously. The increased feeding activity results in physical damage to the newly formed tissues which become necrotic. Intensive damage results in the formation of brown coloured patches. In addition, as the nut grows in size, longitudinal fissures and splits occur on the outer surface of the husk. Occasionally brownish gummy exudate is seen oozing out from the affected surface. Severe infestation results in poor growth, reduced size and copra content and malformed nuts with cracks and hardened husk.

Ecology

The adult coconut mite is able to penetrate between the upper and lower tepals and finally to the fruit surface covered by the perianth within a few weeks to a month after fertilization (Ortega *et al.*, 1965; Mariau and Julia, 1970; Moore and Alexander, 1987a; Howard and Abreu, 1991). Young fruits are almost entirely covered by the perianth. Experiments with marking inks have shown that the tepals are tightly ad-pressed to the fruit during its first month of development (Howard and Abreu, 1991), so that the perianth gives maximal protection at this stage. As the fruit develops, it becomes increasingly larger in relation to the perianth, and within about a month spaces develop between the coconut surface and the preianth which, in the experiments cited, were penetrated by marking inks and thus apparently are sufficiently large to permit the entry of coconut mites. With a development cycle from egg to adult of about ten days (Mariau, 1977) mite numbers can build up rapidly. Apparently, reproductive activities take place beneath the perianth, as spermatophores associated with coconut mite colonies have been observed there. The fruits remain susceptible to mite attack almost throughout the whole development, but on more mature fruits (10 to 13 months), coconut mites are found rarely and in small numbers (Moor and Alexander, 1987).

The coconut mite is found in tropical and subtropical climates, but populations can survive both short periods of frost and periods of temperatures just above zero more prolonged than those normally encountered where coconut palms are grown (Howard *et al.*, 1990). On the basis of observations in particular localities it has been suggested that coconut mite attacks are more severe in relatively dry climates or during the dry season of wetter climates (Zuluaga and Sanchez, 1971; Griffith, 1984). However, in other localities there is not clear relationship between coconut mite populations and wet and dry weather, or, if such a relationship exists, it is obscured by other factors (Doreste, 1968; Mariau, 1969, 1977; Howard *et al.*, 1990).

In a study on the mite population observed under the perianth for one year from November 1999 to October 2000 in a stand of 15 years old coconut palm at Coconut Research Station, Veppankulam revealed that the mite population was found to be very high in 2 to 5 months old buttons. The study conducted also revealed that the percentage of nuts infested with eriophyid mite was maximum (83.8 per cent) in 5 months old bunches as against minimum (14.0 per cent) in 3 months old bunches. The population of mites varied from 15 to 170 per 4.mm^2 area in 5 month old nuts. The observations on the time of entry of mite in the coconut buttons of East Coast Tall variety revealed that the mites were able to enter in to the bracts and colonise the meristematic zone about 34 days after fertilization *i.e.*, 54 days after opening of the inflorescence and thereafter. Though the mites were found throughout the year, the maximum population was recorded during the summer months of March to July when the temperature ranges from 30 to 36°C (Muthiah *et al.*, 2001).

Dispersal/Spread

The mechanisms of dispersal of *A.guerreronis* have not been well studied but it is conjectured that some dispersal may take place by phoresy, either on animals directly attracted to the inflorescences (*e.g.*, pollinating insects such as bees; rodents which feed on the fruits), or on those attracted by such animals (*e.g.*, Predatory lizards, birds, predaceous insects).

However, the principal way in which coconut mites spread and colonize new palms, particularly over long distances, is probably through areial dispersal of inseminated female mites. The coconut palm provides a large target for aerially diersed organisms, and the likelihood of arriving on a fruit is probably increased when air currents carry the mites to recemes, or to the more vertical leaves in the crown, from which they may drop to inflorescences. Coconut mites can walk between touching inflorescences and being negatively geo-tactic, tend to move from older to younger inflorescences (Moore and Alexander, 1987). Coconut mites walk at a rate of 20–100 μms^{-1} but apparently are no efficient in finding feeding sites. As with other eriophyid mites. *A. guerreronis* demonstrates inefficient dispersal and host finding but a high reproductive rate and rapid development.

Yield Loss

The results on the assessment yield losses due to eriophyid mite at Coconut Research Station, Veppankulam revealed that in all the three types of coconut the eriophyid mite infestation caused significant yield loss in percentage terms. An average of 20.27, 16.61 and 18.40 per cent copra yield losses were assessed in East Coast Tall, hybrid (VHC 2) and Chowghat Orange Dwarf varieties respectively due to eriophyid mite. Julia and Mariau (1979) also reported loss of copra varied from 10 per cent in Benin, 16 per cent in Ivory coast (Mariau and Julia, 1979), 30 per cent in Mexico (Hernandez, 1977) and 11–28 per cent in St. Lucia (Moore, 1986). The estimated loss of copra in Tamil Nadu was 27.5 per cent (Ramaraju *et al.*, 2000). The kernel weight/nut and copra yields/nut demonstrated that there was a progressive decline in yield as the mite damage worsens. Even the quite badly damaged nuts (eg. Category 3 nuts) can yield relatively well and that most yield loss was due to nuts from categories 4 and 5. Even the most damaged nuts could yield some copra but dehusking of these nuts was very difficult. There was also a significant reduction in kernel content observed between the eriophyid mite damaged nuts (Categories 2 to 5) and undamaged nuts (Category 1) in the variety Chowghat Orange Dwarf.

The nut size, nut girth, dehusked nut weight and quantity of water also significantly reduced due to eriophyid mite damage in different damage category nuts. The results confirm the findings of Muthiah and Bhaskaran (2000) reported the uneven development of nuts due to mite resulted in the significant reduction in nut size. When the infection occurs in the early stages, development of kernel was affected leading to reduction in copra out turn. If the infection occurs after seven months of fertilization, although lesions develop on the surface of the nuts the nut size and copra content were not much reduced. However, there was no significant difference observed in kernel thickness and shell weight due to eriophyid mite damage in all the three varieties evaluated (Muthiah and Natarajan, 2004).

Management Strategies for the Coconut Eriophyid Mite

Use of biological control, cultural methods, other eco-friendly methods, chemical control and integrated pest management are considered for the management of the eriophyid mite.

Research Efforts in India

Immediately after noticing the incidence of this dreaded pest in Ernakulam district, research efforts were initiated both at Central Plantation Crops Research Institute, Kayamkulam and Kerala Agricultural Univeirsity to evolve control measures against this pest. Noticing the incidence of this pest in Tamil Nadu, Tamil Nadu Agricultural University (TNAU) also initiated the programme for the management of this pest. Alerted by the speedier spread of this pest, focused attention for research and development programmes was laid. Coconut Development Board played a key role in co-ordinating the efforts and got initiated the programme on biological control by sponsoring a research project to the Project Directorate of Biological Control (PDBC). The Directorate identified the '*Hirsutella thampsonii*' a fungus, effective in causing mortality of the population eriophyid mite, and intensified its research further for its commercial formulation. Recognizing the importance of this dreaded pest in the economy of the coconut, a network programme involving all the capable institutions in the coconut growing states was initiated under National institutions in the coconut growing states was initiated under National Agriculture Technology Project (NATP) for coastal agro eco-system, wherein TNAU played a modal role and University of Agricultural Sciences, Bangalore, CPCRI, PDBC, Kerala Agriculture University and other concerned Departments became partners. The National Steering Committee constituted under the chairmanship of Horticulture Commissioner, Government of India took up to monitoring role and facilitated the refinement of technology, based on the discussion for its wider adoption. The steering Committee catalyzed the efforts for collective thinking and speedier adoption of recommendation. The states also played significant role in creating awareness for mitigating the problem. The Government of India extended financial assistance to states for mitigating the problem.

Survey

Survey was conducted from Coconut Research Station, Veppankulam in Thanjavur, Thiruvarur, Nagai and Pudukkottai district to assess the incidence of eriophyid mite and its natural enemies during September 2000. Among the four districts the eriophyid mite population was more in Thanjavur district (10.5 mites/4 sq.mm) and minimum in Nagapattinam district (7.6 mites/4 sq.mm). The percent nut damage by eriophyid mite in harvested nuts was also more in Thanjavur district (87.5 per cent and less in Nagapattinam district (77.0 per cent). The predatory mite *Amblysieus, Paspalivarus* was found to present on the infected nuts. The predatory mite population ranged from 0.2 to 0.8 mites/nut in the four districts surveyed with a maximum of 0.8 mites/nut in Thiruvarur district.

Cultural and Agronomic Practices

Researchers are interested in the possibility of cultural control of the coconut mite. Periods of water deficit result in greater yield losses due to coconut mite, at least under some conditions (Mariau, 1977, 1986), possibly because fruit growth is slower in dry periods resulting in the fruits remaining susceptible over a longer period. Better cultivated trees, with appropriate fertilizer application, may also suffer less coconut mite attack (Mariau, 1977; Romney, 1980).

The recommended dose of fertilizers are needed to keep the coconut trees healthy to resist any pests and diseases. Increased application of potassium @ 3.5 kg/palm recorded the least mite infestation probably due to increased plant resistance to the mite. Since K and neem cake are known to induce resistance to pests they are being recommended at 3.5 and 5 kg/palm respectively. Further, Ca, Mg and B are known to induce some amount of resistance to mites. The application of farm yard manure @ 50 kg/palm is also recommended.

Flood irrigation recorded lesser incidence of the mite in the nuts compared to other irrigations. Abundant water supply in flood irrigation and addition to nutrient through incorporation of green manure might have promoted the uptake of more nutrients resulting in increased vigour of the palms. When high density multiple intercrops was raised, the mite damage was 58–64 per cent (Muthiah and Bhaskaran, 2000). Intercropping of sunnhemp with coconut reduced the mite incidence upto 13.6 per cent and reduced the damaged grade.

The results on the application of calcium (gypsum) and other ameleorants after two years of treatment revealed that application of Borax or calcium alone along with the recommended NPK and FYM registered low damage grade index (3.36 and 3.53) than the nutrients applied in combinations due to mite in the harvested nuts. However, there was no significant difference in the mite damage grade index among the various treatments and control. The mite damage per cent in the young bunches revealed that it was less in application of 'Ca' and 'Mg' along with the recommended NPK and FYM during 2003. However, there was no significant difference in the per cent nut damage in young buttons among the various treatments (Muthiah, Natarajan, 2004). Management of coconut mite with nutrition was attempted earlier by Moore *et al.* (1991). Better cultivated trees, with appropriate fertilizer application, may also suffer less coconut mite attack (Mariau, 1977). Calcium, magnesium and boron are known to induce resistance to mites (Kannaiyan *et al.*, 2002). Through intensive research and observations it has been evident that damage due to mite is negligible in the orchards, which have been appropriately managed, maintaining adequate nutrients and water balance. Periods of water deficit result in greater yield losses due to coconut mite (Mariau, 1986) possibly, because fruit growth is slower in dry periods resulting in the fruits remaining susceptible over a longer period. Natarajan and Rabindra (2003) reported that application of bacterial bio-control agent *Pseudomonas fluorescens* and organic manures in soil against mite was not effective. The present study also confirms that application of macro and micro nutrients reduces the mite damage. However, the effect of different nutrients could not be realized significantly.

Cultivar Resistance

Mariau (1977, 1986) reported that cultivars varied in their susceptibility to coconut mite infestations, and trees of a cultivar from Cambodia suffered no attack. Mariau (1977) suggested that mites could not get under the perianth of the apparently resistant Cambodian cultivar because it adhered tightly to the very rounded form of the fruits. Mite damage was found to be less where the tepals were most tightly addressed to fruits and in rounded fruits compared with elongated ones (Moore, 1986; Moore and Alexander, 1990). Cultivars with improved coconut mite resistance are a future possibility. This may involve breeding of new cultivars or replanting using seed nuts from productive trees which show little mite damage.

Screening of 31 coconut genotypes and 34 hybrids for 4 years at Coconut Research Station, Veppankulam revealed four genotypes Chowghat Orange Dwarf (1.99), Spicata (2.41), Siam (2.50), British Solomon Island (2.5) and two hybrids (JG × ECT (1.61) and CC × PHO (2.34) significantly recorded minimum damage grade index due to mite and found to be moderately resistant. Further screening will be needed for confirmation.

Coconut varieties 'Kenthali', 'Gangabondam' and 'Philippines Ordinary' recorded the lowest nut damage at Coimbatore. West coast tall × Chowghat dwarf orange, West coast tall × Gangabondam, West coast tall × Malaysian dwarf yellow, Gangabondam × Fiji, Gangabondam × Philippines ordinary, Gangabondam × Lakshadweep Ordinary were consistently showing less nut damage on nut surface at Bangalore.

Chemical Status

Spraying

Chemical control of the coconut mite is possible; dicrotophos, monocrotophos or chinomethionate sprayed onto bunches of developing fruits every 20 or 30 days significantly reduced damage (Hernandez, 1977). Similar results were obtained with acaricides applied at 15 day (Mariau and Tehibozo, 1973).

Stem Injection

Julia and Mariau (1979) found that stem injection of monocrotophos every two months was effective on young dwarf plants but was not recommended for mature trees. Stem injection of the pesticide vamidothion was proposed by Griffith (1984) but was ineffective in studies by Moore and Alexander (1987 b). Multi location field experiments revealed that spot application of triazophos 5 ml/l, carbosulfan 5ml/l and monocrotophos 5 ml/l twice on the first six bunches from the top at 60 days interval gave appreciable control of the mite (Rabindra and Karuppusamy 2001).

The results on the management of eriophyid mite with different chemicals at Coconut Research Station, Veppankulam revealed that spraying of carbosulfan 2ml/lit significantly reduced mite population on 15 and 30 days after first treatment followed by Triazophos 5ml/lit. In the second round of treatment Triazophos 5ml/lit significantly reduced the mite population by 70 per cent on 30 DAT and it was on par with carbosulfan 2ml/lit which recorded 68 per cent population reduction (Muthiah *et al.*, 2001). The results confirm the findings of Palanisamy *et al.*, 2000 reported that the spot application of Triazophos 5ml/lit, carbosulfan 5 ml/l and monocrotophos 5ml/l recorded a reduction of 60–76 per cent mite population. The study at Coconut Research Station, Veppankulam also revealed that spray treatment of chemicals are more effective them root feeding treatments. The results confirm the findings of Mohanasundaram *et al.*, 1999; Muthiah and Bhaskaran 1999.

The field trial conducted for the management of mite with chemicals during 1999 revealed that among the different chemicals spraying either methyl demeton 4 ml/lit at 10 days interval or monocrotophos 1.5ml/lit significantly reduced the percentage of nuts damaged to 25 per cent as compared to 53 per cent in control. Bunches sprayed at 10 days interval reduced the mite damage when compared to spraying at 30 days interval.

In general, chemical control is not practicable. A short interval between treatments is required for most acaricides to be effective and chemical treatments would have to be continued indefinitely. This would be economically unfeasible, environmentally hazardous and unacceptable where consumption of fresh coconuts or coconut water is common.

Botanicals and Other Eco-friendly Agents

Among the eco-friendly agents, TNAU Neem oil, Neem azal, Fish oil rosin soap and Garlic + neem oil as psray and neem seed kernel extract as root feeding tested at Somanur of Coimbatore district, spraying of fish oil rosin soap (FORS) @ 4 per cent, garlic + neem oil, TNAU neem oil and neem azal are promising for the control of the coconut eriophyid mite recording a mean population reduction of 44.69 to 59.24 per cent over control. An eco-friendly formulation *viz.*, TNAU–Agro–Biocide containing plant extracts, plant based oils and fungal culture filtrate was developed at azolla laboratory, TNAU based on the biossay studies and it was tested as root feeding @ 30 ml/tree. This recorded a reduction in mite population of 65.1–75.0 per cent, 89.6–94.9 per cent and 64.7–100 per cent over untreated control on 15, 30 and 45 days after treatment (DAT) (Kannaiyan *et al.*, 2000 a). Kerala Agricultural University developed a formulation containing neem oil, garlic and washing soap solution for the management for the coconut mite (Nadarajan *et al.*, 2000).

Biological Control (Natural Agents)

Predators

Coconut mites, as with other mites, are not attacked by parasitoids and their sheltered habitat and biology provide few opportunities for other natural enemies to be effective. Theoretically, predators could attack the coconut mite during dispersal, which occurs regularly (Moore and Alexander, 1987 a), and have been observed occupying the meristematic zone of coconut fruits. These include *Bdellu distincta* Baker and Balock, *Amblyseius largoensis* Muma, *Neoseiulus mumai* Denmark and *N. paspalivorus* DeLeon (Howard *et al.*, 1990), two unidentified phytoseiids and a tarsonemid (Julia and Mariau, 1979). Predaceous mites are observed on infested coconuts only occasionally and in very small populations and there is no evidence that they make a significant impact on coconut mite populations (Hall *et al.*, 1980; Howard *et al.*, 1990).

Pathogens

The acerogenous fungus *Hirsutella thompsonii* (Fisher) has been isolated from samples of coconut mites from tropical America and West Africa (Hall *et al.*, 1980; C.W. McCoy, personal communication, 1990) and from samples of *C.novahebridensis* from New Hebrides, New Guinea and Sri Lanka (Hall *et al.*, 1980). This fungus is specific to mites and has been studied extensively for the control of citrus rust mite, *Phyllocoptruta oleivora* (Ashmead), and a process for a commercial product was developed (Gillespie, 1988). Field use against coconut mite has had variable results. In Mexico up to 75 per cent mortality was achieved using the fungus (Espinosa and Carrillo, 1986) but no success was reported to West Africa (Anon, 1989) or from limited trials in St. Lucia (Moore *et al.*, 1989). In laboratory trials, Sampedro and Rosas (1989) tested seven isolates of *H. thompsonii;* mortality ranged from 88 per cent with an isolate from *A.guerreronis* to 32 per cent with one obtained from *P.oleivora.*

Another acerogenous species attacking *A.guerreronis, H.nodulosa* Petch, has been reported from Cuba (Cabrera and Dominguez, 1987).

Studies conducted at the Department of Plant Pathology, Tamil Nadu Agricultural University, Coimbatore indicated that the entomogenous fungi, *Verticillium suchalosporium* and *Metarhizium anisopliae* were able to reduce the population of *A.guerreronts.* Five entomogenous fungi *viz., Beauveria bassiana, Paecilomyces fumasoroseus. Catenaria* sp., *M. anisopliae* and an unknown culture isolated from dead mite (N/MS) were grown in the PDA broth and sprayed on the eriophyid mite affected nuts at Maruthamalai area, Coimbatore.

Attempts have been made to isolate acarophagous fungi capable of controlling mites. Laboratory and field experiment have indicated that *Fusarium suchlosporium* and *Hirsutella thompsoni* controlled the mites only to an extent of 20 to 40 per cent.

Hirsutella thompsoni var synnematosa the fungal pathogen of CEM was isolated and identified for the first time at Vellayani. A novel entomo-phoresy approach for the delivery of pathogens of the mite to the coconut inflorescence using stingless honebee *Trigona irridipennis* was developed. At PDBC, Bangalore three mycoacaricides *viz.*, Mycohit powder, Mycohit LG 20 and Mycohit OS from a fungal pathogen *Hirsutella thompsoni* has been developed for the biological control of coconut eriophyid mite. The '*Hirsutella thompsoni*', developed and releaed as 'MYCOHIT' in Bangalore, could not prove its efficacy in the other parts of the country indicating the need for improving its formulation. Reformulated MYCOHIT has been tried with varying results. Interestingly, *Hirsutella thompsoni* is observed to be effective in bringing down the population of coconut mite but its impact on reduction of damage is not significant under field condition. In the environment various factors interact which determine the efficacy. *Sporothrix fungorum* has been identified for the first time as a pathogen of coconut mite. A talc based formulations based on *S. fungorum* with a potency of 1 x 10^9 CFU was developed.

Research Strategies/Requirements

1. The coconut mite has not received the research attention that it warrants and many fundamental aspects are poorly understood.
2. *Investigation of basic biology.* A method of rearing and maintaining laboratory colonies of coconut mites is needed so that laboratory studies can be undertaken to elucidate the development, feeding and reproductive behaviour, population dynamics, interactions with natural enemies and other environmental factors, dispersal mechanisms and many other aspects.
3. *Coconut mite-host interactions.* Is the damage purely mechanical or does the mite transmit a toxin which causes yield loss? Other mites are capable of causing similar but lighter damage to coconuts (Hall *et al.*, 1980; Kang, 1981; Howard *et al.*, 1990), usually on a small number of

fruits or trees. The mechanism behind the damage could be of significance in breeding resistant cultivars.

4. *Cultural control of coconut mite*. Are well cultivated trees tolerant of the mite? In Jamaica, the mite is considered a pest only of poorly maintained estates (B.O Been, personal communication, 1985). Research on fertilizer use, spacing, weed control, irrigation and a range of cultural practices could determine if the mite can be managed rather than controlled.
5. *Distribution and status*. In addition to determining the effects of biotic and abiotic environmental factors on coconut mite populations in areas where it is a pest, it should be determined if *A. guerreronis* is endemic in areas where it is not a pest and attains pest status because of some environmental changes. In parts of the Caribbean the coconut mite has rapidly become important in areas affected by hurricanes, or volcanic eruptions, or sprayed with fungicides to protect other crops. When newly detected in some countries, infestations have been found in a number of dispersed localities after which the infestation appeared to spread very quickly (Doreste, 1968; Mariau, 1986).
6. *Testing for resistant cultivars and determination of mechanisms of resistance*. Variability in susceptibility to coconut mite among coconut cultivars (Mariau, 1986) should be thoroughly investigated, and efforts made to identify resistant cultivars or to breed for resistance.
7. *Biological control*. The development of a mycopesticide is probably the most realistic chance for biological control. This would require searches for more species or isolates of pathogenic fungi and the solving of formulation and application problems.
8. *Investigation of different strains of A.guerreronis*. For may of the above it may be helpful to test different strains of the mite, from different areas of its range, to determine variability in behaviour, effects on host, etc. This would cause quarantine problems unless laboratory rearing is achieved and the work is conducted in areas that do not produce coconuts.
9. Efforts should be made for conservation of predatory mite population existing in nature by avoiding indiscriminate use of chemicals which are harmful to the predators and maintain ecological balance.
10. More emphasis should be made to study the natural incidence of *Hirusutella thompsonii* in different regions.
11. Establishment of *H. thompsonii* in nature as a long term strategy to manage the mite population.
12. In INM and IPM technologies for managing mite, emphasis should be laid on cost effective measures like organic recycling of farm waste materials by vermicomposting and using lignin degrading fungi to convert the bio-mass into a acceptable organic manures.
13. Use of Stingless honeybees has to be studied in details for easy deposition of entomogenous fungus formulations in the young buttons during pollination through entomophoresis technique. The safety of such entomogenous fungus formulations to honey bees should also be studied.
14. Collaborative efforts are required for large scale demonstration of existing technologies in farmers fields with the active involvement of Scientists, Department of Officials and local bodies.
15. Research effort should be intensified for further improvement of sprayers developed for carrying out spraying the tall palms.

16. Use of mycelia based formulations in place of spore based formulations may be looked into for infestation of mite by *Hirsutella* fungus.
17. Quality of neem based commercial formulations used for eriophyid mite control should be ensured by registering with the central agency.
18. Usefulness of traditional Panchagavya formulations in control of mite may be tested.
19. Instead of spraying and root feeding a technology may be developed for the feasibilities of soil application in the granular farm may be explored.
20. Community training programmes and large scale demonstrations are need to be taken up on priority. Inputs have to be made available along with credit facilities.
21. Farmers should be educated properly not to use any spurious products for controlling mite which are being sold in market without proper registration from central insecticide board. The state government should insist for registration of products by designated authorities of central/state government.

References

Amrine, Jr. J.W. and Stasny, T.S., 1994. *Catalog of the Eriophyoidea (Acarina: Prostigmata) of the World.* Indira Publishing House, Michigan, USA, 804p.

Anon., 1989. *Eriophyes guerreronis. Oleagineux*, 44: 130–131.

Cabrera, R.I. and Dominguez, D., 1987. Elhongo *Hirsutella nodulosa* nuevo parasito para el acaro del cocotero *Eriophytes guerreronis.* Ciencia Y Technica en la Agricultura, Citricosy Otros Frutales, 10: 45–51.

Doreste, S.E., 1968. El acaro de la flor del cocotero (*Aceria guerreronis*) en Vennezuela. Agronomia Tropical, 18: 379–386.

Espinosa, B.A. and Carrillo, S.J.L., 1986. El hongo *Hirsutella thompsoni* Fisher en el control del eriofido del cocotero, *Eriophyes guerreronis* (Keifer). *Agricultura Tecnica en Mexico,* 12: 319–323.

Gillespie, A.T., 1988. Use of fungi to control pests of agricultural importance. In: *Fungi in Biological Control Systems,* (Ed.) M.N. Burge. Manchester University Press, Manchester, UK, pp. 37–60.

Griffith, R., 1984. The problem of the coconut mite, *Eriophyes guerreronis* (Keifer) in the coconut grooves of Trinidad and Tobago. In: *Proc. 20th Ann. Meeting of the Caribbean Food Crops Soc. East Caribbean Center,* (Eds.) R. Webb, W. Knausenberger and L. Yntema. College of the Virgin Islands and Caribbean Food Crops Soc., St. Croix, Virgin Islands, USA, pp. 128–132.

Hall, R.A., Hussey, N.W. and Mariau, D., 1980. Results of a survey of biological control agents of the coconut mite *Eriophyes guerreronis. Oleagineux,* 35: 395–398.

Hermandez, R.F., 1977. Combate quimico del eriofido del cocotero *Aceria* (*Eriophyes*) *guerreronis* (K.) en la costa de Guerrero. *Agricultura Tacnica en mexico,* 4: 23–28.

Howard, F.W. and Abreu-Rodriguez, E., 1991. Tightness of the perianth of coconuts in relation to infestation by coconut mites. *Fla. Entomol.,* 74: 358–361.

Howard, F.W., Abreu-Rodriguez, E. and Denmark, H.A., 1990. Geographical and seasonal distribution of the coconut mite, *Aceria guerreronis* (Acari: Eriophyidae) in Puerto Rico and Florida, USA. *J. Agric. Univ. Puerto Rico.,* 74: 237–251.

Julia, J.F. and Mariau, D., 1979. Nouvelles recherches en cote'd Ivoire sur *Eriophyes guerreronis* K., acarien ravageur des noix du cocotier. *Oleagineux*, 34: 181–189.

Kannaiyan, S., Kumar, K., Gopalsamy, G., Yasotha, P. and Ramaraju, K., 2000a. Evaluation of a formulation based on plant products, biocides and agrochemicals for its efficacy against eriophyid mite in coconut. Paper presented in *Interactive Workshop on Coconut Eriophyid Mite* held on 19th, May, at Tamil Nadu Agricultural University, Coimbatore–3.

Kannaiyan, S., Rabindra, R.J., Ramaraju, K. and Sabitha Doraiswamy, 2002. Management of eriophyid mite *Aceria guerreronis* (Keifer) on coconut. In: *Integrated Management of Eriophyid Mite on Coconut,* (Eds.) S. Kannaiyan, R.J.Rabindra, K. Ramaraju and Sabitha Doraiswamy. Tamil Nadu Agricultural University, Coimbatore, p. 1–82.

Keifer, H.H., 1965. Eriophyid studies B–14. *Calif. Dept. Agric. Bureau of Entomol.*, 20p.

Mariau, D., 1977. *Aceria (Eriophyes) guerreronis*. un important ravageur des cocoteraies africaines et americaines. *Oleagineux*, 32: 101–111.

Mariau, D., 1986. Comportement de *Eriophyes guerreronis* Keifer a 1, egand de differentes varieties de cocotiers. *Oleagineux*, 41: 499–505.

Mariau, D. and Fulia, J.F., 1970. L' acariose a *Aceria guerreronis* (Keifer), ravageur du cocotier. *Oleagineux*, 25: 459–464.

Mariau, D. and Tchibozo, H.M., 1973. Essais de lutte chimique contre *Aceria guerreronis* (Keifer). *Oleagineux*, 28: 133–135.

Martyn, E.B., 1949. Notes on a visit to Columbia. *Trop. Agric. (Trinidad)*, 26: 48–50.

Moore, D. and Alexander, L., 1990. Resistance of coconuts in St. Lucia to attack by the coconut mite. *Eriophyes guerreronis* Keifer. *Trop Agric*, 67: 33–36.

Moore, D., 1986. Bract arrangement in the coconut fruit in relation to attack by the coconut mite *Eriophyes guerreronis* Keifer. *Trop. Agric. (Trinidad)*, 63: 285–288.

Moore, D. and Alexander, L., 1990. Resistance of coconuts in St. Lucia to attack by the coconut mite, *Eriophyes guerreronis* Keifer. *Trop. Agric. (Trinidad)*, 67: 33–36.

Moore, D. and Alexander, L., 1987. Stem injection of vamidothion for control of coconut mite, *Eriophyes guerreronis* Keifer, in St. Lucia. *Crop Protection*, 6: 329–333.

Moore, D., Ridout, M.S. and Alexander, L., 1991. Nutrition of coconuts in St. Lucia and relationship with attack by the coconut mite *Eriophyes guerreronis* Keifer. *Trop. Agric (Trinidad)*, 68: 41–44.

Muthiah, C. and Bhaskaran, R., 2000. Major outbreak of eriophyid mite of coconut in India. *The Planter*, 76(889): 243–246.

Muthiah, C. and Bhaskaran, R., 1999. Screening of coconut genotypes and management of eriophyid mite *Aceria guerreronis* (Eriophyidae: Acari) in Tamil Nadu. *Indian Coconut Journal*, 30 (6): 10–11.

Muthiah, C. and Natarajan, C., 2004. Varietal reaction and Nutrient management of coconut eriophyid mite. *The Planter*, Kaulalumpur, 80(936): 159–169.

Muthiah, C. and Bhaskaran, R., 2000. Survey, bio-efficacy and management of eriophyid mite on coconut. In: *Paper Presented in Group Meeting on Coconut Eriophyid Mite, Aceria guerreronis* K. on 15th May, at Tamil Nadu Agricultural University, Coimbatore–3.

Muthiah, C., Bhaskaran, R. and Kannaiyan, S., 2001. Bio-ecology and control of eriophyid mite of coconut: An Indian experience. *The Planter*, 77(902): 255–263.

Nadarajan, L., Ranjith, A.M., Jim Thomas, Pathummal Beevi, S. and Nair, Madhavan, 2000. Coconut perianth mite, *Aceria* (*Eriophyes*) *guerreronis* (Keifer) and its management, Kerala Agricultural University, Vellanikkara, pp. 1–10.

Natarajan. K and Rabindra, R.J., 2003. Effect of *Pseudomonas fluorescens* and organic manures for the management of coconut eriophyid mite. *Madras Agric. J.*, 90 (7–9): 389–94 4

Ortega, C.A., Rodriguez, V.J and Garibay, V. C, 1965. Investigaciones preliminares sobre el eriofido del fruto del cocotero *Aceria guerreronis* Keifer, en la Costa Grande de Guerrero. *Agricultura Technica en Mexico*, 2: 222–226.

Ramaraju, K., Natarajan, K., Sundara Babu, P.C. and Palanisamy, S., 2000. Studies on coconut eriophyid mite *Aceria guerreronis* K. in Tamil Nadu, India. In: *Paper Presented at "International Workshop on Coconut Eriophyid Nite"* held at CRI, Sri Lanka, Jan. 5–7, pp. 8–9.

Romney, D.H., 1980. Agronomic performance of "Malayan Dwarf" coconut in Jamaica. *Oleagineux*, 35: 551–554.

Sampedro, L. and Rosas, J.L., 1989. Selection de cepas de *Hirsutella thompsoni* Fisher paracombatir al acaro del cocotero, *Eriophyes guerreronis* Keifer, I. Bioensayos de patogenicidad. *Revista Mexicana de Micologia*, 5: 225–232.

Insect Pest and Disease Management *Pages* ***130–139***
Editor: **Dr. D. Prasad**
Published by: **DAYA PUBLISHING HOUSE (2016)**

Chapter 8

Sustainable Management of Bud Fly, *Dasyneura lini* Barnes in Linseed

Y.P. Malik
Project Coordinating and Germplasm Management Unit (Linseed)
C.S. Azad University of Agriculture and Technology, Kanpur – 208 002, U.P.

Introduction

Linseed (*Linum usitatissimum* L.) is under cultivation since pre-historic times in the world for its immense utility in various industries. India is the largest (21.42 per cent) linseed producing country in the world with its cultivation on 5.26 lakh hectares of land predominantly in Madhya Pradesh, Maharashtra, Chhattisgarh, Uttar Pradesh, Bihar, Orissa and Jharkhand states of the country, besides its limited cultivation in other states. On production front, India ranks third (10.14 per cent) after Canada (36.06 per cent) and China (22.28 per cent) having total production and productivity of 2.12 lakh tonnes and 403 kg/ha, respectively (Anon., 2004). Each and every part of linseed is endowed with commercial and medicinal characteristics, which is utilized either directly or after processing. About 80 per cent of the total oil production is utilized by paint, varnish, oil cloth, linoleum, pad ink and printing ink manufacturing industries, while remaining oil quantity is consumed by the farmers for their domestic purposes. The oil cake is good feed for milch cattle. The stem yields fibre of good quality having high strength and durability. The fibre and woody matter is used for the manufacturing of twines, canvas, rope, currency paper, stationery paper and various indispensable products for defense purposes.

National productivity of linseed being 403 kg/ha is far below than the mean realizable yield of 2050 kg/ha with improved technologies, which have an exploitable yield reservoir of 1643 kg/ha with 408 per cent increase over the national average yield (Husain *et al.*, 2005). Poor crop management practices under low input conditions are the main reasons for such low yield levels. To bridge this vast gap in the productivity of linseed, efficient crop management technologies starting from adoption of proper crop rotation, timely sowing with quality seeds, adequate plant stand through adjustment in seed rate and thinning, timely weed management, life saving irrigation, balanced plant nutrients and need based plant protection have their significant contribution on the productivity front of this crop. Among these components, need based application of plant protection measures have profound influence (18 to 73 per cent) on the productivity of different oilseed crops and in linseed too (Hegde, 2000).

Insect Pest Complex

A total number of 15 insect pests were reported earlier to damage the linseed crop right from seedling to harvest stage in fields (Rai, 1976), but it has touched to 29 insect pests (Table 8.1) with the variety development during recent years (Deshmukh *et al.*,1992 and Malik, 1999). Among these insect pests, bud fly (*Dasyneura lini* Barnes) is designated as the key pest through out the country.

Linseed Bud Fly (*Dasyneura Lini* Barnes)

Identification

Major yield loss in linseed occurs during reproductive phase of the crop by various factors including insects, diseases and physiological disorders. These factors cause unfertilized/infertile flower buds resulting in reduction of the produce as (*i*) Infestation of bud fly (*Dasyneura lini* Barnes) maggots, which feed upon the corolla and androcium, the reproductive parts of the buds interrupting fertilization process, (*ii*) infection of *Alternaria* blight (Bud blight) disease (*Alternaria lini*) at the early age of the floral bud forming a ring on peduncle just below the bud base, which checks the nutrient supply to the upper part of the bud resulting in the bud starved/unfertilized, and (*iii*) the buds, which appear in later stage of the crop growth (secondary/tertiary flowering) and do not receive proper environment for their development and fertilization, remain immature/unfertilized (Malik, 2000). Bud fly infestation and Alternaria blight infection may create a little bit confusion to the research workers/farmers, but bud fly infested buds become hollow due to crumpling of the corolla and remain on the plants till harvest, while Alternaria blight infected buds show brownish spots on sepals with a light yellowish ring just below the bud base on peduncle (Plate 8.1) and can easily be detached from the plant by a slighter jerk.

Extent of Loss

The percentage of bud damage by bud fly at harvest stage reflects the magnitude of yield loss in linseed. Bud fly causes severe yield loss mainly in Gangetic plains, southern plateau, and central and peninsular region of the country due to the favourable agro-ecosystem. The state-wise yield losses have been quantified as high as 89.4 per cent in Maharashtra followed by Uttar Pradesh (80 per cent), Madhya Pradesh (75.5 per cent), Orissa (61.7 per cent), Bihar (60.7 per cent), Himachal Pradesh (50 per cent), Rajasthan (46.2 per cent), West Bengal (35 per cent), Haryana (22.7 per cent), but crop may fail completely under pest epidemics (Anon., 1992-2000 and Malik *et al.*, 1998).

Biology

The female lays whitish, minute, curved eggs, in clusters mostly underneath the sepals of compact floral buds or sometimes on the outer surface of the buds. A single female lays 22 to 103 eggs on 8–17 buds with an average of 5-10 eggs/bud, which hatch in 2-5 days. Neonate maggot is whitish and turns to deep pinkish in full-grown stage, which feed inside the flower bud to the corolla and androecium resulting the buds do not bloom and remain unfertilized. The larval period lasts 5 to 14 days with four larval instars. Pupation occurs generally in soil about 5–7 cm below the soil surface, but occasionally in fallen leaves also, which takes about 5 to 11 days. The adult lives for 1–3 days and the life cycle completes in about 14 to 27 days (Plate 8.2). There are four generations of the pest on linseed in a season. The pest is host specific and aestivates in larval form, which emerges as adult in the ensuing season (Sood and Pathak, 1985 and Singh *et al.*, 1992).

Table 8.1: List of Insect Pests in Linseed Under Field Condition

Sl.No.	*Common Name*	*Scientific Name*	*Nature of Damage*
(A)	**Key/Major Pests**		
1.	Bud fly	*Dasyneura lini* Barnes	Feeds floral buds
2.	Leaf miner	*Chromatomyia horticola* Gour.	Mines leaves
(B)	**Minor/Sporadic Pests**		
3.	Semi looper	*Plusia orichalcea* Fabr.	Defoliation
4.	Lucern caterpillar	*Laphygma exigua* Hubn.	Defoliation
5.		*Euproctis scintillans* Walk.	Defoliation
6.	Termite	*Odentotermes obesus* Rambur	Damages roots
7.	Grass Hopper	*Chrotogonus trachypterous* B.	Defoliation
8.		*Acrida exaltata* Walk.	Defoliation
9.	Cut worm	*Agrotis ypsilon* Hfn.	Cuts seedlings
10.		*Agrotis flammatra* Schiff.	Cuts seedlings
11.	Jassid	*Empoasca kerri* Pruthi.	Sucks sap
12.	White fly	*Bemisia tabaci* Genn.	Sucks sap
13.	Red pumpkin beetle	*Raphidopalpa foveicollis* L.	Feeds on leaves
14.	Grey weevil	*Myllocerus cetulifer* Desb.	Feeds on root and leaf
15.	Tobacco caterpillar	*Spodoptera litura* Fab.	Defoliation
16.	Bihar hairy caterpillar	*Spilarctia obliqua* Walk.	Defoliation
17.	Safflower caterpillar	*Prospalta capensis*	Defoliation
18.	Green bug	*Creontiades pollidifer* Walk.	Suck sap
19.	Army worm	*Mythimna loreyi* Duponch.	Defoliation
20.	Aphid	*Myzus persicae* Sulz.	Sucks sap
21.		*Aphis craccivora* Koch.	Sucks sap
22.	Painted bug	*Bagrada cruciferarum* Kirk.	Sucks sap
23.	Pentatomid bug	*Dollichoris indicus* Stal.	Sucks sap
24.	Cabbage butter fly	*Pieris brassicae* Linn.	Defoliation
25.	Capsule borer	*Helicoverpa armigera* Hubn.	Bore capsules
26.	Red cotton bug	*Dysdercus koenigii* P.	Sucks sap
27.	Green stink bug	*Nezara virudula* Linn.	Sucks sap
28.	Thrips	*Caliothrips indicus* Bagnall	Sucks sap
29.	Anthomyiid fly	*Delia* sp.	Damages root

Plate 8.1

Best multiplication of this pest occurs at 16-20°C temperatures and 60-70 per cent relative humidity, which fall generally during the months of January-February. Temperature plays positively role in the bud fly infestation, whereas the relative humidity negatively (Kumar *et al.*, 1989; Shrivastava *et al.*, 1993 and Malik *et al.*, 2000).

Management Strategy

Injudicious application of pesticides creates the health hazards to human as well as animals, phytotoxicity, insect-pests resistance to insecticides, pest resurgence, danger to pollinators and wild life, pesticide residues and upsetting of ecological balance. To withstand these serious problems and enormous crop losses, Integrated Pest Management (IPM) is the only way out, which is ecologically sound, socially acceptable and economically feasible. A workable IPM schedule especially for bud fly is being given under here, which will also take care of all other insect-pests in different parts of the country under rainfed as well as irrigated situations for sustainability in linseed production.

Host Plant Resistance

Resistant/tolerant variety is one of the prime significance and economic tool of integrated pest management. Although, practically no high yielding resistant variety could be evolved so far, but cultivation of relatively resistant/moderately resistant varieties like Neela, Jawahar-1, Jawahar-7, Kiran, Jawahar-17, Gaurav, Shubhra, Garima, Laxmi-27, Janki, Himalini, Jeevan, Surabhi, Mukta and

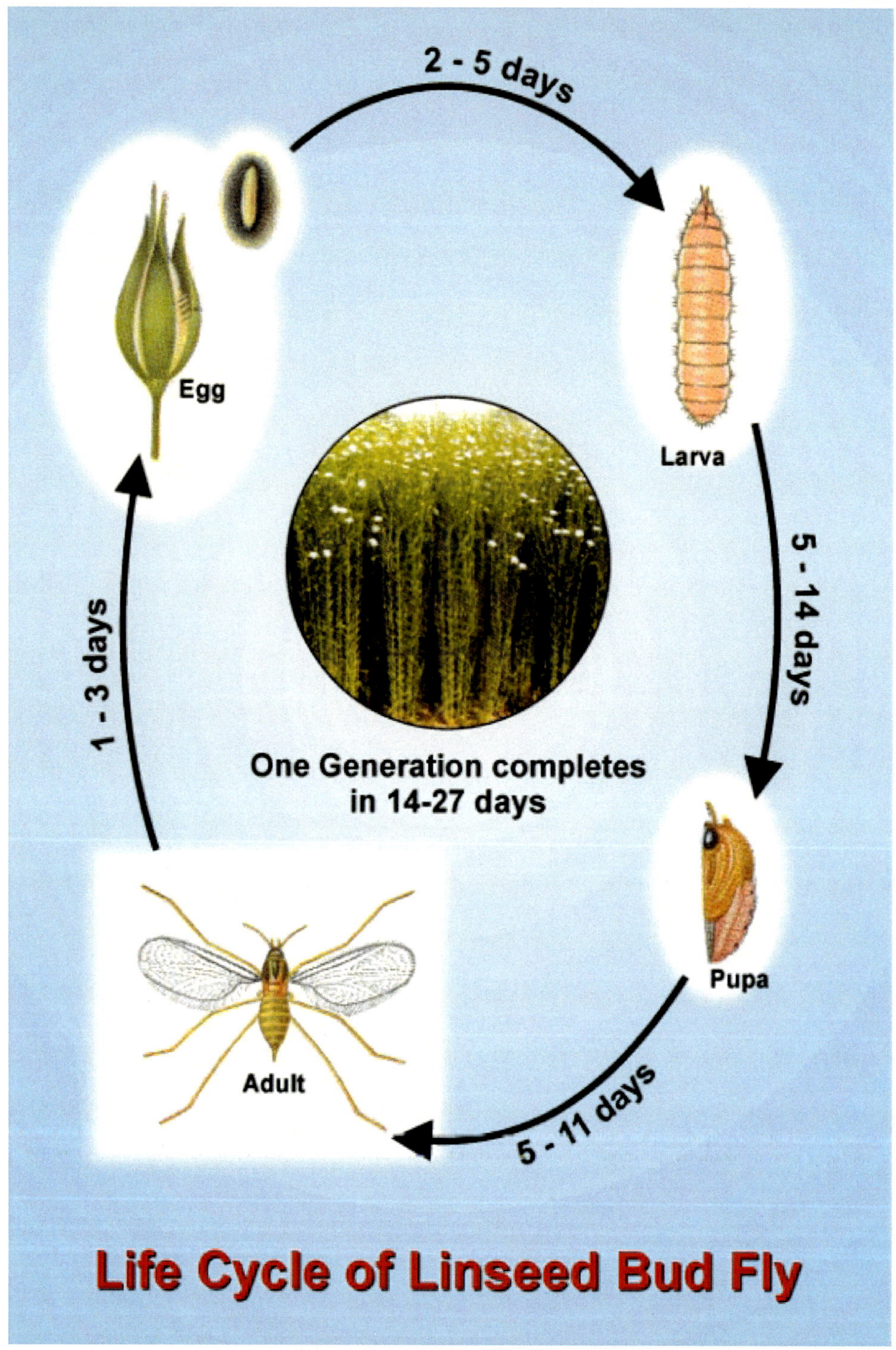
2 - 5 days
Egg
Larva
5 - 14 days
Pupa
5 - 11 days
Adult
1 - 3 days
One Generation completes
in 14-27 days
Life Cycle of Linseed Bud Fly

Plate 8.2

Sweta are recommended for different agro-climatic zones of country, as a principal component of pest management strategy along with suitable cultural, biological and chemical control techniques in an IPM programme (Deshmukh *et al.*, 1982; Jakhmola and Yadav, 1983; Singh *et al.*, 1990 and Malik *et al.*, 1996c).

The resistance in linseed to bud fly is positively associated with short flowering period (Jakhmola and Yadav, 1983; Sood and Pathak, 1983 and Malik *et al.*, 1991) and small size of flower buds (Sood and Pathak, 1990; Katiyar *et al.*, 1992 and Malik *et al.*, 1995). The higher HCN, nitrogen and phosphorus content in leaves at bud initiation stage favoured the bud fly infestation, whereas the polyphenol content affected it inversely (Malik *et al.*, 1995).

Cultural Control

Improved production technology has increased the production of agricultural crops manifold resulting in the self reliance in food production, but continuous planting of same crop/crop variety over large areas, higher cropping intensity, imbalanced use of fertilizers and improper crop geometry have increased the insect pests problems. Following cultural practices would minimize the intensity of linseed pests:

1. Solar treatment of the fields by summer ploughing is helpful in killing of the aestivated larvae of bud fly alongwith other insects (Singh *et al.*, 1992).
2. Crop rotation must be adopted after two to three years of cultivation of linseed in a particular field to avoid the soil borne pests.
3. Optimum time of sowing is a crucial non-monetary input for achieving adequate and uniform plant stand, efficient use of stored moisture, rainfall and nutrient. Optimum planting time takes advantage of the low incidence of key pests on the crop by phenological asynchronization. Sowing of linseed in first fortnight of October in M.P., Maharashtra and Karnataka; second fortnight of October in Himachal Pradesh, Punjab, Gangetic alluvium planes of U.P. and Bihar, while first fortnight of November in West Bengal and Orissa provided higher seed yield with moderate insect pests especially bud fly infestation (Singh *et al.*, 1991b; Malik, 1999 and Husain *et al.*, 2005).
4. Chickpea, lentil, mustard and safflower as companion crops with linseed in various row ratios (Table 8.2) enhanced the net monetary return with reduction in insect pests' infestation in different parts of the country (Malik *et al.*, 1996d; Srivastava *et al.*, 1997 and Husain *et al.*, 2005).
5. Application of optimum dose of 60–90 kg nitrogen and 40 kg phosphorus per hectare and irrigating twice the crop at branching and capsule formation stages provided maximum seed yield with moderate infestation of insect pests (Malik *et al.*, 1996b).

Table 8.2: Intercropping Practices in Different Areas

Sl.No.	*Area*	*Intercropping system*
1.	Bihar, alluvial region of U.P. and M.P.	Linseed + Chickpea (3:1 to 1:3 row ratio) Linseed + Mustard (4:1 to 6:1 row ratio)
2.	Bundelkhand region of U.P., West Bengal and Punjab	Linseed + Chickpea/Lentil (3:1 to 1:3 row ratio)
3.	Maharashtra and Karnataka	Linseed + Chickpea/Safflower (different row ratios)

Biological Control

Different predators and parasites are helpful in reducing the infestation of different insect pests for eco-friendly farming of linseed, which sustain the productivity of this crop with increased profitability. The biological suppression of linseed bud fly occurs by its larval parasites *i.e., Systesis dasyneurae* Mani, *Elasmus* sp., *Eurytoma* sp. and *Tetrastichus* sp. Parasitization of bud fly maggots upto 50 per cent by *S. dasyneurae* is reported during February (Ahmad and Mani, 1939 and Singh and Mathur, 1997). The grub and adult stages of *Coccinella septempunctata* Linn. and *Menochilus sexmaculatus Fabr.* (Plate 8.3) predate upon the maggots of bud fly out side the buds up to 15 per cent before pupation (Malik, 1997 and 2001).

Chemical Control

Rational application of insecticides in crops for the management of insect pests is advocated at economic threshold level for minimizing the yield losses of insect pests. Several insecticides have been recommended for bud fly (*D. lini*) management and their application at different crop growth stages will also reduce the infestation of other insect pests. Economic threshold level for bud fly has been ascertained 7-9 per cent bud infestation under central Uttar Pradesh (Kanpur) condition (Malik, 2004), which may be adopted at other places for efficient and economic management through

Plate 8.3: Linseed Bud Infested by Bud Fly (L) and Alternaria Blight (R)

insecticides.Two fortnightly application of phosphamidon 85 SL (0.03 per cent) or deltamethrin 2.8 EC (0.002 per cent) or endosulfan 35 EC (0.07 per cent) or chlorpyrifos 20 EC (0.05 per cent) or cypermethrin 20 EC (0.005 per cent) have been recommended for the management of linseed bud fly (Jakhmola, 1973; Sood and Pathak, 1984; Singh and Mathur, 1986; Singh *et al.*, 1991a; Singh *et al.*, 1995; Malik *et al.*, 1996a and Malik *et al.*, 2000). Two fortnightly alternate application of phosphamidon 85 SL (0.03 per cent) and deltamethrin 2.8 EC (0.002 per cent) with mancozeb (0.20 per cent) or two sprays of endosulfan 35 EC (0.07 per cent) + mancozeb (0.25 per cent) provide higher seed yield with lowest bud fly as well as Alternaria blight (bud blight) infestation, which are safer to the natural enemies (Coccinellids) of the pest and economic (Malik, 1998b). Application of 2 per cent neem seed kernel extract is also recommended for the management of bud fly infestation (Malik, *et al.*, 1997).

References

Ahmad, T. and Mani, M.S., 1939. Two new chalcidoid parasites of the linseed midge, *Dasyneura lini* Barnes, *Indian Journal of Agricultural Sciences*, 9(3): 531–539.

Anonymous, 1992–2004. *Annual Progress Report, Linseed*. Directorate of Oilseeds Research, Rajendranagar, Hyderabad.

Deshmukh, S.D., Akhare, M.D. and Borle, M.N., 1982. Varietal response of linseed to gall midge, *Dasyneura lini* Barnes, *Indian Journal of Entomology*, 44(1): 97–99.

Deshmukh, S.D., Singh, K.M. and Singh, R.N., 1992. Pest complex and their succession in linseed, *Linum usitatissimum* L., *Indian Journal of Entomology*, 54(2): 168–173.

Hegde, D.M., 2000. Technology for high yields. In: *The Hindu: Survey of Indian Agriculture*, p. 65–71.

Husain, K., Singh, P.K., Singh, J., Malik, Y.P., Dubey, S.D., Chandra, R. and Singh, K., 2005. *Linseed: Technology for Increasing Production*. Publ. Project Coordinating Unit (Linseed), CSAUA&T, Kanpur, p. 1–30.

Jakhmola, S.S., 1973. Chemical control of linseed bud fly, *Dasyneura lini* Barnes (Diptera: cecidomyiidae), *Indian Journal of Agricultural Sciences*, 43(12): 1078–1080.

Jakhmola, S.S. and Yadav, H.S., 1983. Susceptibility of linseed cultivars to bud fly, *Dasyneura lini* Barnes, Indian *Journal of Entomology*, 45(2): 165–170.

Katiyar, O.P., Singh, P. and Srivastava, N., 1992. Bud structure of linseed (*Linum usitatissimum* L.) in relation to incidence of bud fly (*Dasyneura lini*), *Indian Journal of Agricultural Sciences*, 62(1): 783–785.

Kumar, D., Singh, B., Singh, S.V., Katiyar, R.R. and Singh, Y.P., 1989. Incidence of various insect pests of linseed in relation to environmental factors, *Farm Science Journal*, 4(1&2): 87–91.

Malik, Y.P., 1997. New record of two coccinellids as predators of bud fly, *Dasyneura lini* Barnes in linseed. *Journal of Oilseeds Research*, 14(2): 338.

Malik, Y.P., 1998. Compatibility of insecticides and fungicide for the management of bud fly and Alternaria blight infestation on linseed. *Journal of Oilseeds Research*, 15(1): 150–155.

Malik, Y.P., 1999. Major insect pests of linseed and their management. In: *IPM System in Agriculture Vol. 5: Oilseeds*, (Eds.) R.K. Upadhyay, K.G. Mukerji and R.L. Rajak. Aditya Books Pvt. Ltd., 2/37, Sant Vihar Lane, Ansari Road, Darya Ganj, New Delhi, India, pp. 349–364.

Malik, Y.P., 2000. Assessment of yield losses due to host pest interactions in linseed (*Linum usitatissimum*). *Indian J. Agric. Sci.*, 70(1): 53–54.

Malik, Y.P., 2001. Predatory potential of two coccinellids against linseed bud fly (*Dasyneura lini* Barnes). *Journal of Oilseeds Research*, 18(1): 145–146.

Malik, Y.P., 2005. Estimation of economic threshold level for bud fly, *Dasyneura lini* Barnes in linseed, *Linum usitatissimum* L. *Journal of Oilseeds Research*, 22(1): 100–102.

Malik, Y.P., Husain, K., Singh, S.V. and Srivastava, R.L., 2000. Development of management module for bud fly, *Dasyneura lini* in linseed. *Indian Journal of Entomology*, 62(3): 260–269.

Malik, Y.P., Husain, K., Singh, S.V., Singh, K., Srivastava, R.L. and Singh, M., 1998. Development of management module for bud fly, *Dasyneura lini* in linseed. In: *Abst. Nat. Sem. on Entomology in 21st Century: Biodiversity, Sustainability, Environmental Safety and Human Health*, held at Rajasthan College of Agriculture, Udaipur, April 30–May 2, p. 213.

Malik, Y.P., Singh, B. and Singh, S.V., 1997. Efficacy of different plant extracts against bud fly, *Dasyneura lini* Barnes infesting linseed. In: *Abst. Nat. Symposium on Integrated Pest Management in India: Constraints and Opportunities*, held at I.A.R.I., New Delhi, Oct. 23–24, p. 26.

Malik, Y.P., Singh, B. and Pandey, N.D., 1991. Role of blooming period in linseed bud fly (*Dasyneura lini* Barnes) resistance. *Indian Journal of Entomology*, 53(2): 276–279.

Malik, Y.P., Singh, B. and Singh, S.V., 1996d. Infestation of linseed bud fly, *Dasyneura lini* Barnes under different intercropping practices. *Annals of Plant Protection Sciences*, 4(2): 167–169.

Malik, Y.P., Singh, B., Pandey, N.D. and Singh, S.V., 1996a. Bio-efficacy of insecticides against linseed bud fly. *Indian Journal of Entomology*, 58 (2): 115–120.

Malik, Y.P., Singh, B., Pandey, N.D. and Singh, S.V., 1996c. Management of linseed bud fly through non-monetary inputs. *Indian Journal of Entomology*, 58(2): 136–139.

Malik, Y.P., Singh, B., Pandey, N.D. and Singh, S.V., 1996b. Role of fertilizer and irrigation in management of linseed bud fly. *Indian Journal of Entomology*, 58(2): 132–135.

Malik, Y.P., Singh, S., Singh, B., Pandey, N.D. and Singh, S.V., 1995, Determination of physico-chemical basis of resistance in linseed for bud fly. *Indian Journal of Entomology*, 57(3): 267–272.

Rai, B.K., 1976. Linseed. In: *Pests of Oilseed Crops in India and their Management*. Publ. ICAR, New Delhi, pp. 81–88.

Shrivastava, N., Katiyar, O.P., Shrivastava, S. and Das, S.B., 1993. Seasonal abundance of linseed bud fly, *Dasyneura lini* Barnes (Diptera: Cecidomyiidae). *Oil Crops Newsletter*, 10(1): 60–63.

Singh, B. and Mathur, Y.K., 1986. Control of linseed bud fly, *Dasyneura lini* Barnes with modern insecticides. In: *Proc. Nat. Conf. on Key Pests of Agricultural Crops*, held at CSAUA&T, Kanpur, Dec. 21–23, 1985, p. 173–176.

Singh, B., Jha, R.C. and Pandey, N.D., 1990. Relative resistance of different linseed varieties against *Dasyneura lini* Barnes under varied conditions. *Indian Journal of Entomology*, 52(3): 349–353.

Singh, B., Katiyar, R.R. and Malik, Y.P., 1992. Final progress report of adhoc research project of ICAR on Pest complex, seasonal abundance and management of linseed pests particularly gall midge in U.P., Deptt. of Entomology, CSAUA&T, Kanpur.

Singh, B., Katiyar, R.R., Malik, Y.P. and Pandey, N.D., 1991a. Efficacy of some insecticides alone and in combination with urea against linseed bud fly (*Dasyneura lini* Barnes). *Indian Journal of Entomology*, 53(2): 270–275.

Singh, B., Katiyar, R.R., Malik, Y.P. and Pandey, N.D., 1991b. Influence of sowing dates and fertilizer levels on the infestation of linseed bud fly (*Dasyneura lini* Barnes). *Indian Journal of Entomology*, 53(2): 291–297.

Singh, D.R. and Mathur, Y.K., 1997. *Systasis dasyneurae* and coccinellids: Potential bio-control agents of linseed bud fly (*Dasyneura lini* Barnes). In: *Abst. Nat. Symp. on Integrated Pest Management in India: Constraints and Opportunities*, held at IARI, New Delhi, Oct., 23–24, p. 3.

Singh, S.B., Dhakad, N.K., Yadav, H.S. and Katiyar, O.P., 1995. Efficacy and economics of some modern insecticides for the control of linseed bud fly (*Dasyneura lini* Barnes). *Annals of Agricultural Research*, 16(3): 269–272.

Sood, N.K. and Pathak, S.C., 1990. Thickness of sepal: A factor for resistance in linseed against *Dasyneura lini* Barnes. *Indian Journal of Entomology*, 52(1): 28–30.

Sood, N.K. and Pathak, S.C., 1983. Role of flowering duration in linseed resistance to bud fly, *Dasyneura lini* Barnes *Indian Journal of Plant Protection*, 11(2): 141–143.

Sood, N.K. and Pathak, S.C., 1984. Stability of effectiveness of insecticides and varieties in control of linseed bud fly, *Dasyneura lini* Barnes. *Indian Journal of Entomology*, 46(1): 53–59.

Sood, N.K. and Pathak, S.C., 1985. The life cycle of linseed bud fly, *Dasyneura lini* Barnes (Diptera: Cecidomyiidae). *Indian Journal of Zoology*, 13(1): 9–12.

Srivastava, R.L., Singh, J., Husain, K., Malik, Y.P., Dubey, S.D., Rai, J. and Bajpai, M., 1997. Linseed. In: *Efficient Management of Dryland Crops in India: Oilseeds*, (Eds.) Singh, R.P., Reddy P.S. and V. Kiresur. Publ. Indian Society of Oilseeds Research, DOR, Hyderabad, p. 228–256.

Insect Pest and Disease Management *Pages* **140–147**
Editor: **Dr. D. Prasad**
Published by: **DAYA PUBLISHING HOUSE (2016)**

Chapter 9

Ecofriendly Strategies for Management of *Thrips palmi* Karny as Pest and Vector

Anuj Bhatnagar
Central Potato Research Station, Morar, Gwalior – 474 006, M.P.

Introduction

Thrips belong to the order Thysanoptera and more than 4000 species exist in the world (Fenemore, 1984). *Thrips palmi* Karny was first described in 1925 from tobacco plants in Sumatra, Indonesia (Karny, 1925) and was re-described in the early 1980s (Bhatti, 1980), this species was reported as an important pest of many vegetable and agronomic crops in South East Asia (Yoshihara, 1982). It is distributed throughout Asia, Africa and Central and South America as well as the Caribbean (Ananthakrishnan, 1980, Yoshihara, 1982, Nakahara 1984, Wang and Chu, 1989, Anonymous, 1989). It is a polyphagous species feeds on 105 different plants of vegetable, ornamental and agronomic crop as well as weeds. It has become an increasingly important pest of several cucurbitaceae and solanaceae crops in India. Serious injury on cucumber, egg plant and sweet pepper in green houses and fields was reported in Western part of Japan (Kawai, 1986, Matsuzaki *et al*., 1986). Reports from Pueroto Rico, Trinidal and Tobago indicated that *T. palmi* caused 50-90 per cent losses in vegetable crops (Hall *et al*., 1993). A serious out break of *T. palmi* was discovered in Hawaii on cucurbits, egg plants, pepper and Amaranth Spinach (Nakahara, 1984).

Beside causing damage from feeding, *T. palmi* has also been reported to transmit tomato spotted wilt virus (TSWV) that infect watermelon in Taiwan (Yeh *et al*., 1992). *Thrips palmi* have been associated with early potato crop in Central India (Khurana *et al*., 2001). The most predominant and widely occurring species is *Thrips palmi* and reported to cause sever damage on watermelon, muskmelon, cucumber, tomato, egg plant and potato grown in low elevation areas. In some places high populations of *Megalurothrips usitatus* Bagnaill also occur in tomato and potato. The two species some times occur together but they can be differentiated readily by their colour. *Thrips palmi* light orange–brown and *M. usitatus* has a black body. *T. palmi* is distinguished from other species by its bright citron yellow colour with legs having a lighter hue (Dammerman, 1929). In the 1980s, it was discovered that *T. palmi* had become an important pest of potato in the high lands of Java and Sumatara. The damage caused by this insect is very serious, specially in the dry season. Samap *et al.* (1985) stated that high temperature and low relative humidity are favourable for thrips population growth. Kalshoven (1950) mentioned that high relative humidity is prerequisite for thrips development.

Biology and Host Range

Palacio (1978) studied the biology of *T. palmi* on watermelon and muskmelon. He found that the species reproduces parthanogenetically, completing development in 13 days on average. The adults live for 2-18 days and fecundity ranges from 6 to 38 eggs. The host plants are amaranth, castor, corn, egg plant, potato, okra, rice, sweet pepper, soybean, winged bean and tomato. Thrips completed one generation in about 20 days at 30°C, but it was lengthened to 80 days, when cultured at 15°C. Thrips are able to multiply during any season, specially warm whether conditions favoured the development of thrips. When crop was matured their suitability for thrips declines, so thrips growth rate diminishes even in the presence of warm whether conditions (Seal, 1994, Frantz *et al.*, 1995).

Female deposited eggs in leaf tissue by slit cut. The colour of eggs is colorless to pale white and bean shaped. The size of larvae is smaller than adult and resemble with adult except wings. There are two instars during the larval period. The pre pupal and pupal instars are inactive and non feeding stages. The adults are pale yellow or whitish in colour, having dark setae on the body. A black colour line is resulted from the juncture of the wings which runs along the back of the body.

Calilung (1990) reared *T. palmi* on potato, cucumber, egg plant, watermelon, cotton and pepper to determine the possible existence of biotypes. She observed the longest average developmental period (15 days) on pepper and shortest (11.2 days) on watermelon. The highest fecundity (15.6 eggs) and longest adult life (17.4 days) were recorded on watermelon also. This could be one of the reasons for the fast thrips population buildup on and very severe damage on watermelon as reported by many farmers. Adult thrips lived for the shortest period on pepper (3.9 days) and did not reproduce. When thrips were offered a choice among the six crops, they showed a lower preference for cotton and pepper compared with other four crops–potato, egg plant, cucumber and watermelon, which were equally preferred.

T. palmi needs about 15 and 24 days to complete the immature and adult stage. In Southern Tiwan, it takes 20-30 days for one generation and at least 10 generations occur on cucurbits per year (Wen, 1984). The female had shorter pre oviposition duration and laid more eggs than those in sexual reproduction (Wang *et al.*, 1989). Those female that mate immediately after emergence produce more male offspring on the contrary more female offspring were produced by those mated in 4–9 days after emergence (Wang and Chu, 1990). Thrips usually feed on the lower surface of leaves. The cell contents are rasped and sucked from the lower mesophyll. As a result air spaces developed in this tissue and leaf become distorted. Damaged leaves frequently have a silvery sheen and they may bear small spots of fecal matter. Feeding damage on cabbage or potato leads to russeting or bronzing of leaves, young plants are especially susceptible (Kranz et *al.*, 1978). The population of *T. palmi* peaks in December and mid January on vegetable sponge (*Luffa Cylindrica*) and cucumber, respectively and decline during the wet season. The density was higher during July to September on wax gourd, the most preferred cucurbits the high daily activity was found at 1700 -1800 hrs. White traps were most attractive when placed 0.5m above the ground (Huang, 1989). The population increased at late April–May and early October–November. *T. palmi* aggregated on egg plant leaves 75 cm above the ground (Su *et al.*, 1985). They apparently in habit the egg plant flowers throughout the year, with numerous thrips occurring in July.

Transmission of Tospo Viruses

Tospo viruses have received international attention because they cause sever diseases of food, fibre and ornamental crops throughout the world. The symptoms of virus infection vary with plant species and tospo viruses. Several thrips species serve as a vectors of tospo viruses (Sakimura, 1962). Watermelon silver mottle tospo virus (WSMV) to Iwaki *et al.* (1984), groundnut bud necrosis tospo virus (GBNV)–Lakshmi *et al.*, 1985. At the time of Sakimura's review of the status of thrips transmitted viruses, four species–*Thrips tabaci*, *Franklimiella occidententalis*, *F. schultzei* and *F. fusca* had been associated with transmission of TSWV. TSWV was considered the sole member of the tomato spotted wilt virus group of plant viruses and was thought to be the only virus transmitted by thrips. TSWV was still considered the sole member of the tomato spotted wilt virus group of plant viruses, however, several distinct new isolates had been recognized (Law and Moyer, 1989, 1990, Reddy *et al.*, 1990). Taiwanese tospo virus isolate from watermelon and an Indian tospo virus isolate form groundnut (bud necrosis virus). Several new viruses form watermelon (watermelon mottle spot virus and peanuts in India (peanut bud necrosis virus, peanut yellow fan virus thought to be transmitted by *Thrips palmi. F. schultzei* an extensive discussion of the relationships between thrips and tospo virus is given by Ullman (1996) and the relationships of the eight thrips vectors are discussed by Mound (1996b). These vector species are *Thrips palmi*, *T. setosus*, *T. tabaci*, *Frankliniella fusca*, *F. intonsa* and *F. occidentalis*. *Thrips tabaci* was a key vector of TSWV. Some previously reported vector tospo virus associations (*Scirtothrips dorsalis*, *Frankliniella schultzei and Frankliniella tenuicornis*) require experimental validation. *S dorsalis* and *F. schultzei* were initially reported as vectors of peanut bud necrosis virus (Amin *et al.*, 1981). Nine year later it was discovered that this report was probably based on incorrect species identification (Mound, 1996a). Subsequent research finding demonstrated that *Thrips palmi* was the primary vector of GBNV in India, *F. schultzei* transmitted GBNV at a very low rate (2 per cent) and *S. dorsalis* was a non–vector (Lakshmi *et al.*, 1995). Emergence of new tospo viruses and new thrips virus association is the apparently very specific relationship between *Thrip palmi* and GBNV (Lakshmi *et al.*, 1995) and WSMV (Yeh *et al.*, 1992). *Thrips palmi* does not transmit any of the TSWV isolates that have been tested but it commonly feeds upon and completes its life cycle in association with TSWV plant hosts and other tospo virus vectors. The effect of tospo viruses on fecundity and longevity of their thrips is controversial and in need of further study. Transmission by thrips is also known only for a tospo virus causing potato stem necrosis mainly by *Thrips palmi* (Khurana *et al.*, 1997, 2001).

The relationship of thrips with virus is very complex, because a thrips can acquire the virus only through feeding as a first instar larva (Wetering *et al.*, 1996) and the virus is transmitted when the subsequent adult feeds on a suitable plant. But not all populations of these known vector species can transmit tospo viruses (Goldbach and Kuo, 1996).

Current Strategies for Management

Management through prevention is critical to success ding in controlling thrips species and tospo virus spread, because insecticides and natural enemies are seldom efficacious in reducing thrips population and more over plant can not be cured once they are infected. Gaining control of thrips populations with pesticides is extremely difficult because eggs with in leaf tissue, larvae deep with in developing buds and flowers and pupae in the soil/leaf litter are protected from most sprays.

Chemical Control

Rejesus *et al.* (1986) reported that chlorpyrifos + BPMC (Brodan) to which a sticker was added and carbofuran + sticker were the most effective in reducing the thrips population with these pesticides, it was feasible to lengthen the time interval between applications from the usual 7 days to 18 days. Since thrips mainly feed at night, spraying after sunset is recommended. In case of thrips, chemicals is not the answer, if it will be used for the control of thrips at a right time. Satish *et al.* (2004) evaluated the efficacy of monocrotophos, imidacloprid, acetamiprid, thiomethoxam, diafenthiuron and carbosulfan against *Thrips palmi* Karny on sunflower under field and laboratory conditions. Field studies indicated that thiomethoxam was most effective in controlling thrips followed by acetamiprid and monocrotophos. However, laboratory experiments revealed acetamiprid is most toxic to thrips followed by thiomethoxam and monocrotophos. Chaudhari *et al.*(2000) reported that stem necrosis disease incidence/severity and thrips population had a decline with delay in planting. Sprays of insecticides were effective and at par in controlling the disease.

Cultural Control

Rejesus *et al.* (1986) also found that mulching watermelon fields with rice straw or hog manure reduced the thrips population significantly, the ground cover possibly interfering with the process of thrips pupation in the soil. The another cultural practice based upon the farmers experience in potato fields that irrigation can dislodge thrips from the leaves and early planting of potato in Northern part of Philippines where the onset of the dry season is around October, land preparation can be done early to allow planting in late November or early December. This makes possible harvesting of the crop before the occurrence of the peak thrips population which is usually in March, irrigation or flooding can be used to destroy thrips which spend pupal stage in the soil (Lewis, 1973). Aluminum foil mulches may deter flying insects such as aphids and thrips from landing on plants. The use of black plastic mulch has proved effective in suppressing the population of thrips. The plastic mulches consistently delayed thrips damage and thus redused crop losses. The rice straw mulch did not affect the thrips population. A high proportion of nitrogenous compound makes leaf tissue a rich source of nutrients for thrips (Lewis, 1973). Khurana *et al.*, 2001 reported that there is a strong positive correlation between early planting and thrips activity which declines with the postponement of date of planting to end of October or early November. They also reported that high temperature (30 to 35°C) and dry weather during September/October are known to be favourable for thrips activity.

Biological Control

It is consistent with the fact that *T. palmi* is native to South east Asia and thus has effective natural enemies native to this region. A number of arthropods that pray on *T. palmi* were enumerated in South East Asia. Rejesus *et al.* (1986) reported that *O. tantillus* was the most predominant natural enemy preying on nymphs and adult thrips (*T. palmi*) on watermelon. At predatory capacity can consume more than 200 thrips though out its life times. The maximum predatory capacity of the adult bug is 19-20 adult thrips per day. Mituda and Calilung (1989) also studied the life cycle and predation of *Amblyseius* sp. on nymphs of *T. palmi*. The adult predatory mite consumes 2-7 *T. palmi* nymphs per day. They observed prey consumption of the other predators is 1-5 thrips per day for *Campylomma* sp. and 8-25 thrips per day for the spider *Conopistha* sp, Both predators feed on nymphs and adult thrips. During 1987-88, survey of natural enemies of *T. palmi* were made in Thailand, Malaysia and the

Philippines (Hirose, 1990). Among the eight sp. discovered the eulophid larval parasitoid *Ceranisus menes* (Walkar) and the anthocorid predator *Bilia* sp. were effective natural enemies of *T. palmi* in Thailand. The results of survey also showed that the aboundance of *T. palmi* was never higher in unsprayed fields than in those sprayed with insecticide, suggesting the effectiveness of *C. menes* as natural enemy of *T. palmi* in unsprayed field. The predator, *Orius* sp. is known to be effective against *T. palmi* on egg plant in Japan (Nagai *et al.*, 1988). *C. menes* is native to Japan but it was known as a parasitoid of *T. palmi* until 1988, when *C. menes* was observed parasitizing *T. palmi* nymphs on egg plant near Fukuoka.

Host Plant Resistance

The available potato germ plasm at IPB and at the CIP regional station of South East Asia has been evaluated for thirp resistance. Significant differences in the *T. palmi* damage sustained by the germ plasm entries observed (Bernordo and Lit, 1989, Bernordo and Eusebio, 1990), suggesting different tolerance levels. Singh *et al.*, 1997 reported that grow only tolerant potato varities for early planting like Kufri Sutlej, Kufri Badshah and Kufri Jawahar have exhibited lesser infection of thrips while Kufri Bahar and Kufri Chandramukhi shown susceptible reaction against thrips.

Conclusion

Thrips palmi Karny (Potato or melon thrips) is a polyphagous insect species, feeding on many vegetable and ornamental crops as well as weeds. Besides causing damage by direct feeding it has the ability to transmit 3 viruses- Tomato spotted wilt virus (TSWV), Ground bud necrosis virus (GBNV) and Watermelon silver mottle tospo virus (WSMV). Of these viruses, former two viruses are common in India. *Thrips palmi* is an emerging insect pest and vector in many parts of South East Asia and India. The present article is an attempt to review the work of biology, host range, disease transmission by *Thrips palmi* and current management strategies.

References

Amin, P. W., Reddy, D.V.R.and Ghonekar, A.M., 1981. Transimission of tomato spotted wilt virus causal agent of bud necrosis of peanut by *Scirtothrips dorsalis* and *Frankliniella scnultzei. Plant Disease*, 65: 663–665.

Ananthakrishnan, T.N., 1980. Thrips. In: *Vectors of Plant Pathogen*, (Eds.) K.F. Harris and K. Manmarosch. Academic Press, p. 149–164.

Annual Report, 2003. *Annual Report of Central Potato Research Institute*, Shimla, H.P, p. 83.

Anonymous, 1989. European Plant Protection Organization data sheets on quarantine organism *Thrips palmi. EPPO Bull.*, 19(175): 717–720.

Bernado, E.N. and Eusebio, J., 1990. Management of thrips and mites attacking potato in the lowland, study 3, Resistance of potato germplasm to thrips and mites. Progress Report for 1990, submitted to SEARCA, college, Laguna, Philippines.

Bernado, E.N. and Lit, M.C., 1989. Resistance of selected vegetable crops to important arthropod pests, Annual Report for 1987, Institute of Plant Breeding, College of Agriculture, University of the Philippines at Los Banos, Laguna, Philippines.

Bhatti, J.S., 1980. Species of the genus Thrips from order (Thysanoptera). *Syst. Entomol.*, 5: 109–166.

Calilung, V.J., 1990. Management of thrips and mites attacking potato in the lowland. Study 1. Identify biology, host range and natural enemies of potato thrips. *Terminal Research project Report*, 25 p.

Chaudhari, S.M., Paul Khurana, S.M., Patel, P.K. and Patel, R.N., 2000. Management of potato stem necrosis disease through manipulation of planting dates and application of insecticides at Deesa. In: *Potato: Global Research and Development*, (Eds.) S.M. Paul Khurana, G.S. Shekhawat, B.P. Singh and S.K. Pandey. Indian Potato Association, Shimla, 1: 393–397.

Dammerman, K.W., 1929. *The Agricultural Zoology of the Malay Archiplago*. J.H. de Bussy Ltd., Amsterdam, p. 215–217.

Fenemore, P.G., 1984. *Plant Pests and their Control*, Revised Edition. Butterworth, London, 280 p.

Frantz, G., Parks, F. and Mellinger, H.C., 1995. Thrips population trends in peppers in south west Florida. In: *Thrips: Biology Management*, (Eds.) Parper, B.L., Skinner, M. and T. Lewis. Plenum Press, New York, pp. 111–114.

Goldbach, R. and Kuo, G., 1996. Introduction. In: *International Symposium on Tospoviruses and Thrips of Floral and Vegetable Crops. Acta Horticulaturae*, 431: 21–26.

Hall, H.A., Peterkin, D.D. and Pollard, G.V., 1993. A system of caging *Thrips palmi* for laboratory bioassay of pathogen. *Fla. Entomol.*, 76: 171–175.

Hirose, Y., 1990. Prospective use of natural enemies to control Thrips palmi (Thysanoptera: Thripidae). In: *The Use of Natural Enemies to Control Agricultural Crop Pests*, (Ed.) J. Bay Petersen. FFTC, Taipei, pp. 135–141.

Huang, K.C., 1989. The population fluctuation and trapping of *Thrips palimi* in waxgourd. *Bull. Taichung DAIS*, 25: 35–41.

Iwaki, M., Honda, Y., Hanada, K., Tochinara, H., Yonaha, T., Hokama, K. and Yokoyama, T., 1984. Silver mottle disease of watermelon caused by tomato spotted wilt virus. *Plant Disease*, 68: 1006–1008.

Kalshoven, L.G.E., 1950. Deplagen van de cultuurgewassen in Indonesia, Deel, I.N.V.Vitgevi, i.w. Van Hove, Den Haag, Bandung, 512 p.

Kalshoven, L.G.E. 1981. *Pests of Crops in Indonesia*, (Revised and Translated by P. A. Vander Laan). P. T. Ichitar Baru-van Hoeve, Jakarta, p. 83–88.

Karny, H.H., 1925. Die an tabak auf Java and Sumatra angetroffenen Blasenfusser (Thysanoptera). *Bull. Deli Profstation*, 23: 3–55.

Kawai, A., 1986. Studies on population ecology of *Thrips palmi* Karny, 12: Analysis of damage to egg plant and sweet pepper. *Jpn. J. Appl. Entomol. Zool.*, 30: 179–187.

Khurana, S.M. Paul, Pandey, S.K., Singh, R.B. and Bhale, U., 1997. Spread and control of the potato stem necrosis. *Indian J. Virol.*, 13: 23–28.

Khurana, S.M. Paul, Bhale, U. and Garg, I.D., 2001. Stem necrosis disease of potato. *Tech. Bulletin*, 54: 7.

Kranz, J., Schmutterer, H. and Koch, W., 1978. *Diseases, Pests and Weeds in Tropical Crops*. John Wiley and Sons, Chichester, New York, Brisbane and Toronto, p. 280–283.

Lakshmi, K.V., Wightman, J.A., Reddy, D.V.R., Rao, G.V., Ranga, Buiel, A.A.M. and Reddy, D.D.R., 1995. Transmission of peanut bud necrosis virus by Thrips palmi in India. In: *Thrips, Biology and Management*, (Eds.) B.L. Parker, M. Skinner and T. Lewis. Plenum Press, New York, pp. 179–184.

Law, M.D. and Moyer, J.W., 1989. Physio-chemical analysis of a serologically distinct tomato spotted wilt virus strain. *Phytopathology*, 79: 933–938.

Law, M.D. and Moyer, J.W., 1990. A tomato spotted wilt like virus with a serologically distinct N. protein. *Journal of General Virology*, 72: 2597–2601.

Lewis, T., 1973. *Thrips: Their Biology, Ecology and Economic Importance*. Academic Press, London and New York, 349 p.

Matsuzakai, T.K., Ichikawa, K., Kusakawa, H. and Fujimoto, K., 1986. Studies on the control of *Thrips palmi* on egg plant in green house, I: Chemical control. In: *Proc. Associ. Pl. Prot.*, Shikoku, 21: 75–86.

Mituda, E.C. and Calilung, V.J., 1989. Biology of *Orius tantillus* (Motschulsky) (Hemiptera: Anthocoridae) and its predatory capacity against *Thrips palmi* Karney (Thysanoptera: Thripidae) on watermelon. *The Philippine Agriculturist*, 72(2): 165–184.

Mound, L.A., 1996a. The rhysanoptern vector species of tospoviruses. In: *Tospoviruses and Thrips of Floral and Vegetable Crops: An International Symposium*, (Ed.) G. Kuo. Taichung, Taiwan, *Acta Horticulture*, Technical communication of International Society of Horticultural Science, 431: 298–309.

Mound, L.A., 1996b. The Thysanoptera vector species of tospoviruses. *Acta Horticulaturae*, 431: 298–309.

Nagai, K.T., Hiramatsu and Henmi, T., 1988. Predatory effect of *Orius* sp. (Hemiptera: Anthoecridae) on the density of *Thrips palmi* Karney (Thysanoptera: Thripidae) on egg plant. *Jpn. J. Appl. Ent. Zool.*, 32: 300–324.

Nakahara, L.M., 1984. New state record, *Thrips palmi* Karny. *Hawaii Pest Report*, 4(1): 1–5.

Palacio, I.P., 1978. Biological studies of Thrips palmi Karny (Thysanoptera: Thripidae) on muskmelon and watermelon. *Under Graduate Thesis*, Deptt. of Entomology, Collage of Agriculture, UPLB, Laguna, Phillippines.

Reddy, D.V.R., Sudarshana, M.R., Rama, A.S., Reddy, A.S., Amin, P.W., Kumar, L.K. and Murthy, A.K., 1990. The occurrence of yellow spot virus, a member of tomato spotted wilt virus group on peanut (*Arachia hypogaea* L.) in India. In: *Virus-Thrips Plant interactions of Tomato Spotted Wilt Virus*, Proceedings of a USDA Workshop, Beltsvilla, Maryland, Agricultural Research Service, pp. 77–78.

Rejesus, B.M., Malaya, M.T., Javier, P.A., Saavedra, N.T., Palaceo, I.P. and Navasero, M.V., 1986. Management of insect pest of water melon with emphasis on thrips. *Terminal report, UPLBPCARRD Project* No. 34: 264–321.

Sakimura, K., 1962. The present studies of thrips borne viruses, In: *Biological Transmission of Disease Agents*, (Ed.) K. Maramorosch. Academic Press, New York.

Samap, M.M., Nawale, R.N. and Ajri, D.S., 1985. Seasonal incidence of chili thrips and mite. *Journal of Maharastra Agril. Univ.*, 10(3): 345–346.

Satish, M., Reddy, K.D., Basappa, H. and Singh, V.K., 2004. Insecticidal management of *Thrips palmi* Karney on Sunflower. *Indian Journal of Entomology*, 66(2): 119–120.

Seal, D.R., 1994. Field studies in controlling melon thrips, *Thrips palmi* Karny (Thysanoptera: Thripidae) on vegetable crops using insecticides. *Proceedings of the Florida State Horticulture Society*, 107: 159–162.

Singh, R.B., Khurana, S.M. Paul, Pandey, S.K. and Srivastava, K.K., 1997. Screening germplasm for potato stem necrosis resistance. *Indian Phytopath.*, 51: 222–224.

Su, C.Y., Chiu, T.S. and Lin, Y.J., 1985. Study of population fluctuation of *Thrips palmi* and its insecticidal control in the fields on egg plant. *Chinese J. Entomol.*, 5: 119–128.

Ullman, D.E., 1996. Thrips and tospoviruses: Advances and future directions. *Acta Horticulaturae*, 431: 310–324.

Wang, C.L., Chu, Y.I. and Lo, K.C., 1989. The reproductive mechanism of *Thrips palmi* Karny (The female ovipositional behaviour). *Chinese J. Entomol.*, 9: 251–261.

Wang, C.L. and Chu, Y.I., 1990. The reproductive mechanism of *Thrips palmi* Karny. (Development and sex ratio of individuals reproduced parthenogenetically and biosexually. *Chinese J. Entomol.*, 10: 125–132.

Wen, H.C., 1984. The important thrips on cucurbit crops. *Shinung Farm Magazine*, 187: 28–30.

Wetering, F., Vander, R.G. and Peters, D., 1996. Transmission of tomato spotted wilt virus by *Frankliniella occidentalis* after viral acquisition during the first larval stage. *Acta Horticulaturae*, 431: 350–358.

Yeh, I.D., Lin, Y.C., Cheng, Y.H., Jin, C.I., Chen, M.J., and Chen, C.C. 1992. Identification of tomato spotted wilt like virus on watermelon in Taiwan. *Plant Dis.*, 76: 835–840.

Yoshihara, T., 1982. An overview of researches on *Thrips palmi* in Japan. Entomological Laboratory, Kurume Vegetable Experiment Substation, Kurume, Japan.

Insect Pest and Disease Management *Pages* ***148–166***
Editor: **Dr. D. Prasad**
Published by: **DAYA PUBLISHING HOUSE (2016)**

Chapter 10

Spiders: Bio-ecology and Conservation for Insect Pest Management

R.K. Tanwar, O.M. Bambawale and H.R. Sardana
National Centre for Integrated Pest Management, L.B.S. Centre, I.A.R.I. Campus, New Delhi – 110 012

Introduction

Arthropods represent an extremely successful group of animals which live in wide variety of habitats, have a great range of variation and include the greatest number of species. Arthropods constitute 64.5 per cent of the described species compared to plants (14.31 per cent), fungi (4.2 per cent) and vertebrates (2.3 per cent) (Anonymous, 1995). The arachnids constitute the second largest class (7 per cent) of documented arthropods. Included in the arachnids are the spiders, mites, ticks, scorpions, pseudoscorpions, harvestmen, whipscorpions and sundspiders. Out of 90,000 plus presently described arthropods species about 40,000 are spider species (Arachnida: Araneae) which are found all over the world in almost every kind of habitat.

Spiders are important predators of pests of cotton, rice, apple, banana and various other crops and plantations and are of great value to the environment. They have a general blanket or buffer effect, promoting evolution, help keeping insect numbers under control in a broad sense and supplying food for many other animals and therefore, keeping the flow of life moving. Spiders represent more than 90 per cent of the natural enemies of brown planthoppers living in paddy fields in Korea (Lee *et al.*, 1997). They feed almost exclusively on insect but little attention has so far been paid into their possible use in Integrated Pest Management programme because until recently little was known of their ecology, especially as it pertains to interactions with associated prey populations and other arthropod competitors (Riechert and Lockley, 1984). In households, a particular spider known as the giant crab spider has been identified as an effective "tool" in controlling cockroaches and other insect pests found in the domestic environment.

India has a very rich diversity, but the knowledge on diversity and distribution of spiders in India is meager compared to other countries. The first detailed account of spiders in India was given by Pocock (1900) who listed 216 species of spiders under 17 families. The most comprehensive account on the Indian spiders was provided by Tikader (1987). Among the 1066 species of spiders listed under 43 families, Lycosidae, Salticidae, Gnaphosidae, Thomisidae and Araneidae are the predominant ones. Each of these families constitutes about ten percent or more of the spider fauna recorded by Tikader (1987). About 23 families have been reported which include 10 or more species of spiders (Table 10.1).

Table 10.1: Important Families (with ten or more species) of Spiders Occurring in India (Tikader, 1987)

Family	Number of Species	Family	Number of Species
Barychelidae	10	Tetragnathidae	22
Oonopidae	10	Oxyopidae	32
Hersilidae	12	Theridiidae	49
Scytodidae	12	Theraphosidae	50
Linyphidae	12	Heteropodidae	65
Ctenide	13	Clubionidae	69
Agelenidae	13	Lycosidae	97
Pisauridae	13	Salticidae	100
Ctenizidae	14	Gnaphosidae	108
Zodaridae	14	Thomisidae	117
Uloboridae	15	Araneidae	142
Dictynidae	16		

Structure

Spiders are characterized by presence of eight walking legs, anterior appendages bearing fangs and poison glands, and specialized reproductive organs on the second appendages of the male. They commonly make extensive use of silk that they spin. Spiders, like other arachnid species, are terrestrial with the exception of a few species which are adapted to freshwater life by trapping air bubbles underwater and carrying the bubbles with them. Although most of the species are less than 1 cm long but the largest may extend upto 9 cm in body length with much greater leg spans. The body of a spider is divided into an anterior cephalothorax or prosoma and a posterior abdomen or opisthosoma separated by a narrow stalk or pedicel, which gives the animal a flexibility and facilitates in use of silk. The cephalothorax ordinarily bears four pairs of simple eyes that tend to be larger in hunting spiders and smaller in spinners of elaborate webs. Three pairs of eyes are also present in certain spiders such as recluse spiders. Each of the first pair of appendages or chelicerae bears a fang with an opening from a poison gland at the tip. The second pair of appendages are pedipalps, which are leglike and are generally modified into a kind of feeler. In the male the pedipalp bears a peculiar copulatory apparatus called a palpal organ. Also on the cephalothorax are four pairs of walking legs. On the abdomen are located modified appendages, the spinnerets, used in secreting silk. Respiratory openings on the abdomen lead to the so-called book lungs (named for their layered structure) or a system of tubes (tracheae) for carrying air, or both. As the spiders generally digest their prey outside the body, therefore, the digestive system of spiders is adapted exclusively to suck the liquid food. The fairly complex brain is larger or smaller in certain parts, depending on whether the animal locates prey mainly by touch or vision.

Venom

All the spiders have poison glands to inject venom into the body of prey. Most species of the spiders are too small to break human skin, and only a few species mainly the black widow spider and the brown recluse spider are dangerous to humans. The black widow spider is non-aggressive and bites humans only in defense. Their painful bite is followed by faintness, difficulty in breathing, and other symptoms. Although the bite is seldom fatal, especially if it is inflicted on healthy adults, medical attention should be sought at once. Some spiders generally capture their prey and bite to inject venom and digestive fluids, paralyzing their victim. When the prey is immobilized, its liquefied digested tissues are sucked by the spider, leaving only an empty shell. Other spiders crush the body of their prey with processes on the pedipalps, bathe the tissues with digestive fluid, suck up the digested liquefied digested fluid and continue chewing and rolling the body until only a little ball of indigestible material remains.

Spider Silk

Spider silk is a fibrous protein that is secreted by the visible part of silk gland *i.e.* spinnerets as a fluid and forms a polymer, on being stretched, that is much stronger than steel and further resists breakage by its elasticity. It takes web spiders about an hour and a half to create a web (Figure 10.1). Spider web is able to stretch up to 40 per cent of its length without breaking. Seven types of silk are produced by seven glands. A single spider does not posses all seven glands but has at least three: dragline (non sticky), attachment and sticky silk; females have an additional one for egg sac silk. Spiders do not stick to their own web because only the central spiral part of the web is sticky, not the spokes; they know where to step (*Source: Oxford University "The spider" by Ed Nieuwenhuy*). Silk produced by spiders is used for different purposes. Small and, especially, young spiders spin a "parachute" thread that enables them to be carried by the wind, sometimes for hundreds of kilometers. The males use silk in transferring sperm to the palpal organ, and the females make cocoons with it. Silk is also used to make nests and other chambers and to line burrows. The most familiar and amazing use of silk by many species, however, is in making insect traps called spiderwebs. Once prey is caught in such a web, the spider may wrap it in more silk. The diverse webs spun by spiders provide a remarkable example of the evolution of instinctive behaviour. The simplest webs are irregular and generally laid out along the ground. More advanced webs, particularly of orb-weaver spiders, are highly intricate, raised above the ground, and oriented to intercept the paths of flying insects. The spinning itself is a complex process involving the placement and then removal of scaffolding spirals and a combination of sticky and non-sticky strands. In some cases a number of spiders form a kind of communal web, but spiders, in general, are not social. Such spiders rely largely on the sense of touch.

Biology

Spiders have separate sexes. Male spiders are sometimes much smaller than the females. The dwarfing of males is pronounced when the females tend to remain in one place. Males mature earlier and the sooner the male gets to a female the more apt he is to reproduce. The genital openings of both

1. Spider releases stand of silk and allows it to float on the breeze until it anchors itself

2. Spider traverses line and drags a thicker bridging

3. Spider once more crosses the line with a loose silk line behind

4. Spider descends to centre and attaches new silk line to the nearest solid point

5. Spider lays radial spokes and sets off boundary lines

6. Spider begins to spin a loose spiral from the center of web

7. Spider returns to centre and begins to spin a tighter spiral which begins to fill in the web

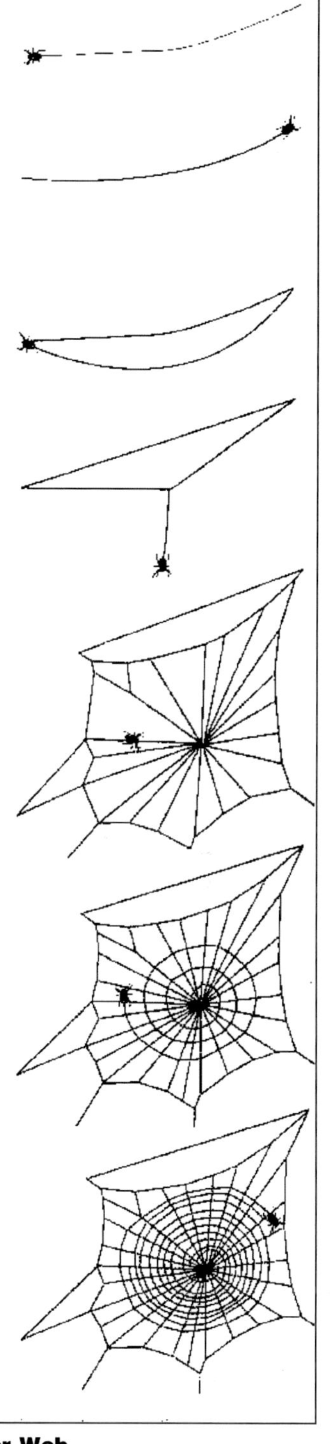

Figure 10.1: Steps (1-7) in the Formation of a Spider Web
(*Source*: Oxford University "The Spider" by Ed. Nieuwenhuy)

Figure 10.2: Wolf Spider Carrying an Egg Sac

Figure 10.3: Spider Guarding the Eggs

Figure 10.4: Egg Mass on Leaf

male and female are located on the abdomen. The male's copulatory organs are complicated structures located on his pedipalps. He spins a little web and deposits sperm in it, then moves the sperm to the palpal organ. Courtship behaviour is often complicated. Males may use draglines to detect and recognize mates, or they may signal their approach by plucking on the female's web. In spiders with well-developed eyes, complex mating displays have evolved that are associated with bright colour patterns. Often the male must avoid having the female treat it as food; even in species where this is common, however, the male often escapes. After sperms are transferred to the female, they can be stored in her body for an extended period.

The number of eggs spiders can lay varies from a few to several hundreds, depending on the species. Some spiders carry the eggs in a little sac until the young spiders hatch from the eggs (*e.g.* wolf spiders–Lycosidae). Others guard the location where the eggs are deposited (*e.g.* Lynx spiders) or place the egg mass in the web or on leaves, covered with fluffy silk. Spiders may live up to 4 months, depending on the species.

Development

Many spiders die after securing the egg cocoon whereas in other species mother guards the eggs until they hatch and secure a safe place for them to learn about the world and even supply food. Eggs usually hatch into spiderlings within three weeks. The young spiders live on yolk-material stored in their abdomens. Generally after their first moult, the young spiders disperse to find food to avoid cannibalism. Some species disperse simply by walking away, others disperse more quickly by ballooning ("Ballooning" is the term used to describe the habit of young spiders of sailing through the wind currents with the help of silk strands). Young Wolf spiders in the genus *Lycosa* spend time clustered on their mother's abdomen after hatching. *Segestria* sp. keeps her young in her burrow with her whereas *Pisaura* sp. builds a special nursery web for her. *Amaurobius terrestris* allows her young to share her burrow until she dies, they then take their last meal from the dead mother's body. Yet even in cases like these the young spiders, usually by their second moult are forced to leave home and find a place to live on their own.

Spiders grow by shedding their old skeleton by new one as their bodies get larger. Spiders stop eating for a few days before moulting, the larger the spider the longer this period can be. During this time the spider's body is secreting a lubricating fluid between the old hard skeleton and the new cuticle and dissolving the inner layers of the old cuticle. This recycles nutrients and makes the whole exoskeleton weaker so it will be easier to escape from. When the spider begins to moult, its heart rate increases and its haemolymph pressure doubles. The spider's body redistributes its internal mass so that the weight of the cephalothorax increases by about 80 per cent and the abdomen decreases by about 30 per cent. The cuticle of the carapace then splits along the front and the sides allowing it to lift away from the body. First the cephalothorax opens up like a trapdoor and swings away from the body, then the sides of the abdomen split and the spider is able to start pulling its body out of the old skin. The legs and palps are always the last parts of the spider's body to be extracted and occasionally one or more of them may fail to come free properly. Most spiders perform this amazing feat hanging upside down from a few strands of web. Moulting can be completed in less than 15 minutes for small spiders but the larger the spider gets the longer the process takes. Adult spiders that live for more than one year may shed their skins once a year even though they are adult already. Moulting is controlled physiologically by a hormone called ecdysone. Like other arthropods, spiders can regenerate lost limbs, if the limb is lost early on after a moult. Spiders have to moult a number of times during their lives, smaller species have to go through less moults than larger ones. Small spiders will normally have less than ten moults to reach maturity while tarantulas have many more. In species where the female is larger than the male the female will also normally have one or more additional moults.

Classification

Spiders make up the order Araneae in the class Arachnida. About 105 families of spiders are known, plus about 10 that are extinct. Two suborders are widely but not universally recognized.

Suborder: *Mesothelae* contains a few primitive, burrowing forms.

Suborder: *Opisthothelae* contains two infraorder:

Mygalomorphae, which consists of the "straight-jawed" forms, usually large, such as the trap-door spiders and the ones called tarantula in the United States.

Araneomorphae, the members of which have the chelicerae somewhat modified and more efficient; it contains the more common and conspicuous forms, such as orb weavers, wolf spiders, and jumping spiders. The cribellate araneomorphs have a specialized organ, the cribellum that helps to produce silk.

The spider families are easily identified by the position and number of eyes (Table 10.2), the overall shape, length of legs and form of the spinnerets. The MontGuide (http://www.montana.edu/wwwpb/pubs/mt9210.html) provides a pictorial guide for determining some important common species and families using conspicuous characteristics.

Table 10.2: Eye Pattern Characteristic to Certain Families of the Spiders

Sl.No.	Family	Eye Pattern
1.	Lyssomanidae	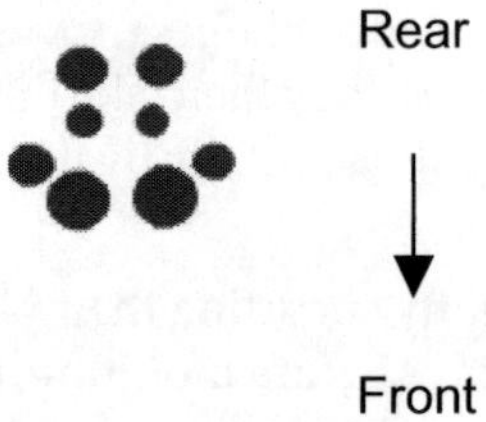 Rear → Front
2.	Hersilidae	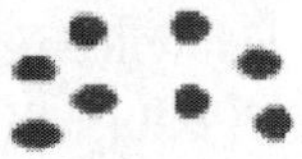
3.	Thomisidae (Crab Spiders)	
4.	Lycosidae (Wolf spider)	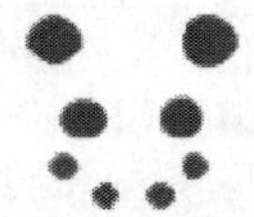
5.	Argyopidae	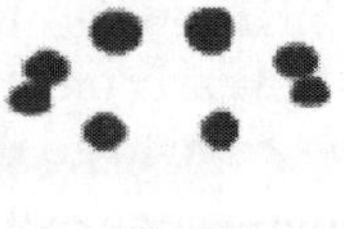
6.	Oxyopidae (Lynx spiders)	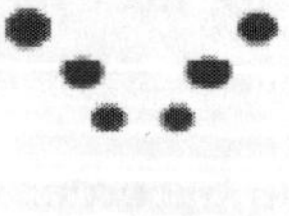
7.	Heteropodidae	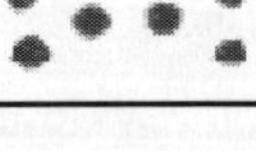

Contd...

Table 10.2–Contd..

Sl.No.	Family	Eye Pattern
8.	Linyphidae	
9.	Tetragnathidae (Long jawed spider)	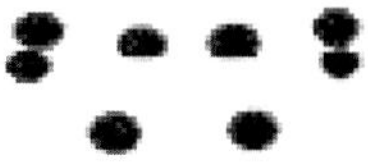
10.	Phoicidae	
12.	Salticidae (Jumping Spiders)	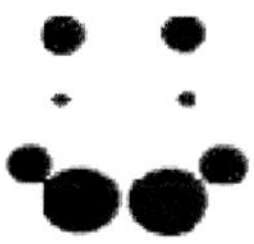
13.	Theridiidae (House or Comb-footed Spiders)	
14.	Araneidae (Orb-weaver spiders)	
15.	Agelenidae (Funnel web spiders)	
16.	Loxoscelidae "Recluse Spiders"	

Source: http://www.montana.edu/wwwpb/pubs/mt9210.html; and http://www.wii.gov.in./envis/rain_forest/chapter7.htm).

Description

Spiders can be divided in two major groups, depending on the method they use to catch their prey. The first group includes the hunting spiders that actively go around in search for their prey and are very mobile and spend a lot of time searching for their prey. The other group of spiders makes webs and then waits for prey to get trapped. They are especially important as predators of flying insects such as adults of moths. Both the hunting spiders and the web-making spiders are very common and important as predators in vegetable and fruit crops as well as in rice fields. Besides the web spinners, many spiders hunt for their food or lie in wait for it. Hunters tend to rely on vision if they feed during the daytime or on touch if they feed at night. Jumping spiders may lurk in ambush for their prey, and a number of them are well camouflaged on flowers by colour or body structure or both.

Web Spinning Spiders

House or Comb-footed Spiders (Family- Theridiidae)

House spiders (Genus *Steatoda*) are medium to small sized, glossy spiders with globular abdomens, thin legs bearing few spines and eight relatively large, protuberant eyes. House spiders, except for the widow spiders, have two of the lateral eyes touching each other on each side and well separated from the other four. Female spiders are1/8 to 3/8 inches long and the abdomen is often covered with dirty whitish or brownish hairs arranged somewhat like chevrons. House spiders spin irregular sheet webs in protected places. Unlike black widow spiders, males and females are often observed living together in the same webs.

Black Widow Spiders

Black widow spiders (Genus *Latrodectus*) get their name from the fact that the female frequently eats the male after mating. These spiders can be easily recognized by reddish or orange hourglass marking on the underside of the abdomen. This marking is lost when preserved in alcohol. The bite of black widow contains a neurotoxin which is very painful. The pain occurs less at the actual bite but rather in the abdomen and limbs. Physiological effects are an accelerated heart beat, increased blood pressure, breathing difficulties, and paralysis. When death occurs, it is due to suffocation.

Orb-weaver Spiders (Family-Araneidae)

Orb-weaver or garden spiders (Figure 10.5) are generally large, with bodies ½ to 1 inch long and stout and spiny legs. They often have conspicuous black, white, yellowish or orange markings. All eight eyes are small, sub-equal, and seemingly grouped into pairs. Web is an orb with a closed central hub from which spokes radiat. Their retreat is often away from the web. The female sits in the hub, waiting to detect movement that signals prey has been captured. After mating, the female seeks a protected place to lay a cluster of 300-800 eggs.

Funnel Web Spiders (Family-Agelenidae)

Funnel web spiders are ½ inch long with two dark longitudinal stripes on a grayish cephalothorax. All eight eyes are small, sub-equal, and arranged in two relatively short rows. The most distinguishing characteristic is a long posterior pair of two segmented spinnerets. Funnel web spiders spin flat webs of silk which extend into funnels into which the spiders retreat for protection. The webs are often found in shrubbery near buildings or in the angles of doors or buildings. The peak population in reached in late summer. Winter is passed in the egg stage.

Spiders that Don't Spin Webs

"Recluse Spiders" (Family-Loxoscelidae)

Brown recluse spiders can be easily recognized by presence of a dark fiddle or violin-shaped marking on the top side of the cephalothorax, and six eyes arranged in three pairs forming a semicircle. The body of female spider is about ½ inch long. Both female and male brown recluse spiders are poisonous and can inject venom in human. A small blister appears and a large swollen area around the bite becomes congested and swollen that can result in a local necrotic lesion.

Figure 10.5: Orb Weaver Spider

Wolf Spiders (Family-Lycosidae)

Wolf spiders (Figure 10.6) are medium to large spiders, and are usually hairy, brown or black, with long legs adapted for running. Wolf spiders have forked or Y-shaped median light bands on their outer coverings. Eyes are arranged in three rows with four large eyes on top and front of head and anterior, slightly curved row of four small eyes. They have good eyesight for chasing down their prey, paralyzing it and feeding. Females carry the globular egg sac attached by the spinnerets. They then carry the young spiderlings about on their backs for a time. The fourth pair of legs is the longest and frequently held stretched out behind the spider. This group is very mobile and may be found moving into houses in the fall in search of prey. Wolf spiders prey on moths and butterflies. They feed on 5 to 15 preys daily and they eat each other when their population is high.

Jumping Spiders (Family-Salticidae)

Jumping spiders (Figure 10.7) are small to medium sized stout bodied spiders with short and stout hairy legs. The first pair of legs is sometimes enlarged. The square-fronted cephalothorax bears four very large anterior eyes. They frequently have contrasting black, reddish, or yellowish markings. Jumping spiders are diurnal and rely on quickness and visual acuity. They are very agile, pouncing and feeding on small insects about the home. They are often seen on screens or near doors or windows.

Crab Spiders (Family-Thomisidae)

Crab spiders (Figure 10.8) are colourful crab-like spiders with the first four legs larger than their hind legs and are able to walk forward, backward, or sideways like a crab. They are medium sized and

Figure 10.6: Wolf Spider

Figure 10.7: Jumping Spider

Figure 10.8: Crab Spider

often brightly coloured, with abdomens that are usually wide at the posterior end. They have eight relatively small, well spaced, light coloured eyes. These are generally found on the blossoms of plants and have the tendency to camouflage in their habitat so as to catch prey, unaware as they pass by.

Rearing Spiders as Biological Pest-Control Agents

The rearing of giant crab spiders has been successfully carried out at village level. The important steps involved in rearing of the spider are available on the website of *"Centre for Indian Knowledge*

Systems". In the rearing method the male and female giant crab spiders are collected from inside houses, where they are found in plenty. A male and a female are placed in a 2-litre container and allowed to mate. After mating, the male is removed. The female constructs an egg sac within two weeks. Young spiderlings, which resemble the adults except for being smaller in size, hatch out from egg sac within 20 days. The newly hatched spiderlings are separated in small containers, provided with lid having a few small holes for aeration and one moderately big hole for feeding. A small strip of paper is inserted into the container to give additional support for the movement of the spiderlings. The bigger hole should be plugged with cotton. The first instar spiderlings should be fed with small larvae, like the second instar larvae of houseflies, *Drosophila* and *Tribolium*. Once these moult to second instar, they can be fed with slightly bigger larvae of *Tribolium* and very small-sized cockroaches. The spiderling can be grown up to the fifth instar stage in this container after which it should be transferred into a 1-litre container where it is kept until the 10th instar. Tenth to fourteenth instar stages and adults should be grown in 2-litre containers. The seventh or eighth instar stage can be fed adult houseflies and medium-sized cockroaches apart from the other prey given earlier. From the 10th and 11th instar onwards, adult cockroaches can be given. In the adult stage, predation is maximum after mating and after the egg sac is dropped once the spiderlings have hatched. During these periods, the spiders should be provided with enough prey. The containers should be numbered properly and a regular register should be maintained to note the date of moulting and other observations. The containers should be cleaned regularly. The spider should first be transferred to another container and the old one should be cleaned with soap water and sun-dried. The spiders should be protected from ants by either putting the containers in water trays or putting a cup of oil or water under the legs of the table on which the containers are kept.

Spider Ecology

Most spiders live in very specific habitats depending on their ability to withstand abiotic variables such as temperature (average as well as the seasonal maximums and minimums), wind presence, moisture, humidity, light intensity and vegetation type thereby avoiding undue competition among them. Spiders can spend long time without food and expend little energy waiting for their prey. Spiders that hunt out their prey obviously use more energy but even these can live for weeks without eating. Microclimate, bark structure as well as height are important features controlling the distribution of spiders. Some spiders change their habitat throughout the year as they grow, thus the orb web spiders *Argiope auranitia* and *Argiope trifasciata* make their webs at different heights in the vegetation during their juvenile stages, but once adult they both use the same vegetation level to build in.

Factors Affecting Population Buildup

Population density of spiders depends on the availability of prey insects. Female spiders lay more eggs when a plenty of prey is available. The population of spider is affected by both biotic as well as abiotic factors. The abiotic causes are weather, storms, floods and droughts and they can all take their toll on spiders directly and indirectly by the loss of prey. The biotic factor includes the predators, particularly the mammals (insectivores, bats and many small primates) and other spiders. One family of spiders, the Mimetidae (Pirate spiders) feed exclusively on other spiders. Another family with a particular liking for other spiders as prey is the Pholcidae. There are also many genera such as *Portia* in the family Salticidae which prefer spiders rather than insects as their prey. Other natural enemies which also affect the population build up are Ichneumons (Pimpline and Cryptine), Wasps (Sphecids, Pompilids, Trypoxylonids, Chalcids and common Yellow Jackets), and various parasitic flies such as the family Acroceridae, beetles, ants, centipedes and even some large crickets. Several species of Nematodes (round worms) also parasitize spiders.

Over Wintering

While spiders are poikilothermic (cold blooded) animals however small size and behavioural adaptations can allow a spider such as *Geolycosa godeffroyi* to maintain its body at a temperature about 4-5 degrees higher than the ambient. Tropical orb-web spiders keep themselves cool by changing their orientation in their webs as the sun moves across the sky. Different spiders use different strategies to survive in temperate climates where the winters are severe and insect food is also short (Table 10.3). Many simply dig down into the leaf little and go under subterranean hibernation during winter. Other species simply die off and leave only their eggs to survive the cold; their lack of cellular development prevents them from damage when they freeze. Once the weather becomes favourable the eggs develop and the young spiders can hatch in the spring when food is again available. Some species in the milder temperate climates will remain active throughout the winter and can survive temperatures down to about minus 5 or 6. Below this they die, but at zero they are still spinning webs.

Table 10.3: Over Wintering Strategies for Known Spiders

As Nymphs only	45 per cent
As Nymphs and Adults	23 per cent
As Adults only	11 per cent
Inside Egg Sac	7 per cent
Winter Active	9 per cent

Source: www.earthlife.net/chelicerata/s-ecology.html.

Camouflage and Mimicry

Many spiders are mimics which helps them to catch their prey, avoid other predators and to live in other peoples homes. Many spiders active during the day are cryptically coloured that make them difficult for their predators to locat *e.g.* Pisaurids, Salticids or Tetragnathans. *Misumena vatia* can be seen to change colour from yellow to white, from white to yellow as it is moved from one flower to the next. The colour change does take a day or two to be complete. Other spiders such as *Cyrtophora cicatrosa* can change their colour rapidly, while dropping from their perch on a thread so that they are different colour on the ground than they were in the vegetation. This undoubtedly confuses some predators. Several hundred species of spider are known to be spider mimics, though some are obviously better than others. Mimics can be either morphological or physiological or both.

Relative Abundance of Spiders in Different Agro-ecosystems

Rain Forests

The Western Ghats regions are characterized by heavy rains averaging 3000 mm between June–August. The survey conducted by Rajashekhar and Raghavendra (2001) in Western Ghats to study the arboreal spider fauna indicated that Oxyopidae, Thomisidae, Araneidae and Salticidae were the major families (Table 10.4). The density of spider remained high during the pre-monsoon season and gradually decreased during monsoon. On some plants like *Areca* and *Ceasalpinia* they were abundant in the month of December and their number decreased by end of January. There was considerable variation in the members of Araneidae during rainy season and winter. In *Areca* plant *Gasteracantha* species were abundant, which suggests that species prefer specfic habitats. Flowering plants such as *Nyctanthus* had high density of spider belonging to Thomisdae

Table 10.4: Relative Abundance of Major Arboreal Spider Families in the Central Western Ghats (After Rajashekhar and Raghavendra, 2001)

Family	*Percentage*
Hersilidae	2.9
Linyphidae	4.2
Oxyopidae	5.4
Thomisidae	11.9
Salticidae	26.2
Araneidae	38.0

Rice Ecosystem

Spiders play an important role in regulating insect pests in the agricultural ecosystem. Spiders represent more than 90 per cent of the natural enemies of brown planthoppers living in paddy fields in Korea (Lee *et al.*, 1997). The spiders (Table 10.5) living in rice fields in Korea comprise 27.96 per cent of all Korean spiders, and can be classified into 22 families, 99 genera and 175 species. *Pirate subpiraticus, Pachygnatha clercki* and *Gnathonarium dentatum* are dominant spider species in rice fields (Paik *et al.*, 1979).

Table 10.5: Important Spider Predators of Rice Pests in Korea (Lee and Kim, 2001)

Season	*Survey Period*	*Site*	*Mean No.*			*Mean Occupancy Rate (%)*	
			Total Spiders	*Web Builders*	*Hunters*	*Web Builders*	*Hunters*
Rice crop	July to Sept.	Paddy field	274	108	166	39.4	60.6
	1994-96	Levee	136	42	166	30.6	69.4
Hibernation	Nov. May 1994-97	Levee	285	70	94	31.6	68.4

During the rice growing period in Korea, spiders generally show two peaks-one when spiders which have over wintered nearby migrate into the paddy and the other is when the second generation of the year appears (Lee and Kim, 2001). After transplanting the young rice plants are too small to provide a suitable habitat for either hunters or web builders, therefore, the immigration of the spiders is delayed (Kim, 1998). Spiders start to build up their populations in the middle of July, 40–45 days after transplanting, (Kim 1992; Yoon, 1997). The different time taken by different spider species to immigrate from levees into paddy fields is the cause of the low initial population densities (Lee *et al.*, 1997). Spider fauna was relatively poor in early period of rice growth (June). *P. subpiraticus* (Lycosidae, hunter) and *G. dentatum* and *Oedothorax insecticeps* (Erigonidae, snares with a hunting habit) were the main early settlers. From July onwards, spider's fauna became rich with the invasion and population growth of *P. clercki* (Tetragnathidae, hunter exceptional in this family) and others (Okuma *et al.*, 1978). Most of the dominant spider species (Table 10.6) immigrated from outside areas into the rice fields by ballooning (habit of young spiders of sailing through the air borne by silk strands on wind currents).

Both web builders and hunters follow a foraging strategy. Song and Lee (1994) suggested that web builders such as *P. clercki* were better able to suppress insect pests than hunters such as *P. subpiraticus*. However, hunters are usually active predators which follow a "pursue and kill" foraging strategy, while web builders follow a passive "sit and wait" strategy (Enders, 1974). Furthermore, the webbing sites of web builders are easily affected by environmental factors. In addition, when the web spaces overlap, there is competition with and between species of web builders. Therefore, hunters probably are more effective predators than web builders.

Table 10.6: Important Spider Predators of Brown Planthopper (*Nilaparvata lugens*) in Paddy Fields in Korea (Lee and Kim, 2001)

Vertical Division (Above Water/ Soil Surface)	*Spiders*				
	P. subpiraticus	*Clubiona kurilensis*	*G. dentatum*	*Ummeliata insecticeps*	*Tetragnatha maxillosa*
<20 cm	+++	++	+++	++	+
20-40 cm	+	++	++	++	+
40-60 cm					++
60-80 cm		++			+++
>80 cm		±			+

±: Very rare; +: Usually present; ++: Fairly common; +++: Abundant.

In India recent study conducted by Kumar and Shiva Kumar (2004) in Gujarat indicated 33 species of spiders in rice under 10 families. The most dominant species were *Lycosa pseudoannulata* (Bosenberg and Strang) and *Oxyopes shweta* (Tiakder) from hunter group and *Tetragnatha* sp., *Argiope aemula* (Walckenaer), *Cytrophora cicatrosa* (Stolixzha), *Neoscona theis* (Walckenaer) and *Neoscona sinhagadensis* (Tikader) from weaver group. The population of *L. pseudoannulata* remained high throughout the cropping season whereas *O. shweta* increased during grain filling stage. Most of the weavers appeared late in the season. No significant correlation was observed with abiotic factors. Rice hispa and phadka grasshoppers indicated a significant correlation with spider density.

Sugarcane Ecosystem

Sugarcane being a long duration crop provides a stable ecosystem favourable for insect pests and their natural enemies. Field survey conducted in three states of southern peninsular India revealed the occurrence of 57 species of spiders belonging to 13 families. Of these, only two species *viz.* wolf spider, *Hippasa greenalliae* Blackwall (Lycosidaae) and *Cyrtophora cicatroswa* Stoliczka (Araneidae) were abundant while six species were less abundant and 49 species were rare in occurrence. Colonizing the sugarcane fields started with hunting spiders and some miscellaneous group after 40-45 days of planting when the sprouting of the crop was completed. After 100 days of planting, when the canopy became dense, the web building spiders also started colonizing the field and the population increased steadily upto 180 days and thereafter it got fairly stabilized. The population reached the peak of 19250 spiders/ha by 270 days. *H. greenalliae* comprised more than 70 per cent of the total population. The population of the spiders was found to be negatively correlated with maximum temperature, minimum temperature and rainfall (Easwaramoorthy *et al.*, 1994).

Conservation of the Natural Spider Fauna

Mulching, especially organic mulch can increase the number of spiders. Spiders can hide in the layer of mulch and they will also find a lot of alternative prey there, such as springtails (Collembolans) and other small insects that feed inside the mulch. The effectiveness of spiders in an agro ecosystem is limited by the disruptive effects of the application of insecticides and the annual harvesting and tilling of the vegetation ground layer. Both the practices represent major sources of mortality to spider populations. The agricultural practices tend to suppress spider diversity and density in this system.

Insecticide Application

Insecticides have a negative effect on the population densities and species diversity of rice field spiders (Kuno, 1968; Kuno and Hokyo, 1970; Kawahara *et al.*, 1971; Choi *et al.*, 1978, Paik *et al.*, 1979, Jang 1981, Ryu *et al.*, 1977, Kim *et al.*, 1984, Paik and Hwang, 1990; Lee *et al.*, 1993; Bae *et al.*, 1994). Carbofuran which is widely used in rice fields is very toxic to spiders (Bae *et al.*, 1994). Even insecticides which generally protect natural enemies often have a negative effect on spiders (Clausen, 1990). Hunting species tend to suffer more damage than web builders (Specht and Dondale, 1960; Legner and Oatman 1964; Bostanian *et al.*, 1984), while active hunters suffer more than less active ones (Bostanian *et al.*, 1984). Some spiders are less sensitive to insecticides than are other predators (Dippenaar, 1979).

By restricting insecticide application during crucial period in the life cycle of the pest species and by limiting the spraying program to midday when the spiders are inactive and in sheltered location, one can conserve spider numbers and diversities in agro ecosystems. The use of selective insecticides applied sparingly in Japan led to the control of leafhopper infestations on rice as the spider populations were not decimated by the chemical (Almquist, 1973). Similar effect has also been reported in cotton (Laster and Brazzel, 1968). Use of selective insecticides might also help in conservation of spider fauna. Spiders are less susceptible to endosulfan (Sharma and Sarup, 1980), dieldrin (Dippenaar, 1979), methyl parathion (Almquist, 1973), emulsifiable concentrates (Pfrimmer, 1964) and carbamates and pyridaphenthion (Laster and Brazzel, 1968; Li,1982) compared to other insecticides.

Cultivation

Various mechanical and cultivation procedure utilized in agricultural systems are perhaps even most disruptive to spider communities due to suppression of spider numbers, loss of egg sacs and destruction of habitat structures. At least once in a year both the habitat and beneficial fauna are destroyed. If small plots are bounded by hedgerows, abandoned fields and other natural habitats, the deleterious effect of destroying the habitat and beneficial fauna each year can be offset by recolonization from adjacent habitats. Some recolonization in large-scale agricultural system is also possible with respect to that portion of the spiders fauna that balloons. The major ballooning spider families are the small sheet line weavers (Linyphiidae-microyphantids), the wolf spiders (Lycosidae) and the orb weavers (Araneidae). One of the reasons is that the Chinese and Japanese have been successful in using spiders for biological control in rice is that of the ballooning linyphiids are the major predators of the leafhopper pests of the system (Riechert and Lockley, 1984).

Augmentation

Recolonization in large scale field by the introduction of egg sacs may be done but it entails large scale collection or rearing programme. Some spiders are being promoted against agricultural insect pests. Linyphiids against aphids in wheat fields in the UK and *Pardosa ramulosa* in California where it has been shown that 20 spiders per square metre can reduce leaf hoppers in rice fields by up to 90 per cent (Source: *www.earthlife.net/chelicerata/s-ecology.html*).The Chinese actually use straw bundles to conserve spiders in paddy (Jones, 1981).These bundles are initially laid down in areas where spiders are numerous and subsequently transported to the field where recolonization is needed. Once the spiders are introduced into a field there is a need to limit their movement outside the field. Spider emigration is associated with unfavourable thermal environments *e.g.* temperature, humidity, low prey availability and disturbance. Prey numbers and local thermal environments can be enhanced by planting beneficial weeds and annual flowers at regular intervals and or by maintaining compost heaps in the garden.

Conclusion

Spiders are potential natural enemies of a diverse array of insect pests and need a helping hand by way be directed research efforts which are lacking at present in India. More emphasis should be made on habitat management tactics in different agro ecosystem to create favourable microclimate for increasing the nesting site for spiders and for increasing the prey populations. Long term experiments are also needed to generate information on the rate of recolonization of the areas following disturbance and the effect of the spider populations on prey population.

References

Anonymous, 1995. *Global Biodiversity Assessment*. Cambridge University Press, pp. 1140.

Almquist, S., 1973. Spiders association in coastal sand dunes. *Oikos*, 24: 444–457.

Bae, Y.H., Lee, J.H. and Hyun, J.S., 1994. A systematic application of insecticides to manage early season insect pests and migratory planthoppers on rice. *Kor. Jour. Appl. Entomol.*, 33(4): 207–215.

Bostanian N.J., Dondale, C.D., Binns, M.R. and Pitre, D., 1984. Effects of pesticide on spiders (Araneae) in Quebec apple orchards. *Canadian Entomologist*, 116: 663–675.

CABI, 2000. *Crop Protection Compendium, Global Module*, 2nd edition. CABI Publishing, Wallingford, UK.

Clausen, I.H.S., 1990. Design of research work based on a pilot study dealing with the effect of pesticides on spiders in a sugar-beet field. *Acta Zool. Fennica*, 190: 69–74.

Choi, S.Y., Lee, H.R. and Ryu, J.K., 1978. Effects of carbofuran root-zone placement on the spider populations in the paddy fields. *Kor. J. Plant Protection*, 17(2): 99–103.

Dippenaar-Schoeman, A.S., 1979. Spider communities in strawberry beds: seasonal changes in numbers and species composition. *Phytophylactica*, 11: 1–4.

Easwaramoorthy, S., David, H., Kurup, B.K. and Santhalakshmi, G., 1994. Studies on the spiders faunal of sugarcane ecosystem in Southern Peninsular India. *J. Biol. Control*, 2(L): 85–93.

Enders, F.A., 1974. Vertical stratification in orb web spiders (Araneae: Araneidae) and a consideration of other methods of coexistence. *Ecology*, 55: 317–328.

Jang, Y.D., 1981. Studies on the protection and utilization of the natural enemy against rice insect pests. *Res. Reports, Agri. Tech., Chungnam Nat. Univ.* 8(1): 19–29.

Jones, R.L., 1981. Report of the USDA Biological Control of Stem borers Study Team's visit to the People's Republic of China, July 1980 submitted to Int. Agric. Prog. Off., Univ., Minnesota, Minneapolis.

Kawahara, S., Kiritani, K. and Sasaba, T., 1971. The selective activity of rice-pest insecticides against the green leafhopper and spiders. *Botyu-Kagaku*, 36: 121–128.

Kim, J.B., Cho, D.J., Kim, C.H., Uhm, K.B. and Chang, S.D., 1984. Studies on the seasonal fluctuation and the timing of control for the brown planthopper (*Nilaparvata lugens* Stål) and white-backed planthopper (*Sogatella furcifera* Horvath) in southern rice cultural area. Res. Reports, ORD (Office of Rural Development, Korea) 26–2 (S.P.M.U): 16–20.

Kim, H.S., 1992. Suppressive effects of wolf spider, *Pirata subpiraticus* (Araneae: Lycosidae) on the Population Density of Brown Planthopper (*Nilaparvata lugens Stål*). Unpublished *Ph.D. Thesis*, Dongkuk Univ, Seoul, Korea, 69 pp.

Kim, S.T., 1998. Studies on the ecological characteristics of the spider community at paddy field and utilization of the *Pirata subpiraticus* (Araneae: Lycosidae) for control of *Nilaparvata lugens* (Homoptera: Delphacidae). *Unpublished Ph.D. Thesis*, Konkuk Univ., Seoul, Korea, 90 pp.

Kumar, Dolly and Shiva Kumar, M.S., 2004. Ecological studies on spiders in rice agroecosystem of Vadodara (Gujarat) with special emphasis on biocontrol aspect. *Indian J. Entomology*, 66: 323–327.

Kuno, E., 1968. Studies on the population dynamics of rice leafhopper in a paddy field. *Bull. Kyushu Agri. Exp. Stn.* 14(2): 131–246.

Kuno, E. and Hokyo, N., 1970. Comparative analysis of the population dynamics of rice leafhopper, *Nephotettix cincticeps* Uhler and *Nilaparvata lugens* Stål, with special reference to natural regulation of their numbers. *Res. Popul. Ecol.*, 12: 154–184.

Laster, M.L and Brazzel, J.H.R., 1968. A comparison of predator populations in cotton under different control programs in Mississippi. *J. Econ. Entomol.*, 61: 714–718.

Lee, H.P., Kim, J.P. and Jun, J.R., 1993. Influences of the insecticidal application on the natural enemies and spider community. *J. Indus. Tech. Grad. Sch. Dongguk Univ.*, 1: 295–307.

Lee, H.P. and Kim, S.T., 2001. Use of spiders as natural enemies to control rice pests in Korea. *FFTC Extension Bulletin*, p. 9–10.

Lee, J.H., Kim, K.H. and Lim, U.T., 1997. Arthropod community in small rice fields associated with different planting methods in Suwon and Icheon. *Kor. J. Appl. Entomol.*, 36(1): 55–66.

Legner, E.F. and Oatman, E.R., 1964. Spiders on apple in Wisconsin and their abundance in a natural and two artificial environments. *Canadian Entomologist*, 96: 1202–1207.

Li, L., 1982. Integrated rice insect pest control in Guangclong Province of China. *Int. Organ. Biol. Controll Gerdat Colloq.*, Paris, April.

Okuma, C., Lee, M.H. and Hokyo, N., 1978. Fauna of spiders in a paddy field in Suweon, Korea. *Esakia* 11: 81–88.

Paik, J.C. and Hwang, J.Y., 1990. Preliminary survey of natural enemies of rice pests for the development of integrated control programs. *Res. Report RDA*, 33: 153–171.

Paik, J.C., Lee, Y.B., Lee, H.R. and Choi, K.M., 1979. Studies on the physiology, ecology and control of the rice insect pests. *Res. Report RDA*, p. 341–367.

Park, J.S., Lee, S.C., Lee, B.H., Kim, Y.I., Park, K.T. and Ahn, K.J., 1972. Influences of the insecticides on the rice insect pests. *Res. Report RDA*, p. 146–169.

Pfrimmer, T.R., 1964 populations of certain insects and spiders on cotton plants following insecticide application. *J. Econ. Entomol.*, 57: 640–644.

Pocock, R.I., 1900. *The Fauna of British India Including Ceylon and Burma.* Secretary of State for India in council, London pp. 279.

Rajashekhar, K.P. and Raghavendra, N., 2001. An overview of spider diversity in India. Soruce: http://www.wii.gov.in./envis/rain_forest/chapter7.htm.

Reichert, S.E. and Lockley, T., 1984. Spiders as biological control agents. *Annual Rev. Entomol.*, 29: 299–320.

Ryu, J.K., Choi, S.Y., Lee, H.R. and Song, Y.H., 1977. Root-zone placement of carbofuran for control of rice insect pests. *Kor. J. Plant Protection*, 16(4): 217–220.

Sharma, V. K. and Sarup, P., 1980. Predatory role of spiders in the integrated control of the maize stalk borer, *Chilo partellus* (Swoinhoe). *Indian J. Entomol.*, 42: 229–231.

Song, Y.H. and Lee, Y.G., 1994. Studies on the spider fauna in the paddy fields of Chinju and Namhae areas. *Korean J. Environ. Agric.*, 13(1): 98–110.

Specht, H.B. and Dondale, C.D., 1960. Spider populations in New Jersey apple orchards. *J. Econ. Ent.*, 53: 810–814.

Tikader, B.K., 1987. *Handbook of Indian Spiders.* Zoological Survey of India, Calcutta, India, pp. 251.

Yoon, J.C., 1997. Arthropod community structure and changing patterns in rice ecosystems of Korea. Unpublished *Ph.D. Thesis*, Seoul National University, Seoul, Korea, 105 pp.

www.earthlife.net/chelicerata/s–ecology.html.

Insect Pest and Disease Management *Pages* **167–170**
Editor: **Dr. D. Prasad**
Published by: **DAYA PUBLISHING HOUSE (2016)**

Chapter 11

Impact of Thiamethoxam on Spiders in Sugarcane Ecosystem

C. Vijayaraghavan* and A. Regupathy
Department of Agricultural Entomology, T.N.A.U., Coimbatore – 641 003, T.N.
***E-mail: vijayaraghavanento@yahoo.co.in**

Introduction

Sugarcane is one of the important commercial crops in the tropics and serve as the main source of sugar in the world. Sugarcane is known to be attacked by about 228 insects and non-insect pests in India (David and Nandagopal, 1986).Vegetative structure and robust nature of the crop limits the insecticidal usage. In India only 2-3 per cent of the pesticides are used on sugarcane. The crop favours the arthropod predators abundance in the sugarcane ecosystem. Use of selective insecticides that are less toxic to natural enemies than to pests should conserve natural enemy populations, and the surviving natural enemies may suppress the pest population, which in turn will reduce the rate of insecticide application (Tanaka *et al.*, 2000). Also the insecticides used in integrated pest management should have minimum effect on beneficial insects and environment and maximum effect on target pests. Information on the effects of pesticides on the natural enemies will be useful to effectively utilize them as bio control agents. Hence, an attempt has been made to assess the effect of thiamethoxam on spiders of sugarcane.

Two season field trials (2004 and 2005) were conducted in two locations *i.e.*, Valaithoppu (Madurai district) and Lalapettai (Karur district) in a completely randomized block design to assess the bioefficacy of thiamethoxam against sucking pests like whitefly, woolly aphid and spider population. In all the trials, imidacloprid, dimethoate and methyl demeton were included as standard checks. Six month old sugarcane crop with natural infestation of the target pests were selected for the study. Insecticides were sprayed @ 1000 liters per ha to run off point using a high volume sprayer. In all experiments an

Table 11.1: Effect of Insecticides on Spiders in Sugarcane Ecosystem (Pooled Mean Data of Three Sprays)

Treatment	*Dose (g.ai./ha)*	*Pre count (Spiders/ Clump)*	*Lalapettai*				*Pre count (Spiders/ Clump)*	*Valaithoppu*			
			Population of Spiders (DAT)					*Population of Spiders (DAT)*			
			1	*3*	*7*	*14*		*1*	*3*	*7*	*14*
Thiamethoxam 25 WG	25	4.7 (2.17)^{b}	4.57 (2.14)^{c}	4.63 (2.15)^{b}	4.7 (2.17)^{b}	4.9 (2.21)^{b}	3.13 (1.77)^{b}	3.13 (1.77)^{b}	3.03 (1.74)^{b}	3.1 (1.76)^{b}	3.2 (1.79)^{b}
Thiamethoxam 25 WG	50	4.7 (2.17)^{b}	4.63 (2.15)^{a}	4.67 (2.16)^{a}	4.83 (2.20)^{a}	5.03 (2.24)^{a}	2.97 (1.72)^{c}	2.97 (1.72)^{c}	2.9 (1.70)^{d}	2.87 (1.69)^{d}	3 (1.73)^{d}
Thiamethoxam 25 WG	100	4.33 (2.08)^{d}	4.2 (2.05)^{e}	4.2 (2.05)^{e}	4.4 (2.10)^{d}	4.5 (2.12)^{e}	2.87 (1.69)^{e}	2.87 (1.69)^{d}	2.8 (1.67)^{e}	2.77 (1.66)^{e}	2.83 (1.68)^{f}
Thiamethoxam 25 WG	200	4.83 (2.20)^{a}	4.6 (2.14)^{b}	4.6 (2.14)^{c}	4.7 (2.17)^{b}	4.87 (2.21)^{c}	2.83 (1.68)^{f}	2.73 (1.65)^{f}	2.7 (1.64)^{f}	2.77 (1.66)^{e}	2.87 (1.69)^{e}
Imidacloprid 200 SL	50	4.03 (2.01)^{e}	3.83 (1.96)^{f}	3.83 (1.96)^{f}	3.93 (1.98)^{e}	4 (2.00)^{f}	3.13 (1.77)^{b}	3.13 (1.77)^{b}	2.93 (1.71)^{c}	3 (1.73)^{c}	3.07 (1.75)^{c}
Dimethoate 30 EC	750	3.13 (1.77)^{f}	2.23 (1.49)^{g}	1.83 (1.35)^{g}	1.93 (1.39)^{f}	2.07 (1.44)^{g}	2.9 (1.70)^{d}	2.8 (1.67)^{e}	2.5 (1.58)^{g}	2.53 (1.59)^{f}	2.67 (1.63)^{g}
Untreated check	-	4.37 (2.09)^{c}	4.37 (2.09)^{d}	4.4 (2.10)^{d}	4.57 (2.14)^{c}	4.6 (2.14)^{d}	3.33 (1.82)^{a}	3.33 (1.82)^{a}	3.43 (1.85)^{a}	3.5 (1.87)^{a}	3.5 (1.87)^{a}

In the table means followed by a common letter are not significantly ($P = 0.05$) different by DMRT.

Figures in parentheses are $\sqrt{x+a}$ values, where x-Population.

untreated check was included. Observations on the spider population were made before and 1, 3, 7 and 14 days after insecticide application on five randomly selected clumps in each plot.

The population of spiders ranged from 3.13 to 4.83 in Lalapettai and 2.87 to 3.33 per clump in Valaithoppu prior to application of insecticides (Table 11.1). The results clearly revealed that the application of thiamethoxam does not influence the spiders in sugarcane ecosystem. However, a short term reduction in the population was noticed with increased rate of application. The effect was noticed up to a short period of one week. The reason may be attributed to the short-term dispersal of spiders to the neighbouring untreated plants. The spider population was more in treated plots rather than untreated plots. The resumption of spider population in treated plots was well indicated by the slower increase of spiders in treated plots. This study was in concordance with findings of Mathirajan (2001) and Senthilkumar (2003); they observed a short-term reduction in spider population in rice and coffee by the application of thiamethoxam. Imidacloprid and chlorpyriphos were found to less hazardous to spiders than the other chemicals, quinalphos and monocrotophos applied in cotton ecosystem (Vanitha, 2000).

A significant short term reduction in the population of foliage dwelling spiders was noticed in plots applied with dimethoate (750 g a.i./ha) which was used as a standard check.This is in conformity with the observations made by Easwaramoorthy *et al.*,1994 in sugarcane and other workers that single applications of an insecticide in various ecosystems resulted in at most only short-term reductions in spider numbers (Dondale, 1972; Culin and Yeargan, 1983; Toft and Jensen, 1998). As a whole, foliar application of neonicotinoids at the recommended levels was found to be less hazardous to the spiders of sugarcane ecosystem.

Conclusion

Impact of new molecule thiamethoxam on spider population in sugarcane ecosystem was studied at various locations in Tamil Nadu. No marked decrease in the population reduction of spiders was noticed by the application of thiamethoxam and imidacloprid. However, significant reduction in the population of spiders was noticed by the application of dimethoate.

References

Culin, J.D. and Yeargan, K.V., 1983. The effects of selected insecticides on spiders in Alfalfa. *Journal of Kansas state Entomological Society*, 52(2): 151–158.

David, H. and Nandagopal, V., 1986. Pests of sugarcane: Distribution symptomatology attack and identification. In: *Sugarcane Entomology in India*, (Eds.) H. David, S. Easwaramoorthy and R. Jayanthi. Sugarcane Breeding Institute, Coimbatore, India, pp. 1–29.

Dondale, C.D., 1972. Effects of carbofuran on arthropod populations and crop yield in hay fields. *Canadian Enomologist*, 104: 1433–1437.

Easwaramoorthy, S., David, H., Kurup, N.K. and Santhalakshmi, G., 1994. Studies on the spider fauna of sugarcane ecosystem in southern peninsular India. *Journal of Bioogical Control*, 8(2): 85–93.

Jepson, P.C., 1989. The temporal and spatial dynamics of pesticide side-effects on non-target invertebrates. In: *Pesticides and Non-target Invertebrates*, (Ed.) P.C. Jepson. Intercept, Wimborne, pp. 95–128.

Mathirajan, V.G., 2001. Diversity and predatory potential of spiders in cotton and rice ecosystems applied with thiamethoxam. *Ph.D. Thesis*, Tamil Nadu Agricultural University, Coimbatore, 140 p.

Senthilkumar, C.M., 2003. Diversity of Lecanicillium (=Verticillium) lecanii (Zimmermann) Viegas and spiders in coffee ecosystem application with thiamethoxam. *Ph.D. Thesis*, Tamil Nadu Agricultural University, Coimbatore.

Tanaka, K., Endo, S. and Kazano, H., 2000. Toxicity of insecticides to predators of rice planthoppers: Spiders, the mirid bug and the dryinid wasp. *Applied Entomology and Zoology*, 35(1): 177–187.

Toft, S. and Jensen, A.P., 1998. No negative sub lethal effects of two insecticides on prey capture and development of a spider. *Pesticide Science*, 52(3): 223–228.

Vanitha, K., 2000. Studies on spiders in cotton pest management. *M.Sc. (Ag.) Thesis*, Tamil Nadu Agricultural University, Coimbatore, 119 p.

Insect Pest and Disease Management *Pages* **171–178**
Editor: **Dr. D. Prasad**
Published by: **DAYA PUBLISHING HOUSE (2016)**

Chapter 12

Life Table and Biotic Potential of *Helicoverpa armigera* (Hubner) on Chickpea

S.K. Singh*[1] and D.K. Yadav[2]
[1]Indian Institute of Pulses Research, Kanpur – 208 024
***E-mail: singhskkanpur@indiatimes.com**
[2]P.C. Unit (Linseed), Chandra Shekhar Azad University Agriculture and Technology, Kanpur – 208 002

Introduction

Gram pod borer, *Helicoverpa armigera* Hubner (Lepidoptera: Noctuidae) is a noxious and polyphagous pest, which ravages 200 cultivated, and wild hosts. This pest become global importance and its seriousness with outbreak situation have been witnessed due to its Polyphagy, multivoltine, migratory, over wintering and diapausing under adverse condition. In Central Uttar Pradesh and most part of country every year realized a heavy toll of losses to chickpea crop during *Rabi* season at most of the reproductive phase. The most damaging stage of the pest is larvae, which bore into the pods, eventually the yield reduction drastically. *Helicoverpa armigera* single larvae can damage 25-30 pods in its lifetime (Singh and Ali, 2005).

Indiscriminate and incessant synthetic chemical pesticide has resulted insect resistance to insecticide, resurgence, and outbreak of secondary pest, side ill effect on non-target organisms including aquatic and terrestrial fauna. So it is imperative to undertake such studies, which could deduce that what is the biotic potential of the pest to thrive and survive under a set of condition. In this perspective, Life tables are one of the most useful tools in the study of population dynamics of insect pest, which insert in quantitative evaluation of host plant. Life table may also help determine relative contribution of immature stages to the population of pest species (Birch 1948).

The net capacity of increase has not been quantified for many insect (Wellick and Pedigo, 1978). Therefore, in the present investigation, the projective potential rate of increase of *Helicoverpa armigera* on chickpea pods were computed considering the fact that the construction of life table is vital statistic to the description and understanding of the population dynamic of the pest. Life table deliver a indication to determine which stages in the life cycle contribute the most population.

Laboratory culture of *H. armigera* was used for this study. Known number of adult pairs was released for egg laying in cages. Eggs laid on white muslin cloth were collected by wet camel hairbrush and such hundred eggs were kept for hatching in the batches of ten each. Immediately after hatching, the larvae were transferred on green pods of chickpea kept in the plastic container (5.0 × 5.5 cm). The fresh chickpea pods were provided daily as food in the morning till all the larvae pupated. To avoid cannibalism, second instars onward larvae were reared individually in the plastic container and foods were provided *ad libitum.* The adults emerged on a particular day were released in separate cages for egg laying. The numbers of eggs laid on subsequent days were recorded daily till the entire female died. The number of eggs laid per female was divided by two (sex ratio 1:1) to get the numbers of female birth (mx). Life table were constructed according to the method of Birch (1948), Howe (1953) and Atwal and Bains (1974).

The intrinsic rate of increase (rm), net reproductive rate (Ro) and mean generation time (Tc) were basic parameters used to assess the population growth. Stable age-distribution (per cent distribution of various stages) was also worked out by calculating the population schedule of birth rate and death rate (mx and lx) when grown-up in a limited space under laboratory condition. The life expectancy table was calculated by methods as those summarized by Southwood (1968), Atwal and Bains (1974), Sridhar and Prasad (1996), Baitha and Ram (2001), Shewale (2003) and Venkatesan *et al.* (2004). Corrected rm (intrinsic rate of natural increase was computed by trial and error method as depicted in Table 12.4.

The maximum duration of egg, larvae and pupa were 3, 8 and 15 days, respectively. The numbers that survived from egg to adult emergence was 78 individuals. It is evident from the table 1 that the immature stages lasted for 33 days. The pre-oviposition period ranged from 36th to 38th days of pivotal age. The first female mortality within the co-hort occurred on 43rd day (lx=0.76) after the emergence of adult and mortality increased subsequently. The female deposited first batch of egg on 39th day, and it was continued up to 48th days of pivotal age. The maximum mean progenies production (mx) per day was 70.33 females per female on the 42nd day of pivotal age, which declined to 19.24 (mx) on 48th day. The net reproductive rate (Ro), representing the total female birth in one generation was 332.43 (Table 12.1).

Thus, the population of *H. armigera* would be able to multiply 332.43 times per generation on chickpea pods under ambient temperature on laboratory condition in February-March (Table 12.1).

The data on intrinsic rate of natural increase (Rm) and generation time are portrayed in Table 12.2. It is apparent from the data documented in the table that the intrinsic rate of natural increase in number (Rm) was 13.54 females per female per day within finite rate of increase (λ) 1.1459 females/ female/day. The population would multiply 2.58 times every week. The mean generation time was 42.88 days. The hypothetical female population in F_2 generation was 110509.709 (Table 12.2). Bilapate *et al.* (1977) reported the rate multiplication of *H. armigera* on Lucerne at 26± 1°C with an infinitesimal rate (rm) of 0.1160 and finite rate (λ) of 1.12 per female/day. The net reproductive rate was of the tune of 185.59 with the mean length of generation 45.03 days. Similar studies were also carried out on pea where the valve of 'Ro' 'rm' and (λ) were reported to 407.39, 0.1346, 1.14, respectively with the mean generation time 44.65 days (Bilapate and Pawar, 1978). Bilapate *et al.* (1981) observed the value of intrinsic rate of increase to 0.1387 females/female/day net reproductive rate (RO) 600.95 and mean generation times 46.13 days on chickpea. More or less similar of generations were also recorded by Dhandapani and Balasubramanium (1979) on cotton against this pests life table in Laboratory; Bilapate *et al.* (1980) on safflower, Bilapate *et al.* (1983) on *tur*, Koshiya (1984) constructed life table of *H. armigera* on chickpea, wheat, cotton and groundnut who found the highest survival of immature stages on chickpea, followed by cotton, where as, it was relatively low on groundnut and wheat. The net reproductive rate was 586.67, 411.84, 318.38 and 355.39 on chickpea, cotton, groundnut and wheat, respectively. The in net capacity for increase in number and finite rate of increase were 0.1478 and 1.159 on chickpea, respectively.

Table 12.1: Life Table (for Females) and Age Specific Fecundity of *H. armigera* on Chickpea (*Cicer arietinum*)

Pivotal Age (in days) (x)	*Survival of ♀ at Different Age Interval (lx)*	*Age Schedule for ♀ Birth at Age x (mx)*	*lx.mx*	*x.lxmx*
0–35	0.78	–	–	Immature stages
36	0.78	–	0.78	28.08
37	0.78	–	0.78	28.86
38	0.78	–	0.78	29.64
39	0.78	19.86	15.49	604.11
40	0.78	58.45	45.59	1823.60
41	0.78	67.24	52.45	2150.45
42	0.78	70.33	54.86	2304.12
43	0.76	65.67	49.91	2146.13
44	0.74	56.23	41.67	1830.84
45	0.65	43.84	28.50	1282.5
46	0.57	34.27	19.53	890.38
47	0.52	25.95	13.49	634.03
48	0.45	19.24	8.66	415.68
49	0.39	–	–	–
			Σlxmx (Ro)=332.43	Σxlxmx= 14168.42

The contribution of different developmental stages towards the stable age, distribution was determined (Table 12.3). It was observed that on reaching the stable age distribution, the population of *H. armigera* in its various stages *viz.*, eggs, larvae, pupae and adult contributed to the extent at 43.24, 53.77, 3.74 and 0.42 per cent, respectively, which clearly indicate that immature stages contributed the maximum towards the stable age distribution. A clear insight of the table further illustrate that the population on reaching the stable distribution, the immature stages *viz.*, eggs and larvae contributed the highest, whereas it was the lowest by pupae and the adults.

Table 12.2: Mean Length of Generation, Innate Capacity for Increase in Number and Finite Rate of Increase in Number of *H. armigera* on Chickpea

Population Growth Statistics	*Formula*	*Calculated Value*
Net reproductive rate	Ro= Σlxmx	332.43
Mean length of generation (Tc)	Tc= Σxlxmx/Ro	42.62 days
Innate capacity for increase in number 'rm'	rm=Log eRo/Tc	0.1362 females/female/day
Arbitrary 'rm' (rc)	0.13 and 0.14	
Corrected 'rm'	$e^{7-rmx.lxmx}$ =1096.6	0.1354Females/female/day
Corrected generation time (T)	T=log eRo/rm	42.88 days
Finite rate of increase in number (λ)	λ=antilog e^{rm}	1.1459Females/Female/day
Weekly multiplication of population	$(F_2)^7$	2.58 times
Hypothetical F_2♀	$(Ro)^2$	110509.7049

On reaching the stage its, distribution population was 52.12, 42.17, 4.68 and 0.38 per cent for eggs, larvae, pupae and adult on chickpea, respectively (Bilapate *et al.*, 1981). Koshiya (1984) reported that *H. armigera* population on reaching the stable egg distribution comprised approximately 97 per cent of immature stages while working out on life table studies of *H. armigera* on pigeonpea *Cajanus cajan*. Bilapate *et al.* (1983) found that on reaching at stable age distribution the population comprised approximately 99 per cent of immature stages.

Table 12.3: Stable Age Distribution of *H. armigera* on Chickpea (rm= 0.1354)

Pivotal Age (in Days) (x)	*Life Table Age Distribution (L_x)*	$e^{-rm(x+1)}$ *Distribution*	$Lxe^{-rm(x+l)}$	*Per cent Contribution*
0	1.0	0.8734	0.8734	13.09324356
1	1.0	0.7628	0.7628	11.43522577
2	1.0	0.6662	0.6662	9.987083652
3	1.0	0.5818	0.5818	8.721833186
				Eggs 43.23738617
4	1.0	0.5081	0.5081	7.616987636
5	0.99	0.4438	0.4394	6.587097803
6	0.99	0.3876	0.3837	5.75209246
7	0.99	0.3385	0.3351	5.023524064
8	0.96	0.2956	0.2838	4.254479646
9	0.95	0.2582	0.2453	3.677321555
10	0.95	0.2255	0.2142	3.211097746
11	0.95	0.1970	0.1871	2.804838414
12	0.95	0.1720	0.1634	2.449548887
13	0.93	0.1502	0.1397	2.094259361
14	0.92	0.1312	0.1207	1.809428095
15	0.90	0.1156	0.1031	1.545584396
16	0.90	0.1001	0.09007	1.350250112

Contd...

Table 12.3–Contd...

Pivotal Age (in Days) (x)	*Life Table Age Distribution (L_x)*	$e^{-rm(x+1)}$ *Distribution*	$Lxe^{-rm\ (x+l)}$	*Per cent Contribution*
17	0.89	0.0874	0.0778	1.166309079
18	0.88	0.0763	0.0672	1.007403214
19	0.88	0.0667	0.0587	0.8799787
20	0.85	0.05823	0.0495	0.742060403
21	0.83	0.0509	0.0422	0.632625232
				Larvae 53.77119588
22	0.83	0.0444	0.0369	0.5531723
23	0.83	0.0388	0.0322	0.48271404
24	0.83	0.0339	0.0281	0.421250451
25	0.83	0.02959	0.0246	0.368781533
26	0.83	0.0258	0.0214	0.3208099502
27	0.82	0.0226	0.0185	0.277335706
28	0.82	0.0197	0.0162	0.242856132
29	0.81	0.0172	0.0139	0.208376557
30	0.81	0.0150	0.0122	0.182891655
31	0.81	0.0131	0.0106	0.158905864
32	0.80	0.0115	0.0092	0.137918297
33	0.79	0.0100	0.0079	0.118429842
34	0.78	0.00875	0.0068	0.101939611
35	0.78	0.0076	0.00596	0.08934707
36	0.78	0.00667	0.0052	0.07795382
				Pupae 3.74268282
37	0.76	0.00583	0.00455	0.068209592
38	0.78	0.00509	0.00397	0.059514743
39	0.78	0.00444	0.00347	0.056066786
40	0.78	0.00388	0.003028	0.045393109
41	0.78	0.00339	0.00264	0.039576554
42	0.78	0.00296	0.00231	0.034629485
43	0.76	0.00259	0.00197	0.029532504
44	0.74	0.00226	0.00167	0.025035169
45	0.65	0.00197	0.00128	0.019188632
46	0.57	0.00172	0.00098	0.014691296
47	0.52	0.001505	0.00078	0.011693073
48	0.45	0.001314	0.00059	0.00884476036
49	0.36	0.001148	0.000448	0.006716021429
			$\Sigma\ Lxe^{-rm\ (x+l)}$ 6.670616	Adults 0.419091711

Table 12.4: Trial and Error Method for the Calculation of Intrinsic Rate of Natural Increases (rm) of *H. armigera*

Pivotal Age (in Days) (x)	rm=0.13			rm=0.14	
	lxmx	e^{7-rmx}	$e^{7-rmx}.1xmx$	e^{-7-rmx}	$e^{7-rmx}.1xmx$
36	0.78	10.176	7.937	7.099	5.537
37	0.78	9.116	7.110	6.179	4.814
38	0.78	7.846	6.1199	5.366	4.185
39	15.49	6.8895	106.72	4.665	72.255
40	45.59	6.0496	275.80	4.055	184.88
41	52.45	5.3122	278.62	3.525	184.908
42	54.86	4.665	255.899	3.065	168.138
43	49.91	4.09596	204.43	2.664	132.98
44	41.57	3.5966	149.51	2.316	96.29
45	28.50	3.158	90.008	2.014	57.39
46	19.53	2.773	54.160	1.751	34.191
47	13.49	2.435	32.8498	1.522	20.53
48	8.66	2.138	18.517	1.323	11.46
49	–	–	–	–	–
			$\Sigma e^{7rmx.1xmx}$ 1487.68		$\Sigma e^{7rmx.1xmx}$ 977.56

Table 12.5: Life Table for Contributing Life Expectancy of *H. armigera* on Chickpea

Pivotal Age (in Day) x	Numbers Surviving to the Beginning of Age Intervals (lx)	Numbers Dying During x (dx)	Mortality Rate/100 Alive at Beginning of Age Interval (dx.100/1x) (100 qx)	Alive Between Age x and x+1 Lx	No. of individual Life Days Beyond x Tx	Expectancy of Further Life ex
0-5	100	1	1.00	99.5	739.0	14.78
5-10	99	4	4.04	97.0	639.5	12.92
10-15	95	5	5.26	92.5	542.5	11.42
15-20	90	5	5.55	87.5	450.0	10.00
20-25	85	2	2.35	84.0	362.5	8.53
25-30	83	2	2.41	82.0	278.5	6.71
30-35	81	3	3.70	79.5	196.5	4.85
35-40	78	0	0.00	78.0	117.0	3.00
40-45	78	22	28.21	39.0	39.0	1.00

While considering the life expectancy of *H. armigera* on chickpea, it was found that at the age of 30-35 days it registered 4.85 days as compared with 14.78 days for newly laid eggs in the present investigation (Table 12.5). Balapate *et al.* (1983) reported the life expectancy at the age of 30-35 days was 7.18 days as compared with 16.26 days for newly laid eggs.

Conclusion

Life table and Biotic Potential of *Helicoverpa armigera* (Hubner) on Chickpea pods was developed and biotic potential was worked out at ambient room temperature. Investigation on life fecundity tables and age specific distribution revealed that the expectancy of newly laid eggs was 332.43. The survival of immature stages was 0.39. The reproductive rate (Ro) of this pest was 332.43. The mean length of generation was 42.62 days. The intrinsic rate of natural increase in number (rm) was 0.1362 females/female days. The number that survived from egg to adults was 78 individuals. Under a limited space and sufficient food supply, the population of *H.armigera* increased with and infinitesimal rate (rm) of 0.1354 and finite rate (λ) 1.1459 females/female/day. A generation was completed in 42.62 days. The population was able to multiply 332.43 times/generation on chickpea pods. The population on reaching a stable age distribution comprised approximately 99 per cent of immature stage. The expectation of further life at the age interval of 34 to 49 days was reduced to 4.85 days from 14.78 days in the beginning. Age distribution of this pest on chickpea manifested that eggs and larvae contributed the highest to the population of stable age, whereas the contribution of pupa and adult was negligible.

References

Atwal, A.S. and Bains, S.S., 1974. *Applied Animal Ecology*. Kalyani Publisher, Ludhiana, pp. 11–35.

Baitha, Arun and Ram, A., 2001. Effect of temperature on *Trichogrammatoidea* sp. Nr. *ormigera* Nagaraja (Trichogrommatoidea: Hymenoptera). *Ann. Pl. Protec. Sci.*, 9: 134–135.

Bilapate, G.G., Pawar, V.M. and Raodeo, A.M., 1977. Growth of population of *Heliothis armigera* Hub. on lucerne. *Indian J. Agric. Sci.*, 47(11): 540–542.

Bilapate, G.G. and Pawar, V.M., 1978. Life tables for the gram pod borer, *Heliothis armigera* Hubner (Lepidoptera: Noctuidae) on pea. *Proc. Indian Acad. Sci., B.*, 87 (6): 119–127.

Bilapate, G.G., Raodeo, A.K. and Pawar, V.M., 1980. Investigation of *Heliothis armigera* Hubner in Marathawada–II. The rate of natural increae when reared on safflower. *Entomon.*, 5(4): 319–326.

Bilapate, G.G., Pawar, V.M. and Raodeo, A.M., 1981. Investigation on *Heliothis armigera* Hubner in Marathawada–III. Life table and intrinsic rate of increase on chickpea. *J. Maharashtra Agric. Univ.*, 6(1): 51–54.

Bilapate, G.G., Raodeo, A.K. and Pawar, V.M., 1983. Investigation on *Heliothis armigera* Hubner in Marathawada–I. The rate of increase of American cotton bollworm on pigeonpea. *Indian J. Ent.*, 45(3): 275–281.

Birch, L.C., 1948. The intrinsic rate of natural increase of an insect population. *J. Anim. Ecol.*, 17: 15–26 [cited by Atwal A.S. and Brains S.S. (1974). *Applied Animal Ecology* pp. 128–138].

Dhandapani, N., Abdul Karrem, A. and Jayraj, S., 1989. Life table studies of *S. litura* on the two variety of cotton. *J. Ent. Res.*, 13(1): 60–63.

Dhandapani, N. and Balasubramanian, M., 1979. Life tables for the gram pod borer, *Heliothis armigera* Hubner (Lepidoptera: Noctuidae) on cotton. *J. Ent. Res.*, 3(2): 186–190.

Howe, R.W., 1953. The rapid determination of intrinaic rate of increase of an insect population. *Ann. Appl. Biol.*, 40: 134–135.

Koshiya, D.J., 1984. Studies on biometrical analysis growth and life tables of *Heliothis armigera* Hubner (Lepidoptera: Noctuidae) on different, hosts. *Ph.D. Thesis*, Gujarat Agricultural University, Sardar Krushinagar (Unpublished).

Shewale, B.S., 2003. Studies on life fecundity table of pomegranate butterfly. *Ann. Pl. Protec. Sci.*, 11: 255–257.

Shridhar, V. and Prasad, V.D., 1996. Life table studies on natural population of *S. litura* on groundnut *Arachis hypogaea*. *Annal. Pl. Pro. Sci.*, 4(2): 142–147.

Singh, Raghvendra and Samshad, Ali, 2005. Efficacy of Biopesticides in the management of *Helicoverpa ornigera* (Hub.) in chickpea. *Ann. Pl. Protec. Sci.*, 13(1): 94–96.

Southwood, T.R.E., 1968. *Ecological Method*. Mathewn and Co. Ltd., London, pp. 391.

Venkatesan, T., Jalani, S.K., Srinivasa Murthy, K., Rabindra, R.J. and Rao, N.S., 2004. Comparative life table studies of *Goniozus nephantidis* (Muesbeck) (Hymenoptera: Bethylidae) on three lepidopteran insect pests. *Ann. Pl. Protec. Sci.*, 12(1): 5–8.

Wellick, M.J. and Pedigo, L.P., 1978. The innate capacity for increase and ovipositional pattern of green clover worm. *Environ. Entomol.*, 7: 171–177.

Insect Pest and Disease Management *Pages* **179–191**
Editor: **Dr. D. Prasad**
Published by: **DAYA PUBLISHING HOUSE (2016)**

Chapter 13

Insect Pathogens and Pest Management

R.K. Murali Baskaran, D.S. Rajavel and K. Suresh
Department of Agricultural Entomology,
Agricultural College and Research Institute,
Tamil Nadu Agricultural University, Madurai – 625 104, T.N.

Introduction

Of late, frequent application and large scale use of insecticides for control of pests led to the endangerment of agro eco-system. Indiscriminate use of insecticides resulted in the destruction of parasitoids and predators of the pests and ultimately led to the resistance of pests to insecticides and insect resurgence. In the light of these problems considerable research has been devoted to the elucidation of the toxic residues in the consumable produce and subtle effects of insecticides on arthropods. Considering the seriousness of insecticidal toxic residues and their impact on human beings, there is an urgent need for developing an effective, economically viable and environmentally safe eco-friendly pest management system. Among the various components involved, biocides like bio-control agents especially microbials contribute substantially to compete with the pests and are considered as most important eco-friendly components in the Integrated Pest Management.

Bio-intensive pest management (BIPM) is one of the recent advances which can clearly demonstrate the scope for biological control in agricultural IPM systems. The successful use of several entomophages and entomopathogens in the control of pests has projected biological control as a promising alternative to ecologically disruptive pest control measures (Jayaraj and Rabindra, 1992). Insect pathogens, virus, bacteria, fungus and protozoans are considered as one of the alternatives for the suppression of key pests attacking several agricultural and horticultural crops. They are severally called as 'Bio-rational insecticides', 'Microbial insecticides' and 'Living insecticides'. Bio-rational insecticides are those, which are naturally occurring or products derived from living organisms or if synthetic, it should resemble a natural product, and their mode of action should be other than direct toxicity.

Microbials

Viruses

Viruses from more than 800 species of insects have so far been isolated and described in the World. Virus in the family Baculoviridae are the best known for all the insect viruses. The specificity accompanied by biological activity and stability have been the main reasons for which they are considered as potential pesticidal agents in biological control (Pawar and Thombre, 1992). They are environmentally safe, species specificity, consequently they are safe to all non-target organisms including humans, animals, plants and even beneficial insects (Miller, 1998). Virus based pesticides are compatible with other methods of pest control including chemical insecticides, resistant host-plants, cultural control practices and parasitoids and predators. Individual population of insects do not acquire resistance, readily up on continual virus pressure. Baculovirus have been recorded from various insect orders like Lepidoptera, Hymenoptera, Diptera and Coleoptera. In India, about 35 insect-virus relationship from the Baculovirus groups have been recorded. Nucleo polyhedron viruses isolated from cotton American boll worm, *Helicoverpa armigera*, tobacco caterpillar, *Spodoptera litura* and groundnut redhairy caterpillar, *Amsacta albistriga* are potent and promising and are being used for bio-suppression of these three insects under field condition in India. However, their relatively slow action, specificity to a single insect pest, high level of UV inactivation as well as glandular secretion of leaves and poor storage capability limit their efficacy and large scale commercial production and use. Efficacy of NPV can be improved through:

1. Identification of virulent desirable strains
2. Suitable UV screen or sunlight protectant and adjuvants
3. Techniques to improve laboratory yield of NPV
4. Suitable formulations
5. Integration with other agents
6. Genetic engineering.

Selection of Desirable Strains

Viruses from members of the same host species from different geographical locations show variability in virulence and biological characteristics (Brown, 1982). Field isolates of NPV collected in diverse geographic locations differ in their efficacy, reported on several pests including *Amsacta albistriga* (Rabindra, 1999; Murali Baskaran, *et al.*, 2001), *Helicoverpa armigera* (Battu and Arora, 1996; Ramesh Arora *et al.*, 1997) *etc.* Rabindra (1992) reported that an isolate of NPV of *H. armigera* obtained from the Nilgris recorded the lowest LC_{50} than those collected from Coimbatore, Aruppukottai and Gujarat. It was also reported that there was no difference in restriction pattern of isolates of NPV of *A. albistriga*, collected from Coimbatore, Madurai, Kancheepuram and Hyderabad. However isolate of NPV of *A. albistriga* collected from Dharwad was found to be 5.6 times more virulent than Madurai, Narikudi and Pavagada isolates with reduced LT_{50} (Murali Baskaran, *et al.*, 2001) (Table 13.1).

Studies are in progress in the Tamil Nadu Agricultural University, Coimbatore to identify suitable cell lines for laboratory production of NPV and NPV with cross infectivity (Anonymous, 2005). The cross infectivity of *Galleria mellonela* NPV is reported on *H. armigera* and *S. litura* (PDBC, Bangalore).

Table 13.1: LC_{50} and LT_{50} of Four Isolates of *AaNPV*

*Isolate**	*LC_{50} (POBs/ml)*	*LT_{50} (h)*
Madurai	2,66,072	163.44
Dharwad	48,621	122.16
Pavagada	2,71,818	189.84
Narikudi	2,68,583	167.52

*Treatment involving NPV contained $1x10^{2-8}$ POBs/ml.

UV Screens for NPV

In the past 25 years, many natural and synthetic organic chemicals were evaluated as sunlight protectants for entomopathogens such as viruses (Podgwaite and Shapiro, 1986), bacteria (Krieg *et al.*, 1980), protozoa Teetor and Kramer, 1977) and nematodes (Gaugler and Bousch, 1979). The most successful materials were aromatics such as para-aminobenzoic acid (Gaugler and Bousch, 1979), 2-hydroxy 4-methoxy-benzophenone (Krieg *et al.*, 1980), 2-phenylbenzimidazole 5-sulfonic acid (Shapiro *et al.*, 1983) and folic acid (Shapiro, 1985) and boric acid (Shapiro and Bell, 1982), Tinopal DCS (Martignoni and Iwai, 1985), chitinase (Shapiro *et al.*, 1987) and congo red (Shapiro, 1989). Among the environmental factors, sunlight especially ultraviolet portion (250 nm to 400 nm) has major impact on the inactivation of NPV (Yendol and Hamlen, 1973; Jaques, 1977). The success of sunscreens or sun blocks was accounted to good absorption in the UV-B region (280 to 310 nm) (Jaques, 1938) rather than absorption in the UV-A region (320 to 400 nm) (Shapiro, 1985; Harms *et al.*, 1986). pH of the NPV + brighteners suspension between 3.5 and 8.7 was safe to NPV.

Optical brighteners were discovered more than 50 years ago (Eggert and Wendt, 1939) and are widely used in the detergent, paper, plastics, fabric softeners and organic coatings industries (Lanter, 1966) and as fluovochromes for microorganisms (Silfkin and Cumbie, 1988). Success of these chemicals was due to good absorption of ultraviolet radiation and transmit light in the blue portion of the visible spectrum (Jaques, 1968). During the past decade, many scientists demonstrated several optical brighteners as successful UV protectants including Douglas-fir tussock moth NPV (Martignoni and Iwai, 1985), fall army worm NPV (Hamm and Shapiro and Dougherty, 1994) etc. Effect of optical brighteners in preserving NPV was concentration dependent. Addition of selected optical brighteners, Ranipal BVN and 2B at a range of 0.001 to 1.00 per cent was able to provide excellent protection (80 per cent original activity remaining) to *S. litura* NPV (Murali Baskaran *et al.*, 1997) and *A. albistriga* (Murali Baskaran *et al.*, 1998) and also increased the virulence of NPV by reducing the LC_{50} and LT_{50} (Table 13.2).

Table 13.2: Per cent OAR of *AaNPV* Suspended in Different Optical Brighteners

Treatment	*% OAR at 8 Day P. I.**			*Mean*
	0.1 per cent	*0.5 per cent*	*1.0 per cent*	
NPV + H_2O (No UV)	93.3	95.5	95.5	94.8
NPV + H_2O (60 UV)	4.6	6.9	7.0	6.2
NPV + Ranipal BVN	90.8	97.6	97.6	95.3
NPV + Ranipal 2B	83.4	95.3	95.4	91.4

* Original activity remaining (OAR) was based up on virus induced mortality before and after UV.

Several brighteners are known to interfere with cellulose (Itoh, *et al.*, 1984) and fibrillogenesis (Herth, 1980). In insects, the peritrophic membrane serves as a barrier for the invasion of micro-organisms including insect viruses (Brandt *et al.*, 1978). Selected optical brighteners may inhibit or alter the peritrophic membrane creating in the lining through which NPV could pass easily.

Some of the 'B' vitamins, P-amino benzoic acid (PABA) (Gaugler and Bousch, 1979) and folic acid and riboflavin (Shapiro, 1985) had been shown to act as excellent ultraviolet protectants for entomogenous nematodes and insect virus, respectively which were attributed to good ultraviolet absorbance at 280-320 nm, most responsible for inactivation (Jaques, 1997). PABA is an effective sunscreen and acted as a UV protectant for insect viruses (Shapiro *et al.*, 1983). Addition of folic acid, pyridoxine and riboflavin provided protection of *S. litura* NPV against the far and near ultraviolet light. Riboflavin and folic acid at one per cent protected the virus by actors of 4.9 and 3.6, respectively (Ramakrishnan and Chaudhari, 1991). Among eight vitamins tested, riboflavin at 1.0 per cent was reported to provide superior protection (51-80 per cent OAR) to *S. litura* NPV (Murali Baskaran *et al.*, 2000). Interestingly, there was a positive correlation between effectiveness and molecular weight of vitamins, which indicated that vitamins with higher molecular weight could perform as good UV protectant to NPV.

Certain dyes and indicators displayed effective protection against sunlight inactivation of entomopathogens (Shapiro, 1989; Shapiro and Robertson, 1992). Congo red in aqueous solution absorbed light intensely in the visible portion of the solar spectrum (absorbance max 496 nm) and the ultraviolet (Gurr, 1971). Congo red absorbs UV over a wide spectrum (320 to 400 nm). Congo red 1.0 per cent has been reported to protect NPV of *A. albistriga* from UV rays and retained 83.3 per cent of viral activity even after exposure to UV for 60 minutes which also decreased the incubation period of NPV (Murali Baskaran *et al.*, 1998). Use of 'Shade' and lignin sulphate as ultraviolet screens resulted in good persistence of *Heliothis* NPV on cotton, soybean and tomato through 48 h (Young and Yearian,1974). Addition of Robin blue and Tinopal at 1 per cent increased the persistence of the *Heliothis* NPV by the higher percentage of original activity remaining (Rabindra and Jayaraj, 1989; Rabindra *et al.*, 1989).

Adjuvants for NPV

Adjuvants may be organic or inorganic substances, which are added ot NPV to enhance its activity (Murali Baskaran *et al.*, 1999). Rabindra and Jayaraj (1988) tested various adjuvants and found that crude sugar 0.5 per cent, cotton seed kernel powder 1 per cent, groundnut oilcake 3 per cent and chickpea flour 1 per cent acting as phagostimulants increased the efficacy of NPV. Laboratory experiments have revealed that 4 per cent larval extract of *H. armigera, S. litura* and *Corcyra cephalonica,* 20 per cent whole milk, 10 per cent whole egg homogenate, 10 per cent yellow of egg, 10 per cent egg white, 20 per cent tender coconut water and 20 per cent crude sugar increased the efficacy of NPV of *H. armigera* (Rabindra and Jayaraj, 1988). Results of the field experiments on cotton revealed that four rounds of ULV application of NPV at 450 LE (larval equivalent)/ha with endosulfan 350 g a.i./ha with adjuvants like larval extract of *H. armigera* 4 per cent or whole milk 10 per cent + crude sugar 15 per cent or cotton seed kernel powder 2.5 per cent + crude sugar 17.5 per cent were as effective as endosulfan 700 g a. i./ha in reducing the larval population of *H. armigera* and damage caused to flowers, squares and bolls. NPV applied with adjuvants significantly increased the seed cotton yield (Dhandapani *et al.*, 1987).

Application NPV 450 LE + endosulfan 500 ml/ha with the adjuvants jaggery 15 per cent and milk 10 per cent or larval extract of *H. armigera* 4 per cent as ULV spray significantly reduced the damage to squares, flowers and bolls. The larval population was also significantly reduced after three rounds of treatment (Dhandapani *et al.*, 1987). Results of the field experiments with different NPV dosages and spraying system revealed that the larval population and damage in cotton leaf, square, flower and bolls were lower and seed cotton yield was higher in plots applied with NPV at 450 LE + endosulfan 350 g a. i./ha with the adjuvants crude sugar + whole milk or crude sugar + *H. armigera* larval extract or crude sugar + cotton seed kernel powder as high volume or ULV application (Dhandpani, 1990).

Lab and field experiments demonstrated the usefulness of several phagostimulants and UV protectant adjuvants like cotton seed kernel extract and crude sugar in increasing the efficacy of HaNPV. The possibility of incorporating these adjuvants in the basic formulation was shown (Muthuswamy *et al.*, 1995). Results of field experiments on the control of *H. armigera* on winter and summer cotton revealed that *H. armiger* could be successfully controlled by CDA of NPV along with the cotton seed kernel extract based adjuvant mix. Seed cotton yield were also increased significantly (Rabindra and Jayaraj, 1996).

Application of urea through foliage along with insecticide is a common practice in fiber crops, which increased the yield by potentiating the activity of insecticides (Raman *et al.*, 1975a; Raman *et al.*, 1975b; Raju *et al.*, 1982). Joint effect of fertilizer with NPV was studied by Chaudhari (1987) who reported that combination of *Diacrisia obliqua* NPV with fertilizer, urea, muriate of potash and superphosphate against Bihar hairy caterpillar resulted in increased mortalities of 100.0, 91.9 and 70.0 per cent, respectively as compared to 59.9 per cent in NPV alone. In all these combinations, reduced LT_{50} values between 135 and 144 h was observed. No death due to fertilizers at tested concentrations was observed. Addition of urea 1 per cent potentiated the *S. litura* NPV by reducing the LT_{50} by 1.29 folds to 107.04 h (Murali Baskaran, 1955). The adjuvant mix, NPV + urea 0.5 per cent + crude sugar 0.5 per cent + Ranipal BVN 0.1 per cent was reported to enhance the virulence of *A. albistriga* NPV, resulting 96.7, 93.3 and 81.7 per cent mortality to II, III and IV instar, respectively while NPV alone caused 40.0, 41.7 and 40.0 per cent mortality at 8 days post inoculation (Table 13.3).

Table 13.3: Influence of Adjuvants on the Mortality of *A. albistriga* Larvae Due to AaNPV

Treatment	*% Mortality at 8 Day P. I.*	*LT_{50} (h)*
NPV + urea 0.5 per cent + Ranipal BVN 0.1 per cent + crude sugar 0.5 per cent	81.7	142.08
NPV + boric acid 0.5 per cent + Ranipal BVN 0.1 per cent + crude sugar 0.5 per cent	70.0	174.96
NPV alone (1×10^4 POBs/ml)	40.0	240.00

Infectivity and *in vivo* Production of NPV

Influence of Host Plants

Susceptibility of insects to microbial pathogens vary according to the type of host plants, pH and constituents of the leaf tissues which either alter acquisition rate of pathogens or midgut pH of insects (Benz, 1987). Infecting *S. litura* with NPV through castor leaves increased the acquisition of polyhedra, which in turn caused the highest virus induced mortality (Mahadevan, 1978). The pH of leaf materials or extracts altered the activity of NPV either through additive or antagonistic effect (Kunimi and Aruga, 1974). Host plant foliar constituents and larval midgut conditions were responsible for the virus induced mortality of many lepidopteran insects (Keating, *et al.*, 1990; Forschler, *et al.*, 1992). The mortality of *A. albistriga* larvae fed on calotropis, castor and gingelly was maximum in the order of 91.1, 86.7 and 82.2 per cent, respectively, followed by groundnut (80.0 per cent), cotton (77.8 per cent), soybean (77.8 per cent) and cowpea (77.8 per cent) while it was least on bajra (57.8 per cent). Fourth instar of *A. albistriga* infected through cotton and calotropis gained maximum weight which in trun yielded more polyhedra. High preference of *A. albistriga* to NPV treated calotropis and cotton might have increased the yield of NPV. Cotton leaf has been reported to have tannin content which in turn strongly influences both POBs distribution rates and virion survival in many lepidopteran pests (Gudauaskas and Cannerday, 1968; Murali Baskaran *et al.*1996).

NPV + Juvenile Hormone

Juvenile hormone analogues and plant extracts with juvenilizing effect have been reported to interfere with the endogenous juvenile hormone synthesis of insects and increased larval weight and supernumerary instars (Staal, 1975). Methoprene altered insect susceptibility to NPV (Ben-Shaked and Harpaz, 1996; Mohamed *et al.*, 1984) and increased the yield of NPV *in vivo* (Im *et al.*, 1989; Hirsch, *et al.*, 1995). Similarly, Gunasekaran and Chelliah (1985) and Murali Baskaran *et al.* (1987) reported that certain plant extracts like *Tribulus terrestris* L. produced supernumerary larval of *S. litura* and *Heliothis armiger* (Hubner) by increasing the larval biomass. Administration of *Aa*NPV along with *T. terrestris* 10 per cent or either 24 h later or earlier to fourth instar of *A. albistriga* significantly increased the yield of polyhedra (Table 13.4). Im *et al.* (1989) and Hirsch *et al.* (1995) observed increased yield of NPV *in vivo* through methoprene in *S. litura* and *L. dispar* by 15 and 19 per cent, respectively. Treatment of *T. terrestris* simultaneously and 24 h earlier to NPV triggered the larvae resulting in increase in size of larva and yield of NPV, but 24 h later to NPV treatment could not influence much as *T. terrestris* did not find binding place at cellular receptors of insect midgut (Riddiford, 1995).

Table 13.4: Yield of *Aa*NPV as Influenced by *T. terrestris*

Treatment	POBs/larva (x 10^9)	POBs/g of larva (x 10^9)
NPV + *T. terrestris* 10 per cent simultaneously	3.272	5.260
T. terrestris 10 per cent + NPV 24 ha later	3.396	5.524
NPV + *T. terrestris* 10 per cent 24 h later	3.285	5.280
NPV + *T. terrestris* 10 per cent 24 h later	2.053	3.300

Interaction with Botanicals

There are several reports to potentiate the NPV with botanical pesticides in which neem products play vital role. Nicotine sulphate and pyrethrum were found to produce additive to potentiation effect on the NPV in *S. litura* larvae (Chaudhari and Ramakrishnan, 1983). While screening botanicals for increasing the efficacy of *H. armigera* NPV, Rabindra and Jayaraj (unpublished) found that aqueous extracts of *Vitex negundo* and *Prosopis julifera* enhanced the larval mortality of *H. armigera.* Muthiah (1988) reported that ULV formulation of NPV with one per cent neem seed kernel extract or neem seed oils was more effective against second instar larvae of *H. armigera* than ULV formulation individually. Devaprasad (1989) found that the addition of methanolic fractions of *Ocimum sanctum* and *Acorus calamus* at 0.01 per cent increased the efficacy of NPV against third instar larvae of *H. armigera* and *S. litura*. Combination of *H. armigera* NPV with 10 per cent aqueous extract of leaves of *V. negundo, P. julifera, Tagetes patula* and *Argemone mexicana* did not increase the efficacy of the virus.

The combination of one per cent of neem oil and NSKE and neem cake extract 5 per cent with *S. litura* NPV (2 x 10^4 POBs/ml) enhanced the virus induced mortality from 31.1 to 81.5, 31.1 to 81.5 and 31.1 to 57.6 per cent, respectively (Murali Baskaran *et al.*, 1999). LC_{50} and LT_{50} of *S. litura* also got reduced. *V. negundo* 10 per cent had no additive effect with NPV while *P. julifera* and *Ipomoea carnea* had antagonistic effect with *S. litura* NPV. Muthuswamy *et al.* (1993) stated *V. negundo* alone could not increase the efficacy of NPV however, the efficacy of NPV was increased when *V. negundo* was used along with French bean extract based adjuvant mix. Compounds of *V. negundo* like terpenes, cinole, 1 sabinene and sesquiterpenes (Manalo, 1982) might have weakened and predisposed the larvae of *S. litura* to action of NPV.

Neem products with NPV reduced the leaf consumed by larva, which in turn affected the acquisition of virus and further infection. Though larvae fed on neem products acquired relatively low quantity of virus, presence of antifeedant effect of neem products, such larvae got pre-disposed to infection by starvation. Use of neem products as adjuvant is a promising approach because larval feeding and subsequent defoliation would be reduced greatly without interference with larval activity.

Integration with *Bacillus thuringiensis*

The bacterium, *B. thuringiensis* ssp. kurstaki is used widely to control insect pests belonging to Lepidoptera, Coleoptera and Diptera. Muthuswamy *et al.* (1993) emphasized the possibility of integrating *B. t.* with virus mixture on groundnut and cotton so that other Lepidopterous pests like the leaf miner, *Aproerema modicella* may also be successfully managed along with *H. armigera* and *S. litura.*

Interaction with Chemical Insecticides

Viruses have been found to be compatible with chemical insecticides and a few agrochemicals (Benz, 1971; Shapiro and Bell, 1982; Chaudhari and Ramakrishnan, 1983; Mohamed *et al.*, 1983). Ignoffo and Montoya (1966) stated that methyl parathion decreased the activity of the virus while endrin, DDT, toxaphene and carbaryl had no effect on the activity of virus. Komolpith and Ramakrishnan (1978) studied the joint action of NPV of *S. litura* with insecticides and found potentiation with pyrethrin and antagonism with malathion. Chaudhari and Ramakrishnan (1980) found that the effectiveness of the NPV of *S. litura* on cauliflower was improved when sub-lethal doses of DDT and endosulfan were added. Effectiveness of the combination spray of NPV and methamidaphos in the reduction f field population of *S. litura* on cotton was reported by Jayaraj and Santharam (1985). NPV induced susceptibility of larvae to insecticides may be possible reason for the superior control achieved by the application of *H. armigera* NPV 250 LE/ha followed by endosulfan 262 g a. i./ha three and five days later on sunflower (Rabindra *et al.*, 1991). A combination of *H. armigera* NPV 250 LE/ha + endosulfan 175 g a. i./ha effectively controlled *H. armigera* larvae and increased the yield of chickpea (Rabindra and Jayaraj, 1987; Rabindra and Jayaraj, 1988). Murali Baskaran (2002) also reported the compatibility of endosulfan 0.035 per cent and NPV of the management of *A. albistriga* both under field and laboratory conditions.

Interaction with IGR

Diafenthiuron (Polo) 0.05 per cent has been reported to be compatible with *S. litura* NPV, resulting 33.3 per cent mortality at day two and cent per cent at five day after treatment, in combination with NPV (Murali Baskaran, 1995). Reduction in leaf consumption by *S. litura* on IGRs as well as IGRs + NPV treated leaves indicted that IGRs had antifeedant and deterrent effect as reported by Rao (1982), Sundaramurthy (1977), Natesan (1979), Natarajan (1982) and Aguilane (1986) which weakened and predisposed the larvae to NPV.

Interaction with Other Bio-control Agents

Dhandapani *et al.* (1991) reported that three releases of biocontrol agents, *Trichogramma chilonis* (1,00,000/ha/release) and *Brinkochrysa scelestes* (50,000/ha/release) on 90, 105 and 120 days after sowing and two rounds of application of *Spodoptera* NPV at 250 LE/ha with adjuvants effectively checked the population of *H. armigera* and *S. litura* on cotton which was equal to that of insecticides. NPV in combination with a sequential release of *T. chilonis* and *B. scelestes* was superior in reducing the larval population of *H. armigera* on cotton (Sathiah, 1991) and *S. litura* on groundnut (Kalyanasundaram *et al.*, 1991). Manjunath (1993) managed *H. armigera* on cotton in Andhra Pradesh through the use of sex pheromone trap, *Trichogramma* spp., *C. carnea* and NPV. The use of biocontrol agents, *T. chilonis*, *C. carnea* and NPV individually and in different combinations was as effective as untreated check in controlling *S. litura* on cotton. NPV 250 LE/ha or release of either *T. chilonis* 1,25,000 parasitoid/ha or *C. carnea* 1,00,000 eggs/ha followed by NPV treatment was effective and comparable with endosulfan 0.07 per cent. Release of either *T. chilonis* or *C. carnea* separately was less effective (Murali Baskaran, 1995). Significant reduction in larval population and leaf damage was observed whenever the release of either *T. chilonis* or *C. carnea* followed by NPV. The larvae hatched out from unparasitized or unpredated eggs could be infected by NPV as suggested by Dhandapani *et al.* (1991) on cotton and Kalyanasundaram *et al.* (1991) on groundnut. Two releases of *Trichogramma chilonis* (1 lakh/ha) followed by two rounds of application of *Helicoverpa* NPV (500 LE/ha) recorded less mean larval population of *H. armigera* and damage on fruiting parts of rainfed cotton (Balakrishnan *et al.*, 2004). Similarly, lowest mean larval population of *H. armigera* and highest yield was recorded in the rainfed cotton received with two releases of *Chrysoperla carnea* (50,000 eggs/ha) followed by two rounds of application of *Ha*NPV (500 LE/ha) (Balakrishnan *et al.*, 2004).

NPV Formulation

In India, the NPVs of *S. litura* (Chaudhari and Ramakrishnan, 1980; Dhandapani and Kumaraswami, 1982; Ethiraju, 1986) and *H. armigera* (Ethiraju, 1986; Rabindra *et al.*, 1988, Muthiah, 1988) have been formulated as basic formulations like dust, wettable powders and ultra low volume products. Under field conditions on chickpea against *H. armigera*, Ethiraju (1986) reported that application of NPV formulations, dust and wettable powder were as effective as the ULV application @ 250 LE/ha. The persistence of virus was notice in the field even up to seventeen days. Rabindra *et al.* (1988) and Muthiah (1988) found that ULV formulation of NPV was as effective as the unformulated virus against *H. armigera* on chickpea. Among the ULV formulations NPV-ULV (Lactose) had the lowest LT_{50} value (54.11 h) and all the ULV formulations were as effective as the unformulated virus (Muthiah, 1988). In another laboratory experiment, the ULV formulation with one per cent neem seed kernel extract or neem oil was more effective against second instar larvae of *H. armigera* than ULV formulation individually.

Pot culture studies on the efficacy of NPV formulations against second instar larvae of *H. armigera* on sunflower showed 78 per cent mortality on plants sprayed with wettable powder (Dedenol) formulation. Dust formulation was not effective (Ethirajau, 1986). ULV formulation of NPV (125 LE/ha) in combination with endosulfan (175 g a. i./ha) was effective against *H. armigera* on pigeonpea as ULV application of endosulfan 350 g a. i./ha. Addition of boric acid to ULV formulation of NPV had not increased the mortality. The flower and pod damage in NPV treatments were on a par with ULV application of endosulfan (Muthiah, 1988). In laboratory and field trials, the wettable powder formulation of *S. litura* NPV at a higher dose of 450 LE/ha was able to prevent the damage by the larvae on groundnut.

It was found that the wettable powder formulation of *H. armigera* NPV was not very stable under room temperature. Within three months, the wettable powder formulation involving acetone lost 31-58 per cent of its activity. When water was used in the process of formulation the resultant product was found to be slightly more stable with only 17-25 per cent loss of original activity. The loss of virulence upon storage was found to be lesser in the case of *S. litura* NPV formulation (involving water) with about 16-29 per cent loss of original activity after six months (Ethiraju, 1986). Even after seven months of storage the NPV-ULV formulation retained 86.2 per cent of the original activity. However, the LT_{50} values progressively increased indicating loss of virulence (Muthiah, 1988).

Field Efficacy of NPV

Among the several alternative methods of pest management tried for *H. armigera*, entomopathogens like nucleo polyhedron virus is the most promising and its efficacy has been tested against the pest on a number of crops (Rabindra and Jayaraj, 1990). The use of NPV for the control of *Heliothis* spp. has been well documented (Narayanan, 1979; Jayaraj, 1989). Several attempts have been made to control the field populations of *H. armigera* on different crops with the NPV in Tamil Nadu and elsewhere in India (Jayaraj, 1989). The virus at the rate of 1.5 x 1012 POBs/ha sprayed three to four times in the evening hours at an interval of 4-10 days, when the pest was in the very early stages, effectively controlled the pest on chickpea (Narayanan, 1979; Rabindra and Jayaraj, 1988a, 1988b), cotton (Dhandpani *et al.*, 1987), pigeonpea (Santharam *et al.*, 1981; Sanap and Deshmukh, 1988), tomato (Mistry *et al.*, 1984), sunflower (Rabindra *et al.*, 1985) and lablab bean (Jayaraj *et al.*, 1987).

NPV + endosulfan was more effective than NPV alone in the control of *H. armigera* (Jayaraj *et al.*, 1985). Dhandapani *et al.* (1987) reported that four rounds of ULV application of the virus at 450 LE/ha along with endosulfan 350 g a. i./ha with adjuvants like larval extract of *H. armigera* 4 per cent or whole milk 10 per cent + crude sugar 15 per cent or cotton seed kernel powder 2.5 per cent + crude sugar 17.5 per cent were as effective as endosulfan 700 g a. i./ha. Boric acid was known to increase the efficacy of NPV of *S. litura* (Rao *et al.*, 1987). Efficacy of combination of NPV with insecticide in reducing larval population of *S. litura* in other crops has been reported earlier (Ramakrishnan, 1976; Jayaraj *et al.*, 1980; Ramakrishnan *et al.*, 1981). The effectiveness of chlorpyriphos and fenitrothion combination against *S. litura* on cotton was reported by Jayaraj *et al.* (1980). Rabindra and Jayaraj (1990) demonstrated the increased susceptibility of virus infected final instar of *H. armigera* and *S. litura* to fenvalerate, cypermethrin, endosulfan and monocrotophos and the enhanced susceptibility was maximum in cypermethrin followed by endosulfan. Two rounds of application of NPV 125 LE + endosuflan 0.035 per cent/ha + Ranipal 1.0 per cent + crude sugar 0.5 per cent was able to suppress the population of *S. litura*, appreciably on cotton (Murali Baskaran, 1995). A tank mix formulation, NPV + urea 0.5 per cent + Ranipal 2B 0.1 per cent + crude sugar 0.5 per cent has been developed for the management of *A. albistriga* (Murali Baskaran and Muthulakshmi, 2005).

Compatibility of NPV with IPM Modules

NPV of *H. armigera* and *S. litura* are considered as one of the components in several IPM modules like Dharwad, Surat, TNAU, Akola and Guntur developed in India for the management of early season sucking pests and boll worm in rainfed cotton. Among the modules tested, one round of application of NPV of *H. armigera* and *S. litura* during 80 DAS along with 27 components in TNAU module recorded the lowest population of early season sucking pests and bollworms with highest cost benefit ratio of 1:2.20 (Balakrishnan *et al.*, 2005). This was possible because of inclusion of intercrops and border crop in TNAU module while other modules were suggested without intercrops or border crops.

Genetically Engineered Viral Pesticides

Improvement of biological pesticides through genetic modification has enormous potential and the insect baculoviruses are particularly amenable to this approach. A key aim of genetic engineering is to increase their speed of kill. The innovations of modern recombinant DNA techniques, knowledge of molecular biology and genetics of baculoviruses have accelerated a new phase of research *i.e.* to modify the Baculovirus so as to develop a potent effective viral pesticide. Using this approach, Carbonell *et al.* (1988) successfully demonstrated genetically modified viral pesticide by introducing insect toxins from the scorpion, *Buthus eupeus*. Subsequently, Merryweather *et al.* (1990) introduced the *B. t.* kurstaki toxic gene into *Ac*MNPV genome and showed biologically active pro-toxin expression in recombinant virus infected larvae. Maeda (1989) reported that the recombinant virus with diuretic hormone under the polyhedron gene promoter showed larval mortality one day ahead of their wild type virus infection. The first field trial with improved recombinant NPV was carried out in UK during 1993 (Cory *et al.*, 1994). This genetically engineered virus contained an insect specific neurotoxin gene from the scorpion.

The pesticidal efficiency of baculoviruses could be enhanced by inactivation or deletion of growth regulating genes present in Baculovirus genome (O'Reilly and Miller, 1991). Several NPV genes such as Ecdysteriod UDP-Glucosyl transferase (egt) (O'Reilly and Miller, 1989), p74 (Kuzio *et al.*, 1989) and Enhancin (Granados and Gorsaro, 1990), have been identified as potential candidate genes to enhance the pesticidal activity of baculoviruses. All these genes are known to interfere in the normal host developmental process. Enhancin increases the efficacy of virus infection by disrupting the integrity of peritrophic membrane of the host, which acts as a barrier for viral infection (Derksen and Granados, 1988; Gallo *et al.*, 1991). p74 has been identified as an anti-apoptosis gene, which reduces virus, induced cell death (Clem *et al.*, 1991). EGT lengthens the life of the infected larvae by inactivating the moulting hormone. Inactivation or elimination of these types of genes may enhance the pesticidal efficacy or may result in cessation of feeding prematurely.

Bacteria

The important bacterial pathogens of insects belong to Bacillaceae and Enterobacteraceae. Out of 100 species of pathogenic bacteria described from insects, *Bacillus thruingiensis* has been found to be the most successful one. It was isolated by Berliner in 1915 in Germany from the diseased larvae of the Mediterranean flour moth, *Ephestia kuhniella*.

B. thuringiensis is a gram-positive spore forming bacteria. Each bacterial cell forms a spore at one end and a crystal at the other end. At the time of sporulation, they produce few insecticidal toxins, alpha-exotoxin (heat labile), beta-exotoxin (heat stable) and delta-endotoxin (crystalline toxin or crystal).

Mode of Action

B. thuringiensis produces a protenaceious parasporal crystalline inclusion during sporulation. Upon ingestion by the insects, this crystalline inclusion is solubilized in the midgut, releasing proteins called delta-endotoxin. Theses proteins are activated by midgut proteases and the activated toxins diffuse through the peritrophic membrane of midgut causing a disruption in membrane integrity and ultimately leading to lethal septicemia and death.

Symptoms

The first sign of poisoning in insect is paralysis of the gut and mouthparts, leading to a cessation of feeding. Subsequent to the onset of gut paralysis, there is a swelling and destruction of microvilli of the midgut which leads to disruption of ion and glucose separation of midgut cells from the basement membrane and bursting of separated cells in the midgut lumen. Disruption of midgut structure and function lead to ion and pH imbalances in the haemolymph, total body paralysis and death.

Table 13.5: Varieties of *B.t.*

1.	*B. thuringiensis* var. kurstaki	:	Lepidoptera
2.	*B. t.* var. israeliensis	:	Diptera
3.	*B. t.* var. tenebrionis	:	Coleoptera
4.	*B. popilliae*	:	Scarabaeid beetles
5.	*B. sphericus*	:	Mosquitoes

Table 13.6: Classification of Insecticidal Protein from *B. thuringiensis*

Host	*Protein*
Lepidoptera	Cry 1 A a to A c, Cry 1 B a, C a, C b, D a, F a, Cry 9A a, Cry 2 A b
Coleoptera	Cry 3 A a, B a, Cry 7 A a, Cry 3 C a
Diptera	Cry 4 A a, B a, Cry 10 A a
Coleoptera/Lepidoptera	Cry 116

Table 13.7: *B. thuringiensis* Products Available in India

Trade Name	*Target Pest*	*Crop*	*Manufacturer*
Halt	Diamond back moth, lepidopterous caterpillars	Cabbage	Wockhardt Ltd., Bombay
Biolet	Cotton bollworms	Cotton	Biotech International Ltd., New Delhi
Bioasap	Cotton bollworms, tobacco caterpillar	Cotton	Biotech International Ltd., New Delhi
Delfin WG	Cotton bollworms	Cotton, castor	M/s Sandoz India Ltd., Bombay
Dipel 8L	Cotton bollworms, tobacco caterpillar	Cotton, tomato, brinjal, okra	Lupin Agrochemicals Pvt. Ltd., Bombay
Spicturin	American bollworm, leaf-folder, tobacco caterpillar, diamond back moth	Cotton, rice, cabbage, cauliflower, chillies	Tuticorin Alkali Chemicals and Fertilizers Ltd., Chennai
Biobit	Lepidopterous caterpillar	several	Rallis India Ltd., Bangalore

The transgenic *B. t.* cotton has been developed which resist the attack of bollworms particularly *H. armigera* which alone cause extensive damage to cotton in India, Pakistan and Australia.

Vaeck *et al.* (1987) was the first who developed *B. t.* cotton and it is found resistant to lepidopterous pests. India being the very important cotton growing country recognized the importance of *B. t.* cotton which was first developed at USA by M/s Monsanto. In India, Monsanto Mahyco have developed *B. t.* cotton which contain Cry 1 Ac gene which is resistant against bollworms. Results conducted with *B. t.* cotton across the country revealed that *B. t.* hybrids produced 30-40 per cent higher yield with reduced insecticide application by 40-50 per cent than the non *B. t.* cotton. During March 2003, Genetic Engineering Approval Committee (GEAC), New Delhi has permitted Mahyco company to sell

B. t. cotton seeds in India. Farmers of certain cotton growing pockets in India are permitted to grow *B. t.* cotton. Thus the *B. t.* forms another important IPM tool in reducing pesticide load and related problems without affecting their yield. *B. t.* rice has been developed by International Rice Research Institute, Philippines during 1996. It is under testing elsewhere in the country.

Fungus

Among the insect pathogens, fungi constituted the largest group with more than 750 species causing mycoses in insects. The classes Phycomycetes and Deuteromycetes (Fungi imperfecti) include most of the fungal pathogens and among them, the general *Entomophthora, Metarhizium, Beauveria, Nomuraea* and *Verticillium* are noteworthy. Fungus diseases are commonly seen in insect orders such as Homoptera, Lepidoptera, Coleoptera, Hymenoptera and Diptera. Fungus based biopesticides are popularly called as Myco-insecticides.

Mode of Action

The fungus invades into the host insect through the body wall and spiracles primarily through invasion through the alimentary canal also occurs by the contamination of mouthparts. Fungus conidia germinate on the insect cuticle by producing germ tube which penetrate the body wall. Once the fungus gains entry into the insect body, it produces small hyphal bodies, which are fragments of mycelia. The hyphal bodies multiply rapidly in the body cavity. In some cases, the fungus produces toxins at this stage to cause death before invading into major organs. It invades different tissues primarily the fat bodies. The developing mycelium ramifies all the internal organs. When the insect is completely filled with fungus, the conidiophores are formed protruding through the cuticle and spores are produced outside the insect.

Fungal Toxins

Fungi are known to produce substances that are toxic to insects. Toxins produced by entomopathogenic fungi are

Cordyceops miliaris	:	*Cordycepin*
Beauveria bassiana	:	Beauvericin
B. brongnihartii	:	Beauverolides
		Bassinolidae
Metarhizium anisopliae	:	Isarolides A, B, C, D
Paecilomyces spp.	:	Beauvericin
Isaria spp.	:	Isarin
Verticillium	:	Similar to Bassinolidae

Field Application

The three pathogens, *B. bassiana, M. anisopliae* and *N. rileyi* belong to order Moniliales and class Deutereomycotina. *B. bassiana and B. brongniartii* are pathogenic to a wide range of insects belonging to Lepidoptera, Coleoptera, Hemiptera and Diptera. The fungi cause the white muscardine disease in insects. *Metarhizium* spp. is responsible for the green muscardine disease known in more than 200 species of insects. *Nomuraea* spp. also cause green muscardine disease in insects. At 1 x 10^7 conidia/ ml, *B. bassiana* caused mortality *of H. armigera* larvae (I-V) inflicting 60-100 per cent and 100 per cent on the eggs. Also to 1 x 10^9 conidia/ml, 60 and 80 per cent mortality was observed in pre-pupae and adults, respectively with an incubation period ranging from 2-14 days (Gopalakrishnan and Narayanan, 1990). Natural occurrence of *B. bassiana* on *H. armigera* ranged from 60.8 to 19.2 per cent and the temperature had a negative correlation with fungal infection (Gowda and Ravi Prasad, 1992). *Verticillium leecanii* was reported to be effective on coffee green scale, *Coccus viridis* (Eswaramoorthy, 1979).

Table 12.8: Fungal Formulations: Commercial Bio-pesticides Based on Pathogenic Fungi

Beauveria bassiana	Colorado potato beetle, corn root worm, citrus root weevil	Bovericin
Hirsutella thompsoni	Citrus rust mite, coconut perianth mite	Mycar
Verticillium lecanii	Aphids, whiteflies	Vertilac, Mycotal
Metarrhizium anisopliae	Spittle bug, sugar hopper	Metaquino
Phytophthora palmivora	Milk weed vine	Devine
Alternaria cassiae	Sickle pod	Cast
Colletotricum gloeosporoides	Northern joint vetch	College
Fusarium lateritium	Velvet leaf	Vego

A biocide formulation based on *Fusarium* sp. isolated from the dead mite has been released by Tamil Nadu Agricultural University, Coimbatore under the brand name 'TNAU Agrobiocide' to control mite menace in coconut.

Protozoans

More than 1000 species of protozoa pathogenic to insects have been described. *Vairimorpha necatrix* infects a number of important noctuid pests and appears quite promising for development as a biopesticide (Battu *et al.*, 1993). Four species of microsporidia have been recorded from acridids, *Nosema locustae, N. acridophagus, N. cuneatum* and *Perezia dichloroplusae* in India. The pathogenicity of *Farinocystis tribolii* on *Tribolium castaneum* was recorded by Rabindra (1979).

Conclusion

Insect pathogens are found to be one of the best alternatives to overcome the menace due to indiscriminate use of pesticides. Virulent isolates, UV protectants and adjuvants identified can be better utilized for enhancing the efficacy of NPV. NPV formulations developed so far in India are not in use due to poor stability and shelf-life, which is one of the thrust areas to be investigated. Research on insect cell lines should be concentrated. There are enormous scopes for recombinant NPV which may be extended to field condition. In India, Monsanto Mahyco *B. t.* cotton has been introduced recently which forms an important IPM tool in reducing pesticide load. However, proper monitoring on the field performance of *B. t.* cotton by entomologists should be done periodically. Several entomopathogenic myco-insecticides available in India can be exploited for the pest management of cotton during rainy and winter season. Separate system should be there to monitor the quality of the preparations of the insect pathogens. Registration procedure for bio-pesticides should be developed to sell in open market.

Insect Pest and Disease Management *Pages* ***192–208***
Editor: **Dr. D. Prasad**
Published by: **DAYA PUBLISHING HOUSE (2016)**

Chapter 14

Rice Diseases and their Management through Biocontrol Agents

Ashraf Ali Khan[1] and D. Prasad[2]
[1]Division of Plant Pathology, [2]Division of Nematology,
Indian Agricultural Research Institute, New Delhi – 110012

Introduction

Rice (*Oryza sativa* L.) is one of the world's most important cereals for human consumption, grown in 113 countries and the staple food for over half the world's population. It is the globe's most popular food that shapes religious festivals, customs, cuisine and celebrations. World rice production in 2004 was just under 610 million tons. Asian farmers produce about 90 per cent of the total production, with two countries, China and India, growing more than half the total crop. Rice is the staple food of 65 per cent of the total population in India. It constitutes about 52 per cent of the total food grain production and 55 per cent of total cereal production (Anonymous, 2004).

In recent years, the increasing use of pesticides in agriculture has been the subject of growing concern for both environmentalists and public health authorities. Besides their non-target effects and hazardous nature, they are becoming more expensive and some are losing their effectiveness due to development of resistant strains. Biological control has emerged as an alternative and most promising means of the management of plant pathogens. Apart from being effective and economical, biological control offers several advantages over chemical control like (*i*) there is no pollution problems and health hazards (*ii*) it is more stable without any problem of resistance development in pathogen (*iii*) biological control agents are not phytotoxic, (*iv*) it may have the growth promoting effect and (*v*) it causes little disturbance in ecological balance. So, it is now widely recognized that use of eco-friendly bio-pesticides is a distinct possibility for the future and can be successfully exploited in modern agriculture especially with in the frame work of integrated disease management system without affecting our specious eco-system.

Biological Control

Biological control is the reduction of the amount of inoculum or disease producing activity of a pathogen accomplished by or through one or more organisms other than man (Cook and Baker, 1983). Biological control as a strategy and philosophy for reduction crop loss from plant disease is not new to agriculture (Cook, 1982). Even the earliest farmers practiced biological control by rotating their crops, burrowing disease infested crop residues and fertilizing with organic manures. These and many other traditional practices give effective disease control by allowing time and opportunity for biological destruction of disease organisms. The challenge today is how to achieve effective biological control of pests in modern agriculture where farmers must specialize with one or two crops, use minimal or no-tillage to save fuel and stop soil erosion and where organic fertilizers are unavailable, inconvenient or too expensive. Pesticides are necessary at present but are not a long-term solution to crop health. Besides, their non-target effects, petroleum based pesticides will become even more expensive and some are now losing their effectiveness (Cook, 1982).

Biological control of plant pathogens can be achieved by either promoting the native antagonists to reach a density sufficient to suppress a pathogen(s) or by introducing alien antagonists. Though some of the earlier works related to promoting the native antagonist was by using organic amendments, the recent trend is to isolate, multiply and introduce the antagonists to soil or specific court of infection to achieve a successful biological suppression of a disease. The one time skepticism that biological control by introduced antagonist may not work stands invalid in the present days as evidenced by enormous literature in favour of successful biological control research has been reached a state where we have commercial formulations of biocontrol agents sold in the market of several countries. There remains no doubt that biological control has come to stay and in near future it will be an integral part of disease management strategies for many crops. Nothing in biological control should be thought of as impossible at the present state of our technology (Papavizas, 1985; Baker, 1986; Mukhopadhyay, 1987; Gabriel and Cook, 1990).

Production and Formulations of Biocontrol Agents

It has been realized that the greatest obstacle to biological control by introduced antagonists is the lack or scarcity of methods for mass multiplication of the biocontrol agents (Papavizas and Lewis, 1982). A fermentation system providing high quality biomass is one of the main factor for the success of the biocontrol agents. The first step in production of biocontrol agent is development of a suitable medium using inexpensive, readily available agricultural bio-products with the appropriate nutrient balance (Latze and Soper, 1977). Many attempts have been made so far to standardize growth media for mass multiplication of the antagonists. Now a days most of the research on mass production of antagonists continue to emphasize solid substrate fermentation consisting of molasses, wheat bran and yeast for the large scale production of the *T. harzianum* biomass. A variety of media have been used for various researchers for production of *Trichoderma* spp. in stationary flasks. Molasses–brewer's yeast medium is being used widely for commercial production of *Trichoderma* by fermentation process. In India, many commercial *Trichoderma* producers are using molasses yeast medium for mass production. Brencer's yeast which serves as nitrogen sources is expensive and there is a need to identify inexpensive and more commonly available alternative to yeast and the molasses soy medium could serve as a better alternative to FYM for large scale production of *Trichoderma* spp. (Prasad *et al.*, 2000). Fravel, *et al.*, 1985 reported that encapsulation of biocontrol agents in alginate pellets have potential to control plant diseases. Sabaratnam and Traquair (2002) demonstrated that vegetative propagules from actively growing filaments are acceptable inoculants for *Streptomyces* sp. formulations.

The talc based formulations of *Pseudomonas fluorescence* and *Trichoderma viride* can be recommended as one of the crop protection strategies for the management of sheath blight of rice. Bacterial antagonists have been formulated in a variety of ways to control plant pathogens. Tests revealed that formulations of antagonistic bacterial strains *Pp*V14I and *Pf*7-14 with combination of methyl cellulose and talc at 1:4 and $CaCO_3$ are most satisfactory. In this formulation, bacteria can survive up to 10 months. Formulated *Pp*V14I applied as seed treatment, root dip or foliar sprays can suppress sheath blight up to 60 per cent. While formulation of *Pf*7-14 applied as seed and multiple foliar sprays can suppress 60 per cent and 72 per cent of leaf and neck blast, respectively (Mathivan *et al*,.2005). Talc, $CaCO_3$, CMC and talc + CMC (4:1) based formulations of *Trichoderma* and *Pseudomonas* were tested, and it was observed that talc + CMC (4:1) formulation was more effective as compared to other formulations (Singh and Sinha, 2005a; Khan and Sinha, 2006).

System of Delivery of Biocontrol Agents

Based on studies, Papavizas and Lewis (1981) concluded that the methods of inoculum preparation of biocontrol agents, their mode of application, use of different food bases, and an integrated application of other chemicals or cultural practices can improve the effectiveness of biocontrol agents.

Seed Treatment

A more common and often effective method of biological control is to introduce the antagonist with the planting material. This approach may be a practical way to introduce antagonists for biological control of externally seed borne and soil borne pathogens with introduced antagonists (Cook and Baker, 1983). Inoculation of seeds with fast growing antagonists prevents seed decay and seedling blight by pre-emptying leaked nutrients and their restricting pre-penetration increase of pathogen biomass, as well as by providing a protective covering, producing pathogen inhibiting enzyme, antibiotics and occasionally by parasitism (Baker, 1987). Spore suspension as well as dry powder have been used to coat the seeds with potential antagonists (Harman *et al.*, 1981; Mukhopadhyay *et al.*, 1992). The general procedure in biological seed treatment is to harvest the conidia from a well sporulated potato dextrose agar culture with the help of camel hair brush, making a spore suspension of desired concentration (with or without a sticker), soaking the seeds in the suspension and air-drying them before sowing, when the treated seeds germinate, the antagonist proliferates on the seed coat without affecting germination, thus creating a physical barrier against invasion by the pathogen. The application of *Trichoderma viride* and *Bacillus subtilis* as seed treatment reduced sheath blight infection of rice in pots (Das *et al.*, 1998). Khan *et al.*, 2005 reported that rice seeds treated with *Trichoderma* spp. enhanced the plant biomass.

Several strategies can be adopted for enhancing efficacy of biological seed treatments. The important and feasible ones are:

1. Permitting the bio-protectant to colonize the seed surface before planting.
2. Applying the bio-protectant in a thin layer that permits it to proliferate for few hours after planting in the absence of competitive microflora.
3. Inclusion of adjuvant to seed treatments that may
   (*a*) Control pH to level favourable to the biocontrol agents but unfavourable to competitive microflora.
   (*b*) Add food bases selectively to the bio-protectants.

(*c*) Use selective toxicants which did not restrict the growth of bioagents.

(*d*) To alter the timing *i.e.* period when biocontrol agent is active in relation to the pathogen between the bioprotectant and the pathogen.

Soil Treatment

The application of biocontrol agents by broadcast, sprays, or by direct application in seed furrow at the time of sowing, has the greatest potential for use in commercial glasshouse operations or small scale field experiments, for biological control of root pathogens (Cook and Baker, 1983). The beneficial effects of soil treatments with biocontrol agents in plant disease management lie in their ability to suppress pathogen populations. Manian and Paulsamy (1987) reported that there was significant control in sheath blight incidence and severity, when the soil was amended with *Trichoderma aurioviride*. Das *et al.* (1996) observed that a 2 per cent application of *Trichoderma viride* in *R. solani* inoculated soils reduced sheath blight infection and increased rice plants growth and yield. Soil application of *A. terreus* were made and found that the effective suppression of disease was greater when *A. terreus* was applied to soil. *T. harzianum* and *T. virens* increased germination of seeds, length of root, shoot and weight of fresh root and shoot and dry root and shoot in rice plants as well controlled the sheath blight effectively when applied to soil (Khan *et al.*,2005; Khan and Sinha 2005b).

Foliar Application

Spraying the spore suspension of biocontrol agents for controlling sheath blight disease of rice is reported by several workers. Rosales and Mew (1982) reported that there was significant reduction in lesion development when spore suspension of fungal biocontrol agent of *Trichoderma* spp. were sprayed on the pathogen (*R. solani*) inoculated rice plants. Spraying with conidial suspension of *Aspergillus terreus* on inoculated rice plants showed significant reduction in sheath blight severity. Spraying with spore suspension of *Trichoderma koningii, T. harzianum, T. viride* and *Aspergillus niger* for controlling the rice sheath blight has been reported by several workers (Das *et al.*, 1996). The application of talc-formulation through seed, seed treatment plus foliar spray and foliar spray alone significantly reduced the leaf blight incidence both under greenhouse and field conditions.

Maximum reduction in disease severity and incidence and increasing the grain yield and 1000-weight was observed with the combination of the treatment viz seed + soil + root dip + foliar sprays (Meena and Muthusami 1998; Commare *et al.*, 2002; Singh and Sinha, 2005b; Khan and Sinha 2005c).

Management of Rice Diseases through Biocontrol Agents

Research on biological control of rice pathogens started recently, mainly in the 1980s, and is still concentrated on the identification, evaluation and formulation of potential biocontrol agents for deployment. A number of fungus, bacteria, virus, nematode and mycoplasma-like organisms cause disease to rice plants. Among these the fungal diseases *viz.* blast (*Pyricularia grisea*), brown spot (*Bipolaris oryzae*), stem rot (*Sclerotium oryzae*), sheath blight (*Rhizoctonia solani*), sheath rot (*Sarocladium oryzae*), bacterial disease such as bacterial blight (*Xanthomonas oryzae* pv. *oryzae*) and viral disease such as tungro (rice tungro virus) are most important. These diseases are considered as a serious constraint for rice production. Rice disease management strategies mainly aim at prevention of outbreak or epidemics through the use of host plant resistance and chemical pesticides. However, research during the previous two decades indicates another potential option for rice disease management. That is, biological control of rice diseases. The effects of some of the biocontrol agents on the pathogens and/or plant are summarized in the following:

Rice Blast [*Pyricularia grisea* (Cooke) Sacc.]

The Rice blast is among the most well-known rice diseases. It occurs in most of the humid rice-producing areas of the world. In India occurrence of this disease was first reported from the Tamil Nadu and subsequently several other parts of India. In severe cases loses amounting to 70-80 per cent of grain yield are reported. The disease attacks the leaves, culms, branches of the panicle and the floral structures. Although leaves and neck of the panicle are more commonly affected.

Different species of *Bacillus* applied to rice plants as a seed treatment before sowing, root dip prior to transplanting, and two foliar sprays prior to inoculation can suppress rice blast pathogen by up to 59 per cent, resulting in a two-fold increase in plant height and grain yield. In *in vitro* tests isolates of *Chaetomium cochlioides* was antagonistic to the rice blast pathogen. When rice seeds, cv. IR 442-2-58, were coated with a spore suspension or culture filtrate of *C. cochlioides*, early infection by blast was controlled. The treatment was as effective as that with captan. The protective capacity of the culture filtrate suggested that *C. cochlioides* may produce an antagonistic substance inhibiting the development of *P. oryzae* (Soytong, 1990). Antagonism by *Pseudomonas aureofaciens* and *P. putida* was highest at 20-30°C in dual culture assay. The number of rice blast lesions/plant was lower when seed was treated with *P. fluorescens* and *P. putida* than in soil incorporation treatment (Lee *et al.*,1990). Antibiotics secreted by *B. subtilis* inhibited mycelial growth and conidial germination of *P. oryzae in vitro*. The antagonistic substances produced by *B. subtilis* were thermostable and effective at low concentrations against rice blast pathogen (Bettiol and Kimati, 1990; Peng *et al.*,2002; Liu *et al.*, 2005). Foliar spray of *P. fluorescens*, in addition to LFA and 45 kg potash/ha significantly reduced the rice blast incidence and increased crop yield (Karpagavalli *et al.*, 2001). Fungicidally active secondary metabolites of micro organism have been used either directly as microbially produced fungicides (*e.g.* kasugamycin) or as lead molecules for chemical synthesis of novel chemistry (*e.g.* pyrrolnitrin) used as a fungicidal seed dressing against rice blast (Lange *et al.*, 1993). Antifungal substances isolated from *F. graminearum* completely inhibited the growth of *P. oryzae*, when the substance was sprayed directly to plants, it effectively controlled rice blast, rice brown spot and barley powdery mildew with >80 per cent disease control (Kim *et al.*, 1995). Salem, 1999 reported that isolates belong to *Bacillus* spp. and *Pseudomonas* spp. are inhibitory against *P. oryzae* the causal organism of blast disease in both laboratory and greenhouse tests, these isolates showed the mycelial reduction ranging from 53.6 to 40.5 per cent *in vitro*. In greenhouse tests, all isolates were effective compared with the control. Krishnamurthy and Gnanamanickam, 1998, reported that strains of rice-associated pseudomonads *Pseudomonas fluorescens* 7-14 and *P. putida* V14i caused an induced systemic resistance (ISR) in cultivars of rice. ISR response coincided with increased levels of salicylic acid and biological suppression (25 per cent) of rice blast disease. They suggested that the ISR triggered by these bacteria which were spatially separated from the rice blast pathogen, is an important mechanism for the biological suppression of rice blast. Rice plants preinoculated with either 200 ml culture filtrates or 200 ml conidial suspension of *T. harzianum* prior to inoculation of blast pathogen led to a decrease in blast severity (Ngueko, *et al.*, 2002). Han, *et al.*, 2005 revealed that a novel, plant growth promoting bacterium *Delftia tsuruhatensis*, strain HR4, antagonistic and this strain could suppress the growth of various plant pathogens effectively, especially the three main rice pathogens (*Xanthomonas oryzae* pv. *oryzae*, *R. solani* and *P. oryzae in vitro*. In the greenhouse also this strain, evidently controlled rice blast, rice bacterial blight and rice sheath blight. *P. fluorescens strain* Pf 7-14, is a potential biological control agent, against rice blast (Li *et al.*, 2005). Saravanakumar, *et al.*, 2005 reported that antifungal protein produced by *B. polymyxa* strain VLB16 inhibited the growth of *P. grisea* and *Rhizoctonia*.

Sheath Blight (*Rhizoctonia solani* Kuhn)

Sheath blight (Sh.B) of rice hitherto regarded a minor disease has assumed the status of major one in most of the rice growing areas. Since the introduction of high yielding rice cultivars, it has become one of the most important diseases in rice production in both temperate and tropical rice growing countries causing serious yield losses. The disease was first reported from Japan in 1910 by Miyake (1910) who named the causal organism as *Sclerotium irregulare.* Subsequently, in Sri Lanka, China and Philippines the pathogen was identified as *Rhizoctonia solani.* In India, first report of its occurrence was from Gurdaspur, Punjab and later from Uttar Pradesh. Sheath blight of rice occurs commonly in tropical areas of the world, resulting in severe yield losses. Under climatic conditions favourable for its development, the disease can kill the leaves of an affected rice plant. Grain yield of different varieties of rice is chiefly reduced by sheath blight disease. The symptoms of this disease, initially appear as water soaked lesions on leaf sheath, that are circular to oblong, ellipsoid to ovoid or even irregularly elongated 3 cm x 1 cm discoloration that later turn into discrete lesions with pale greenish-grey to grayish white center with narrow blackish to dark brown margins. Finally, 4-5 such lesions coalesce and girdle the whole leaf sheath, culm, boot and flag leaf whereby the tiller is encircled to death.

The fluorescent and nonfluorescent strains of a number of antagonistic bacteria associated with upland and lowland rice rhizosphere soils have been found effective *in vitro*, greenhouse and the field against *R. solani* (sheath blight). Strains of *Bacillus* spp., inhibited growth and sclerotial germination of *Rhizoctonia solani in vitro*. The strains were also effective in controlling rice sheath blight in field tests (Mew *et al.*, 1998; Li *et al.*, 2000; Chen *et al.*, 2000; Yi *et al.*, 2000). *In vitro* antagonistic assay showed that plant growth-promoting bacterium *Delftia tsuruhatensis* strain HR4 could suppress the growth of *R. solani*. Culture of this strain also, controlled the rice sheath blight in the greenhouse (Han *et al.*, 2005). Application of *Serratia marcescens* effectively reduced the incidence of sheath blight. It survived in soil under glasshouse conditions at 108 colony forming units/g of soil for 4 weeks after application. These results suggest that *S. marcescens* has potential, as an effective and persistent biological control agent for rice sheath blight (Someya *et al.*, 2000). Nandakumar *et al.*, 2001 reported that the mixture of plant growth promoting rhizobacterial (PGPR) strains of *P. fluorescens,* strains significantly reduced the sheath blight incidence when applied as either bacterial suspension through seed, root, foliar and soil application in glasshouse conditions, or as talc-based formulation under field conditions, compared to the respective individual strains. In addition to disease suppression, treatment with mixture of PGPR strains promoted plant growth in terms of increased plant height and number of tillers, and ultimately grain yield. Antagonistic fungi *T. harzianum*, *T. viride* and *Gliocladium virens* are antagonistic to *R. solani* and reduces Sheath blight, under green house conditions. In field test control effect of fungal strains mixture was 32.25 per cent, and it also had good effects on seed setting rate and 1000-grain weight of rice plants. (Wang *et al.*, 2000; Tang *et al.*, 2002, Kazempour *et al.*, 2003). *In vitro* antibiosis of *P. fluorescens* isolates resulted in reduction of mycelial growth of *R. solani* and the inhibition zone ranged from 1.3 to 22.5 mm in different isolates on Kings-B medium. In a greenhouse study, there was a 17.7-100 per cent reduction in rice sheath blight infection due to pretreatment of sclerotia for 4 weeks with suspension of *P. fluorescens* isolates (Pande *et al.*, 2003, Li *et al.*, 2003). *B. subtilis* showed strong antagonistic activities to several fungal pathogens, *e.g. R. solani, Magnaporthe grisea in vitro*. Its culture showed good control efficiency to rice sheath blight in the field (He *et al.*, 2002). Rangaraj *et al.*, 2002, tested antifungal activity of *P. fluorescens* against the rice sheath blight pathogen, *R. solani*. They reported that strains of *P. fluorescens* produced antibiotic 2,4-diacetyl phloroglucinol, iron chelating siderophore, hydrogen cyanide, lytic enzymes, such as chitinase and beta-1,3-glucanase in culture, and induced chitinase in plants. The soaking of *R. solani* sclerotia in bacterial culture filtrates or in sap extracted from plants treated with *P. fluorescens* strains significantly reduced the germination capacity of the sclerotia. Application of *P. fluorescens* and *T. viride* resulted in a significant reduction of sheath blight incidence and was comparable to the systemic fungicide, Carbendazim. The number of productive tillers, grains per panicle and grain test weight, straw and grain yields were also significantly increased when compared with control (Mathivan *et al.*, 2005). The sclerotial germination of *R. solani* decreased when the sclerotia were incubated in the culture filtrate of the *Serratia marcescens* strain B2. The antifungal activity of the culture filtrate was enhanced by the treatment in combination with low concentrations of the chemical fungicides flutolanil, pencycuron and validamycin (Someya *et al.*, 2005). Nagarajkumar *et al.*, 2005 reported that seed treatment followed by soil application of rice with *P. fluorescens*, reduced the severity of sheath blight by 75 per cent compared with the control, under green house conditions. They also reported that *P. fluorescens* was capable of detoxifying Oxalic Acid which is responsible for virulence in *R. solani*. Several biocontrol agents, *i.e. Trichoderma, Gliocladium, Aspergillus, Pseudomonas* etc. have ability to control Sheath blight and widely using against this disease. (Khan, 2003; Khan and Sinha, 2005a; Khan and Sinha, 2005c; Singh and Sinha, 2005a; Singh and Sinha, 2005b). Spraying of Talc + CMC formulations of bioagents *viz., T. harzianum, T.virens, T. viride* and *A. niger* were found quite effective against ShB (Khan, 2003; Khan and Sinha, 2006).

Bacterial Blight [*Xanthomonas oryzae* pv. *oryzae* (Ishiyama) Swing *et al.*]

Bacterial blight of rice first time reported from Japan and Philippine about 65 years ago. It is described in our country from Pune in 1959. Now the disease occurs all over the country in the exotic and indigenous rice varieties and is considered a severe menace to rice production in India. The bacterium includes either wilting of plants or leaf blight. The earliest symptom of the blight phase is the appearance of dull greenish water soaked or yellowish spots 5-10 mm in length on the leaf towards the tip or margins, leading to tip and marginal drying. The infection soon extends along one or both margins, some times to the leaf sheath. In the severe infection, all the leaves are attacked and premature drying results. The losses in grain yield due to this disease may range up to 60 per cent. The control of rice bacterial leaf blight in greenhouse and field plot tests with an avirulent mutant Du728 strain showed that, the rate of control with one application of Du728 spray (106 c.f.u./ml) was 48.5 per cent. Higher dosages did not significantly increase control. When Du728 was mixed with salicylic acid (10 micro g/ml), control was increased to 60 per cent. In field tests, the control effect (up to 80.1 per cent) obtained from 3 applications of Du728 and salicylic acid combined was better than when the chemicals were applied singly (Liu and Wang, 1998). *Bacillus subtilis*, an antagonistic bacteria isolated from the leaves of rice, strongly inhibits growth of *X. o.* pv. *oryzae*, the causal agent of rice bacterial blight. Crude extract was obtained by precipitation of the cell-free culture of *B. subtilis* with 70 per cent saturated $(NH_4)2SO_4$. The suspension of the precipitate strongly inhibited the growth of *X. o.* pv. *oryzae.* (Xu *et al.*, 1998; Tong *et al.*, 1999) *In vitro* antagonistic assay conducted by Han *et al.*, 2005 showed that *Delftia tsuruhatensis* HR4 strain could suppress the growth of various plant pathogens effectively, especially the *X. oryzae* pv. *oryzae*, they reported that it also controlled rice blast in the greenhouse, evidently.

Brown Spot or Blight Disease [*Bipolaris oryzae* (Breda de Haan) Shoemaker]

The disease appears at first as small, purplish brown spots, on the leaves and glumes. Later the spots grow larger, and have a grey center and brown border. The fungus causes damage by infecting the seed directly, which then germinates poorly and produces weak seedlings, but primarily by attacking the leaves during the seedling stage either in the seed beds or in the field. In such infections the young foliage is severely damaged, the plants are weakened, and the yield drastically reduced.

Ray *et al.* (1990) reported that rice brown spot was reduced to some extent when the plants were treated with the cell suspensions of Fluorescent pseudomonads before inoculation of *B. oryzae* but post inoculation treatment failed completely. Various fungi isolated by Kumar and Mishra (1994) from the rice leaf surface and tested against *D. oryzae* on Czapek Dox Agar, they reported that, most effective antagonist was *T. viride*, which could be exploited as a possible biocontrol agent for the brown spot pathogen. Antimicrobial substances produced by *F. graminearum* was sprayed directly to plants, it effectively controlled rice brown spot with >80 per cent disease control (Kim *et al.*,1995). Kamala and Vidhyasekaran (1998) reported that, *P. fluorescens* strain P1 caused the highest inhibition of growth of foliar pathogens of rice (*P. oryzae, Helminthosporium oryzae, X. o.* pv. *oryzae* and *R. solani*) *in vitro.* A talcum powder-based formulation of the bacterial strain was developed and its efficacy in controlling blast, brown spot, bacterial blight and sheath blight diseases caused by these pathogens was evaluated under greenhouse conditions. Seed inoculation with the product at 4 g/kg (1.4 x 109 colony forming units (c.f.u.) was effective in controlling these diseases. As the dosage of the product increased the protection due to p1 also increased. Seedling root dipping in 0.5 per cent (1.8 x 109 c.f.u./l) of the product was effective in controlling all these diseases. Appreciable control was obtained with foliar spray at 0.1 per cent (3.6 x 109 cfu/l). Combined application of the product by seed treatment, seedling root dipping and foliar spray, induced 60-70 per cent control in the field. There was also a significant increase in yield due to the treatment. The protective effect of the product against these diseases varied from 15-60 days depending on the mode of application.

Sheath Rot [*Sarocladium oryzae* (Sawada) W. Gams and D. Hawksworth]

Vishwanathan and Narayanasamy (1990) reported that, isolates of *Bipolaris zeicola* inhibited growth of *S. oryzae* in culture. The undiluted culture filtrate of *C. carbonum* was also inhibitory to *S. oryzae*. Dried leaf extract of Ipomoea spp. effectively controlled the growth of the rice sheath rot pathogen, *Sarocladium oryzae,* while the growth of beneficial biocontrol agents, *Bacillus subtilis, Pseudomonas fluorescens* and *Trichoderma viride,* was not affected at 4 per cent concentration of leaf extract in *in vitro* conditions (Rajappan *et al.*, 1997).

Native strains of rhizobacteria of *P. fluorescens* type showed *in vitro* inhibitory activity against *S. oryzae*. Bacteriarization resulted in the effective biological control of sheath rot of rice and stem rot of groundnut and enhanced plant growth/yield (Sakthivel *et al.*, 1998).

Rice Tungro Virus

Tungro is characterized by stunting of the plant and discoloration of leaves, ranging from various shades of yellow to orange and rusty blotches spreading downwards from the leaf tip. The young leaves show a mottled appearance and slightly twisted, whereas the older leaves appear rusty coloured. In less susceptible varieties viral infection delayed flowering. The Tungro virus is transmitted by green leaf-hopper *Nephotettix verescens.*

Crude culture filtrate from *A. flavus* reduce the RTV infection by 87 per cent followed by *Neurospora* sp. (75 per cent) and sonicated suspension of yeast also reduced the RTV infection by 70 per cent culture filtrates from *Bacillus subtilis* reduced the RTV infection by 72.9 and 65.9 per cent, respectively. Muthulakshmi (2002) reported that. The culture filtrates obtained after shorter incubation periods of these bacteria (3 and 6 days) were more effective than that obtained after longer periods of incubation. There is no significant difference in the antiviral activity of protein and non-protein fractions of culture filtrates of these fungi and bacteria in reducing RTV infection. Suzuki *et al.*,1996 analyzed the effect of natural enemies on the population dynamics of *Nephotettix virescens* in rice fields. They reported that Egg parasitoids, lycosids and other predators attacking nymphs and adults, and pipunculids were important natural enemies of *N. virescens*, but none of them caused a density-dependent increase in the mortality rate of the 1st generation. Cultivation of resistant varieties was considered the most feasible and effective method for controlling the virus.

Commercially Available Biocontrol Products for the Management of Plant Pathogens

A number of commercial, bacterial biofungicides and mycofungicides for disease control of different crops are available in the markets of different countries, but some important ones are given below with complete address of their Manufacturer/Distributor.

Table 14.1: Biocontrol Products Commercially Available, Listed Alphabetically

Actinovate	
Biocontrol Organism	*Streptomyces lydicus*
Formulation	Water-dispersible granule
Manufacturer/Distributor	Natural Industries, Inc. 6223 Theall, Houston, TX.77066 Phone: 888-261-4731, Fax: 281-440-9206 www.naturalindustries.com/index.htm
Bio-save 10LP, 110	
Biocontrol Organism	*Pseudomonas syringae*
Formulation	Lyophilized product, frozen cell concentrated pellets
Manufacturer/Distributor	Village Farms LLC153 Sable Palm Dr., Longwood, FL 32779 Phone 1-407-682-8542 www.villagefarms.com/biosave/index.html
BlightBan A506	
Biocontrol Organism	*Pseudomonas fluorescens* A506
Formulation	Wettable powder
Manufacturer/Distributor	NuFarm Inc., 1-708-754-3330. www.nufarm.com
Cedomon	
Biocontrol Organism	*Pseudomonas chlororaphis* strain
Formulation	Seed treatment
Application Method	Seed dressing
Manufacturer/Distributor	BioAgri AB P.O. Box 914; visiting address: Dag Hammarskjolds 180 SE-751 09 Uppsala, Sweden; Phone: 46(0) 18-67 49 00; Fax: 46(0) 18-67 49 01. www.bioagri.se
Companion	
Biocontrol Organism	*Bacillus subtilis*GB03, *other B. subtilis,B. lichenformis, B. megaterium*
Formulation	Liquid
Manufacturer/Distributor	Growth Products PO Box 1259, Westmoreland Ave., White Plains, NY 10602 USA; Phone: 1-800-648-7626. www.growthproducts.com
HiStick *N/T*	
Biocontrol Organism	*Bacillus subtilis* MBI600.
Manufacturer/Distributor	Becker Underwood Inc.MicroBio Group, Ltd. Unit 2 Centro, Boundary Way, Hemel Hempstead, Herts HP2, 7SU, UK; Phone 01442 399123 (Production); 014420399124 (R&D); FAX 01442-399128, US www.microbiogroup.com Distributor: Helena Chemicals, Memphis TN www.helenachemical.com

Contd...

Table 14.1–Contd...

Kodiak (several formulations)	
Biocontrol Agent	*Bacillus subtilis* GB03
Formulation	Dry powder; usually applied with chemical fungicides
Manufacturer/Distributor	Gustafson, Inc. 1400 Preston Rd, Plano TX 75093 USA; Phone 1-800-248-6907 or 1-972-985-8877; FAX 1-972-985-1696. www.gustafson.com
Mycostop	
Biocontrol Agent	*Streptomyces griseoviridis* strain K61
Formulation	Powder
Manufacturer/Distributor	Kemira Agro Oy, Porkkalankatu 3, P.O. Box 330, 00101 Helsinki, Finland; Phone 358-0-13-211; FAX 358-0 694-1375. U.S. Distributor: AgBio Development Inc., 9915 Raleigh St., Westminster, CO 80031; Phone 877-268-2020, 303-469-9221; Fax 303-469-9598. www.agbio-inc.com
Primastop	
Biocontrol Agent	*Gliocladium catenulatum*
Formulation	Powder
Manufacturer/Distributor	Kemira Agro Oy, Porkkalankatu 3, P.O. Box 330, 00101 Helsinki, Finland; Phone 358-0-13-211; FAX 358-0 694-1375. U.S.
Distributor	AgBio Development Inc., 9915 Raleigh St., Westminster, CO 80031; Phone 877-268-2020, 303-469-9221; Fax 303-469-9598. www.agbio-inc.com
RootShield, Plant Shield, T-22 Planter box	
Biocontrol Agent	*Trichoderma harzianum* Rifai strain KRL-AG2 (T-22)
Manufacturer/Distributor	Bioworks, Inc. 122 North Genesee St., Geneva, NY 14456 USA; Phone 1-315-781-1703; FAX 1-315-781 1793. www.bioworksbiocontrol.com
Serenade	
Biocontrol Agent	*Bacillus subtilis* QST716.
Manufacturer/Distributor	AgraQuest, Inc. 1530 Drew Avenue, Davis, CA 95616 USA; Phone 1-530-750-0150; FAX 1-530-750-0153 www.agraquest.com
SoilGard	
Biocontrol Agent	*Gliocladium virens* (a.k.a. *Trichoderma virens*) GL-21
Formulation	Granules
Manufacturer/Distributor	Certis, Inc.9145 Guilford Road, Suite 175, Columbia, MD 21046 USA; Phone 1-301-604-7340; FAX 1 301-604-7015. www.certisusa.com
YiedShield	
Biocontrol Agent	*Bacullus pumillus* GB34
Manufacturer/Distributor	Gustafson, Inc. 1400 Preston Rd, Plano TX 75093 USA; Phone 1-800-248-6907 or 1-972-985-8877; FAX 1-972-985-1696. www.gustafson.com

Mechanism of Disease Suppression by Biocontrol Agents Against Rice Pathogens

Understanding the mechanism of disease pathogen suppression is essential for the successful deployment of biological control as a disease management strategy. Rice disease suppression by biocontrol agents is governed by a multitude of factors. The influence of these factors varies with the type of biocontrol agent and the nature of the pathogen targeted for control. Currently research is underway to improve understanding of the mechanism of rice disease suppression by various agents. The available information is summarized below. Biocontrol agents exert disease suppression by different modes of action including competition, direct parasitism, antibiosis and induction of plant resistance mechanism (Whipps, 2001).

Antibiosis

The antagonistic organism releases an antibiotic or other chemicals, which are harmful to the pathogen and inhibits its growth. One of the most frequently used methods to begin selecting fungi and bacteria as potential biocontrol agents (mostly for bacteria) involves screening of biocontrol agents for antibiosis *in vitro* (Fravel, 1988). The mechanism by which *P. fluorescens* exerts its antagonism against *Pythium* sp. has been extensively studied and appears to involve the production of a variety of antibiotic compounds, competition and induced resistance (Gravel, *et al.*, 2005). Presence of thirmonematic principle in the culture filtrate of *Trichoderma lignorum* was first evidenced by Weindling (1934, 1941). Production of antibiotic by *T. harzianum* has been reported by Upadhyay and Mukhopadhyay (1985); Mukhopadhyay and Kaur (1990). Gliotoxin a metabolic product of *T. viride* was highly fungistatic and finally named as "Viridin" but later on Webstar and Lomas (1964) reported that gliotoxin/viridin was produced by *T. virens* rather than *T. viride.* In addition to Gliotoxin *Trichoderma* spp. also produced 1.5-2.1 KDa antibiotic like polypeptides such as Alamethicin, Paracelsin and Trichotoxin. These antibiotics are correlated with the disease suppression caused by *Rhizoctonia solani* and *Pythium ultemun* (Lumsden *et al.*, 1991).

Mycoparasitism

Mycoparasitism (= hyperparasitism) is defined as parasitism of one organism by another. Several hyperparasites of soil borne plant pathogens have been exploited in biocontrol. The mechanism of hyperparasitism includes different kinds of interactions like coiling of hyphae around the pathogen, penetration production of haustoria and lysis of hyphae. Recent investigations have revealed that necrotrophic parasites kills the susceptible host by the action of toxins, antibiotics and enzymes (Mukhopadhyay, 1987). Coiling of hyphae by *Trichoderma* spp. followed by vacuolation, coagulation of cytoplasm and lysis have been recorded (Rosales and Mew, 1982; Khan, 2003).

Competition

Competition play a role in microbial interaction, it may be an independent phenomenon, or it may be connected with antibiosis or parasitism. Competition as a mechanism of biological control of plant diseases by biocontrol agents includes, competition for infection site and nutrients (Weller, 1988). Competition for infection site was implicated as a mechanism by biological control by Osburn *et al.* (1989). O'sulliran and O'Gara (1992) suggested that nutrients competition varies at different rhizosphere depending on the viable sources of carbon, nitrogen, sulphur, phosphate and other micronutrients.

Induced Resistance

Some nonpathogenic microbial agents induce a sustainable change in the plant, resulting in an increased tolerance to subsequent infection by a pathogen. This phenomenon is known as induced systemic resistance. This is different from systemic acquired resistance, which refers to the host reaction in response to localized infection by pathogens, manifested as broad range of protection against other pathogens.

Future Prospects

We are very interested in developing new methods of biocontrol. A systemic research is required in biological control of rice diseases that would about bring significant control of diseases in field conditions. The development of strains those able to withstand with the fungicidal treatments, where fungicidal compound applied in combination with biocontrol agents to suppress the disease. The production and formulations of bioagents remains a challenge for growers. Effective production and formulations protocols must be easily available. Before any formulated product is marketed, it must first be thoroughly tested by growers whose comments and suggestions must be considered. Some enzymes have and the genes encoding them identified, from *Trichoderma* and other microbes. These enzymes degrade chitin, which is a structural component of pathogenic fungi and herbivorous insects. We are interested in large-scale production of the enzymes themselves, and other uses of genes and gene products for the control of plant pests.

Conclusion

The success of potential biocontrol agents in disease suppression depends on a suitable form for application to the plant system. Biocontrol agents can be applied either by direct inoculation (dipping seeds in culture, aerial spraying or spreading it in sowing furrows by a drip system or by the use of various solid-phase inoculants. A diverse group of biocontrol agents such as bacteria, fungi, viruses exist in nature. Among them bacterial antagonist are considered ideal candidates because of their rapid growth, ease of handling and aggressive colonizing character. Bacterial antagonists, *Pseudomonas*, and *Bacillus* in particular, are good candidates for biological control. *Bacilli* are gram-positive endospore-producing bacteria that are tolerant to heat and desiccation; a very good feature required for field application. The pseudomonads are gram-negative rods and have simple nutritional requirements; they are excellent colonizers and widely prevalent in rice rhizosphere. As many as 23 bacterial antagonists belonging to the genera *Bacillus, Pseudomonas, Serratia,* and *Erwinia* have been found to inhibit mycelial growth of *R. solani*, while a few of them also inhibit growth of other fungal pathogens like *Sclerotium oryzae* (stem rot), *B. oryzae* (brown spot), *P. grisea* (blast), *Sarocladium oryzae* (sheath rot) and *Fusarium fuijkuroi* (bakanae). Laboratory studies also revealed that a large number of bacterial strains possess the ability to protect rice plants from diseases such as blast, sheath blight, sheath rot, and stem rot. The fungal antagonists include *Trichoderma* spp., *Penicillium, Myrothecium verrucaria, Chaetomium globosum* and *Laerisaria arvalis* are also important biocontrol agents of rice diseases.

References

Anonymous, 2004. *FAOSTAT online electronic database.*

Baker, R., 1986. Biological control: An overview. *Can. J. Plant Path.*, 8: 218–221.

Bettiol, W. and Kimati, H. 1990. Effects of *Bacillus subtilis* on *Pyricularia oryzae*, causal agent of rice blast. *Pesquisa Agropec. Brasil*, 25(8): 1165–1174.

Chen, Z.Y., Xu, Z.G., Gao, T.D., Ni, S.K., Yan, D.F., Lu, F. and Liu, Y.F., 2000. Evaluation and utilization of antagonistic bacteria against rice sheath blight. *Chinese J. Rice Sci.*, 14(2): 98–102.

Commare, R.R., Nanadakumar, R., Kandan, A., Suresh, S., Bharathi, M. and Raguchander, T., 2002. *Pseudomonas fluorescens* based bio-formulation for the management of sheath blight and leaf folder insect in rice. *Crop Prot.*, 21: 671–677.

Cook, R.J., 1982. Biological control of plant pathogens: Overview. In: *Biological Control in Crop Production*, (Ed.) G.C. Papavizas, All an held, Osmum, Granada, U.S.A., pp. 23–44.

Cook, R.J. and Baker, K.F., 1983. The nature and practice of biological control of plant pathogens. *Am. Phytopath. Soc.*, St. Paul, 539 pp.

Das, B.C., Bora, L.C., Phookan and Bhagabati, A.K., 1996. Antagonistic effects of *Aspergillus terreus, Trichoderma harzianum* and *Trichoderma viride* on sheath blight of rice. *Oryza*, 33: 62–65.

Das, B.C., Khairuzzaman, A.S.M. and Bora, L.C., 1998. Biological seed treatment for management of sheath blight of rice. *Indian J. Mycol. Plant Path.*, 28: 45–47.

Fravel, D.R., Marois, J.J., Lumsden, R.D. and Connick, W.J., 1985. Encapsulation of potential biocontrol agents in an alginate-clay matrix. *Phytopath.*, 75: 774–777.

Fravel, D.R., 1988. Role of antibiosis in the biocontrol of plant diseases. *Ann. Rev. Phytopath.*, 26: 75–91.

Gabriel, C.J. and Cook, R.J., 1990. Biological control: The need for new scientific frame work. *Biol. Sci.*, 40: 204–206.

Gravel, V.C., H., Martinez, Antoun and Russell, J. T., 2005. Antagonist microorganisms with the ability to control pythium damping-off of tomato seeds in rockwool. *Biocontrol*, 50(5): 771–786.

Han, J.G., Sun, L., Dong, X.Z., Cai, Z.Q., Sun, X.L., Yang, H.L., Wang, Y.S. and Song, W., 2005. Characterization of a novel plant growth-promoting bacteria strain *Delftia tsuruhatensis* HR4 both as a diazotroph and a potential biocontrol agent against various plant pathogens. *System. Applied Microbiol.*, 28(1): 66–76.

Harman, G.E., Chet, I. and Baker, R., 1981. Factors affecting *Trichoderma harzianum* applied to seed as biocontrol agents. *Phytopath.*, 71: 569–572.

He, Q.F., Chen, W.L. and Ma, Z.C., 2002. Purification and properties of antagonistic peptide produced by *Bacillus subtilis* A30. *Chinese J. Rice Sci.*, 16(4): 361–365.

Hitoshi, K., Kazuhiko, N. and Teruyoshi, H., 2004. Suppression of rice blast by phylloplane fungi isolated from rice plants. *J. Gen. Plant Pathol.*, 70(2): 131–138.

Kamala, N. and Vidhyasekaran, P., 1998. Evaluation of a new biopesticide formulation for the management of some foliar diseases of rice. In: *Proceedings of International–Conference on Ecological Agriculture Towards Sustainable Development*, Chandigarh, India.

Karpagavalli, S., Marimuthu, T., Jayaraj, J. and Ramabadran, R., 2001. An integrated approach to control rice blast through nutrients and biocontrol agent. *Res. on Crops*, 2(2): 197–202.

Kazempour, M.N., Pedramfar, H. and Elahinia, S.A. 2003. Effect of certain fungicides and isolates of antagonistic fungi on *Rhizoctonia solani*, the causal agent of rice sheath blight. *J. Sci. Technol. Agri. Natural Resour.*, 6(4): 151–158.

Khan, A. A. and Sinha, A.P., 2006. Influence of formulations, doses and time of application of fungal bio-agents on rice sheath blight. *Annl. Pl. Protec. Sci.*, 14: 157–161.

Khan, A.A., Sinha, A.P. and Rathi, Y.P.S., 2005. Plant growth promoting activity of *Trichoderma harzinaum* on rice seed germination and seedling vigour. *Indian J. Agric. Res.*, 39: 256–262.

Khan, A.A. and Sinha, A.P., 2005a. Comparative antagonistic potential of some biocontrol agents against sheath blight of rice. *Indian Phytopath.*, 58: 41–45.

Khan, A.A. and Sinha, A.P., 2005b. Influence of different factors on the effectivity of fungal bioagents to manage sheath blight of rice, in nursery. *Indian Phytopath.*, 58: 289–293.

Khan, A.A. and Sinha, A.P., 2005c. Influence of soil and nutritional factors on the effectivity of *Trichoderma harzianum* against sheath blight of rice. *Indian Phytopath.*, 58: 276–281.

Khan, A.A., 2003. Factors affecting efficacy of biocontrol agents against sheath blight of rice. *Ph.D. Thesis*, G.B. Pant University of Agriculture and Technology, Pantnagar, 255 p.

Kim, B.S., Kim, K.W., Lee, J.K., Lee Y.W. and Cho, K.Y., 1995. Isolation and purification of several substances produced by *Fusarium graminearum* and their antimicrobial activities. *Korean J. Plant-Pathol.*, 11(2): 158–164.

Krishnamurthy, K. and Gnanamanickam, S.S., 1998.Induction of systemic resistance and salicylic acid accumulation in *Oryza sativa*, L. in the biological suppression of rice blast caused by treatments with *Pseudomonas* spp. *World J. Microbiol. Biotechnol.*, 14(6): 935–937.

Kumar, R.N. and Mishra, R.R., 1994. Interaction studies *in vitro* between brown spot pathogen of paddy and certain epiphytic phylloplane fungi. *Plant-Cell-Incompat. Newsl.*, (26): 40–47.

Lange, L., Breinholt, J., Rasmussen, F.W., and Nielsen, R.I., 1993. Microbial fungicides-the natural choice. *Pesticide Sci.*, 39(2): 155–160.

Latge, J.P. and Soper, R.S., 1977. Media suitable for industrial production of *Entomophthora virulenta* Sygospores. *Biotech. Bioeng.*, 19: 1269–1284.

Lee, Y.H., Shim, G.Y., Lee, E.J., and Mew, T.W., 1990. Evaluation of biocontrol activity of fluorescent pseudomonads against some rice fungal diseases *in vitro* and greenhouse. *Korean J. Plant-Pathol.*, 6(1): 73–80.

Li, Q.Q., Lin, W., Chen, Y.N., Peng, H.W., Meng, J.R. and Wei, S.X., 2000. Biological control of sheath blight of rice by *Bacillus* sp. *Chinese J. Biol. Cont.*, 16(4): 160–162.

Li, X.M., Hu, B.S., Xu, Z.G. and Mew, T.W., 2003. Spatial-temporal colonization patterns of *Pseudomonas fluorescens* strain 7-14 on rice plants. *Acta-Phytopath. Sinica*, 33(5): 468–473.

Liu, B.W., Peng, H.X., Chen, S.Q. and Liu, H.Y., 2005. Study on control effect of *Bacillus* spp. against pathogen of economic crops and stability test. *J. Yunn. Agric. Uni.*, 20(1): 16–19.

Liu, F.Q. and Wang, J.S., 1998. A preliminary study on controlling rice bacterial leaf blight with a virulent gene deleted strain DU728. *Chinese J. Biol. Cont.*, 14(3): 115–118.

Lumsdon, R.D., Locke, J.C. and Walter, J.F., 1991. Approval of *Gliocladium virens* by the US Environmental Protection Agency for biological control of *Pythium ultimum* and *Rhizoctonia solani* damping off. Forth Int. *Trichoderma* and *Gliocladium* Workshop. Belgivate, Italy, July 17–20, Petria, 1: 138.

Manian, S. and Paulsamy, S., 1987. Biological control of sheath blight disease of rice. *J. Bio. Control*, 1: 57–59.

Mathivanan, N., Prabavathy, V. R. and Vijayanandraj, V. R., 2005. Application of talc formulations of *Pseudomonas fluorescens* Migula and *Trichoderma viride* Pers. ex S.F. Gray decrease the sheath blight disease and enhance the plant growth and yield in rice. *J. Phytopath.*, 153(11–12): 697–701.

Meena, B. and Muthusami, M., 1998. Efficacy of different mode of application of *Pseudomonas fluorescens* against rice sheath blight disease. *Oryza*, 35: 293–294.

Mew, T.W, Chen, Z.Y., Lu, F., Yin, S.Z. and Abao, S., 1998. Field experiments on biocontrol using beneficial bacteria in rice. In: *Integrating Science and People in Rice Pest management proceedings of the rice integrated pest management IPM conference*, Kuala Lumpar, Malaysia, p. 99–101.

Mukhopadhyay, A.N., 1987. Biological control of soil borne plant pathogens by *Trichoderma* sp. *Indian J. Mycol. Plant Pathol.*, 16: 1–8.

Mukhopadhyay, A.N. and Kaur, M.P., 1990. Biological control of chickpea nut complex by *Trichoderma harzianum*. In: *Proc. Third Int. Conference on Plant Protection in the Tropics*, Malaysia, March 20–23.

Mukhopadhyay, A.N., Patel G.J. and Brahmbhatt, A., 1986. *Trichoderma harzianum*: A potential biocontrol agent for tobacco damping-off. *Tobacco Res.*, 12: 26–35.

Mukhopadhyay, A.N., Shrestha, S.M. and Mukharjee, P.K., 1992. Biological seed treatments for control of soilborne plant pathogens. *FAO Plant Protect Bull.*, 40: 1–40.

Muthulakshmi, P., 2002. Effect of bacterial and fungal metabolites on rice tungro virus (RTV) infection. *J. Ecobiol.*, 14(2): 99–103.

Nagarajkumar, M., Jayaraj, J., Muthukrishnan, S., Bhaskaran, R. and Velazhahan, R., 2005. Detoxification of oxalic acid by *Pseudomonas fluorescens* strain PfMDU2: implications for the biological control of rice sheath blight caused by *Rhizoctonia solani*. *Microbiol. Res.*, 160(3): 291–298.

Nandakumar, R., Babu, S., Viswanathan, R., Sheela, J., Raguchander, T. and Samiyappan, R., 2001. A new bio-formulation containing plant growth promoting rhizobacterial mixture for the management of sheath blight and enhanced grain yield in rice. *Bio. Cont.*, 46(4): 493–510.

Ngueko, R.B., Shen, Y. and Xu, T., 2002. Induced resistance in rice plants by *Trichoderma harzianum* P1 and its efficiency to *Magnaporthe grisea*. *CRRN, Chinese Rice Res. Newsl.*, 10(3/4): 16–18

Osburn, R.M., Schroth, M.N., Hancock and Hendson, M., 1989. Dynamics of sugarbeet seed colonization by *Pythium ultimum* and *Pseudomonas* spp. effect on seed rot and damping off. *Phytopath.*, 76(6): 709–716.

O'Sullivan, D.J. and O'Gara, F., 1992. Traits of fluorescent *Psuedomonas* spp. Involved in suspension of plant pathogen. *Microbiol. Rev.*, 56: 662–676.

Pande, V. S. and Chaube, H.S., 2003. Effect of *Pseudomonas fluorescens* isolates on sclerotial viability of *Rhizoctonia solani* (Kuhn). *Annls. Plant Prot. Sci.*, 11(1): 57–60.

Papavizas, G.C., 1985. *Trichoderma and Gliocladium*: Biology, ecology and potential for biocontrol. *Annu. Rev. Phytopath.*, p. 23–54.

Papavizas, G.C. and Lewis, J.A., 1981. Introduction of new biotypes of *Trichoderma harzianum* resistant to Benomyl and other fungicides. *Phytopath.*, 71: 247–248.

Papavizas, G.C. and Lewis, J.A., 1982. Introduction and augmentation of microbial antagonists for the control of soil borne plant pathogens. In: *Biological Control in Crop Production*, (Ed.) G.C. Papavizas. Allanheld, Osmum, Granada, U.S.A., pp. 305–322.

Peng, H.X., Liu, B.W., Chen, X.J. and Liu, H.Y., 2002 Biological control of rice blast with *Bacillus* spp. *Chinese J. Biol. Control*, 18(1): 25–27.

Prasad, R.D., Rangeshwaran, R. and Sreerama, P. K., 2000. Mass production of *Trichoderma harzianum* and its shelf-life in different formulations. In: *International Conference on Integrated Plant Disease Management for Sustainable Agriculture Proceedings. Indian Phytopath Society*, 1281 pp.

Rajappan, K., Mariappan V. and Kareem, A.A., 1997. Effect of dried leaf extract of Ipomoea on rice sheath rot pathogen and beneficial microorganisms. *Indian Phytopath*, 50(3): 329–331.

Rangaraj, N., Subramanian B. and Ramalingam, R., 2002. *Pseudomonas fluorescens* mediated antifungal activity against *Rhizoctonia solani* causing sheath blight in rice. *Phytopath. Mediter.*, 41(2): 109–119.

Ray, S., Ghosh, M. and Mukherjee, N., 1990. *Fluorescent pseudomonads* for plant disease control. *J. Mycopathol. Res.*, 28(2): 135–140.

Rosales, A.M. and Mew, T.W., 1982. Antagonistic effects of soil microorganisms on rice sheath blight pathogen. *Int. Rice Res. Newsl.*, 7(6): 12–13.

Sabratnam, S. and Traqair, J.A., 2002. Formulations of a *Streptomuces* biocontrol agent for the suppression of Rhizoctona damping-off in tomato transplants. *Biol. Control*, 23: 245–253.

Sakthivel, N., Sivamani, A., Savithiry S. and Gnanamanickam, S., 1998. Beneficial bacteria for plant disease management. In: *Advances in Research on Plant Pathogenic Bacteria Based on the Proceedings of the National Symposium on Phytobacteriology* held at the University of Madras, 213–220 pp.

Salem, E.A., 1999. Biological control of rice blast disease with antagonistic bacteria under Egyptian conditions. *Egyptian J. Agric. Res.*, 77(3): 995–105.

Saravanakumar, K., Sivanesan, S., Samuel, G., Mohammed, I. and Rajadas, J., 2005. Isolation and partial characterization of antifungal protein from *Bacillus polymyxa* strain VLB16. *Proc. Biochem.*, 40(10): 3236–3243.

Singh, Rajbir and Sinha, A.P., 2005a. Effect of *Pseudomonas fluorescens* formulations on sheath blight and yield component of rice. *Annl. Pl. Protec.*, 13:159–162.

Singh, Rajbir and Sinha, A.P., 2005b. Influence of application methods of *Pseudomonas fluorescens* on rice sheath blight. *Indian Phytopath.*, 58: 474–476.

Someya, N., Nakajima, M., Watanabe, K., Hibi T. and Akutsu, K., 2000. Potential of *Serratia marcescens* strain B2 for biological control of rice sheath blight. *Biocont. Sci. Technol.*, 15(1): 105–109.

Someya, N., Nakajima, M., Watanabe, K. and Akutsu, K., 2005. Synergistic antifungal activity of the culture filtrates of *Serratia marcescens* strain B2 and chemical fungicides against the sclerotial viability of the rice sheath blight pathogen, *Rhizoctonia solani*. *Biocont. Sci.*, 10(3): 97–100.

Soytong, K., 1990. Biological control of rice blast pathogen by coating seeds with *Chaetomium cochlioides* and *C. cuniculorum*. *Khon Kaen Agric. J.*, 18(2): 89–96.

Suzuki, Y., Aryawan, I.G.N., Astika, I.N.S., Suwela, I.N., Widiarta, and Soeroto, I.N., 1996. Effect of major biocontrol agents on the population dynamics of the green rice leafhopper *Nephotettix virescens*, the most efficient vector of rice tungro disease. In: *Biological Pest Control in Systems of Integrated Pest Management, Proceedings of the International Symposium on The use of Biological Control Agents under Integrated Pest Management*, pp. 252–260.

Tang., J.B., Ma, B.T., Li, L.X., Wang, P., Zheng, A.P. and Chen, H., 2002. Biological control of rice sheath blight with *Trichoderma* and *Trichoderma* like species. *Chinese J. Rice Sci.*, 16(1): 63–66.

Tong, Y.R., Ma, Z.C., Chen, W.L. and Li, D.B., 1999. Purification and partial characterization of antagonistic proteins from *Bacillus subtilis* B034. *Acta Microbiol. Sinic.*, 39(4): 339–343.

Upadhyay, J.P. and Mukhopadhyay, A.N., 1985. Effect of non-volatile and volatile antibiotic of *Trichoderma harzianum* on growth of *Scleroetium rolfrii*. *Indian J. Mycol. Plant Path.*, 13: 232–233.

Viswanathan, R.P. and Narayanasamy, 1990. Potential antagonist of rice sheath rot (ShR) pathogen. *International Rice Res. Newsl.*, 15(2): 26–27.

Wang, Y., Shen, Y., and Xu, T., 2000. Study on *Trichoderma harzianum* strains to control rice sheath blight. *Acta Phytophylac. Sinic.*, 27(2): 97–101.

Webster, J. and Lomas, N., 1964. Does *Trichoderma viride* produce gliotoxin and viridin. *Trans Br. Mycol. Soc.*, 47: 537–540.

Weindling, R., 1934. Studies on a lethal principle effective in the parasitic action of *Trichoderma lignorum* on *Rhizoctonia solani* and other soil fungi. *Phytopath.*, 24: 1153–1179.

Weindling, R., 1941. Experimental consideration of the mold toxins of *Gliocladium* and *Trichoderma. Phytopath.*, 31: 991–1003.

Weller, D.M., 1988. Biological control of soil-borne plant pathogens in the rhizosphere with bacteria. *Annu. Rev. Phytopath.*, 26: 379–407.

Whipps, J.M., 2001. Microbial interactions and biocontrol in the rhizosphere. *J. Exp. Bot.*, 52: 487–511.

Xu, J.P., Guan, J.Z., Chen, W.L., Gong, H.F. and Li, D.B., 1998. Diethyl sulfate induced mutation and screening of *Bacillus subtilis* B826. *J. Zhejian. Agric. Univ.*, 24(6): 595–596.

Yi, T.Y., Gao, B.D., He, K. and Chen, X.Y., 2000. Studies on the screening of biocontrol bacteria to rice sheath blight disease. *J. Hunan Agric. Uni.*, 26(2): 116–118.

Insect Pest and Disease Management *Pages* ***209–226***
Editor: **Dr. D. Prasad**
Published by: **DAYA PUBLISHING HOUSE (2016)**

Chapter 15

Eco-friendly Approaches for Sclerotinia Disease Management in Vegetable Crops

Ramesh Singh[1] Udit Narain[2], Alka[3] and Sandhya Kant[4]
[1]Department of Plant Pathology, T.D.P.G. College, Jaunpur – 222 002, U.P.
[2]Department of Plant Pathology,
C.S. Azad University of Agriculture and Technology, Kanpur – 208 002, U.P.
[3]Dayanand Girls College, Kanpur – 208 001, U.P.
[4]D.A.V. College, Kanpur – 208 001, U.P.

Introduction

Vegetables have got an important place in agriculture development, human diet and economy of the country. All the developed and developing countries realized the importance of vegetables as an essential diet due to medicinal and nutritional value for human health. There is a steady upward trend in vegetable production. China is ranking first in the world and currently produces 237 million tonnes of vegetable. India has made a quantum jump in vegetable production securing the second position in the world. The total production of vegetable is more than 91 million tonnes in the country. Diseases and pests are the most important factors for low vegetable production. Various fungi, bacteria, viruses, phytoplasma and nematodes are responsible for causing disease in vegetables resulting heavy reduction in yield and quality. Due to the wide range of agroclimatic conditions prevailing in our country, maximum diseases like late blight, wilt, white rot, fruit rot, root rot, black rot, sclerotinia blight, alternaria leaf spot and damping off cause serious damage to cole crops, tomato, chilli and brinjal. A total of 65 million tonne pre harvest losses in vegetables disease alone in the world have been estimated that account about 10.5 per cent. *Sclerotinia sclerotiorum* de Bary is a distructive soil borne plant pathogen infecting wide range of host crops in various parts of world. It is likely that the population of *S. sclerotiorum* in our soils has gradually built up high under the changed conditions of intensive cultivation and is producing new disease problem in vegetable crops. The original concept that *S. sclerotiorum* inhibits only soil and plant roots and causes stem rot, sclerotinia blight, white rust, fruit rot and black rot has been modified. Pesticides indiscriminatory kill the pathogen and beneficial micro-organisms both. They are becoming expensive year after year and loosing their efficacy also. Development of pesticide resistant, new strain and secondary pests is the result of indiscriminate use of pesticides. At the present time, the systemic fungicides are being frequently used by the vegetable growers. It is always resulting certain level of toxicity in the plant system and thus health hazardous. It is also disturbs complete ecology of the soil microbial diversity of the ecosystem. Effective management packages can not be advocated in absence of details symptomatology and pathological information. Considering the necessity of environmental safety, pesticide residue problem and cost effective production of the crops, it is desired to develop ecofriendly management approaches for the whole year.

Geographic Distribution

Sclerotinia sclerotiorum occurs in relatively cool and moist areas of the world, as the generalized statement made by many individuals. However, *Sclerotinia sclerotiorum* also occurs in localities generally considered to be hot and dry. When the temperature approaches the freezing point (0°C) or when temperature more than 32°C prevails, *S. sclerotiorum* is much less active than at temperature between these generalized extremes. The precise location and the environmental condition prevailing at the time the diseases or fungus observed were not specified. *S. sclerotiorum* has been reported from many countries located in all continents. It is probable that the fungus occurs somewhere in almost every country.

Symptomatology

The disease is observed when low temperature, cloudy weather and high atmospheric humidity accompanied by soil moisture prevail during December to January. Infection starts during flowering stage as soft, wet, water-soaked, rotting of stem, leaves, fruits, inflorescence and flowers and are observed with white fungal growth over it. The mycelium develops honeydew stage soon after colonization and later on after drying, the entire portion is converted into compact mycelial mat followed by the formation of hard sclerotial body. Fruits are severely affected where sclerotic are formed both in and out side of the fruits. It produced symptoms on lettuce as soft watery decay (Purdy, 1952), on tomato as water-soaked rotting of stem, leaves, fruits and flowers (Pandey *et al.*, 2002), on cauliflower as water-soaked rotting of curd, petiole and stalk (Shyam *et al.*, 1994) and was recorded severe wilting and stunting of *Capsicum anuum* (Srivastava and Diwakar, 1987). The symptom was observed as water-soaked of stem and petiole on brinjal (Singh, 2003).

Taxonomy and Nomenclature

Pezziza sclerotiorum of (Libert, 1937) was transferred by Fuckel to his new genus, *Sclerotinia* and the specific epithet was changed by him to *libertiana* resulting in the combination of *Sclerotinia libertiana* (Fuckel, 1970). However, it is preferred to use the original specific epithet in the combination of *Sclerotinia sclerotiorum* Fuckel (De-Bary, 1984).

Sclerotinia sclerotiorum is the typical representative of the family Sclerotiniaceae of sub-division Ascomycotinia (Ainsworth, 1973). The taxonomic position of *Sclerotinia sclerotiorum* (Lib.) deBary, *S. trifoliorum* and *S. minor* Jagger has not yet been resolved. Many authors (Jagger, 1920; Keay, 1939;

Whetzel, 1945; Bjorling, 1951; Williams and Western, 1965) considered them completely different, while others (Purdy, 1955; Walker, 1969; Morrall, 1976) recognized only one species *i.e. Sclerotinia sclerotiorum.*

Morphological Characters of the Pathogen

The morphological structure of *Sclerotinia sclerotiorum* deBary was given by Kolte (1985), the mycelium was thin 9 to 8 µm in diameter with lateral branches of smaller diameter than the main hyphae. The vegetative hyphae were multinucleate with the haploid eight chromosome and also described the sclerotia were black, round to semispherical in shape and measured 3 to 4mm in diameter. The sclerotic could be easily detached from staked apothecia and several apothecia grew form a single sclerotium (Willet and Wong, 1980). While describing the structure of apothecia of the fungus (Kosasih and Willet, 1975) it was found that hymenium is made of asci and paraphysis, which measured 119–162 × 6.5–10.9 µm in size, and are inoperculate, cylindrical, narrow, round at the apex with 8 ascospores in each ascus. Each ascospore is hyaline, ellipsoid, smooth wall and measure 10.2–14.0 × 6.4–7.7 µm in size. Studies of apothecial morphogensis of *Sclerotinia sclerotiorum* (Singh *et al.*, 1998) formed several anomalies like apical proliferation of stipes, development of secondary normal apothecia and rudimentary stipes from an aborted apothecium, dichotomous branching of stipes bearing normal apothecia apically, dwarf stipes each bearing several apothecia (Abd-El Moity and Shatla, 1981; Abdou *et al.*, 1982; Achbani *et al.*, 1995; Adams, 1989; Adams, 1990; Adams and Ayers, 1979; Adams and Ayers, 1981; Adams and Ayers, 1982; Adams and Ayers, 1983; Adams and Fravel, 1990) on its swollen head.

Survival and Spread of Disease

Sclerotia are the structure which allow species of *Sclerotinia* to survive for long periods of time under adverse conditions. The sclerotia of *Sclerotinia trifoliorum*, (Dillon-Wester *et al.*, 1946) estimated that sclerotia could survive for 6-8 years. Tribe (Tribe, 1957) reported that sclerotia of *Sclerotinia trifoliorum* buried at a depth greater than 44 cm persisted for 8 years. There is opinion (Davis, 1925) that sclerotia of *Sclerotinia sclerotiorum* near the soil surface do not retain viability for more than 1 year. However after 4 years rotation the sunflower could be grown on a field with a history of Sclerotinia wilt (Young and Morris, 1927). Under favourable dry weather condition, sclerotia of *Sclerotinia sclerotiorum* remain viable for at least 10 years and this is in contrast to the estimated 4-5 years of survival for sclerotia in field soil. *Sclerotinia sclerotiorum* could survive through seed in the form of mycelium infecting the testa of tomato seed (Neergaard, 1958). The seed borne nature of *Sclerotinia sclerotiorum* was established (Kappor, 1988) and found that more than 40 per cent seeds of brinjal were severely affected by the pathogen.

Sclerotinia spp. became established and spread from field to field, and from one geographical area to another, by several means in which windborne ascospores can be the major means of field to field spread (Backhouse and Stewart, 1987). *Sclerotinia* spp. also may be spread from field to field in soil adhering to seedling, farm equipments, animals or man (Dillon-Wester *et al.*, 1946; Starr *et al.*, 1953) in the form of sclerotia or as mycelium in infected host tissues. Irrigation also has been shown to be involved in the spread of *Sclerotinia* spp. from field to field (Steadman *et al.*, 1975). Probably the greatest potential for long distance spread of *Sclerotinia* spp. is either by seed infected with mycelia or by seed contamination with sclerotia. Sclerotinia infected or infested seed has been reported for cabbage and cauliflower (Neergaard, 1958) and bean (Starr *et al.*, 1953).

Host Range

Sclerotinia sclerotiorum is a cosmopolitan species with a host range of over 350 species in 60 families (Schwartz, 1974). Twenty four plant species reported as the hosts of *Sclerotinia sclerotiorum* (28) and some of them were *Brassica* spp., *Capsicum* spp., sesame, coriander, chickpea, sunflower and tobbaco plants. Roy (Roy, 1973) studied the host-range of *Sclerotinia sclerotiorum* and listed 27 hosts of which 13 were new records in India (Roy, 1973). The number of plants species were reported as the hosts of *Sclerotinia sclerotiorum* from time to time by many workers (Morrall, 1976; Purdy, 1979; Bilgrami *et al.*, 1979; Cristea *et al.*, 1985; Shin and Lee, 1987; Rathore *et al.*, 1993; Satter *et al.*, 1995; Ghasolia *et al.*, 2004).

Epidemiology

Species of the genus *Sclerotinia* can function either as soil borne or airborne pathogens. Infection of aboveground plant parts result from ascospore inoculum, where as soil line infection may result either from ascospore or sclerotia; below ground infection, however result from mycelial germination of soilborne sclerotia. Accordingly, the epidemiology of two type of infections incited by *Sclerotinia* spp., is quite different and the effect of weather factors on their incidence and development differs considerably.

The infection of beans occurs on aboveground parts and it seems unlikely that the large sclerotia of *Sclerotinia sclerotiorum* are an important form of inoculum and they function essentially to produce airborne ascospores (Abawi and Grogan, 1975; Cooke *et al.*, 1975; Saiio, 1977; Schwartz and Steadman, 1978). Apothecia formation was observed at mean temperature of 9-17°C for 14 days periods. Number of apothecia and level of aerial infection, were not related to the amount of total monthly rainfall (Letham *et al.*, 1967). The outbreak of white blight of mustard caused by *Sclerotinia sclerotiorum* of all the above ground parts of the plants was in cool and wet environment. A temperature of 20-25°C and relative humidity of 90-10 per cent was congenial for stalk rot *(Sclerotinia sclerotiorum)* development in cauliflower (Sharma and Sharma, 1985). Low temperature and high moisture favoured the development of sclerotia of *Sclerotinia sclerotiorum* on the surface of the pea under natural conditions (Sharma, 1985). Apothecia production in case of *Sclerotinia* spp. was governed by temperature between 0-30°C followed by intermediate R. H. 98 per cent (Achbani *et al.*, 1995).

Disease Management

Management mainly aims to prevent epidemic and widespread out breaks of disease in host plant population. This may be achieved through reducing the initial population of the pathogens and of them limiting the rate of increase in reduced population in most diseases of caused by *Sclerotinia sclerotiorum*, *S. minor* and *S. trifoliorum* (Vander, 1968).

The explosive pathogenicity of those fungi under favourable condition and the ability of their sclerotia to withstand adverse conditions allow them to be successful pathogen on many crops. The use of protectant chemicals materials that inhibit and destroy sclerotia, biological and modified cultural practices may help in achieving ecofriendly management of the Sclerotinia disease.

Biological Control

Control of plant diseases particularly those which are soil borne in nature, through biological agents has now become an integral part of plant disease control measure, mainly due to hazardous of chemicals in soil and environment. Only recently, the importance of biological control of *Sclerotinia* spp. in vegetable crops has been realized.

Isolation and Evaluation of Potential Bio-agents

Isolation and recovery of numerous potentially antagonistic fungi and bacteria have been reported from naturally decomposing sclerotia of *Sclerotinia* spp. The following organisms which have been isolated from sclerotia in nature have been the most well studied in recent years.

Trichoderma spp. (Adams and Ayers, 1979; Sariah and Cheng, 1999), *Coniothyrium minitans* (Adams and Ayers, 1979; Huang, 1981; Huang and Kokko, 1987; Trutmann *et al.*, 1982), *Gliocladium virens* from sclerotia of *Sclerotinia sclerotiorum* (McCredie and Sivasithamparam, 1985; Phillips, 1986; Tu, 1980) and of *S. rolfsii* (Papavizas, 1985), *Sporidesmium sclerotivorum* from sclerotia of *Sclerotinia sclerotiorum* and *S. minor* (Adams and Ayers, 1979; Ayers and Adams, 1979), *Penicillium* spp. from sclerotia of *Sclerotinia sclerotiorum* (Merriman, 1976; Rai and Saxena, 1975) and on *S. cepivorum* (Harrison and Stewart, 1988; Oliveira de *et al.*, 1984; Utkhede and Rahe, 1980), *Bacillus subtilis* from sclerotia of *Sclerotinia cepivorum* (Utkhede and Rahe, 1980; Wong and Hughes, 1986) and several other organisms have recently been described and showed to also have potential as antagonists either *in vitro* or *in vivo*. These organisms include *Alternaria* and *Cladosporium* sp. (Boland and Hunter, 1988; Boland and Inglis, 1989), *Chetomium globosum* (Harrison and Stewart, 1988), *Streptomyces diastaticus* (Singh and Dwivedi, 1987) and *Teratosperma oligocladium* (Adams, 1989; Adams and Ayers, 1983; Uecker *et al.*, 1980).

To conduct the evaluation of isolated microorganisms for antagonistic potential under laboratory conditions, a number of widely different technique have been utilized and these are described below:

1. Pairing of the micro-organisms with the pathogen on agar medium in dual culture and evaluating pathogen growth inhibition or extent of growth of the antagonist over the pathogen or reduction in sclerotial formation production (Deb and Dutta, 1991; Harrison and Stewart, 1988; Henis *et al.*, 1984b; Jackson *et al.*, 1991a; Lynch, 1987; Phillips, 1986; Phillips, 1989; Zhou and Reeleder, 1990; Zazzerini *et al.*, 1987). This method is very commonly used and can provide the most rapid results, but is not necessarily the most appropriate for extrapolation to how the antagonist may behave under natural field conditions.
2. Inoculation of sclerotia of the pathogen with propagules of the antagonist and evaluating the degree of decomposition and reduction in germinability or survival (Adams, 1989; Ayers and Adams, 1979; Backhouse and Stewart, 1989; Hadar and Gorodecki, 1991; Henis *et al.*, 1984a; Mueller *et al.*, 1985; Budge *et al.*, 1995; Srivastava and Diwakar, 1987; Whipps and Budge, 1990). This method is rapid and can provide a much better indication of the suitability of the bio-control agents for reducing sclerotial viability in soil.
3. Microscopic evaluation of mycoparasitism expressed as hyphal infection or sclerotial colonization (Abd-El Moity and Shatla, 1981; Elad *et al.*, 1983; Oliveira de *et al.*, 1984; Stewart and Harrison, 1988; Tong-Kwee and Keng, 1990; Trutmann and Keane, 1990; Trutmann *et al.*, 1982), which may provide an indication of one of the potential mode of actions of the antagonist.
4. Screening of antagonists for production of antibiotics, lytic enzymes, or other toxins (Abd-El Moity and Shatla, 1981; Harrison and Stewart, 1988; Jackson *et al.*, 1991a; Oliveira de *et al.*, 1984; Tong-Kwee and Keng, 1990), which may not be indicative of their involvement in bio control activity.
5. The inoculum of host tissue with antagonist preyer to or simultaneously with the pathogen and assessing pathogen growth, sclerotial formation and disease development (Boland and Inglis, 1989; Lynch, 1987; Trutmann *et al.*, 1982; Whipps, 1987). This method is less frequently

used but can provide useful preliminary indication of bio-control potential under control environment conditions.

The antagonists which have been the most well studied for biological control of all these sclerotium forming plant pathogens are described in more detail below:

Trichoderma

Trichoderma spp. are common soil saprophytes which have prolific growth rate and capability for sporulation when in culture. *Trichoderma* spp. have been studied extensively and showed promising results (Chet, 1987; Papavizas and Collins, 1990; Papavizas and Lumsden, 1980). Many *Trichoderma* isolate was antagonistic to a range of fungi producing volatile and non-volatile antibiotics (Dennis and Webster, 1971; Ghisalberti and Sivasithamparam, 1991; Upadhyay and Mukhopadhyay, 1983). The general mechanism believed to be involved in bio-control is a combination of competition, lysis, antibiosis and mycoparasitism.

Trichoderma spp. have also been reported to parasitize hyphae of *S. rolfsii* and *S. cepivorum* and induce lysis (Chet, 1987; Deb and Dutta, 1988; Elad *et al.*, 1983; Elad *et al.*, 1980; Oliveira de *et al.*, 1984; Tong-Kwee and Keng, 1990), although the importance of this mechanism in reducing pathogen survival under natural field condition is difficult to determine. Sclerotia of *Sclerotinia* spp. have also been reported to be colonized by species of *Trichoderma*, which resulted in lower germination and reduced survival in soil (dos Santos and Dhingra, 1982; Budge and Whipps, 1991a; Abawi *et al.*, 1985; Huang, 1980; Imolehin and Grogan, 1980; Knudsen *et al.*, 1991; Trutmann and Keane, 1990; Zazzerini and Tosi, 1985). The various events which occur during parasitism of sclerotia and hyphae of *Sclerotinia sclerotiorum* by *Trichoderma* spp. have also been described in detail (Trutmann and Keane, 1990). Application of *Trichoderma* inoculum into soil under field conditions was shown to reduce survival of sclerotia of *Sclerotinia sclerotiorum* (Huang, 1980; Knudsen *et al.*, 1991; Trutmann and Keane, 1990). *Trichoderma* spp. have been successfully used for management of disease caused by *Sclerotinia sclerotiorum* in Ajowan (Singh *et al.*, 2003a), celery and lettuce (Budge and Whipps, 1991a), rape and mustard (Singh, 1998), tomato (Kokalis, 2002) and egg plant (Sariah and Cheng, 1999) and *Sclerotinia cepivorum* on onion and garlic (Abd-El Moity and Shatla, 1981; Oliveira de *et al.*, 1984). The reduction in disease which achieved ranged from 36–83 per cent.

Coniothyrium

C. minitans, a pycnidial forming fungus, has been well described as a mycoparasite of *Sclerotinia* spp. under laboratory and natural conditions (Huang, 1980, Huang, 1981; Huang and Hoes, 1976; Huang and Kokko, 1987; McCredie and Sivasithamparam, 1985; 90. Phillips, 1989; Phillips and Price, 1983; Whipps and Budge, 1990). From *in vitro* studies, the fungus was shown to parasitize the sclerotia and subsequently formed pycnida on and within the sclerotia (McCredie and Sivasithamparam, 1985; Phillips and Price, 1983, Tu, 1984; Cael *et al.*, 2001, Whipps and Budge, 1992; Whipps *et al.*, 1993). Sclerotia formed in natural field soil and on infected host tissue were also found to be parasitize by *C. minitans* (Huang, 1981; Zazzerini and Tosi, 1985; Budge *et al.*, 1995; Mc-Laren *et al.*, 1996; Singh *et al.*, 2003a).

Application of *C. minitans* inoculum to soil has been reported to reduce the survival of sclerotia of *Sclerotinia sclerotiorum* (Bullock *et al.*, 1986; Huang, 1980; Mc-Quiken *et al.*, 1995; Cael *et al.*, 2001). In addition, development of Sclerotinia wilt of sunflower was reduced and yield increased under field conditions (Huang, 1980) primarily due to reduction in initial infection from sclerotia. In greenhouse, *C. minitans* was shown to reduce disease on lettuce and to also reduce survival of sclerotia of *Sclerotinia sclerotiorum* (Budge and Whipps, 1991). Application of inoculum of *C. minitans* as soil and seed dressing against *Sclerotinia cepivorum* on onion seedling reduced the growth of *Sclerotinia cepivorum* (Ahmed and Tribe, 1977). The ability of *C. minitans* to parasitize sclerotia of *Sclerotinia sclerotiorum* inside host plant root and stems as well as those on root surface of infected sunflower plants was demonstrated (Huang and Hoes, 1976). Up to 65 per cent of sclerotia of *S. trifoliorum* were destroyed in field soil by application of pycnidal dust preparation of *C. minitans* (Turner and Tribe, 1975).

Gliocladium

Gliocladium virens has been found to be associated with the sclerotia of *Sclerotinia sclerotiorum* (Liu, 1989; McCredie and Sivasithamparam, 1985; Phillips, 1986; Phillips, 1989; Tu, 1980; Singh, 2000) in natural soil. *In vitro* studies have been shown that fungus is a mycoparasite which can parasitize both mycelia and sclerotia of *Sclerotinia sclerotiorum* and can inhibit sclerotium formation in culture (Tu, 1980; Singh *et al.*, 2000; Palat *et al.*, 2004; Singh, 2003). When evaluated for antagonistic capability against various sclerotium forming pathogens, the fungus was shown to reduce the viability and to destroy the sclerotia of *Sclerotinia sclerotiorum, S. minor, S. cepivorum* and *Sclerotium rolfsii* (Singh *et al.*, 2003a; McCredie and Sivasithamparam, 1985; Phillips, 1986; Stewart and Harrison, 1988; Whipps and Budge, 1990). Inhibition of carpogenic germination of sclerotia of *Sclerotinia sclerotiorum* has also been described (Mueller *et al.*, 1985; Phillips, 1989; Tu, 1980). Application of *Gliocladium virens* inoculum to soil was shown to reduce the disease due to *Sclerotinia sclerotiorum* on Ajowan under wirehouse condition (Singh *et al.*, 2003a), on rajmash (Palat *et al.*, 2000), on brinjal (Singh, 2003) and on carrot and tomato (Ristanio *et al.*, 1994).

Sporidesmium

Sporidesmium sclerotivorum is a slow growing Dematiaceous Hypomycetes which forms conidia (both macro and micro), chalmydospores and microsclerotia in culture (Adams, 1990). The fungus has been isolated from soil in numerous area and has been shown to be an effective, near obligate mycoparasite against *Sclerotinia sclerotiorum, S. minor, S. cepivorum* (Adams and Ayers, 1981; Adams and Ayers, 1983; Ayers and Adams, 1979; Bullock *et al.*, 1986). It has been suggested the *Sclerotinia sclerotiorum* may be involved in natural decline of population of those pathogen in many soils (Adams and Ayers, 1981). Germination of the macro-conidia is stimulated by sclerotial exudates, and the germ tubes can grow towards and infect the sclerotia (Ayers and Adams, 1979). Growth of the antagonist from infected sclerotia into soil can occur over a distance of up to 3 cm (Adams and Ayers, 1983). When applied to natural soil, *Sclerotinia sclerotiorum* reduced the survival of sclerotia of *Sclerotinia sclerotiorum* and *S. minor* within 6-10 weeks. Under field conditions, the inoculum density of *S. minor* was reduced to a level at which significant disease reduction was observed. Futhermore, the mycoparasite was able to survive and propagate itself (Adams and Ayers, 1982; Adams and Fravel, 1990).

Penicillium

Penicillium spp. isolated from decaying sclerotia were shown to be antagonistic to *Sclerotinia sclerotiorum in vitro* (Liu, 1989; Rai and Saxena, 1975), when paired in dual culture with these pathogen, the inhibition zone developed (Harrison and Stewart, 1988; Oliveira de *et al.*, 1984; Rai and Saxena, 1975).

Bacillus

Bacillus spp. are the promising bio-control agents as they produce endospore that are tolerant to heat and desiccation. Bacterium is isolated from sclerotia of *S. cepivorum* from a number of different areas (Backhouse and Stewart, 1989; Utkhede and Rahe, 1980; Utkhede and Rahe, 1983a; Wong and

Hughes, 1986; Zazzerini *et al.*, 1987). Antagonism to *Sclerotinia sclerotiorum* can be demonstrated by marked inhibition zones which develop in dual culture, or by inhibition of sclerotial germination (Backhouse and Stewart, 1989; Utkhede and Rahe, 1980; Singh *et al.*, 2003a; Palat *et al.*, 2000). The bacterium may be associated in nature only the sclerotial surface (Backhouse and Stewart, 1989) or also occur internally (Utkhede and Rahe, 1983b). When applied as soil amendments, *Bacillus sublitis* significantly reduced the development of Sclerotinia blight of brinjal under green house condition (Singh, 2003).

Cultural Management

We know the importance of cultural practices from ancient time. Suitable adjustment in cultural practices can modify the environment in such a manner that becomes unfavourable for the pathogen and disease development. It is scientifically proved that if we have done slight change in common cultural practices, the amount of inoculum and disease can be reduced to a great extent there by increasing the production of vegetable in many folds.

Organic Amendments

Soil amendments is one of the effective methods of bio-control. Amendments in the form of plant debris, green manuring, farm yard manure, compost, oil cakes and fertilizer are known to improve crop productivity by improving nutrient status and soil tilth. The disease may be reduced by certain amendments due to suppressiveness. The disease caused by *S. sclerotiorum* was a serious problem in cauliflower seed crop and the application of farm yard manure at 10-40 tonns/acre gave effective control and reduced number of sclerotia (Sandhu, 1992). The sclerotinia soft rot of lettuce and survival rate of sclerotia of *S. sclerotiorum* were reduced by addition of organic amendments in field (Asirif *et al.*, 1994). The stalk rot of cauliflower caused by *S. sclerotiorum* in field condition reduced the disease intensity by the application of different soil amendments (Aujala *et al.*, 2001). The application of different soil amendments against *S. sclerotiorum* on Ajowan (Singh *et al.*, 2000), rajmash (Palat, 2001), brinjal (Singh *et al.*, 2005a) were reduced the disease incidence in field conditions.

Adjustment of Sowing Dates

The choice of sowing dates in relation to crop diseases has one principal aims *viz*: to reduce to a minimum the period over which infective agent (propagules, vectors) meets susceptible host tissue. Since most crops grow under ranges of temperature and humidity, wider than those favouring pathogen development, sowing or planting at seasons which gave the crop and advantage is obviously strategy. The incidence of a disease in infested *S. sclerotiorum* field, was minimum in early sown crop (Singh *et al.*, 1979; Singh and Singh, 1984). The disease of *Trifolium alexandrium* was effectively controlled by sowing the crop in November rather than in late September–October (Singh and Singh, 1995). The different dates of sowing on the incidence of white rot of pea caused by *S. sclerotiorum*, early sown (21 October) crop resulted in the highest, where as sowing in the last week of November, resulted in minimum disease incidence (Singh, 1996). The adjustment of sowing date on the incidence of stem rot of Ajowan (Singh *et al.*, 2002) and of rajmash (Palat, 2001) and Sclerotinia blight of brinjal (Singh, 2003) caused by *S. sclerotiorum* was recorded minimum in field condition. The effect of sowing date on the severity of Sclerotinia rot was determined. A significant reduction in disease severity in late season crop was observed (Singh and Tripathi, 1995).

Moisture Regulation

Flooding of soil has been effective means for suppression of soil born plant pathogen (Cooke and Barker, 1983). Flooding of soil during warm periods can be very effective in biological suppression of the pathogen (Cooke and Barker, 1983). Flooding a field continuously for 23-45 days of cycles of alternate flooding and drying led to destruction of sclerotia of *S. sclerotiorum* and reduced disease in Florida. The sclerotia of *S. sclerotiorum* and *S. minor*, when dried for short periods and remoistened in soil, leaked nutrient, were rapidly colonized by micro-organisms were delayed in 2-3 week (Smith, 1972), although the implication for disease control were obvious. Sclerotia had to loss gol of their moisture before survival was affected. It would be difficult to reduce moisture in field soil enough to allow sclerotia to loose that amount of water.

Crop Rotation

Crop rotation is a disease control recommendation and often has been advocated for control of Sclerotinia disease. However, sclerotia survive in the soil at least 3 years and tillage operation generally assure the presence of sclerotia at or near the soil surface (Cooke *et al.*, 1975). In Nebraska, sclerotial population was comparable in all field sampled in various crop rotations despite differences in occurrence of the host in the previous crop history (Schwartz and Steadman, 1978). In addition, apothecia were found in field of host crop. Thus, it would assure that air-borne or irrigation water-borne ascospores dissemination could occur. Although not effective for control of Sclerotinia diseases caused by the large sclerotia types, crop rotation is a necessary practice for minimizing the many other plant disease and should be used also.

Soil Texture

The incidence of stem rot of Ajowan caused by *S. sclerotiorum* was tested in different types of soil and the disease intensity was maximum in sandy soil whereas, it was minimum in black soil (Singh, 1999; Singh, 2000). The stem rot of rajmash and Sclerotinia blight of brinjal was noticed the maximum disease in sandy and minimum in black soil (Palat, 2001; Singh, 2003). During the investigation regarding sclerotia germination, it was observed that sclerotia germinate better in sandy soil than other soil which are compact in nature (Singh, 2000).

Chemical Control

The endeavours made by several workers to evaluate efficacy of host extract, biopesticides and chemical in inhibition of growth of *S. sclerotiorum in vitro* gave premising results. The biopesticide (Nimbidine) gave about 94 per cent growth inhibition over control (Singh *et al.*, 2005b; Singh *et al.*, 2003b). The fungicides like Bavistin, Vitavax, Vapam and Topsin-M completely inhibited the growth of the fungus *in vitro* (Singh *et al.*, 2003b; Palat *et al.*, 2004; Roy and Saikia, 1976; Abdou *et al.*, 1982; Singh and Gangopadhay, 1984). Germination of sclerotia was prevented by Vitavax, Benomyl and Thiram at 5ppm and caused about 60 per cent inhibition of mycelial growth of *S. sclerotiorum*, Captan and Thiram completely inhibited ascospore germination at 3-4 ppm. Benomyl at 1000 ppm reduced the viability of sclerotia (Howthorne and Jarvis, 1973). In glasshouse condition against soft rot of cucumber it was found that use of Benomyl reduced the disease incidence (Bycheno and Simmul, 1974). The effect of an aqueous leaf extract of garlic on *S. sclerotiorum* was studied and it was found that the growth of the pathogen reduced greatly (Singh *et al.*, 1979). Germination of sclerotia was also prevented by the use of Vitavax at 10 ppm concentration (Abdou *et al.*, 1982). The garlic extract has fungicidal and inhibitory action against fungal spore (Moore and Atkins, 1977). Antifungal properties of neem product has also been noticed (Murty and Sirsi, 1957).

The efficacy of Bavistin, Vitavax, Vapam and Topsin -M under field condition against *S. sclerotiorum* in *Tranchyspermum ammi* and *Phaseolus vulgare* was found to be effective in controlling the disease (Singh *et al.*, 2003b; Palat *et al.*, 2004). The Sclerotinia stem rot of French bean caused by *S. sclerotiorum* was effectively controlled by spraying of Benomyl at the rate of 1 g per litre at full bloom stage (Peg, 1983). The spraying of Benomyl (0.1 per cent) or Fundazol (0.1 per cent) gave best result against white rot of cucumber (Dujickova, 1977). Good control of carrot leaf blight due to *S. sclerotiorum* was achieved the application of Ronlin @2.0 kg/ha or Benomyl @4.0 kg/ha four times at an interval of two weeks (Tate, 1980). In tomato and cauliflower crops the excellent control was found with application of Benomyl @ 100g/100 liter water sprayed at 14 days intervals (Letham *et al.*, 1967). The spraying of Vitavax (0.1 per cent) at 10 days intervals gave almost complete control of white rot of seed crop of cauliflower caused by *S. sclerotiorum* (Singh and Gangopadhay, 1984). Combined application of Benoyml and Thiram improved control of *S. sclerotiorum* causing stem rot of bean, when applied in soil. They also effectively reduced the sclerotia germination, apothecia production by germinating sclerotia, rate of apothecia production and delayed apothecial emergence (Yarden *et al.*, 1986). The disease could be minimized by the spraying of Vitavax, Topsin-M, Indofil M-45, and Thriam (Fayslle, 1995). Foliar spray of Carbendazim @ 0.1 per cent was effective at flowering stage alternating with one spray of Mancozeb @ 0.25 per cent at 10 days intervals, during cool, cloudy and high humidity weather against *S. sclerotiorum* on tomato, chilli and cauliflower (Pandey *et al.*, 2002).

References

Abawi, G.S., Grogan, R.G and Duniway, J.M., 1985. Effect of water potential on survival of sclerotia of *Sclerotinia minor* in two California soils. *Phytopathology*, 75: 217–221.

Abawi, G.S. and Grogan, R.G., 1975. Source of primary inoculum and effect of temperature and moisture on infection of bean by *Sclerotinia sclerotiorum*. *Phytopathology*, 65: 300–309.

Abd-El Moity, T.H. and Shatla, M.N., 1981. Biological control of white rot disease (*Sclerotium cepivorum*) of onion by *Trichoderma harzianum*. *Phytopath. Z.*, 100: 29–35.

Abdou, Y.A.M., El. Sharkawy, T.A., Osman, A.R. and Rajab, M.M., 1982. *In vitro* test for control of *Sclerotinia sclerotiorum*. *Phytopathology*, 14: 87–92.

Achbani, E.N., Tourvieille, D. and Lenormand, M., 1995. Apothecia production in *Sclerotinia* spp. *Alawamina*, 90: 31–50.

Adams, P.B., 1989. Comparison of antagonists of *Sclerotinia* spp. *Phytopathology*, 79: 1345–1347.

Adams, P.B., 1990. The potential of mycoparasites for biological control of plant diseases. *Annu. Rev. Phytopathol.*, 28: 59–72.

Adams, P.B. and Fravel, D.R., 1990. Economical biological control of sclerotinia lettuce drop by *Sporidesmium sclerotivorum*. *Phytopathology*, 80: 1120–1124.

Adams, P.B. and Ayers, W.A., 1979. Ecology of *Sclerotinia* spp. *Phytopathology*, 69: 896–899.

Adams, P.B. and Ayers, W.A., 1981. *Sporidesmium sclerotivorum*: distribution and function in natural biological control of sclerotial fungi. *Phytopathology*, 71. 90–93.

Adams, P.B. and Ayers, W.A., 1982. Biological control of sclerotinia lettuce drop in the field by *Sporidesmium sclerotivorum*. *Phytopathology*, 72: 485–488.

Adams, P.B. and Ayers, W.A., 1983. Histological and physiological aspect of infection of sclerotia of two *Sclerotinia* spp. by mycoparasites. *Phytopathology*, 73: 1072–1076.

Ahmed, A.H. and Tribe, H.T., 1977. Biological control of white rot of onion (*Sclerotium cepivorum*) by *Coniothyrium minitans*. *Plant Pathol.*, 26: 75–78.

Ainsworth, G.C., 1973. Introduction and key to higher taxa. In: *The Fungi: An Advanced Treatise Vol. IV(A)*, (Eds.) Ainsworth, G.C., Sparrow, P.K. and A.S. Susman. Academic Press, New York.

Asirif, K.N., Morgan, W.C. and Parbey, D.G., 1994. Suppression of sclerotinia soft rot of lettuce with organic soil amendment. *Augst. J. Expt. Agric.*, 34: 131–136.

Aujala, I.S., Sandhu, K.S., Sohal, B.S. and Singh, P.P., 2001. Bio-chemical basis of efficacy of soil amendments in managing stalk rot of cauliflower. *Plant Disease Research*, 16: 220–224.

Ayers, W.A. and Adams, P.B., 1979. Mycoparasitism of *Sclerotinia* and *Sclerotium* spp. by *Sporidesmium sclerotivorum*. *Can. J. Microbiol.*, 25: 17–23.

Backhouse, D. and Stewart, A., 1987. Anatomy and histochemistry of resting and germinating sclerotia of *Sclerotium cepivorum*. *Trans. Br. Mycol. Soc.*, 83: 561–567.

Backhouse, D. and Stewart, A., 1989. Interaction between *Bacillus* spp. and sclerotia of *Sclerotium cepivorum*. *Soil Biol. Biochem.*, 21, 173–176.

Bilgrami, K.S., Jamaluddin and Rizwi, M.A., 1979. *Fungi of India: List and References*. Today and Tomorrows Printers, New Delhi.

Bjorling, K., 1951. *Underdie Entwicklungs. Geschichte varialtitate and pathogenitate von Sclerotinia trifoliorum Eriksson. Phytopath. Z.*, 18: 129–156.

Boland, G.J. and Inglis, G.D., 1989. Antagonism of white mould *Sclerotinia sclerotiorum* of bean by fungi from bean and rapeseed flowers. *Can. J. Bot.*, 67: 1775–1781.

Boland, G.J. and Hunter, J. E., 1988. Influence of *Alternaria alternata* and *Cladosporium cladosporioides* on white mould of bean caused by *Sclerotinia sclerotiorum*. *Can. J. Plant Pathol.*, 10: 172–177.

Budge, S.P and Whipps, J.M., 1991. Glasshouse trials of *Coniothyrium minitans* and *Trichoderma* spp. for the biological control of *Sclerotinia sclerotiorum* in celery and lettuce. *Plant Pathol.*, 40: 59–66.

Budge, S.P and Whipps, J.M., 1991a. Without any permanent carpogenic Ugermination of sclerotia. *Mycological Res.*, 95: 195–198.

Budge, S.P., Mc-Quilken, M.P., Fenlon, J.S. and Whipps, J.M., 1995. Use of *Coniothyrium minitans* and *Gliocladium virens* for biological control of *Sclerotinia sclerotiorum* in glasshouse lettuce. *Biological Control*, 5: 513–522.

Bullock, S., Adams, P.B.,Willets, H.J. and Ayers, W.A., 1986. Production of haustoria by *Sporidesmium sclerotiorum* in sclerotia of *Sclerotinia minor*. *Phytopathology*, 76: 101–103.

Butler, E.J. and Bibsy, G.R., 1931. *The Fungi of India.* Imp. Coun. Agric. India Sci. Mono., 18: 257.

Bycheno, N.I. and Simmul, G.E., 1974. The effectiveness of Benomyl applied by different methods against cucumber disease. *Khimiya-V-Sel.-Skam-Khozyaistve*, 12: 14–43.

Cael, N., Penaud, A. and Decroos, Y., 2001. New element in the blight against Sclerotinia on oilseed crops. A new biological control fungicide studied in France. Champignon antagoniste *Coniothyrium minitans. Phytoma*, 539: 12–16.

Chet, I., 1987. *Trichoderma*: Application mode of action and potential as a biocontrol agent of soil borne plant pathogenic fungi. In: *Innovative Approaches to Plant Disease Control*, (Ed.) I. Chet. John Wiley, New York, p. 137–160.

Cooke, G.E., Steadman, J.R. and Boosalis, M.G., 1975. Survival of *Whetzelinia sclerotiorum* and initial infection of dry ediable bean in Western Nebraska. *Phytopathology*, 65: 250–255.

Cooke, R.J. and Barker, K.F., 1983. *The Nature and Practice of Biological Control of Plant Pathogen.* American Phytopathological Society, St. Paul, Minnsota, U.S.A., p. 539.

Cristea, G., Rimer, S.R. and Pritchard, M.K., 1985. *Sclerotinia sclerotiorum* pathogen of sunflower white rot parasitic on some medicinal and aromatic plants. *Bulitin ul de Ptotecta Plantelor, No.* 1–2.

Davis, W.H., 1925. Drop of Chinese cabbage and common cabbage caused by *Sclerotinia sclerotiorum* (Lib.) Masse. *Phytopathology,* 15: 249–259.

Deb, P.R. and Dutta, B.K., 1988. *Trichoderma* as biocontrolling agent against *Sclerotium rolfsii. Microbios Lett.,* 37: 107–113.

Deb, P.R. and Dutta, B.K., 1991. Studies on biological control of foot rot disease of soybean caused by *Sclerotium rolfsii* Sacc. *Z. Pflkrankr. Pflanzenschutz.,* 98: 539–546.

de-Bary, A., 1984. Ueper eirige sclerotinen and sclerotinkarank heiten. *Batanzeit,* 44: 374–377.

Dennis, C. and Webster, J., 1971. Antagonistic properties of species group of *Trichoderma* I Production of non-volatile antibiotics. *Trans. Mycol. Soc.,* 67: 25–39.

Dillon-Wester, W.A.R., Loveless, R.A. and Taylor, E.R., 1946. Clover rot. *J. Agric. Sci.,* 36: 18–28.

Dujickova, H., 1977. Preparation against white rot of cucumber. *Agrochemia Brastislava,* 17: 228–230.

Elad, Y., Chet, I., Boyle, P. and Henis, Y., 1983. Parasitism of *Trichoderma* spp. on *Rhizoctonia solani* and *Sclerotium rolfsii,* scanning electron microscopy and fluorescence microscopy. *Phytopathology,* 73: 85–88.

Elad, Y., Chet, I. and Katan, J., 1980. *Trichoderma harzianum*: As biocontrol agent effective against *Sclerotium rolfsii* and *Rhizoctonia solani. Phytopathology,* 70: 119–121.

Fayslle, C., 1995. (Control of sclerotinia) cultivar (Rvell-Malmasion), 395: 44–46.

Fuckel, L., 1970. Symbloic Pilze Jahrb. *Nassauischen vercins Naturk,* 23–24: 1–459.

Ganesan, P. and Gnanamanikam, S.S., 1987. Biological control of *Sclerotium rolfsii* Sacc. in peanut by inoculation with *Pseudomonas fluorescens. Soil Biol. Biochem.,* 19: 3538.

Ghasolia, R.P., Asha, Shivpuri and Bhargava, A.K., 2004. Some common weed species as hosts for *Sclerotinia. sclerotiorum. J. Mycol. Pl. Pathol.,* 34(2): 670.

Ghisalberti, E.L. and Sivasithamparam, K., 1991. Antifungal antibiotics produced by *Trichoderma* spp. *Soil Biol. Biochem.,* 23: 1011–1020.

Hadar, Y. and Gorodecki, B., 1991. Suppression of germination of a sclerotia of *Sclerotium rolfsii* in compost. *Soil Biol. Biochem.,* 23: 303–306.

Harrison, Y.A. and Stewart, A., 1988. Selection of fungal antagonists for biological control of onion white rot in New Zealand. *N.Z. J. Expt. Agric.,* 16: 249–256.

Henis, Y., Adams, P.B., Lewis, J.A. and Papavizas, G.C., 1984a. Penetration of sclerotia of *Sclerotium rolfsii* by *Trichoderma* spp. *Phytopathology,* 73: 1043–1046.

Henis, Y., Adams, P.B., Lewis, J.A. and Papavizas, G.C., 1984b. Interaction between *Sclerotium rolfsii* and *Trichoderma* spp: relationship between antagonism and disease control. *Soil Biol. Biochem.,* 16: 391–395.

Howthorne, B.T. and Jarvis, R., 1973. Differential activity of fungicides on various stages of life cycle of *Sclerotinia* spp., *Newzealand J. Agric. Res.,* 16: 551–557.

Huang, H.C., 1980. Control of sclerotinia wilt of sunflower by hyperparasites. *Can. J. Plant Pathol.* 2: 32.

Huang, H.C., 1981. Distribution of *Coniothyrium minitans* in Manitoba sunflower field. *Can. J. Plant Pathol.,* 3: 319–322.

Huang, H.C. and Kokko, E.G., 1987. Ultrastructure of hyperparasitism of *Coniothyrium minitans* on sclerotia of *Sclerotinia sclerotiorum. Can. J. Bot.,* 65: 2483–2489.

Huang, H.C. and Hoes, J.A., 1976. Penetration and infection of *Sclerotinia sclerotiorum* by *Coniothyrium minitans. Can. J. Bot.,* 54: 406–410.

Imolehin, E.D. and Grogan, R.G., 1980. Factor affecting survival of sclerotia and effect on inoculum density, relative position and distance of sclerotia from the host on infection of lettuce by *Sclerotinia minor. Phytopathology,* 70: 1162–1167.

Jackson, A.M., Whipps, J.M. and Lynch, J.M., 1991a. *In vitro* screening for the identification of potential biocontrol of *Allium* white rot. *Mycol. Res.* 95,430–434.

Jagger, I.C., 1920. *Sclerotinia minor,* new spp. the cause of a decay of lettuce, celery and other crops. *J. Agric. Dis.,* 20: 331–334.

Kappor, K.S., 1988. *Sclerotinia sclerotiorum,* a threat to seed production of egg plant. *Capsicum Newsletter, No.* 7: 52.

Keay, M.A., 1939. A study of certain species of genus Sclerotinia. *Annl. Appl. Biol.,* 26: 227–246.

Knudsen, G.R., Eschen, D.J., Dandurand, L.M. and Bin, L., 1991. Potential for biocontrol of *Sclerotinia sclerotiorum* through colonization of sclerotia by *Trichoderma harzianum. Plant Dis.,* 75: 466–470.

Kokalis, B.N., 2002. Biological control of tomato disease. In: *Biological Control of Crop Disease,* (Ed.) Gnanmanickam *(Sclerotinia sclerotiorum).* New York, U.S.A. Marcel Dekker Inc., p. 252–262.

Kolte, S.J., 1985. *Diseases of Annual Ediable Oilseed Crops, Vol. II: Rapeseed, Mustard and Sesame Disease.* CRC Inc., Boca Raton, Florida, 135 pp.

Kosasih, B.D. and Willet, H.S., 1975. Types of normal apothecia produced by *Sclerotinia sclerotiorum. Mycologia,* 67: 89–97.

Letham, D.B., Huett, D.O. and Trimboli, D.S., 1967. Biology and control of *Sclerotinia sclerotiorum* in cauliflower and tomato crops in New South Wales. *Plant Dis. Reportr.,* 60: 286–289.

Libert, M.A., 1937. Plante cryptogamicae arduennae (Exsiceati) No. 326. Publshed by M.A. Libert.

Liu, H.Y., 1989. Hyperparasitism of *Gliocladium deliquescens* and *Penicillium vermiculatum* on *Sclerotinia sclerotiorum. Plant Prot. Bull.,* 31: 175–184.

Lynch, J.M., 1987. *In vitro* identification of *Trichoderma harzianum*: As a Potential antagonists of plant pathogen. *Curr. Microbiol.,* 16: 49–53.

McCredie, T.A. and Sivasithamparam, K., 1985. Fungi mycoparasite on sclerotia of *Sclerotinia sclerotiorum* in some Western Australian soils. *Trans. Br. Mycol. Soc.,* 84: 736–739.

Mc-Laren, D.L., Huang, H.C. and Rimmer, S.R., 1996. Control of apothecia production of *Sclerotinia sclerotiorum* by *Coniothyrium minitans. Plant Dis.,* 80: 1373–1378.

Mc-Quiken, M.P., Mitchell, S.J., Budge, S.P., Whipps, J.M., Felon, J.S. and Archer, S.A., 1995. Effect of *Coniothyrium minitans* on sclerotial survival, and apothecia production of *Sclerotinia sclerotiorum* in field grown oilseed rape. *Plant Pathology,* 49: 883–896.

Merriman, P.R., 1976. Survival of sclerotia of *Sclerotinia sclerotiorum* in soil. *Soil Biol. Biochem.*, 8: 385–389.

Moore, G.S. and Atkins, R.D., 1977. The fungicidal and fungistatic effect of an aqueous garlic extract on medically important yeast like fungi. *Mycologia*, 69: 341–348

Morrall, R.A.A., 1976. list of host records of *Sclerotinia sclerotiorum* (Sensu Purdy, 1955) in Canada by family.

Mueller, J.D., Cline, M.V., Sinclair, J.B. and Jacobsen, B.J., 1985. An *in vitro* test for evaluating efficacy of mycoparasites on sclerotia of *Sclerotinia sclerotiorum. Plant Dis.*, 69: 584–587.

Murty, S.P. and Sirsi, M., 1957. Pharmocological studies on *Melia azadirachta*. Part I, Antibacterical, antifungal and antitubeveuler activity of neemoil and its fractions. *Symp. Utilization Indian Med. Pl.* Lucknow, pp. 55.

Neergaard, P., 1958. Mycelial seeds infection of certain crucifers by *Sclerotinia sclerotiorum* (Lib.) de Bary. *Plant Dis. Reptr.*, 42: 1105.

Oliveira de, V. L., de M. Bellei, M. and Borges, A.C., 1984. Control of white rot of garlic by antagonistic efficacy of mycoparasites of on sclerotia of *Sclerotinia sclerotiorum. Can. J. Microbiol.*, 30: 884–889.

Palat, Ram, 2001. Pathological studies on sclerotinia stem rot of Franchbean. Ph D. Thesis C.S.A. University of Agric. &Tech. Kanpur.

Palat, Ram, Singh, P.N. and Ramesh, Singh, 2000. Biological control of stem rot of Rajmash. In: *Paper Presented in National Symposium on Plant Disease on North Eastern India and their Management* of Zonal chapter Meeting on Dec., 6–7at N. D. U. of Agric. & Tech. Kumargang, Faizabad.

Palat, Ram, Singh, P.N. and Narain, Udit, 2004. Management of sclerotinia stem rot of Franchbean through fungicides and biopesticides. *Ann. Pl. Protec. Sci.*, 12(2) : 447.

Pandey, K.K., Pandey, P.K. and Satpathy, S., 2002. Integrated management of disease and insets of tomato, chilli and cole crops. *Technical Bulletin* No. 9. I.I.V.R., Varanasi.

Papavizas, G.C and Collins, D.J., 1990. Influence of *Gliocladium virens* on germi nation and infectivity of sclerotia of *Sclerotium rolfsii. Phytopathology*, 80: 627–630.

Papavizas, G.C. and Lumsden, R.D., 1980. Biological control of soil borne fungal propagules. *Ann. Rev. Phytopathol.*, 18: 389–413.

Papavizas, G.C., 1985. *Trichoderma* and *Gliocladium:* Biology, ecology and potential for biocontrol. *Ann. Rev. Phytopathol.*, 23: 23–54.

Peg, K.G., 1983. Control of Sclerotinia rot in bean with Benomyl. *Australian J. Experimental Agric. and Animal Husbandry*, 12: 81–82.

Phillips, A.J.L. and Price, K., 1983. Structure aspects of the parasitism of sclerotia of *Sclerotinia sclerotiorum* (Lib.) deBary by *Coniothyrium minitans. Phytopath. Z.*, 107: 193–203.

Phillips, A.J.L., 1986. Factor affecting the parasitic activity of *Gliocladium virens* on sclerotia of *Sclerotinia sclerotiorum* and a note on its host range. *J. Phytopathol.*, 116: 212–220.

Phillips, A.J.L., 1989. Fungi associated with sclerotia of *Sclerotinia sclerotiorum* in South Africa and their effects on the pathogen. *Phytophylactia*, 21: 135–139.

Purdy, L.H., 1952. Infection of lettuce by *Sclerotinia sclerotiorum* (Lib.) deBary. *Phytopathology*, 42: 518.

Purdy, L.H., 1955. A broader concept of the species, *Sclerotinia sclerotiorum* based on variability. *Phytopathology*, 54: 421–427.

Purdy, L.H., 1979. *Sclerotinia sclerotiorum*: History, disease and symptomatology, host range geographic distribution and impact. *Phytopathology*, 69: 875–880.

Rai, J.M. and Saxena, V.C., 1975. Sclerotial mycoflora and its role in natural biological control of White rot disease. *Plant and Soil*, 43: 509–513.

Rathore, R.S., Rajpurohit, T.S., Solanki, J.S. and Bishoni, H.R., 1993. Wild onion a new record of *Sclerotinia sclerotiorum*. *Indian Phytopath.*, 46: 261–262.

Ristanio, J.B., Lewis, J.A. and Lumsden, R.D., 1994. Influence of isolates of *Gliocladium virens* and delivery system in biological control of southern blight on carrot and tomato in field. *Plant Dis.*, 78: 153–156.

Roy, A.K., 1973. Host range of *Sclerotinia sclerotiorum* and *Sclerotium rolfsii* in Jorhat. *Assam Sci. and Culture*, 39: 319–320.

Roy, A.K. and Saikia, U.N., 1976. White blight of mustard and its control. *Indian J. Agric. Sci.*, 46: 274–277.

Saiio, I., 1977. Studies on the maturation and germination of sclerotia of *Sclerotinia sclerotiorum* (Lib.) deBary, a causal fungus of bean stem rot. *Agric. Exp. Stn.*, 26: 106.

Sandhu, K.S., 1992. Control of stalk rot disease of cauliflower with farmyard manure. *Plant Dis. Res.*, 7: 11–18.

Santos, A.F. and Dhingra, O.D., 1982. Pathogenicity of *Trichoderma* spp. on the sclerotia of *Sclerotinia sclerotiorum*. *Can. J. Bot.*, 60: 472–475.

Sariah, M. and Cheng, S.F., 1999. Effect of *Trichoderma* inoculants on growth and collar rot of soilles. *Mix. Gforct CABI. Publishing*, p. 71–73.

Sattar, A., Saini, H.B. and Alam, M., 1995. Occurrence of *Sclerotinia sclerotiorum* on important medicinal and aromatic plants. *Indian J. Plant Pathol.*, 13: 5–6.

Schwartz, H.F., 1974. Host range of *Whetzelinia sclerotiorum* (Lib.) Korf and Dumont [*Sclerotinia sclerotiorum* (Lib.) deBary]. Unpublished Manuscript 17pp.

Schwartz, H.F. and Steadman, J.R., 1978. Factor affecting sclerotium population and apothecium production by *Sclerotinia sclerotiorum*. *Phytopathology*, 68: 383–388.

Sharma, A.K., 1985. Epidemiological studies on white rot *(Sclerotinia sclerotiorum)* of pea. *Indian Phytopath.*, 38: 319–321.

Sharma, R.C. and Sharma, S.L., 1985. Correlation of meteorological factors with stalk rot of cauliflower caused by *Sclerotinia sclerotiorum*. *Indian J. Ecol.*, 12: 327–330.

Shin, D.B. and Lee, J.T., 1987. Ecological studies on lettuce crop disease occurring under controlled cultivation condition in drained paddy fields. *Korean J. Plant Pathol.*, 3: 252–260.

Shyam, R.K., Gupta, S.K. and Mandradia, R.K., 1994. Prevalence of different types of curd rots and extent of yield loss. *Indian J. Mycol. Pl. Pathol.*, 24: 172–175.

Singh, Dharmbir, 1996. Effect of sowing dates on white rot (*Sclerotinia sclerotiorum*) development in pea. *Indian J. Agric. Soc.*, 65: 621–623.

Singh, Hardeep and Singh, Harnam, 1995. Cultural control of stem rot of berseem caused by *Sclerotinia sclerotiorum* Eri. *Plant Dis. Res.*, 10: 2832.

Singh, L.B., 2003. Management of Sclerotinia blight of brinjal. *M.Sc. (Ag.) Thesis*. T.D.P.G. College, Jaunpur (U.P.).

Singh, R.K. and Dwivedi, R.S., 1987. Studies on biological control of *Sclerotium rolfsii* Sacc. causing foot rot of barly. *Acta. Bot. Indica*, 15: 160–164.

Singh, Rajender and Tripathi, N.N., 1995. Effect of sowing dates and stage of crop growth on severity of Sclerotinia rot of sunflower. *Haryana Agric. J. Res.*, 25: 276–281.

Singh, Ramesh, 2000. Pathological studies on Sclerotinia stem rot of Bishop weed *(Trachyspermum ammi*. *Ph.D. Thesis,* C.S.A. University of Agric. and Tech., Kanpur.

Singh, Ramesh, Palat, Ram and Sachan, N.S., 2001. Biological control of sclerotinia stem rot of Ajowan. In: *Paper Presented in National Symposium Plant Protection Strategies for Sustainable Agric. Horticulture* on 12–13 Oct. at S. K. U. of Agric. & Tech. R.S. Pura, Jammu.

Singh, Ramesh, Singh, S.B. and Palat, Ram, 2000. Management of Sclerotinia stem rot of Ajowan through soil amendments. In: *Paper Presented in National Symposium on Plant Disease on North Eastern India and their Management*, Zonal Chapter Meeting on Dec., 6–7at N. D. U. of Agric. & Tech. Kumargang, Faizabad.

Singh, Ramesh, Singh, S.B. and Palat, Ram, 2002. Management of Sclerotinia stem rot of Bishop weed by adjustment of sowing dates. In: *Paper Presented in Annual Meet and National Symposium on Integrated Management of Plant Disease of Mid Eastern India with a Cropping System Perspective*, Dec. 5–7 at N. D. U. of Agric. & Tech. Kumargang, Fizabad.

Singh, Ramesh, Singh, S.B. and Palat, Ram, 2003b. Management of Sclerotinia stem rot of Ajowan through fungicides and biopesticides. *Ann. Pl. Protec. Sci.*, 11(1): 164–165.

Singh, Ramesh, Narain, Udit and Palat, Ram, 2003a. Evaluation of bio-agents against *Sclerotinia* stem rot of Ajowan. *Ann. Pl. Protec. Sci.*, 11(2): 386.

Singh, Ramesh, Singh, L.B. and Singh, D., 2005a. Evaluation of different soil amendments against Sclerotinia blight of brinjal. In: *Paper presented in 6th Nat. Symp. on Sustainable Plant Protection Strategies Health and Environment Concern,* Oct. 15–17, 2005 Dr. Balasheb Sawant Konkan Krshi Vidyapeeth. Dapoli. Ratnagiri Maharashtra.

Singh, Ramesh, Singh, D., Singh, P.C. and Kumar, Dinesh, 2005b. Effect of some plant extract against *Sclerotinia sclerotiorum.* In: *Paper Presented in 15th Annual Conference of PAS,* Nov., 19–20 at T.D.P.G. College, Jaunpur (U.P.).

Singh, U.P. and Gangopadhyay, 1984. Control of white rot of seed cauliflower. *Pesticides,* 18: 23–24.

Singh, U.P. and Singh, R.B., 1984. Effect of date on sowing of the incidence of Sclerotinia blight and wilt of gram [*Cicer arietium* (L.)]. *Phytopath. Z.*, 109: 254–260.

Singh, U.P., Pathak, K.K., Khare, M.N. and Singh, R.B., 1979. Effect of leaf extract of garlic on *Sclerotinia sclerotiorum*, *Fusarium oxysporium* f. sp. *ciceri* on gram seed. *Mycologia*, 71: 556–564.

Singh, Yogendra, 1998. Biological control of Sclerotinia rot of rape and mustard caused by *Sclerotinia sclerotiorum*. *Plant Dis. Res.*, 13: 144–146.

Singh, Gurbaj, 1999. Cultural and pathological studies on Sclerotinia stem rot of Ajowan. *M.Sc. (Ag.) Thesis,* C.S.A. University of Agric. and Tech., Kanpur.

Singh, R.B., Mishra, K.K., Singh, J. and Chand, Ramesh, 1998. Influence of nutritional status of growth substrate on hyphal branching growth and sclerotia formation in *Sclerotinia sclerotiorum. Indian Pl. Pathol.,* 16: 28–37.

Smith, A.M., 1972. Biological control of fungal sclerotia in soil. *Soil Biol. Biochem.,* 4: 131–134.

Srivastava, B.K. and Diwakar, M.C., 1987. Note on stem rot and wilt of chillies in Harayan (India). *Plant Protection Bull., India,* 39: 35.

Starr, G.C., Walters, H.G. and Bridgmon, G.H., 1953. White mould of bean. *Agric. Exp. Stn. Bull.,* 322.

Steadman, J.R., Maier, C.R., Schwartz, H.F. and Keer, E.D., 1975. Pollution of surface irrigation waters by plant pathogenic organisms. *Water Resour. Bull.,* 11: 796–804.

Stewart, A. and Harrison, Y.A., 1988. Mycoparasitism of sclerotia of *Sclerotinia sclerotiorum. Australasian Plant Pathol.,* 18: 1014.

Tate, K.G., 1980. Sclerotinia in baby carrots. *Newzealand Commercial Grower,* 35: 24–25.

Tong-Kwee, L. and Keng, T.B., 1990. Antagonism *in vitro* of *Trichoderma* species against several Basidiomycetous soil borne pathogens and *Sclerotium rolfsii. Z. Pflanzenkr. Pflanzenschutz,* 97: 1014.

Tribe, H.T., 1957. On the parasitism of *Sclerotinia trifoliorum by Coniothyroum minitans. Trans. Br. Mycol. Soc.,* 40: 489–499.

Trutmann, P. and Keane, P.J., 1990. *Trichoderma koningii* as a biological agents for *Sclerotinia sclerotiorum. Australasian Soil Biol. Biochem.,* 22: 43–50.

Trutmann, P., Keane, P.J. and Merriman, P.R., 1982. Biological control of *Sclerotinia sclerotiorum* on aerial part of plants by the hyperparasite *Coniothyrum minitans. Trans. Br. Mycol. Soc.,* 78: 521–529.

Tu, J.C., 1980. *Gliocladium virens,* a destructive mycoparasite of *Sclerotinia sclerotiorum. Phytopathology,* 70: 670–674.

Tu, J.C., 1984. Mycoparasitism by *Coniothyrum minitans* on *Sclerotinia sclerotiorum* and its effects on sclerotial germination. *Phytopath. Z.,* 109: 261–268.

Turner, G.J. and Tribe, H.T., 1975. Preliminary field plot trial on biological control of *Sclerotinia trifoliorum by Coniothyrum minitans. Plant Pathol.,* 24: 109–113.

Uecker, F.A, Ayers, W.A. and Adams, P.B., 1980. *Teratosperma oligicladium* hyphomcetes mycoparasite on *Sclerotinia trifoliorum* and *S. minor. Mycotaxon,* 10: 421–427.

Upadhyay, J.P. and Mukhopadhyay, A.N., 1983. Effect of non-volatile and volatile antibiotic of *Trichoderma harzianum* on growth of *Sclerotium rolfsii. Indian J. Mycol. Pl. Pathol.,* 13: 232–233.

Utkhede, R.S. and Rahe, J.E., 1980. Biological control of onion white rot. *Soil Biol. Biochem.,* 12: 101–104.

Utkhede, R.S. and Rahe, J.E., 1983a. Intraction of antagonist and pathogen in biological control of onion white rot. *Phytopathology,* 73: 890–893.

Utkhede, R.S. and Rahe, J.E., 1983b. Effect of *Bacillus subtilis* on growth and protection of onion against white rot. *Phytopathol. Z.,* 106: 199–203.

Vander, Plank, 1968. *Disease Resistance in Plants.* Academic Press, New York, pp. 206.

Walker, J.C., 1969. *Plant Pathology*, 3rd edn. McGraw Hill, New York and London, pp. 849.

Whetzel, H.H., 1945. A synopsis of the genera and species of the Sclerotiniaceae Family of stromatic inoperculate Discomycetes. *Mycologia*, 39: 648–714.

Whipps, J.M., Mc-Quilken, M.P. and Budge, S.P., 1993. Use of fungal antagonists for biocontrol of damping off and Sclerotinia disease. *Pesticide Science*, 37: 309–303.

Whipps, J.M., 1987. Behaviour of fungi antagonistic to *Sclerotinia sclerotiorum* of plant tissue segments. *J. Gen. Microbiol.*, 133: 1495–1501.

Whipps, J.M. and Budge, S.P., 1990. Screening for sclerotial mycoparasites of *Sclerotinia sclerotiorum. Mycol. Res.*, 94: 607–612.

Whipps, J.M. and Budge, S.P., 1992. Biological control of *Sclerotinia sclerotiorum* in glasshouse crops. *Brighton Crop Protection Conference Pests Disease,* 1: 127–132.

Willet, H.J. and Wong, A.L., 1980. The biology of *Sclerotinia sclerotiorum* with emphasis on specific nomenaclature. *Bot. Rev.*, 46: 101–165.

Williams, G.H. and Western, J.H., 1965. A biology of *Sclerotinia trifoliorum* Evikks and other species of sclerotia forming fungi-I: Apothecium formation from sclerotia. *Ann. Appl. Biol.*, 56: 253–260.

Wong, W.C. and Hughes, I.K., 1986. *Sclerotium cepivorum* Berk. in onion (*Allium cepa* L.) crops: Isolation and characterization of bacteria antagonistic to the fungus in Queensland. *J. Appl. Bact.*, 60: 57–60.

Yarden, O., Ben, Y.Y., Katan, J. and Aharonson, N., 1986. Fungal control of *Sclerotinia sclerotiorum* in soil with a combination of Benomyl and Thiram. *Pl. Dis.*, 70: 738–748.

Young, P.A. and Morris, H.E., 1927. Sclerotinia wilt of sunflower, Montana. *Agric. Exp. Stn. Bull.*, 208.

Zazzerini, A. and Tosi, L., 1985. Antagonistic activity of fungi isolated from sclerotia of *Sclerotinia sclerotiorum. Plant Pathol.*, 34: 415–421.

Zazzerini, A., Tosi, L. and Rossi, J., 1987. Antagonistic effects of *Baccilus* spp. on *Sclerotinia sclerotiorum. Phytopathologica,* Mediterranean, 26: 185–187.

Zhou, T. and Reeleder, R.D., 1990. Selection of strains of *Epicoccum purpurascens* for tolerance to fungicides and improved biocontrol of *Sclerotinia sclerotiorum*. *Can. J. Microbiol.*, 36: 754–759.

Insect Pest and Disease Management *Pages 227–238*
Editor: **Dr. D. Prasad**
Published by: **DAYA PUBLISHING HOUSE (2016)**

Chapter 16

Integrated Disease Management in Pulses

Jameel Akhtar[1], V.B. Nargund[2] and Abdul Khalid[1]
[1]Department of Plant Pathology, B.A.U., Ranchi – 834 006, Jharkhand
[2]Department of Plant Pathology, College of Agriculture, Gulbarga – 585 287, Karnataka

Introduction

The pulses are called "poor man's meat", due to their protein value. Pulses are also known as short duration crops. Pulse crops provide about 10 per cent of the worlds dietary protein India is the largest producer of pulses (13 million tonnes) but the average productivity is very low (614 kg/ha) (Ali, 1998). All pulses require very little amount of water, hence are termed as dry crops. Another advantage in the cultivation of pulses is that, all pulses play an important role in crop rotation, due to help in maintaining soil fertility, through atmospheric nitrogen fixing bacteria. The major pulse crops grown in India are chickpea, pigeonpea, blackgram, greengram, lentil field pea and red gram are the major pulse crops grown in different systems as rain fed and irrigated crop in different seasons.

Pulses, the food legumes, have been grown by farmers since millennia, and these have contributed in providing nutritionally balanced food to the people of India. While pigeon pea, black gram, green gram, lablab bean, moth bean, and horse gram have definitely originated and domesticated in the Indian subcontinent, there is a probability that chickpea and lentil (Indian type) were also domesticated in the Indian subcontinent. Pea, grass pea, and cowpea were introduced in India millennia ago. Only faba bean was introduced in medieval times. This paper briefly describes how the pulses constituted an important item of food, how these were cultivated, and how food preparations evolved.

There are many constrain to the pulse above pulse crops, including the heavy losses, caused by a number of diseases affecting the increasing productivity of pulses apart from ecological and biological constraints. In red gram, sterility mosaic and in black gram and green gram, yellow mosaic caused by Mungbean Yellow Mosaic virus (MYMV), leaf crinkle caused by Urdbean leaf crinkle virus (ULCV), leaf spot caused by Cercospora canasence, web blight caused by *Rhizoctonia solani*, powdery mildew and dry root rot and in gram, wilt and Ascochyta blight are most important one. These are the potentially dangerous diseases. The losses due to the diseases to pulse crops are estimated from 4 to 44 per cent, depending upon the crop variety. The diseases attracted attention after severe epidemic years, during which the diseases drastically reduced the total production. The diseases therefore, have the capacity to cause heavy losses both in volume in commodity and to farmer's income.

Only possible way to reduce the yield loss due to these diseases are adopting the integrated plant disease management (IPDM) practice employing many strategies like use of resistant varieties, use of disease free seeds, manipulation of cultural practices, management of vectors, application of bio-control agents and chemicals.

A brief account of the most important diseases of pulse crops in India, including the causes, symptoms, preventive and curative control measures of very destructive diseases, are crop wise discussed, which are responsible for reducing overall production to a great extent in some areas, in some years, and generate great concern because of their effects on the quality and/or quantity of the crop produce. It must be kept in mind that wilt and blight diseases are reported as most destructive diseases in all pulse crops. These diseases are discussed in gram, only to avoid the repetition.

The four basic requirements for management of plant diseases are clean and healthy seeds, clean field or pathogen free soil, prevention of entry of and infection by a pathogen in a standing crop and precautions during harvesting and storage of the produce. An ideal schedule for controlling a disease is to integrate measures covering these four requirements. In an integrated disease management programme in an annual crop the first step is the choice of healthy seed of a resistant variety. The healthy clean seed can be obtained as a certified seed from a reliable agency. The next step is to treat the seed chemically or with bioagents if the supplier has not done the same. The healthy, pathogen free seed is then planted in relatively clean field. Crop rotation is another precaution. During growth of the crop periodical rouging of diseased plants, if any, should be done. The inoculum of a pathogen may arrive from external sources. This inoculum should be prevented from causing infection by sprays of suitable chemicals. At the time of harvest, care should be taken, to discard diseased plants, do the harvesting in a clean weather, and store the produce in proper storage conditions.

A successful disease management program depends on a crop production system which is closely aligned with the goals of pest management. Integrated plant disease management means using all the tactics available to the growers (cultural, biological, host-plant resistance and chemical) that provides acceptable yield and quality at the least cost and is compatible with the tenets of environmental stewardship.

The five components of an IPDM programmes are prevention, monitoring, correct disease diagnosis, development and use of optimum selection of management tools. The management strategies available include genetic control, cultural control, biological control, and chemical control. What management strategy is most relevant depends in part on the particular pathogen. In general, weeds are most effectively controlled with cultural and chemical practices. Diseases are controlled most by host-plant resistance, cultural practices, and chemicals. For disease management, it is important to understand the potential of a pathogen to infest and spread in the crop. The three parameters of disease progress are:

1. The initial amount of inoculum
2. The rate of disease increase, and
3. The time the crop is grown.

These parameters interact in ways to produce a rapid pathogen population increase, manifested as exponential growth in many production systems. The rate of disease increase over time is dependent upon the interactions of the pathogen, plant host, and the environment. For disease management purposes, we are most concerned with the interactions of the pathogen and host at the population level of organization. However, environmental conditions play a critical role in determining the nature of plant disease epidemics.

Understanding the biology of the pathogen, host-pathogen interactions, and the effect of environmental factors on this dynamic process in time and space, is critical for planning and implementing effective and efficient management strategies. These strategies can affect particular aspects of the growth of the pathogen population. Host-plant resistance can affect all three parameters by:

1. Reducing the amount of inoculum via resistance to particular strains of the pest
2. Slowing the rate of pathogen buildup by reducing its reproductive capacity, and
3. Reducing the total period of exposure in short-season varieties.

Cultural control is aimed at reducing the primary inoculum (sanitation) or reducing the rate of disease increase by modifying the crop environment. A good example of the later is the use of drip irrigation rather than overhead to reduce free water on foliage. The total time that a crop is exposed to a pathogen may be minimized by optimizing plant growth, thus reducing the time to harvest. Biological control usually affects the rate of pathogen buildup. Finally, chemical control can affect the amount of inoculum available at the beginning of the season (*i.e.* soil fumigation) and/or reduce the rate of disease development by killing a portion of the pathogen involved in later stages of the epidemics.

Cultural Practices

Cultural practices serve an important role in prevention and management of plant diseases. The benefits of cultural control begin with the establishment of a growing environment that favors the crop over the pathogen. Reducing plant stress through environmental modification promotes good plant health and aids in reducing damage from some plant diseases Sanitation practices aimed at excluding, reducing, or eliminating pathogen populations are critical for management of infectious plant diseases. It is important to use only pathogen-free transplants, especially for late blight, bacterial spot, viral diseases and early blight. In order to reduce dispersal of soil-borne pathogens between fields, stakes and farm equipment should be decontaminated before moving from one field to the next. Reduction of pathogen survival from one season to another may be achieved by destruction of volunteer plants and crop rotation. Removal of cull piles and prompt destruction of crops should be applied as a general practice. Avoid movement of soil from one site to another to reduce the risk of moving pathogens. For example, sclerotia of *Sclerotinia sclerotiorum* and *Sclerotium rolfsii*, are transported primarily in contaminated soil. Minimizing wounds during harvest and packing will reduce post harvest disease problems. Depending on crops and other factors, sanitation of soil can be achieved to some degree by solarization. Crop rotation is a very important practice, especially for soil-borne disease control. For many soil-borne diseases, at least a 3-year-rotation using a non-host crop will greatly reduce pathogen populations. This practice is beneficial for Fusarium wilt, but longer rotation periods (up to 5-7 years) may be needed. Land previously cropped to alternate and reservoir hosts should be avoided whenever possible.

Fields should be located as far away as possible from inoculum and insect vector sources. Weed control is important for the management of viral diseases. Weeds may be alternate hosts for several important vegetable viruses and their vectors. Elimination of weeds might reduce primary inoculum. Cover crops help to reduce weed populations that may harbor pathogens between seasons. For this purpose use cover crops that grow fast and provide maximum biomass. Non-host cover crops will help to reduce weed populations and primary inoculum for soil-borne pathogens. Excessive handling of plants such as in thinning, pruning and tying may be involved in spread of pathogens, particularly bacteria. It is advisable to handle plants in the field when plants are driest. Because some pathogens can only enter the host through wounds, situations which promote plant injury should be avoided. Soil aeration and drying can be enhanced through incorporation of composted organic amendments in the soil. Build up of inoculum can be reduced by removing all plant materials after harvest. Between-row cover crops reduce plant injury from blowing sand. Polyethylene mulch can be used as a physical barrier between soil and above-ground parts of plants.

Host Resistance

It is very important to choose cultivars with multiple pathogen resistance whenever possible. Practical management of many diseases of pulses (Fusarium wilt, leaf spot and powdery mildew) is achieved primarily by this method. Recently, multiple resistant varieties/genotypes of lentil (KLB 327, Pl 023) to Fusarium wilt, rust, stem rot and Ascochyta blight, resistant genotypes/varieties (Pant P 31) of field pea to powdery mildew, genotype JG 62 of gram resistant against wilt have been identified.

Biological Control

Use of biocontrol agents in disease management of pulses are increasing. These products are considered safer for the environment and the applicator, than conventional chemicals and are mainly used against soil-borne diseases. Examples of commercially available biocontrol agents include the fungi *Trichoderma harzianum* and *T. virens*, an actinomycete *Streptomyces griseoviridis*, and a bacterium *Pseudomonas fluorescens* have been found as an effective biocontrol agent for the management of soil-borne fungal and bacterial diseases.

Chemical Control

Pesticides including fungicides and bactericides are an important component of integrated disease management programme. It is important to remember that chemical use should be integrated with all other appropriate tactics mentioned in this chapter. Information regarding physical mode of action of a fungicide will help producers improve timing of fungicide applications. Always read the pesticide labels and follow the instructions carefully. Remember, the label is the law. Fumigants can be used to manage soil-borne pathogens. Before applying, it is important to review the disease history of the specific site when choosing fumigant materials. Effective management of whiteflies, thrips, and aphids should be practiced to reduce the incidence and secondary infections of viral diseases of pulses vectored by these insects.

The other virus diseases of black gram and green gram are bean common mosaic, alfalfa mosaic, bean yellow mosaic, Cowpea aphid-borne mosaic and etc. The major fungal diseases of black gram and green gram are the powdery mildew, dry root rot, and leaf spot diseases like Cercospora, Alternaria leaf spots and rust (Singh and Bhan, 1998: Muthukrishnan *et al.*, 1995).

Diseases of Chickpea

The gram (Chickpea, or Bangal gram) ranks first as a pulse crop in the country followed up by green gram only in area and production. Actually, the gram is almost only the *rabi* pulse crop, that is successfully grown just after the harvest of the rice crop, without irrigation water. In case of irrigated crop, only one irrigation found enough at flowering stage. The crop suffers from a number of diseases, but wilt and blight reduces considerably the yield as well as the quality.

1. Wilt
2. Ascochyta blight

Wilt (*Fusarium oxysporum* f. sp. *ciceri*)

Identification

The disease symptoms can be observed within 3 weeks of sowing by discolouration of leaf leading to yellow colouration. Whole seedling collapse and lie flat on ground. These seedlings retain their dull green colour. The plants start wilting and ultimately dry up at seedling and/or at flowering stage (Figure 16.1). The diseased plants are pulled out easily, due to the loss of rigidity and show uneven shrinking of the stem above and below the colour region. Affected plants show external rotting, drying or root discolouration. Sometimes, sudden drooping of leaves and/or only a few branches of a single plant is affected. The disease mostly spreads in patches, but entire field may also be affected in severe conditions.

Figure 16.1: Wilt of Chickpea

Disease Management

Cultivation of disease resistant variety *e.g.* JG 62, Gaurav, Pant G114, is only the most easy, economical and safe method. However, following measures could be beneficial if applied.

1. Avoid fields having history of wilt in previous years as well as with water logging problems.
2. Avoid the cultivation of gram in the diseased area for three years.
3. Avoid the making of *bhusa* stacks in the field.
4. Improvements of soil condition with good drainage minimize the disease incidence.
5. Deep ploughing of chickpea fields to bury infested debris and resting spores/propagules.
6. Application of peat culture mixed with organic manure or sand along with the rows at the time of sowing and at 30 and 60 days after sowing.
7. Diseased plants should be uprooted from the field and burnt.
8. Late and deep sowings reduce the incidence of the disease.
9. Mixed cropping of gram with wheat, barley, rape, mustard, safflower, sorghum and millet help to check the disease.
10. Soil application with *P. fluorescens* peat culture @ 2.5 kg/ha
11. Sweeping the threshing floor and burning or burying all plant debris.
12. Use clean certified seeds treated with *Tricoderma viride* with a mixture of Bavistin or Thiram (1:1) @ 2.5 gm/kg of seeds.

Ascochyta Blight (*Ascochyta rabiei*)

Identification

Figure 16.2: Ascochyta Blight

Very early symptoms of Ascochyta blight are small (1-2 mm), reddish brown spots on the leaves and stems. The spots enlarge gradually and turn light or dark brown and necrotic. These lesions often have a darker margin surrounded by a yellow (chlorotic) area. Severely affected leaves eventually wilt. On stems, the lesions are dark brown and elongated. Small, black fruiting bodies (pycnidia) are produced in the stem lesions and can be seen with a hand lens. When lesions girdle a stem the portion of the plant above the lesion wilts or the stem becomes weak and breaks easily. If the main stem is girdled at the stem base, the whole plant dies. On pods, the lesions are usually circular with a dark margin (Figure 16.2). The small, black fruiting bodies (pycnidia) are arranged in concentric rings. Chickpea seed infected by Ascochyta blight are shriveled, spotted and dark discoloured. In 1999, another kind of fruiting bodies (pseudothecia) containing ascospores was found on naturally infected chickpea debris. It was identified as the sexual spore stage *Didymella rabiei* of the same pathogen which causes Ascochyta blight.

Disease Management

Follow Crop Rotation

Since only chickpea is susceptible to Ascochyta blight caused by *A. rabiei*, inclusion of non-host crops in the rotation will bring about a reduction in the inoculum level. Chickpeas should not be grown more frequently than every 3-4 years. Furthermore, new crops should not be grown near fields that were infested with blight in the previous year due to the risk of infection by air borne ascospores.

Tillage

Under most conditions, deep ploughing will hasten decomposition of infected chickpea straw and remove it as a source of inoculum. Studies showed that the pathogen survived in naturally infected plant material for more than 2 years when situated on the soil surface, but lost its viability rapidly at soil depths of 10-40 cm. All infested crop residues and volunteer plants should therefore be destroyed by thorough tillage and proper crop rotation.

Use Resistant Cultivars

Practical management of this disease is achieved primarily by this method. Two cultivars Pant G114 and C-235 have a good level of resistance until flowering when lesions may start to develop on leaves and stems. If the weather conditions are favourable, the disease will increase rapidly in these cultivars and fungicide application should be considered.

Chemicals

Growers should take special care to protect green and healthy foliage during pod filling, especially if blight symptoms appear and increase in severity. Seed treatment with bavistin + thiram (1:2) @ 3.0 g/kg seed is very effective. Two spray of carbendazim or chlorothalonil @ 0.2 per cent at different crop stages have shown that spraying at early poding reduced pod damage and markedly increased seed yields. Foliar fungicide applications may not prove cost effective when disease pressure is low.

Field Pea Diseases

Field Peas are subjected to a number of diseases, but following are considered very common and destructive.

1. Powdery mildew
2. Root rot
3. Sclerotinia stem and pod rot

Powdery Mildew (*Erysiphe polygoni*)

Identification

Powdery mildew is probably the most commonly observed disease in field peas, develops just before or at harvest. As the name implies, a white powdery coating covers the leaves (Figure 16.3). Eventually, the speckles turn into spots and the plant turns purple, bluish-green and, lastly, brown. Warm, dry days and cool nights are critical to this organism's development. Late-seeded fields are typically at higher risk. Frequent heavy dews in the absence of rain, accelerate the disease's progression.

Figure 16.3: Powdery Mildew of Field Peas

Root Rot

Identification

Root rot is caused by several fungi: *Fusarium solani* f sp. *pisi, Pythium* species and *Rhizoctonia solani*. Affected plants are stunted and yellow, lose lower leaves and die prematurely. The lower stem turns black or brown. Roots are pinched off and rotted (Figure 16.4). *Fusarium solani* f sp. *pisi* infections may be distinguished from the others by the reddish brown external colour of the roots. High moisture and warm soil temperatures (18 to 25°C) favour root rot development of all of these fungi.

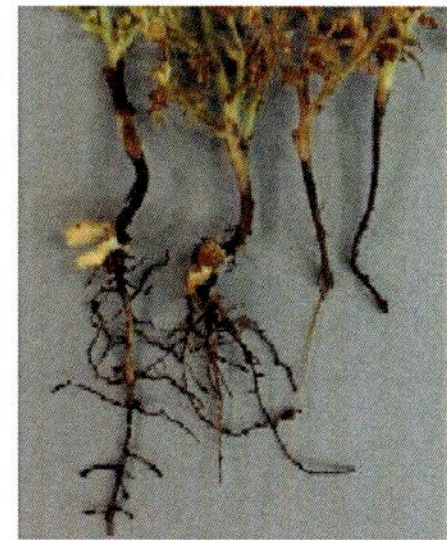

Figure 16.4: Root Rot of Field Peas

Sclerotinia Stem and Pod rot

Identification: Symptoms of this disease are soft stems and pods with a cottony white mould and hard charcoal-like resting bodies, called sclerotia, on the inside or outside of the stems. During periods of high soil moisture, the sclerotia grow golf tee-like structures that disperse spores into the air and onto the plant. Researches on *Sclerotinia*, revealed the burying of residue decreased the survival rate of the organism causing disease by 6 to 32 per cent.

Disease Management

There are a number of factors to consider for controlling diseases in field peas.

1. Crop rotation, using a diversity of broad-leaved and cereal crops, will ensure lower levels of all diseases.
2. Avoid growing beans, fababeans, and sunflower before or following field peas to lower the chances of attack by *Sclerotinia sclerotiorum*.
3. For diseases such as Ascochyta blight, Mycosphaerella blight and foot rot, a chemical treatment with Chlorothalonil fungicide is available.
4. Using field pea varieties with an open canopy structure, like the semi-leafless types, will allow better air flow and should reduce disease levels.
5. Powdery mildew resistance will be incorporated into all new field pea varieties.

Diseases of Urd and Mungbean

Mung and urd bean are also subjected to a number of diseases, but following are considered very common and destructive.

1. Mungbean Yellow Mosaic Virus
2. Cercospora leaf spot

Mungbean Yellow Mosaic (Mungbean Yellow Mosaic Virus)

Identification

It is the most serious problem of mungbean in the Indian subcontinent. Infected plants become chlorotic and stunted. Initially mild yellow spots appear on young leaves. The next trifoliate leaves emerging from the growing apex shows irregular yellow and green patches altering with each other. Spots gradually increase in size and ultimately some leaves turn completely yellow (Figure 16.5). Diseased plants are stunted, mature late and produce very few flowers and pods. The pods are deformed and contains shriveled under sized seeds.

Figure 16.5: Mungbean Yellow Mosaic

Disease Management

Planting of MYMV-tolerant/resistant varieties is the best control measure. Use of resistant varieties such as pant Mung 2, Narendra Mung 1, Vamban 1, Vamban 2, and Vamban 3 are fairly resistant to yellow mosaic virus. In addition,

1. Removal of yellow mosaic infected plants.
2. Removal of weeds.
3. Vector management through spray of Monocrotophos @ 500 ml/ha or Methyl demeton @ 500 ml/ha and repeat after 15 days.
4. Collection of the seeds from disease free plants, are important to mange MYMV.

Cercospora Leaf Spot (*Cercospora canescens*)

Identification

Cercospora canescens commonly attacks mung bean in the tropics. Cercospora leaf spot (CLS) is recognized by the appearance of leaf spots that are circular to irregularly shaped with grayish white centers and reddish brown to dark brown margins (Figure 16.6). These spots increase in number and size, which turn into lesions of a reddish brown margin. The size of pods and seed is reduced, hence yield decreased considerably. It can cause yield losses of up to 58 per cent.

Figure 16.6: Cercospora Leaf Spot

Disease Management

The losses due to CLS disease can be avoided by planting tolerant/resistant cultivars. Crop debris and weed hosts should also be removed at the time of planting. When susceptible cultivars are planted, spray with fungicides such as carbendazim or chlorothalonil @ 2 kg a.i./ha at two-week intervals. Seed treatment with talc formulation of *Trichoderma virde* @ 4g/kg of seed (or) *Pseudomonas fluorescens* @ 10 g/kg of seed (or) carbendazim or thiram @ 2 g/kg of seeds.

Diseases of Lentil or Masoor

Lentil wilt disease is common; his symptoms, perpetuation and control measures are similar as that of wilt of gram.

Wilt (*Fusarium oxysporum* f. sp. *lentis*)

Identification

The disease symptoms appear in the field in patches at both seedling and adult stages. Seedling wilt is characterized by sudden drooping followed by drying of leaves and the whole seedling. The wilt symptoms in the adult plants first appear at flowering to late pod-filling stage by sudden drooping of top leaflets of the affected plants (Figure 16.7).

Figure 16.7: Wilt of Lentil

Disease Management

Use of disease resistant variety *e.g.* Pant L 234, Pant L 639, Pusa 3 etc. is economical and safe method. However, following measures could be beneficial if applied:

1. Use clean certified seeds treated with *Tricoderma viride* with a mixture of bavistin and thiram (1:1) @ 2.5 gm/kg of seeds or treat the seeds with benlate T @ 3 g/kg seed.
2. Spot drench around the base of the affected plants as well as healthy plants surrounding them with carbendazim @ 1 g/litre for wilt.
3. Soil application of *Pseudomonas fluorescens* @ 2.5 kg/ha mixed with 50 kg of well decomposed FYM at 30 days after sowing.
4. Avoid fields having history of wilt in previous years as well as with water logging problems.
5. Avoid the cultivation of lentil in the diseased area for three years.
6. Improvements of soil condition with good drainage minimize the disease incidence.
7. Diseased plants should be uprooted from the field and burnt.
8. Soil application with *P. fluorescens* peat culture @ 2.5 kg/ha

Diseases of Pigeon Pea

The cultivation of pigeon pea (arhar, or red gram) is very common after gram in the country. Actually, the pigeon pea is mostly taken as *kharif* pulse crop. The crop suffers from a number of diseases, but sterility mosaic and wilt diseases cause serious damage thereby reduce considerably the yield as well as the quality.

1. Sterility Mosaic
2. Wilt

Sterility Mosaic (Pigeon Pea Sterility Mosaic Virus)

Identification

Diseased plants can be identified from a distance as patches of bushy pale green plants without flowers or pods (Figure 16.8). The leaves of these plants are small and show a light and dark green mosaic pattern and vein clearing on young leaves. Eriophyid mite vectors the disease.

Disease Management

1. Grow resistant varieties like Bahar, Pant A 3, UPAS-120, Rampu Rahar etc.
2. Rouge out the infected plants in the early stages of growth.
3. Seed dressing with Furadan 3 G or 10 per cent aldicarb @ 3 g/kg seed.
4. Spraying of acaricide like kelthane, morestan, metasystox @ 0.1 per cent on noticing the initial symptoms and repeat after a fortnight.

Figure 16.8: Sterility Mosaic of Pigeon Pea

Wilt (*Fusarium udam*)

Identification

Patches of infected plants in the field can be seen when the crop is at flowering or poding stage (Figure 16.9). Symptoms of wilt is browning of the stem tissues in the region of the purple band and blackening of xylem bundle when split.

Disease Management

1. Select a field site with no previous record of wilt disease for at least three years.
2. Grow resistant varieties like Birsa Arhar, Mukta Prabhat etc.
3. Grow pigeon pea mixed with cereal crops such *e.g.* sorghum.
4. Seed dressing with benlate T @ 3 g/kg seed.
5. Spot drench around the base of the affected plants as well as healthy plants surrounding them with carbendazim @ 1 g/litre for wilt.
6. Soil application of *Pseudomonas fluorescens* @ 2.5 kg/ha mixed with 50 kg of well decomposed FYM at 30 days after sowing.

Figure 16.9: Wilt of Pigeon Pea

Conclusion

IPDM embodies an ecological approach to the pathogen problem with the objective of minimizing use of chemical pesticides while sustaining production levels. The gains are reduction in costs of production, more economic access of food and conservation of the resilience and integrity of the ecosystem. The modules developed and validated have to be either crop or location specific and therefore, lack universal application.

References

Agrios, G.N., 1997. *Plant Pathology*. Academic Press, New York.

Ali, M., 1998. Research, development and management for production of pulses. In: *IPM System in Agriculture. Volume 4. Pulses*, (Eds.) R.K.Upadhyay, K.G. Mukerji and R.L. Rajak. Aditya Books Private Limited, New Delhi, pp. 1–40.

Allen, G. and Bath, J.E., 1980. The conceptual and institutional aspects of integrated pest management, *Bioscience*, 30: 658.

Anonymous, 1999. *Crop Production Guide*. Directorate of Agriculture, Chennai, pp. 73–102.

Banto, S.M. and Sanchez, F.F., 1972. The biology and chemical control of Callasobruchus chinensis (Linn) (Coleoptera: Bruchidae). *Philipp. Entomol.*, 2: 167–182.

Bolkan, H.A. and Reinert, W.R., 1994, Developing and implementing IPM strategies to assist farmers: An industry approach. *Plant Dis.*, 78: 545.

Chaube, H.S. and Pundhir, V.S., 2005. *Crop Diseases and their Management*. Prentice Hall of India, New Delhi, pp. 703.

Chaube, H.S. and Singh, U.S., 1990. *Plant Disease Management: Principles and Practices*, CRC, Press, Boca Raton, USA.

Eversmeyer, M.G. and Kramer, C.L., 1987. Components and techniques of integrated pest management threshold determination for pathogens. *Plant Dis.*, 71: 456.

Ferris, H., 1987. Components and techniques of integrated pest management threshold determination for soil borne pathogens. *Plant Dis.*, 71: 452.

Lal, S.S., Yadava, C.P. and Dias, C. A.R., 1985. Assesment of crop losses caused by *Helicoverpa armigera. FAO Plant Protection Bulletin*, 33: 27–35.

Muthukrishnan, K., Arjunan, G. and Raguchander, T., 1995. Some pathological studies on Macrophomina root rot of urbean. *Indian Journal of Pulses Research*, 8: 162–165.

Raguchander, T., Rajappan, K. and Prabakar, K., 1995. Evaluation of tale based product of *Trichoderma viride* of the control of blackgram root rot. *Journal of Biological Control*, 9: 63–64.

Singh, Gurdip and Livinder Kaur Bhan, 1998. Disease of mungbean and urbean and their management. In: *IPM System in Agriculture. Volume 4. Pulses*, (Eds.) R.K.Upadhyay, K.G. Mukerji and R.L. Rajak. Aditya Books Private Limited, New Delhi, pp. 311–371.

Singh, R.S., 2000. *Plant Disease Management*. Oxford and IBH Publishing Co. Pvt. Ltd., New Delhi.

Talekar, N.S., 1990. *Agromyzid Flies of Food Legumes in Tropics.* Wiley Eastern Limited, New Delhi.

Thurston, H.D., 1991. *Sustainable Practices for Plant Disease Management in Traditional Farming Systems*. Westview, Boulder, Colorado.

Vidhyasekaran, P. and Muthamilan, M., 1995. Development of formulation of *Pseudomonas fluorescens* for the control of chickpea wilt. *Plant Dis.*, 79: 782–786.

Yadava, C.P. and Lal, S.S., 1998. Major insect pests of chickpea and their management. In: *IPM System in Agriculture. Volume 4. Pulses*, (Eds.) R.K.Upadhyay, K.G. Mukerji and R.L. Rajak. Aditya Books Private Limited, New Delhi, pp. 197–231.

Insect Pest and Disease Management *Pages* **239–256**
Editor: **Dr. D. Prasad**
Published by: **DAYA PUBLISHING HOUSE (2016)**

Chapter 17

Eco-Friendly Approaches: Combat for Rice Diseases

Ali Anwar, G.N. Bhat, K.A. Bhat, M. Shahjahan Dar and F.A. Khan*
Section of Plant Pathology, *Crop Physiology, Rice Research and Regional Station, SKUAST-K, Khudwani, Anantnag – 192 102, J&K
E-mail: zaman04@rediffmail.com

Introduction

The production of rice in India, however, increased gradually because of significant development in agricultural research, education extension and infrastructure. Farmers adoption of high-yielding varieties and other improved techniques as well as their use of chemicals, fertilizers have undoubtly contributed to increase rice production, while Kashmir Division of Jammu and Kashmir would require stupendous production to feed its increasing population because Kashmir is reaching a stage at which further expansion in rice area will not be possible. The alternative is to give more emphasis to increase the production potential per unit area and simultaneously finding ways to achieve this farm through appropriate technology. In India, the productivity is less than those in agriculturally advanced countries because poor agronomic practices followed in many remote areas and partly because a huge amount of crop being damaged by insects, fungi, bacteria, viruses and nematodes. However, the problems of rice cultivation in Kashmir are essentially those of temperate regions associated with conditions of low temperature, high altitude, short growing seasons and long day conditions with continuous and extensive monocropping of rice, which induces certain fungal diseases (Plates 17.1 and 17.2) occurring sporadically every year but may appear epidemic form in such a season when cloudy weather resulting high humidity prevails for long periods during growing season resulting tremendous losses in rice production. Among the diseases, blast (*Magnaporthe grisea*) and sheath blight (*Rhizoctonia solani*) diseases have emerged as most devasting and damaging factor of rice crop. Moreover, the pathogens have very high degree of pathogenic variability thus development of disease

Plate 17.1: Symptoms of Fungal Diseases on Rice in Kashmir Valley

(A) Leaf blast; (B) Neck blast and (C) Nodal blast (*Magnaporthe grisea*); (D) Disease complex in the field; (E) Brown spot disease (*Helminthosporium oryzae*); (F) False smut (*Ustilaginoidea virens*); (G) Glume discolouration caused by fungal complex

Contd...

Plate 17.1–Contd...

Contd...

Plate 17.1–Contd...

F

Contd...

Plate 17.1–Contd...

resistant varieties has always been considered to be of paramount importance in rice breeding which makes an economic, effective and eco-friendly method of disease management.

Use of resistant cultivars alone has not achieved disease control, because the inherent level of resistance is low. Despite continuous efforts of International Rice Research Institute (IRRI) and in various rice improvement programmes, diseases have not been satisfactorily controlled by host resistance (Mew and Rosales, 1986).

The sub terranean nature of the pathogens, its wide host range, competitive saprophytic ability

Plate 17.2: Peculiar Symptoms of Sheath Blight Disease (*Rhizoctonia solani*)

and prolonged survival in soil made the classical methods of disease management such as cultural and chemical methods in effective. Further, the adverse effects of agrochemicals on non target organisms, accumulation of toxic residues in soil environment and ground water pollution are other constraints (Baby and Manibhushanrao, 1996). Moreover, use of such chemicals entails a substantial cost to the nation and developing country like India cannot afford it. With the concern shown for hazardous nature of chemical pesticides towards environment it is imperative that biological or herbal chemical control agents are explored to combat the diseases. Biological methods, on the other hand is economical long lasting and free from residual side effects. Fungi in the genus *Trichoderma* are of increasing interest as bioprotectants against diseases of rice, besides plant extracts have also assumed special significance in the present day strategy of developing ecologically safe methods of plant disease management.

Effects of Bioagents Against Pathogens

In vitro

Chowdhury *et al*. (2003) worked on 3 bioagents *viz.*, *Trichoderma viride*, *T. harzianum* and *T. virens* and *antagonistic rhizospheric* bacteria in vitro and found significant inhibition of mycelial growth of *R. solani* by wheat rhizospheric bacteria (WRb-8) and *T. virens*. Bhat (2005) placed agar disc of 3mm dia of 7-days-old culture of *R. solani* in the centre of petridishes contained PDA medium and after 24 hours

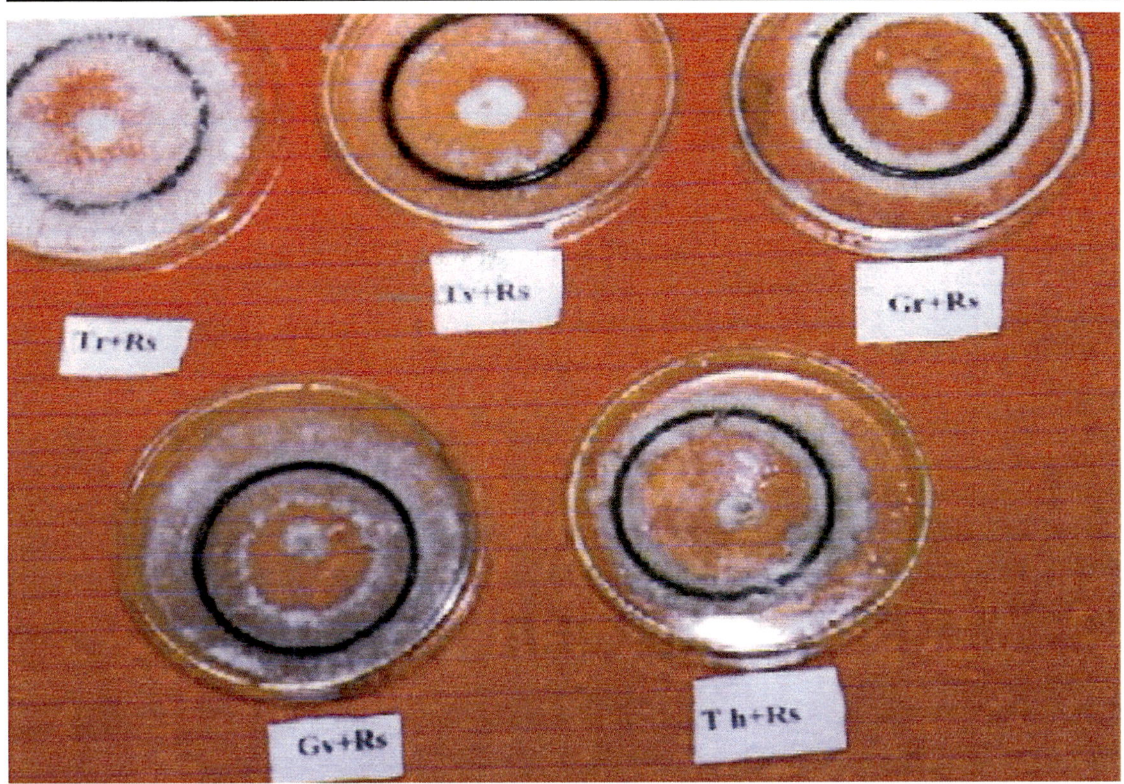

Plate 17.3: Antagonism Between Bioagents and *Rhizoctonia solani* by Dual Culture Bangle Method
Rs: *Rhizoctonia solani*; Tr: *Trichothecium* sp.; Tv: *Trichoderma viride*; Gr: *Gliocladeum roseum*; Gv; *Gliocladeum virens, Trichoderma harzianum*

of incubation at 25 ± 2°C glass bangles dipped in spore suspension of different bioagents put around them (Plate 17.3) which clearly exhibited that *T. harzianum* was the most effective bioagent showing highest inhibitory and antagonistic action followed by *T. viride, G. virens, Trichothecium* sp. and *G. roseum*. In dual culture, *T. viride* and *G. virens* produced distinct inhibition zones as compared to *T. harzianum, Trichothecium* and *G. roseum*. After 10 days of incubation *T. harzianum, Trichothecium* sp. and *G. roseum* completely over grew the test fungus while *T. viride* and *G. virens* were seen over growing the test fungus after 12 days. Similarly, Dubey (1998) and Chowdhary *et al*. (2003) have also reported highest mycelial inhibition by *T. harzianum* followed by *T. viride* and *G. virens.*

Mycoparasitism

Wen-Shi *et al*. (1985) reported the *Aspergillus* spp. and Penicillium spp. generally expressed loose parasitic association with *R. solani*, while *T. harzianum* performed most drastic hyper-parasitic activity to *R. solani* on potato dextrose agar medium and in soil. Hyphae of *T. harzianum* formed pegs then penetrated and eroded the hyphae of *R. solani*. The cytoplasm of injected hyphae was translucent and became shrinked. Sreenivasaprasad and Manibhushanrao (1990) observed that *Gliocladium virens* produced varying levels of wall lytic enzymes such as chitinase, glucanase and cellulose, capable of degrading, the cell wall of *R. solani* and also exhibited strong antibiosis as well as parasitism against the pathogen. They further reported that in *T. longibrachatum* parasitism and lytic enzymes seem to play an important role in the biocontrol of *R. solani*. Dubey (1998) apparently observed that hyphae of *G. virens* coiled the hyphae of *R. solani* and also penetrated at the point of coiling ultimately killed the host hyphae. Hand-cut sections of sclerotia showed the mycelium of G. *vriens* being inter and intra cellular. Hyphae of *T. harzianum* either directly penetrated the host hyphae or formed a granular mycelium bunch around the entry point.

The remaining process was similar to that of *G. virens* while *T. viride* formed either hooks or bunch around the host hyphae from where penetration took place. Similar observations were already found by Elad *et al.* (1980) and Tu (1980). Tarek and Moussa (2002) found most effective biocontrol agent. *T. harzianum* against *R. solani*. Hyphal interactions between *T. harzianum* and *R. solani* were observed by scanning electron microscopy and *T. harzianum* was found attached to the host by hyphal coils.

Bhat (2005) studied and showed in (Plates 17.4a–f) of parasitism that hyphae of *T. harzianum* grew on the surface of *R. solani* hyphae and parasitized throughout the coiling and also penetrated host hyphae. Parasitism between *T. viride* and *R. solani* revealed that *T. viride* hyphae only formed peg like structures and ran along the host hyphae and remained adhered to the host hyphae. The host hyphae also showed the stress conditions due to the parasitic activity of the *T. viride* which was apparent by disintegrated appearance of invaded hyphae (Plates 17.5a–f). However, no coiling was observed in case of *T. viride*. Similarly, the hyphal interaction between *G. virens* and *R. solani* revealed that *G. virens* hyphae also ran along the host hyphae and formed appressoria like intrusions which penetrate the host hypahe (Plates 17.6a–e). Further at later stages of observation, the disintegrated appearances of host hyphae were also exhibited (Plate 17.6d). Study between *Trichothecium* sp. and *R. solani* showed loose parasitism and parastic hyphae was not frequently found running along the host hyphae. The bioagents produced branches which fused with the hyphal cells of the host (Plates 17.7a–b) whileas, not interaction was observed between *G. roseum* and *R. solani* the mycelium of both the fungi were seen intermingling with each other showing no parasitism (Plates 17.8).

In vivo

Bhat and Anwar (2002) while studying the potential effect of antagonists (*T. harzianum*, *T. viride* and *Pseudomonas fluorescens*) as seed dressing and found most efficient *Pseudomonas fluorescens* strain afforded 60 per cent sheath blight suppression. Among the *Trichoderma* spp., *T. harzianum* isolate exhibited highest antagonistic effect to the sheath blight disease followed by *T. viride*. Resulted significant increase in biomass and grain yield of rice. Chowdhury *et al.* (2003) treated soil with bioagents exhibited maximum reduction in sheath blight incidence by WRP alone or in combination with other bioagents or soil amendments. But a higher reduction in sheath blight incidence as well as its severity was observed when soil treated with *T. virens* + WRPF Gliricidia. Bhat (2005) applied bioagents @ 160 kg ha^{-1} of sand maize medium in soil and found highest effect of *T. harzianum* followed by *T. viride*, *G. viren* and *Trichothecium* sp.

Biological control is the reduction in inoculum intensity on disease producing activities of pathogen in its active or dormant state, by one or more organisms, accomplished naturally or through manipulation of the environment host or antagonist or by mass introduction of one or more antagonist. *Pseudomonas flourescence* caused the highest inhibition of growth of blast pathogen in rice (*P. grisea*) (Krishnamurthry and Gnamanichalam, 1997) by seed treatment with talcum powder based formulation of the bacteria @ 4g/kg or seed (Bhat and Anwar, 2002). Seedling root dipping in 0.5 per cent of the product was effective in minimizing the disease incidence. Concomitent application of the product *P. flourescence*, *T. viride* and *T. harzianium* by seed treatment and seedling root dipping induced 40-60 per cent control of blast in field. There was also a significant increase in yield due to the treatment of bioinoculants (Anwar, 2000, 2001).

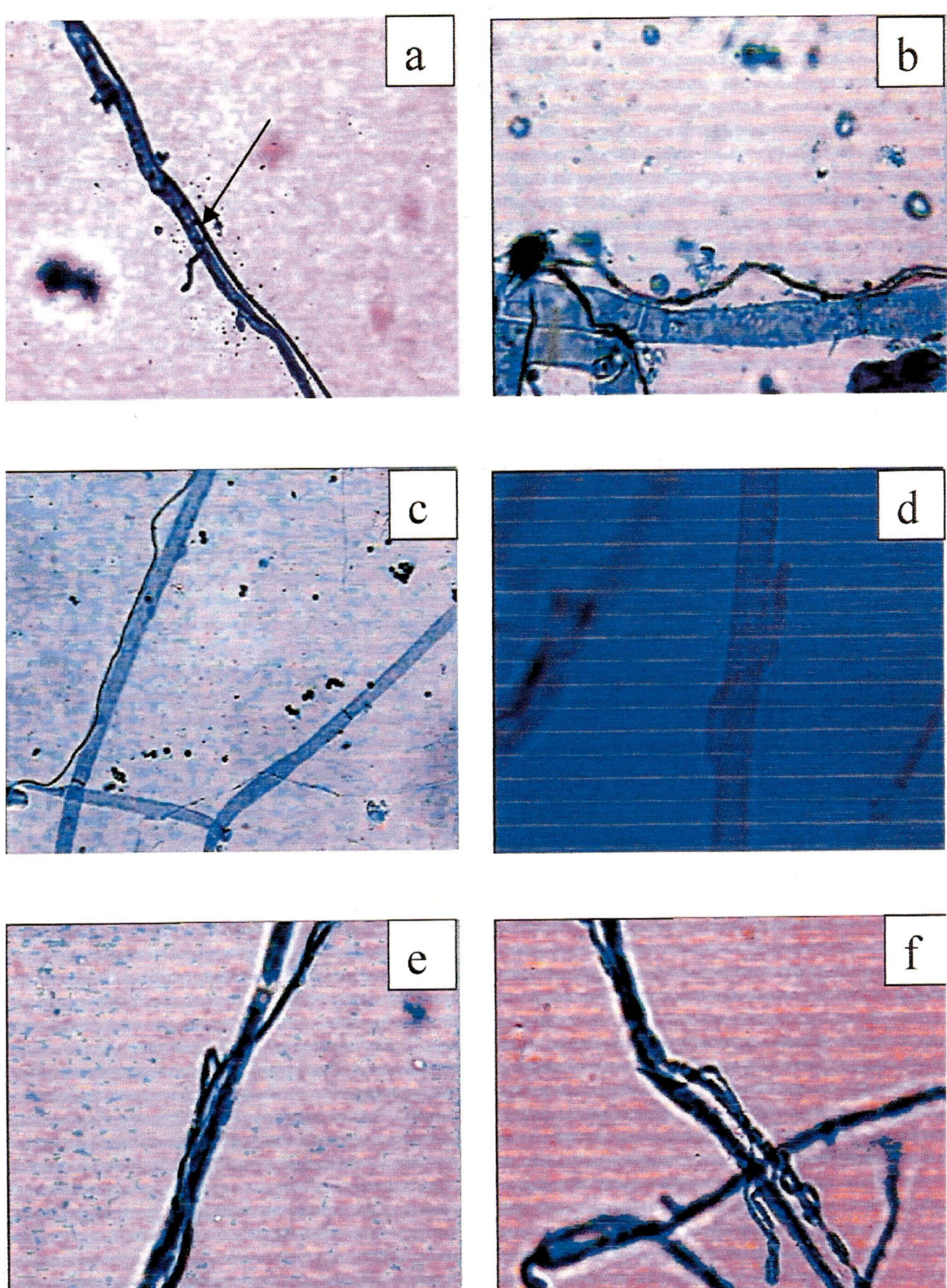

Plate 17.4: (a) ***T. harzianum*** **attacking** ***R. solani*** **by running along the host and penetrating pegs (x 450); (b)** ***T. harzianum*** **starting colonization on hyphae of** ***R. solani*** **(x 1000); (c) Hyphal attachment of** ***T. harzianum*** **on** ***R. solani*** **(x 450); (d)** ***T. harzianum*** **attacking** ***R. solani*** **by hyphal coils (x 1000); (e) Coiling of** ***T. harzianum*** **around** ***R. solani*** **(x 450); (f) Coiling of** ***T. harzianum*** **around** ***R. solani*** **causing disintegration of host (x 450)**

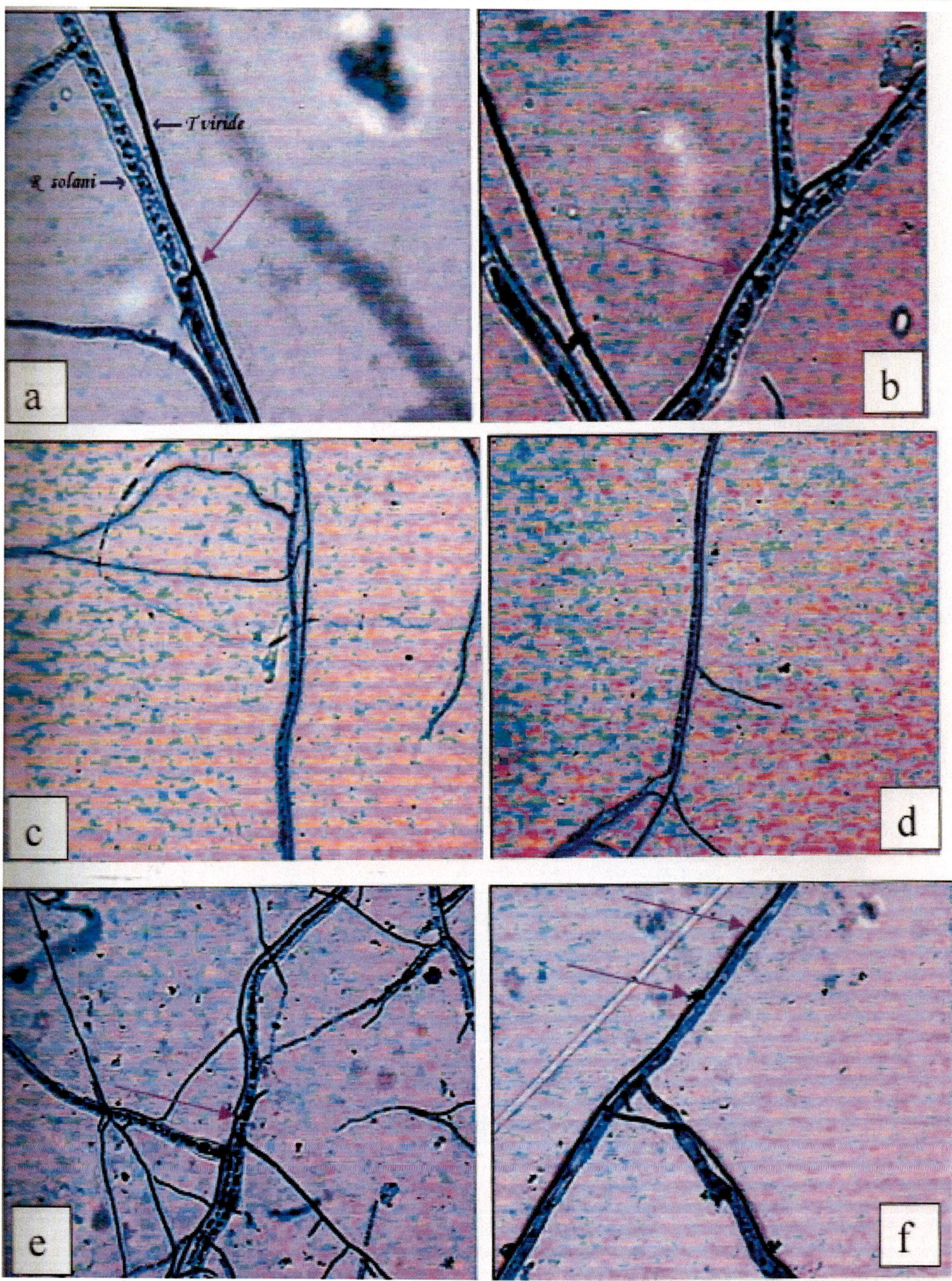

Plate 17.5: (a) Hyphal interaction between *T. viride* and *R. solani* bioagent attached to host by forming pegs penetrating the host (x 450); (b) *T. viride* attached to host hyphae by pegs. The invaded hyphae appearing disintegrated (x 1000); (c) *T. viride* hyphae attached to host (x 100); (d) *T. viride* attached to *R. solani* (x 100); (e) *T. viride* adhered to *R. solani* with penetrating pegs causing disintegration to host hyphae (x 450); (f) *T. viride* invading *R. solani* with pegs and running along the host surface

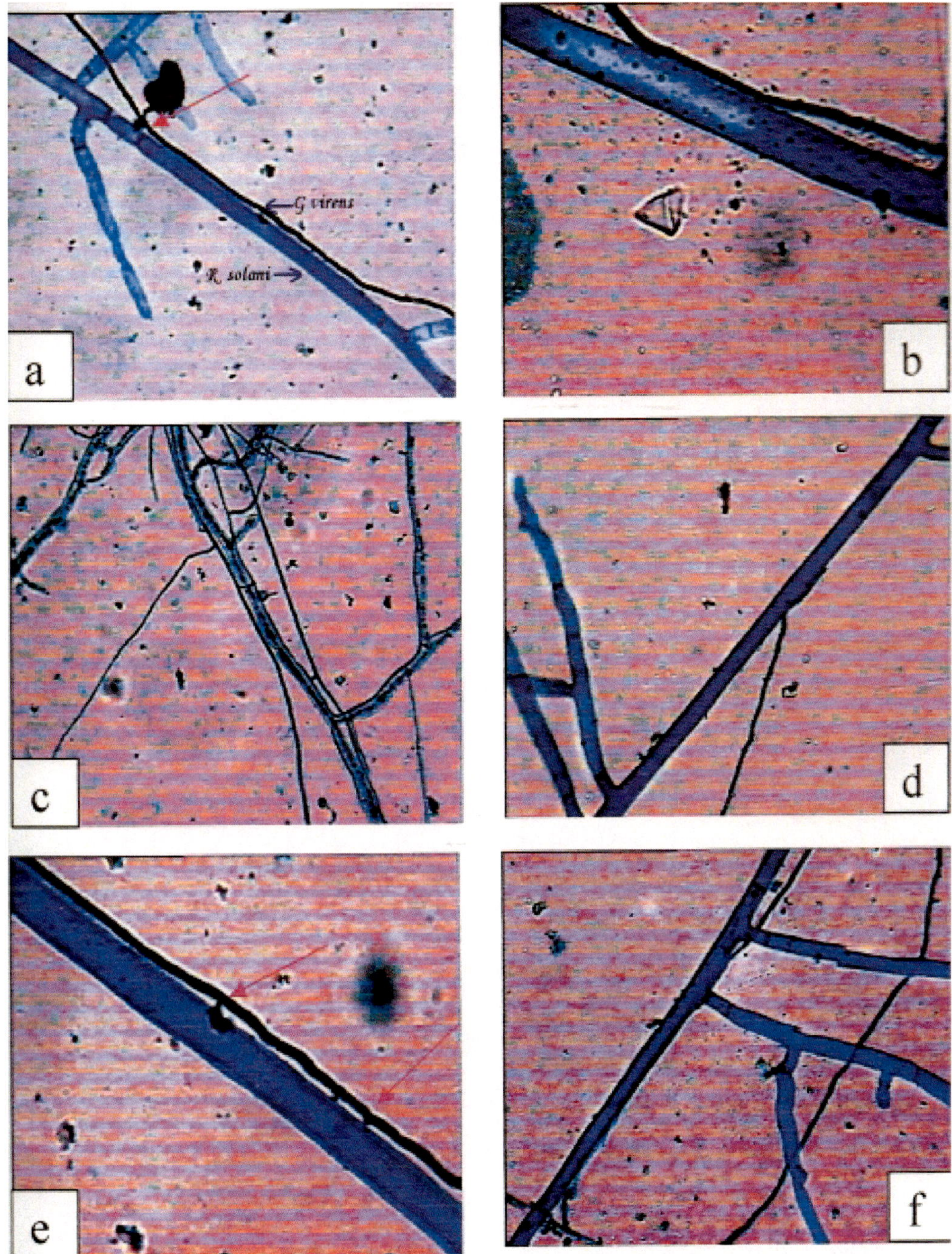

Plate 17.6: (a) *G. virens* hyphae running along *R. solani* forming appressoria like intrusions (x 450); (b) *G. virens* hyphae adhered to host hyphae (x 1000); (c) *R. solani* invaded by *G. virens* and causing disintegration of host cells (x 450); (d) *G. virens* hyphae reaching the host hyphae and initiating parasitism *R. solani* (x 450); (e) *G. virens* hyphae running along *R. solani* forming appressoria and pegs (x 1000); (f) *G. virens* hyphae adhered and running along the hyphal surface of *R. solani* (x 450)

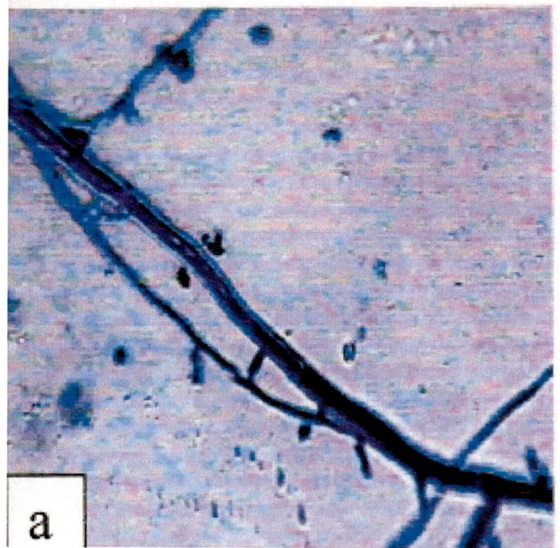

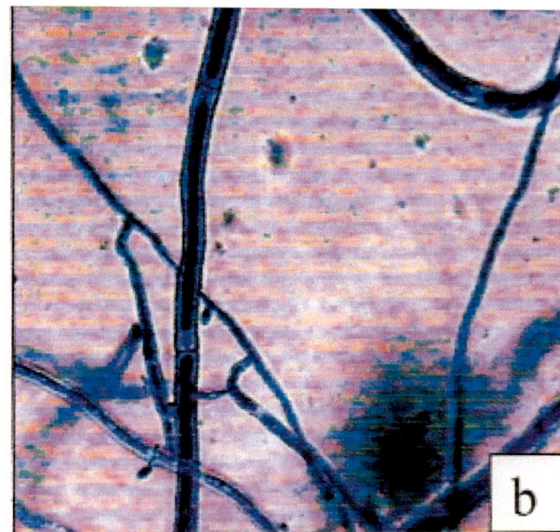

Plate 17.7: (a) *Trichothecium* sp. hyphae fusing with cells of *R. solani* (x 450), (b) *Trichothecium* sp. producing hyphae which fuse with host hyphal cells (x 450)

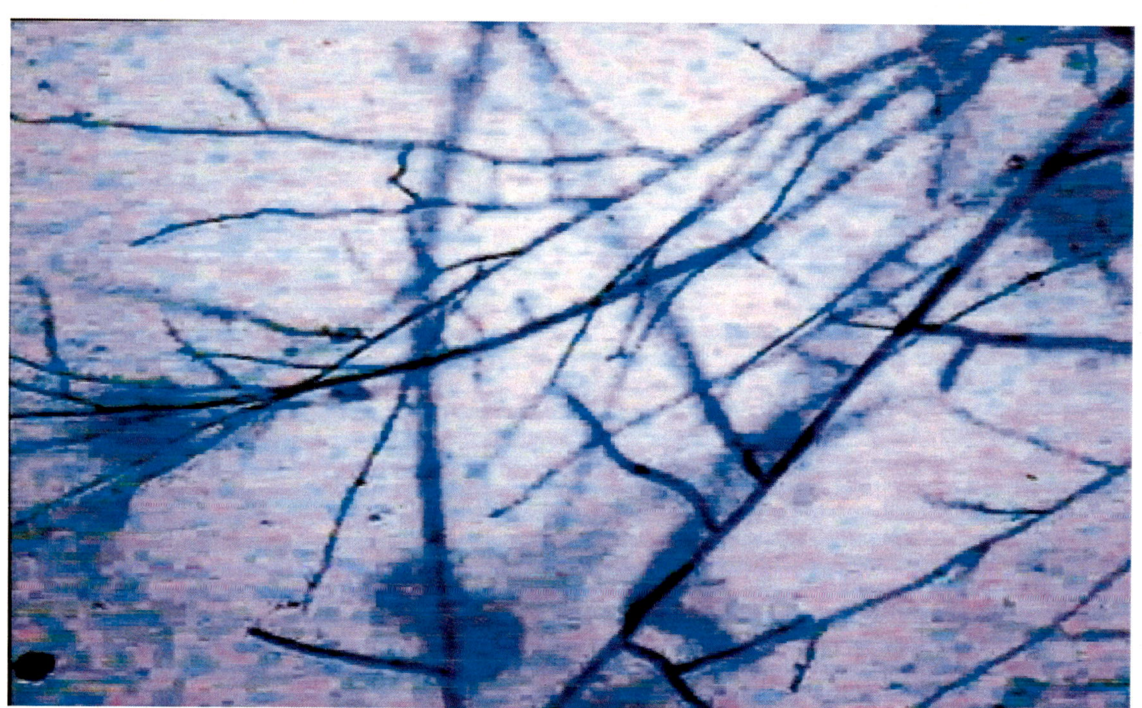

Plate 17.8: The Mycelium of *R. solani* and *G. roseum* Intermingled with each other showing no parasitism (x 100)

Moreover, the extracts of marigold leaves exhibited more effective followed by the extract of *Mentha* sp. leaves, when seed treatment and seedling root dipping were carried out in field conditions (Anwar and Bhat 2003; Kamala *et al.*, 2001). As several plant extracts have been demonstrated to possess excellent fungicidal properties (Anwar and Khan, 2001, Kandhari and Singh, 2000). Recently excellent fungitoxicants of extracts of *Vitex negundo* and calotropis sp. has been reported (Dev *et al.*, 2002).

Commercial Production of Bio-control Agents and their Shelf Life

Trichoderma spp., have in the recent years come out as a viable and feasible alternative bio-pesticide. As a result seven commercial products of *Trichoderma* have appeared in Europe and USA (Whipps, 1992). The product *Trichoderma* is extensively used in Russia to cover 4.4 x 10 ha/year. Tamil Nadu Agricultural University has developed talc based formulation for seed treatment which has been commercially produced and distributed to farmers. So far about 1.5 tone, have been distributed to cover about 13000 ha of oil seeds and pulse crop (Jayarajan and Nakkeeran, 1996). Vermiculite formulation for soil application was developed by Lewis *et al.* (1991). This product was tested and was found promising effect against diseases. Granular formulation of *Trichoderma* and *Gliocladium* spp. was also reported by Prasad and Rangeshwaran (1999) which was tested successfully against *R. solani*.

Mass production of bio-control fungus has done in a variety of media in stationary flasks (Patel, *et al.*, 1984). Molasses yeast medium (MYM) is being used widely for commercial production of *Trichoderma* by fermentation process. However, an improved molasses soy medium is reported by Prasad and Rangeshwaran (2002) where brewers is yeast which serve as nitrogen source is replaced by cheap soy flour. Besides vermiculite formulation, talc based formulations was also prepared by Jayarajan and Narkkeeran (1996). They reported that using wheat bran as food base, *T. viride* continues to multiply during storage. They further confirmed that highest population of *T. viride* continues to multiply during storage. They further confirmed that highest population of *T. viride* was observed in milky white bags and produce can be stored at 20-30°C. At this range of temperature even after 75 days product contained 230-242 x 10^6 cfu/g and it was much above the minimum of 20 x 10^6. Adams (1990) has indicated that the bio-control efficiency of *Trichoderma* is 5 x 10^4 cfu for every cfu of *R. solani.* Bhat (2005) proposed talc based formulations of *T. harzianum* and added carboxyl methyle-cellulose as sticker (Plate 17.9a) and stored in white poly-propylene bags then the initial population was estimated by serial dilution and agar method (Plate 17.9b) that significant difference in population was observed at 30-180 days. The number of cfu/g declined to less than 10^6 cfu/g by 6 months.

Cultural Practices

This approach has an excellent promise and future cultural practices are found to affect the incidence and intensity of diseases. Successful manipulation in cultural practices for disease combat can be made when the farmers eager to get on apprehension about the nature of pathogen and environment to disease development. However, irrigated rice cultivation with the malaise peculiar practices is being followed by the farmers of Kashmir therefore, Anwar and Bhat (2004) often suggested cultural practices are the only feasible means or method of control in rice crop as a brief outlines of selected cultural practices are given in reaction to diseases as follows:

Seed Selection

1. Healthy and stout seed should be selected.
2. Seeds should be free from the disease.

Plate 17.9(a): Talc Based Formulation of *Trichoderma harzianum*

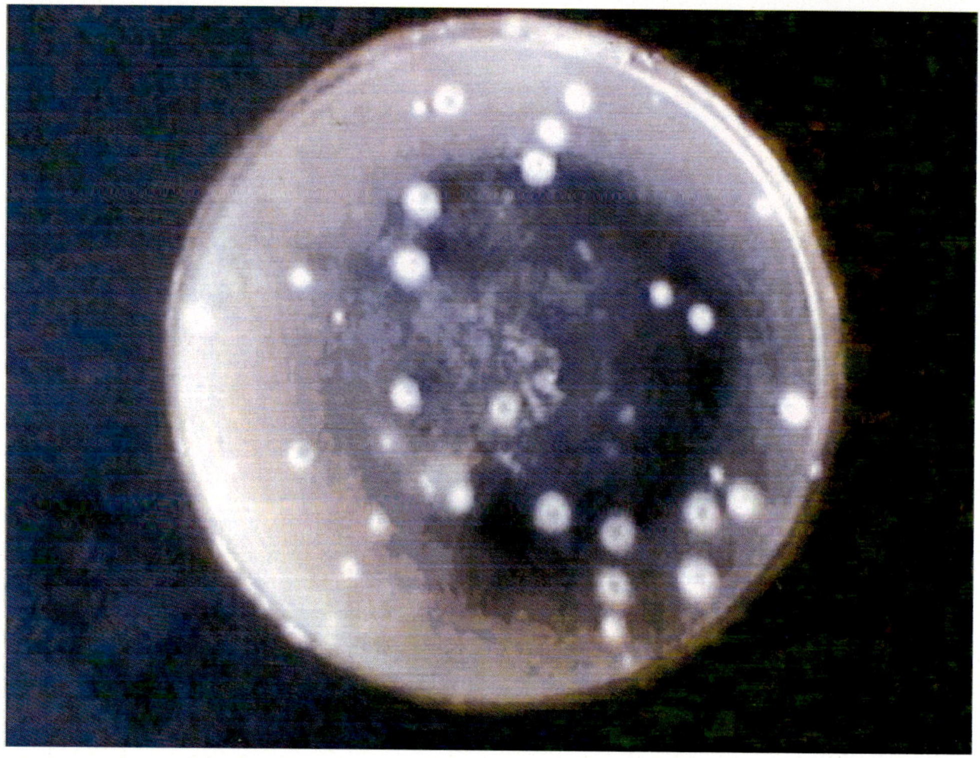

Plate 17.9(b): *Trichoderma* Colonies Developing in Serial Dilution and Plating Method (x 10^7)

3. Do not use the chaffy or partially filled seeds.
4. Obtain the seeds from a reliable source.
5. Collect the seeds of disease tolerant varieties or free from the disease.
6. Choose the variety which suitable for the particular rice growing area.
7. Avoid mechanical injury in seeds during the threshing and seed processing.

Ejection of Chaffy or Diseased Seeds

1. Prepare a solution of salt @ 2 per cent
2. Pour the seeds into the prepared double quantity of salt solution.
3. Stir the seeds slowly by hand or stick
4. Remove the floated seeds and decant the rest of the seeds (Plate 17.10).
5. Sun dry the seeds for 4-6 hours before chemical treatment.

Plate 17.10: (A) Floated chaffy and diseased seeds; (B) Discarding of chaffy and diseased seeds

Seed Soaking

1. Put the treated seeds into fresh cotton bags with the capacity up to 50 kg.
2. Water for soaking should no be contaminated.
3. Stagnant decomposed water of rural drainage should not be used for soaking of seeds.
4. After 24 hours the seed bags should be kept under the temperature of 20–30°C.
5. Do not use the paddy straw or cow dung for increasing the temperature.
6. Polythene sheet can be cover the seed bags for sustaining the temperature.
7. Either put the seed bags for one day in sunny heat.
8. After 4-6 days the sprouts emerged, should be sown.
9. Sprouted seeds should not be treated with any fungicide because of plumule/radicle can be damaged.

Preparation of Nursery

1. Select 1/10 area of the total cultivated land for nursery.
2. Eradicate the stubbles and grasses from the fields by burying deep in soil in known to reduce build up of pathogens.
3. Do adequate puddling and pulverization of nursery field because soil toxicant can not make any layer in upper soil horizon.
4. After decantation of the sand particles remove the floated debris by outletting of the water.
5. Immediately incorporate the sprouted seeds into prepared nursery with an optimum seed rate.
6. If there is any delay in seed broadcasting then the water moulds develop which resist to establish the seeds in soil and accelerate the rotting and damping off by *Achlya* spp.
7. After 4-6 days of seed broadcasting, drain water in alternate days to avoid the mould development.
8. The maximum seed rate also influence the seedlings due to overlapping of sprouted seeds resulting rotting and weakening in seedling vigour.
9. Use the recommended basal dose of NPK in nursery.
10. Do not broadcast urea in nursery.
11. Avoid scarcity of water in nursery.
12. Seedlings become susceptible when air and water temperature are different, particularly under cool air and warm water therefore, avoid the supply of cool water.
13. Avoid the consequent flow of the cold water into the field because rotting can be started in seedlings.

References

Adams, P.B., 1990. The potential of mycoparasites for biological control of plant disease. *Annual Review of Phytopathology*, 28: 59–72.

Anwar, A. and Bhat, M.S., 2001. Management of rice blast disease through the cultural practices. *Agric. Biol. Res.*, 20(1): 85–95.

Anwar, A. and Bhat, M.S., 2003. Bio-inoculants and herbal effect in management of blast disease *Manoporthe grisea* in rice. In: *National Seminar on Organic Production and their Future Prospects.* October 21–22, SKUAST-K, Srinagar, pp. 117–118.

Anwar, A. and Khan, F.U., 2001. Effect of aqueous leaf extracts of medicinal plants on the growth of rhizosphere fungi of tomato cv. Pusaruby *in vitro. SKUAST-J. Res.,* 3: 60–63.

Baby, U.I. and Manibhushanrao, K., 1996. Fungal antagonists and VA mycorrhizal fungi for biocontrol of *Rhizoctonia solani,* the rice sheath blight pathogen In: *Recent Developments in Bio-control of Plant Pathogens,* (Eds.) K.M. Roa and A. Mahadevan. Today and Tomorrows Printers and Publishers, New Delhi, pp. 1–9.

Bhat, G.N. and Anwar, A., 2002. Effect of bio-inoculants, seed dreassing and spray schedule of carbendazium against sheath blight (*Rhizoctonia solani*) of rice. *SKUAST,* 4: 73–76.

Bhat, K.A., 2005. Studies on biological control of *Rhizoctonia solani* Kuhn causing sheath blight of rice. *M.Sc. Thesis,* Division of Plant Pathology, SKUAST-K, Shalimar Srinagar.

Chowdhury, A., Dutta, S., Chowdhury, A.K. and Laha, S.K., 2003. Effects of biotic/abiotic elicitors on the management of sheath blight of rice under terai agro-ecological region of West Bengal. *Indian Mycol. Pl. Pathol.,* 33: 378–386.

Dev, U., Devakumar, C., Aggarwa, P.C., Jitendra, M., Joshi, K.d. and Indra Rani, 2002. Antifungal effect of *Vitex negundo, Calotropis* spp. and other plant extracts against seed-borne fungi. *Pesticide Res. J.,* 14: 229–233.

Dubey, S.C., 1998. Evaluation of different fungal antagonists, plant extracts and oil cakes against *Thanatephorus cucumeris.* Causing banded blight of rice. *Indian J. of Mycol. Pl. Pathol.,* 28(3): 266–269.

Elad, Y., Chet, I. and Katan, J., 1980. *Trichoderma harzianum* and biocontrol agent effective against *Sclerotium rolfsii* and *Rhizoctonia solani. Phytopathology,* 70: 119–121.

Jayarajan, R. and Nakkeeran, S., 1996. Exploitation of biocontrol potential of *Trichoderma* for field use. In: *Recent Development in Biocontrol of Plant Pathogens,* (Eds.) K.M. Rao and M. Mahadevan. Today and Tomorrows Printers and Publishers, New Delhi, pp. 61–66.

Kandhari, J. and Singh, R.P., 2000. Efficacy of different extracts of certain indigenous plants against sheath blight pathogen of rice. *Indian J. Mycol. P. Pathol. Res.,* 38: 125–127.

Krishnamurthy, K. and Gnamnichalam, S.S., 1997. Biological control of sheath blight of rice. Induction of systematic resistance in rice by plant associated *Pseudomonas* spp. *Current Science,* 72(5): 331–334.

Lewis, J.A., Papaviza, G.C. and Lumsden, R.D., 1991. A new formulation system for the application of biocontrol fungi to soil. *Biocontrol Science and Technology,* 1: 59–69.

Patel, S.T., Misra, Ashok and Misra, A., 1994. Production of *Trichoderma* in stationary flasks. *Gujarat Agricultural University Research Journal,* 19: 53–56.

Prasad, R.D. and Rangeshwaran, R., 1999. Granular formulation of *Trichoderma* and *Gliocladium* spp. In biocontrol of *Rhizoctonia solani* of chikpea. *Indian J. of Mycol. Pl. Pathol.,* 29(2): 222–226.

Prasad, R.D. and Rangeshwaran, R., 2002. An improved medium for mass production of the biocontrol fungus *Trichoderma harzianum. Indian J. of Mycol. Pl. Pathol.,* 30(2): 233–235.

Sreenivasaprasad, S. and Manibhushanrao, K., 1990. Antagonistic potential of *Gliocladium virens* and *Trichoderma longibrachiatum* to Phytopathogenic fungi. *Mycopathologica,* 109: 19–26.

Tarek, A. and Moussa, A., 2002. Studies on biological control of sugarbeet pathogen *Rhizoctonia solani* Kuhn. *Online Journal of Biological Sciences,* 2(12): 800–804.

Tu, J.C., 1980. *Gliocladium virens* a destructive myco–parasitic of *Sclerotinia sclerotium*. *Phytopathology,* 70: 670–674.

Wen-Shi, W., Shan-Da, L., Chang, Y.C. and Johannes, T., 1985. Hyper-parasitic relationship between antagonists and *Rhizoctonia solani. Plant Protection Bulletin,* 28: 91–100.

Whipps, J.M., 1992. Status of biological disease control in horticulture. *Biocontrol Science and Technology,* 2: 3–24.

Insect Pest and Disease Management *Pages* **257–267**
Editor: **Dr. D. Prasad**
Published by: **DAYA PUBLISHING HOUSE (2016)**

Chapter 18

Active Oxygen in Plant Disease Control: Possible Role and Future Scope

Chinmay Biswas, S.K. Biswas and S.S.L. Srivastava
Department of Plant Pathology, C.S.A. Univ. of Agri. and Tech., Kanpur – 208 002, U.P.

Introduction

The active oxygen species are toxic intermediates that result from the reduction of molecular oxygen (O_2). The predominant species present in the plant system are superoxide radical (O_2^-), hydrogen peroxide (H_2O_2) and hydroxyl radical (OH^-). In healthy plant cells under normal metabolic conditions, these active oxygen (AO) species produced at low levels are not allowed to accumulate and get detoxified through various antioxidant mechanisms. However, striking release of active oxygen which is termed as oxidative burst has been reported in many plant-pathogen interactions. The AO produced in response to pathogens or elicitors are believed to play key roles in signal transduction, antimicrobial activity, cell wall modification, induced resistance and hypersensitive cell death (Baker and Orlandi, 1995). Therefore, active oxygen could be an important element contributing in plant disease control. Although use of AO directly as antimicrobial agent is limited due to phytotoxicity, treatment with low concentration of H_2O_2 leads to development of induced resistance against many diseases. Development of transgenic crops expressing gene(s) responsible for elevated levels of AO would be a novel approach for plant disease control. Studies on the occurrence of the oxidative burst in various plant-pathogen interactions, proper understanding of the regulation and properties of the molecular components involved in the generation of AO will be the right step towards that direction.

Active Oxygen Species in Plants

Molecular oxygen is nontoxic and unreactive, but it becomes reactive when its electron structure is altered. Its successive reduction yields different AO species. The first one electron reduction of O_2 forms superoxide radical (O_2^-). This step requires a slight input of energy provided by NADPH in plant cells. However, the subsequent steps are exothermic and can occur spontaneously, either catalyzed or not (Dat *et al.*, 2000). The superoxide radical (O_2^-) is highly reactive and relatively shortlived. In aqueous solution, it undergoes spontaneous dismutation to get reduced to hydrogen peroxide (H_2O_2).

$$2\,O_2^- + 2\,H^+ \longrightarrow H_2O_2 + O_2$$

This reaction occurs at neutral or slightly acidic pH or with superoxide dismutase as a catalyst.

Hydrogen peroxide is relatively stable and longlived molecule. Its reaction with biological molecules is slower than that of O_2. H_2O_2 can oxidize transition metals such as Fe^{2+}. The last species generated by this series of reductions is hydroxyl radical (OH^-). It is formed through Fenton or Haber-Weiss reactions.

$$O_2^- + Fe^{3+} \longrightarrow Fe^{2+} + O_2$$

$$H_2O_2 + Fe^{2+} \longrightarrow Fe^{3+} + OH^- + OH^-$$

$$\text{Summary: } H_2O_2 + O_2 \longrightarrow OH^- + OH^- + O_2$$

OH^- has a very strong potential and short a half-life. Therefore it shows very high affinity for biological molecules at its site of production. As the transition metals act as a catalyst in its formation, the site of production is determined by the location of metal ion complexes in the plant cells.

Sources of AO in Plant System

The AO species are routinely generated at low levels by plant cells due to electron transport in chloroplasts, mitochondria and enzymes in other cell organelles.

Photosynthesis

In the noncyclic photophosphorylation water molecule is utilized as a source of electron. The electron from Photosystem (PS)-II passes through plastoquinone, cyt. b_6, cyt.f, and plastocyanin, then through PS-I eventually reaches ferredoxin which in turn donates to reduce NADP. The reduced NADP (NADPH + H^+) is utilized for reduction of CO_2 to carbohydrate. However, under conditions limiting the availability of $NADP^+$ the electron is donated to alternative acceptor–molecular oxygen (O_2) leading to formation of superoxide radical (O_2^-). Leakage of electrons to O_2 from electron transport chain is also inevitable because very few plants have CO_2 fixation rate high enough to use more than 50 per cent of the light energy absorbed (Baker, 1991). Thus, conditions limiting CO_2 fixation lead to enhanced generation of AO species, as more O_2 molecules will be used as electron acceptors. Superoxide can initiate a chain reaction to form other AO Species. The sum of the reaction is transfer of electrons from H_2O to O_2, resulting in production of O_2 at PS-II, O_2^- at PS-I (Polle, 1996)

Photorespiration

Under normal conditions during carbon assimilation the enzyme ribulose–1, 5–bisphosphate carboxylase catalyses fixation of CO_2. But under conditions of high O_2/CO_2 ratio, O_2 successfully competes with CO_2 and ribulose–1, 5-bisphosphate carboxylase switches to oxygenase activity. Oxygenation of RuDP yields two glycolates which are then transported from chloroplasts to peroxisomes. There oxidation of glycolate catalysed by glycolate oxidase yields H_2O_2 and glyoxylate. (Foyer, 1996)

Respiration

The electron transport system in the mitochondria is made up of several dehydrogenase complexes that reduce a common pool of ubiquinone. Then the latter is oxidized either by the cytochrome or the alternative pathway. Ubiquinone radicals and NADH dehydrogenases are the main O_2^- generators in mitochondria. The superoxide is formed through auto oxidation of reduced components of respiratory chain. (Dat *et al.*, 2000)

Regulation of Active Oxygen

Active oxygen species generated in healthy plant cells would be toxic, if they are allowed to accumulate. Therefore, the cell has several antioxidant mechanisms for the detoxification of these species. The capacity of AO to serve as signals adds to the importance of antioxidants for specifically regulating different active Oxygen species at different cullular sites.

Superoxide Dismutases (SOD)

SOD is an important antioxidant group which is basically a family of metalloenzymes and are found in the cytosol, chloroplasts and mitochondria. It catalyzes the dismutation of O_2^- and HO_2^- to H_2O_2.

$$HO_2^- + O_2^- + H^+ \longrightarrow H_2O_2 + O_2$$

$$O_2^- + O_2^- + 2H \longrightarrow H_2O_2 + O_2$$

Yim *et al.* (1990) reported the capacity of SOD to catalyze the generation of OH^- from H_2O_2.

Catalase (CAT)

Catalases are tetrameric heme containing enzymes that can convert H_2O_2 to O_2 and H_2O (Sanchez-Casas and Klessig, 1994).

$$H_2O_2 + H_2O_2 \longrightarrow 2H_2O + O_2$$

Catalases are present in peroxisomes, glyoxisomes and other organelles where H_2O_2 generating enzymes, such as glycolate oxidase, are found. It plays significant role in reducing high levels of H_2O_2. Due to its high K_m, catalase is very inefficient in scavenging low levels of H_2O_2 produced in cells.

Ascorbate

Ascorbate is a common antioxidant found in plant cells. It directly scavenges O_2^- and OH^- (Halliwell, 1974) and detorifies H_2O_2 via ascorbate peroxidase (Foyer *et al.*, 1994). Ascorbate is regenerated by dehydroascorbate reductase using glutathione which is in turn regenerated by the enzymc glutathione reductase at the expense of NADPH.

Vitamin E (α–tocopherol) is another important scavenger molecule for oxygen detoxification.

Active Oxygen Production During Plant Defense Response

Doke (1983) for the first time reported that upon inoculation with an incompatible race of *Phytophthora infestans* eliciting hypersensitive response (HR) in potato tubers produced AO species O_2^-. It was suggested that O_2^- may be involved in host cell hypersensitivity. Subsequently, many workers have reported AO generation in various plant-pathogen interactions.

Fungal Pathogens

AO generation was reported in race specific interaction between tomato and the fungus, *Cladosporium fulvum.* It was also observed that apoplastic (extracellular) fluid from infected leaves exhibiting HR stimulated AO generation. However, apoplastic fluid from uninfected plants or plants infected with a race not eliciting HR had little effects on AO levels. (Vera-Estrella *et al.*, 1992). Borden and Higgins (2002) further reported that H_2O_2 generation in tomato against the pathogen. *C. fulvum* was also associated with a variety of defense responses including protein cross-linking, callose deposition and accumulation of phenolics.

Enhanced levels of H_2O_2 was recorded in tomato cultivars less susceptible to the necrotrophic pathogen, *Botrytis Einerea* indicating important role of AO in resistance. (Patykowski and Urbanek, 2003)

Nonspecific fungal elicitors can also cause oxidative burst in plant cells. Soybean and bean cell suspensions treated with non specific elicitors rapidly produced AO species (Apostol *et al.*, 1989, Anderson *et al.*, 1991).

Sundar *et al.* (2002) studied the effect of a glycoprotein elicitor isolated form the mycilial walls of red rot pathogen of sugarcane, *Glomerella tucumanensis* on AO generation in suspension-cultured cells of sugarcane. The elicitor rapidly induced generation of H_2O_2 and O_2^- within 10 min after treatment in suspension-cultured cells of both resistant and susceptible cultivars. However, the generation of AO was more rapid in resistant cultivar compared to that of susceptible one.

Changing levels of AO were noticed in apple fruits of susceptible and less susceptible cultivars upon wounding or inoculation with *Penicillium expansum* after harvest. Immediately after wounding or inoculation, the less susceptible fruit showed a significant increase in H_2O_2 levels concomitantly with higher activity of SOD. No significant changes in CAT and POX were found. In susceptible fruit, both H_2O_2 levels and SOD activity in wounded/inoculated fruit remained similar to control. In contrast, CAT and POX (Peroxidase) activity significantly increased as a consequence of wounding and inoculation. (Torres *et al.*, 2003).

Donato *et al.* (1997) reported induction of AO production in cell suspension and callus homogenates of a melon cultivar (Jerac) resistant to *Fusarium oxyporum* f.sp. *melonis* only when hyphal wall components of a specific race (O) of the pathogen was applied.

Bacterial Pathogens

AO production has also been reported in the plant cells in interaction with specific plant pathogenic bacteria. Tobacco plants infected with *Pseudomonas syringae* pv. *syringae* exhibited HR and it was accompanied by enhanced O_2^- generation (Adam *et al.*, 1989).

Harpin, a bacterial protein has been reported as an elicitor of HR in tobacco. Harpin producing *Erwinia amylovora* induced AO generation in tobacco cell suspensions. (Baker *et al.*, 1993).

Bestwick *et al.* (1997) reported localized accumulation of H_2O_2 during HR in lettuce leaves to *Pseudomonas syringae* pv *phaseolicola*. Inoculation with wild type cells of *P. syringae* pv. *phaseolicola* induced localized accumulation of H_2O_2 in plant cell walls adjacent to attached bacteria. Increased levels of AO was reported also in a bean plant (*Phaseolus vulgaris*) in incompatible interaction with *P. syringae* pv. *phaseolicola* (Milosevic and Slusarenko, 1996).

AO production was reported in pepper (*Capsicum annuum*) leaves infected with *Xanthomonas campestris* pv. *vesicatoria* and it went uncontrolled at advanced stages of infection. (Buonaurio and Kumar, 1995).

Inoculation of *Arabidopsis thaliana* leaves with avirulent *Pseudomonas syringae* pv. *tomato* induced secondary oxidative bursts in discrete cells at distant tissues leading to low frequency systemic microhypersensitive responses which are not confined to the inoculation site. The resulting AO species were O_2^- and H_2O_2 (Alvarez *et al.*, 1998).

Viruses

AO generation was detected in tobacco leaf discs undergoing HR due to interaction with tobacco masaic virus as early as in 1988. (Doke and Ohashi, 1988).

Oxidative burst was reported in *Capsicum annuum* leaves due to infection of cucumber mosaic virus. Increased soluble peroxidase activity through its involvement in lignification process might be considered as a part of the defense mechanism of *C.annuum* against viral infection. (Candela *et al.*, 1994).

Hernandez *et al.* (2001) examined the response of antioxidative enzymes to plum pox virus (PPV) in apricot (*Prunus armeniaca*). In the inoculated resistant cultivar a decrease in catalase (CAT) as well as an increase in total SOD and dehydroascrobate reductase (DHAR) activities were observed. Ascorbate peroxidase (APX), glutathione reductase (GR) and mono dehydroascarbate reductase (MDHAR) did not change significantly in relation to non-inoculated (control) plants. In the susceptible cultivar inoculation with PPV brought about a decrease in CAT, SOD and GR, whereas a rise in APX, MDHAR and DHAR activities was found in comparison to non-inoculated plants.

Two Phased Production of AO

The production of AO occurs in two distinct phases: phase-I and II. Phase-I is the initial relatively short lived, non specific phase which follows within minutes of pathogen addition and Phase-II is the relatively long-lived secondary phase, dependent on recognition of incompatible pathogens by the host which begins 1-3 h after he intial burst (Baker *et al.*, 1991) The phase-II AO production occurs only in nonhost-pathogen interaction and incompatible race-cultivar interations. For example, *Pseudomonas syringae* pv. *tabaci*, a tobacco pathogen, does not elicit a Phase-II AO generation in tobacco but does elicit the response in non host soybean. The race 4 and 6 of *P.syringae* pv. *glycinea* both elicited phase-II AO response in tobacco, a non host but in their host soybean only the incompatible race 6 could do so, not the race 4. (Baker and Orlandi, 1995)

It was reported that increases in the level of bacterial inoculum increased phase-I AO production, but the phase-II AO production was lower when higher amounts of bacteria were used. This discrepancy could be attributed to induction of increased AO-scavenging activity by higher inoculum levels of bacteria (Baker *et al.*, 1995).

Staskawicz *et al.* (1984) reported that phase-II AO response in race-cultivar specific incompatible interactions in soybean was due to interactions between a virulence gene avrA in *Pseudomonas syringae* pv. *glycinea* race 6 and corresponding resistance gene in soybean. Incorparation of avrA gene transformed the compatible race 4 into and incompatible one that induced Phase-II AO response in soybean cells (Orlandi *et al.*, 1992). Phase-II AO production due to non-host pathogen incompatible interaction has been reported to be controlled by the genes located in the bacterial hrp/hrm region. Mutation of these hrp genes blocked the Phase-II AO response. However, insertion of hrp/hrm region of *P. syringae* pv. *syringae* into saprophyte, *P. fluorescens* transformed it into an incompatible bacterium which could induce phase-II AO response in suspension cells (Huang *et al.*, 1988., Glazener *et al.*, 1991).

Mechanism of AO Generation During Plant Pathogen Interaction

Several potential sources of AO have been identified in plant cells. But still the exact mechanism of AO production during plant-pathogen interaction is not clear. Various enzyme systems have been proposed as the sources of AO generation such as NADPH oxidases, peroxidases etc. Based on the available evidences it is hypothesized that:

1. Hyphal wall or plant cell wall derived glycoprotein or carbohydrate elicitor triggers oxidative burst. (Apostol *et al.*, 1989)
2. The elicitor binds to the plasmamembrane bound specific receptors. (Horn *et al.*, 1989)
3. The elicitor-receptor complex is likely to be associated with GTP binding or G-proteins as found in animal systems. (Legendre *et al.*, 1992)
4. AO generation depends on increased intracellular Ca and protein kinase activation. The elicitor receptor coupled with a G-protein leads to Ca influx, which then activates a Ca^{++} dependent protein kinase and ultimately the O_2^- generating oxidase. (Mehdy, 1994).
5. The oxidase that catalyzes the formation of O_2^-, NADPH dependent oxidase thus gets activated. Then it transfers electrons from NADPH on the inside of the membrane to molecular oxygen on the outside of the membrane leading to the generation of O_2^-.
6. Another oxidase that could be the sources of O_2^- generation is a NADH-dependent peroxidase associated with the external surface of plasma membrane. (Sutherland, 1991).
7. In addition to their role in AO scavenging peroxides can also act as NAD(P)H oxidases leading to formation of AO. Oxidation of NAD(P)H yields O_2^- and H_2O_2. The source of extracellular NAD(P)H is cell wall bound malate dehydrogenase which uses malate coming through the cell membrane. (Baker and Orlandi, 1995)

Possible Role of AO in Plant Defense

Strengthening of Plant Cell Wall

The active oxygen appears to play a vital role in strengthening plant cell wall against pathogen attack. In the cell wall, H_2O_2 is used in lignin formation. H_2O_2 mediated oxidative cross-linking of specific structural proteins was demonstrated in bean and soybean suspension cells after treatment with elicitors. (Bardley *et al.*, 1992). This cross-linking could discourage or slow down the pathogenic advance.

Role in Phytoalexin Production

There are several reports correlating AO production and phytoalexin accumulation. Direct application of superoxide generating systems or H_2O_2 to potato tuber slices was reported to elicit phytoalexin production (Chai and Doke, 1987). Inhibitory effects of antioxidant enzymes or AO scavengers on phytoalexin production (Apostol *et al.*, 1989) suggest that AO may serve as signal intermediates that induce phytoalexin biosynthesis. However, the mechanisms by which the AO regulate phytoalexin production/accumulation are not known. As the accumulation of phytoalexins has been shown due to transcriptional activation it appears that AO affects the gene expression.

Antimicrobial Effects of AO

The antimicrobial activity of AO in animals was recognized long back. In recent years, the antimicrobial effects of exogenously applied O_2^- and H_2O_2 have been demonstrated against plant fungi and bacteria (Kiraly *et al.*, 1993). Spore germination of a member of fungal pathogens has been reported to be inhibited by micromolar concentration of H_2O_2 (Ping and Kuc, 1992). However, it is still not clear whether the antimicrobial effects of AO play a role *in vivo.* (Kiraly *et al.*, 1993)

Role in Hypersensitive Cell Death

Levine *et al.* (1994) reported that exogenous application of H_2O_2 (2-10 mM) caused cell death in soybean. The toxicity of AO may contribute to host cell death during HR as well. The hypothesis that plant cell death during HR is at least partly due to elevated AO levels is supported by studies showing that addition of antioxidants reduces death of hypersensitively responding cells (Keppler and Novacky, 1987). But based on the recent studies it was concluded that although exogenous application of AO in non physiological concentration (much above that found *in vivo*) elicits defense responses and cell death, the AO production found during initial hours of bacterial treatment may be one of a series of events leading to HR but is not itself the direct cause of hypersensitive cell death. (Baker and Orlandi, 1995).

AO in Plant Disease Management

Since the role of AO in plant defense has been recognized various attempts have been made to provide AO based plant disease control. Plant diseases could be managed in two ways: by exogenous application of AO/AO generating substances and developing transgenic plants with elevated AO levels.

Exogenous Application of AO

The use of AO in plant disease management is due to antimicrobial effects and the capacity of some AO species to induce resistance. Lu *et al.* (1999) reported that application of 4-5 mM H_2O_2 retarded *Cladosporium fulvum* germ tube growth and killed germtubes at higher concentrations.

Oxycom TM that contains active oxygen species has under commercial field conditions improved the performance of a variety of plant species exposed to a range of naturally occurring pathogens, including Pythium, downy mildew and powdery mildew (Kim *et al.*, 2001).

Iriti and Faoro (2003) reported that a single spray of 0.3 mM Benzo-(1,2,3)-thia diazole-7 carbothioic acid S-methyl ester (BTH) 7 days before inoculation was sufficient to control the pathogen, *Uromyces appendiculatus* in bean. BTH causes H_2O_2 accumulation in treated tissues. The acquired resistance against bean rust is induced by H_2O_2 which promotes the activity of anionic peroxidases and strengthens cell wall through lignification.

The treatment of rice seeds with 1.0 ppm BTH or 0.5 per cent H_2O_2 reduced blast severity on seedling leaves. Foliar sprays with BTH (10 ppm) reduced the blast severity by 77-80 per cent (Mosa, 2002).

Transgenic Plants Expressing Elevated AO Levels

Transgenic potato plants expressing the *Aspergillus niger* gene encoding glucose oxidase, which generates H_2O_2 when glucose is oxidized were found to have elevated H_2O_2 levels in both leaf and tuber tissues. The transgenic potato tubers exhibited strong resistance to bacterial soft rot caused by *Erwinia carotovora* sub sp. *carotovora*. The transgenic plants also exhibited enhanced resistance to potato late blight (Wu *et al.*, 1995).

Wu *et al.* (1997) further studied H_2O_2 mediated disease resistance in transgenic potato plants expressing glucose oxidase gene infected with *Verticillium dahliae.* The constitutively elevated sublethal levels of H_2O_2 induced an array of host defense mechanisms including accumulation of salicylic acid in the leaf tissue, increased accmuluation of extracellular peroxidases and chitinase accompanied by a significant increase in the lignin content of stem and root tissues of the transgenic plants.

Lee *et al.* (2002) developed transgenic tobacco and cabbage by transferring glucose oxidase (GO) gene from *A. niger* (strain 9642). The transgenic tobacco plants showed delayed symptoms to *Phytophthora nicotianae* infection. The development of leaf lesions by *Xanthomonas campestris* pv. *campestris* was significantly reduced in transgenic cabbage and the disease tolerance was positively associated with the GO activity in leaves.

The involvement of AO in plant-pathogen interactions is an emerging field of disease physiology. Although there has been a flurry of reports regarding AO generation in various plant pathogen interactions, AO studies have not yet been initiated in many plant-pathogen combinations. Studies on the mechanism of AO generation, involvement of elicitor-receptors, regulation of defense response by AO etc. are only the beginning unravelling the subtle interactions among diverse plant pathogens. Further study is required for proper understanding of chemical and molecular basis of AO generation and their role in plant defense. Use of AO in plant disease management is a very recent approach. Studies on induced resistance may widen the scope of exogenous use of AO in plant disease control. Transgenic plants expressing elevated AO levels have been developed only in very few crops. Identification of new genes responsible for AO generation will help in developing transgenics in many other crops.

References

Adam, A., Farkas, T., Somlyai, G., Hevesi, M. and Kiraly, Z., 1989. Consequence of O_2^- generation during a bacterially induced hypersensitive reaction in tobacco: deterioration of membrane lipids. *Physiol. Mol. Plant. Pathol.*, 34: 13–26.

Alvarez, M.E., Pennell, R.I., Meijer, P.J., Ishikawa, A., Dixon, R.A. and Lamb, C., 1998. Reactive oxygen intermediates mediate a systemic signal network in the establishment of plant immunity. *Cell Cambridge*, 92 (6): 773–784.

Apostol, I., Heinstein, P.F. and Low, P.S., 1989. Rapid stimulation of an oxidative burst during elicitation of cultured plant cells. Role in defense and signal transduction. *Plant Physiol.*, 90: 109–116.

Baker C.J., O'Neill N.R., Keppler, L.D., and Orlandi, E.W., 1991. Early responses during plant bacteria interactions in tobacco cell suspensions. *Phytopathology*, 81: 1504–7.

Baker, C.J. and Orlandi, E.W., 1995. Active oxygen in plant pathogenesis. *Annu. Rev. Phytopathol.*, 33: 229–321.

Baker, C.J., Orlandi, E.W. and Mock, N.M., 1993. Harpin, an elicitor of the hypersensitive response in tobacco caused by *Erwinia amylovora*, elicits active oxygen production in suspension cells. *Plant Physiol.*, 102: 1341–1344.

Baker, N.R., 1991. A possible role for photosystem II in environmental perturbations of photosynthesis. *Physiol. Plant*, 81: 563–570.

Bestwick, C.S., Brown, I.R., Bennett, M.H.R. and Mansfield, J.W., 1997. Localization of hydrogen peroxide accumulation during the hypersensitive reaction of lettuce cells to *Pseudomonas syringae* pv. *phaseolicola*. *Plant Cell*, 9(2): 209–221.

Borden, S. and Higgins, V.J., 2002. Hydrogen peroxide plays a critical role in the defense response of tomato to *Cladosporium fulvum*. *Physiol. Mol. Plant Pathol.*, 61 (4): 227–236.

Bradley, D.J., Kjellbom, P. and Lamb C.J., 1992. Elicitor and wound induced oxidative cross linking of a proline rich plant cell wall protein: a novel, rapid defense response. *Cell*, 70: 21–30.

Buonaurio, R. and Kumar, N.N.U., 1995. Active oxygen producing and scavenging enzymes in pepper leaves infected with *Xanthomonas campestris* pv. *vesicatoria. J. Phytopathol.*, 143(3): 165–168.

Candela, M.E., Munoz, R., Alcazar, M.D. and Espin, A., 1994. Isoperoxidase involvement in the resistance of *Capsicum annuum* to infection by cucumber masaic virus. *J. Plant Physiol.*, 143(2): 213–217.

Chai H.B. and Doke N., 1987. Activation of the potential of potato leaf tissue to react hypersensitively to *Phytophthora infestans* by cytospore germination fluid and the enhancement of this potential by calcium ions. *Physiol. Mol. Plant. Pathol.*, 30: 27–37.

Dat, J., Vandenabeele, S., Vranova, M., Montagu, M.V., Inze, D. and Breusegen, F.V., 2000. Dual action of the active oxygen species during plant stress responses. *Cell. Mol. Life Sci.*, 57: 777–795.

Doke N. and Ohashi Y., 1988. Involvement of an O_2^- generating system in the induction of necrotic lesions on tobacco leaves infected with tobacco mosaic virus. *Physiol. Mol. Plant Pathol.*, 32: 163–175.

Doke, N., 1983. Involvement of superoxide anion generation in the hypersensitive response of potato tuber tissues to infection with an incompatible race of *Phytophthora infestans* and to the hyphal wall components. *Physiol. Plant Pathol.*, 23: 345–57.

Donato, M.D., Mozzetti, C., Chiavazza, P., Rubino, R., Matta, A. and De-Donato, M., 1997. Active oxygen production in cell–free melon homogenates elicited with hyphal wall components of *Fusarium oxysporum* f. sp. *melonis*. 1. A response correlated with race-specific resistance. *J. Plant Pathol.*, 78(1): 35–43.

Foyer, C.H., Descourvieres, P. and Kunert, K.J., 1994 Protection against oxygen radicals an important defense mechanism studied in transgenic plants. *Plant Cell Environ.*, 17: 507–523.

Foyer, C.N., 1996. Oxygen processing in photosynthesis. *Biochem. Soc. Trans.* 24: 427–433.

Glazener, J.A., Huang H.C. and Baker C.J., 1991) Active oxygen induction in tobacco cell suspensions treated with *Pseudomonas fluorescens* containing the cosmid pHIR11 and with strains containing TnPhoA mutations in the hrp cluster. *Phytopathology*, 81: 1196 (Abstr.).

Halliwell B., 1974. Superoxide dismutase, catalase and glutathione peroxidase: Solutions to the problems of living with oxygen. *New Phytol.*, 73: 1075–1086.

Hernandez, J.A., Talavera, J.M., Martinez, G.P., Dicenta, F. and Sevilla, F., 2001. Response of antioxidative enzymes to plumpox virus in two apricot cultivars. *Physiologia Plantarum*, 111(3): 313–321.

Horn, M.A., Heinstein, P.F. and Low P.S., 1989. Receptor mediated endocytosis in plant cells. *Plant Cell,* 1: 1003–1009.

Huang, H.C., Schuurink, R., Denny, T.P., Atkinson, M.M. and Baker, C.J., 1988. Molecular cloning of a *Pseudomonas syringae* pv. *syringae* gene cluster that enables *Pseudomonas fluorescens* to elicit the hypersensitive response in tobacco. *J. Bacteriol.*, 170: 4747–4756.

Iriti, M. and Faoro, F., 2003. Benzothiadiazole (BTH) induces cell death independent resistance in *Phaseolus vulgaris* against *Uromyces appendiculatus. J. Phytopathol.*, 151 (3): 171–180.

Keppler, L.D. and Novacky, A., 1987. The initiation of membrane lipid peroxidation during bacteria-induced hypersensitive reaction. *Physiol. Mol. Plant Pathol.*, 30: 233–245.

Kim, Y.C., Blee, K.A., Robins, J. and Anderson, A.J., 2001. Oxycom TM under field and laboratory conditions increases resistance responses in plants. *European J. Plant Pathol.*, 107(1): 129–136.

Kiraly, Z., El-Zahaby, H., Galal, A., Abdou, S. and Adma, A., 1993. Effect of oxyfree radicals on plant pathogenic bacteria and fungi and on some plant diseases. In: *Oxygen Free Radicals and Scavengers in the Natural Sciences,* (Eds.) Mozsik, G., Emerit, I., Feher, J., Matkovics, B. and Vincze, A., Budapest: Akademiai Kiado, 356. pp.

Lee, Y.H., Yoon, I.S., Suh, S.C. and Kim, H.I., 2002. Enhanced disease resistance in transgenic cabbage and tobacco expressing a glucose oxidase gene from *Aspergillus niger. Plant Cell Reports,* 20(9): 857–863.

Legendre, L., Heinstein P.F. and Low, P.S., 1992. Evidence for participation of GTP binding proteins in elicitation of the rapid oxidative burst in cultured soyabean cells. *J. Biol. Chem.*, 267: 20140–20147.

Levine, A. Tenhaken, R., Dixon, R. and Lamb, C., 1994. H_2O_2 from the oxidative burst orchestrates the plant hypersensitive disease resistance response. *Cell,* 79: 583–593.

Lu, H., Higgins, V.J. and Lu, H.G., 1999. The effect of hydrogen peroxide on the viability of tomato cells and of the fungal pathogen *Cladosporium fulvum Physiol. Mol. Plant Pathol.,* 54(5–6): 131–143.

Mehdy, M.C., 1994. Active oxygen species in plant defense against pathogens. *Plant Physiol.,* 105: 467–472.

Milosevic, N. and Slusarenko, A.J., 1996. Active oxygen metabolism and lignification in the hypersensitive response in bean. *Physiol. Mol. Plant. Pathol.,* 49 (3): 143–158.

Mosa, A.A., 2002. Induced resistance in rice against blast disease using abiotic and biotic agents. *Ann. Agric. Sci., Cairo,* 47(3): 993–1008.

Orlandi, E.W., Hutcheson, S.W. and Baker, C.J., 1992. Early physiological responses associated with race–specific recognition in soybean leaf tissue and cell suspensions treated with *Pseudomonas syringae* pv. *glycinea. Physiol. Mol. Plant Pathol.,* 40: 173–180.

Patykowski, J. and Urbanek, H., 2003. Activity of enzymes related to H_2O_2 generation and metabolism in leaf apoplastic fraction of tomato leaves infected with *Botrytis cinerea. J. Phytopathol.,* 151(3): 153–161.

Peng, M. and Kuc J., 1992. Peroxidase-generated hydrogen peroxide as a source of antifungal activity *in vitro* and on tobacco leaf disks. *Phytopathology,* 82: 696–699.

Polle, A., 1996. Mehler reaction: Friend or foe in photosynthesis? *Bot. Acta,* 109: 84–89.

Sanchez-Casas, P. and Klessig, D.F., 1994. A salicylic acid binding activity and a salicylic acid inhibitable cartalase activity are present in a variety of plant species. *Plant Physiol.,* 106: 1675–1779.

Staskawicz, B.J., Dahlbeck, D. and Keen, N.T., 1984. Cloned avirulence gene of *Pseudomonas syringae* pv. *glycinea* determines race-specific incompatibility on *Glycine max* (L.) Merr. *Proc. Natl. Acad. Sci.* USA, 81: 6024–6028.

Sundar, A.R., Velazhahan, R., Viswanathan, R. and Vidhyasekaran P., 2002. Induction of active oxygen species (AOS), lipoxygenase and lipid peroxidation in suspension cultured sugarcane cells by a glycoprotein elicitor isolated from *Colletotrichum falcatum. Zeitschrift fur Pflanzenkrankheiten und Pflanzenschutz,* 109 (5): 441–451.

Sutherland, M.W., 1991. The generation of oxygen radicals during host plant responses to infection. *Physiol. Mol. Plant Pathol.*, 39: 79–93.

Torres, R., Valentines, M.C., Usall, J., Vinas, I. and Larrigaudiere, C., 2003. Possible involvement of hydrogen peroxide in the development of resistance mechanisms in 'Golden Delicious' apple fruit. *Postharvest Boil. Tech.*, 27(3): 235–242.

Vera-Estrella R., Blumwald E., and Higgins, V.J., 1992. Effect of specific elicitors of *Cladosporium fulvum* on tomato suspension cells. *Plant Physiol.*, 99: 1208–1215.

Wu, G., Short, B.J., Lawrence, E.B., Levine, E.B., Fitzsimmons, K.C., Shah, D.M. and Wu, G.S., 1995. Disease resistance conferred by expression of a gene encoding H_2O_2 generating glucose oxidase in transgenic potato plants. *Plant Cell*, 7(9): 1357–1368.

Wu, G., Shortt, B.J., Lawrence, E.B., Leon, J., Fitzsimmons, K.C., Levine, E.B., Raskin, I., Shah, D.M. and Wu, G.S., 1997. Activation of host defense mechanisms by elevated production of H_2O_2 in transgenic plants. *Plant Physiol.*, 115(2): 427–435.

Yim, M.B., Chock, P.B. and Stadtman, E.R., 1990. Copper, zinc superoxide dismutase catalyzes hydroxyl radical production from hydrogen peroxide. *Proc. Natl. Acad. Sci..*, USA, 87: 5006–5010.

Insect Pest and Disease Management *Pages* ***268–281***
Editor: **Dr. D. Prasad**
Published by: **DAYA PUBLISHING HOUSE (2016)**

Chapter 19

Sclerotinia Stem Rot of Mustard and its Management

Rajendra Prasad and Saroj Kumar
Department of Plant Pathology,
C.S.A. University of Agriculture and Technology, Kanpur – 208 002, U.P.

Introduction

Presently, India is playing a vital role in global demand spectrum of oilseeds through accounting for about 12-13 per cent of world oilseeds area, 6-7 per cent of world oilseed output, 6-7 per cent of world oil meal production, 5-6 per cent of world oil meal export, 6-7 per cent of world vegetable oil production, 15-17 per cent of world vegetable oil import and 10 per cent of the world edible oil consumption. Among oilseed crops, rapeseed- mustard contributes 19.2 per cent and 22.7 per cent in area and production, respectively. These crops occupy an area of 6.85 million ha. yielding 8.64 million tonnes with an average productivity of 1220 kg/ha. It is estimated that 58 million tonnes of oil will be required by the year 2020 wherein the share of rapeseed mustard will be around 24.2 million tonnes. Rapeseed and mustard are grown chiefly for their oil. The oil is utilized for human consumption throughout the country and particularly in North India, where it is difficult to be replaced by any other oil. The oil content ranges between 37-49 per cent.

More than twenty diseases affect the crop with great economic bearings. Among the diseases, stem rot caused by *Sclerotinia sclerotiorum* (Lib.) de Bary is next only to Alternaria blight in importance. Stem rot caused by *Sclerotinia sclerotiorum* was considered to be the disease of minor importance of rapeseed and mustard but due to newer agriculture technology and extensive cultivation of new varieties of crops susceptible to this fungus, it has assume the greater importance in recent past. Of late, it has assumed serious proportion and causing considerable damage to the crop in rapeseed and mustard growing areas of U.P. Haryana, Punjab and Rajasthan.

Economic Importance and Distribution

A survey was conducted during the crop season 2002-03 in different rapeseed and mustard growing areas of U.P. It was observed that stem rot caused by *S. sclerotiorum* was prevalent in all the mustard growing areas causing substantial damage to the crop. The incidence of disease varied from 25 per cent at Kumarganj, Faizabad (U.P.) to 38 per cent in Aligarh (Saroj Kumar, 2006). The disease was observed in severe form in Gujarat during 1987-88 and 1988-89 causing losses up to 50 per cent (Dang *et al.*, 1992). Chauhan *et al.* (1992) reported that stem rot of mustard caused yield loss up to 72 per cent in severe cases in Uttar Pradesh. Yield losses in susceptible crops due to stem rot may be as high as 100 per cent (Purdy, 1979). Six hundred and forty fields were surveyed in 24 district of Rajasthan by Ghasolia *et al.* (2004) and 40-60 per cent disease incidence was recorded in 31 fields of 10 districts and 20-39 per cent disease incidence of *Sclerotinia sclerotiorum* in 71 fields and in rest of the 538 fields were found less than 20 per cent disease incidence. In Rajasthan, more than 60 per cent yield loss due to Stem rot was reported by Shivpuri *et al.* (2000) in mustard crop. However, Sharma *et al.* (2001) reported about 40 per cent yield loss due to this disease in Rajasthan. The shattering of siliquae which mature prematurely due to the disease may be an additional yield loss factor. At the time of harvesting and threshing, sclerotia of the fungus get mixed with seed, which affects the marketability of the crop.

Stem rot of rapeseed and mustard is wide spread and worldwide in occurrence and has been reported from other countries like Canada (Platford and Breiner, 1975; Morall, 1976; Dueck, 1977), Holland (Hubbeling, 1955), Germany (Kirchner and Pluschkell, 1979), U.K. (Rawlinson and Muthgalu 1979), Nepal (Chaudhary, 1993), and China. In India, Shaw and Ajrekar (1915) reported this disease first time from Pusa, Bihar on mustard crop. Since then it was found to occur in other parts of the country. While Mehta *et al.* (1994) also reported stem rot diseases on *Eruca sativa*. Roy (1973) reported it from different parts of Assam and Uttar Pradesh. Rai and Saxena (2004) noticed it on *B. campestris* var. *toria* in serious form. Recently, stem rot of rapeseed and mustard emerged in serious form in many parts of the country *viz.*, Haryana, Himachal Pradesh, Punjab, Delhi, U.P., Bihar, West Bengal, N E Hill states, Andman Nikobar (Kumar and Thakur, 2000), Rajasthan (Lodha *et al.*, 1992). During the course of survey undertaken in 1986 for the fungal diseases of crucifers, showed that *Sclerotinia sclerotiorum* attacked a number of crucifers in various areas of Bihar and U.P.

Symptoms

The pathogen attacked all the parts of mustard plant and produces different kind of symptoms in severe conditions. The sclerotia of the fungus may be formed on the outside of the root where they are quite apparent and causing root rot (Figure 19.1). The stem of the mustard plants shows the symptom of the early, as the stem is the primary target of *Sclerotinia sclerotiorum*. On the stem, the disease appears as elongated water-soaked area. Later on this water-soaked areas are covered with a cottony mycelial growth of the fungus. The lesions gradually increase in size and in favourable conditions girdle the stem completely resulting in wilting and death of the plants. In many cases, the infection normally not spreads much and gets restricted to the smaller area of the pith and thus resulted in stunting and premature ripening rather than sudden collapse and death of plants. The affected area become white and tends to shred and numerous grayish-white to black sclerotia of various forms and sizes appear on the surface of the lesions and also inside the pith (Figures 19.2–19.4).

Water-soaked circular to oval whitish lesions are also frequently noticed on leaves, which later on are covered with cottony mycelial growth of the fungus. The lesions were first appeared at the tip or margin of the leaves as faded green patches which soon turned to whitish and enlarging rapidly down word and in ward in favourable weather conditions. Ultimately, complete leaves were covered with cottony growth of mycelium on the surface of the lesions in four to five days (Figure 19.5). The growth of mycelium was also observed to spread on petioles and stem causing premature defoliation of leaves and death of plant. Grayish white to black sclerotia were also notice on the surface of the lesions. Similar symptoms were also noticed on siliquae of affected plants, as a result few or some time no seed formation takes place. Grayish white coloured mycelial growth of the fungus tends to form various size of sclerotia on the surface of the pods (Figure 19.6). Chahal (1982) has also observed similar types of symptoms on stem.

Taxonomy and Morphology of the Pathogen

The fungus was first described as *Peziza sclerotiorum* (Libert 1837), that binomial for the fungus stood until Fuckel (1870) described the genus *Sclerotinia* and renamed *Peziza sclerotiorum* with newly coined binomial *Sclerotinia sclerotiorum*. Wakefield (1924) cited G.E. Massee as the proper authority for *Sclerotinia libertiana* (Lib.) Massee, because he had used that binomial in 1895, but de Bary used it in his 1884. (de Bary 1884). Thus, the accepted name and authority for this fungus is *Sclerotinia sclerotiorum* (Lib.) de Bary (Purdy 1979).

The genus *Sclerotinia* was designated type genus of the family Sclerotiniaceae which was erected in 1945 by Whetzel to accommodate inoperculate discomycetes that produce stromata, stipitate apothecia, ellipsoidal ascospores and globose spermatia. He provided keys and diagnosis to the genera assigned to Sclerotiniaceae. Korf and Dumont (1972) had proposed *Whetzelinia* as a new generic name of *Sclerotinia sclerotiorum*. Purdy (1979) opined that eventually *Whetzelinia* Korf and Dumont would be considered a superfluous name, and *Sclerotinia* Fuckel would be retained for the generic epithet and that type species would continue to be *S. sclerotiorum* (Lib.) de Bary as proposed by Dennis (1974) and Buchwald and Neergaard (1973).

The morphological and cultural characters of the fungus studied on Potato dextrose agar medium (in culture) and host (in nature for the identification of fungus revealed that the colonies in culture were white, compact, circular, fast growing and formed numerous sclerotia. Mycelium septate, branched, hyaline to light brown in colour and measure 10 to 19 mm in size. Sclerotia–superficial, formed in one or two concentric rings on Potato dextrose agar medium within 15 days. Sclerotia are grayish white to black in colour, spherical to cylindrical or irregular in shape, measuring 6.4-4.8 mm in size with smooth to rough and wrinkled surface (Figure 19.7). The sclerotia of fungus germinated by mycelium (myceliogenic) or by formation of apothecia (Figures 19.8 and 19.9). Apothecia–on germination, the sclerotia formed stalked apothecia, which were disc shaped and yellowish brown in colour and 3-6 mm in diameter. Asci-inoperculate, cylindrical, narrow, rounded at the appex, eight ascospores in each ascus and measuring 117-107 x 5.6-11 mm in size. Ascospores- hyaline, ellipsoidal, smooth wall, uniform in size and measuring 9.5-12.5 x 6.1-7.9 mm in size (Figure 19.10). The morphological characters of the pathogen above agree very closely with those described by Kosasih and Willet (1975), Willet and Wong (1980) and Kolte (1985) for *Sclerotinia sclerotiorum* (Lib.) de Bary and thus, the fungus was identified as such.

Epidemiological Factors

Weather parameters such as prevalent temperature, relative humidity (RH) and rainfall influenced the stem rot incidence. The disease was noticed in the first week of January in the crop season 2002-03 when maximum and minimum temperature was 15.8 and 4.6°C, respectively and relative humidity 84.8 per cent and 12.1 mm rainfall recorded in a day of the week. In the crop season 2003-2004 disease was noticed in last week of December when maximum temperature was 14.36°C and minimum 7.5°C, respectively and relative humidity was 88.2 per cent and 18 mm rainfall. It was observed that the

Figure 19.1: Formation of Sclerotia on Root

Figure 19.2: Water Soaked Lesion on Stem

Figure 19.3: Symptom on Collar Region

Figure 19.4: Sclerotia Inside the Pith

Figure 19.5: Symptoms on Stem and Leaves

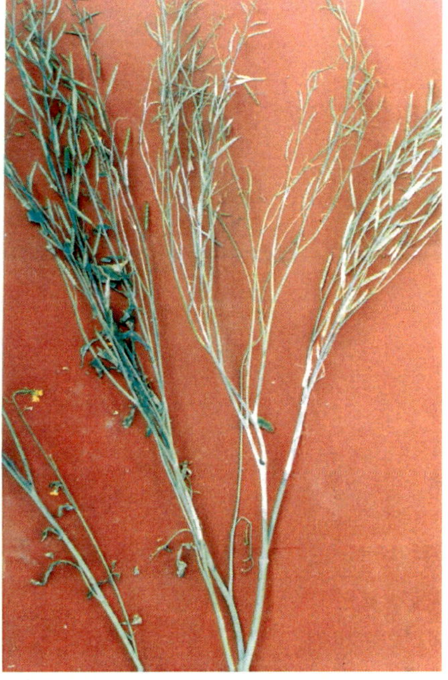

Figure 19.6: Partial Infection on Siliquae

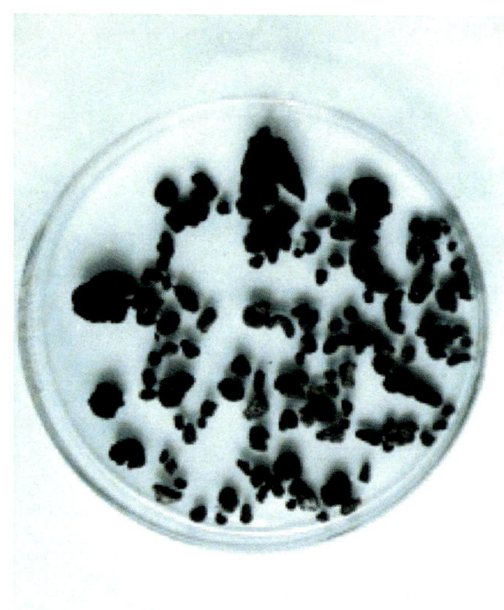

Figure 19.7: Different Size of Sclerotia

Figure 19.8: Germination of Sclerotium

Figure 19.9: Carpogenic Germination of Sclerotia

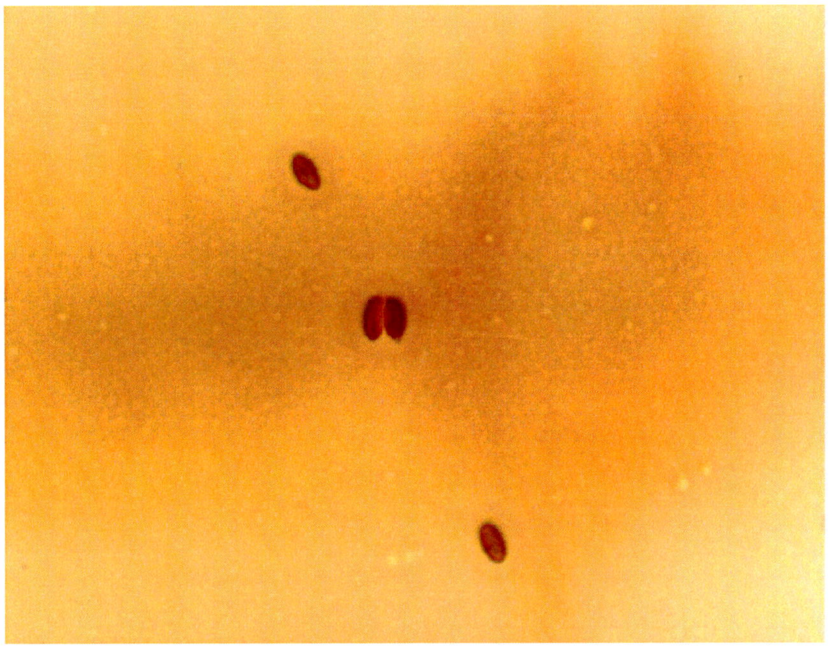

Figure 19.10: Ascospores

disease incidence increased with the decrease in temperature (25°C) and increase in relative humidity (up to 80 per cent). There after, there was no further increase in disease as the temperature rose though the relative humidity sufficient enough. Correlation analysis of the data showed that there was a positive correlation of maximum and minimum temperature with disease appearance in both crop seasons Saroj Kumar (2006). Various workers (Moore *et al.*, 1949; Hass and Bolwyn 1972) reported 20-25°C temperature favourable for ascospore germination, production of sclerotia, mycelial growth and lesion initiation and development. Abawi and Grogan (1975) reported that moisture is a limiting factor in the development of disease of Frenchbean caused by *S. sclerotiorum*. Apothecia were produced only in saturated or nearly saturated soils and infection occurred only when free moisture was maintained for a relatively long period on the surface of bean tissues. Ascospore germination was not greatly affected by temperature but 25°C was optimum for germ tube growth. Mycelial growth, sclerotia production and lesion initiation and development were observed maximum at 20-25°C.

The temperature range for apothecial production was 10-25°C. The optimum temperature 10°C did not appear to be a limiting factor in the development of white mould in Narvey. Carpogenic germination of sclerotia was observed at mean temperatures of 9-17°C for 14 day periods. Number of apothecia and level of aerial infection were not related to the amount of total monthly rainfall (Letham *et al.*, 1976). Roy and Saikia (1976) found an outbreak of white blight of mustard in cool and wet environment. Weiss *et al.* (1980) reported that maximum incidence of the disease was observed at 10-25°C and the lowest at 30°C in northern bean crop. However, Zazzerini and Tosi (1985) reported that the stem rot of rape caused by *S. sclerotiorum* was favoured by low temperature and heavy rainfall. According to Nodin *et al.* (1992), rainfall in the month of June or first week of July favoured the developement of apothecia of Sclerotinia stem rot on spring sown rapeseed. While, Sigvald and Wearn

(1992) reported the low temperature during spring and summer and considerable rainfall in early summer favoured the Sclerotinia stem rot of rape.

Integrated Disease Management

To bridge the gap in our understanding of the disease and to formulate effective strategies to manage this emerging problem, following management practices are reviewed here in brief.

Seed Treatment

Seed treatment is one of the most effective and economic methods of disease management. Among different seed treatment combinations, *T. viride* + Carbendazim and *T. Viride* + garlic extract proved best and both treatments reduced the disease incidence up to 90.91 per cent followed by Companion + *T. viride* (81.83 per cent). Seed coating either alone with fungicides or in combination with bio-control agent significantly controlled stem rot. Seed coating with Benomyl and Carbendazim has been reported most effective against *S. sclerotiorum* in mustard (Singh *et al.*, 1994). Seed treatment with garlic bulb extract, *Trichoderma viride*, Mancozeb and Carbendazim was found effective in reducing disease incidence and better plant stand in mustard crop (Chattopadhyay *et al.*, 2002 and 2004). Pathak *et al.* (2001) evaluated six fungicides for seed dressing against *Sclerotinia sclerotiorum* and reported that all the fungicides performed better in comparison to check. However, it was found that Ridomil gave maximum germination (44 per cent) with zero post emergence mortality and Carbendazim gave 40 per cent germination with 6 per cent post emergence mortality. Singh *et al.* (1997) found that the growth of pathogen greatly reduced at 7000 and 5000 ppm respectively, seeds treated with garlic bulb extract before sowing produced wilt free seedlings.

Source of Resistance

Resistant varieties are the most simple, practical, effective and economical method of plant disease management. *Sclerotinia sclerotiorum* has an extremely wide host-range and has been reported to affect more than 408 species of plants. The dearth of reports before 1968 indicates that many researchers formally accepted the ideas that resistance to *S. sclerotiorum* did not exists (Steadman, 1979). Brun *et al.* (1979,). Studied the reaction of 150 rapeseed genotypes against *S. sclerotiorum.* Rapeseed genotypes of Japanese origin showed some resistance to *S. Sclerotiorum*. Sharma and Sharma (1986) screened eight genotypes of *Brassica oleracea* against *S. Sclerotiorum*, and found that "Orena" and "Avans" were more tolerant to the pathogen. Singh *et al.* (1994) also, screened 12 *Brassica* spp. under natural and artificial conditions. They found *Brassica napus* and *B. juncea* cv. Rugosa posse's resistance in the field as well as under green house conditions. The genotypes PCR 10, RW 8410, RW 9401 and RGH 8006 were observed resistant against stem rot (Pathak and Godika, 2002). Varieties namely Hayola 401, PBN 9501, RH 492, PAB 9511, PWR 954 and Kiran also showed resistant reaction against this disease. Saroj Kumar (2006) observed that out of tested 45 genotypes, none of the genotypes was found resistant/ free. However, seven genotypes *viz.*, DHR-9618, BAUR-9802, TKG-180-1, JTC-18, TKG-G-16, NUDB-09 and IIUM-0004 showed moderately resistant reaction against Sclerotinia stem rot of mustard.

Effect of Sowing Date

Alternation in sowing dates affect stem rot incidence in mustard and several other crops. The mustard crop sown on 15th November showed highest disease intensity (33.50 per cent) and 10.15 q/ ha yield. Maximum crop yield of 15.02 q/ha and the minimum disease intensity (5.83 per cent) was recorded in Ist October sown crop. However, disease intensity after 15th November sowing decreased gradually and the crop yield was also decreased in that sowing dates (Saroj Kumar, 2006). Sharma *et al.* (2001) reported that Indian mustard (T-59) sown between 4th week of October and the Ist week of November recorded lowest (9.5 per cent) disease severity while crop sown in second and third week of October recorded the highest disease severity of 49.2 per cent.

Gupta *et al.* (2004) studied the effects of dates (21st October, 5th Nov., 5 Dec., 20 Dec., 5 Jan.) of sowing on disease incidence and crop yield. They concluded that sowing dates significantly affected Sclerotinia stem rot incidence and crop yield. Early sowing (21st October) resulted greatest disease incidence (10.5 per cent on average), disease incidence decreased with the delay in sowing. The crop established at 5th January sowing was not affected with Sclerotinia stem rot. Sowing on 21st October gave the highest yield (1585.7 Kg/ha) further delay in sowing significantly reduced the yield. Ghasolia *et al.* (2004) in Rajsthan observed that mustard crop sown in last week of October and Ist week of November showed minimum disease incidence (2.9 per cent) of stem rot. However, it was 28.70 per cent in crop sown from second to third week of October.

Effect of Antagonists

Antagonism is the balance wheel of nature, where there is life there is antagonism. In biological control, one exploits the natural enemies of the pathogen to one's benefit. One introduces the known enemy of the pathogen in the environment, so that it can put a check on the pathogen. In China Luo *et al.* (1987) isolated *Trichoderma viride, T. harzianum, T. hamatum, Gliocladium virens, Coniothyrium minitans* and *Fusarium* spp. from the rotted sclerotia of *S. sclerotiorum*. Out of which, the frequency of *T. viride* was about 10 per cent to the total isolates. Seven isolates of *T. viride* caused more than 90 per cent rotting of sclerotia when inoculated at 100 spores/g soil. In winter oilseed rape, *Coniothyrium minitans* on soil application infected sclerotia and decreased sclerotial survival, carpogenic germination and production of apothecia. Good results were obtained when *C. minitans* was applied in autmn, at the time of sowing, rather than when applied in spring (Mc- quilken *et al.*, 1995).

Sharma *et al.* (1982) detected eight fungal antagonistic species antagonistic to *S. sclerotiorum* from rotted sclerotia and infested soil. Isolates of *T. viride, T. harzianum* and *Gliocladium roseum* were most effective in inhibiting mycelial growth. Culture filtrate of these three antagonists was also effective in inhibiting germination of sclerotia. Singh and Singh (1998) studied the effect of *Trichoderma viride* and *T. harzianum* culture filtrate on the growth of *Sclerotinia sclerotiorum in vitro* and *in vivo*. They have also reported that *T. viride* was better than *T. harzianum* in checking the diseases. The germination of *C. minitans* conidia could be stimulated by *S. sclerotiorum* on lesions by 51 per cent and 95 per cent, after 48 hrs and 72 hrs, respectively. All the test microorganisms *viz., T. harziamum, T. viride, Gliocladium virens, Penicillium citrinum, Aspergillus niger* and *A. flavus* caused significant inhibition in mycelial growth of the pathogen over control. *Trichoderna viride* isolate of Pantnagar showed maximum inhibition (69.07 per cent) in colony growth of the test pathogen followed by *T. harzianum* isolate (67.22 per cent) of Kanpur (Saroj Kumar, 2006).

Effect of Plant Extracts

In recent years, an increasing consciousness about environmental pollution due to pesticides has forced Plant Pathologists to search for eco-friendly methods for disease management. Earlier bio-agents were widely used against various groups of pathogens including *S. Sclerotiorum*. The host plant extracts can be alternative for disease management. Recently, garlic extract has shown to have fungicidal and inhibitory action against several fungal species including some that cause mycoses (Moore and Atkins, 1977). Saroj Kumar (2006) observed that the botanicals tested against the pathogen *in vitro*, bulb extracts of garlic (*Allium sativum*) provided the best performance showing 72.22 per cent inhibition followed by *Azadirachata indica* (48.15 per cent) and *Ocimum sanctum* (37.4 per cent). *Azadirachata indica* commonly known as *neem*, has reputed value for medicinal (Shah *et al.*, 1959), antiviral (Rao *et al.*, 1969) antibacterial (Chopra *et al.*, 1952) and antifungal (Murthy and Sirsi, 1957) properties.

Shivpuri and Gupta (2001) tested 18 plant extracts against *Sclerotinia sclerotiorum* by poison food technique and reported that extracts of *Allium cepa, Azadirachta indica, Datura stramonium, Ocimum sanctum, Polyalthia longifolia, Tagetes erecta, Catharanthus roseus, Withania somnifera* and *Allium sativum* inhibited sclerotial formation (0.71 to 1.77 sclerotia per plate). Other treatments enhanced the sclerotial production from 2.08 to 4.10 sclerotia per plate, against 2.18 of the control. Chattopadhyay *et al.* (2002) evaluated leaf extract of *Ocimum sanctum, Azadirachata indica, Datura stramonium, Eucalyptus longifolia, Calotropis gigantea, Acacia arabica, Bougainvillea spectabilis* and bulb extract of *Allium sativum* @ 1 per cent w/v. They have found that bulb extract of *A. sativum* checked the mycelial growth of *S. sclerotiorum*, 94.4 per cent over control and rest of the extracts were ineffective against the pathogen except *Acacia arabica* leaf extract which inhibited 5.5 per cent growth of the pathogen. Seed treatment and foliar spray of bulb extract of *Allium sativum* was also found effective in response to higher initial plant stand, reduction in disease incidence and seed yield. Zewain *et al.* (2004) studied the effect of various fungicides and *neem* extract on the mycelial growth and myceliogenic germination of sclerotia of *S. sclerotiorum*. They reported that mycelial growth of fungus was completely checked at 5000 ppm of neem extract. However, sclerotia of the fungus germinated at all the concentrations tested. Various *neem* based pesticides were found effective against *S. sclerotiorum*. Neem extract (25 per cent) at 6 ml/litre was found effective against stem rot of cauliflower and reduced the disease incidence. (Zewain *et al.*, 2005).

Organic Amendment

Soil amendment is an effective methods of bio-control. Sandhu (1992) reported that the application of Farm Yard Manure @ 10-40 tonnes/acre gave effective control and reduced number of sclerotia.

Chemical Control

Effect of various group of fungicides are well known against the pathogen. In order to check the secondary spread of the disease, the possibility to manage the disease through foliar spray have been described. Noon *et al.* (1988) evaluated flutriafol based formulation containing either Carbendazim or Chlorothalonil on oilseed rape against stem rot and reported good control of disease together with significant yield. Morall *et al.* (1989) showed that Benomyl gave better control of *S. sclerotiorum* in rapeseed than Ipridione. Other experiments proved that spraying of Benomyl at late flowering stage effectively managed the disease. Fayslle (1995) suggested that stem rot of sunflower caused by *S. sclerotiorum* could be effectively minimized by spraying of Vitavax, Topsin-M, Indofil M-45 and Thiram. Sharma and Bansandrai (1997) evaluated Carbendazim, Triadimephon, Trimorphine (Tridemorph), Ipriodione Mancozeb, Dinocap and elemental sulphur fungicides *in vitro* against *S. sclerotiorum* and reported that the most effective being Iprodione followed by Triadimephon in sclerotial germination. All the fungicides were effective in inhibiting ascospore germination, the maximum inhibition being at 0.5 per cent concentration of Iprodione, Triademephon and Dinocap. Ipridione and Triademephon also significantly reduced the number of sclerotia.

Singh *et al.* (1998) tested ten fungicides against rape stem rot in field and reported that all the fungicides were effective with Iprodione Vnclozolin, Benomyl, Metalaxyl and Carbendazim gave best control. In cauliflower and cabbage presowing application of Carbendazim (5 per cent) granules also controlled stem rot disease. Shivpuri *et al.* (2000) evaluated several fungicide and plant extract against

S. sclerotiorum using poisoned food technique, Carbendazim, Thiophanate methyl and Phenyl pyrrole completely inhibited the growth of pathogen at all concentrations tested. Copper oxychloride was the least effective one. Saroj Kumar (2006) has recorded that spraying of Carbendazim (0.1 per cent) proved best among all the fungicidal treatments and reduced disease incidence 75.40 and 72.67 per cent in 2002-03 and 2003-04 crop seasons, respectively. The next best effective fungicide was Topsin-M, resulting 71.42 and 71.97 per cent disease control, respectively. Garlic bulb extract was also found effective under field conditions in both crop seasons and gave 49.21 and 52.51 per cent disease control over check plots. He also found that spraying of fungicides and plant extracts enhanced seed yield and oil content. The highest seed yield (13.25 q/ha) and oil content (40.3 per cent) was obtained by spraying of Carbendazim (0.2 per cent). However, the lowest seed yield and oil content was obtained in sulphur dust spraying. It was observed that oil content was positively correlated with the crop yield. All the treatments gave significantly better yield and oil content over check plots, but there was no effect of spraying of fungicides and botanicals on the oil quality (fatty acid composition).

Singh *et al.* (2003) studied the effect of various fungicides on myceliogenic and carpogenic germination of sclerotial bodies of *S. sclerotiorum.* They observed that Carbendazim (25 and 50 mg/ml) and Ridomil MZ-72 (50 mg/ml) completely inhibited mycelial growth. All the fungicides tested, significantly reduced the carpogenic germination of sclerotia. However, the most effective was Carbendazim (100 mg/ml/20 minute). Ridomil and Kitazin were effective at higher concentrations.

The yield and oil content in mustard crop improved with the management practices. Noon *et al.* (1988) reported that the significant seed yield was increased with the spraying of Carbendazim against *S. sclerotiorum* in rapeseed. Agrawal *et al.* (1997) reported that oil content and quality of oil decreased in mustard seeds due to *Sclerotinia* infection. Chattopadhyay *et al.* (2002 and 2004) observed significantly better seed yield and oil content with the spraying of Carbendazim, garlic extract and *Trichoderma viride.*

Cultural Control

Since the disease is carried over through the sclerotia and crop debris so that collect and burn the debris of crop along with the sclerotia after harvest, if the disease has occurred during the crop season. Cleaned and healthy seeds should be used for the sowing. Deep ploughing the land during summer season and crop rotation with resistant host are essential for preventing the sclerotia population in the soil. Soil solarization through coverage of transparent polythene (0.1 mm) for 45 days has been observed a good management practice. Sharma *et al.* (2001) reported that the irrigation at pre sowing and supplemented with three irrigation at branching, flowering and at pod formation minimize the disease incidence. EPTC treatment significantly increase the degree of control. The crop rotation with non host crops *viz.* wheat, barley, rice, maize and gram helps in minimizing the inoculm buildup and prefer to grow mustard in alternate year, if disease has occurred during the crop season.

References

Abawi, G.S. and Grogan, R.G., 1975. Source of primary inoculum and effect of temperature and moisture on infection of bean by *Sclerotinia sclerotiorum. Phytopath.*, 65: 300–309.

Agarwal, R.A.K., Kumar, A. and Thakur, H.L., 1997. Effect of Sclerotinia rot on oil quality in low erueic acid cultivars of rapeseed. *Cruciferae Newsletter*, 19: 103–104.

Brun, H., Renard, M., Joun, B., Tangue, X. and Lamargue, C., 1979. Preliminary observations on some rape diseases in France. *Sclerotinia sclerotiorum, Cylindrosporium concentrium, Ramularia armoraciae. Sciences Agronomiques, Rennes*, 7: 77.

Buchwald, N.F. and Neergaard, P., 1973. A plea for the retention of *Sclerotinia sclerotiorum* as type species for the genus Sclerotinia Fckl. Emend. *Friesia*, 10: 96–99.

Chahal, S.S., 1982. Diseases of rapeseed and mustard. *Indian Farming*, 32(6): 27–32.

Chattopadhyay, C., Meena, P.D. and Kumar, Sudhir, 2002. Management of Sclerotinia rot of Indian mustard using ecofriendly strategies. *J. Mycol. Pl. Pathol.*, 32(2): 194–200.

Chattopadhyay, C., Meena, P.D. and Meena, R.L., 2004. Integrated Management of sclerotinia rot of Indian mustard. *Indian J. Pl. Protection.*, 32(1) 88–92.

Chaudhari, B.N., 1993. Yield loss estimations in rape and mustard by *Sclerotinia sclerotiorum* (Lib.) de Bary. *J. Institute of Agri. and Animal Sciences*, 14: 113.

Chauhan, L.S., Singh, J. and Chandra, D.R., 1992. Proceeding of National Symposium on management of microbes in service of mankind. Nov. 19–21 at Univ. of Allahabad, Allahabad, pp. 65–66 (Abs.)

Chopra, I.C., Gupta, K.C. and Nazir, B.N., 1952. Preliminary study of antibacterial substances from *Melia azadirachta*. *J. Med. Res.*, 40: 511.

Dang, S.R.S., Pandey, R.N. and Desai, B.G., 1992. Disease in oilseed crops in India: A current status. In: *Progress of Plant Pathological Research*, (Eds.) Sokhi, S.S., Chahal, S.S. and P.P. Singh.

de Bary, A., 1884. Verglaeich and Morphologie und Biology der Pilze, Mycetozen and Bacterian (comparative morphology and Biology of the fungi, mycetozoa and bacteria. Translated by H.E.F. Garnasey revised by I.B. Balfour, the Claredom Press, Oxford. 525 (*cf. Phytopathology*, 69: 875–880).

Dennis, R.W.G., 1974. Whetzelinia korf and Dumont, a superflouous name. *Kew Bull.*, 29: 89–91.

Dueck, J., 1977. Sclerotinia in Rapessed. *Can. Agri.*, 22: 7

Fayslle, C., 1995. Control of Sclerotinia cultivar (Rveil–Malmasion) No. 394: 44–46.

Fuckel, L., 1870. Symbolic Pilze. Jahrb. Nassauisschen vereins Naturk. 23: 1–459.

Ghasolia, R.P. and Asha Shivpuri, 2005. Screening of rapeseed-mustard genotypes for resistance against Sclerotinia rot. *Indian Phytopath.*, 58(2): 242.

Ghasolia, R.P. Shivpuri, Asha and Bhargava A.K., 2004. Sclerotinia rot of Indian mustard in Rajasthan. *Indian Phytopath.*, 57(1): 76–79.

Gupta, R., Awasthi, R.P., Kolte, S.J. and Gupta, R., 2004., Influence of sowing dates on the incidence of sclerotinia stem root of rapeseed-mustard. *Annals of Pl. Prot. Sci.*, 12(1): 223–224.

Hass, J.H. and Bolwyn, 1972. Ecology and epidemiology of Sclerotinia wilt of white beans in Ontario. *Can. J. Pl. Sci.*, 52(4): 525–533.

Hubbeling, N., 1955. Inst.Plzicktk onduz, Wageniangen, 83: 80.

Kirchner, H.A. and Pluschkell, H.J., 1979. On the occurrence of rape stalk break [*Sclerotinia sclerotiorum* (Lib.) de Bary] in root stock. *Rev. Pl. Pathol.*, 52: 3862.

Kolte, S.J., 1985. *Diseases of Annual Edible Oilseed Crops, Vol. II: Rapeseed, Mustard and Sesame Disease.* CRC Inc. Boca Raton, Florida, pp. 135.

Korf, R.P. and Dumant, K.P., 1972. Whetzelinia: A none genetic name ofr *Sclerotinia sclerotiorum* and *S. tuberosa*. *Mycologia*, 64: 248–251.

Kosasih, B.D. and Willetts, H.J., 1975. Types of abnormal apothecia produced by *Sclerotinia sclerotiorum*. *Mycologia*, 67: 89–97.

Kumar, A. and Thakur, K.S., 2000. *National Seminar on Oilseeds and Oil Research and Development Needs in the Millenium*, Feb. 2–4, DOR Hyderabad, pp. 224–225.

Kumar, Saroj, 2006. Diagnosis and management of Sclerotinia stem rot of Mustard [*Brassica juncea* (L.) Czern and Coss. *Ph.D. Thesis*, Department of Plant Pathology, C.S. Azad University of Agriculture and Technology, Kanpur (India).

Leetham, D.B., Huett, D.O. and Trimboli, D.S., 1976. Biology and control of *S. sclerotiorum* in cauliflower and tomato crops in Costal New South Wales. *Plant Dis. Reptr.*, 60: 286–289.

Libert, M.A., (1837. Plantae cryptogamicae arduennac (Exsiccati) No. 326. Published by M.A. Libert (*cf. Phytopathology*) 69: 875–880.

Lodha, B.C., Bhatnagar, M.K., Mathur, K., Doshi, A., Mathur, S., Bairwa, L.N., Sharma, D. and Trivedi, A., 1992. *Plant Pathological Thoughts and News*, Department of Plant Pathology, Raj. Coll. Agric., Udaipur (India).

Luo, K., Ren, X.G. Zhou, B.W., Chen D.Y. and Yang, J., 1987. Study of the parasitic fungi of sclerotia of *Sclerotinia sclerotiorum* in rape. *Oil Crops of China*, 3: 40–44.

Mc–Quilken, M.P., Mitchell, S.J., Budge, S.P., Whipps, J.M., Fenlon, J.S. and Archer, S.A., 1995. Effect of *Coniothyrium minitans* on sclerotial survival, and apothecia production of *Sclerotinia sclerotiorum* in field grown oilseed rape. *Pl. Pathol.*, 44: 883–896.

Mehta, R.L., Zayas, J.F., Yang, S. and Yang, S.S., 1994. Ajowan as a source of natural lipid antitoxidant. *J. Agric. and Food Chemistry*, 42: 1420–1422.

Moore, W.D., Robert, A.C. and David, L.S., 1949. The Sclerotiniaceae Diseases of Vegetable Crops. *Bull. Fla. Agri. Exp. Sta.*, p. 457.

Morall, R.A.A., 1976. List of host records of *Sclerotinia sclerotiorum* (Sensu Purdy, 1955) in Canada by family.

Morall, R.A.A., Rogers, R.B. and Rude, S.V., 1989. Improved techniques of controlling Sclerotinia stem rot of Canola (oilseed rape) with fungicide in western Canada. *Medielinger Vau de Facultul Landbouwenschap. EnRijKsumiversitet. Crent.*, 54: 643–649.

Murthy, S.P. and Sirsi, M., 1957. Pharmacological studies on *Melia azadirachata* Part–1; Antibacterial, Antifungal and antitubecular activity of neem oil and its fractions. Symp. Utilization. *Indian Med. Pl.*, Lucknow, pp. 55.

Nodin, K., Sigvald, R. and Sevenson, C., 1992. Forecasting the incidence of Sclerotinia stem rot of spring sown rapeseed. *Zeitschriftfu-rpflanzenkrankheiten-Und pflanzenschutz*, 99: 245– 255.

Noon, R.A., Northwood P.J., Brown, M.C. Mortuary, A. and Charlet, C., 1988. Elutria fol based formulations for control of oilseed rape diseases. In: *Brighton Crop Protection Conference Pests and Diseases, Vol. 3: Thornton Health*, U.K., British Crop Protection Council, p. 947–952.

Pathak, A.K. and Godika, S., 2002. Efficiency of seed dressing fungicides in management of stem rot disease of mustard caused by *Sclerotinia sclerotiorum* (Lib.) de Bary. *Cruciferae News Letter*, 24: 93–94.

Pathak, A.K., Godika, S., Jain, J.P. and Muralia, Suresh, 2001. Effect of antagonistic fungi and seed dressing fungicides on the incidence of stem rot of mustard. *J. Mycol. Pl.Pathol.*, 31(3): 327–329.

Platford, R.G. and Brainer, C.C., 1975. Diseases of rapeseeds in Manitoba. 1973–74. *Can. Pl. Dis. Surv.*, pp. 55.

Purdy, L.H., 1979. *Sclerotinia sclerotiorum:* History disease and symptomatology, host range, geographic distribution and impact. *Phytopathology*, 69: 875–880.

Rai, J.N. and Saxena, V.C., 2004. Influence of sowing dates on the incidence of Sclerotinia stem rot of rapeseed-mustard. *Annals of Plant Protection Science,* 12(1): 223–224.

Rao, A. R., Sukumar, S. Paramisivam, T. B., Kamalwanshi, S. and Shantha, 1969. Study of antiviral activity of tender leaves of margosa tree (*Melia azadirchata*) vaccinia variala virus: A preliminary repport. *Ind. J. Med. Res.*, 57: 495.

Rawlinson, C.J. and Muthyalu, G., 1979. Diseases of winter oilseed rape: Occurrence, effect and control. *J. Agric. Sci. Comb.*, 93: 593.

Roy, A.K., 1973. Host range of *Sclerotinia sclerotiorum* and *Sclerotium rolfsi* in Jorhat, Assam. *Science and Cult.*, 39: 319–320.

Roy, A.K. and Saikia, U.N., 1976. White blight of mustard and its control. *Indian J. Agric. Sci.*, 6: 274–277.

Shah, M.P.,Shah, M.J. Seth, U.K. and Bhide, N.K., 1959. Clinical trials with new indigenous diuretisodium nimbidinate. *J. Assoç. Physician India*, 7: 235.

Sandhu, K.S., 1992. Control of stalk rot disease of Cauliflower with Farm Yard Manure. *Plant Dis. Res.*, 7: 11–18.

Sharma, B.K., and Basandrai, B.K., 1997. Effect of biocontrol agents, fungicides, and plant extracts on selerotial viability of *Sclerotinia sclerotiorum. Ind. J. Agric. Sci.* 67: (3) 132–133.

Sharma, R.C and Sharma, S.L., 1986. Survival of *Sclerotinia sclerotiorum* and initial infection of cauliflower in Himanchal Pradesh. *Indian. J. Eco.*, 13: 367–369.

Sharma, R.C., Sharma, S.L. and Sharma, L., 1982. Varietal reaction of cauliflower against stalkrot caused by *S. sclerotiorum. Crop Improvement*, 9(2): 167–168.

Sharma, S. Yadav, J.L., Sharma, G.R. and Sharma, S., 2001. Effect of various agronomic practices on the incidence of white rot of Indian mustard caused by *S. sclerotiorum. J. Mycol. Pl. Pathol.*, 31(1): 83–84.

Shaw, F.J.W. and Ajrekar, S.L., 1915. The genus Rhizoctonia in India. *Mem. Dep. Agric. India Bot. Ser.*, 7: 177–194

Shivpuri, Asha, Sharma, K.B. and Chhipa, H.P., 2000. Some studies on stem rot disease of mustard in Rajasthan. *Paper Presented in 22nd Conference of ISMPP* at Solan on 3–4 May.

Shivpuri, Asha, and Gupta, R.B.L., 2001. Evaluation of different fungicides and plant extracts against *Sclerotinia sclerotiorum* causing stem rot of mustard. *Indian Phytopath.*, 54(2): 272–274.

Sigvald, R. and Wearn, P., 1992. *Plant Protection Year 1991 Agriculture.* vaxtskyddsnotiser, 56(1): 7–12.

Singh R.B., Chand, Ramesh, Singh, J. and Singh, R.B., 1998. Anomalous apothecia formation in *Sclerotinia sclerotiorum Indian. J. Pl. Pathol.*, 16: 24–27.

Singh, A.K., Singh, D.B. and Chauhan, L.S., 1997. Screening for sources of resistance in rapeseed (*Brassica napus* Sp. Oliefera) and mustard for rot caused by *S. sclerotiorum. Indian J. of Agricultural Sciences*, 67 (7): 324.

Singh, R., Narain, Udit and Ram Palat, 2003. Evaluation of bio-agents against Sclerotinia stem rot of Ajowan. *Ann. Pl. Protec. Sci.*, 11(2): 53–56.

Singh, R., Tripathi, N.N., Kaushik, C.D., and Singh. R., 1994. Management of Sclerotinia rot of Indian mustard by fungicides. *Crop Research Hisar*, 7: 2, 276–281.

Singh, R., Tripthi, N.N. and Singh, R., 1994. Assured methods of inoculation and screening of Brassica spp. against *Sclerotinia sclerotiorum* (Lib.) de Bary. *Crop Res. Hisar*, 8: 570–574.

Singh, Yogendra and Singh, Y., 1998. Management of Sclerotinia rot of rapeseed and mustard through chemicals. *Pl. Dis. Res.*, 13 (2) 149–150.

Steadman, J.R., 1979. Control of plant diseases caused by *Sclerotinia* species. *Phytopathology*, 69(8): 904–907.

Wakefield, F.N., 1924. On the names of *Sclerotinia sclerotiorum* (LIb.) massee, and *S. libertiana* Fuckl. *Phytopathology*, 14: 126–127.

Weiss, A., Kerr, E.D. and Steadman J.R., 1980. Temperatures and moisture influence on development of white mold disease (*Sclerotinia sclerotiorum*) of great northetn beans. *Pl. Dis.*, 8: 757–759.

Whetzel, H.H., 1945. A synopsis of the genera and species of the sclerotiniaceae, a family of stromatic in operculate Discomyeetes. *Mycologia*, 37: 648–714.

Willet, H.J. and Wong, A.L., 1980. The biology of *Sclerotinia sclerotiorum* with emphasis on specific nomenclature, *Bot. Rev.*, 46: 101–165.

Zazzerini, A. and Tosi, L., 1985. *Sclerotinia sclerotiorum* (Lib.) de Bary a new parasite of Rape (*Brassica napus* L. *oleifera* D C) in Italy. *Informatore Fitopathologico*, 35(2): 31–33.

Zewain, Q.K., Bahadur, P. and Sharma, P., 2004. Effect of fungicides and neem extracts on mycelial growth and myceliogenic germination of *Sclerotinia sclerotiorum. Indian Phytopath.*, 57(1): 101–103.

Zewain, Q.K., Bahadur, P. and Sharma, P., 2005. Integrated disease management of stalk rot of cauliflower. *Indian Phytopath.*, 58(2): 167–173.

Insect Pest and Disease Management *Pages 282–290*
Editor: **Dr. D. Prasad**
Published by: **DAYA PUBLISHING HOUSE (2016)**

Chapter 20

Spot Blotch of Wheat: Management Options with Special Reference to Biological Control

S.K. Biswas[1], Chinmay Biswas[1], Biswajit Bhowmik[2] and S.S.L. Srivastava[1]
[1]Department of Plant Pathology, C.S.A. University of Agriculture and Technology, Kanpur – 208 002
[2]Indira Gandhi National Open University, Siliguri Sub-regional Centre, West Bengal

Introduction

Spot blotch of wheat caused by *Drechslera sorokiniana* (Sacc.) Subram. and Jain (Syn. *Helminthosporium sativum* P.B.K.; perfect stage *Cochliobolus sativus* Ito and Kuribay) is a predominant disease in South-east Asia including the Indian subcontinent. It is one of the serious diseases of wheat prevalent in Bolivia, Brazil, Paraguay, Zambia, Pakistan, Bangladesh and Nepal. In India, the disease was first recorded in Bihar as early as in 1914 by Mohy (HCIO No. 12508). In India, warm-humid conditions occur in North eastern region, therefore, it is considered a serious problem of Bihar and West Bengal. Of late, spot blotch has spread to North-Western region comprising Punjab, Haryana, Delhi and western U.P., central and peninsular areas of Maharashtra and Karnataka particularly, under rice-wheat cropping system.

Economic Losses

The Spot blotch disease causes considerable losses in wheat. In India, an epidemic of *D. sorokiniana* was reported by McRae (1930) in the vicinity of Pusa (Bihar). Nema and Joshi (1971) estimated 3-20

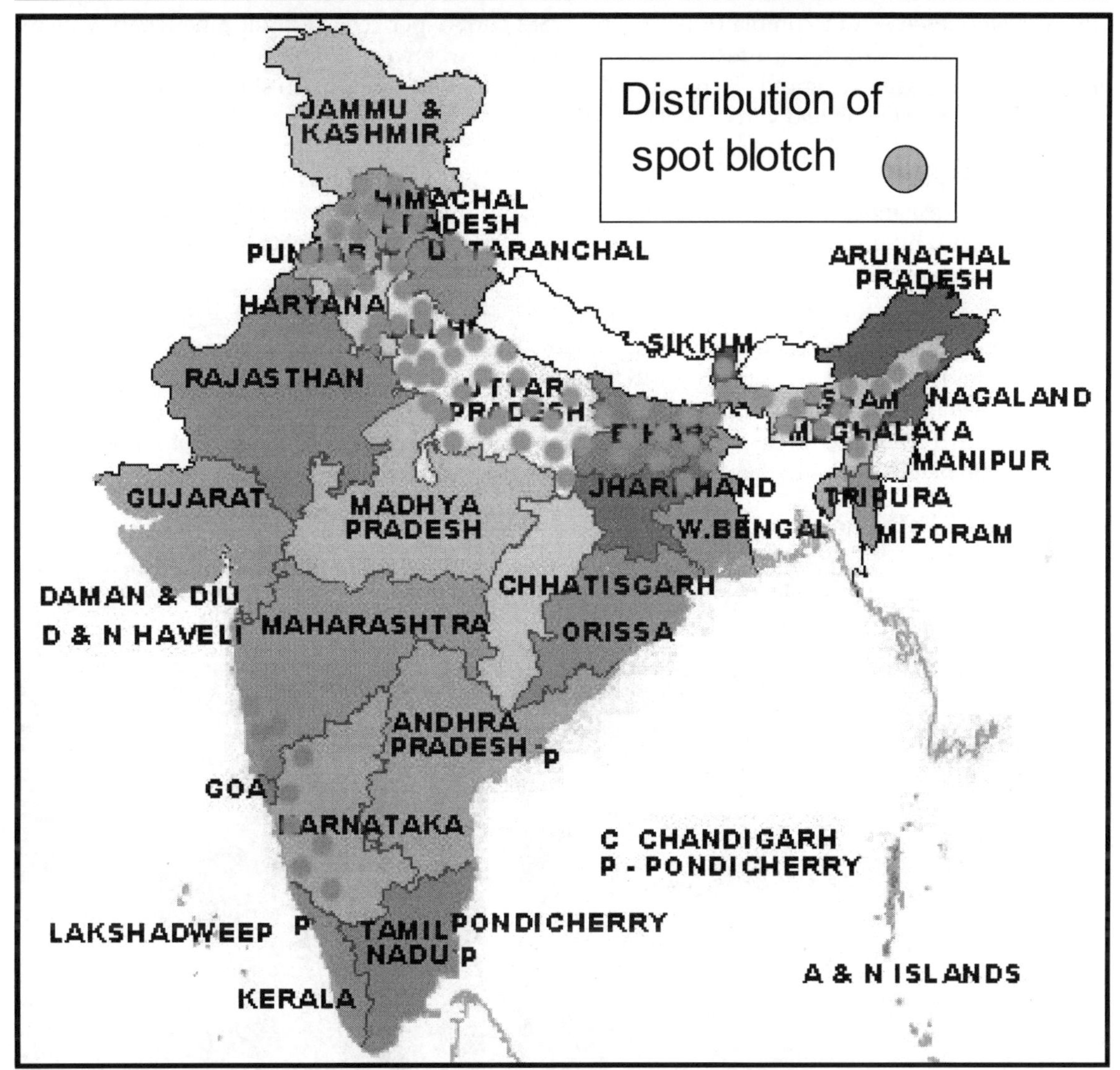

Figure 20.1: Distribution of Spot Blotch

per cent yield loss under Indian conditions. The experimental data on yield losses caused by this pathogen accounted for 2.72 to 36.24 per cent under different agroclimatic conditions (Anon., 1995).

In Bangladesh, the disease caused 4-21 per cent loss in grain yield in commercial varieties like Sonalika, Akber, Kanchan and Ashrani (Razzaque and Hossain, 1991) while Badaruddin *et al.* (1994) recorded about 71 per cent yield loss. Malaker *et al.* (1994) reported average yield loss in Bangladesh to the tune of 25 to 29 per cent due to Helminthosporium leaf blight at farmers field. Yield loss has commonly been 20 to 30 per cent and in some cases above 75 per cent in China (Cheng *et al.*, 1998). Mahto (1999) reported 16.19 to 28.99 per cent wheat yield loss and 6.97 and 15.78 per cent thousand grain weight loss in commercial cultivar UP 262 at farmer's field in Nepal. De Milliano and Zadoks (1985) found 38 per cent yield loss in growth chamber studies in Netherlands. Losses to the tune of 85 per cent were assessed in Zambia (Raemakers, 1988) and 40 per cent in Philippines (Lapis, 1985). During favourable years losses due to spot blotch alone were between 30 and 86 per cent in Brazil and in some fields loss reached upto 100 per cent (Hetzler *et al.*, 1991). In Mexico about 25 per cent losses were estimated by Dubin and van Ginkel (1991).

Disease Symptoms

The disease manifests itself in different types of symptoms. Seed rot, seedling blight, foot rot, leaf spot/blight and stunting of plants develop due to *D. sorokiniana.* However, foliar infections via airborne conidia occur more frequently in India.

Initially, the leaf symptoms are most frequent on lower leaves. They appear as distinct, elongate brown lesions that rarely exceed one cm in diameter. The dark lesions contrast sharply with green or straw coloured leaf tissues. Gradually, with the advancement of the plant growth, the spots become irregular in shape and size having brown necrotic centre which is surrounded by chlorotic margin. The spots further coalesce with each other causing extensive drying of foliage. Consequently, all leaves dry up before the milk stage and the whole field looks blighted.

The Causal Pathogen

The genus *Helminthosporium*, a Dematiaceous Hyphomyceteous fungus was established by Link in 1809 and it was validated in 1821 by S.F. Grays. Subsequently, Nisikado (1928) erected two sub-

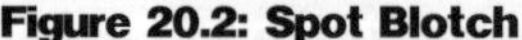

Figure 20.2: Spot Blotch

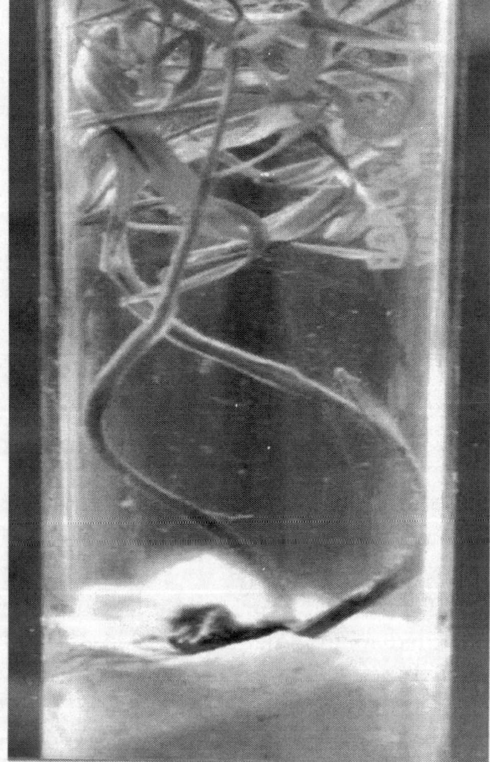

Figure 20.3: Foot Rot

genera on the basis of shape and germination of conidia, *viz.* (i) *Cylindro-Helminthosporium* having cylindrical conidia that germinate from any cell and (ii) *Eu–Helminthosporium* which has fusoid or curved conidia and germinate from the end cells. Ito (1930) changed the sub-genus *Cylindro Helminthosporium* to *Drechslera* who characterised the members of new genus as having "cylindrical not curved conidia germinating from every cell and associated with "Pyrenophora" as perfect stage". Luttrell (1951) recognised three major groups. (i) *H. sigmoideum* group where conidia are borne on sterigmata. (ii) *H. gramineum* group having cylidrical conidia with slightly tapering ends which are sessile and germinate from any or all cells. (iii). *H. sativum* group with ellipsoidal to fusoid conidia germinating by polar germ tube.

Shoemaker (1959) differentiated the graminicolous species into two: the species that germinated from any or all the cells were included under *Drechslera,* and those with fusoid pharagmospore germinating by one germ tube from each end (bipolar germination) were designated *Bipolaris.* Later, Subramanian and Jain (1966) amended *Drechslera* to include all the graminicolous species. A new genus *Exserohilum* was erected for species with a protuberant hilum by Leonard and Suggs (1974). In the monograph "Dematiaceous Hypomycetes", Ellis (1976) has accepted *Drechslera* in the wider sense and listed *Bipolaris* and *Exserohilum* as synonyms. The teleomorph of spot blotch pathogen has been identified as *Cochliobolus sativus* (Ito and Kuribay) Drechs. ex Dastur and has been reported to occur in nature in Zambia (Raemaekers, 1988).

Sorokin collected infected spikes of wheat and rye in Ussuria (Russia) in 1890 but the causal fungus was identified by Saccardo as *H. sorokinianum* Sacc. It was Drechsler (1923) who suggested that the organism which Saccardo identified was synonymous to *H. sativum.* Since then nomenclature complexion continued and the name of the causal organism has been used differently according to individual person concerned. The causal agent is presently termed as *D. sorokiniana* (Syn. *B. sorokiniana*).

Mycelium of the *D. sorokiniana* is typically deep olive brown. Conidiophores are single or clustered, simple, erect and 6-8 x 110-150 µm, with septations. Conidia develop terminally and then remain laterally attached at pores beneath each newly produced conidiophore septum. They are olive brown and ovate to oblong with tapered ends and a prominant basal scar. Conidia are 15-20 x 60-120 µm, 3 to 10-septate, and slightly curved. Their walls are smooth and thickened noticeably at septa.

Attempts were made to identify races of *D. sorokiniana* as early as in 1922 and at least 37 races were identified by Christensen (1922). The isolates of *D. sorokiniana* differ in their parasitic capabilities irrespective of host or the geographical distribution. Recently, Mahto (2000) has shown the existence of pathogenically distinct pathotypes of *D. sorokiniana* in North-western region of India. However, constant mutation and saltation make race identification still more difficult. Therefore, chances of prevalence of more pathotypes can not be ruled out.

Disease Cycle

Drechslera sorokiniana is ubiquitous on grasses and cereals including wheat. Dormant conidia of the pathogen survive freely in the soil unless hosts are available and cause local infection. Reis *et al.* (1998) observed that crop residue is also the main source of infection. Kumar and Arya (1973) studied the role of seed, soil and plant debris on the development of the disease and observed that the pathogen perpetuates on the infected seed and causes local infection. Disease transmission rate from seeds was found to be very high. High sporulation on coleoptile tips confirmed the potential of seed inoculum for disease development. Once inoculum is introduced by seed, disease develops on above ground plant parts where pathogen sporulates. Conidia are transported to new infection sites and to the soil by wind and rain splashes.

The environmental factors particularly temperature and humidity influence the disease intensity, sporulation and spore deposition. A temperature around 28°C and relative humidity 92 per cent has been found to favour the maximum development of foliar blight under field condition (Chaurasia, 1997). Rahman *et al.* (1996) reported 25°C to be the best temperature and pH 6.5 ideal for the growth of the fungus. Mehta (1998) expressed the view that higher predisposition at advanced growth stage like DC-56 can be very problematic under favourable weather conditions *viz.*, continuous rains for 5-6 days followed by daily average temperatures of 20-22°C. Such conditions can lead to a spot blotch epidemic.

Management Options

Efforts have been made to manage this disease in various ways. The management options include developing host resistance, chemical control, cultural practices, biological control and induced resistance.

Host resistance: Considering the high yield losses, breeding for resistance commands high priority. However, resistance to foliar blight in India is not yet at high level. Therefore, identification of sources of resistance and knowledge of inheritance of resistance is considered essential to transfer resistance in high yielding wheat genotypes.

Efforts are being made all over the world to identify sources of resistance to *D. sorokiniana.* However, at present, there is no immune cultivar so far in the world. Sources of spot blotch resistance which have been identified over the years, fall into three categories *i.e.* Latin American, China and wild relatives of wheat or alien species. The Latin American sources are mainly derived from Brazil and may be traced back to Italian ancestry. Resistant Brazillian commercial varieties are BH 1146 and CNT1 (Mehta, 1985). Alien resistance sources are many and being used in crosses involving *Thinopyrum curvifolium* (Mujeeb *et al.*, 1996). CIMMYT has used this source in developing some outstanding lines *i.e.* Mayoor and Chirya series.

Studies on genetics of resistance to spot blotch pathogen revealed that the resistance in Sharbati Sonora and E4858 is conditioned by two complimentary genes. Adlakha *et al.* (1984) opined that the inheritance of resistance to spot blotch is simple and it is governed by one or two dominant factors. According to Bailey *et al.* (1988) resistance was controlled by three genes in different crosses.

Attempts were made to transfer resistance from genotypes that showed higher levels of resistance in agronomically desirable cultivars of wider adaptation using two back crosses (Mehta *et al.*, 1993). However resistance to *D. sorokiniana* still remains low among commercial wheat cultivars. Because of this situation, Rajaram (1988) emphasised that genes for resistance from several sources other than wheat have to be combined to achieve sufficient resistance through pyramiding of genes. Transferring resistance genes from alien species is problematic because of the ploidy problem. Nonetheless, some progress in transferring resistance genes from alien species *Thinopyrum curvifolium, Elymus curvifolium* and *Thinopyrum tauschii* to wheat germplasm has been achieved (Mujeeb *et al.*, 1996). Induced variation for high levels of resistance is also suggested through somatic embryogenesis (Vasil and Vasil, 1986).

Chemical Control

The effective control of the disease has been achieved by seed treatment with chemicals such as Arasan, Agrosan G.N., N.I. Ceresan, Parzate and foliar spray with Dithane Z-78, propiconazole, iprodione and imazalil. However, their use under environment conducive to high disease pressure is not economical. In pot experiments, Brestan, Brestanol, Duter and Thiram reduced spot blotch intensity on wheat plants but the spray of former three fungicides proved phytotoxic (Singh and Virk, 1980).

Cultural Management

Cultural practices like, crop rotation with legumes, clean cultivation, wider line sowing and restricted irrigation are also effective for lowering the disease level from heavily infested soil and saprophytic growth of the pathogen in crop residues. Soil application of organic amendments like neem cake can also check the disease development to some extent.

Biological Control

Biological control, was first used in relation to plant pathogen by Von Tubeff, 1914. Garrette (1965) defined the biological control as any condition under which or practice whereby survival and activity of a pathogen is reduced through the agency of any other organism (excluding the role of the host and man) with the result that there is a reduction in the incidence of the disease caused by pathogen. Keeping in view the functional relationship between host, pathogen and environmental factors for development of disease, Baker and Cook (1974) have given a broader definition of biological control. According to them "biological control is the reduction of inoculumn density or disease producing activities of a pathogen or parasite in its active or dormant state by one or more organism (s) accomplished naturally or through manipulation of the environment, host or antagonist or by mass introduction of one or more antagonist (s).

Biological control of spot blotch of wheat was for the first time attempted by Butler in 1953. The bio-agents was *Trichoderma viride*. Prasad *et al.* (1978) also reported bio-control of *D. sorokiniana* by *T. viride*. Fernandez (1992) noticed *T. harzianum* to be equally effective against the spot blotch pathogen. Mandal (1995) reported antagonism of *T. reesei, T. harzianum T. hamatum, Gliocladim virens and Talaromyces flavus* to *D. sorokiniana*. *In vitro* and *in vivo* experiments conducted by Biswas (2001) revealed that *Chaetomium globosum* could be successfully used for the management of spot blotch disease in wheat. Application of crude extract of C. *globosum* @ 0.2 per cent reduced the disease severity by 94.9 per cent.

Culture filtrates of these bio-agents inhibited conidial germination of the pathogen upto 92 per cent. SEM micrographs showed lysis of host mycelium due to over growth and penetration through hyphal pegs and tight coiling produced by the antagonists (Mandal, 1995). Foliar spray of culture filtrate of *C. globosum* on *D. sorokiniana* infected wheat leaves revealed distortion of conidial wall and non-germination of conidia of the pathogen resulting in the reduced spot blotching (Biswas *et al.*, 2000).

Table 20.1: Mechanism of Bioagents Against Spot Blotch Pathogen

Bioagents	*Mechanism of Control*
Chaetomium globosum	Antibiosis and induced resistance
T. viride	Inhibition
T. reesei	Over growth and profuse sporulation
T. pseudokoningii	Over growth and profuse sporulation
Trichothecium roseum	Inhibition

(Mandal, 1995, Biswas,1997).

Induced resistance: It is an innovative approach for management of spot blotch of wheat. Besides antagonism to *D. sorokiniana, Chaetomium globosum* could be used for induced resistance against spot blotch in wheat. Pre-treatment of wheat seedling with chaetomium crude extract provided protection for considerable period of time. Mildly virulent strain of *D. sorokiniana* can also be used to induced resistance against the virulent race. Pre-treatment of wheat seedling with conidial suspension of mild virulent starin of *D. sorokiniana* @ 10^4/ml reduced the PDI by 49.9 per cent (Biswas, 2001).

Conclusion: The complexity of these non traditional blights in the main wheat belt and of the conditions in which they occur, makes it essential to seek a holistic approach for the control of the disease. Integration of promising control measures would be the right step to that direction. At the present level of technological development seed treatment with chemical fungicides, soil application of organics amendments and foliar application of bioagents/inducers would constitute an eco-friendly IDM package for the management of the spot blotch disease in wheat.

References

Adlakha, K.L., Wilcoxson, R.D. and Raychaudhuri, S.P., 1984. Resistance of wheat to leaf spot caused by *Bipolaris sorokiniana. Plant Dis.*, 68: 320–321.

Anonymous, 1995. Report of the Co-ordinated Experiments, 1994–95, Vol. 1, Crop Protection (Pathology) (Eds.) Prasher, M., Nagarajan, S., Goel, L.B. and J. Kumar. AICWIP, Directorate of Wheat Research Karnal, pp. 43–54.

Badaruddin, M., Saunders, D.A., Siddique, A.B., Hossain, M.A., Ahmed, M.U., Rahman, M.M., and Parveen, S., 1994. Determining yield constraints for wheat production in Bangladesh. In: *Wheat in Heat Stressed Environments: Irrigated, Dry Areas and Rice-Wheat Farming Systems*, (Eds.) Saunders D.A. and G.P. Hettel. CIMMYT, Mexico, pp. 265–271.

Bailey, K.L., Knott, D.R. and Harding, H., 1988. Heritability and inheritance of resistance to common root rot (*Cochliobolus sativus*) in wheat (*Triticum aestivum*). *Can. J. Pl. Pathol.*, 10: 207–214.

Baker, K.F. and Cook, R.J., 1974. *Biological Control of Plant Pathogens.* W.H. Free Man, San Francisco, 433 pp.

Biswas, S.K., 1997. Biocontrol potential of *Chaetomium globosum* against *Drechslera sorokiniana. M.Sc. Thesis*, IARI, New Delhi, pp. 62.

Biswas, S.K., 2001. Biochemical changes in wheat induced by Chaetormium globosum against spot blotch pathogen. *Ph.D. Thesis*, IARI, New Delhi, pp. 91.

Biswas, S.K., Srivastava, K.D., Aggarwal, R., Prem Duneja and Singh D.V., 2000. Antagonism of *Chaetomium globosum* to *Drechslera sorokiniana,* the spot blotch pathogen of wheat. *Indian Phytopath.*, 53: 436–440.

Butler, F.C., 1953. Saprophytic behaviour of some cereal root-rot fungi. II. Factors influencing saprophytic colonization of wheat straw. *Ann. Appl. Biol.*, 40: 298–304.

Chaurasia, S.K., Chand, R. and Joshi, A.K., 1997. Relative dominance of *Alternaria triticina* and *Helminthosporium sativum* in different growth stage of wheat. In: *Proc. Intern. Conf. on Integrated Plant Disease Management for Sustainable Agriculture.* Indian Phytopathological Society, New Delhi (2000), pp. 604–605.

Cheng, Naitao, Zenggui, Gao., Xianlu, Jia., Wu, Wei and Wu Yousano, 1998. *In vitro* selection for spot blotch resistance in wheat: In: *Helminthosporium Blights of Wheat: Spot Blotch and Tan Spot, Proceedings of an International Workshop,* (Eds.) Duvesller, E., Dubin, H.J., Reeves, J. and A. McNab. CIMMYT, Mexico, pp. 253–262.

Christensen, C.M., Nelson, G.H., Mirocha, C.J., Bates, F. and Dorworth, C.E., 1966. Toxicity to rats corn invaded by *Chaetomium globosum*. *Appl. Microbiol.*, 14: 774–777.

De-Milliano, W.A.J. and Zadoks, J.C., 1985. The effect of early foliar infection by *Helminthosporium sativum* on some yield components of two African wheats. In: *Wheat for More Tropical Environments*, (Eds.) Vallaredal, R.L. and A.R. Klatt. *A proceedings of the International Symposium*, CIMMYT, Mexico, pp. 154–157.

Dreshsler, C., 1923. Some graminicolous species of *Helminthosporium*. *J. Agric. Res.*, 24: 641–739.

Dubin, H.J. and Ginkel, M., 1991. The status of wheat diseases and disease research in warmer areas. In: *Wheat for the Nontraditional Warm Areas, A Proceedings of the International Conference*, (Eds.) D.A. Saunders. CIMMYT, Mexico, pp. 125–145.

Ellis, M.B., 1976. *More Dematiaceous Hyphomycetes*, C.M.I., Kew.

Fernandez, M.R., 1992. The effect of *Trichoderma harzianum* on fungal pathogens infesting soybean residues. *Soil Biology and Biochemistry*, 10: 1027–1029.

Garrette, S.D., 1965. *Biology of Root Infecting Fungi*. Cambridge University Press, London and New York.

Hetzler, J., Eyal, Z., Mehta, Y.R., Campos, L.A., Febrmann, H., Kushnir, V., Zekaria-Oren, J., and Cohen, L., 1991. Interactions between *Cochliobolus sativus* and wheat cultivars. In: *Wheat for the Nontraditional Warm Areas*, (Ed.) D.A. Saunders. CIMMYT, Mexico, pp. 146–164.

Ito, S., 1930. On some new ascigerous stages of the species of *Helminthosporium* parasitic on cereals. *Proc. Imper. Acad., Tokyo*, 6: 352–355.

Kumar, V.R. and Arya, H.C., 1973. Taxonomy of fungus causing leaf blight of wheat. In: *Proc. 62nd Sci. Cong. Part III*, pp. 40–41.

Lapis, D.B., 1985. Insect pests and diseases of wheat in the Philippines. In: *Wheat for More Tropical Environments*, (Eds.) Vallareal R.L. and A.R. Klatt. *Proceedings of the International Symposium*. CIMMYT, Mexico, pp. 152–153.

Leonard, K.J. and Suggs, E.G., 1974. *Setosphaeria prolatam*, the ascigerous state of *Exserohilum prolatum*. *Mycologia*, 66: 281–297.

Luttrell, E.S., 1951. A key to the species of *Helminthosporium* reported on grasses in the United States. *Plant Dis. Reptr. Suppl.*, 201: 59–67.

Mahto, B.N., 1999. Management of Helminthosporium leaf blight of wheat in Nepal. *Indian Phytopath.*, 52: 408–413.

Mahto, B.N., 2000. Characterisation of variatibility in *Drechslera sorokiniana* causing spot blotch of wheat. *Ph.D. Thesis*, IARI, New Delhi, pp. 148.

Malakar, P.K., Saha, N.K., Rahman, Hossain A.B.S., Hug, M.I. and Kabir, K.H., 1994. Yield loss assessment of wheat due to Helminthosporium leaf blight of farmers field. *Bangladesh J. Sci. Indul. Res.*, 29: 49–57.

Mandal, S., 1995. Effect of some antagonists on *Dreschelera sorokiniana*, causal agent of spot blotch of wheat. *M.Sc. Thesis*, I.A.R.I., New Delhi, 74 pp.

Mc Rae, W., 1930. Report of the Imperial Mycologist. Sci.Rep. Imp. Agric. Res. Inst., Pusa, 1928–29 pp. 51–66.

Mehta, Y.R., 1985. Breeding for wheat resistance to Helminthosporium spot blotch. In: *Wheats for Non Tropical Environments, A proceeding of the International Symposium*. CIMMYT, Mexico, pp. 135–144.

Mehta, Y.R., 1993. Spot blotch (*Bipolaris sorokiniana*). In: *Seed Borne Diseases and Seed Health Testing of Wheat*, (Eds.) Mathur, S.B. and M. Cunfer, B.B. Institute of Seed Pathology for Developing Countries Denmark, 168 pp.

Mehta, Y.R., 1998. Constraints in the integrated management of spot blotch of wheat. In: *Helminthosporium Blight of Wheat: Spot Blotch and Tan Spot Proceeding of and International Workshop*, (Eds.) Duveller, E., Dubin, H.S., Reeves, J. and A. McNab. CIMMYT, Mexico, pp 18–27.

Mujeeb, K.A., Villareal, R.L., Gilchrist, L.A. and Rajaram, S., 1996. Registration of five wheat germplasm lines resistant to Helminthosporium leaf blight. *Crop. Sci.*, 36: 216–217.

Nema, K.G. and Joshi, L.M., 1971. The spot blotch disease of wheat caused by *Helminthosporium sativum*. *Proc. Second Int. Symp. Pl. Path.*, IARI, New Delhi pp 42.

Nisikado, Y., 1928. Studies on Helminthosporium disease of Gramineae in Japan. *Ohara Inst. Agri. Res. Spec. Reptr.*, 4: 384.

Prasad, K.S.K., Kulkarni, S. and Siddarmaiah, A.L., 1978. Antagonistic action of *Trichoderma* sp. and *Streptomyces* sp. on *Drechslera sativum* (*Helminthosporium sativum* Pam. King and Bakke). Subram and Jain. *Curr. Res.*, 12: 202–213.

Raemaekers, R.H., 1988. *Helminthosporium sativum*; disease complex on wheat and sources of resistance in Zambia. In: *Wheat Production Constraints in Troical Envoronments*, (Ed.) A.R. Klatt. CIMMYT, Mexico, 175–186 pp.

Rahman, M.L., Rahman, M.Z., Main, I.H. and Alam, K.B., 1996. Effect of different level of temperature and pH on growth of *Bipolaris sorokiniana* and method of its inoculation into wheat. *Bangladesh J. Plant Pathol.*, 12: 51–58.

Rajaram, S., 1988. Breeding and strategies developed wheat for rice wheat rotation areas, In: *Wheat Production Constraints in Tropical Environments*, (Ed.) A.R. Klatt. CIMMYT, Mexico, pp. 187–196.

Razzaque, M.A. and Hossain, A.B.S., 1991. The wheat development programme in Bangladesh, In: *Wheat for the Non Traditional Warm Areas*, (Ed.) D.A. Saunders. CIMMYT, Mexico, pp. 34–43.

Reis, E.M., Medeiros, C.A. and Casa, R.T., 1998. Control of leaf blights of wheat by elimination of the inoculum sources. In: *Helminthosporium Blight of Wheats Spot Blotch and Tan Spot, Proceedings of International Workshop*, (Eds.) Duveillar E., Dubin, H.J., Reeves, J. and A. McNab. CIMMYT, Mexico. pp. 327–332.

Sharma, A.K., Kumar, J. and Nagarajan, S., 1996. Report of the coordinated experiments 1995–96: *Crop Protection* (*Patholgoy and Nematology*) All India Coordinated Wheat Improvement Project, Directorate of Wheat Research, Karnal, India 185 pp.

Shoemaker, R.A., 1959. Nomenclature of *Drechslera and Bipolaris* grass parasites segregated from *Helminthosporium*. *Canad. J. Bot.*, 37: 879–887.

Singh, A. and Virk, S.K., 1980. Chemical inhibition of *Helminthosporium sativum*. *Indian J. Mycol. Pl. Path.*, 10: 115–116.

Subramanian, C.V. and Jain, B.L., 1966. A revision of some graminicolous *Helminthosporia*. *Curr. Sci.*, 35: 350–355.

Vasil, I. and Vasil, V., 1986. Regeneration in cereal and other grass species In: *Cell Cultural and Somatic Cell Genetics of Plants* (Eds.) I. Vasil and V. Vasil, p. 121–150.

Insect Pest and Disease Management *Pages* **291–315**
Editor: **Dr. D. Prasad**
Published by: **DAYA PUBLISHING HOUSE (2016)**

Chapter 21

Ecologically Sustainable Management of Sheath Blight Disease of Rice

Rajbir Singh[1], A.P. Sinha[2], Ashraf Ali Khan[3], G.P. Gangwar[2] and D. Prasad[4]

[1]Department of Plant Pathology, Gochar Mahavidyalaya, Rampur Maniharan, Saharanpur – 247 451, U.P., E-mail: rajbir25805@yahoo.com

[2]Department of Plant Pathology, G.B. Pant University of Agriculture and Technology, Pantnagar – 263 145

[3]Division of Plant Pathology, [4]Division of Nematology, Indian Agricultural Research Institute, New Delhi – 110 012

Introduction

Rice continues to be the most dynamic sector in world grain production and India is the second largest world rice producer after China with the production of about 90 million tones annually. Increased and sustained production of rice is fundamental to food security in our country. In India, rice is grown in 45 million hectares in four major ecosystems: irrigated, rainfed, low lands, flood prone, and rainfed upland ecosystems. No other country in the world has such diversity in rice ecosystems. With the rice area getting stabilized and the population growing at the rate of 1.9 per cent annually, about 134 million tones of rice will be required by 2020 AD to sustain the share of rice in total food production (Mishra, 2005). There is no scope for increasing rice production through as increase in rice area; hence productivity of rice is of great concern. An important constraint to achieving higher rice yields is losses caused by diseases. The crop is vulnerable to diseases as it is mainly grown during Kharif season when the weather is hot and humid, which is congenial for rapid development of pathogens. Cultivation of high yielding dwarf rice cultivars since mid sixties, that necessitates the high inputs use including excess nitrogenous fertilizers and consequent altered microclimate accentuated the insect-pest, disease and nematode problems. The extent of losses due to these pests fluctuates widely depending on the prevailing factors of abundance of these pests in a particular year/season and the local macroclimatic conditions.

Sheath blight of rice (ShB) caused by *Rhizoctonia solani* is a potential threat to rice cultivation, causing severe yield losses (up to 69 per cent) to the crop (Naidu, 1992). Due to intensive and changed cultivation practices and use of higher quantity of chemical fertilizers the minor disease like ShB has occupied a prominent place in the disease scenario of the country. The disease was first reported from Japan in 1910 by Miyake (1910). In Philippines, ReinKing (1918) and Polo (1926) found a similar disease, which they considered to be due to a fungus of *R. solani* group. In India, first report of the occurrence of ShB of rice was made by Paracer and Chahal (1963) from Gurdaspur (Punjab) and later from Uttar Pradesh (Kohli, 1966). Since then, it has become one of the most common and destructive disease in Kerala and in many rice growing tracts of Tamil Nadu (Kannaiyan and Prasad, 1979). The disease has caused endemic in various parts of M.P. and Maruteru, East Godawari (A.P.) and Cuttack (Orissa) (Gangopadhyay and Chakrabarti, 1982) and Manipur (Mathur, 1983). The organism was identified by Saksena and Chaube (1972) as *Thanatephorus cucumeris* (Frank) Donk (Ou, 1985). Higher ShB incidence has been reported from those rice growing areas where heavy dose of fertilizers are applied to high yielding cultivars which have large number of tillers, resulting in an increase in the humidity near the base of tillers of rice plants (Ou, 1985).

The pathogen is soil borne and remains viable in soil for several months (Ou, 1985). The sclerotia of *R. solani* in soil play a significant role in perpetuation and spread of the disease in rice fields under water logged conditions (Kannaiyan and Prasad, 1978). Use of chemicals in the management of ShB is being discouraged because it is known to cause serious thread to environment, imbalance in the ecosystem and human health hazards. Though the development of disease resistant cultivars forms an effective long term strategy, this approach has certain obvious limitations. Hence, other disease management methods have to be integrated to realize the maximum yield output from the available rice cultivars.

Therefore, in the present disease management strategies the options lie in the manipulation of resistant genes in the rice cultivars, suitably tailored cultured practices and need based use of biological, plant biocides and for chemicals on the basis of surveillance. The movement of pathogens is checked through plant quarantine regulatory measures. The possible use of fungal and bacterial antagonist(s) against the pathogen has been viewed as an alternate disease management strategy. Although bioagent(s) have been found effective in reducing disease under *in vitro* conditions, but their application in field has given inconsistent and erratic results. The predisposing factors responsible for the out break of the disease vary from one location to other. Keeping these factors in mind one can make strict survey and surveillance of diseases for adopting appropriate control measures. In this article an attempt has been made to compile all the information available in the literature pertaining to ecologically sustainable management of ShB of rice.

Economic Importance

Sheath blight of rice occurs commonly in tropical and sub tropical areas of the world and causes severe yield losses, which varies from negligible to 69 per cent depending on the degree of severity of disease, crop stage and environmental conditions (Naidu, 1992; Sharma, 1995). Grain yield of different varieties of rice is chiefly reduced by ShB (Ou, 1985). In, East Asia 0.8 per cent (IRRI, 1983), East China 0.8 per cent (IRRI,1993), Japan and Republic of Korea 5.2 per cent (IRRI,1973), Vietnam and Philippines

25-50 per cent (Mizuta, 1956), Japan 20-25 per cent (Mizuta, 1956), Philippines 24 per cent (IRRI,1975), United states (Arkansas) 2 per cent (Boyetto and Lee, 1979) and Untied States (Southern) 7-50 per cent (Lee and Rush, 1983) losses have been estimated. A close correlation between the percentage of diseased hills in a field and the yield loss, with 0-50 per cent hills inoculated, low and high nitrogen applied, susceptible and moderately resistant cultivars showed 7.5-22.7, 8.6-23.7, 9.4-10.0, 2.5-13.2 per cent yield losses, respectively, with fungicidal treatment at 50 and 100 per cent hills infected, yield increase were 1.6, 6.4 and 8.0-10.1 per cent respectively (Ou and Bandong, 1976).

Symptoms

The symptoms of the disease first appear in the form of lesions on the leaf sheath. The lesions are at first ellipsoid or ovoid somewhat irregular, greenish grey, varying from 1 to 3 cm in length. The centre of the lesion becomes greenish-white with a brown margin. Sclerotia are formed on or near these lesions which are easily detached. The size and colour of lesion and formation of sclerotia depends upon environmental conditions (Ou, 1985). The mycelium of the fungus may grow over the surface of the leaf sheath and can spread to a considerable distance in 24 hours under humid conditions. In the field, the lesions usually first appear near the water line. When conditions are favourable to pathogen, they are also visible on upper leaf sheaths and the leaf blades as well. The presence of several large lesions on the leaf sheath usually causes death of the whole leaf and in severe cases, all the leaves of a plant may be blighted. The diseases also called as "Oriental leaf and sheath blight" (Kozaka, 1975) and "brown-bordered leaf and sheath spot" (Templetion and Johnston, 1969).

Under favourable microclimatic conditions of low sunlight, humidity near 95 per cent and high temperature (28-32°C) infection spreads rapidly by mean of runner hypae to upper plant parts including leaf blades, and to adjacent plants (Lee and Rush, 1983). Dasgupta (1992) stated that under favourable conditions disease may progress inward, vertically and horizontally, (*a*) in ward spread–the symptoms of disease spread from outer to inner sheaths with bleached centre and irregular purple brown border; (*b*) vertical spread–upward rapidly invading the lamina lossening the sheath from the culm, even causing blight of boot, flag leaf, panicle and lodging and grains become chaffy or partially filled in the lower parts of the panicle, (c) horizontal spread- laterally the disease spreads from tiller to tiller and hill to hill apparently by physical contact in densely crowded planting (Singh *et al.*, 1988, 1989).

When the conditions are not favourable, the lesions remain restricted to lower leaf sheath following some advancement of disease, as subsequently opening of the plant canopy due to leaf death permits increased penetrations of sunlight and reduced relative humidity, lesions become dry, oblong, well defined, white tan or grey with brown to reddish brown border. They coalesce, overlap and form a typical cobra skin pattern (Dasgupta, 1992). The disease normally attacks the leaf sheath and leaf blade but symptoms were found on emerging panicle which were chaffy, grayish brown and matted together by fungal mycelium. Numerous symptoms were reported when inoculum from 5 days culture was spread on newly emerged panicles, leaves and leaf sheath of susceptible cultivar IR-50 (Singh *et al.*, 1988). The sclerotia are formed after 6 days near the spots and are easily detached. Under humid conditions the mycelium of the fungus may grow over the surface of the leaf sheath and can spread to considerable distance (several cm) in 24 hrs (Ou, 1985).

Disease Management

Several attempts have been made in the past to manage ShB of rice by adopting different methods including cultural, biological, chemical control and through host resistance but sheath blight continued to be a serious threat to rice cultivation.

Host Resistance

Host resistance is the cheapest and direct method for the control of diseases and free from all the adverse effects posed by the fungicides. Further, it is compatible with all other methods of disease control. Most of the modern cultivars released and grown widely in insect-pest/disease prone areas possess resistance to at least one insect-pest or disease. In case of ShB also, the use of resistant varieties is the cheapest, safest and most effective method to mange the disease. The efforts through conventional breeding so far made are not successful in developing commercial rice cultivars, resistant to ShB. Thousands of entries have been evaluated under high disease pressure obtained by various inoculation techniques in greenhouse and field test (Lee and Rush, 1983). However, a few moderately resistant cvs. such as Jaya, Triveni, Jyoti, Pankaj, IET 1491 (Dasgupta, 1992), Zhongzu 1 resistance and Well is moderately susceptible to ShB. (Lee *et al.*, 1999) are available. Among them hybrid rice (PBCMS 2A × IET 10016, PBCMS 2A × Pankaj, PBCMS 2A × Ranjeit, PBCMS 2A × Vajram, PBCMS 3A × Bahdur, PBCMS 3A × Kushal, PBCMS 3A × Pankaj, PBCMS 3A × Ranjit and PBCMS 10A × Vijram) were found to be moderately resistant to the disease. Pattnayak *et al.* (2000) tested two anther culture derived double haploid lines (DH 7 and DH 21) against ShB in the field conditions at Barapani, India. DH 7 was resistant while DH21 was moderately resistant to the disease. Of 120 rice varieties/cultivars screened for their relative resistance to ShB, none was found to be resistant. AVT-M (H) 1808, AVT-M (H) 1810, AVT-M (H) 1814, IVT-J 2503, IVT-J 2507, IR 51673-50-2 and ACM-56 were moderately resistant. Among the remaining cultivars 13, 98 and 2 were moderately susceptible and highly susceptible to the disease, respectively (Meena *et al.*, 2000). One hundred and eleven high yielding and traditional rice cultivars screened with a highly virulent strain of *R. solani* by Biswas (2001). Eight cultivars (IR 54, Palman 579, Dinesh, Scetabhog, Dee-Geo-Woo-Gen, Matangini, Malta and Sabita) scored 3 (resistant), 25 scored 5 (moderatly resistant) and the rest scored 7 or 9 (susceptible). The disease incidence was heavy in most of the very early, early and mid early duration cultivars where all plants inoculated with the pathogen died. A number of resistant and moderately resistant cultivars observed in the medium and late maturity groups. The disease was more severe in high yielding than in traditional cultivars. Kalita *et al.* (2000) screened 11 rice cultivars against ShB two of the cultivars, Adrewsali and Monoharsali showed resistant reaction against this disease. The newly introduced Basmati cultivar gave a highly susceptible reaction. Biswas (2000) screened 161 entries of NSN-1 against ShB, none of the test entries recorded a resistant reaction, 6 showed moderately reaction and the other 155 a susceptible reaction. A high yielding indica-japonica hybrid rice variety Hosiyutaka, developed from Chugo Ju 55 × 89, was selected for long slender, non-glutinous grains with high amylase content. This variety escapes damage by ShB due to its late maturity (Sakai *et al.*, 1989).

In China, 17 early rice cultivars (*Oryza sativa*) were evaluated for their resistance level. Among these Jiayu 948 and Zhongfu 955, with disease index (DT) of 7.86 and 15.99, respectively, expressed a much higher resistance to ShB of rice as compared with other cultivars (Chen *et al.*, 2000). Sudhakar *et al.* (2000) tested 51 medium duration rice cultivars (120-125 days) to ShB in Andra Pradesh. Among them lowest disease index was recorded in Bhadra, ADT 40, WGL 48684 and WGL 47970. Lincombe *et al.* (2001) developed a variety in USA 'Earl' which is moderately resistant to ShB. The resistance of 64 *Oryza sativa* sub sp. indica cultivar/lines from China, lemont and SY 63/LH (susceptible controls), Jasmine 85 and Teging (moderately-resistant controls), to rice ShB was studied. Fourteen of the cultivars were resistant to highly resistant. The fourth-type cultivars, which were bred for accumulation of resistance from the cross Teging/91-3, were distributed in moderately resistant and resistant groups (60 per cent of these were in the resistant group) (Pan *et al.*, 2001). Dixiebelle a highly resistant variety to ShB was developed in USA by McClung *et al.* (1998).

The indica rice cultivars, Chinsurah Bora II, IR 72 and IR 51500 were transformed with the tip gene construct by PEG mediated direct gene transfer to protoplast and biolistic transformation using immature embryos. Bioassays of transgenic plants challenged with the ShB pathogen indicated that over expression of tip resulted in enhanced resistance compared to control (Datta *et al.*, 1999). Krisnamurthy *et al.* (2001) produced transgenic rice (cv. M202) expressing puroindolines (PINs) genes, PINA and/or PINB throughout the plants. PIN extracts of leaves from the transgenic plants reduced *in vitro* growth of *Magnaporthe grisea* and *R. solani* by 35 to 50 per cent and an 11 to 22 per cent reduction in symptoms. Velazhahan *et al.* (2001) identified protien with antifungal properties in the seeds of the pearl millet. The purified non-specific lipid transfer protiens (LTPs) inhibited mycelial growth of rice ShB fungus *in vitro*. T4 transgenic rice plants containing the *Seratia marcescens* chitinase gene showed resistance to ShB (Li *et al.*, 2003). Kim *et al.* (2003) developed rice plants containing MOD 1 and RCH 10 under the control of the rice rbcS and Act 1 promoters, respectively. Analysis of the response of transformed plants to *R. solani*, indicated significantly symptoms reduction. The Xa 21 gene (resistance to bacterial blight), the Bt fusion gene (for insect resistance) and the chitinase gene (for tolerance to ShB) were combined in a single rice line by reciprocal crossing of two transgenic homozygous IR 72 lines. In F4 the identified lines pyramid with different genes to protect against yield loss showed high tolerance of ShB disease (Datta *et al.*, 2003). *Agrobacterium tumifaciens* mediated transformation of cultivar Pusa Basmati 1 that harbours a rice chitinase gene (Chi 11) was performed using LBA 4404 (pSB 1, pMKU-RF2) under the control of the maize ubiquitin (Ubi 1) promoter intron. T1 plants revealed constitutive expression of chitinase at high levels. Bioassays of T1 plant indicated enhanced resistance to the ShB pathogen. A homozygous transgenic line was established from one to line, which exhibited the maximum resistance to *R. solani*. Karabi *et al.* (2001) introduced a PR-3 rice chitinase gene (RC 7), isolated from *R. solani* infested rice plants, into indica rice cultivars IR 72, IR 64, IR 68899B, MH 63 and Chinsura Boro II. The transformants synthesized different levels of chitinase protiens constitutisely and progeny from the plants containing the chitinase gene showed different level of enhanced levels of resistance when challenged with the ShB pathogen.

Chemical Control

Control of ShB disease with fungicides has been attempted many years in India as well as in other countries. The fungicidal control of ShB disease in India was well established (Kannaiyan and Prasad, 1978). The growth of *R. solani* in Czapeks pectin medium was markedly inhibited by the fungicides Hinosan and Kitazin. The production of polygalacturonase (PG), polygalacturonate trans eliminase (PGTE) and pectin transeliminase (PET) were inhibited by Kitazin. Treatment of rice seeds with Bavistin, Demosan, Daconil, Vitavax, Benlate and Agrosan reduced the seed-borne infection of *R. solani* and improved the seed germination, shoot and root growth, seedling vigour and preserved the viability upto 8 months. In field trial, the best curative and protective effects against ShB of rice were given by Validamycin through the highest grain yield was obtained from IBH plots (Devi *et al.*, 1987). In Indonesia, Rovral (iprodione) and folicur gave better control of the disease than Validamycin (Harman *et al.*, 1984).

In vitro, Benomyl, Epridione + carbendazim and Mepronil at 50 ppm, Carboxin + Thiram and carbendazim at 100 ppm, Edifenphos at 200 ppm and Zineb at 400 ppm inhibited mycial growth of *R. solani*. In field trials, Iprodione + Carbendazim, Benomyl and Validamycin decreased disease and increased yield, but Mepronil, Zineb, Carboxin + Thiram and Edifenphos were less effective (Torabi and Benesh, 1987). In Louisiana, Tilt (Propiconazole) has given better yields than Benomyl (Mukherjee, 1988, Groth and Rush, 1988). In inoculated field plots the disease caused by *R. solani*, was best controlled by soil application of kitazin 17 G (IBP) @ 40 kg/ha. However Topsin-M 70 WP (thiophanate methyl) applied as 2 foliar sprays at 50 and 65 days after sowing, respectively gave the best net return (Mishra *et al.*, 1989). Das and Mishra (1990) tested the effectiveness of some fungicides and found seed treatment with Topsin M effectively controlled disease and increased yield significantly. It is followed by seed treatment with Brassicol @ 5 g/kg seed + 2 sprays of Hinosan. Groth (1996) found that Benlate (Benomyl), Tilt 248 (propiconazole) Rovral (iprodione), Moncut (flutolanil) and Quadris reduced the ShB disease and increased grain yields upto 1400 pounds/acre and milling yield by 12 per cent. In field trial six fungicides namely, Opus, Baynate, Atemi, Tilt 25 per cent EC, Antracol 70 WP and Octave 50 WP significantly superior in reducing disease severity. Amongst the six fungicides opus reduced the ShB to maximum level. It was followed by Baynate and Atemi (Kumar *et al.*, 1997). Sheath blight was reduced when foliar sprays of Probenfos (as ketazin) were applied before the inoculation (Bera and Purkayastha, 1997). Ahmed *et al.* (1998) tested 7 fungicides in field experiment in Bangladesh and found that Propiconazole and Thiophanate methyl + Thiram gave best control of rice ShB disease and increased yield. Thiabendazole, Iprodione, Thiabendazole + Maneb and Edifenphos increased the yield, but did not significantly decrease the disease. Contaf reduce disease severity and incidence and increased test weight and 1000- grain yield (Singh and Sinha, 2004). Sudhakar *et al.* (2005) tested hexaconazole, propiconazole, cyperoconazole, terbuconazole, mencozeb, chlorothalonil, pencycuron, iprodione and found that carbendazim and Carbendazim, hexaconazole and propiconazole were most effective. Carbendazim, hexaconazole and propiconazole were most effective. Validamycine reduce disease severity when applied as foliar sprays (Sharma and Trimurty, 2005).

Cultural Control

Historically, cultural practices were the farmer's most important methods of preventing crop losses. By manipulating suitably cultural practices taking in to account of all the predisposing factors, the disease intensity can be reduced to a great extent. Maintenance of field sanitation, *viz.*, burning of diseased straw, stubbles, chaff, eradication of disease–harboring weeds and deep summer ploughing of the field are some of the visual practices adopted. Judicious level of fertilization should be used in order to reduce disease incidence. Heavy application of fertilizers particularly nitrogen tend to increase disease incidence. Besides, top dressing of N fertilizers should be done split doses. Basal application of urea in the form of neem cake caoated urea and mussoorie rock phosphated urea reduced disease incidence and gave higher yields as compared to three application of prilled urea (Baruah 1991). Since shades of trees enhance the development of infection crop should not be grown under shades. Destruction or burning of crop residues could not eliminate sclerotia from soil but has been suggested to be a part of integrated management practices (Lee and Courtney, 1981). Destruction of collateral hosts in the field or bunds was effective (Kurschner, 1988). Lee and Rush (1983) reported that crop rotation may be ineffective because of wide host range of the pathogen. Roy (1978) showed that by transplanting rice seedlings at spacing of 25 × 25 cm against 15 × 15 cm, the favourable effect of high N application in disease development could be reduced. Split application of potassium in a NPK combination of 40: 20: 20 kg/ha reduced the incidence of ShB and increased grain yield of rice (Baruah, 1995). Sarkar and Chowdhary (2003) determined the role of spacing on disease incidence, different spacing between plant to plant and row to row were used *i.e.* 15 × 15, 20 × 15, 20 × 20, 25 × 20 and 25 × 25 cm. Increasing of spacing reduced the infection rate. Planting of rice seedling far from the bund reduced the disease incidence. The use of potash fertilizer improved the pest resistance of rice plants and reduces the damage. Crop density degree affected pest occurrence in late season directly sown rice field Hung *et al.* (1999). Incorporation of oil cakes (Rajan and Menon, 1975) and some green manuring crops particularly *Sesbania aculeta*, sunhemp and green gram reduced survival of *R. solani*. Prasad *et al.* (1998) tested the effect of oil cakes (neem cake, ground nut cake, sesame cake) and 3 leaf manures (*Calotropis* sp., *Crotalaria* sp. and *Phaseolus* sp.) on sclerotial survival of *R. solani* was studied. Neem cake and Calotropis were most effective in affecting the survival of sclerotia in the soil and the combined effect of *Calotropis* with groundnut cake was most effective combination in reducing sclerotial viability in soil. The effect of minimal or traditional tillage, crop residues incorporated or not incorporated into the soil or burned, sowing by manual broadcasting or in row and crop densities at 100 and 150 kg seed/ha on disease was studied by Rodriquez *et al.* (1999). With 100 kg seed/ha, traditional tillage and sowing in rows, the incidence of ShB was significantly reduced (about 50 per cent). Lakpala *et al.* (1997) reported that ShB severity was maximum in the early transplanting cultivars followed by the medium and then late duration cultivars. Slight occurrence of ShB in direct sowing on a ridged plot without ploughing was reported by Wang *et al.* (1999). Introduction of minimum tillage which merely stirs the soil rather than inverting improves condition for survival of sclerotia. Incorporation of oil cakes and some green manuring crops particularly *sesbania aculeata*, sunnhemp and green gram reduced survival of *R. solani.* Roy (1993) reported loss of survival of sclerotia in the root zone of *S. aculeate.* Inorganic fertilizers can also influence saprophytic survival of the pathogen, K and P reduced it (Roy, 1993).

Studies may be under taken in tarping soil with polythene sheet in upland situation (Gangopadhayay and Chakrabarti, 1982) which reduces *R. solani* infection in some other crops. Trapping soil with clear polyethylene plastic during summer months has been found to control *R. solani* in other crops and it is worth attempting to control ShB in upland and dry farming areas by this method. Before or during trapping treatments the soil should be irrigated with 5-10 cm water. High soil moisture is a necessary part of soil heating and the control of soil borne fungi because it increases the conductivity of the soil to heat and lowers the ability of many soil borne organisms to withstand high temperatures. Eradication of *R. solani* from the top 15 cm soil takes longer in pre-irrigated trapped soil. Soil temperature at 5-15 cm depths increase greatly and kills the sclerotia and mycelial fragments (Gangopadhayay and chakrabarti, 1982). Application of micronutrients such as borax, zinc sulphate and ferrous sulphate at 0.05 per cent with two foliar sprays reduced the disease and increased the grain yield during the Thaladi (June to September) and Kuruvai (October to January) season in Tamil Nadu (Kannaiyan and Prasad, 1979). Although high nitrogen applications enhance disease development, this effect may be reduced with wider spacing (Roy 1993). Soil type may also influence the incidence of disease, especially if this is related to moisture content. In Japan, sheath blight is more prevalent in drained rice field than the upland fields and is less frequent in light clay soil than in and loam (Gangopadhayay and Chakrabarti, 1982).

Botanical Control

Sunderraj *et al.* (1996) observed that the plant extracts of *Allium sativum, Prosopis juliflora, Gyanandropsis pentaphylla, Leucos aspera* and *Vitex nequndo* and animal faeces such as litter and pig dung have inhibitory effect towards *R. solani*. Seed extract of *Trachispermum ammi* and leaf extract of *Ocimum* sp. have fungicidal activity against sheath blight pathogen (Ansari, 1995). Leaf extracts of *Clerodendron viscosum* and *Lantana camera* have inhibitory effect to *R. solani* causal agent of ShB of rice (Shyalaja *et al.*, 1997). Different plant extract namely Ajwain (seed), lemon grass (leaves), lawsone (leaves), Tulsi (leaves), mentha (leaves), rauwolfia (leaves) and samhalu (leaves) were tested against *R. solani* and it was observed that maximum reduction in growth and sclerotia of fungus was in Ajwain and Lawsone treated plates (Anon, 1976). Meena *et al.* (2002) evaluate the efficacy of oils, oil cakes, bulb and seed extracts were tested for their ability to suppress *R. solani* in vitro and rice ShB. Plasmarosa oil completely inhibited mycelial growth and gave good disease control. Extracts of 15 common weed species were tested against *R. solani* under laboratory conditions by Ranjana *et al.*

(1999). All the weed extracts (50 per cent concentration) significantly inhibited the mycelial growth. Based on *in vitro* results, pot and field grown rice plants (CVIR-50) were used to study the efficacy of the weed extracts as pre and post inoculation sprays. Sheath blight infection could be considerably suppressed by bracken fern extract followed by goat weed and sickle pod senna. *Eucalyptus citridora* and its major constituent citronellal at 10 and 20 ppm were tested against *R. solani*. Citronellal was found more effective in reducing the radial growth and dry weight of fungus (Ramezani *et al.*, 2002). Among the neem formulations tested in glasshouse, neem fungicide (0.5 per cent) and neemark (extracts of *Azadirachta indica*) (0.5 per cent) were found best in controlling ShB disease in rice. *In vitro* screening of the plant extracts against *R. solani* revealed that 0.1 and 0.5 per cent palmarosa oil completely inhibited mycelial growth while there was 8.87 cm growth in the control (Meena and Muthusamy, 1998). Kandhari and Gupta (2000) evaluated extracts of eight environmental friendly plants (*Ocimum basilicum, Citrus reticulata, Syzygium aromaticum, Jasminum officinale, Tegetes tenuifolia, T. erecta, Pyrus pashia* and *Gladiolus* sp.) and were evaluated against *R. solani*. *In vitro* complitly inhibition of fungal growth was observed at 100 ppm concentration by *S. aromaticum* and methyl anthranilate showed 77.7, 88.3 and 83.3 per cent antifungal spectrum at the same concentration. The benzene and acetone extracts of *T. erecta* also showed completely inhibition at higher concentration (1000ppm). Meena *et al.* (2002) evaluate the effecacy of leaf extracts of *Acacia concinna, Capsicum annuum, Lycopersicon esculantum, Mimosa pudica, Pithecellobium dulce, Prosopis juliflora, Solanum indicum, S. nigrum* and *S. xanthocarpum* against *R. solani in vitro.* The leaf extract of *Pithecellobium dulce* was highly effective in inhibiting the mycelial growth (2.5 cm) compared to the control (8.9 cm). The leaf extracts of *Pithecellobium dulce* and *Prosopis juliflora* were at par in inhibiting the sclerotial number, dry weight and germination of the pathogen. The leaf extract of *S. xanthocarpum* was not effective. The extracts of *Prosopis juliflora* and *Pithecellobium dulce* were effective in controlling the disease, reducing 32.3 and 33.3 per cent, respectively, compared to 76.2 per cent in the control. Biswas and Raychoudhury (2003) tested achook (neem formulation), neemazal, neemgold, spictaf, tricure and wanis (neem formulation). All the botanicals reduce disease severity. Among all these spictaf performed best. Of the various plant extracts, the leaf extract of zimmu (*Allium cepa* × *Allium sativum*) showed the maximum antifungal activity against *R. solani* and other agronomically important fungal and bacterial pathogens (Satya *et al.*, 2005).

Biological Control

Biological control of plant pathogens can be achieved by either promoting the native antagonists to reach a density sufficient to suppress a pathogen(s) or by introducing alien antagonists. Though some of the earlier works, related to promoting the native antagonists was by using organic amendments, the recent trend is to isolate, multiply and introduce the antagonists to soil or specific court of infection to achieve a successful biological suppression of a disease. The possible use of fungal and bacterial antagonists of rice pathogen(s) has been viewed as an alternative disease strategy. Although bioagent(s) have been found effective in inhibiting the growth of pathogen(s) under *in vitro* conditions, they fail to control the pathogen in field in most of the cases due to various reasons. Therefore, biocontrol may hold promise under a certain set of conditions only (especially under controlled conditions). Soil and nutritional factors, time of application, the inoculum level and potential of pathogen as well as bioagents, mode and form of application or delivery system of the bioagent etc. play vital role in biocontrol strategy.

Fungal Bio-control Agents

Several fungal biocontrol agents, *i.e. Trichoderma, Gliocladium, Aspergillus* etc. have gained wide attention due to their ability to control ShB (Khan and Sinha, 2005 a; Teweri and Singh, 2005).

Trichoderma sp. colonizing the sclerotia of *Thanetephorus cucumeris* in a dry land rice field was isolated and tested for antagonism *in vitro*. Pathogen growth ceased after contact with the *Trichoderma sp.* which eventually covered the whole plate (Rosales and Mew, 1982). Coiling of hyphae by *Trichoderma spp.* followed by vacuolation, coagulation of cytoplasm and lysis have been recorded (Roy, 1977, 1989; Rosales and Mew, 1982; Khan, 2003). *In vitro* test by using different biocontrol agents isolated from soil and irrigated water from paddy fields indicated that *A. niger* and *T. viride* were most antagonistic fungi, inhibiting linear growth of *R. solani* (Gokulapalan and Nair, 1984). When, *T. harzianum* was introduced under rainfed conditions, it decomposed rice straw and by replenishing the substrates reduced survival of *R. solani* (Mew and Rosales, 1984).

Manibhushanrao *et al.* (1989) reported that in the antagonistic potential of *G. virens* and *T. longibrachiatum* there was no specific connection between antibiotic production and mycoparasitic potential of either of biocontrol agents. However, considerable variation in mycoparasitic and antibiotic potential was found between various strains both were able to utilize various components of fungal cell as carbon sources and their mycoparasitic potential was reduced by cycloheximide. Fungi from rice phylloplane were screened for activity against *R. solani* the cause of ShB. Inhibition zone were formed when *R. solani* was grown with *T. harzianum, T. viride* and *Chaetomium globosum*. After seven days, the *Trichoderma* spp. completely over grew and parasitize *R. solani* (Gokulapalan and Nair, 1991).

Dubey (1995) evaluated some fungal biocontrol agents against *R. solani* and showed that *T. harzianum, T. viride* and *G. virens* significantly inhibited the mycelial growth and sclerotial production of *R. solani*. Of these, *G. virens* found to be most effective. The spraying with *T. viride* on artificially inoculated rice plants had given significant control of disease (Das *et al.*, 1996). The application of *T. viride* and *Bacillus subtilis* as seed treatment reduced sheath infection (*R. solani*) of rice in pot experiment in Assam, India. *T. viride* was more effective than *B. subtilis* in reducing infection (Das *et al.*, 1998). It has been definitely shown, however, that *T. harzianum* destroys the rice stubbles/straw on which the pathogen, *R. solani* survives and also parasitizes the sclerotia. So reducing the viable *R. solani* in straw from 100 per cent occurrence in freshly incorporated straw to 20 per cent after two weeks and causing complete removal of pathogen by 16 weeks (Mew and Rosales, 1986). Spraying of Talc + CMC formulation of bioagents *viz., T. harzianum, T.virens, T. viride* and *A. niger* were found quite effective against ShB (Khan and Sinha, 2006). A number of commercial formulations of fungal biocontrol agents are available in the market viz Soil gard/Gliogard *G. virens* based product for controlling many soil-borne diseases (Cook 1993), (WRC-AP1) *G. virens* based product for control of Pythium and Rhizoctonia root rot on ornamental plants (Ristaino and Thomas 1997), Trichostar *T. harzianum* based bio-pesticide for controlling *R. solani, Fusarium, sclerotenia* disease (Mukhopadhyay 1996).

Bacterial Biocontrol Agents

Bacterial antagonists have emerged as an effective most promising means for ShB management. Several bacterial species have so far been tried as biocontrol agents viz *Agrobacterium, Actnoplanes, Alcaligenes, Amorphosprangium, Arthobacter, Azatobacter, Bacillus, Bradyrhizobium, Cellulomonas, Enterobacter, Erwinia, Flavobacterium, Micromonaspora, Hafnia, Pseudomonas, Pasteuria, Rhizobium, Serratia, Streptomyces* and *Xanthomonas* (Weller, 1988). Reports are on hand indicating effectiveness of *Pseudomonas fluorescens* against *R. solani* (ShB pathogen). *P. fluorescens* has been attempted in biocontrol against various diseases because of the properties like its ability to colonize the rhizosphere of plants. short generation time, ability of utilize a large number of organic substrates commonly found in root and seed exudates (Rao and Johri, 1999), relatively easy cultivation under laboratory conditions, mobility as well as potential to produce an array of compounds inhibiting growth of fungal pathogens, compatibility with commonly used pesticides and other biocontrol agents (Hass *et al.*, 1991 and Tripathi, 1999). They can be manipulated by current genetic techniques. Some strains of fluorescent pseudomonads induce systemic resistance and reduce plant diseases (Van Peer *et al.*, 1991). It is well known that different *Pseudomonads* (fluorescent) have different abilities to colonize a particular root-niche (Loper *et al.*, 1985; deWeger *et al.*, 1987). There was good colonization of roots of rice seedlings, when *P. fluorescens* was applied as seed treatment (Kamala *et al.*, 1998; Vidhyasekaran and Muthamilan, 1999; Trane *et al.*, 2001). A number of commercial formulations *viz.* Biomass, Biospriny, Biosheld, Cingaurd, Dagger-G, Deny, Epic-Gus 376, Mycostop, Pant Biocontrol Agent-1, Pant Biocontrol Agent-3 and PfALR2 of bacterial biocontrol agents are available in the market for controlling the diseases caused by *R. solani.*

Mass Production of Fungal/Bacterial Biocontrol Agents

It has been realized that the greatest obstacle to biological control by introduced antagonists is the lack or scarcity of methods for mass multiplication of the biocontrol agents (Papavizas and Lewis, 1982). A fermentation system providing high quality biomass is one of the main factors for the success of the biocontrol agents. The first step in production of biocontrol agent is development of a suitable medium using inexpensive, readily available agricultural bio-products with the appropriate nutrient balance (Latze and Soper, 1977). Many attempts have been made so far to standardize growth media for mass multiplication of the antagonists. There are numerous reports for mass multiplication of *Trichoderma* and *Gliocladium* spp. Substrate used for the mass production of the fungal antagonists includes wheat bran and sterile bran sand water 1:1:2 w/w/v (Lewis and Papavizas, 1984), soil (Chet, 1987), wheat bran and saw dust (Singh, 1994; Malik *et al.*, 2005), rice and sorghum grains (Malik *et al.*, 2005), italian millet (Tewari and Singh 2005), groundnut shell + tap water (Dubey and Patel 2002), wheat bran, rice husk, oat bran, eucalyptus leaf, maize meal and sodium alginate (Mafia *et al.*, 2003) and FYM (Sangeeta *et al.*, 1995). Lactose, microcrystalline cellulose and sodium carboxy methyl cellulose (Wiwattanapatapee *et al.*, 2004). The unique problem in developing biopesticides is that it represents a living system, which must be able to stand the process of formulation and should remain sufficiently viable for a period until it reaches the farmers (Vidhyasekaran and Muthamilan, 1995).

The bacterial biocontrol agent *Pseudomonas* has been used as cell-suspension for experimental purpose. For commercial production of these antagonists' different technologies has been adopted on industrial scale (Mukhopadhyay and Mukherjee, 1998; Vidhyasekaran and Muthamilan, 1999). Commercial formulation of *P. fluorescens* was developed using a mixture of carboxy methyl cellulose and talc (1:100, w/w) (Vidhyasekaran *et al.*, 1997; Mishra *et al.*, 2001). Singh and Sinha, (2005a) tested talc, $CaCO_3$, CMC and talc + CMC (4:1) and found that talc + CMC (4:1) was more effective against ShB. Different base materials *viz.* peat (Vidhyasekaran and Muthamilian 1995; Rabindran and Vidhyasekaran 1996; Shivakumar *et al.*, 2000), hydrogenated vegetables oil, monohydrate, lactose polyvinyl pyrrolidone and cross linked sodium carboxymethyl cellulose (Kanajanmaneesathian *et al.*, 1998), dried xanthin gum (Demir *et al.*, 1999), methyal cellulose: talc (1:4) (Krishanamurthy and Gnanamanickam 1998) have been used against ShB.

Mode of Application

Based on experiments, Papavizas and Lewis (1981) concluded that the methods of inoculum preparation of biocontrol agents, their mode of application, use of different food bases, and an integrated application of other chemicals or cultural practices can improve the effectiveness of biocontrol agents. The application of biocontrol agents by broadcast, sprays, or by direct application in seed furrow at the time of sowing, has the greatest potential for use in commercial glasshouse operations or small scale field experiments, for biological control of root pathogens Cook and Baker (1983).This method gives the antagonist an opportunity to possess a great "Competitive advantage" when introduced on seed rather than soil because in the soil the roots must be within convenient reach of the antagonist in order for the antagonist to colonize on the roots before the pathogen. Foliar spray of *T. harzinaum* significantly reduced the disease severity and incidence and increasing the grain yield and 1000-weight as compared to soil treatment or seedling root dip against ShB disease of rice (Tewari and Singh, 2005). Different workers have applied biocontrol agents by different methods. All methods of application of *P. fluorescans viz.* seed treatment, soil treatment, root dip and foliar spray reduce the diseases severity and incidence of ShB but combinations of more than two methods of application were more effective in reducing ShB as compared to single method of application. Maximum reduction in disease severity and incidence and increasing the grain yield and 1000-weight was observed with the combination of the treatment viz seed + soil + root dip + foliar sprays (Meena and Muthusami 1998; Mishra 1998; Nandkumar *et al.*, 2001; Commare *et al.*, 2002; Singh and Sinha, 2005b).

Mechanism of Action of Antagonists

Starting from the classical work of articles from 1932 to 1941 various mechanism involved in biological control have been proposed by different authors. The entire basis of biological control is the exploitation of the antagonists potential of the biocontrol agents. Antagonism is the balanced wheel of nature and is the basis of biological control (Snyder, 1960).

Mechanism involved in biological disease control comprise of the following:

1. Antibiosis, which results from the liberation of an antibiotic or other chemical by one microorganism and harmful to the pathogen.
2. Competition for same nutrient or other commodity in limited supply but needed by the pathogen.
3. Predation, hyperparasitism, mycoparasitism or other forms of direct exploitation of a pathogen by another organism.
4. Siderophore production, resulting in iron unavailability to plant pathogen.
5. Systemic resistance, bioagents are inducer of systemic resistance.

Direct exploitation is the only form of antagonism where the beneficial organism may exhibit a host-dependent population trend. Competition and antibiosis are usually fortuitous. With these mechanisms, the antagonist is unlikely even to make physical contact with the pathogen (Cook, 1982). An antagonist may use more than one form of antagonism and the action of same antagonist may fit under more than one mechanism (Cook and Baker, 1983). The antagonists act by (*a*) reducing the population of pathogens (*b*) preventing the pathogen from infecting the plant and (*c*) hunting disease development after infection (Mukherjee *et al.*, 1995).

Factors Affecting Efficacy of Fungal/Bacterial Bioagents

Scanty information is available in the literature on the factors which may affect the efficacy of biocontrol agent against ShB. Isolate, time, rate mode of application and formulation of bioagent, pathogen inoculum level, plant growth, soil factors, nutrition availability etc. play important role in biocontrol strategy (Singh, 2003). The amendments protected plants better than the antagonists alone. The highest grain yield was obtained with a combination of organic manures and *G. virens*. Plant growth and disease incidence were recorded for 22-days old plant of rice cultivar IR 50 inoculated with 2 per cent *R. solani* and grown in pots in acidic (pH 5.2) or neutral (pH 6.8) soil. Inoculation with

2 per cent *T. harzianum* significantly reduced infection in all treatments in both seasons, but was more effective in reducing ShB incidence in acidic soils than in neutral soils. Gamlie and Katan (1991) reported that regression analysis showed a significant, inverse relationship between soil pH and increased growth and between soil pH and population densities of fluorescent Pseudomonads in rhizosphere. Disease incidence was also significantly lower when it was applied simultaneously of 7 days before inoculation with *R. solani*. Grain yield increased significantly in the presence of *T. harzianum* when it was applied either simultaneously or 7 days before *R. solani* (Das *et al.*, 1995; Khan and Sinha 2005 b; Singh and Sinha, 2005c). Effectivity of *T. harzianum* against ShB was more in soil having low pH (5.2) as compared to neutral and high pH (Khan 2003).

Harman *et al.* (1991) reported that field application of biocontrol agent is hampered due to absence of suitable technique and carrier/substrates. Farmyard manure (FYM) has been found to be an effective substrate for growth and mass multiplication of *T. harzianum* (Kausalya and Jeyarajan, 1988). Baby and Manibhushanrao (1993) observed that both the antagonists (*T. longibrachaetum* and *G. virens*) and the organic manures, alone or together, reduced ShB. Organic amendment such as green manures, stable manures, and compost can provide the food base and has long been recognized to facilitate the biological control if applied well ahead of planting. *Trichoderma* spp. is reported to be a predominant effective biocontrol agent of *R. solani* in compost prepared from lignocelluloses wastes. Incorporation of *T.harzianum* in soil having soil nutrition P_{180} and K_{40} showed maximum reduction of the disease. Among different combinations of NPK fertilizers tested on the effectivity of *T. harzianum* on ShB, maximum reduction in ShB was observed with $N_{60}P_{60}K_{40}$ (Khan, 2005c). Babich and Stotzky (1978) reported that concentrations of Zn^{2+} upto 1 mM did not significantly inhibit the growth of fungi. However, 10 mM Zn^{2+} significantly decreased the mycelial growth of *F. solani*, *A niger* and *T. viride* and completely inhibited the growth of *R. solani*. The addition of Zn to zinc deficient soils resulted in reduced yield loss in the presence of *R. solani*, a reduction in disease score and no changes in the concentration of nutrients in the shoots. However, under zinc deficiency, increasing levels of added *R. solani* resulted in significant yield loss, and increase in disease score (Streeter *et al.*, 2001). Soil application of the bioagents in zinc amended soil showed 15.96 per cent disease severity and 30.17 per cent incidence. However, application of bioagents in zinc deficient soil resulted in 20.36 and 32.29 per cent disease severity and incidence, respectively. Soil application of neem cake enhanced the effectivity of the bioagent in increasing seedling emergence and reducing ShB severity and incidence (Singh, 2003).

(Singh and Sinha 2004) tested different isolates of *P. fluorescens* against ShB and found that isolate Pf1 exhibited maximum reduction in disease severity and incidence and increase the grain yield and 1000-grain weight. Talc+ CMC based formulation exhibited maximum reduction of disease severity and incidence than others formulations. (Singh and Sinha, 2005 a). They also tested different combinations of NPK fertilizers on the effectivity of *P. fluorescens* on ShB, maximum reduction in disease was observed with $N_{60}P_{60}K_{40}$ however, the effectivity was minimum with treatment $N_0P_0K_0$. Application of *P. fluorescans* 7 days before inoculation of *R. solani* was found highly effective in increasing seedling emergence, reducing disease severity and incidence than simultaneous application (Singh and Sinha 2006). Of the three rates *viz.*, 2, 4 and 8 g/litre of the *P. fluorescens* tested against ShB on transplanted rice, higher rate (8 g/litre) was found highly effective in reducing disease severity and incidence and increasing grain yield and 1000-grain weight (Singh and Sinha 2005a).

Integrated Management of Sheath Blight of Rice

Economic control of plant diseases is rarely achieved by a single method. It must be supported by different control measures. An integrated pest management system involves simultaneous or sequential use of an optimal combination of various means of disease control eco-friendly. Biological control is of particular interest as a component of IPM and can best be exploited within the framework of an IPM system (Klassen, 1982). Species of *Trichoderma* and *Gliocladium* have distinct advantage as biocontrol agents, as they are compatible with most of the agricultural practices and hence, can be successfully utilized as a part of total crop management practices or broadly as a component of agro-ecosystem management strategies. *Gliocladium* spp. and *Trichoderma* spp. have been integrated with cultural practices, physical soil treatment (solarization), fungicide and disease resistant varieties for the management of different plant diseases.

Combining the antagonist *T. harzianum* with soil solarization in *R. solani* infested soil improved disease control and delayed inoculum build up (Elad *et al.*, 1980, Chet *et al.*, 1982). One of the greatest advantage in using *Trichoderma* and *Gliocladium* is that these two antagonists are insensitive to many fungicides of common use for controlling plant diseases (Mukhopadhyay *et al.*, 1992). Gogoi and Ali (2005) found that *T. harzianum* was more compitable with captan, as it inhibited the growth of the rice sheath blight pathogen more than *T. viride* and *G. virens*. Single or combination doses of iprodione (Rovral 50) and carbendazim (Bavistin 50 WP), both at 0.5, 1.0, 2.0, or 4.0 ppm, and the compatibility of these fungicides with *T. harzianum* and *G. virens* were tested. Single and combination doses of the fungicides were highly effective in controlling the disease. The plant pathogen was more tolerant to the fungicides than the biological control agents. Single dose of carbendazim exibited better results than iprodione and the doses of carbendazim (Sitansu *et al.*, 2001). Pathak *et al.* (2002) conducted experiment to validate an integrated pest management (IPM) model for pest and pathogens of rice cv. Basmati-1. Seed treatment, spraying of chlorpyrifos, streptocycline and carbandazim application of methyl-parathion, butachlor or amilofos and hand weeding as well as release of *Trichogramma japonicum* in the experimental fields were included in the IPM programme. Sheath blight recorded in the fields was low in IPM (60 per cent) than FP subjected fields (13.5 per cent).

Many positive results obtained with fungal biocontrol agents and organic amendments in various crops suggest possibility of combining these two systems for effective crop protection. The beneficial effects of soil amendments in plant disease management lie in their ability to suppress pathogen populations through the stimulation of soil saprophytes and also by toxic substances produced during their decomposition. Soil amended with organic amendments such as rice chalf, mustard cake, sawdust, neem cake, composed coir with and farmyard manure significantly reduced the seedling infections, decrease severity (Kanniyan and Prasad, 1981; Meena and Muthusamy, 1999). Baby and Manibhushanrao (1993) reported that there was significant decrease in ShB severity when two fungal biocontrol agents (*G. virens* and *T. longibrachaetum*) were integrated with two organic soil amendments (gliricidia leaf and neem cake). In addition to disease suppression, the growth of rice plants as well as grain yield were also increased.

Mathivanum *et al.* (1997) tested the efficacy of some fungicides with integration of neem oil against sheath blight disease and reported that integration of carbendazim + neem oil was most effective followed by contaf + neem oil. In the pots experiment lower incidence of sheath blight of rice were observed when these pots treated with sub lethal dose of (0.05 per cent) thiram + *T. harzianum* than treatment with its lethal dose (0.3 per cent) alone (Bora *et al.*, 1999). Surulirajan *et al.* (2003) evaluated *T. viridi*, carbendazim and soil amendments (farmyard mannure and sawdust (1 per cent)), singly or in combination in pot experiment. Among the treatments, the application of *T. viridi* + carbendazim 50 WP + FYM + sawdust resulted in the lowest disease incidence and highest grain and straw yields. PF-9 (fluorescent pseudomonas sp.), B-44 (*Bacillus* sp.) and Chb-1 (chitinolytic bacterium) were found to be compatible with carbendazim (bavistin) at 500 and 1000ppm concentrations. Among them, PF-9 was the most effective in reducing disease severity either alone or in combination with one spray of 0.1 per cent Bavistin, followed by combination of PF-9 and B-44 (Laha and Venkataraman, 2001). Mishra and Sinha (1997) observed effective control of ShB with the application of *P. fluorescens* and Tilt. Goel *et al.* (2003) recommended of N_{120} P_{30} K_{30} Zn_{3}, N_{120} P_{30} K_{30} or FYM+N_{60} P_{30} K_{30} kg/ha as a component of integrated disease management to minimize the occurrence of ShB and attain maximum grain yield in rice.

Plant Growth Promotion by Bio-agents

As a result of rather intensive work carried out with *Trichoderma*, several independent research groups have noticed that the addition of the antagonists *Trichoderma* spp. induced the growth of various crops. Activity of biocontrol agents could also reduce the concentration of substances in soil that are inhibitory to plant growth (Windhan *et al.*, 1986). Baby and Manibhushanrao (1993) observed 25 per cent increases in growth of rice plants, when antagonists were applied to soil. Mishra and Sinha (1998) observed seed treatment or soil application of fungal biocontrol agents (*T. harzianum, G. virens* and *A. niger*) significantly increased root and shoot growth of rice plants. Direct growth promotion occurs when a biocontrol agent produces metabolites that directly promote plant growth without producing any harmful effect on plant (Schiper *et al.*, 1987). In rice *T. longibrachiatum* was found to induce 25 per cent increase in plant growth. *T. harzianum* and *T. virens* were found to double the root length (Sankar and Jeyarajan, 1996). (Khan and Gupta, 1998) reported growth promoting activity of *T. harzianum, T. virens, A. niger, G. virens* and *P. fluorescens* in rice. Khan *et al.* (2005) reported that *T. harzianum* and *T. virens* increased germination of seeds root and shoot length and fresh root and shoot weight and dry root and shoot weight in rice plants.

Reports are on hand which indicate that bacterial biocontrol agents also induce the growth of rice plants. Singh (2003) reported that *P. fluorescens* increased germination, root and shoot length and fresh root and shoot weight and dry root and shoot weight in rice when applied as seed treatment and soil treatment. Promotion of plant growth, increase of plant height, numbers of tillers and yield have been observed in *P. fluorescens* treated rice seedlings by Mishra and Sinha (1999, 2000) and Nandkumar *et al.* (2001). These responses may be due to altered root growth because of N production (Baker and Schippers, 1987; Voisard *et al.*, 1989), hormonal effect on plant growth (Imam *et al.*, 1977), increased root exudation (Prikryl and Vancura, 1980), induced host resistance (Kempe and Sequeria, 1983). Activity of biocontrol agents could also reduce the concentration of substances in soil that are inhibitory to plant growth (Windham *et al.*, 1986). Enhancement of nutrient uptake by the plant and enzyme synthesis to regulate the level of plant ethylene have been observed (Krishamurthy and Gnanamanickam, 1997). There is production of plant hormone like substances or increase in the uptake of nutrient such as N, P, Zn, Fe, Mn, etc. either by changing physiology of plant or by converting nutrients into available form as well as synergetic effect on plant growth, yield and dinitrogen (Vidyasekaran and Muthamilan, 1995). These bacteria are ideally suited as soil inoculants because of their potential of rapid and aggressive colonization. Direct growth promotion occurs when a rhizobacterium produces metabolites that directly promote plant growth without interactions with native soil micro-flora (Dube, 2001). In contrast, antibiotics, siderophores and HCN, which decreased activities of pathogen or deleterious microorganisms and thereby, increased plant growth, are examples of indirect growth promotion by biocontrol agents.

Conclusion

Sheath blight caused by a fungus *Rhizoctonia solani* is one of the important diseases of rice the world over including India. It may cause serious yield losses up to 69 per cent. The disease cause more loss in dwarf, high tillering, high doses of nitrogen fertilizer, metrological factors, time, incidence and disease severity. Plant is more susceptible at the actively tillering stage and susceptibility booting and heading stage. High relative humidity (> 85 per cent) and moderate temperature (23°-35°C) favours the disease. The lesion development is rapid and medium at normal sunshine and least in complete shading. Soil-borne sclerotia and a lesser degree mycelium in plant debris are the means of pathogen survival and primary inoculum. Once infection occurs, secondary spread takes place through direct contact. Sclerotia may move from one field to another through irrigation water and during movement they may produce mycelia and secondary or tertiary sclerotia. Effective management of this disease can be achieved by adopting host resistance, cultural practices biological control and chemicals in an integrated manner.

References

Ahmad, H.U., Shahjahan, A.K.M. and Miah, S.A., 1997. Fungicides to control rice sheath blight (ShB.). *Int. Rice Res. Newsl.*, 13: 37–38.

Anon., 1996. The world rice production during 1995–1996. *Rice J.*, 99: 30.

Ansari, M., 1995. Control of sheath blight of rice by plant extracts. *Indian Phytopath.*, 48: 268–270.

Babich, H. and Stotzky, G., 1978. Toxicity of zinc to fungi, bacteria and coli phages: Influence of chloride ions. *Am. Soc. Microbiol.*, 36: 906–914.

Baby, U.I. and Manibhushnrao, K., 1993. Control of rice sheath blight through the integration of fungal antagonists and organic amendments. *Trop. Agric.*, 30: 240–244.

Baker, R., 1986. Biological Control: An Overview. *Can. J. Pl. Path.*, 8: 218–221.

Baker, R. and Scher, F.M., 1987. Enhancing the activity of biological control agents. In: *Innovative Approaches to Plant Disease Control*, (Ed.) I. Chet. Wiley Intersci. New York, U.S.A., pp. 1–18.

Baruah, B.P., Saikia, L., Ali, L. and Senapoty, D., 1991. Nitrogen, management for wet land rice in bacterial leaf blight endemic areas. *Oryza*, 28: 507– 508.

Baruah, R.P., 1991. Nutrient management of sheath blight disease of rice. *Indian J. Hill Farm.*, 8: 64–67.

Bera, S. and Purkayastha, R.P., 1999. Multicomponent coordinated defence response of rice to *Rhizoctonia solani* causing sheath blight. *Curr. Sci.*, 76: 1376–1384.

Biswas, A. and Roychoudhury, U.K., 2003. Relative efficacy of some botanicals against sheath blight disease of rice. *J. Mycopat. Res.*, 41: 163–165.

Bora, K., Das, B.C. and Roy, A.K., 1999. Integrated management of sheath blight disease of rice with *Trichoderma harzianum* and chemicals. *Oryza*, 36: 238–240.

Boyette, C.D. and Lee, F.N., 1979. Reduction in rice yield caused by sheath blight. *Arakansas Farm Res.*, 28: 5–7.

Callan, N.W., Mathre, D.E., Miller, J.B. and Vavrina, C.S., 1997. Biological seed treatment: Factors involved in efficacy. *Hort. Sci.*, 32: 179–183.

Chen, Z.Y., Xu, Z.G., Gao, T.D., Yan, D.F., Lu, F. and Liu, Y.F., 2000. Evaluation and utilization of antagonistic bacteria against rice sheath blight. *Chinese J. Rice Science*, 14: 98–102.

Chet, I., Elad, Y., Kalfon, A., Hadar, Y. and Katan, J., 1982. Integrated control of soil-borne and bulb-borne pathogens in iris. *Phytoparasitica*, 10: 229–231.

Chet, I., 1987. *Innovative Approaches to Plant Disease Control.* John Wiley and Sons, New York, 372 p.

Commare, R.R., Nanadakumar, R., Kandan, A., Suresh, S., Bharathi, M. and Raguchander, T., 2002. *Pseudomonas fluorescens* based bio-formulation for the management of sheath blight and leaf folder insect in rice. *Crop Prot.,* 21: 671–677.

Cook, R.J., 1982. Biological control of plant pathogens.: Overview. In: *Biological Control in Crop Production,* (Ed.) G.C. Papavizas. Allanheld, Osmum, Granada, U.S.A., pp. 23–44.

Cook, R.J. and Baker, K.F., 1983. *The Nature and Practices of Biological Control of Plant Pathogens.* APS Books, St. Paul, USA, 599 p.

Das, B.C., Khairuzzaman, A.S.M. and Bora, L.C., (998. Biological seed treatment for management of sheath blight of rice. *Indian J. Mycol. Plant Path,* 28: 45–47.

Das, B.C., Bora, L.C., Phookan, A.K. and Bhagabati, 1996. Antagonistic effects of *Aspergillus terreus, Trichoderma harzianum* and *Trichoderma viride* on sheath blight of rice. *Oryza,* 33: 62–65.

Das, R.C., Gohain, A.M. and Talukdar, M.C., 1995. Effect of soil reaction on Aspergillus terrius in suppressing rice sheath blight. *J. Agril. Sci.,* 9: 181–191.

Dasgupta, M.K., 1992. Rice sheath blight: The challenge continues. In: *Plant Diseases of International Importance: Disease of Cereals and Pulses,* Vol. I., (Eds.) U.S. Singh, A.N. Mukhopadhyay, J. Kumar, H.S. Chaube. Prentice Hall Englewood Chiffs, New Jersey, pp. 130–157.

Datta, K., Baisakh, N., Thet, K.M. and Tu, J., 2003. Pyramiding transgenes for multiple resistance in rice against bacterial blight, yellow stem borer and sheath blight. *Theor. Appl. Genet.,* 106: 1–8.

Datta, K., Velazhahan, R.N., Oliva, N., Ona, I., Mew, T., Khush, G.S., Muthukrishnan, S. and Datta, S.K., 1999. Over-expression of the cloned rice thaumatin-like protein (PR–5) gene in transgenic rice plants enhances environmental friendly resistance to *Rhizoctonia solani* causing sheath blight disease. *Theor. Appl. Genet.,* 98: 1138–1145.

Datta, S.K., Datta, K., Tu, J., Velazhahan, R., Mew, T.W. and Muthukrishnan, S., 2001. Enhanced resistance to sheath blight by constitutive expression of infection-related rice chitinase in transgenic elite indica rice cultivars. *Plant Sci.,* 160: 405–414.

De'fago, G. and Hass, D., 1990. Pseudomonads as antagonists of soil borne plant pathogens: Modes of action and genetic analysis. In: *Soil Biochemistry,* (Eds.) J.M. Bokag and G. Statzky. Marcel Dekker, New York, 6: 249–292.

Demir, G., Karcilioglu, A. and Onane, A., 1999. Protection of cotton plants against damming–off disease with rhizobacteria. *J. Turkish Phytopath.,* 28: 111–118.

Devi, R.R., Paul, T.S. and Gokulapalan, C., 1987. Efficacy of different fungicides in the control of sheath blight of rice. *Indian J. Plant Prot.,* 15: 69–70.

DeWeger, L.A., Jann, B., Jann, K. and Lugenberg, B., 1987. Lipopolysaccharides of *Pseudomonas* spp. that stimulate plant growth. Composition and use for strain identification. *J. Bactriol.,* 169: 1441–1446.

Dube, H.C., 2001. Rhizobacteria in biological control and plant growth promotion. *J. Mycol. Pl. Path.,* 31: 9–21.

Dubey, S.C., 1995. Evaluation of fungal antagonistic against *Thanatephorus cucumeris* causing banded blight of rice. In: *Paper Presented at the First International Symposium on Rhizoctonia, ISR'95* at Leeuwenhorst Congress Centrum Noordwijkerhout, Netherland, June 27–30.

Dubey, S.C. and Patel,B., 2002. Mass multiplication of antagonists and standardization of effective dose for management of web blight of urd and mungbean. *Indian Phytopath.*, 55: 338–341.

Elad, Y. and Chet, I., 1987. Possible role of competition for nutrients in biocontrol of Pythium damping–off by bacteria. *Phytopathology*, 77: 190–195.

Elad, Y., Katan, J. and Chet, I., 1980. Physical, biological and chemical control integrated for soil borne diseases of potato. *Phytopathology*, 70: 418–522.

Elad, Y., Barak, R., Chet, I. and Henis, Y., 1983. Ultra structural studies of the interaction between *Trichoderma* spp. and plant pathogenic fungi. *Phytopathology*, 107: 168–175.

Elad, Y., Hadar, Y., Chet, I. and Henis, Y., 1982. Prevention with *Trichoderma harzianum* Rf. of reinfestation of *Sclerotium rolfsii* Sacc. and *Rhizoctonia solani* Kuhn of soil fumigated with methyl bromide and improvements of disease control in tomatoes and peanuts. *Crop Prot.*, 1: 199–211.

Gamiliel, A. and Katan, J., 1991. Involvement of fluorescent pseudomonads and other microorganisms in increased growth response of plants in solarized soils. *Phytopathology*, 81: 494–502.

Gangopadhyay, S. and Chakrabarti, N.K., 1982. Sheath blight of rice. *Rev. Plant Path.*, 61: 451–460.

Goel, R.K., Rekhi, R.S. and Singh, P., 2003. Integrated nutrient management in relation to the occurrence of some major diseases and grain yield in rice. *Plant Dis. Res.*, 18: 34–36.

Gogoi, N.K. and Ali, M.S., 2005. Integrated management of sheath blight of winter rice involving *Tricoderma harzinum*, few soil amendments and captan. *Crop Res., Hissar*, 30: 423–427.

Gokulapalan, C. and Nair, M.C., 1992. Phylloplane fungi of rice as mycoparasites of *Rhizoctonia solani. Indian Phytopath.*, 45: 367–368.

Gokulapalan, C. and Nair, M.C., 1984. Antagonism of few fungi and bacteria against *R. solani* Kuhn. *Indian J. Microbiol.*, 24: 57–58.

Groth, D.E. and Rush, M.C., 1988. Foliar fungicides for use in the management of rice disease. *Bulletin Lousiana Agric. Expt. St.*, 840: 43.

Growth, D.E., 1996. Two new fungicides to control rice diseases. *Louisiana Agric.*, 39: 31–33.

Harman, G.E., 2000. Myths and dogmas of biocontrol: changes in the perceptions derived from research on *Trichoderma harzianum* T-22. *Plant Dis.*, 84: 377–393.

Hass, D., Keel, C., Laville, T., Maurhofer, M., Oberhansli, T., Schnider, U., Voisard, C., Wuthrich, B. and Defago, G., 1991. Secondary metabolites of *Pseudomonas fluorescens* strain CHAO involved in the suppression of root diseases. *Adv. Molec. Genet. Plant Microbe Int.*, 1: 450–456.

Huang, C.W, Qian, J.Q., Ge, W.B, Jin, M.S. and Feng, B.C., 1999. Occurrence patterns of rice pests in the rice fields with a simple and labour saving cultivation. *Acta-Agriculturae-Zhejiangensis*, 11: 287–292.

Imam, M.K., Badawy, F.H. and El-Hudairy, A., 1977. Response of tomato and two other vegetable crops to inoculation with Azotobacter. *The Libyan J. Agric.*, 6: 156–167.

IRRI, 1986. *Annual Report for 1985.*

Kim, J.K., Jang, I.C., Lee, Y.H., Ahn, I.P. and Nahm, B.H., 2003. Co-expression of a modified maize ribosome-inactivating protein and a rice basic chitinase gene in transgenic rice plants confers enhanced resistance to sheath blight. *Transgenic Res.*, 12: 475–484.

Klassen, W., 1982. The role of biological control in integrated pest management systems. In: *Biological Control in Crop Production*, (Ed.) G.C. Papavizas. Allanheld, Osum, Granda, U.S.A., pp. 433–466.

Kloepper, J.W., Zablotowicz, R.M., Tripping, E.M. and Lifshitz, R., 1991. Plant growth promotion mediated by bacterial rhizosphere colonizers. In: *The Rhizosphere and Plant Growth*, (Eds.) D.L. Keister and P.B. Cregan. Kluwer Academic Publ., Dordrecht, 315–326 pp.

Kohli, C.K., 1966. Pathogenicity and host range studies on paddy sheath blight pathogen (*R. solani* Kuhn). *J. Res.*, 3: 37–40.

Kousalya, G. and Jeyarajanm, R., 1988. Techniques for mass multiplication of *Trichoderma viride* Pers. Fr. And *T. harzianum* Rafai. In: *Abstracts of Papers Presented in National Seminar on Nanagement of Crop Diseases with Plant Products/Biological Agents*. 10–12 January. Agricultural College and research Institute, Madurai, pp. 32– 33.

Kozaka, T., 1975. Sheath bight in rice plants and its control. *Rev. Plant Prot. Res.*, 8: 69–80.

Krishnamurthy, K. and Gnanamanickam, S.S., 1997. Biological control of sheath blight of rice. Induction of systemic resistance in rice plant associated *Pseudomonas* spp. *Curr. Sci.*, 72: 331–334.

Krishnamurthy, K. and Gnanamanickam, S.S., 1998. Biocontrol of rice sheath blight with formulated *Pseudomonas putida*. *Indian Phytopath.*, 51: 233–236.

Krishnamurthy, K., Balconi, C., Sherwood, J.E. and Giroux, M.J., 2000. Wheat puroindolines enhance fungal disease resistance in transgenic rice. *Mol. Pl. Microbe Inter.*, 14: 1255–1260.

Kumar, B.S.D., Balamani, B. and Bezbaruah, B., 1997. Plant growt promotion and fungal pest control through an antibiotic and siderophore producing fluorescent pseudomonas strain from tea [*Cammellia sinensis* (L.) O'Kuntze] plantations. *Indian J. Expt. Biol.*, 35: 289–292.

Kurschner, E., 1988. Effect of weed control on the development of crop plants and infection with rice blast in upland rice. *Mitteil Biol. Bundes Land Forest. Berlin Dahlem*, 245: 201.

Laha, G.S. and Venkataraman, S., 2001. Sheath blight management in rice with biocontrol agents. *Indian Phytopath.*, 54: 461–464.

Lakpale, N., Thrimurty, V. S. and Kumar, R., 1997. Influence of crop duration on development of sheath blight of rice. *J. Mycol Pl. Pathol.*, 27: 343–344.

Latge, J.P. and Soper, R.S., 1977. Media suitable for industrial production of *Entomophthora virulenta* Sygospores. *Biotech. Bioeng.*, 19: 1269–1284.

Lee, F.N. and Rush, M.C., 1983. Rice sheath blight: a major rice disease. *Plant Dis.*, 67: 829–832.

Lee, F.N. and Courtney, M.L., 1981. Foliar fungicide testing for rice sheath blight control. *Arkansas Farm Sec.*, 30: 11.

Lee, F.N., Dilday, R.H., Moldenhauer, K.A.K., Rutger, J.N. and Yan, W., 1999. Sheath blight and rice blast resistance in recently introduced rice germplasm. *Res. Series Arkansas Agril. Expt Station*, 468: 195–210.

Leeman, M., den Ouden, F. M., van Pelt, J.A., Cornellissen, C., Matamallagarros, A., Baker, P.A.H.M. and Schippers, C., 1996. Suppression of Fusarium wilt of radish by co–inoculation of fluorescent *Pseudomonas* spp. and root colonizing fungi. *Eur. J. Plant Path.*, 102: 21–31.

Lewis, J.A. and Papavizas, G.C., 1984. A new approach to stimulate population of *Trichoderma* spp. and other potential biocontrol fungi introduced into natural soils. *Phytopathology*, 74: 1240–1244.

Li, X.X., He, Y.C. and Gao, B.D., 2003. Studies on target genes transferred by rice floral organ-mediated method. *Chinese J. Rice Sci.*, 17: 196–200.

Linscombe, S.D., Jodar, F., Bollich, P.K., Groth, D.E., White, L.M., Chou, O.R., Dunand, R.T. and Sanders, D.E., 2001. Resistration of 'Earl' rice. *Crop Sci.*, 41: 2003–2004.

Loper, J.E., Hoak, C. and Schroth, M.N., 1985. Population dynamics of pseudomonads in rhizosphere of potato (*Solanum tuberosum* L.). *Appl. Environ. Microbiol.*, 49: 416–422.

Mafia, R.G., Alfenas, A.C., Mafia, L.A., Ventura, G.M. and Sanfuentes, E.A., 2003. Encapsulation of *Trichoderma hamitum* for the biocontrol of *Rhizoctonia solani* in colonal propagation of Eucalyptus. *Fitopathologia Brasileira*, 28: 101–105.

Malik, G., Dawar, S., Sattar, A. and Dawar, A., 2005. Efficacy of Trichoderma after multiplication on different substrate in the control of root rot fungi. *Int. J. Biol. Biotech.*, 2: 237–242.

Manibhushanrao, R., Prasad, S., Baby, U.I. and Jay, Y., 1989. Susceptibility of rice sheath blight pathogen to mycoparasites. *Curr. Sci.*, 58: 515–518.

Mathivanum, N., Srinivasan, K. and Chelliah, S., 1997. Evaluation of fungicides and their combinations with neem oil for the management of rice diseases. *Pestology*, 21: 16–17.

Mathur, S.C., 1983. Fungal diseases of rice in India. In: *Recent Advance in Plant Pathology*. Print House, Lucknow (India), 368p.

McClung, A.M., Marchetti, M.A., Webb, B.D. and Bollich, C.N., 1998. Regisration of 'Dixiebelle' rice. *Crop Sci.*, 38: 898.

Meena, B. and Muthusami, M., 1998. Efficacy of different mode of application of *Pseudomonas fluorescens* against rice sheath blight disease. *Oryza*, 35: 293–294.

Meena, B., Ramamurthy, V., Marimutghu, T. and Velazhahan, R., 2000. *Pseudomonas fluorescens* mediated systemic resistance against late leaf spot of groundnut. *J. Mycol. Pl. Path.*, 38: 898.

Meena, B., Ramamoorthy, V. and Muthusamy, M., 2002. Effect of some plant extracts on sheath blight of rice. *Curr. Res.*, 31: 49–50.

Meena, B. and Muthusamy, M., 1999. Effect of organic soil amendments against rice sheath blight. *Indian Phytopath.*, 52: 92–93.

Mew, T.W. and Rosales, A.M., 1986. Bacterization of rice plants for control of sheath blight caused by *Rhizoctonia solani*. *Phytopathology*, 76: 1260–1264.

Mishra, D.S. and Sinha, A.P., 1999. Laboratory evaluation of fungicides against *Rhizoctonia solani* Kuhn, the cause of sheath blight of rice. *Agric. Sci. Digest*, 19: 211–213.

Mishra, D.S. and Sinha, A.P., 2000. Comparative efficacy of some biocontrol agents applied as seed or soil treatment on seedling infection of rice caused by *Rhizoctonia solani*. In: *52nd Annual Meeting of Indian Phytopathological Society and National Symposium on "Role of Resistance in Intensive Agriculture"* held at Karnal, Haryana, Feb. 15–17, (Abstr.).

Mishra, D.S. and Sinha, A.P., 1997. Biocontrol potential of some biocontrol agents against *Rhizoctonia solani* Kuhn, the cause of sheath blight of rice. In: *International Conference on Integrated Plant Disease Management for Sustainable Agriculture*, New Delhi, 10–15, November, Abstr. pp. 209.

Mishra, D.S. and Sinha, A.P., 1998. Plant growth promotion by some biocontrol agents. In: *Fiftieth Annual Meeting of Indian Phytopathological Society and National Symposium on "Present Scenario in Diseases of Oilseeds and Pulses"* held at Aurangabad, Maharastra, 17–19 Feb. (Abstr.).

Mishra, D.S., Singh, U.S. and Dwivedi, T.S., 2001. Comparative efficacy of normal seed treatment and seed bio–priming with commercial formulations of *Trichoderma* spp. In: *53rd annual meeting of Indian Phytopathological Society and National Symposium on "Ecofriendly Approaches for Plant Disease Management"* held at Chennai, Jan. 21–23.

Mishra, S.K., Karmakar, S.K., Naik, R.P. and Dash, B.K., 1989. Chemical control of sheath blight of rice. *Orissa J. Agric. Res.*, 2: 199–201.

Misra, B., 2005. More crop per drop. *Survey of Indian Agriculture, The Hindu*, p. 41–46.

Miyake, I., 1990. Studies-uber die reispflance in Japan. *J. Cell. Agric. Imp. Univ.*, 2: 237–276.

Mizuta, H., 1956. On the relation between yield and inoculation times of sheath blight *Corticum sasakii* in the earlier planted paddy rice. *Assoc. Plant Prot.*, Kyushu, 2: 100–102.

Mukherjee, P.K., 1991. Biological control of chickpea wilt complex. *Ph.D. Thesis*, G.B. Pant University of Agriculture and Technology, Pantnagar, India. 188 p.

Mukherjee, P.K. and Mukhopadhyay, A.N., 1995. *In situ* mycoparasitism of *Gliocladium virens* on *Rhizoctonia solani*. *Indian Phytopath.*, 48: 101–102.

Mukhopadhyay, A.N. and Mukherjee, P.K., 1998. Biological control of plant diseases: Status in India. In: *Biological Suppression of Phytopathogens, Nematodes and Weeds*, (Eds.) S.P. Singh and S.S. Husain. Project Directorate and Biological Control, Bangalore, pp. 107–135.

Mukhopadhyay, A.N., 1996. Recent innovation in plant disease control by ecofriendly biopesticides (Presidential Address). In: *83rd Annual meeting. Indian Science Congress*, Patiala, January 1–8.

Mukhopadhyay, A.N., Shrestha, S.M. and Mukherjee, P.K., 1992. Biological seed treatments for control of soil borne plant pathogens. *FAO Plant Prot. Bull.*, 40: 1–10.

Naidu, V.D. and Venkatrao, G., 1997. Comparative evaluation of different fungicides against sheath blight of rice. *Pestology*, 21: 16–20.

Naidu, V.D., 1992. Influence of sheath blight of rice on grain and straw yield in some popular local varieties. *J. Res., Assam Agric. Univ.*, 10: 78–80.

Nanadkumar, R., Babu, S., Viswanathan, R., Raguchander, T. and Samiyappan, R., 2001. A new bio-formulation containing plant growth promoting rhizobacterial mixture for the management of sheath blight and enhanced grain yield in rice. *Biocotnrol*, 46: 493–510.

Neilands, J.B. and Leong, J., 1986. Siderophores in relation to plant growth and diseases. *Ann. Rev. Plant Physiol.*, 37: 187–208.

Ou, S.H. and Bandong, J.M., 1976. Yield losses due to sheath blight of rice. *Int. Rice Res. Newslt.*, 6: 14.

Ou, S.H., 1985. *Rice Diseases*, 2nd edn. Common Wealth Mycological Institute, Kew, 272 p.

Papavizas, G.C. and Lewis, J. A., 1982. Introduction and augmentation of microbial antagonists for the control of soil borne plant pathogens. In: *Biological Control in Crop Production*, (Ed.) G.C. Papavizas. Allanheld, Osmum, Granada, U.S.A., pp. 305–322.

Paracer, C.S. and Chahal, D.S., 1963. Sheath blight of rice caused by *R. solani* Kuhn. A new record in India. *Curr. Sci.*, 32: 328–329.

Pathak, M., Garg, D.K., Jeswani, M.D. and Singh, R.N., 2002. Farmers participatory IPM in Basmati rice, a case study. *Annal Pl. Protec. Sci.*, 10: 386–388.

Pattanayak, A., Bhuyan, R.N., Gupta, H.S., Sreedhar, M. and Prasad, M.S., 2000. Performance of cold-tolerant rice lines developed through anther culture for mid-altitude areas of Meghalaya, India. *Int. Rice Res Notes*, 25: 11–12.

Polo, M.A., 1926. Rhizoctonia disease of rice. I. A study of the influence of certain conditions upon the viability of the sclerotial bodies of causal funguss. *Phillippines Agriculturists*, 15: 361–375.

Prasad, M.S., Lakshmi, B.S., Shaik, M. and Mohiddin, S., 1998. Survival of sclerotia of *Rhizoctonia solani* in rice soils amended with oil cakes and green leaf manures. *Annals of Agric. Res.*, 19: 44–48.

Prikryl, Z. and Vancura, V., 1980. Root exudates of plants. IV wheat root exudation as dependent on growth concentration gradient of exudates and the presence of bacteria. *Plant and Soil*, 57: 69–83.

Rabindran, R. and Vidhysekaran, P., 1996. Development of a formulation of *Pseudomonas fluorescens* of PfALR2 for management of rice sheath blight. *Crop Prot.*, 15: 715–721.

Rajan, K.M. and Menon, M.R., 1975. Effect of organic soil amendment on plant growth and intensity of sheath blight of rice. *Agric. Res. J. Kerala*, 13: 179–181.

Ramaezani, H., Singh, H.P. , Batish, D.R., Kohli, R.K., Dargan, J.S. and Zydenbos, S.M., 2002. Fungicidal effect of volatiles oils from *Eucalyptus citriodora* and its constitutuents citronellal. In: *Proc. of a Conference* held at Rotoura, New Zealand, pp. 327–330.

Ranjana, S., Phookan, A.K., Bhagabati, K.N. and Sarma, R., 1999. Efficacy of some plant extracts in the management of sheath blight disease or rice. *J. Mycol. Pl. Path.*, 29: 336–339.

Rao, C.V.S. and Johri, B.N., 1999. Seed and root extracts in chemotaxis, agglutination, adherence and root colonization of soybean (*Glycine max*) by fluorescent pseudomonads. *Indian J. Microbiol.*, 39: 31–38.

Ravindran, R. and Vidhyasekaran, P., 1996. Development of a formulation of *Pseudomonas fluorescens* PfA1R2 for management of rice sheath blight. *Crop Prot.*, 15: 715–721.

Raymond, R.N., Muller, G. and Matzanke, B.F., 1984. Complexation of iron by siderophores: A review of their solution and structural chemistry and biological function. *Trop. Curr. Chem.*, 123: 49–102.

Reddy, C.S., Sudhakar, R. and Manoharachary, C., 2002. Efficacy of plant products and other chemicals in the management of sheath blight of rice. In: *Frontiers in Microbial Biotechnology and Plant Pathology*, (Eds.) D.K. Purohit, S. Reddy, M.A Singaracharya and S.S. Girisham. Scientific Publishers (India), Jodhpur, India, pp. 263–267.

Reinking, O.A., 1918. Philippines economic pant disease. *Philippines J. Sci.*, 843: 165–274.

Ristaino, R. and Thomas, P., 1997. Agriculture methyl bromide and ozone hole. Can we fill the gaps? *Plant Dis.*, 81: 964–977.

Rodriguez, H.A., Nass, H., Cardona, R. and Aleman, L., 1999. Alternatives to control sheath blight caused by *Rhizoctonia solani* in rice. *Fitopatologia-Venezolana*, 12: 18–21

Rosales, A.M. and Mew, T.W., 1982. Antagonestic effects of soil microoganisms on rice sheath blight pathogen. *Int. Rice Res. Newsl.*, 7: 12–13.

Rosales, A.M., Vantomma, R., Swing, J.D.E. and Mew, J.W., 1993. Identification of some bacteria from paddy antagonistic to several rice pathogens. *J. Phytopath.*, 138: 189–208.

Roy, A.K., 1993. Sheath blight of rice in India. *Indian Phytopath.*, 46: 45–49.

Roy, A.K., 1989. Source of seed borne infection of sheath blight in rice. *Oryza*, 26:111–112.

Saikai, P. and Baruah, P., 1990. Linear model for predicating yield loss in rice (*Oryza sativa*) due to sheath blight (*Rhizoctonia solani*) incidence. *Indian Agric. Sci.*, 60: 360–361.

Sakai, M., Shinanda, H., Hoshina, T. and Okamotos, M., 1989. A high yielding *indica-japonica* hybrid rice variety Hoshiyutaka. *Japan Agric. Res. Quart.*, 23: 81–85.

Saksena, H.K. and Chaube, R.D., 1972. Banded blight disease of paddy. In: *Paper presented in the All Indian Co-operative Rice Research Workshop*, 1972, Hyderabad (India).

Sankar, P. and Jayarajan, R., 1996. Seed treatment with *Trichoderma* and Gliocladium for biological control of *Macrophomena phaseolina* in seasamum. *Indian Phytopath.*, 49: 148–151.

Sarkar, S.C. and Chowdhury, A.K., 2003. Spread of rice sheath blight under field situation. *Environ. Ecol.*, 21: 632–638.

Satya, V.K., Radhajayalakshmi, R., Kavitha, K., Paranidharan, V., Bhaskaran, R. and Velazhahan, R., 2005. *In vitro* antimicrobial activities of zimmu (*Allium sativum, Allium cepa* L) leaf extract. *Arch. Phytopath. Pl. Protec.*, 38: 185–192.

Schiper, B., Baker, A.W., and Baker, P.A.H. M., 1987. Interaction of detererious and beneficial rhizosphere microorganism and the effect of cropping practice. *Ann. Rev. Phytopath.*, 25: 339–358.

Sharma, D. and Thrimurty, V.S., 2005. Protection and therapeutant ability of validamycin against sheath blight of rice. *Env. Ecol.*, 23: 933–934.

Sharma, M.K., 1996. Assessment of yield losses and chemical control of sheath blight of rice caused by *Rhizoctonia solani. M.Sc. Thesis*, G.B. Pant University of Agriculture and Technology, Pantnagar. pp. 72.

Shivakumar, G. and Narayanaswamy, N.T., 1998. Biological control of sheath blight of rice with *Pseudomonas fluorescens. Oryza*, 35: 57–60.

Shylaja, M., Ranakausar, S.M. Reddy, Shrivastava, H.P., Purohit, D.K., Reddy, S. Ram, 1997. Antifungal activity of some leaf extracts on the growth and sclerotial production of *R. solani. Microbiol. Biotech.*, 143–144.

Singh, Rajbir and Sinha, A.P., 2005b. Influence of application methods of *Pseudomonas fluorescens* on rice sheath blight. *Indian Phytopath.*, 58: 474–476.

Singh, N.I., Devi, R.K.T. and Singh, K.V., 1988. Occurrence of rice sheath blight (ShB) *Rhizoctonia solani* Kuhn on rice panicles in India. *Int. Rice Res. Newslt.*, 13: 29.

Singh, N.I., Devi, R.K.T. and Singh, K.V., 1989. *Rhizoctonia solani:* An agent of rice boot blight. *Int. Rice Res. Newslt.*, 14: 22.

Singh, Rajbir and Sinha, A.P., 2005c. Influence of time of application of *Pseudomonas fluorescens* in suppressing sheath blight of rice. *Indian Phytopath.*, 58: 30–34.

Singh, Rajbir and Sinha, A.P., 2006. Effect of *Pseudomonas fluorescens* on sheath blight of rice in nursery. *Annl. Pl. Protec. Sci.*, 14: 264–265.

Singh, Rajbir, 2003. Factors affecting biocontrol potential of *Pseudomonas fluorescens* against sheath blight of rice. *Ph.D. Thesis*, G.B. Pant University of Agriculture and Technology, Pantnagar, 262 p.

Singh, Rajbir and Sinha, A.P., 2002. Effect of different methods of application of *fluorescent pseudomonads* against sheath blight of rice. In: *54th Annual Meeting of IPS and National Symposium on CropsProtection and WTO An Indian Prospective* held at Kassargad January, 22–25.

Singh, Rajbir and Sinha, A.P., 2005e. Influence of soil nutrition on the effectivity of *Pseudomonas fluorescens* on rice sheath blight. *Indian Phytopath.*, 58: 477–479.

Singh, Rajbir and Sinha, A.P., 2003. Plant growth promoting activity of *Pseudomonas fluorescens* in rice seedlings. In: *Paper Presented in 5th National Symposium on Biocontrol Agents for Sustainable Management of Pests* held at G.B. Pant University of Agriculture and Technology, Pantnagar, Dec. 18–20.

Singh, Rajbir and Sinha, A.P., 2004. Comparative efficacy of local bioagents, commercial bio-formulations and fungicide for the management of sheath blight of rice, under glass house condition. *Indian Phytopath.*, 57: 494–496.

Singh, Rajbir and Sinha, A.P., 2005 a. Effect of *Pseudomonas fluorescens* formulations on sheath blight and yield component of rice. *Annl. Pl. Protec.*, 13: 159–162.

Sinha, A.P., 2000. Evaluation of new fungicidal formulations for sheath blight control in kharif 2000. *Progress Report Directorate of Rice Research*, Hyderabad, pp. 381.

Sitansu, P., Hazra, S., Roy, A. and Pan, S., 2001. Efficacy of iprodione and carbendazim against rice sheath blight and their compatibility with biocontrol agents. *J. Interacademicia*, 5: 11–18.

Sivakumar, G., Sharma, R.C. and Rai, S.N., 2000. Biocontrol of banded leaf and sheath blight of maize by peat based *Pseudomonas fluorescens* formulation. *Indian Phytopath.*, 53: 190–192.

Snyder, W.C., 1960. In: *Biological Control of Plant Pathogens*, (Eds.) K.F. Baker and R.J. Cook. W.H. Freeman and Co., San Francesco, 433 pp.

Streeter, T.C., Rengel, Z., Meate, S.M. and Graham, R.D., 2001. Zinc fertilization increases tolerance to *Rhizoctonia solani* (AG8) in Medicago trucalula. *Plant and Soil*, 228: 233–242.

Sudhakar, R., Rao, K.C. and Reddy, C.S., 2000. Evaluation of medium duration rice varieties against sheath blight disease. *Indian J. Pl. Protc.*, 28: 215–217.

Sudhakar, R., Rao, K.C. and Reddy, C.S., 2005. Chemical control of rice sheath blight incited by *Rhizoctonia solani* Kuhn. *Res. Crops*, 6: 343–348.

Sunderraj, T., Kurucheve, V. and Jayaraj, J., 1996. Screening of higher plant and animal faces for the fungi toxicity against *Rhizoctonia solani*. *Indian Phytopath.*, 49: 398–403.

Surulirajan, M and Kandhari, J., 2003. Effects of integrated treatments against rice sheath blight severity, disease incidence and yield parameters. *Annl. Pl. Protec Sci.*, 11: 284–288.

Suslow, T.V. and Schroth, M.N., 1982. Rhizobacteria of sugarbeet: Effect of seed application and root colonization on yield. *Phytopathology*, 72: 199–206.

Templeton, G.E. and Johnston, T.H., 1969. Brown-bordered leaf and sheath spot on rice. *Arkansas Farm Res.*, 18: 5.

Tewari, A.K., 1996. Biological control of chickpea wilt-complex using different formulations of *Gliocladium virens* through seed treatment. *Ph.D. Thesis*, G.B. Pant University of Agriculture and Technology, Pantnagar, India, 167 p.

Tewari, L., Chaube, H.S., Singh, Rajbir, Kaur, P. and Singh, S., 2003. Optimization of conditions for conidial yield of *Trichoderma harzianum*. In: *Paper Presented in 5th National Symposium on Biocontrol Agents for Sustainable Management of Pests* held at G.B. Pant University of Agriculture and Technology, Pantnagar, Dec. 18–20.

Torabi, M. and Binesh, H., 1987. Effect of some fungicides on growth of R. solani causal organism of sheath blight disease of rice, media and in the field. *Entomologie et Phytopathologie Appliques*, 54: 51–52.

Trane, C., Nielsen, M.N., Sorensen, J. and Olsson, S., 2001. *Pseudomonas fluorescens* DR54 reduces sclerotia formulation, biomass development and incidence of *Rhizoctonia solani* causing damping-off in sugarbeet. *Microbial Ecol.*, 42: 438–445.

Trichoderma spp. In: *54th Annual Meeting of IPS and National Symposium on Crop Protection and WTO an Indian Prospective* held at Kassargad Jan. 22–25.

Tripathi, M., 1999. Assessment of biocontrol potential of *fluorescent pseudomonads* for some fungal pathogens. *Ph.D. Thesis*, G.B. Pant University of Agriculture and Technology, Pantnagar, 171 p.

Van Peer, R., Niemann, G.J. and Schippers, B., 1991. Induced resistance and phytoalexin accumulation in biological control of Fusarium wilt of Carnation by *Pseudomonas* spp. strain WCS 417r. *Phytopathology*, 81: 728–734.

Velazhahan, R, Radhajeyalakshmi, R., Thangavelu, R. and Muthukrishnan, S., 2001. An antifungal protein purified from pearl millet seeds shows sequence homology to lipid transfer proteins. *Biologia Plantarum*, 44: 417–421.

Vidhyasekaran, P. and Muthamilan, M., 1995. Development of formulations of *Pseudomonas fluorescens* for control of chickpea wilt. *Plant Dis.*, 79: 782–786.

Vidhyasekaran, P. and Muthamilan, M., 1999. Evaluation of powder formulation of *Pseudomonas fluorescens* Pf l for control of rice sheath blight. *Biocontrol Sci. Technol.*, 9: 67–74.

Vidhyasekaran, P., Rabindran, R., Muthamilan, M., Nayar, K., Rajappan, K., Subramaniam, N. and Vasumathi, K., 1997. Development of powder formulation of *Pseudomonas fluorescens* for control of rice blast. *Plant Path.*, 46: 291–297.

Voisard, C., Keel, C., Hars, D. and Defago, G., 1989. Cyanide production by *Pseudomonas fluorescens* helps suppress black root rot of tobacco under genotobiotic conditions. *EMBO J.*, 8: 351–358.

Wang, T.C., Deng, C.R., Du, Z.L., Ding, X.T., Shi, R.Q. and Huang, H.M., 1999. occurrence of major pests in rice fields with a new cultivation pattern, direct seedling on ridged plots without ploughing and the optimal control technique of rice pests in the southwest mountain area of Zheijiang province. *Acta Agriculture Zhe Jiangensis*, 11: 356–359.

Wei, G., Kloepper, J.W. and Tuzum, S., 1996. Induced systemic resistance to cucumber diseases and increased plant growth by plant growth promoting rhizobacteria under field conditions. *Phytotathology*, 86: 221–224.

Weller, J.M., 1988. Biological control of soil borne plant pathogens in the rhizosphere with bacteria. *Ann. Rev. Phytopath.*, 26: 379–407.

Windham, M.T., Elad, Y. and Baker, R., 1986. A mechanism for increased plant growth induced by *Trichoderma* spp. *Phytopathology*, 76: 518–521.

Wiwattanapatapee, R.A., Pengnoo, M., Kanjanamaneesathian, W., Matchavanich, Nilratana and Jantharangsri, A., 2004. Floating pellets containing bacterial antagonist for control sheath blight of rice: formulations, viability and bacterial release studies. *J. Controlled Release*, 95: 455–462.

Insect Pest and Disease Management *Pages* **316–330**
Editor: **Dr. D. Prasad**
Published by: **DAYA PUBLISHING HOUSE (2016)**

Chapter 22

Integrated Disease Management on Maize

Shahid Ahamad
Regional Agricultural Research Station, SKUAST-J, Rajouri – 185 131, J&K
E-mail: shahid_2056 @yahoo.com

Introduction

In Jammu and Kashmir, maize being the staple food of intermediate zone is grown in 3 m ha with an acreage of approximately 2 m ha in Jammu division. The district wise area in Jammu division under this crop in Udhampur, Doda, Rajouri, Jammu, Poonch and Kathua is 55.58, 48.19, 44.89, 26.77, 23.67 and 15.89 thousand hectares, respectively (Directorate of Economics and Statistics J&K, 1999).

In recent years, the average yield of maize has declined and this is accounted by biotic, abiotic and nutritional stresses. Among the different stresses, biotic factors are the major constraints in the production and productivity of maize (Desai and Hegde, 1990). Due to the attack of several plant diseases caused by fungi, bacteria, viruses and possibly nematodes, yields are often reduced considerably (Rangarajan and Chakrabarty, 1969). Diseases are important constraints to maize production, wherever the crop is grown. The importance and magnitude of damage depend on interaction between the host plant and the pathogen under a given set of environmental conditions. Losses due to corn diseases are estimated at 9.4 per cent per year worldwide, 12 per cent in Asia and 14 per cent in Africa (Cramer, 1967). On a country basis, losses are estimated at 12 per cent in the USA and 13.2 per cent in India (Payak and Sharma, 1985). Several diseases, under favourable conditions, have the potential to completely destroy the crop or to render the produce unfit as human food or animal feed. The present findings are in accordance with Shurtleff (1980), Ullstrup (1977), De Leon (1984), Payak and Sharma (1982).

The incidence and economic impact of corn diseases in India vary from state to state (Table 22.1). A large number of pathogens infest maize in the tropical lowland, subtropical, intermediate, transition zone and high land maize growing ecological zones covered by CIMMYT's mandate. In 1988, a study was conducted by CIMMYT where information was compiled on maize production along with perceived constraints for each of these defined zone (Tables 22.2 and 22.3). An attempt has been made to estimate losses caused by disease outbreaks and incidence in the crop in terms of shelled grain. Potential loss that can be caused by diseases has been worked out individually for 10 diseases (Table 22.4.) Based on experimental work and other observations, estimates have been prepared for India as a whole (Table 22.5).

Table 22.1: Distribution of Maize Diseases in India

Diseases	*India*	*J&K*
	Viruses	
SCMV/MDMV	+	
MStpV	+	
	Downy mildew	
Sorghum (*P. sorghi*)	+++	
Java (*P. maydis*)	+	
Philippines (*P. phillipinensis*)	+	
Sugarcane (*P.s acchari*)	++	
Brown stripe (*P. rayssiae*)	+++	+++
	Foliar diseases	
Northern leaf blight	+++	+++
Southern leaf blight	+++	+++
Diplodia leaf blight	+	
Brown spot	+++	+++
Curvularia leaf spot	++	
Gray leaf spot	+	
Banded leaf and sheath blight	+++	+++
	Rust and smuts	
South rust	++	
Common rust	+++	+
Common smut	+	+++
Head smut	+	+++
	Stalk/Root rots	
Bacterial	+++	
Diplodia	+	
Fusarium	+++	+++
Charcoal	+++	
Phythium	+++	+++
Black bundle/late wilt	++	
	Ear and Kernel diseases	
Rhizoctonia	+++	
Fusarium	+++	+++
Aspergillus	++	++

+: Not important; ++: Moderately important; +++: Important.

Table 22.2: Percentage of Total Area in Tropical Lowland Maize with Reported Economic Losses to Diseases (Worldwide)

Diseases	*% Losses*
Maydis leaf blight (*B.maydis*)	25.5
Polysora rust (*P.polisora*)	33.3
Downy mildew (all genera and spp.)	30.5
Maize streak virus	17.6
Corn stunt complex (CSS,MBS,MRFV)	17.6
Ear rots (*Fusarium* spp., *Stenocarpella* spp., *M. phaseoli, C. maydis, A. strictus*)	26.7

Source: CIMMYT Maize Program, 1988. Maize production region in developing countries, Mexico.D.F.

Table 22.3: Percentage of Total Area in Sub-tropical, Mid Altitude, Transition Zone and High Land Maize with Reported Economic Losses to Diseases (Worldwide)

Diseases	*% Losses*
Turcicum leaf blight (*E.turcicum*)	58.6
Common rust (*P. sorghi*)	34.2
Downy mildew (all genera and spp.)	29.4
Maize streak virus	15.5
Corn stunt complex (CSS, MBS, MRFV)	1.8
Ear rots (*Fusarium* spp., *Stenocarpella* spp., *Aspergillus* spp.	26.7
Stalk rots (*Fusarium* spp., *Stenocarpella* spp., *M. phaseolina, C. maydis, A.s trictus, Pythium* spp. and *E. chrysanthemi*)	32.1

Source: CIMMYT Maize Program, 1988. Maize production region in developing countries, Mexico.D.F.

Table 22.4: Yield Losses by 10 Diseases Under Experimental Conditions in Susceptible Cultivars of Maize

Diseases	*% Losses*
Bacterial stalk rot	100
Pythium stalk rot	100
Late wilt	50.9
Charcoal rot	39.5
Maydis leaf blight	30.3
Turcicum leaf blight	66.0
Common rust	32.0
Brown spot	27.0
Brown stripe downy mildew	63.0
Banded leaf and sheath blight	40.5

Table 22.5: Estimation of Losses Due to Major Diseases of Maize in India

Diseases Group	*% Losses*
Seed and seedling blight	0.1
Downy mildews	2.1
Foliar diseases (Turcicum leaf blight, Maydis leaf blight, Phaseospharia leaf spot, rust, Brown stripe downy mildew)	5.0
Stalk rots, root rots, ear rots	5.0
sheath blight, smut	1.0
Viruses and other	13.2

Major Diseases and their Distribution

About 72 pathogens have so far been reported causing substantial damage in maize, 16 have been identified as major ones and are given here. These includes diseases which occur in both tropical and temperate environments.

1. Turcicum leaf blight (*Exherohilum turcicum*) occurs in Jammu and Kashmir (J&K), Himanchal Pradesh (H.P.), Sikkim, West Bengal (W.B.), Meghalaya, Tripura, Assam, Rajasthan, Uttar Pradesh (U.P.), Bihar, Madhya Pradesh (M.P.), Gujrat, Rajasthan, Andhra Pradesh (A.P.), Karnataka, Tamil Nadu (T.N).
2. Maydis leaf blight (*Drechslera maydis*), occurs in J&K, H.P., Sikkim, Meghalaya, Rajasthan, Punjab, Haryana, Delhi, U.P., Bihar, M.P., Gujrat, Rajasthan, A.P., Maharastra.
3. Common rust (*P.sorghi* II,III stage only) occurs in J&K, H.P., Sikkim,W.B., Rajasthan, Punjab (winter), Haryana (winter), Delhi, U.P., Bihar, M.P., Gujarat, A.P., Maharastra, TN.
4. Phaseospharia leaf spot (*Phaseospharia maydis*) is found in Sikkim, W.B. U.P., Karnataka, Maharastra.
5. Erwinia stalk rot (*Erwinia chrysanthemi* pv. zeae) occurs in H.P., Sikkim, W.B. Rajasthan, Punjab, Haryana, Delhi, U.P., Bihar, M.P., Rajasthan, A.P.
6. Pythium stalk rot (*Pythium aphanidermatum = P.butleri*) is found in Sikkim, W.B., Rajasthan, Punjab, Haryana, Delhi, U.P., Bihar.
7. Fusarium wilt and stalk rot (*Fusarium moniliforme*) occurs in Rajasthan, U.P., Bihar, M.P., A.P., Karnataka,T.N.
8. Late wilt (*Cephalosporium maydis*) is reported from Rajasthan, U.P., Bihar, A.P.
9. Charcoal rot (*Macrophomina phaseolina*) occurs in J&K, Punjab, Haryana, Delhi, U.P., M.P., A.P., Karnataka, T.N.
10. Brown stripte downy mildew (*Sclerophthora rayssiae* var. *zeae*) occurs in J&K, H.P., Sikkim, W.B., Rajasthan, Punjab, Haryana, Delhi, U.P., Bihar, Meghalaya, M.P., Gujarat.
11. Phillipine downy mildew (*Peronosclerospora phillipinensis*) occurs in Punjab, Haryana, Delhi, U.P., M.P., Bihar.
12. Sorghum downy mildew (*Peronosclerospora sorghi*) occurs in Rajasthan, A.P., Maharashtra., Gujarat, Karnataka, T.N.
13. Sugarcane downy mildew (*Peronosclerospora sacchri*) is found in U.P., Bihar.

14. Brown spot (*Physoderma maydis*) occurs in J&K, H.P., Sikkim, Rajasthan, Punjab, M.P., W.B., Rajasthan, Maharastra, Karnataka,.
15. Banded leaf and sheath blight (*Rhizoctona solani* f.sp. *sasakii*) occurs in J&K, Sikkim, Rajasthan, Punjab, M.P., Haryana, Rajasthan, U.P. and Delhi.
16. Cyst nematode (*Heterodeara zeae*) occurs in Rajasthan, M.P., Bihar, U.P.

Some Major Epidemics

Six major disease out breaks have been recorded in India since the sixties.

1. Bacterial stalk rot (*Erwinia chrysanthemi* pv. *zeae*) appeared in the foothills (*tarai*) area of U.P. in 1959 and in 1969 in H.P. on Hybrid Ganga.
2. Brown stripe downy mildew (*Sclerophthora rayssiae* var. *zeae*) appeared in 1962 and 1964 in WB hills, 1964 in western MP and south-eastern Gujrat,1964 in north Bihar and in 1965 in the *tarai* region of UP and parts of Himanchal Pradesh.
3. There was an out break of Sorghum downy mildew (*Peronosclerospora sorghi*) in Rajasthan, Karnataka and Tamil Nadu from 1968 to 1973.
4. Common rust appeared from 1973 onwards in the winter (Rabi) season on hybrids GS-2 and Hi-starch in Bihar.
5. Turcicum leaf blight (*Exherohilum turcicum*) virulent strain of *E. turcicum* appeared in 1974 on hybrid Ganga5, Deccan, Piox104, X114 and Kisan and in 1989-90 in Bihar.
6. Banded leaf and sheath blight (*Rhizoctona solani* f.sp. *sasakii*) In the late sixties there was an outbreak of this disease in HP, and other areas of western Himalaya.

One of the cheapest and best method of disease control is the use of resistant varieties but the break down of resistance year after year resulting problem in gene pyramiding has restricted this approach.Chemical control is practically feasible but non judicious and untimely application has made this approach ecologically and economically unsuitable on the other hand, biological control is ecologically and socially acceptable but difficulty in having an effective bio-control agent as well as lack of awareness among the farmers has limited its inclusion in regular disease management schedule of crop. Considering all these constraints it is imperative formulate an IDM programme incorporating all the conventional disease management approaches in such a way so that the limitation of one approaches in such a way so that the limitation of one approach can be neutralized with other one.

Integrated Disease Management (IDM): Concepts and Principles

The Integrated Disease Management (IDM) involves the need based used of pesticide only when the disease incidence reach economic threshhold levels and this will promote the build up of many bio-control agent in the crop ecosystems. Thus IDM, a greener alternative to the conventional use of chemicals, is an attempt to promote natural, economic and sociological farming methods through the most effective combination of farming techniques and judicious and limited use of fungicide. In other words, IDM programme implies all the available disease management approaches including cultural, biological and chemical control with the main objective to keep the disease incidence below economic threshold level.

IDM: Components

Broadly IDM involves four components namely:

1. Host resistance
2. Biological control
3. Cultural control and
4. Chemical control

Host Resistance

Resistance is the character of a plant which suppress pathogen and disease development. The use of resistant genotypes is a highly effective approach to suppress disease to tolerable levels. Resistance alone is sometimes sufficient to suppress disease to tolerable levels. In resistant genotypes, disease appearts late, build up slowly and results in little damage to the crop.

Biological Control

In this method the pathogen actively is reduced through the use of other living organisms *e.g.* hyper-parasites, resulting in a reduction of disease incidence and severity.

Cultural Control

This involves deliberate manipulation of the crop environment to make it less favourable to harmful organism *e.g.* by disrupting their reproduction cycles, eliminating their food sources, or encouraging their natural enemies. Included in this method are such practices as intercropping, crop rotation, field sanitation, manipulation of sowing dates etc. some of these techniques provide only small benefits when integrated with other techniques, they significantly improve disease management.

Chemical Control

In IDM programme chemical control implies the judicious and need based use of pesticides (fungicides, insecticides and herbicides). Chemical control is essential in areas where diseases appear in the early stage of plant growth and environmental conditions are likely to spread them fast.

Foliar Diseases

Of the many foliar diseases occurring in India, three may be considered important based on geographical distribution and yield loss potential. Northern leaf blight (*Helminthosporium turcicum=Exserohilum turcicum*) is a major problem in the region with cooler environment like J&K, H.P. etc. under conditions favourable for disease development, susceptible materials suffer heavy blightening resulting in premature drying of leaves and lightweight kernels. It is a major disease in the mountainous areas and in the deep hillside areas of Rajouri and Poonch districts, especially in the cool and shady environment of intermediate zone of J&K. The highest disease intensity *i.e.* 66 per cent was recorded in intermediate zone of Jammu region (Ahamad, 2004, 2005).

The foliar blight disease complex in cool conditions favoured for turcicum leaf blight. At the Regional Agricultural Research Station, Rajouri and MBRSS, Poonch, it was moderate to high. The natural disease pressure every winter cycle allowed for the selection of disease resistant germplasm. It is important that, where a disease outbreak is feared, spraying should be undertaken at an early stage of crop growth, *i.e.* when it is not more than 30 days old (knee-high stage). For large scale high volume sprays 1.8-2.6 kg of fungicide in 900-1200 litre of water/ha is recommended, if the plant population is around 56,000/ha.

Control of foliar blight disease can be achieved using host resistance which has been quite successful, chemical control is also feasible, it becoming profitable on high volume materials like popcorn, sweet corn, baby corn or seed crop.

Leaf Rust: *Puccina sorghi*

Rust pustules appear on the above ground plant parts specially on leaves. Round to elongated uredo-pustules are on both the surfaces of the leaf. They may coalesce and thereby produce scorching or drying of the leaves. Later on these pustules turn black due to formation of teleutospores. This disease is air born. The secondary infection take place by means of uredopustules which are wind born disseminated to the neighboring plants. The major approach adopted has been the utilization of host resistance. Both hybrids and composites possessing resistance have been released.

Downy Mildews

Ten Downy mildew (DM) diseases are identified on maize. Of these, seven are caused by *Peronoslerospora* species two by *Sclerospora* and one by *Ensclerospora*. All the *Peronosclerospora* incited diseases are of old world origin, three of which namely, *P. sorghi, P. sacchari,* and *P. phillipinensis* are of common occurrence in India.

Sugarcane Downy Mildew (*Sclerospora sacchari* Miyake)

The characteristic symptom is the development of long, rather broad chlorotic stripes along almost entire length of the leaves. The colour of the stripe is whitish in early stage which changes to dark brownish yellow in very late stages. The growth of the fungus can be seen on both the surfaces of the stripes. The affected plants may be malformed with undeveloped tussles and ears. The cobs are poorly filled.

Source of primary inoculum includes collateral cultivated and wild hosts, infected volunteer plants, kitchen gardens and certainly oospores, where formed. Infection usually begins with a few plants in early sown fields. Subsequently adjacent planted fields can be severely damaged.

The *Peronosclerospora* spp. are seed borne in nature. However, this does not appear to present a problem. Oospores have not been implicated as being carried out on kernels surface. If they were they could be shaken off or killed chemically. Internal mycelium is inactivated by drying seed to 20 per cent or less moisture and storing for about 3 weeks.

Philippines Downy Mildew (*Sclerospora philippinensis*)

The symptoms are very old similar to that of sugarcane downy mildew except for intensity of colour of stripes. The leaf infection results in long, chlorotic stripes with downy fungal growth. Affected plants produce malformed tassels or aborted ears which may appear at any time of silking, but the plants affected early are stunted and often die. The downy mildew fungi are mostly on alternate hosts and perenate through oospores which falls down in the soil and secondary infection takes place by means of conidia which are wind disseminated (Saha,1990).

Peronosclerospora Downy Mildew

Greatest loss results from systemic infection. This results in death or bareness. Losses from India and several SE Asian countries have been reported to be as high as 40-60 per cent. In India, significant loss in maize yield is usually localized to late planted areas. South India especially in TN and Karnataka have been reported epidemic at various times.

Brown Stripe Downy Mildew (BSDM) (*Sclerophthora rayssiae* var. *zeae* Payak and Renfro)

It has been most severe in UP, HP, southern Rajasthan, Punjab, hilly part of WB, J&K especially in areas that receive 100-200 cm of rains. A yield loss of 63 per cent was recorded in the *tarai* areas of UP. As the leaves are infected by the pathogen, it drastically reduces the total photosynthetic area of the diseased plants, causing reduction in grain yield. The yield losses depend on susceptibility of the host, spread of the disease and environmental conditions prevailing in the growing season. This is more destructive disease in the intermediate zone of Jammu region.

Disease symptoms have been observed only on leaves. In the initial stages, the lesions start developing on lower leaves as narrow chlorotic or yellowish stripes, 3-7 mm wide but variable in length. The stripes extend in parallel fashion, have well defined margins and are delimited by veins. The presence of downy whitish to creamy growth usually on the ventral surface of the infected leaves corresponding to stripes is the most characteristic symptom. In cloudy weather, the growth is profuse and can be seen. Later, these chlorotic stripes turn brown and give a burnt appearance to be leaves.

The Pathogen is an obligate parasite and can not be grown in artificial media. It is mainly soil borne. Its seed borne nature is controversial. Some worker reported this pathogen as seed-borne but it was found later that *S. rayssiae* var. *zeae* is not seed borne because no disease symptoms were observed in spite of providing ideal conditions to the pathogen grown from infected seeds. There seems to be no controversy about soil borne of the disease. Oospores remain viable for at least three years and this viable oospores material constitutes the main source of inoculums. Primary infection invariably occurs on lower proximate to the ground level. Once the primary infection becomes established, its spread is possible through the agency of sporangia.

A definite relationship of disease development with rainfall pattern in different regions has been mapped. In areas where rainfall varies from 40-60 cm, disease was observed from trace to low, in 60 to 100 cm rainfall areas it was from low to moderate, but in the areas having more than 100 cm rainfall, maximum disease ratings have been made. However, in areas of low rainfall, disease may be severe only when continuous spell of shower and cloudiness for a few days accompanied by about 25°C temperature are prevalent. The severity of the disease is influenced by temperature and moisture. Sporangia are produced at low temperature (20-22°C), while oospores are formed at high temperature. Soil temperature of 28-32°C favour disease development. Sporangial production and infection require a film of moisture for 12-96 hours. Zoospore germination occurs in the temperatures (15-30°C), with an optimum at 22-25°C. Sporangia and spore of the pathogen are disseminated by wind, rain and animals. The pathogen infects crab grass. Zinc deficiency predisposes plants to infection.

Integrated Disease Management

1. *Cultural practices*: Several cultural practices reduce severity of diseases. These are designed to eliminate the causal pathogens from a particular area, to significantly reduce primary inoculum or to stimulate growth during the first month after planting.
2. Sow the seed before rainy season begins.
3. *Fungicidal management*: Metalaxyl (Ridomil), can be applied to seed, in furrows as granules or sprayed on foliage. Seed application (Apron 35 W.P., @ 0.25 per cent (2.5 g/kg seed) in a slurry is most economical and total control can be obtained. Systemic fungicides usually act on but one site in the fungus, whereas protective (non-systemic) fungicides act on several sites. Thus in the former case only one change in the fungus is required to develop tolerance. Treatment of seed with apron 35 w.p. @ 0.25 per cent is a good safeguard against any possible seed-borne infection. Spraying of fungicides such as mancozeb @ 2.5-3.0 per cent as soon as disease symptoms appear to protect valuable and breeding materials. Sun drying of seeds lead to inactivation of mycelium present in seed and also reduce moisture levels. Seeds with less than 15 per cent moisture content produce healthy plants.
4. Use only disease resistant hybrids and composites in disease prone areas.

Sheath Blight

It affects the sheaths although they do infect leaves, husks and other parts. Two have been determined to cause significant damage, brown spot and banded leaf and sheath blight.

Brown Spot

This disease mainly occurs in sub-tropical and intermediate areas. It is distributed in the entire Himalayan area as also in states of Rajasthan and Karnataka. Resistance has been identified on the basis of natural incidence only and methods to induce artificial disease development in the field through resisting sporangia are to be worked out. A study conducted at Rajasthan, indicates that the yield loss in hybrid Ganga 5 was 27.0, while in the local open-pollinated variety it was Malan 24.5 per cent.

Banded Leaf and Sheath Blight (BLSB) (*Thanatophorus sasakii* Tu and Kimbrough= *Rhizoctonia solani* f.sp. *saakii* Exner)

It was first reported from Sri Lanka (Bertus, 1927). In India, it was first recorded by Ullstrup in the *Tarai* areas of U.P. (Payak and Renfro, 1966). Now it is present in J&K, HP, UP, MP, Haryana, Punjab, Rajasthan, WB, Meghalaya and Assam. This disease has been considered as one of the major diseases of maize (Payak and Sharma, 1981). Grain losses have been estimated in the range of 23.9-31.9 per cent in ten cultivars (Sangam Lal *et al.*, 1980). Singh and Sharma (1976) estimated 40.5 per cent loss in grain yield with 71 per cent disease index. The magnitude of grain loss may reach as high as 100 per cent if the ear rot phase of the disease predominates.

The disease appears on plants at pre-flowering stage (40 to 50-day old plant) and within a period of 15 to 20 days spreads under favourable conditions from the lower most sheath to the ear shoot.

Rhizoctonia solani Kunh is morphologically characterized by features like pale to brown colour of mycelium, branching near the basal septum in young growing hyphal, presence of a construction and formation of a septum in branch near the point of origin, absence of clamp connections, sclerotia of un-differentiated texture, young multinucleate hyphal cells with a prominent septal pore apparatus and rapid growth rate.

The primary source of inoculums are sclerotia in the soil and grass hosts that grow in the vicinity of maize crop. Secondary spread of disease is by contact of infected leaves with parts of adjoining healthy plants. Secondary spread by basidiospores has not been observed in maize crop.

The optimum temperature for *in vitro* growth of the pathogens is 30°C and the highest level of disease is induced when RH is in the range of 90-100 per cent. At RH of 70 per cent or lower, the disease development is negligible or absent.

Integrated Disease Management

Ahuja and Payak (1978, 82, 86) found under laboratory tests that the growth of the pathogen was completely inhibited by carbendazim, benodanil, thiobendazol and validamycin while field condition benodanil, validamycin and carbendazim were most effective in reducing disease severity. Lal *et al.* (1985) observed that foliar spray with thiobendazole (60 W.P. @ 0.05 per cent) followed by carboxin (75W.P. @ 0.1 per cent) and Duter (20 w -0.05 per cent) were most effective in reducing disease.

Sharma and Hembram (1990) have worked out a simple and economical method to control this disease. The methodology comprised of mechanical stripping of third leaf sheath from the gound level which is effective in checking disease developing further up. This method was also found to be effective in minimizing the yield loss.

Stalk Rot

Stalk rots are the most serious and widespread group of the diseases in maize. The term "stalk rot" is often used to include stalk breakage, stalk lodging, premature death of plants and occasionally root lodging. In true sense, it is decay of the internal pith tissues of the stalk. Typically, the first sign of stalk rot is plant wilting. In initial stage, leaves become gray, ears drop and outer rind of lower stalk may turn brown. When outer stalk tissue is brown, pith tissue in the lowest internodes is rotted and pulled away from the rind, which may result in lodging. Plants are weak and those with rotted stalks always have rotted roots.

There nature is often complex as a number of fungi, nematodes and sometimes bacteria are involved in causation of the disease(s). This group of disease is broadly divisible into two categories, *viz.* pre-flowering and post-flowering types. Of the 12 diseases recorded so far, Pythium stalk rot (*P. aphanidermatum*) and bacterial stalk rot (*Erwinia chrysanthemi* pv. *zeae*) belong to pre-flowering type, while others such as late wilt, charcoal rot, *Fusarium* wilt and stalk rot, *Acremonium* stalk rot, *Botryodiplodia* stalk rot belong to the post–flowering type.

Losses due to stalk rot may occur in the following three ways: (*i*) yield loss may result from premature plant death, thus stopping normal grain fill. Total grain weight on stalk rot affected plants is less than the weight on healthy plant. (*ii*) another component of yield loss is that plants with stalk rot may lodge and not be harvested with mechanical equipment. Harvest is slowed if stalk rot is severe and losses also occur due to time loss during harvest, (*iii*) losses also occur with ear rot as a results of the ear on lodged plant coming in contact with soil. This results in reduced grain quality and potential dockage when the grain is marketed. Because losses due to stalk rot may occur in several ways, yield loss estimates are difficult to obtain.

Technique used to estimate losses are: (*i*) yield of hybrids with practically nil natural stalk rot incidence have been compared to yields in years when stalk rot is severe (Koehler, 1960). These types of studies are confounded by year-to-year in average yield (*ii*) Another technique used is comparing inoculated vs. un-inoculated plots. These studies, however, are confounded by natural infection in un inoculated plots, and by the fact that inoculation does not exactly duplicate the natural stalk rot condition, (*iii*) in paired plant technique, grain yields of adjacent diseased and healthy plants are compared. In this, the problem in diseased plants have much less kernels than on healthy plants. The final problem with yield loss studies is that the loses are estimated on the basis of hand-harvested yield and do not take into account those losses due to lodging or ear rots.

Pre-flowering Stalk Rot

Erwinia Stalk Rot (*Erwinia chrysanthemi* pv. *zeae* Victoria, Arboleda and Munoj)

It is one of the most important disease of maize in tropical countries. At first, the upper leaves show signs of wilting. The symptom appear at the lower node and may remain confined at one or two internodes only. The inner tissues also get infected in severe cases. The infected tissues are at first soft but later on they turn into dry mass of shredded fibers. The plants may topple down at this stage. In some cases the infection (light to dark brown rotting) may spread rapidly throughout the leaf sheath and cause withering of leaf sheath and leaf under favourable environmental conditions. The stalks and ears may also be affected which emit an offensive odour. High disease incidence is linked with irrigation by sewage water, it is particularly favoured by high temperature (28°C and above) and the high ambient moisture which commonly prevails in the most of the maize growing areas 3 to 4 week after sowing.

The disease can be minimized, using indigenously formulated compound bleaching powder containing 33 per cent chlorine @ 10 kg/ha as soil drench at pre-flowering in standing crop. Avoiding of water logging and proper drainage also helps in reducing disease incidence. Planting the crop on ridges rather than in flat soil is recommended. Spraying mixture solution containing 50 g Agrimycin and 2 kg of ceresan wet/ha will also control the disease.

Pythium Stalk Rot [*P. aphanidermatum* (Eds.) Fitzp.]

The disease is common during rainy season under hot and high soil moisture conditions. The most diagnostic feature of this disease is that the rot is contained mostly to a single basal internode leading to destruction of pith parenchyma and consequent weakening to the stalk. The affected plants topple over but do not die for upto 2 weeks after attack. The general symptoms of the disease are poor emergence or patchy growth of seedlings or the seedlings become pale yellow or dry up just after germination. On pulling out the affected plants, rotting of the seeds may be clearly seen. Stalk rot is first recognized when the plants first lodge but do not break over. Stalks often rot at the first internode above the soil level and become soft and brown.

This disease has been reported in *Kharif* season in areas having temperature ranging 30-35°C and RH (relative humidity) 80-100 per cent. When RH drop 50 per cent or below, no rot develop in the stalks. This explain why high disease incidence occurs when the field is water logged, low lying or poorly drained, plant age (pre-flowering stage) and high plant population (< 60,000/ha). The tooth pick method as described for post flowering stalk rot development can be used to inoculate plants at knee high stage (30-35 days) for screening genotypes (Payak (1975), Payak *et al.* (1971), Payak and Sharma (1985).

Post-flowering Stalk Rot

Fusarium Stalk Rot [*Fusarium moniliforme* sheld.=*Gibberella moniliforme* (Sheld.) Wineland]

The disease appears both on young and adult plants. In young seedlings the plants dry up in the early stage of the growth. In the adult plants the symptoms are not visible till maturity. At this stage drying of top leaves from tips begin showing water scarcity. Ultimately the entire plant dries up. The cob formation either does not take place or the grain formation is particularly or wholly affected. The characteristic reddening of the internal tissues are discernible when affected plants are longitudinally split open. In some cases the red streak may also be observed. The pith becomes hollow and white to pale pink mycelial growth may also seen in the cavity thus formed in advance stage of disease development.

Disease Management

1. Sow the resistant varieties like GS-2 and Ganga-101 which are comparatively resistant.
2. Follow 2-3 years crop rotation.

Charcoal Rot [*Macrophomina phaseolina* (Manbl.) Ashby]

This is a common disease in warm and dry areas. A characteristic sign of disease is the presence of numerous, minute, black sclerotia, particularly on the vascular bundles and inside the rind of the stalk. This may cause the stalk to appear gray-black. The disease symptoms appear only near maturity.

Water soaked, brown lesions appear on roots which turn brown later on. The disease is favoured by high temperature *i.e.* 30-42°C and low soil moisture. The pathogen over winters as sclerotia and may penetrate roots and lower stems during growing season.

Control

Avoid water stress after flowering of the plant.

Black Bundle Disease (*Cephalosporium acremonium* Corda)

The characteristic symptoms is the blackening of the vascular bundles as black dots on the cut ends of the stalk. Other important symptoms of this disease are a reddening or purpling of the leaves and stalks, lesions on basal portion of the stalk, multiple ear formation at node and excessive tillering. In severe cases, leaves dry and plants may wilt. Ears may rot or even the ears may not be formed.

Control

Seed treatment with carbendazim or benlate @ 2 g/kg seed.

Smuts

Two smut diseases are occuring in maize, the common smut (*Ustilago maydis*) and head smut (*Sphacelothica reilina*).

Common Smut (*Ustilago maydis* (DC) Cda

The characteristic galls are formed on the infected tissues. The gall may appear on the stem, leaves, axillary's buds and parts of the male flower. When formed on the cob they cause extensive damage. The developing mycelium between the thin walled cells induces hypertrophy. The galls are light coloured in the beginning and later on become dark. As the galls are enlarging they appear light coloured or almost white. The individual flowers may be infected on tassel and even through individual flower parts may be transformed into galls. The light, shiny membrane ruptures and dry black spores masses are exposed. The black powdery masses are the chlamydospores of the fungus. Stem galls result in loss in yield and bending of the stalk.

Head Smut (*Sphacelothica reilina*) (Kunh.) Clinton

This attacks maize, sorghum and other grasses and is moderately destructive disease in the sub-temperate, intermediate zone of hilli areas of Rajouri and Poonch districts of J&K. It has occurred in severe form on Pioneer H-3054 and caused 25-35 per cent yield losses during Kharif-2005 in Manialdara region of Rajouri district (Ahamad, 2005). It appear with the formation of tassels and ears where partial or complete transformation of the inflorescence into sori may be observed. The tassel is partly or wholly converted into smut spores. Galls are first covered with a membrane that soon breaks open to expose a powdery mass and the vascular bundle of the host. If only the parts of the tassels and ears are converted into galls, the floral bracts turn into leafy structures. Ears frequently abort and are replaced by leafy buds. It is externally seed borne and soil borne. Galls on the ear usually destroy it to a large extent while large galls above the ear cause much greater reduction in yield than do galls below the ear. Losses from corn smut are highly variable from one location to another and may range from a trace upto 35 per cent or more in localized areas (Ahamad, 2005). The spore retain viability for two years. Although, the disease may be externally seed-borne, the major source of infection is soil–borne inoculums. Only young plants are susceptible. Low temperature is favourable for infection.

Control Measures

1. Field should be cleaned off the trash after harvest. Since only few plants are affected in a field it is possible to locate and destroy the infected ears before they shed the ears.
2. Follow up crop rotation.
3. Seed treatment with captan or thiram @ 3 g/kg seed.
4. Grow resistant/moderately resistant var. like KH-517, KH-612, GS-2 etc.

Virus Diseases

Viruses are the most mysterious disease agents affecting both plants and animals, and hence the limited amount of studies and where there is the least information on their incidence and economic effects. The lack of information is more acute in the developing countries, where there is a lack of trained personnel and appropriate equipment, and mostly to the concept that when these viral diseases are present, nothing can be done to control them. The problem with such mentality, the incidence of these diseases continues to increase, especially under the present actively changing circumstances in agriculture when new germplasm is almost freely moving throughout all environments, agronomic practices are changing, and there is a conspicuous expansion of area where maize is being grown due to an increased demand. These changes in technology are bringing along an increase in problems, specially insects and diseases, that were not previously considered problems of any significance.

Three viruses occurs on maize in India. These are maize mosaic virus I, maize mosaic (a strain of sugarcane mosaic virus) and vein enotion. The information is required to both pathologists and breeders to understand the complexity of the viral pathotypes, both in their genome and their evolution, to be considered when selecting for stable resistance to their pathogens.

Mosaic Virus (Mosaic Virus Transmitted by Leaf Hopper)

Small white fleck may occur on one side of the midrib near the base of the young leaf and is usually associated with whitening of veins. The specks elongated to form fine discontinuous stripes. Sometimes the stripes coalesce and form yellow bands of the leaves. Stripes may also be formed on sheaths, husks, and the stalks. The plants are pale green, stunted and show mosaic mottleling on the leaves. Cobs are not filled well.

Control

Grow resistant hybrids.

Maize Dwarf Mosaic Virus (Sugarcane Mosaic Virus, Transmitted by Aphids, *Myzus persicae*)

Early infested plants produce long chlorotic streaks along the veins at the base of the leaf lamina. With the growing age of plant, leaves in the whorl become chlorotic and in turn the chlorotic areas in leaves turn red or purple.

Control

Sow resistant varieties.

Maize Mosaic

This has been found to be prevalent in the several states and the incidence ranges from 2.2 to 10.6 per cent. Resistance of a high order even immunity (CM 103 and CM 104) has been identified and the improved released cultivars have not been reported to suffer much from virus diseases.

References

Ahamad, S., Kushwaha, K.P.S. and Manzer, A., 1998. Economics in plant disease control. *Indian Farmers Times*, 2: 35–39.

Ahamad, S., Narain, U. and Yasmeen, 2000. Plant biostress control: Approach for the century. *Pesticide World*, 12: 58–60.

Ahamad, S., 2005. Maize diseases and their management. *RARS Bulletin* No. 6.

Ahamad, S., 2005. Incidence of maize diseases in two districts of intermediate zone of Jammu region. *Annals. Agric. Res.* (In Press).

Ahamad, S. and Kalha, C., 2005. Fungal diseases of maize in the intermediate zone of Jammu region. In: *6th Nat. Symposium on Sustainable Plant Protection Strategies: Health and Environment Concerns* held on 15–17 October, at Dr. Balasaheb Sawant Konkan Krishi Vidyapeeth, Dapoli, Ratnagiri, Maharastra.

Ahuja, S.C. and Payak, M.M., 1978. A field inoculation technique for evaluating maize germplasm to banded leaf and sheath blight. *Indian Phytopath.*, 31: 517–520.

Ahuja, S.C. and Payak, M.M., 1982. Symptoms and signs of banded leaf and sheath blight of maize. *Phytoparacitica*, 10: 41–49.

Ahuja, S.C. and Payak, M.M., 1986. *In vitro* response of maize and rice isolates of rhizoctonia solani to antibiotics and fungitoxicants. *International Rice Research Newsletter*, 11: 16.

AICMIP, 1983. All India Coordinated Maize Improvement Project. Techniques of screening for resistance to important diseases of maize. All India coordinated Maize Improvement Project, IARI, New Delhi, pp. 133.

Bertus, I.S., 1927. *Yearbook*. Deptt of Agriculture, Cylone, p. 44–46.

CABI, 1997. *Crop Protection Compendium:* CD Rom Module I: South-East Asia and Pacific. Wallingford, Oxon, U.K.

CIMMYT, 1988. *Maize Production Region in Developing Countries*, 137 pp.

Cramer, H.H., 1967. *Plant Protection and World Crop Production*. Bayer, Leverkusem, Germany.

De Leon, C., 1984. *Maize Diseases: A Guide for Field Identification*. Centro International Mejoramiento de Maize Trigo (CIMMYT) El Batan, Mexico.

Desai, S. and Hegde, R.K., 1990. A preliminary survey of incidence of stalk rot complex of maize in two districts of Karnataka. *Indian Phytopath.*, 43: 575–576.

D.M.R., 1999. *Maize Protection Technology*. Directorate of Maize Research, pp. 123.

Directorate of Economics and Statistics, 1999. *Digest of Statistics 1997–98*. Planning and development Department, Govt. Jammu and Kashmir, pp. 59–62.

Koehler, B., 1960. Corn stalk rot in Illinois. III. *Agric. Experiment Stn. Bull.*, 658: 90.

Lal, S., Barauh, P. and Butchaiah, K., 1980. Assessment of yield losses in maize cultivars due to banded sclerotial diseases. *Indian Phytopath.*, 33: 440–443.

Payak, M.M., Lal, S., Thind, B.S., Renfro, M.C., Diwaker, R.C., Sharma, R.C., Lilaramani, L. and Laxminarayan, C., 1971. Stalk rot of maize in India: Germplasm evaluation and assessment of losses. In: *7Th Inter–asian Corn Improvement Workshop*, LosBanos, Philippines.

Payak, M.M. and Sharma, R.C., 1979. Resistance to diseases in maize. The Indian experience. *Crop Improv.*, 6: 75–85.

Payak, M.M. and Sharma, R.C., 1982. Disease Resistance in maize in India. In: *Advancing Frontier of Mycology and Plant Pathology*. Today and Tommorrow's Printers and Publishers, New Delhi. pp. 229–237.

Payak, M.M., 1975. Epidemiology of maize downy mildews with special reference to those occurring in Asia. *Trop. Agric. Res. Ser.*, 8: 81–91.

Payak, M.M. and Sharma, R.C., 1985. Maize disease and approaches to their management in India. *Trop. Pest Management*, 31: 302–310.

Renfro, B.L. and Ullstrup, A.J., 1976. A comparison of maize diseases in temperate and tropical environments. *PANS*, 22: 491–498.

Rangrajan, M. and Chakravarti, B.P., 1969. Bacterial stalk rot of maize in Rajasthan, effect on seed germination and varietal susceptibility. *Indian Phytopath.*, 23: 470–477.

Saha, L.R., 1990. *Handbook of Plant Protection*. Kalyani Publishers, 928 pp.

Sharma, R.C., De Leon, C. and Payak, M.M., 1993. Diseases of maize in South and South-East Asia: Problems and progress. *Crop. Protect.*, 12: 414–422.

Shurtleff, M.C., 1980. *Compendium of Corn Diseases*. The American Phytopathological Society St. Paul Menneseta, U.S.A., pp. 105.

Ullstrup, A.J., 1977. Diseases of corn. In: *Corn and Corn Improvement*, (Ed.) G.F. Sprague. American Society of Agronomy Inc. Publishers, Madison, Wisconsin, USA, 8: 391–500

Insect Pest and Disease Management *Pages* **331–341**
Editor: **Dr. D. Prasad**
Published by: **DAYA PUBLISHING HOUSE (2016)**

Chapter 23

Present Scenario of Management Strategies of Plant Viral Diseases

K.K. Biswas, Sumita Kumari and Avijit Tarafdar
Plant Virology Unit, Division of Plant Pathology,
Indian Agricultural Research Institute, New Delhi – 110 012

Introduction

Increasing quantities of plant protection chemicals employed and their wide and uncritical application at a time by modern spraying and dusting machineries has had some unfortunate effects. The ultimate aim of plant pathologists is to evolve suitable control measures so that the loss due to diseases may be prevented and the economic values of the crops rose but it is true that the excessive and indiscriminate use of chemicals has led to various environmental and health hazards. One example in this respect was that a notorious large scale campaign in the USA to eliminate the fire ant by the use of chloro-organic insecticides had a limited effect on the pest but resulted in the death of the several wild, domestic and other non-target organisms and ultimately illness of the man. Matter came to a lead in 1962 with the publication of Rachel Carson's "Silent Spring", which drew attention to the abuses in the use of the agro-chemicals to control pests, diseases and weeds. Then, the global community has expressed a willingness to reduce or stop its reliance on chemical pesticides and to move recently towards "organic farming".

Organic Farming

Presently the rapid growth of organic farming has been among the most remarkable changes in global agriculture. The organic movement began as a reaction of agricultural scientists and farmers against the industrialization of agriculture. Advances in biochemistry, plant breeding and engineering in the early 20th century led to profound changes in farming. By the use of hybrid seeds, large-scale irrigation, fertilization, and pesticides fields grew in size and cropping became specialized to make efficient use of machinery and reap the benefits of the green revolution. The organic farming refers to that promotes and enhances biodiversity, biological cycles and soil biological activity based on minimal use of off-farm inputs and management practices that restores, maintains and enhances ecological harmony. Enhancing soil health is the corner stone of organic farming. A variety of methods are employed, including crop rotation, green manure, cover cropping, application of compost, and mulching. Organic farmers also use certain processed fertilizers such as seed meal, and various mineral powders such as rock phosphate and greensand, a naturally occurring form of potash.

Organic pest control against target pests, weeds and diseases involves the cumulative effect of many techniques allowing for an acceptable level of pest damage encouraging or even introducing beneficial organisms These techniques generally provide benefits in addition to pest control, soil protection and improvement, fertilization, water conservation etc. and these benefits are both complementary and cumulative in overall effect on farm health. Many farm studies concluded that organic farming produces the same as conventional methods, but consumes less energy and contains no pesticide residues. In other cases the farm studies showed an average of 20 per cent lower organic yields over conventional, along with 50 per cent lower expenditure on fertilizer and energy, and 97 per cent less pesticides. In some cases it has been observed that organic yields were 95-100 per cent of conventional yields. Organic methods often require more labour, providing rural jobs but increasing costs to urban consumers.

All aspects of organic farming and organic food are under debate by environmentalists, food safety advocates, various consumer protection, social justice and labor groups, small independent farmers to agribusiness, food consumers and government agricultural policies. Controversy centers on the overall value and safety of chemical agriculture, with organic farming popularly regarded as the "opposite" of modern, large-scale, chemical-based farming, but also of the value of organic farming's solutions. The following topics may be argued in organic farming *viz.*, pesticides, environment, food contamination and quality, human health, soil conservation, crop rotation, genetically modified organisms, shifting from chemical to ecofriendly methods of pest control, multiple diseases and host resistance.

Cultural Control

There are many ways to control diseases caused by plant viruses. However, unlike control of fungal diseases, use of chemical pesticides is not available in practice for viral pathogens. Numerous chemicals have been reported to control viral multiplication, movement and formation of diseases symptoms caused by viruses, but none of them has been adopted for practical purpose in crop production systems. The single case where chemical treatment of virus infections have a practical value, is the preparation of healthy propagative materials from totally infected lines of crops such as garlic, potato and other herbaceous ornamental crops. The chemicals are applied to meristem tip or callus cultures and often heat treatment leading to virus free lines.

Virus diseases may also be avoided or controlled by planting certified virus free material or vegetatively propagated crops such as potatoes and ornamental crops, perennial trees, by testing seed lots for the absence of seed transmitted viruses, by using cultural practices designed to avoid transmission or mechanically transmitted viruses, and by seeking to prevent transmission of vector transmitted viruses by controlling the vector of its access to the host plant. All these methods can make a valuable contribution to practical disease control in specific favourable situations, but they do not offer the whole for all situations.

Cross Protection

Cross protection was first demonstrated by McKinney in 1929 inoculating of tobacco plant with a "green" strain of *Tobacco Mosaic Virus* (TMV) to protect systemic infection by another severe strain of TMV inducing yellow mosaic symptoms. It was also of potential interest for protecting plants against viruses in the field. Cross protection has not only been observed to occur between viruses, it was also demonstrated to occur between viroids or plant virus satellites. Many mild strains have been utilized as cross protecting strains in many crops in field level experimentally or commercially. An ideal mild strains to be used in the field for cross protection should have some features; it should induce milder symptoms than isolates commonly encountered in then field, it should be fully systemic and invade all host tissues, should be genetically stable and not evolve towards more severe forms at times, should not easily disseminated by means of vectors, should provide protection towards wider range of challenging severe isolates and the protective inoculum should be easy to produce, store and provide to farmers etc.

Table 23.1: Virus for which Mild Strains have been Used for Cross Protection

Virus	*Origin of Mild Strains*	*Crops Controlled*
Cocoa swollen shoot	Field isolates	Cocoa*
Citrus tristeza	Field isolates	Citrus**
Arabis mosaic	Field isolates	Grapevine*
Papya ring spot	Mutagenesis	Papaya**, Squash*
Zucchini yellow mosaic	Greenhouse variants	Squash**, cucumber**
Soybean mosaic	Cold treatment	Soybean*
Tomato mosaic virus	Mutagenesis, field isolates and Heat treatment	Tomato**

*: Experimental; **: Commercial.

Cross protection proved to be a very efficient and popular method to control *Tomato mosaic virus* (ToMV). Mild strains were obtained either from naturally occurring strains or by culturing a severe strain in tissue held above 34°C. The mild strain MII-16, which was extensively used in Western Europe was obtained by nitrous acid mutagenesis followed by single local lesion selection. Cross protection against ToMV in tomatoes has widely used in Europe, Canada, New Zealand and Japan. In France, a subculture of MII-16 was used commercially from 1972 to 1983. Preliminary experiments and pilot tests demonstrated that it did not reduce the yields but only slightly delayed fruits setting. It provided very efficient protection against ToMV severe strain; however, the risk like synergism with other viruses limited the use of ToMV crosses protection.

Citrus tristeza virus (CTV) is prevalent in many regions of citrus growing countries in the world. It causes very important losses, particularly in certain scion root combination where it reduces fruit sizes and yield causing stem-pitting, declining and premature death of the tree. It has been currently used for grapefruit in South Africa and Australia and for Perra Sweet oranges in Brazil. More than 8 million Pera Orange trees were cross protected in 1980 and more than 50 millions trees in 1993. Mild isolates in Brazil, Australia and South Africa were isolated from the trees noticed as showing no or very discrete symptoms in areas where all others trees were severely infected. In the areas where server stem pitting isolates are widespread and the brown citrus aphid is epidemic cross protection has been used as a control strategy. Cross protection presumably requires a certain minimum degree of relationships between the protecting and challenge isolates at least in certain areas of viral genome. Better knowledge to more accurately select the effective and mild protecting isolates is needed. Defective interfering RNAs have recently been associated with CTV infection may also have applications for cross protection.

Cross protection has been successfully used in the bio-control of several cases of *Cucumber mosaic* (CMV) infection. Cross protection can be obtained using a sat-RNA free mild strain. Deploying this technique, a CMV pseudo recombinant was assembled, characterized and used as a cross protection agent in Italy and called CMV-ST. The CMV-ST was tested worldwide and had no deleterious effect on growing plants or environment. Cross protection has been favourably exploited in the management of viroids. When tomato plants with a prior infection by a mild strain of *Potato spindle tuber viroids* (PSTVd) are challenged-inoculated with a severe strain, symptom development by the challenged strain is delayed by 48 days. More recently it has been demonstrated that tomato plant pre-inoculated with non-necrogenic satellite RNA of CMV (CARNA-5) produced milder symptoms, when challenged inoculated with a severe strain of PSTVd.

Disease Resistance

The recent advances made in molecular biology with plant transformation and regeneration technologies have widened the scope of conferring resistance to viruses in plants, although, for an effective viral disease control the breeding for disease resistance through conventional breeding methods always have most important role. Extensive work has been done on breeding for resistance against plant viruses. The resistant varieties if available are an effective strategy for minimizing the losses caused by the plant viruses. This is largely due to intrinsic properties of viruses, which do not permit their control by any simple physical and chemical control. The molecular biology has facilitated different steps of conventional methods of breeding for disease resistance and also helped in incorporating novel resistance genes. The development of various molecular markers in breeding resistance has also shortened the selection process.

Genetic Engineering

Genetic engineering is the branch of biotechnology that modify genetic structure of the crop or pest genome which is accomplished either by gene deletion or blocking or by transferring gene from one organism to another. Successful genetic improvement of crop plants by methods developed through biotechnology has resulted in introducing new genes for virus resistance into existing commercial crop varieties. These new genes are derived from the viruses themselves in a concept referred to as pathogen derived resistance (PDR). The organisms/crops that have been modified by one of this gene modifying technique are referred to as genetically modifying organisms/crops (GMOs/GM crops). Sometime it also called as genetically engineered organisms/crops (GEOs/GE crops). The term transgenic crop is applied when the genetic modification involves transfer of genes from one organism to another. Those genes which are transferred called transgenes. Thus, identification of genes conferring pest resistance, and then transfer these genes to crop plants are performed by the genetic engineering techniques. Since the first report of this strategy to protect tobacco and tomato plants against TMV infection by engineering the viral coat protein gene into plants, there have been many report of transgenic plants have been resistant to viral diseases reviewed. Success has been obtained in field level in crops like wheat against *Wheat streak mosaic* (WSMV), citrus against *Citrus tristeza* (CTV) and *Citrus psorosis*

(CPV), Squash against CMV, *Zucchini yellow mosaic* (ZYMV) and *Watermelon mosaic* II (WMV II) and Papaya against *Papaya ring spot* (PRSV) virus.

Table 23.2: Transgenic Crop Resistance to Viruses Using Pathogen Derived Genes

Crop	*Transformation with the Gene*	*Against the Virus*
Citrus	CP gene P23 gene, RdRp gene, Replicase	CTV
Squash	Replicase gene	CMV
Potato	CP gene, Replicase gene	PVX
	CP gene	PVX, PVY, PLRV
Sugarcane	CP gene	SCMV
Maize	CP gene	MDMV, MCMV
Rice	CP gene	RSV, RTSV
Wheat	CP gene	WSMV
Apricot	CP gene	PPV
Cantaloupe	CP gene	ZYMV*,WMV2*,CMV*
Grape	CP gene	GCMV, GFLV, ToRSV
Muskmelon	CP gene	ZYMV*
Papaya	CP gene	PRV*
Plum	CP gene	PPV
Squash	CP gene	ZYMV*, WMV2*
Pepper	CP gene	TSWV
Tomato	CP gene	ToMV* YMV* CMV*, TYLCV*
Potato	CP gene	Yes
Lettuce	CP gene	LMV, TSWV
Pea	CP gene	PEMV
Cucumber	CP gene	CMV*
Sugarbeet	CP gene	BNYVV
Peanut	CP gene	TSWV
Soybean	CP gene	BPMV
Bean	CP gene	BPMV

*: Field tested.

Coat Protein Gene

Use of viral coat protein gene as a transgenes for obtaining virus resistant plants is one of the most important successes achieved in the plant biotechnology. Coat protein-mediated resistance (CP-MR) was demonstrated for the first time a decade ago (Powell-Abel *et al.*, 1986) by demonstrating of virus resistance in transgenic plants utilizing TMV coat protein (CP) gene expressed in tobacco plants.

Replicase Gene

In addition to CP genes, other genetic elements from plant viruses are also being tested in transgenic plants for the ability to confer resistance to virus infection or disease. The expression of replicase

(RNA dependent RNA polymerase, RdRp) native genes and modified version has been shown to confer resistance to virus infection. The modified proteins are thought to be defective in functions critical for the normal replication or spread of virus in the plant. Both, the expression of native viral replicase genes and modified replicase genes have resulted in transgenic plants demonstrating high levels of resistance to virus infection or disease.

Table 23.3: Comparative Performance of Transgenic Virus Resistance Plants

Host	*Transgene*	*Yield Increase (%)*
Tomato	TMV CP	40
Tomato	CMV satellite	14
Potato	PVX + PVY CP	38
Squash	CMV + ZYMV + WMV2 CP	97
Squash	ZYMV + WMV2 CP	90
Squash	ZYMV CP	77
Papaya	PRSV CP	90

Movement Protein Gene

These are the virus encoded proteins essential for cell-to-cell movement of plant viruses. These proteins have been shown to modify the gating function of plasmodesmata, thereby allowing the viral particles or their nucleoprotein derivatives to spread to adjacent cells. This phenomenon was first used to engineer resistance against TMV in tobacco by producing modified movement protein (MP) gene which is partially active as a transgenes. The conferred resistance is believed to be based on the competition between wild-type viruses encoded MP gene and the preformed dysfunctional MP to bind to the plasmodesmatal sites. The above resistance was also effective against distantly related or unrelated viruses. As for example, resistance against TMV can be achieved in tobacco by using the MP gene derived from *Brome mosaic virus*, suggesting the functional conservation of this protein among several viruses. This protein can be made from single MP gene as in *Tobamovirus* where as it is made by a set of three overlapping genes, known as triple-gene-block (TGB), in case of potex- Carla virus. The resistance is dependent upon the interaction of viral derived and the transgene derived MPs gene.

Satellite RNA

Satellite RNAs are sub viral RNA. They can be distinguished from the viral genome by their dispensability for normal viral propagation. The classical definition of satellite is a linear or circular RNA that requires a helper virus to supply proteins for replication but shared little sequence relatedness with the helper virus and not required for the accumulation of the helper virus. The replications of satellite viruses are completely dependent on enzymes encoded by the helper virus. Thus their amplification is limited to confined cells. It is nearly exclusively associated to plant viruses. Satellites are sequence that is either mostly or completely unrelated to any large contiguous segment of the helper virus. They can cause large alteration to normal disease progression induced by their helper virus, thus are fascinating tool to control viral diseases. Satellite RNAs are divided into two groups, ones that generally encode a single non structural protein are known as large satellite RNAs, and other with no functional coding capacity called small satellite RNAs. Satellite viruses are those, which are capable to encode their own coat protein.

Defective Interfering Viral Nucleic Acid

Defective interfering (DI) viral nucleic acid is also a type of sub viral RNA. It is derived mainly from the genome of the helper virus by one or more premature termination and reinitiating events that infrequently results in a functional template capable of further amplification. DI RNAs are associated with many plants and nearly all animal RNA viruses. These viruses are missing some or all the viral genes required for the movement, replication and encapsidation, but contain all cis acting element necessary for replication by RdRp of the parental virus. Most DI RNAs moderate the symptoms of their helper viruses, but some of them exacerbate the severity of symptoms as in DI RNAs of *Broad bean mottle virus* and *Turnip crinkle virus* (TCV). No single mechanism is accepted to account for the DI RNA interference with the accumulation of their helper virus and corresponding symptom attenuation. Many mechanisms of DI RNA symptom modulation have been proposed.

1. Most logical and widely used explanation for the reduction of helper viral genomic RNA levels when associated with DI RNAs is a competition between the DI RNA and the viral genome for limited factors required for replication; reduced virus levels in turn lead to attenuated symptoms.
2. Alternative model proposed suggesting that a link between DI RNA accumulation and viral gene expression leads to symptom attenuation.
3. Activation of PTGS is also believed to be mechanism of symptoms modulation.

Post Transcriptional Gene Silencing

Post transcriptional gene silencing (PTGS) is an adaptive sequence specific RNA degradation system of any organism that plays a role in control of aberrant, unwanted excess of foreign RNA

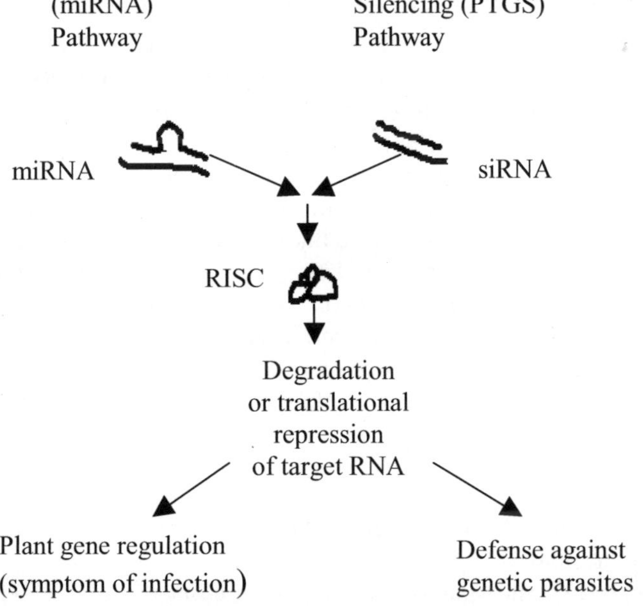

Figure 23.1: Schematic Diagram Showing Micro RNA (miRNA) and Post Transcriptional Gene Silencing (PTGS) Pathway

intracellularly in a homology dependent manner. It is prevalent in various forms of life of plant, fungus and invertebrate animals. This activity could be present constitutively to help normal development or it may be induced in response to cellular defense mechanisms against pathogens. According to recent models PTGS is activated by double stranded (ds) RNA and results in cleavage of the RNA into 21-26 nucleotides long double stranded (ds) small interfering (si) RNAs by an Rnase III like enzyme called DICER. The generated siRNA are associated with an enzyme complex (RNA-induced silencing complex; RISC) and guide the RISC to degrade any RNA having sequence homology with the inducer dsRNA. This complex (RISC) is made of different cellular factors, namely RNA dependent RNA polymerase (RdRp), RNA-helicase, translation elongation factor, RNAse, etc. This degradation process initiating from a concerned cell having the elicitor RNA, spreads later within the entire organism in a systemic fashion. This process is generally regarded to evolve as a plant defense mechanism against invading viruses containing either DNA or RNA genomes. In addition to the cell-autonomous defense function, PTGS in plants generates mobile signals that instruct target RNA degradation at a distance (systemic PTGS). When the viral RNA is either elicitor or target of PTGS, the degradation mechanism is known as virus induced gene silencing (VIGS). The VIGS come into play when plant recovered from initial viral infection.

Conclusion

The development of transgenic plants resistant to viral diseases requires an effective expression cassette for the expression of viral transgenes. The expression cassette consists of a transcriptional promoter and terminator, which control the production of viral gene messenger RNA in transgenic plants. Generally constitutive promoters that results in high levels of gene expression have achieved greatest success. The *Cauliflower mosaic virus* (CaMV) 35S promoter is a single or double copy version (enhanced 35S) is used most frequently although several other promoters have also been used successfully. In cases where a virus is restricted to certain cells or tissues, the use of tissue specific promoters may be considered.

Wherever possible, major crop varieties should be considered for transformation to allow for immediate use of transgenic plants. In cases where target varieties cannot be transformed, other varieties can be used to introduce genes of interest, following by breeding into commercial varieties. The CP-MR and replicase-MR are the two PDR strategies most likely to be commercially available in the near future, especially in vegetable and grain crops. Viral genes should be selected based on analogy with previous performance experience. In most cases CP would be considered first because there are many examples in which CP has conferred higher resistance to viruses from various groups. The coat protein gene is generally well characterized and therefore easily cloned. Following expression in plants, the coat protein can be detected using readily available antisera generated to virions. In the case of luteoviruses, CP has conferred only weak to moderate resistance, so the use of other genes such as replicase should be considered. For virus groups where high resistance has not yet been achieved, basic research must to be done to select the best resistance genes or modifications of commonly used genes.

References

Antignus, Y., Vunsh, R., Lachman, O., Pearlsman, M., Maslenin, L., Hananya, U. and Rosner, A., 2004. Truncated Rep gene originated from tomato yellow leaf curl virus-Israel [Mild] confers strain specific resistance in transgenic tomato. *Annals of Applied Biology*, 144: 39–44.

Asad, S., Hars, W.A.A., Bashir, A., Zafar, Y., Malik, K.A., Malik, N.N. and Lichtenstein, C.P., 2003. Transgenic tobacco expressing Gemini viral RNAs are resistant to the serious viral pathogen causing cotton leaf curl disease. *Archives of Virology*, 148: 2341–2352.

Biswas, K.K. and Varma, A., 2001. Evaluation of resistance in blackgram (*Phaseolus mungo*) to variants of mungbean yellow mosaic geminivirus. *Indian Agril. Sci.*, 71: 215–218.

Biswas, K.K., Das, D.R. and Ahlawat, Y.S., 2005. Viral diseases of fruits, vegetables and spices in North Eastern Himalayan region of India and their management. In: *Crop Protection: Management Strategy*, (Ed.) D. Prasad. Daya Publishing House, New Delhi, pp. 225–226.

Biswas, K.K. and Gupta, A., 2007. Management of important plant viruses in Indian agriculture. In: *Sustainable Pest Management*, (Ed.) D. Prasad. Daya Publishing House, New Delhi, pp. 208–238.

Boogaart, T., Maule, A.J., Davies, J.W. and Lomonossoff, G.P., 2004. Source of target specificity associated with the recovery against *Pea seed–borne mosaic virus*, infection mediated by RNA silencing in pea. *Molecular Plant Pathology*, 5: 37–43.

Dasgupta, I., Malathi, V.G. and Mukharjee, S.K., 2003. Genetic engineering for virus resistance. *Current Science*, 84: 341–354.

Domínguez, A, Guerri, J., Cambra, M., Navario, L., Moreno, P. and Pena, L., 2000. Efficient production of transgenic citrus plants expresseing the coat protein gene of citrus tristeza virus. *Plant Cell Reporter*, 19: 427–433.

Febes, V.J., Niblet, C.L, Lee, R.F. and Moore, G.A., 2003. Characterization of grapefruit plants (*Citrus paradisi* Macf.) transformed with *Citrus tristeza* clostero virus genes. *Plant Cell Rep.*, 21: 421–428.

Fitch, Maureen M.M., Manshardt, R.M., Gonsalves, D., Slightom, J.L. and Sanford, J.C., 1992. Virus resistant papaya plants derived from tissues bombarded with the coat protein gene of papaya ringspot virus. *Bio/Technology*, 10: 1466–1472.

Ghorbel, R., Lopez, C., Fagoaga, C., Moreno, P., Navaro, L., Flores, R and Pena, L., 2000. Transgenic citrus expressing the citrus tristeza virus p23 protein exhibit viral like symptoms. *Mol. Plant Pathol.*, 2: 27–36.

Golemboski, D.B., Lomonssoff, G.P. and Zatlin, M., 1990. Plant transformed with tobacco mosaic virus no structural gene sequence are resistant to virus. *Proc. Nat. Aca. Sci., USA*, 87: 6311–6315.

Gonsalves, D. and Garsney, S.M., 1989. Cross protection techniques for control of plant virus diseases in the tropics. *Plant Disease*, 73: 592–597.

Gonsalves, D., Chee, P., Providenti, R., Seem, R. and Slightom, J.L., 1992. Comparison of coat protein-mediated and gentically-derived resistance in cucumbers to infection by cucumber mosaic virus under field conditions with natural challenge inoculations by vectors. *BioTechnology*, 10: 1562–1570.

Grumet, R., 1994. Development of virus resistant plant via genetic engineering. *Plant Breeding Reviews*, 12: 47–49.

Gutierrez, M.A., Luth, D. and Moore, G A., 1997. Factor affecting *Agrobacterium* mediated transformation in Citrus and production of sour orange (*Citrus aurantium* L.) plants expressing the coat protein gene of citrus tristeza virus. *Plant Cell Reporter*, 16: 745–753.

Hafner, G.J., Harding, R.M. and Dale, J.L., 1995. Movement and transmission of banana bunchy top virus DNA component one in bananas. *Journal of General Virology*, 76: 2279–2285.

Harriron, B.D., Mayo, B.D. and Boulcombe, D.C., 1987. Virus resistance in transgenic plant that expresses Cucumber mosaic virus satellite RNA. *Nature*, 328: 799–801.

Hull, R., 2002. *Matthews' Plant Virology*, 4th edn. Academic Press. A Harcourt Science and Technology Company, Harcourt Place, 32 Jameston Road. London, NW1 7BY, UK.

Khoury, J., Singh, R.P., Boucher, A. and Coombs, D.H., 1988. Concentration and distribution of mild and severe strains of potato spindle tuber viroid in cross-protected tomato plants. *Phytopathology*, 78: 1331–1336.

Khetrapal, R.K., Maisonneuve, B., Maury, Y., Maury, Y., Chalhoub, B., Dinant, S., Lecoq, H. and Varma, A., 1998. Breeding for resistance to plant viruses. In: *Plant Virus Disese Control*, (Eds.) Hadidi *et al.* APS Press, pp. 14–32.

Lawson, E.C., Kaniewski, W., Haley, L., Rosman, R., Newell, C., Sanders, P. and Turner, N., 1990. Engineering resistance to mixed virus infection in a commercial potato cultivar: Resistance to potato virus X and potato virus Y in transgenic Russet Burbank. *Bio/Technology*, 8: 127–134.

MacDiarmid, R., 2005. RNA silencing in productive virus infections. *Ann. Rev. Phytopathol.*, 43: 523–44.

May, G.D., Afza, R., Mason, H.S., Wiecko, A., Novak, F.J. and Arntzen, C.J., 1995. Generations of transgenic Banana (*Musa acuminata*) plants via Agrobacterium-mediated transformation. *Biotechnology*, 13: 486–492.

Miller, N., Sulistyowati, E. and Dietzgen, R.G., 2003 Cucumber mosaic virus infection transiently breaks DsRNA–induced transgenic immunity to potato virus Y in tobacco. *Molecular Plant Microbe Interactions*, 16: 936–944.

Montasser, M.S., Tousignant, M.E. and Kaper, J.M., 1991. Satellite-mediated protection of tomato against cucumber mosaic virus: I. Greenhouse experiments and simulated conditions in the field. *Plant Disease*, 75: 86–92.

Powel-Abel, P., Nelson, R.S., De, B., Hoffman, N., Rogers, S.G. Fraley, R.T. and Beachy, 1986. Delay of disease development in transgenic plant exprees the tobacco mosaic virus coat protein. *Gen. Science*, 232: 738–743.

Scorza, R., Ravelonandro, M., Malinowski, T., Cambra, M. and Minoiu, N., 2003. Potential use of transgenic plums resistant to Plum pox virus field infection. *Acta Horticulture*, 622: 119–122.

Simon, A.E., Rossinck, M.J. and Havelda, Z., 2004. Plant virus satellite and defective interfering RNAs: new paradigms for a new sentury. *Ann. Rev. Phytopathol.*, 42: 415–37.

Singh, S.J., 2003. *Virus and Phytoplasma Diseases of Papaya, Passion Fruit and Pineapple*. Kalyani Publishers, Noida.

Smith, G.R., Ford, R., Frenkel, M.J., Shikla, D.D. and Dale, J.L., 1992. Transient expression of the coat protein of sugarcane mosaic virus in sugarcane protoplasts and expression in *E. coli*. *Arch. Virol.*, 125: 15–23.

Tapfer, M., 2002. Risk assessment of virus-resistant transgenic plants. *Ann. Rev. Phytopathol.*, 40: 467–91.

Vigne, E., Komar, V. and Fuchs, M., 2004. Field safty assessment of recombination in transgenic grape veins expressing the coat protein gene of *Grapevine fanleaf virus*. *Transgenic Research*,. 13: 165–179.

Yang, Z.N., Ingelbrecht, I.L., Louzada, E., Skaria, M. and Mirkov, T.E., 2000. Agrobacterium-mediated transformation of the commercially important grapefruit cultivar Rio Red (*Citrus paradisi* Maca.). *Plant Cell Reports*, 19: 1203–1211.

Young, Y., Sherwood, T.A., Patte, C.P., Hiebert, E. and Polston, J.E., 2004. Use of *Tomato yellow leaf curl virus* (TYLCV) *Rep* gene sequences to engineer TYLCV resistance in tomato. *Phytopathology*, 94: 490–496.

Insect Pest and Disease Management *Pages 342–359*
Editor: **Dr. D. Prasad**
Published by: **DAYA PUBLISHING HOUSE (2016)**

Chapter 24

Bacterial Endophytes of Plants and their Uses in Agriculture

Biswajit Bhowmik[1], Tusar Kanti Bag[2] and S.K. Biswas[3]
[1]Indira Gandhi National Open University, Siliguri Sub-regional Centre, Raja Ram Mohan Roy Road, East Vivekananda Pally, Siliguri – 734 006, West Bengal
[2]National Research Centre for Orchids (ICAR), Pakyong – 737 106, Sikkim
[3]C.S. Azad University of Agriculture and Technology, Kanpur – 208 002, U.P.

Introduction

There are several definitions of endophytes, which in general include fungi and bacteria (Chanway 1996). Endophytes were originally defined as those organisms that live within a plant tissue (de Bary 1866). Bacterial endophytes have been defined by Kado (1992) as "bacteria that reside within living plant tissues without doing substantive harm or gaining benefit other than securing residency." Quispel (1992) considered endophytes only as bacteria that establish an endosymbiosis with the plant, whereby the plant receives an ecological benefit from the presence of the symbiont, such as increased stress tolerance or plant growth promotion. Beattie and Lindow (1995) viewed the term epiphytic and endophytic as two ends of one spectrum reflecting the growth patterns of leaf bacteria, rather than as two distinct groups of organisms. These authors suggested that an active exchange occurs between the internal and external populations of one bacterial strain. Hallmann *et al.* (1997) defined "endophytes as those bacteria that can be isolated from surface disinfested plant tissue or extracted from within the plant, and that do not visibly harm the plant." This definition includes internal colonists with apparently neutral behaviour, as well as symbionts and bacteria that, during their endophytic phase, fluctuate between endophytic and epiphytic colonization. According to Petrini (1998), "endophytes include all organisms that, at some time of their life cycle, live symptomlessly within plant tissues." Endophytes can be exploited in various beneficial aspects such as growth promotion, suppression or ecofriendly management of fungal or bacterial diseases of agriculturally important crop plants.

Distribution of Endophytic Bacteria in Plants

The theory that nonpathogenic bacteria reside in plant tissues dates to Perotti (1926), but research on bacteria residing in the internal tissues of non-symptomatic plants dates back to the 1870s with work by Pasteur and others (Smith 1911). Since 1940 there have been numerous reports on indigenous endophytic bacteria in various plant tissues, including seeds and ovules (Mundt and Hinkle 1976), tubers (Trevet and Hollis 1948), roots (Philipson and Blair 1957; Agarwal and Shende 1987), stems and leaves (Henning and Villforth 1940) and fruits (Samish *et al.*, 1961; Sharrock *et al.*, 1991).

Endophytic bacteria have been detected in different tissues of various symptomless plants. Samish *et al.* (1961) isolated *Xanthomonas, Pseudomonas, Flavobacterium, Aerobacter, Corynebactrium* etc. from surface sterilized tomatoes. Endophytic *Bacillus subtilis* was isolated from peanut kernels (Pattit *et al.*, 1968). The internal tissues of healthy fruits and vegetables were found to contain a diverse bacterial flora (Samish *et al.*, 1963; De Boer and Copeman 1974). The presence of endophytic *Erwinia carotovora* in healthy cucumber fruit was demonstrated by Meneley and Stanghellini (1975). Mundt and Hinkle (1976) isolated a large number of endophytic bacteria from ovules and seeds of various plants. The bacteria were distributed among 19 genera and 46 species. The species isolated in greatest number were *Bacillus megaterium, B. cereus, Erwinia herbicola, Flavobacterium devorans* and *Pseudomonas fluorescens*. Gardner *et al.* (1982) identified 556 endophytic bacteria from lemon root xylem belonging to 13 different genera, with Gram-negative, rod shaped bacteria predominating. Jacobs (1985) obtained 7 most commonly observed bacterial genera (*Bacillus subtilis, Erwinia herbicola, Pseudomonas aeruginosa, P. fluorescens, Lactobacillus* sp., *Xanthomonas* sp. and *Corynebacterium* sp.) associated with apparently healthy sugar beet root tissue. Isolates of *Erwinia* sp., *Bacillus* sp., *Clavibacter* sp. and *Xanthomonas* sp. were recovered from surface sterilized radicles, roots, stems, unopened flowers and bolls of green house and field grown cotton (*Gossypium hirsutum*) plants (Misaghi and Donndelinger 1990); *Erwinia* sp. was the most prevalent bacterium. Whitesides and Spotts (1991) isolated endophytic *Pseudomonas syringae* from roots and stems of pear trees. The diversity of indigenous bacterial endophytes within stems and roots of sweet corn (*Zea mays*) and cotton (*G. hirsutum*) was enumerated by McInroy and Kloepper (1995a). They observed that Gram-negative bacteria comprised 70.50 per cent of the endophytic bacteria and 27 of the genera were identified. Bhowmik *et al.* (2002) isolated fluorescent and non-fluorescent *Pseudomonas* from cotton (*G. hirsutum*).

Endophytic *Pseudomonas fluorescens* has been isolated from root cortex of cotton by Hallmann *et al.* (1997a). Germida *et al.* (1998) randomly selected over 300 rhizoplane and 220 endophytic bacteria from roots of canola plants on agar solidified trypticase soy broth and identified using fatty acid methyl ester (FAME) profiles. Seventy three per cent of the identified isolates belonged to four genera *viz. Bacillus* (29 per cent), *Flavobacterium* (12 per cent), *Micrococcus* (20 per cent) and *Rathayibacter* (12 per cent). Genera identified in the endophytic bacteria were also found in the rhizoplane isolates. They concluded that endophytes were a subset of the rhizoplane community. Christiansen-Weniger (1998) reported that in the induced tumor tissues of cereal crops (rice, wheat and maize), introduced diazotrophs (*Azospirillum* sp., *Azorhizobium caulinodans, Rhizobium* sp.) potentially colonize endophytically the tissues of stem and root tumors with high cell numbers. Agarwal and Shende (1987) microscopically observed the presence of diazotrophs within the root tissue of mustard plant. In cross sections of roots, a number of coccoid, spiral and bacillus shaped deeply stained cellular bodies of various sizes were observed. Some of these appeared to be motile. However, they could not identify the bacteria. Chanway (1998) isolated an endophytic *Bacillus* strain (Pw 2) from inside surface sterilized pine root tissues. This strain possessed nitrogenase activity and could colonize pine seedlings systemically after soil inoculation. Marti *et al.* (1999) isolated endophytic *Agrobacterium tumefaciens* from stem tumors of several rose cultivars.

In general, Gram-negative species were found to be dominant compared with Gram-positive strains, accounting for 78 per cent of the strains isolated from grapevine xylem (Bell *et al.*, 1995), 84 per cent from xylem of citrus roots (Gardner *et al.*, 1982), 94 per cent from alfalfa roots (Gagne *et al.*, 1987) and 75 per cent from corn and cotton tissue (Mclnroy and Kloepper 1994). However, these results were contradictory to Lalande *et al.* (1989), who found 88 per cent Gram-positive bacteria in 2 month old roots of corn and to Leifert *et al.* (1989), who reported an average of 70 per cent Gram-positive bacteria from nine micropropagated plant species after 12 months.

Table 2.1 lists the most common taxa of endophytic bacteria recovered from internal plant tissue. The main taxa belong to the former *Pseudomonas* group (*Pseudomonas, Burkholderia, Phyllobacterium*) and Enterobactriaceae (*Enterobacter, Erwinia, Klebsiella*) (Gardner *et al.*, 1982; Mclnroy and Kloepper 1995a, Bell *et al.*, 1995; Jacobs 1985). A comparison of the bacterial spectrum of monocotyledonous (*Zea mays*) and a dicotyledonous (*Gossypium hirsutum*) host yielded several endophytic species in cotton that did not occur in corn and vice versa (Mclnroy and Kloepper 1995a, 1995b). In addition, the bacterial spectrum isolated from corn stems and roots was similar with only a few exceptions, whereas for cotton, 14 taxonomic groups present in roots were not found in stems.

Isolation of Endophytic Bacteria

The isolation procedure for endophytic bacteria is of key importance for research with endophytes. Several techniques have been employed for the isolation of endophytic bacteria with each having advantages and disadvantages.

Trituration and Surface Disinfestations

The most common procedures for the isolation of endophytic bacteria were based on trituration of surface disinfested plant tissue using various disinfectants such as sodium hypochlorite (Fisher *et al.*, 1992; Gardner *et al.*, 1982; Hinton and Bacon 1995; Hallmann *et al.*, 1997a, 1997b), ethanol (Dong *et al.*, 1994; Fisher *et al.*, 1992; Gagne *et al.*, 1987; Marti *et al.*, 1999), hydrogen peroxide (Mclnroy and Kloepper 1994; Misaghi and Donndelinger 1990), mercuric chloride (Gagne *et al.*, 1987; Hollis 1951; Sriskandarajah *et al.*, 1993) or combination of two or more of these (Caetano-Anolles *et al.*, 1990; Pleban *et al.*, 1995; Germida *et al.*, 1998), followed by several washes in sterile water or buffer solutions. Surface detergents like Tween 20 (Mahaffee *et al.*, 1997a; Pleban *et al.*, 1995), Tween 80 (Sturz 1995) or Triton X-100 (Misaghi and Donndelinger 1990; Musson *et al.*, 1995) may be used to reduce surface tension of the solvent, thereby allowing the disinfestant to reach protected sites and grooves beyond collapsed epidermal cells on the plant surface. Following surface disinfestations, the plant tissue is triturated with a mortar and pestle or mechanical device (Mahaffee *et al.*, 1997a) in sterile water, buffer solutions or liquid media. The triturate is then processed for bacterial enumeration. To decrease the possibility of recovery of surface contaminants, sterility checks are included to monitor the efficiency of the disinfestation procedure for each sample. Only if no bacterial growth occurs in the sterility check after a certain time, the recovered bacteria should be considered as endophytes (Hallmann *et al.*, 1997). The following sterility checks have been successfully applied: dipping roots in nutrient broth (Gagne *et al.*, 1987), transferring 0.1 ml of the final washing solution to a test tube with bacterial growth medium (McInroy and Kloepper 1994) or imprinting the surface disinfested plant tissue on nutrient medium (Pleban *et al.*, 1995; Shishido *et al.*, 1995).

Table 24.1: Most Common Indigenous Endophytic Bacterial Genera of Plant Parts of Different Crops

Plant Part	*Bacterial Taxa*	*Plant Species*	*Reference*
Seed	*Bacillus, Erwinia, Flavobacterium, Pseudomonas*	27 plant Species including cereals, vegetables and woody plants	Mundt and Hinkle, 1976
Tuber	*Bacillus*	Potato	Hollis 1951
Tuber	*Micrococcus, Pseudomonas, Bacillus, Flavobacterium, Xanthomonas, Agrobacterium, Coryneforms*	Potato (*S. tuberosum*)	de Boer and Copeman, 1974
Root	*Pseudomonas, Erwinia*-like bacteria, *Bacillus, Pseudomonas, Corynebacterium*	Alfalfa (*M. sativa*), Corn (*Z. mays*)	Gagne *et al.*, 1987; Lalande *et al.*, 1989
Root	*Agrobacterium, Burkholderia, Serratia*	Cotton (*G. hirsutum*)	McInroy and Kloepper 1995a
Root	*Burkholderia, Enterobacter*	Corn (*Z. mays*)	McInroy and Kloepper 1995a
Root	*Pseudomonas, Bacillus, Enterobacter, Agrobacterium, Chryseobacterium, Burkholderia, Arthrobacter, Stenotrophomonas*	Cucumber (*C. sativa*)	Mahaffee *et al.*, 1997a
Root	*Pseudomonas, Enterobacter, Bacillus, Corynebacterium* and Other Gram-positive bacteria (*e.g. Serratia*)	Rough lemon (*C. jambhiri*)	Gardner *et al.*, 1982
Root	*Bacillus, Erwinia, Pseudomonas, Chryseobacterium, Lactobacillus, Xanthomonas*	Sugar beet (*B. vulgaris*)	Jacobs 1985
Root	*Azospirillum* spp.	Rice and wheat	Christiansen-Weniger 1998
Root	*Bacillus* sp.	Pine	Chanway 1998
Stem	*Enterobacter, Klebsiella, Pseudomonas*	Corn (*Z. mays*)	Fisher *et al.*, 1992
Stem	*Bacillus*	Corn (*Z. mays*)	McInroy and Kloepper 1995a
Stem	*Bacillus*	Cotton (*G. hirsutum*)	McInroy and Kloepper 1995a
Stem	*Enterobacter, Pseudomonas, Pantoea, Rhodococcus*	Grapevine	Bell *et al.*, 1995
Stem	*Agrobacterium tumefaciens*	Rose	Marti *et al.*, 1999
Root, radicle, stem, unopened flowers, boll	*Erwinia, Bacillus, Clavibacter, Xanthomonas*	Cotton (*G. hirsutum*)	Misaghi and Donndelinger 1990
Root, Stem and Leaf	*Pseudomonas fluorescens*	Cotton (*G. hirsutum*)	Bhowmik *et al.*, 2002
Fruit	Pseudomonadaceae, Micrococca-ceae, Enterobactriaceae, Achromobacteriaceae	Tomato, Cucumber 1963	Samish *et al.*, 1961,

Vacuum and Pressure Extraction

A vacuum technique was successfully used to extract xylem sap from the roots of perennial plants like grapevine (Bell *et al.*, 1995) and citrus (Gardner *et al.*, 1982) and the scholander pressure bomb was used for extracting root sap from cotton and other agricultural crops (Hallmann *et al.*, 1997a).

A comparison of the bacterial community recovered from cotton roots by the trituration and pressure bomb technique appears to indicate that only a subset of the bacterial endophytes are recovered by the pressure bomb technique. *Bacillus* species, a predominant group of the soil rhizosphere and endorhiza communities of cucumber, were only recovered by the trituration technique, whereas *Pseudomonas* species, common dominated endophytic bacterial community, were recovered with pressure bomb technique (Hallmann *et al.*, 1997). The trituration technique consistently gave higher number of endophytic bacteria compared with either the pressure bomb (Hallmann *et al.*, 1997a) or vacuum extraction technique (Bell *et al.*, 1995).

Population Dynamics of Endophytic Bacteria

The simplest technique for monitoring population of bacterial endophytes is their culturing on artificial media. Several markers are available to differentiate the introduced microorganism from the indigenous microflora. A common marker has been the selection of spontaneous antibiotic resistant derivatives of the wild type strains (Kluepfel 1993; Mahaffee *et al.*, 1997a). Although multiplication of endophytic bacteria in plant tissue is difficult to demonstrate, the fact that several endophytes approach and maintain certain plateau population densities indicates bacterial multiplication. Hurek *et al.* (1994) reported inter- and intra-cellular multiplication of *Azoarcus* sp. in the root cortex of rice and kaller grass as shown by electron microscopy. Chen *et al.* (1995) reported that *Pseudomonas corrugata* CC-471, introduced at concentrations of 1.0×10^3 and 1.0×10^6 cfu/ml into cotton stems, first increased within 3 days to 1.8×10^6 and 4.3×10^6 cfu/g stem tissue respectively, before the population finally decreased to 1.0×10^3 cfu/g at 28th day, regardless of the inoculums densities. According to Frommel *et al.* (1991), bacterization of potato plantlets with *Pseudomonas* sp. strain PsJN at 3.7×10^5, 9.3×10^6 and 2.8×10^8 cfu/ml, all resulted in a mean final root population of 1.1×10^5 to 1.0×10^6 cfu/cm root tissue after 23 days. The stem population decreased from an initial concentration of 3.3×10^5 to 1.9×10^3 cfu/cm.

Gardner *et al.* (1982) reported that endophytic bacterial population in rough lemon roots of Florida citrus trees ranged from 0.1 to 22 per mm^3 of root tissue (about 1×10^2 to 2×10^4 cfu/g of xylem tissue) when bacterial counts were made on vascular fluid, but these numbers were 10 to 1000 fold greater when aseptically homogenized xylem tissue was examined. Gagne *et al.* (1987) reported that xylem residing bacterial population in alfalfa roots ranged from 6.0×10^3 to 4.3×10^4 cfu/g of fresh xylem tissue and was not affected by plant age or cultivar or by the sampling locations. Misaghi and Donndelinger (1990) studied the population levels of endophytic bacteria in symptom free cotton plants. They reported that average population sizes of *Erwinia* sp. throughout plants were relatively low *i.e.* 1.6×10^4 to 2.3×10^4 cfu/g fresh weight after its introduction into germinated cotton seeds. Population levels for *Clavibacter xyli* subsp. *cynodontis* in different tissues of mature corn varied from 1.0×10^7 to 1.0×10^9 cfu/g fresh weight (Reeser and Kostka 1988). Pleban *et al.* (1995) succeeded in recovering the endophytic *Bacillus cereus*, even after 72 days from the roots and stems of cotton plants. They used bacterial suspension containing 5.0×10^8 cfu/ml of endophyte in water. During that time, there was a decrease in the number of bacteria in both stems and roots after an initial increase. However, from day 30 and onward, the remaining bacteria began to multiply, causing a second increase in their numbers within the roots. After 24 h, the bacteria have already spread to all plant parts. The number of bacteria never decreased below 5×10^4 cfu/g fresh weight of stems and roots. After 72 days the population level in roots was 2.8×10^5 cfu/g fresh weight. Mclnroy and Kloepper (1994) reported that endophytic bacteria could approach population densities up to 1.0×10^9 to 1.0×10^{10} cfu/g fresh weight under certain conditions without affecting plant growth. Mclnroy and Kloepper (1995b) studied the population dynamics of endophytic bacteria in field grown cotton and sweet corn. Endophytic populations in cotton roots ranged from 4-6 log cfu/g fresh weight for most of the growing season in 1990 and 1991. In cotton petioles, mean populations generally fluctuated between 3-7 log cfu/g fresh weight during both the seasons. In case of sweet corn they reported that the endophyte population at seeding emergence ranged from 4-6 log cfu/g fresh weight in 1990 and 4-7 log cfu/g fresh weight in 1991.

Hallmann *et al.* (1997b) observed that following seed treatment endophytic *Pseudomonas fluorescens* 89B-61 colonized cotton roots endophytically at mean population density of 1.1×10^3 cfu/g fresh weight. However, bacteria were not detected in cotyledons, leaves or stems. After inoculation onto leaves, *Pseudomonas fluorescens* 89 B-61 established a mean internal population density of 1.6×10^4 cfu/g leaf tissue. Bhowmik *et al.* (2002) studied the population level of the two antagonistic endophytic bacterial isolates (*viz.* Endo PR4 and Endo PR8) which were isolated from root and stem of green house grown healthy cotton seedlings in two cotton cultivars (*viz.* Pusa 8-6 and Suvin) following seed bacterization. Results showed that the population of endophytes kept on increasing up to 15 days after seedling emergence, which is considered the active phase during which the plants/seedlings are infected by *Rhizoctonia solani*, *Sclerotium rolfsii* or by *Xanthomonas campestris* pv. *malvacearum*. The results obtained by Hallmann *et al.* (1997a) showed that following inoculation of cotton seedlings with endophytic *Enterobacter arburiae* JM22 suspension (1.0×10^{10} cfu/ml), the mean internal bacterial population density for seedling was 3.8×10^3 cfu/g radicle tissue.

Agricultural Uses of Endophytic Bacteria

Biological Control of Plant Diseases

Endophytic bacteria colonize an ecological niche similar to that of plant pathogens, especially vascular wilt pathogens, which might favour them as candidates for biocontrol agents. Their unique ability to survive inside plant tissues with little or no microbial competition make them potential candidates for biological control (Misaghi and Donndelinger 1990).

Biological Control of Fungal Diseases

In targeting fungal diseases, endophytic bacteria have shown significant control in model systems of *Fusarium oxysporum* f. sp. *vasinfectum* in cotton (Chen *et al.*, 1995), *Verticillium alboatrum* and *Rhizoctonia solani* in potato (Nowak *et al.*, 1995). Pleban *et al.* (1995) recovered isolates of different endophytic bacteria from surface disinfested seeds from various sources. *Bacillus cereus* from *Sinapis* (isolate no. 65) inhibited growth of *Rhizoctonia solani*, *Pythium ultimum* and *Sclerotium rolfsii*. *Bacillus subtilis* from onion tissue culture (isolate no. 72) inhibited *R. solani* and *P. ultimum* growth. *B. cereus* from cauliflower (isolate no 78) inhibited growth of *R. solani*. *B. pumilus* from sunflower (isolate no. 85) inhibited growth of *R. solani* and *S. rolfsii*. They also reported that when cotton seedlings were inoculated with *B. cereus* (isolate no. 65), *B. subtilis* (isolate no. 72) or *B. pumilus* (isolate no. 85), disease incidence caused by *R. solani* was reduced in the greenhouse by 51 per cent, 46 per cent and 56 per cent respectively. In bean seedlings inoculated with *B. subtilis* (isolate no 72), *B. cereus* (isolate no 78) or *B. pumilus* (isolate no. 65), disease incidence caused by *S. rolfsii* was reduced by 72 per cent, 79 per cent and 26 per cent respectively as compared to control. In cotton and bean seedlings, these endophytes reduced the disease index by more that 50 per cent. The endophytic bacteria *Enterobacter cloacae*, strain EcCT 501 suppressed *Pythium ultimum* that causes damping-off of many important vegetable and turf crop species (Nelson and Craft 1991; Wilson *et al.*, 1987; Howell *et al.*, 1988).

Mukhopadhyay *et al.* (1996) isolated seven endophytic bacteria from two varieties, IR-42 and BKNFR of rice seedlings grown from intact and dehulled seeds. The bacteria were tested against four fungi, *viz. Rhizoctonia solani*, *Pythium myriotylum*, *Gaeumannomyces graminis* and *Heterobasidium annosum* for their antifungal activity. The most effective bacteria against all the four fungi were *Enterobacter cloacae* and *E. agglomerans*. In Greenhouse tests, sclerotia of *R. solani* soaked for 7 and 14 days in cultures and culture filtrates of *E. cloacae* and *E. agglomerans* were unable to infect rice leaves or grow on PDA plates. Bhowmik *et al.* (2002) reported that five pseudomonads (non-fluorescent and fluorescent) out of fifty isolates of endophytic bacteria isolated from root and stem of green house grown healthy cotton seedlings were found to be antagonistic to *Rhizoctonia solani* and *Sclerotium rolfsii* which cause damping-off disease in cotton. Seed bacterization with one of the endophytes (Endo PR 8) reduced damping-off disease incidence caused by *R. solani* and *S. rolfsii* by 84 per cent and 76 per cent respectively. Hinton and Bacon (1995) reported that *Enterobacter cloacae*, an endophyte isolated form corn exhibited *in vitro* antibiosis against *Fusarium moniliforme*.

Biological Control of Bacterial Diseases

Antagonistic activity of endophytic bacteria against bacterial pathogens has been reported for *Clavibacter michiganensis* subsp. *sepedonicum*, the causal agent of bacterial ring rot of potato (Van Buren *et al.*, 1993). In addition, *Pseudomonas fluorescens* 89 B-27 and *Serratia marcescens* 90–166 were observed to induce resistance in cucumber against *Pseudomonas syringae* pv. *lachrymans* as well as against the fungal pathogens *F. oxysporum* f. sp. *cucumerinum* and *Colletotrichum arbiculae* (Liu *et al.*, 1995). Bacterial endophytes are also thought to play a role in the resistance observed in the multiple adversity resistant (MAR) cotton lines (Bird *et al.*, 1980; Bird 1982).

Bhowmik *et al.* (2002) reported that five pseudomonads (non-fluorescent and fluorescent) out of fifty isolates of endophytic bacteria isolated from root and stem of green house grown healthy cotton seedlings were found to be antagonistic to *Xanthomonas campestris* pv. *malvacearum* (*Xcm*) which causes bacterial blight in cotton. Seed bacterization with endophytes showed that Endo PR 8 was the most effective isolate to reduce the cotyledonary infection by *Xcm*, the percent protection of cotyledonary infection varied from 53.3 (in unsterilised soil) and 60.0 (in sterilized soil).

Plant Growth Promotion by Endophytic Bacteria

Endophytic bacteria have been associated with the growth promotion of several crops, including tomato, lettuce (Bashan *et al.*, 1989; Nowak *et al.*, 1995; Van Peer and Schippers 1989), potato (Frommel *et al.*, 1991; Sturz 1995; Van Peer and Schippers 1989), corn (Hinton and Bacon 1995; Lalande *et al.*, 1989), cucumber (Van Peer and Schippers 1989; Kloepper *et al.*, 1992b; Nowak *et al.*, 1995), rice (Hurek *et al.*, 1994), and cotton (Bashan *et al.*, 1989). According to Sturz (1995) approximately 10 per cent of bacterial isolates recovered from within potato tubers were shown to promote plant growth. The nonfluorescent *Pseudomonas* sp. strain PsJN induced significant increases in potato root numbers (24–196 per cent), root dry weight (44-201 per cent) and stem length (26-28 per cent), as well as enhanced leaf hair formation (55-110 per cent), secondary root branching and total plant lignin content (43 per cent) (Frommel *et al.*, 1991).

Inoculation of rice with the diazotrophic endophyte *Azoarcus* sp. strain BH 72 significantly promoted plant growth (Hurek *et al.*, 1994). In this particular case, growth promotion also occurred with Nif mutants indicating that nitrogen fixation by *Azoarcus* sp. was apparently not involved in plant growth promotion. The observed plant growth promotion might have been caused by enhanced plant mineral uptake and improved plant water relationship associated with the colonization by *Azoarcus* (Hallmann *et al.*, 1997). Some strains of *Pseudomonas, Enterobacter, Staphylococcus, Azotobacter* and *Azospirillum* produce plant growth regulators such as ethylene, auxins or cytokinins and have, therefore, been considered as casual agents for altering plant growth and development (Arshad and Frankenberger 1991; Bashan and Holguin 1997; Leifert *et al.*, 1994). In addition to a direct mechanism for growth promotion, plant growth promotion is also thought to be due to the suppression of a deleterious microflora by the introduced endophyte (Kloepper *et al.*, 1991). Van Peer and Schippers (1989) reported that tomato seedlings inoculated with *Pseudomonas* sp. strain WCS417r showed increased plant growth accompanied by dense colonization of the applied strain in internal root tissues and a consequent displacement of indigenous endophytic pseudomonads that were deleterious to plant growth.

Resistance Induction in Plants Against Bacterial Diseases through Pre-inoculation of Bacterial Endophytes

Pre- or simultaneous inoculation of the plant with saprophytic bacteria can protect plants against infection by a virulent bacterium (Goodman 1967; Verma *et al.*, 1980). Pre-inoculative protective ability of certain phylloplane bacteria (Plb) against bacterial blight of cotton has been well documented (Verma and Singh 1976; Verma *et al.*, 1978; Verma *et al.*, 1983). Habish (1968a, 1968b) observed that *Flavobacterium* sp. isolated from wet cotton trash infected by *Xanthomonas axonopodis* pv. *malvacearum* (*Xam*) caused delay in symptom appearance and decrease in disease grade when cotton leaves were sprayed along with *Xam.*

Hsieh and Buddenhagen (1971) studied several saprophytic bacteria from bacterial blight infected as well as healthy rice leaves, which were found to be inhibitory to *Xanthomonas oryzae in vitro. In vivo* symptom development was much reduced when low concentration of *X. oryzae* (10^6 cells/ml) was combined with high concentration (10^8 cells/ml) of saprophytes in the inoculum. Later, Hsieh and Buddenhagen (1974) found that though *Erwinia herbicola* was not antagonistic to *X. oryzae in vitro,* it reduced symptom appearance when inoculated into rice plant mixed with *X. oryzae.* Watanabe *et al.* (1976) observed that lesion length was markedly inhibited by preliminary inoculation with incompatible strain of *X. oryzae.*

Verma *et al.* (1978, 1980, 1982, 1983) investigated the possibility of biological control of bacterial blight of cotton with the use of phylloplane bacteria (Plbs), using high disease pressure *i.e.* with highly susceptible cultivars of cotton and aggressive strains of *Xanthomonas campestris* pv. *malvacearum* (*Xcm*), both under field and laboratory conditions. A pre-inoculative protective effect of avirulent strain, heat killed cells and Plbs in susceptible cotton cultivars was found on challenge inoculation of virulent strains of *Xcm* given at 8-10 h. No such protection was expresses in mixed inoculation (Verma and Singh 1974, 1976; Verma *et al.*, 1978). Specific phylloplane bacteria like *Flavobacterium* (Plb 5 and Plb 6), *Aeromonas* (Plb12) and *Pseudomonas* (Plb 51) reduced the disease intensity when applied as pre-inoculative spray (Verma *et al.*, 1983). It was observed that application of Plbs resulted in significant reduction of the number of lesions per leaf while the number of infected leaves per plant remained same as untreated control. Randhawa *et al.* (1987) identified *Erwinia herbicola, Enterobacter cloacae* and *Pseudomonas* spp. as microflora associated with cotton seeds of cultivar F 414 of *G. hirsutum* and LD 133 of *G. arboreum. Erwinia herbicola* strains were found effective in reducing seed borne infection by *Xcm* (Randhawa and Singh 1981).

Besides pre-inoculation, the inoculation of mixtures of microbial agents have also been used successfully to increase the level of biocontrol ability above that achieved with individual strain of the mixtures (Kloepper 1983; Pierson and Weller 1994; Duffy and Weller 1995; Duffy *et al.*, 1996; Janisiewicz 1996; Leeman *et al.*, 1996). Schister *et al.* (1997) used combinations of antagonists at 1: 100 concentrations of the individual strains making up pair. Sixteen such pairs of antagonists controlled dry rot development in potato better than predicted based on the performance of the individual strains making up the pairs.

Disease Suppression with Endophytic Bacterial Metabolites

Many bacteria afforded biological control by producing one or more of a variety of secondary metabolites such as siderophores, antibiotics, cyanide etc. (Kloepper *et al.*, 1980; Voisard *et al.*, 1989; Mondal and Verma 1998).

Fluorescent Pseudomonads are known to produce a variety of anti-microbial metabolites such as phenazine 1-carboxylate (Pierson and Thomashow 1992; Powell 1995), pyrrolnitrin (Pfender *et al.*, 1993), 2-4 diacetylphloroglucinol (Keel *et al.*, 1992; Vincent *et al.*, 1991; Defago 1993; Nowak-Thomson *et al.*, 1994; Mondal *et al.*, 2000), pyoluteorin (Kraus and Loper 1990, Maurhofer *et al.*, 1992) and hydrogen cyanide (Freeman *et al.*, 1975; Voisard *et al.*, 1989). All these metabolites have been reported as key factors contributing to the suppression of plant disease. The role of phenazine 1-carboxylic acid (PCA) produced by *Pseudomonas fluorescens* 2-79 in biological control of take-all pathogen, *Gaeumannomyces graminis* var. *tritici* has been well established (Hamdan *et al.*, 1991; Thomashow and Weller 1990). Mutants of strain 2-79 deficient in production of the PCA showed reduced level of suppression than that of the parental strains whereas the siderophore, pyoverdin deficient mutants controlled take-all as effectively as their respective parental strains. Thus, it has been suggested that the PCA is the dominant factor in take-all suppression while fluorescent siderophore, pyoverdin plays little or no role (Hamdan *et al.*, 1991).

The production of antibiotic, 2,4-diacetylphloroglucinol (Phl) by *Pseudomonas fluorescens* strain CHAO has been proved to be an important factor in the suppression of black root rot of tobacco caused by *Thielaviopsis basicola* (Keel *et al.*, 1990, 1992). A mutant, CHA 625, which is defective in Phl production, suppressed black root rot to a distinctly smaller extent that did wild type CHAO. A cosmid constructed from a genomic library of strain CHAO when introduced into the mutant strain, restored the ability of strain CHA 625 to produce Phl and to suppress the disease. Further, addition of synthetic 2, 4-diacetylphloroglucinol to soil reduced disease severity in the absence of bacteria suggesting the dominant role of Phl in the suppression of black root rot of tobacco by *P. fluorescens* strain CHAO. Prevalence of Phl production was also detected in other strains like *Pseudomonas fluorescens* pf-5, a biocontrol agent of diseases caused by *Pythium ultimum*, *Rhizoctonia solani* and *Erwinia carotovora* subsp. *atroseptica* (Nowak-Thompson *et al.*, 1994) and in *Pseudomonas fluorescens* strain PFM2, an antagonist of *Septoria tritici* (Levy *et al.*, 1992). Maurhofer *et al.* (1995) developed an over producing strain, CHAO/pME3090 by introduction of recombinant cosmid pME3090 into *P. fluorescens* strain CHAO that overproduced antibiotic Phl and Ptl by 3-5 folds and showed an increased capacity to protect wheat against *Pythium ultimum* and *Gaeumannomyces graminis* var. *tritici* and cucumber against *Fusarium oxysporum* f. sp. *cucumerinum* and *Phomopsis scleroboides* compared with the wild type strain CHAO.

Mulya *et al.* (1996) suggested that the ability of *Pseudomonas fluorescens* pfG 32 to produce antibiotics and siderophores is the important factor in determining its biocontrol activity against bacterial wilt disease of tomato caused by *Ralstonia solanacearum*. Howie and Suslow (1991) confirmed the role of antibiotic of *Pseudomonas fluorescens* in the inhabitation of *Pythium ultimum* in cotton spermosphere and rhizosphere. Hydrogen cyanide producing strain CHAO of *Pseudomonas fluorescens* has been found to be involved in the suppression of black root rot of tobacco caused by *Thielaviopsis basicola* (Ahl *et al.*, 1986; Voisard *et al.*, 1989). Piotrowska-seget (1995) investigated the hydrocyanic acid (HCN) production by stains P11 and S241 of *Pseudomonas fluorescens* known to suppress *Thielaviopsis basicola* on wheat and tobacco and its effect on plant growth. When grown on King's B broth supplemented with glycine, strain S241 produced 483 n mol/ml and P11, 1109 n mol/ml HCN over a 48 hours period. Strain P11 inhibited the growth of *T. basicola* strongly as compared to S241. It was suggested that stronger growth inhibition caused by P11 was related to higher levels of cyanide production in soil compared to S241. Howell et al. (1988) reported the possible role of ammonia produced by the endophytic bacteria, *Enterobacter cloacae* in the biological control of *Pythium* pre-emergence damping-off of pea, beet, cotton and cucumber. According to Loper *et al.* (1993), the bacterium produces siderophores that inhibit the growth of *Pythium ultimum* through iron competition. Costa and Loper (1994) reported that *Enterobacter cloacae* EcCT-501, which suppressed *Pythium* damping-off of cucumber and other plant hosts, produced the hydroxamate siderophore aerobactin and a catechol siderophore tentatively identified as enterobactin. However, aerobactin and enterobactin production by *E. cloacae* did not contribute to its activity in biological control of *Pythium* damping-off of cotton or cucumber. They concluded that these two siderophores might contribute to the ecological fitness of *E. cloacae* and its interactions with other rhizosphere and spermosphere microorganisms. Mukhopadhyay *et al.* (1996) isolated several endophytic bacteria from rice seeds, which exhibited strong anti-fungal activity against *Rhizoctonia solani*, *Pythium myriotylum*, *Gaeumannomyces graminis* and *Heterobasidium annosum*. They concluded that the anti-fungal activity of *Enterobacter* spp. was either due to ammonia or due to peptides. Anith *et al.* (1999) reported that *Pseudomonas* sp. strain EM 85, a rhizosphere isolate, showed *in vitro* antifungal activity against a number of soil-borne plant pathogens and produced multiple anti-fungal compounds such as HCN, siderophores and fluorescent pigments. *In vivo* experiment established that antifungal antibiotic was responsible for suppression of damping-off of cotton caused by *Rhizoctonia solani*. Recently Mondal *et al.* (2000) reported that *Pseudomonas fluorescens* (CRb-26) produced four phenolic compounds, tentatively designated as I, II, III and IV. Compound I was hydroxy type (fluorescent), compounds II and III were polyhydroxy type (non-fluorescent) and compound IV was of dihydroxy type (fluorescent). Compounds II and IV completely inhibited the growth of *Xanthomonas campestris* pv. *malvacearum* Race 32 *in vitro* at a concentration of 200 and 100 mg/ml respectively. These two compounds also protected cotton plants from bacterial blight disease. Bhowmik *et al.* (2002) isolated three metabolites from endophytic *Pseudomonas fluorescens* (isolate no. Endo PR8) tentatively referred to as compound I, II and III. Compound I was of polyhydroxy phenol group while compound II and II belonged to hydroxy phenol. Only compound I could inhibit the growth of *Xanthomonas campestris* pv. *malvacearum*, *Rhizoctonia solani* and *Sclerotium rolfsii.*

Conclusions

Endophytic bacteria could be exploited in the successful biocontrol of important fungal or bacterial diseases of plants such as damping-off caused by *Rhizoctonia solani*, *Sclerotium rolfsii*, *Pythium* sp. and bacterial blight of cotton caused by *Xanthomonas campestris* pv. *malvacearum*. These disease suppression mechanisms by endophytic bacteria could also be integrated into the Integrated Disease Management (IDM) practices. Plant growth promotion is another aspect, which can be achieved with the exploitation of endophytic bacteria, thus reducing the gestation period of many agriculturally important crops. The genes responsible for metabolite production in endophytic bacteria can be identified and may be incorporated in transgenic production for better crop yield. Although the endophytes were found effective in growth promotion and plant disease management, their applications remained restricted only to a few important field crops against a few selected plant pathogens. Further comprehensive researches are needed for successful application of the endophytes with wider adaptability and acceptability involving many other agri horticultural crops for growth promotion and management of many other destructive crop diseases.

Acknowledgements

The authors are grateful to Dr. J.P. Verma, Former Head, Dr. R. P. Singh, Sr. Scientist, and Dr. Jayashree Jayaraman, Sr. Scientist, Division of Plant Pathology, Indian Agricultural Research Institute, New Delhi, India for their constant support, encouragement and guidance in carrying out research on endophytic bacteria.

References

Agarwal, S. and Shende, S.T., 1987. Tetrazolium reducing microorganisms inside the root of *Brassica* species. *Curr. Sci.,* 56: 187–188.

Ahl, P., Voisard, C. and Defago, G., 1986. Iron bound siderophore, cyanide and antibiotic involved in suppression of *Thielaviopsis basicola* by a *Pseudomonas fluorescens* strain. *J. Phytopathol.,* 116: 121–134.

Anith, K.N., Tilak, K.V.B.R. and Manomohandas, T.P., 1999. Analysis of mutation affecting antifungal property of *fluorescent Pseudomonas* sp. during cotton-*Rhizoctonia* interaction. *Indian Phytopath.,* 52(4): 366–369.

Arshad, M. and Frankenberger, W.T., 1991. Microbial Production of Plant Hormones. In: *The Rhizosphere and Plant Growth,* (Eds.) D.L. Keister and P.B. Cregan. Kluwer Academic Publishers, Dordrecht, The Netherlands, pp. 327–334.

Bashan, Y. and Holguin, G., 1997. *Azospirillum*–plant relationships: environmental and physiological advances, 1990–1996. Can. J. Microbiol., 43: 103–121.

Bashan, Y., Ream, Y., Levanony, H. and Sade, A., 1989. Non-specific responses in plant growth, yield, and root colonization of non cereal crop plants to inoculation with *Azospirillum brasilense. Can. J. Bot.,* 67: 1317–1324.

Beattie, G.A. and Lindow, S.E., 1995. The secret life of foliar bacterial pathogens on leaves. *Ann. Rev. Phytopathol.,* 33: 145–172.

Bell, C.R., Dickie, G.A., Harvey, W.L.G. and Chan, J.W.Y.F., 1995. Endophytic bacteria in grapevine. *Can. J. Microbiol.,* 41: 46–53.

Bhowmik, B., Singh, R.P., Jayaraman, J. and Verma, J.P., 2002. Population dynamics of cotton endophytic *Pseudomonas,* their antagonism and protective action against the major pathogens of cotton. *Indian Phytopath.,* 55(2): 124–132.

Bird, LS., 1982. The MAR (multi-adversity resistance) system of genetic improvement of cotton. *Plant Dis.,* 66: 172–176.

Bird, L.S., Leverman, C., Thaxaton, P. and Percy, R.G., 1980. Evidence that microorganisms in and on tissues have a role in a mechanism of multi-adversity resistance in cotton. In: *Proceedings of Belt wide Cotton Production Research Conferences,* (Ed.) J.M. Brown. National Cotton Council Memphis Tenn., 40: 283–285.

Caetano-Anolles, G., Favelukes, G. and Bauer, W.D., 1990. Optimization of surface sterilization for legume seed. *Crop Sci.*, 30: 708–712.

Chanway, C.P., 1996. Endophytes: They're not just fungi! *Can. J. Bot.*, 74: 321–322.

Chanway, C.P., 1998. Bacterial endophytes: Ecological and practical implications. *Proc. the 7th ICPP*, Edinburgh, U.K.

Chen, C., Bauske, E.M., Musson, G., Rodriguez Kabana, R. and Kloepper, J.W., 1995. Biological control of *Fusarium* wilt on cotton by use of endophytic bacteria. *Biol. Control*, 5: 83–91.

Christiansen-Weniger, C., 1998. Endophytic establishment of diazotrophic bacteria in auxin induced tumors of cereal crops. *Crit. Rev. Plant Sci.*, 17(1): 55–76.

Costa, J.M. and Loper, J.E., 1994. Characterization of siderophore production by the biological control agent *Enterobacter Cloacae. Mol. Plant Microbe. Interact.*, 4: 440–448.

de Bary, A., 1866. Morphologie Und Physiologie der Pilze, Flechten Und Myxomyceten. Engelmann, Leipzig, pp. 316.

De Boer, S.H. and Copeman, R.J., 1974. Endophytic bacterial flora in *Solanum tuberosum* and its significance in bacterial ring rot disease. *Can. J. Plant Sci.*, 54: 115–122.

Defago, G., 1993. 2,4-diacetylphloroglucinol, a promising compound in biocontrol. *Plant Pathol.*, 42: 311–312.

Dong, Z, Canny, M.J., McCully, M.E., Robordo, M.R., Cabadilla, C.F., Ortega, E. and Rodes, R., 1994. A nitrogen fixing endophyte of sugarcane stems. *Plant Physiol.*, 105: 1139–1147.

Duffy, B.K. and Weller, D.M., 1995. Use of *Gaeumannomyces graminis* var. *graminis* alone and in combination with fluorescent *Pseudomonas* species to suppress take-all of wheat. *Plant Dis.*, 79: 907–911.

Duffy, B.K., Simon, A. and Weller, D.M., 1996. Combination of *Trichoderma koningii* with *fluorescent pseudomonads* for control of take-all of wheat. *Phytopathol.*, 86: 188–194.

Fisher, P.J., Petrini, O. and Scott, H.M.L., 1992. The distribution of some fungal and bacterial endophytes in maize (*Zea mays*). *New Phytol.*, 122: 299–305.

Freeman, L.R., Angelini, R.P., Silverman, G.J. and Merrett, C., 1975. Production of hydrogen cyanide by *Pseudomonas fluorescens. Appl. Microbial.*, 29: 560–561.

Frommel, M.I., Nowak, J. and Lazarovits, G., 1991. Growth enhancement and developmental modifications of *in vitro* grown potato (*Solanum tuberosum* ssp.*tuberosum*) as affected by a nonfluorescent *Pseudomonas* sp. *Plant Physiol.*, 96: 928–936.

Gagne, S., Richard, C., Rousseau, H. and Antoun, H., 1987. Xylem residing bacteria in alfalfa roots. *Can. J. Microbiol.*, 33: 996–1000.

Gardner, J.M., Feldman, A.W. and Zablotowicz, R.M., 1982. Identity and behaviour of xylem–residing bacteria in rough lemon roots of Florida citrus trees. *Appl. Environ. Microbiol.*, 43(6): 1335–1342.

Germida, J.J., Siciliano, S.D., De Freitas, J.R. and Seib, A.M., 1998. Diversity of root–associated bacteria associated with field–grown canola (*Brassica napus* L.) and wheat (*Triticum aestivum* L.). *FEMS Microbiol. Ecol.*, 26: 43–50.

Goodman, R.N., 1967. Protection by apple stem tissues against *Erwinia amylovora* infection by avirulent strains and three other bacterial species. *Phytopathol.*, 57: 22–24.

Habish, H.A., 1968a. Role of antagonistic bacterium in the decline of *Xanthomonas malvacearum* in wet cotton debris. *Cott. Gr. Rev.*, 45:36–41.

Habish, H.A., 1968b. Effect of an antagonistic bacterium on degree of infection of cotton with *Xanthomonas malvacearum*. *Cott. Gr. Rev.*, 45: 137–140.

Hallmann, A.Q., Benhamou, N. and Kloepper, J.W., 1997a. Bacterial endophytes in cotton: Mechanisms of entering the plant. *Can. J. Microbiol.*, 43: 577–582.

Hallmann, A.Q., Hallmann, J. and Kloepper, J.W., 1997b. Bacterial endophytes in cotton: location and interaction with other plant associated bacteria. *Can. J. Microbiol.*, 43: 254–259.

Hallmann, J., Hallmann, A.Q., Mahaffee, W.F. and Kloepper, J.W., 1997. Bacterial endophytes in agricultural crops. *Can. J. Microbiol.*, 43: 895–914.

Hamdan, H., Weller, D.M. and Thomashow, L.S., 1991. Relative importance of fluorescent siderophores and other factors in biological control of *Gaeumannomyces graminis* var. *tritici* by *Pseudomonas fluorescens* 2–79 and M4–80R. *Appl. Environ. Microbiol.*, 57(11): 3270–3277.

Henning, K. and Villforth, F., 1940. Experimentelle untersuchungen zur Frage der Bakteriensymbiose in hoheren Pflanzen und ihrer BeeinfluBung durch Leitelemente. *Biochem Z.*, 305: 299–309.

Hinton, D.M. and Bacon, C.W., 1995. *Enterobacter cloacae* is an endophytic symbiont of corn. *Mycopathologia*, 129: 117–125.

Hollis, J.P., 1951. Bacteria in healthy potato tissue. *Phytopathol.*, 41: 350–366.

Howell, C.R., Beier, R.C. and Stipanovic, R.D., 1988. Production of ammonia by *Enterobacter cloacae* and its possible role in the biological control of *Pythium* pre-emergence damping-off by the bacterium. *Phytopathol.*, 78: 1075–1078.

Howie, W.J. and Suslow, T.V., 1991. Role of antibiotic biosynthesis in the inhibition of *Pythium ultimum* in the cotton spermosphere and rhizosphere by *Pseudomonas fluorescens*. *Mol. Plant Microbe. Interact.*, 4: 393–399.

Hsieh, S.P.Y. and Buddenhagen, I.W., 1971. Interactions between *Xanthomonas oryzae* and saprophytic bacteria isolated from rice leaves. *Phytopathol.*, 61: 1321.

Hsieh, S.P.Y. and Buddenhagen, I.W., 1974. Suppressing effect of *Erwinia herbiolcola* on infection by *Xanthomonas oryzae* and on symptom development in rice. *Phytopathol.*, 64: 1182–1185.

Hurek, T., Hurek, B.R., Montagu, M.V. and Kellenberger, E., 1994. Root colonization and systemic spreading of *Azoarcus* sp. strain BH in grasses. *J. Bacterial.*, 176(7): 1913–1923.

Jacobs, M.J., 1985. Enumeration, location and characterization of endophytic bacteria within sugar beet roots. *Can. J. Bot.*, 63: 1262–1265.

Janisiewicz, W., 1996. Ecological diversity, niche overlap and coexistence of antagonists used in developing mixtures of biocontrol of post-harvest disease of apples. *Phytopathol.*, 86: 473–479.

Kado, C.I., 1992. Plant pathogenic bacteria. In: *The Prokaryotes*, (Eds.) A. Balows, H.G. Truper, W. Dworkin, W. Harder and K.H. Schleifer. Springer-Verlag, New York, pp. 660–662.

Keel, C., Schnider, U., Maurhofer, M., Voisard, C., Laville, J., Burger, U., Wirthner, P., Haas, D. and Defago, G., 1992. Suppression of rot diseases by *Pseudomonas fluorescens* CHAO: importance of bacterial secondary metabolite, 2,4-diacetyphloroglucinol. *Mol. Plant Microbe. Interact.*, 5(1): 4–13.

Keel, C., Wirthner, P., Oberhansli, T., Voisard, C., Burger, U., Haas, D. and Defago, G., 1990. *Pseudomonads* as antagonists of plant pathogen in rhizosphere: Role of the antibiotic 2,4-di acetyl phloroglucinol in the suppression of black root-rot of tobacco. *Symbiosis*, 9: 327–341.

Kloepper, J.W., 1983. Effect of seed piece inoculation with plant growth promoting rhizobacteria on population of *Erwinia carotovora* on potato roots and daughter tubers. *Phytopathol.*, 73: 217–219

Kloepper, J.W., Leong, J., Teintze, T. and Schroth, M., 1980. *Pseudomonas* siderophore: A mechanism explaining disease suppressive soils. *Curr. Microbiol.*, 4: 317–320.

Kloepper, J.W., Wei, G. and Tuzun, S., 1992b. Rhizosphere population dynamics and internal colonization of *Cucumber* by plant growth: Promoting rhizobacteria, which induce systemic resistance of *Colletotrichum orbiculare*. In: *Biological Control of Plant Diseases*, (Ed.) E.S. Tjamos. Plenum Press, New York, p. 185–191.

Kloepper, J.W., Zablotowicz, R.M., Tipping, E.M. and Lifshitz, R., 1991. Plant growth promotion mediated by bacterial rhizosphere colonizers. In: *The Rhizosphere and Plant Growth*, (Eds.) D.L. Keister and P.B. Cregan. Kluwer Academic Publishers, Dordrecht, The Netherlands, pp. 315–326.

Kluepfel, D.A., 1993. The behaviour and tracking of bacteria in the rhizosphere. *Ann. Rev. Phytopathol.*, 31: 441–472.

Kraus, J. and Loper, J.E., 1990. The role of pyoluteorin and fluorescent siderophore production by *Pseudomonas fluorescens* Pf-5 in biocontrol of *Pythium* damping-off of cotton. *Phytopathol.*, 80: 1048–1054.

Lalande, R., Bissonnette, N., Coutlee, D. and Antoun, H., 1989. Identification of rhizobacteria from maize and determination of their plant growth promoting potential. *Plant Soil* 115: 7–11.

Leeman, M., den Ouden, F.M., Pelt, J.A., Coruelissen, C., Matamala, A.G., Bakker, P.A.H.M. and Schippers, B., 1996. Suppression of *Fusarium* Wilt of radish by co-inoculation of fluorescent *Pseudomonas* spp. and root-colonizing fungi. *European J. Plant Pathol.*, 102: 21–31.

Leifert, C., Morris, C.E. and Waites, W.M., 1994. Ecology of microbial saprophytes and pathogens in tissue culture and field-grown plants: Reasons for contamination problems *in vitro*. *Crit. Rev. Plant Sci.*, 13: 139–183.

Leifert, C., Waites, W.M. and Nicholas, J.R., 1989. Bacterial contaminants of micropropagated plant cultures. *J. Appl. Bactreiol.*, 67: 353–361.

Levy, E., Gough, F.J., Berlin, K.D., Guiana, P.W. and , J.T., 1992. Inhibition of *Septoria tritici* and other phytopathogenic fungi and bacteria by *Pseudomonas fluorescens* and its antibiotics. *Plant Pathol.*, 41: 335–341.

Liu, L., Kloepper, J.W. and Tuzun, S., 1995. Induction of systemic resistance in cucumber against *Fusarium* wilt by plant growth promoting rhizobacteria. *Phytopathol.*, 85: 843–847.

Loper, J.E., Ishimaru, C.A., Carnegie, S.R. and Vanavichit, A., 1993. Cloning and characterization of the aerobactin biosynthesis genes of the biological control agent *Enterobacter cloacae*. *Appl. Environ. Microbiol.*, 59: 4189–4197.

Mahaffee, W.F., Bauske, E.M., Van Vuurde, J.W.L., Van der Wolf, J.M., Van den Brink, M. and Kloepper, J.W., 1997a. Comparative analysis of antibiotic resistance, immunofluorescent colony staining and a transgenic marker (bioluminescence) for monitoring the environmental fate of a rhizobacterium. *Appl. Environ. Microbiol.*, 63: 1617–1622.

Marti, R., Cubero, J., Daza, A., Piquer, J., Salcedo, C.I., Morente, C. and Lopez, M.M.., 1999. Evidence of migration and endophytic presence of *Agrobacterium tumefaciens* in rose plants. *European J. Plant Pathol.*, 105: 39–50.

Maurhofer, M., Keel, C., Hass, D. and Defago, G., 1995. Influence of plant species on disease suppression by *Pseudomonas fluorescens* CHAO with enhanced antibiotic production. *Plant Pathol.*, 44(1): 40–50.

Maurhofer, M., Keel, C., Schnider, U., Voisard, C., Haas, D. and Defago, G., 1992. Influence of enhanced antibiotic production in *Pseudomonas fluorescens* strains CHAO on its disease suppressive capacity. *Phytopathol.*, 82: 190–195.

Mclnroy, J.A. and Kloepper, J.W., 1995a. Survey of indigenous bacterial endophytes from cotton and sweet corn. *Plant Soil*, 173: 337–342.

Mclnroy, J.A. and Kloepper, J.W., 1995b. Population dynamics of endophytic bacteria in field grown sweet corn and cotton. *Can. J. Microbiol.*, 41: 895–901.

Mclnroy, J.A., and Kloepper, J.W., 1994. Studies on indigenous endophytic bacteria of sweet corn and cotton. In: *Molecular Ecology of Rhizosphere Microorganisms*, (Eds.) F.O. Gara, D.N. Dowling, B. Boesten and V.C.H. Verlagsgesellschaft. Germany, Weinheim, pp. 19–28.

Meneley, J.C. and Stanghellini, M.E., 1975. Establishment of an inactive population of *Erwinia carotovora* in healthy cucumber fruit. *Phytopathol.*, 65: 670–673.

Misaghi, I.J. and Donndelinger, C.R., 1990. Endophytic bacteria in symptom-free cotton plants. *Phytopathol.*, 80(9): 808–811.

Mondal, K.K. and Verma, J.P., 1998. Role of secondary metabolites of *Pseudomonas fluorescens* in the biocontrol of plant pathogens. In: *Biological Suppression of Plant Diseases, Phytoparasitic Nematodes and Weeds*, (Eds.) S.P. Singh and S.S. Hussaini. Project Directorate of Biological Control, Bangalore, pp. 70–80.

Mondal, K.K., Singh, R.P., , P. and Verma, J.P., 2000. Secondary metabolites of cotton rhizobacteria in the suppression of bacterial blight of cotton. *Indian Phytopath.*, 53(1): 22–27.

Mukhopadhyay, N.K., Garrison, N.K., Hinton, D.M., Bacon, C.W., Khush, G.S., Peck, H.D. and Datta, N., 1996. Identification and characterization of bacterial endophytes of rice. *Mycopathologia*, 134: 151–159.

Mulya, K., Watanabe, M., Goto, M., Takikawa, Y. and Tsuyumi, S., 1996. Suppression of bacterial wilt disease of tomato by root dipping with *Pseudomonas fluorescens* PfG32: The role of antibiotic substances and siderophore production. *Ann. Phytopath. Soc., Japan*, 62(2): 134–140.

Mundt, J.O. and Hinkle, N.F., 1976. Bacteria within ovules and seeds. *Appl. Environ. Microbiol.*, 32: 694–698.

Musson, G., Mclnroy, J.A. and Kloepper, J.W., 1995. Development of delivery systems for introducing endophytic bacteria into cotton. *Biocontrol Sci. Techno.*, 5: 407–416.

Nelson, E.B. and , C.M., 1991. Introduction and establishment of strains of *Enterobacter cloacae* in golf course turf for the biological control of dollar spot. *Plant Dis.*, 75: 510–514.

Nowak, J., Asiedu, S.K., Lazarovits, G., Pillay, V., Stewart, A., Smith, C. and Liu, Z., 1995. Enhancement of *in vitro* growth and transplant stress tolerance of potato and vegetable plantlets co-cultured with a plant growth promoting *Pseudomonad* bacterium. In: *Ecophysiology and Photosynthetic in Vitro Cultures,* (Eds.) F. Carre and P. Chagvardieff. Commissariat a l'engergie, France, pp. 173–179.

Nowak-Thompson, B., Gould, S.J., , J. and Loper, J.E., 1994. Production of 2,4-diacetylphloroglucinol by the biocontrol agent *Pseudomonas fluorescens* Pf-5. *Can. J. Microbiol.,* 40(12): 1064–1066.

Pattit, R.E., Taber, R.A. and Foster, B.G., 1968. Occurrence of *Bacillus subtilis* in peanut kernels. *Phytopathol.,* 58: 254–255.

Perotti, R., 1926. On the limits of biological enquiry in soil science. *Proc. Int. Soc. Soil Sci.,* 2: 146–161.

Petrini, O., 1998. What are endophytes any way? *Proceeding of the 7th ICCP,* Edinburgh, U.K.

Pfender, W.F., Kraus, J. and Loper, J.E., 1993. A genomic region from *Pseudomonas fluorescens* Pf-5 required for pyrrolnitrin production and inhibition of *Pyrenophora tritici repentis* in wheat straw. *Phytopathol.,* 83: 1223–1228.

Philipson, M.N. and Blair, I.D., 1957. Bacteria in clover root tissue. *Can. J. Microbiol.,* 3: 125–129.

Pierson, E.R. and Weller, D.M., 1994. Use of mixture of fluorescent pseudomonads to suppress take-all and improve the growth of wheat. *Phytopathol.,* 84: 940–947.

Pierson, L.S. and Thomashow, L.S., 1992. Cloning and heterologous expression of the phenazine biosynthetic locus from *Pseudomonas aureofaciens. Mol. Plant Microbe. Interact.,* 5: 330–339.

Piotrowska-seget, Z., 1995. The effect of cyanogenic strains of *Pseudomonas fluorescens* on plant growth. *Acta Microbiol. Polon,* 44(2): 161–170.

Pleban, S., Ingel, F. and Chet, I., 1995. Control of *Rhizoctonia solani* and *Sclerotium rolfsii* in the greenhouse using endophytic *Bacillus* spp. *European J. Plant Pathol.,* 101: 665–672.

Powell, K.A., 1995. The production of chemicals by biological control agents. *Pesticide Science,* 44(4): 395–397.

Quispel, A., 1992. A search for signals in endophytic microorganisms. In: *Molecular Signals in Plant Microbe Communications,* (Ed.) D.P.S. Verma. CRC Press, Boca Raton, Florida, pp. 471–490.

Randhawa, P.S. and Singh, N.J., 1981. Detection of *Xanthomonas campestris* pv. *malvacearum* from cotton seed by a modified phage multiplication method. In: *Proc. of IIIrd Int. Symp on Plant Disease Problems,* Indian Phytopathological Society, New Delhi.

Randhawa, P.S., Singh, N.J. and Schaad, N.W., 1987. Bacterial flora of cotton seeds and biocontrol of seeding blight caused by *Xanthomonas campestris* pv. *malvacearum. Seed Sci. Techno.,* 15: 65–71.

Reeser, P.W. and Kostka, S.J., 1988. Population dynamics of *Clavibacter xyli* subsp. *cynodontis* (CXC) and a CXC/*Bacillus thuringiensis* subsp. *kurstaki* (BT) recombinant in corn (*Zea mays*). (Abst.) *Phytopathol.,* 78: 1540.

Samish, Z, Etinger-Tulczynska, R. and Bick, M., 1961. Micro flora within healthy tomatoes. *Appl. Microbiol.,* 9: 20–25.

Samish, Z., Etinger-Tulczynska, R. and Bick, M., 1963. The micro-flora within the tissue of fruits and vegetables. *J. Food Sci.,* 28: 259–266.

Schister, D.A., Slininger, P.J. and Bothast, R.J., 1997. Effect of antagonist cell concentration and two strain mixtures on biocontrol of Fusarium dry rot of potatoes. *Phytopathol.,* 87: 17–183.

Sharrock, K.R., Parkes, S.L, Jack, H.K, Rees-George, J. and Hawthorne, B.T., 1991. Involvement of bacterial endophytes in storage rots of buttercup squash (*Cucurbita maxima* D.) hybrid Delica. *N.Z. J .Hort. Sci.,* 19: 157–165.

Shishido, M., Loeb, B.M. and Chanway, C.P., 1995. External and internal root colonization of lodgepole pine seedlings by two growth-promoting *Bacillus* strains originated from different root microsites. *Can. J. Microbiol.,* 41: 707–713.

Smith, E.F., 1911. *Bacteria in Relation to Plant Disease,* Vol. 2. Carnegic Institution of Washington, Washington, D.C.

Sriskandarajah, S., Kennedy, T.R., Yu, D. and Tehan, Y.T., 1993. Effects of plant growth regulators on acetylene-reducing associations between *Azospirillum brasilense* and wheat. *Plant Soil,* 153: 165–177.

Sturz, A.V., 1995. The role of endophytic bacteria during seed piece decay and potato tuberization. *Plant Soil,* 175: 257–263.

Thomashow, L.S. and Weller, D.M., 1990. Role of antibiotic and siderophores in biocontrol of take-all disease of wheat. *Plant Soil,* 129: 93–99.

Trevet, I.W. and Hollis, J.P., 1948. Bacteria in the storage organs of healthy plants. *Phytopathol.,* 38: 960–967.

Van Buren, A.M., Andre, C. and Ishimaru, C.A., 1993. Biological control of the bacterial ring-rot pathogen by endophytic bacteria isolated from potato (Abst.). *Phytopathol.,* 83: 1406.

Van Peer, R. and Schippers, B., 1989. Plant growth responses to bacterization with selected *Pseudomonas* spp. strains and rhizosphere microbial development in hydroponic cultures. *Can. J. Microbiol.,* 35: 456–463.

Verma, J.P. and Singh, R.P., 1974. Recent studies on bacterial diseases of fibre and oilseed crops in India. In: *Current Trends in Plant Pathology,* (Eds.) S.P. Raychaudhuri and J.P. Verma. University of Lucknow, India, pp. 134–145.

Verma, J.P. and Singh, R.P., 1976. Races of *Xanthomonas malvacearum,* loss in their virulence and the protective effect of avirulent strain, heat killed cells and phylloplane bacteria. *Z. Pfkrankh. Pflschutz,* 83: 748–757.

Verma, J.P., Chowdhury, H.D. and Singh, R.P., 1978. Interaction between phylloplane bacteria and *Xanthomonas malvacearum.* In: *Proc. IVth Int. Conf. Pl. Pathogenic Bacteria,* Angers, France, pp. 795–802.

Verma, J.P., Chowdhury, H.D. and Singh, R.P., 1982. Biological control of bacterial blight of cotton. In: *Advances in Mycology and Plant Pathology,* (Eds.) S.B. Chattopadhyay and N. Samajpati. Oxford and IBH Publishing, Kolkata, pp. 170–178.

Verma, J.P., Singh, R.P. and Chowdhury, H.D., 1980. Relationship of epiphytic bacteria to bacterial diseases of plants. In: *Recent Advances in Biology of Microorganisms,* (Eds.) K.S. Bilgrami and K.M. Vyas. Bishen Singh, Mahendra Pal Singh, Dehra Dun, India, pp. 95–106.

Verma, J.P., Singh, R.P., Chowdhury, H.D. and Sinha, PP., 1983. Usefulness of phylloplane bacteria in the control of bacterial blight of cotton. *Indian Phytopath.,* 36: 574–577.

Vincent, M.N., Harrison, L.A., Brackin, J.M., Kovacevich, P.A., Mukerji, P., Weller, D.M. and Pierson, E.A.., 1991. Genetic analysis of the antifungal activity of a soil-borne *Pseudomonas aureofaciens* strain. *Appl. Environ. Microbiol.*, 57: 2928–2934.

Voisard, C., Haas, D. and Defago, G., 1989. Cyanide production by *Pseudomonas fluorescens* helps suppress black root-rot of tobacco under gnotobiotic conditions. *EMBO J.*, 8: 351–358.

Watanabe, M., Nakanishi, K., Kumakura, K., Shimoyama, N. and Furukawa, M., 1976. Studies on the mechanism of resistance in the rice plants infected with *Xanthomonas oryzae*. Inhibition of lesion length enlargement by pre inoculation with the non-pathogenic bacteria. *Ann. Phytopath. Soc., Japan*, 42: 295–303.

Whitesides, S.K. and Spotts, R.A., 1991. Frequency, distribution and characteristics of endophytic *Pseudomonas syringae* in pear trees. *Phytopathol.*, 81(4): 453–457.

Wilson, C.L, Franklin, J.D. and Pusey, P.L., 1987. Biological control of *Rhizopus* rot of peach with *Enterobacter cloacae*. Phytopathol., 77: 303–305.

Insect Pest and Disease Management *Pages 360–391*
Editor: **Dr. D. Prasad**
Published by: **DAYA PUBLISHING HOUSE (2016)**

Chapter 25

Major Diseases of Medicinal Plants

P.K. Gupta[1], N.D. Sharma[2] and Yogita Gharde[3]
[1]Division of Plant Pathology, I.A.R.I., New Delhi – 110 012
E-mail: pkgtaxo@rediffmail.com
[2]D.E.S. ATIC, J.N.K.V.V., Jabalpur – 482 004
[3]Ph.D. Student, IASRI, New Delhi – 110 012

Introduction

Medicinal plants have been used by man from prehistoric times. The Aryans of the Indus valley wrote three treatises *viz.*, the Regveda (2000 BC), Atharvaveda (2000-1000 BC) and Ayurveda (100-600 BC) which mentioned several medicinal plants and their uses (Chopra, *et al.*, 1956). These plants inspite of tremendous developments in the field of allopathy during the 20th century, still remain one of the major source of drugs in modern as well as traditional systems of medicine throughout the world. Approximately over 60 per cent of all pharmaceuticals are plant based. Chemically, depending on their active principles, plant may have alkaloids, glycosides, steroids or other groups of compound which may have marked pharmaceutical actions as anticancerous, antemalerial, antehelminthic or antidycentric (Husain, 1993). These drugs are derived from trees, shrubs and herbs and even for primitive kinds of plants which do not fall into the above categories. They are made from, flower, fruits, leaves, stem, roots and seeds (Maheshwari, 2000).

The annual export of vegetable drugs, from India has been on the increase and during the year 1995-96 it has gone up to Rs 189.30 crores, beside the export of Rs 70.03 crores worth of alkaloids and the total import of these commodities during the year 1995-96 was little over Rs 43.39 crores (Monthly Statistics of Foreign trade of India, 1996). Globally, the plant diseases cause enormous losses of the crops. A large quantity of these valuable plant get spoiled due to diseases while they are in the time of seed sowing in the field harvesting or while in storage before processing (Butler, 1918). Causes of Plant Diseases the pathogen also associated with medicinal plants therefore, disease is a symptom caused by the invasion of a pathogen that is able to survive, perpetuate and spread. Further, the word "pathogen" can be broadly defined as any agent or factor that incites pathos or disease in an organism or host. In strict sense, the cause of plant diseases are grouped under following categories:

Animate or biotic causes Fungi, bacteria, Algae, Phanerogams, Phytoplasma, Protozoa, Rickettsia-like organisms, Nematodes, *Mesobiotic causes* Viruses and Viroides are the cause of serious diseases of medicinal/aromatic plants some times *Inanimate or abiotic causes* these factors are also cause damages ie. Deficiencies or excess of nutrients, Light, Moisture, Air pollutants, Temperature, Toxicity of pesticides, Improper cultural practices, Lack of oxygen, Abnormal soil conditions (acidity, alkalinity). These factor are also spoil medicinal property and even leave many toxic metabolites in various plant parts including leaves, stem, fruits/seed, and roots (Mukherji and Bhasin, 1986) after harvest or in storage.

History of Major Diseases of Medicinal Plants with Special Reference to Indian and Abroad Works

Medicinal plants have been associated with the health of mankind from the immemorial. The Aryans of the Indus Valley wrote three treatises, *viz.*, the *Rigveda* (2000 BC), *Atharvaveda* (2000–1000 BC) *Ayurveda* (100-600 BC) which mention several medicinal plants and their uses, including the hallucinogenic mushrooms *Amantia muscaria* and *Rauvolfia serpentina* used to treat snakebite, epilepsy, mental disorders and other illnesses. This plant is the source of resepine, a tranquillizer and hypertensive agent widely used in modern pharmacy. The *Charaka Samhita*, an encyclopaedia of Indian medicine, published at Varanasi between 1000 BC and 100 AD, is a comprehensive record of medicinal plants and their uses.

Medicinal plants have been one of the important source of medicines ever since the dawn of human civilization. In spite of tremendous developments in the field of allopathy during the 20th, century, plants still remain one of the major source of drugs in modern as well as traditional systems of medicine throughout the world. Approximately one-third of all pharmaceuticals are of plant origin, where in fungi and bacteria are also included. Over 60 per cent of all pharmaceuticals are plant-based.

Chemically, depending on their active principles, plants may have alkaloids, glycosides, steroids or other groups of compounds which may have marked pharmaceutical actions are anticancerous, antimalarial, antihelminthic or antidysentric, etc. Many of the essential oils, dyes, lattices and even tannis and vegetable oils are also widely used as medicines (Farooqi and Sreeramu, 2001).

Out of nearly 30000 species of higher plants available, only a small proportion have been investigated for medicinal properties, and a still smaller number yield well-defined drugs. The same is the case with lower plants and with plants of the sea. Thus, the knowledge of plant constituents gained so far is still merge, considering the huge number of species available in the world. Approximately, only 10 per cent of the organic constituents of plants are reported to be known and the remaining 90 per cent are yet to be explored.

A very small proportion of Indian medicinal plants are lower plants like lichens, ferns, algae, etc. The majority of medicinal plants are higher plants. The major families in which medicinal plants occur are *Fabaceae, Euphorbiaceae, Asteraceae, Poaceae, Rubiaceae, Cucurbitaceae, Apiaceae, Convolvulaceae, Malvaceae* and *Solanaceae*.

Out of the same period came yet another great figure of medicine, Galen, a Greek physician who practised in Rome during the 2nd century AD. Galen's fame is primarily associated with the fauna, not the flora, of his time and place. He revolutionized medicine by performing animal experiments, from which he developed the first performing animal experiments, from which he developed the first medical theory based on scientific investigations. Although many of Galen's theories were proved wrong, because he assumed animal studies would directly apply to human beings, his place as the founder of 'Experimental Medicine' is unchallenged. He is credited with the divergent theories of Allopathic (cure a disease by its opposite) and Homeopathic (like cure like disease) medicine, both of which have grown out of Galen's doctrines.

A vast majority of the world population today are finding themselves unable to afford the products of the western pharmaceutical industry, and they have to depend mainly upon the use of traditional medicines. This reality has been recognized, documented and compiled by the WHO in an inventory of medicinal plants numbering over 20000 species. The world population is expected to touch 750 crores by the turn of century, and this will further escalate the health budget especially those of the developing countries. The developing countries spend roughly 40-50 per cent of their total health budget on drugs and, as a strategy to reduce the financial burden on developing countries, the WHO encourages, recommends and provides for the inclusion of herbal medicines in natural health care programmes. Such herbal medicines are easily available at a cheaper price for the common man, they are time-tested and considered safer than some of the modern synthetic drugs. Of late, there has been a resurgence of interest in herbal medicines in the Western European countries. This is evident from the fact that 2 volumes of *British pharmacopoela* were published and 33 million US dollars worth of literature on herbal products was sold in the USA in 1990. Japan has recently adopted a health policy based on phytopharmaceuticals to be implemented in the next five years (Bhakuni and Jain, 1995).

It is estimated that in 1980, the sales of herbal medicines in European community accounted for US$ 2.25 billions (Rs 7000 crores). This is only 3 per cent of the European pharma market estimated at US$ 66 billion (Rs 211000 crores). In the EC market of phytomedicines Germany has the largest share at US$ 1.5 billion (Rs 4800 crores), *i.e.* 23.2 per cent of the total. The size of the French market for herbal medicine is $ 0.21 billion (Rs 670 crores) which amounts to less than 1 per cent of the EC market. The UK market is $ 425 million (Rs 1300 crores).

The share of the Indian market is negligible, notwithstanding the fact that we have a varied emporium of medicinal plants and are the exporters of over 200 major drugs and pharmaceutical.

The total import of these commodities during the year 1995-96 was little over Rs 43.39 crores. Among the major items imported were vegetables saps and extracts, ginseng, pyrethrum flowers, liquorice roots, agar agar, and chirata.

A number of opium alkaloids such as morphine, codeine, papaverine and nascopine are extensively used in medicine. Except papaverine, which is also obtained from synthetic sources, all other alkaloids are still obtained from opium poppy (*Papaver somniferum*) either from latex or from direct extraction of the poppy husk or straw.

The crude drug Senna is obtained from the leaves and pods of *Cassia angustilofia*, often referred as Indian Senna, and *C. acutifolia* designated as Alexandrian Senna. Sennosidas are important laxative used in medicine.

Plants suffer from diseases caused of which are similar to those affecting animals and man. Plant growth and yield depend on protecting the plants from diseases. Anything that affects the health of plants is likely to affect their growth and yield and also seriously reduce their usefulness to mankind. Plant diseases are thus important to man because they not only damage plants but also the plant products which are the source of our traditional medicines.

Taxonomy of the fungi disease can be identified by the symptoms and the presence of the pathogen especially on the site of infection.

Coleus forskohlii associated with a few fungus till now *Rhizoctonia solani* Kuhn (Shukla *et al.*, 1993).

Digitalis lanata is annual, biennial or perennial herb. Nearly 70 glycosides have been reported from the leaves of *D. lanata*, of these, lanatoside C and digoxin are most important glycosides used in medicine. Both lanatoside C and digoxin are most commonly used drug in congestive heart failure and various other heart ailments (Husain, 1993). *Digitalis* species have been reported to associated with a number of fungi.

The species this fungus *Colletotrichum fuscum* Laub. is reported from USA, UK, Japan, Czechoslovakia and other European countries and Asia (Goto, 1938) *Septoria digitalis* Pass. Cause more damage to *Digitalis purpuria* and *D. lanata* at the cotyledonery stage of the seedling and young plants perpetuation of the disease has been known through pycnidia, stroma and partly as mycelium (Kowalski, 1968).

Panax ginseng have been reported to be affected by a number of disease. Only a few fungi are found to be associated and damage or deteriorate the storage medicinal value like *Colletotrichum dematium* (Pers.) Grove (McPartland and Hosoya, 1998), *Erysiphe panax*. Bai and Wang (Sholberg *et al.*, 1966) and the flower and seed stalk of *Panax ginseg* were also attacked by the pathogen *Alternaria panax* Whetzel (Nair, 1991) and is attacked by a number of diseases like *Phytophthora cactorum* (Lebert and Cohn) Schroter (Darmono and Parke, 1990), *Fusarium solani* (Mart.) Sacc. (Cheon and Kim, 1999), *Cylindrocarpon destructans* (Zinssm) Scholten (Shin *et al.*, 1986, Parke and Shotwell, 1989), *Sclerotinia panacis* (Parke and Shotwell, 1989). *Hyoscymus niger* and *H. muticus* are reported to be affected by a number of disease caused by fungi and the association of fungi was first reported in *Hyoscymus niger* by Padwick and Merch (1943). Ganguly and Pandotra (1962), *Alternaria alternata* (Fr.) Keissler. (Kumar *et al.*, 1982, 1984) prevalent in Jammu and Kashmir, *Pythium butleri* Subramaniam (Shukla *et al.*, 1988) and *Thielavia basicola* Zopf. (Pandey, 1995; Sastry *et al.*, 1977). Opium poppy (*Papaver somniferum* L.) is one of the most important medicinal plants, its latex contain more than two dozen alkaloids. To a limited extent crude opium is still used in modern medicine as well as traditional medicine. A number of fungi cause damage to crop (*Papaver somnifera*) like *Pythium ultimum* Trow, *Pythium mamillatum* Meurs (Angell, 1950) *Pythium dissotocum* Drechsler. (Alam *et al.*, 1996) *Entyloma fuscum*, Schorter (Savulescu, 1932, Kovachevsky, 1936, Kochman, 1936). *Cercospora papaveri*, *C. purpurea* and *Alternaira brassicae* have also been reported from India, Pakistan and Germany in opium poppy (Pavgi and Upadhyay, 1964, Zaman and Khan, 1969, Grummer, 1953), *Helminthosporium papaveris* Saw. (P.S. *Pleospora papaveraceae* (denot. Sacc.) was reported destructive in Bulgaria, Switzerland, Czechoslovakia. The Netherlands, Romania and Yugoslavia, Girzitska (1928), Darpous (1945), Meffert (1949), Ballarin (1950), Miltovic (1952), Bogarda *et al.* (1971) observed the associated fungi cause severely damage to opium poppy.

A new leaf blight disease of opium poppy was observed in an epidemic form in the poppy plantation in U.P. caused by *Alternaria puragmospora* Yan Emden. (Gupta and Janardhanan, 1989) Downy mildew associated fungi *Perenospora arborescens* (Berk.) deBery is one of the serious and wide spread disease of opium poppy. The disease was observed epiphytotic in Argentina, Germany, Iran, Hungary, Austria and India (Fokin, 1922; Yossifovitch, 1928; Moesz, 1938; Masurat and Stephan, 1961; Kothari and Prasad, 1970). *Catharanthus roseus* of an accidental discovery of the anticancer properties of its leaf alkaloids, A large number of fungi associated and deteriorate the medicinal value. Fungi like *Colletotrichum dematium* (Pers ex Fr.) Grove (Janaradhanan *et al.*, 1977) and *Rhizopus stolonifer* (Ehren.) Lind. causes serious damage to the commercial plantations (Saini *et al.*, 1996). *Rhizotonia solani* Kuhn. (*Thanatephorus cucumeris* (Frank) Donk.) associated on the lower leaves of Periwinkle as water soaked area by the side of midrib near the leaf based which gradually develop to both the side.

Numerous small brown to black sclerotia were found (Khatua and Maiti, 1982) Balasubramanian and Bhama, 1977). Leaf blight *Alternaria alternata* (Fr.) Keissler. (Goyal and Pathak, 1982). Target spot *Corynespora cassiicola* (Berk and Curt.) Wei (McGovern, 1994), *Phytophthora nicotianae* Breda de Haan (Gill *et al.*, 1977) are the major fungi associated with Periwinkle, *Rauvolfia serpentina* used for the treatment of snake bite, insect sting and mental disorder and also used for epilepsy, diarrhoea and dysentery based on the Ayurvedic literature and folk medicine. Most of the supply of this plant is from wild sources. The plant is cultivated large scale in India and Bangladesh, *Rauvolfia* is susceptible to several fungi like *Fusarium oxysporum* F. sp. *rouvolfii* (Schlecht.) (Janaradhanan *et al.*, 1964), *Leveillula taurica* (Lev) Atz. (Pandey, 1995), *Alternaria tenuis* Nees. (Ganguly and Pandotra, 1962). The leaf spot disease of *R. serpentina* associated fungi *Cercospora rauvolfiae* Chupp and Muller was first reported by Mohanthy and Addy (1957) and then Chandra (1956). Pandotra and Husain (1966) reported a similar associated fungus on *Rauvolfia serpentina* caused by *Cercospora serpentina* Pandotra and Husain. Another leaf blight associated fungus *Phomopsis sethii* Mehrotra of *R. serpentina* was found prevalent in the plantations in Madhya Pradesh (Mehrotra, 1976) and it was also suffer heavily due to *Corynespora cassiicola* (Berk and Curt.) Wei (Mohanthy and Addy, 1958, Reddy *et al.*, 1971).

Withania somnifera Dunal cultivated on commercial scale and the important compounds isolated from different parts of the plant are steroidal lactones called Withanolids, The roots are also demonstrated to have significant Psychotropic effects and antitumor activity. A number of associated fungi which cause diseases decay of roots, seeds, stem and leaves and lesser than value have been reported like *Alternaria alternata* (Fr.) Keissler (Pandey, 1988; Nigam, 1984; Gupta *et al.*, 1993) and *Myrothecium roridum* Tode ex Fries (Mahrishi, 1986).

Medicinal plant may be associated with a broad variety of microbial contaminants (Kneifel, 2002) and Taxonomy of Some Fungi Associated with Medicinal Plants (Gupta, 2005) therefore, become an important subject in the establishment of modern hazard analysis and critical point scheme with methodological aspects and by considering the influence of different commonly used pharmaceutical preparation technique on the microbiological status of the products.

Janardhanan (2002) presented a comprehensive knowledge of diseases and control measures of 16 medicinal plants.

Much of the information on major disease of medicinal plants had been collected by Janardhanan (2002).

Most of the fungi/pathogens/symbionts have been reported from main substrate/host *i.e.* flowering plants being single source of edible natural medicinal source for all biological from/food chain.

In spite of the tremendous progress in the development of modern medicine, plants continue to be an important source of drugs throughout the world, particularly in the developing countries. The last two decades registered significant revival in the use of higher plants for health care and treatment of several diseases for which modern medicine was not enhanced enormously. This has necessitated in the large-scale cultivation of a number of medicinal plants in the recent years in order to maintain sustainable supply to the pharmaceutical industries. Extensive cultivation of these plants and their industry have increased to great extent.

A number of books on the chemistry, pharmacognosy and utilization of medicinal plants have been brought out during the last two decades. However, no comprehensive publication on the diseases of these plants and their management has been published.

The diseases of these plants and their intensity have multiplied. A number of books on the chemistry, pharmacognosy, agrotechnology and utilization of medicinal plants have published in recent years.

Gupta and Masood (2003) first reported occurrence of a disease on *Aloe vera* from Uttar Pradesh. The leaves of *Aloe vera* were found to be infected with leaf spot disease. The fungus responsible for the disease has been identified as *Alternaria alternata* (Fr.) Keissler. Mandal and Maiti (2005) first reported Bacterial soft rot of aloe caused by *Pectobacterium chrysanthemi*: a new report from India. (Jin *et al.*, 1994) Identified the bacterial isolates obtained from diseased orchid and aloe plants caused by *Erwinia chrysanthemi* (Laat, PCA *et al.*, 1994). Identified the bacterial leaf rot of *Aloe vera* L., caused by *Erwinia chrysanthemi.*

Kumar and Masood (2003) first reported the occurrence of *Alternaria alternata* from Uttar Pradesh on *Casearia tomentosa* as leaf spot pathogen.

Belladana is reported to be affected by two leaf spot diseases, leaf necrosis and leaf spot (Ganguly and Pandotra, 1962; Singh and Singh, 1984). Leaf necrosis was reported from India and also from Argentina. Seedling blight of *Cinchona* caused by *Phytophthora* was first observed in Darjeeling district of West Bengal and species of *Phytophthora* was found to be the causal agent (McRae, 1933). Leaf blight and pod blight caused by *Alternaria crassa* were reported on *Datura nutel* from Israel (Halfan-Meir, 1973). Similar disease was found on *Datura stramonium* in Lucknow, India (Sattar *et al.*, 1986).

Dioscorea deltoidae is found to be affected by a leaf spot disease, caused by *Cercospora dioscoreae* Ellis and Mart. (Ganguly and Pandotra, 1963).

Glycyrrhiza glabra Linn. is found to be affected by several diseases. The major diseases are leaf spots (Khan and Kamul, 1982) and rust (Casulli and Ippolito, 1995).

Commercial cultivar of *Plantago ovata* Forsk. has been reported to suffer from a number of diseases like downy mildew (Kapoor and Chowdhary, 1976), damping-off and wilt (Chastagner *et al.*, 1979) powdery mildew (Shakhela *et al.*, 1985) and leaf blight (Patel and Desai, 1987). Senna (*Cassia angustifolia* Vahl.) is attacked by a number of diseases damping off (Patel and Patel, 1984), leaf blight (Patel and Pillai, 1979) and leaf spot [*Colletotrichum gloeosporioides* (Penz.) Sacc.] (Perfect State: *Glomerella cingulata* (Gupta *et al.*, 1997).

Parasitism and Pathogenesis

An organism, which lives in or on the other living organisms and derives their nutrients from the latter is called *parasite*. The relationship between a parasite and its hosts is known as parasitism. Many fungi and bacteria grow on a non-living substrate within a living plant. The organism of this type of mode of nutrition is called *saprophyte*. Based on the different types of modes, nature and nutrition, the relationship between the host and parasite or saprophyte is termed in many ways *viz.*, obligate parasite (biotroph), obligate saprophyte, facultative parasite, facultative saprophyte, hemibiotroph and necrotroph (perthotrophs or perthophyte).

Parasitism in cultivated crops are in medicinal plants is a common phenomenon. Therefore, any agent that can cause suffering or damage or disease is called a *pathogen*. In plant pathology, the term 'pathogen' is usually used to the living or infectious organisms. The ability of a pathogen or parasite to cause disease is known as *pathogenicity*.

It is clear that a medicinal plant becomes diseased when it is attacked by a pathogen or parasite. The ultimate condition *i.e.* disease occurs by passing through some distinct events. Thus, the genesis or chain of events or stages of disease development is called *pathogenesis*. This is also called as *disease cycle*. The events that occur in specific order are namely inoculation, penetration, and establishment of infection, invasion or colonization, growth and reproduction, dissemination and survival of the pathogen. The events will continue to repeat in the same order in presence of both the host and pathogen/parasite that may lead to severe disease condition.

Causal Agents of Plant Diseases

Fungi

Fungi are characterized by non-motile bodies (thalli) constructed of apically elongating walled filaments (hyphae), a life cycle with sexual and asexual reproduction, usually from a common thallus, haploid thalli resulting from zygotic meiosis, and heterotrophic nutrition. Spindle pole bodies, not centrioles, usually are associated with the nuclear envelope during cell division. The characteristic wall components are chitin (beta-1, 4-linked homopolymers of N-acetylglucosamine in microcrystalline state) and glucans primarily alpha-glucans (alpha-1, 3- and alpha-1, 6- linkages) (Griffin, 1993).

Bacteria

Bacteria are perhaps the simplest and oldest form of life, found throughout nature in the soil, in the water and riding on dust in the air. Bacteria are so unlike plants and animals–they're just DNA/ RNA encased in a hard cover–that they get their own category. They are able to reproduce, but they are extremely small, none bigger than the smallest cell. There are thousands of species of bacteria, in all sizes and shapes such as cones, spirals, rods. A gram of soil contains about 2.5 billion bacteria.

Nematode

Nematodes belong to the Animal kingdom, and phylum Nematoda. Most of the important parasitic genera belong to the order Tylenchida and few under Dorylaimida. Important plant parasite species are *Meloidogyne incognita* (root knot nematodes), *Heterodera* and *Globodera* (cyst nematodes), *Tylenchulus semipenetrans* (citrus nematode), *Pratylenchus* (lesion neamtode), *Ditylenchus* (stem and bulb nematode), *Anguina* (seed-gall nematode), *Aphelenchoides* (foliar nematode) etc.

Virus

Virus as "a set of one or more nucleic acid template molecules, normally encased in a protective coat, or coats of protein or lipoprotein, which is able to organize its own replication within suitable host cells. Within such cells, virus production is (a) dependent on the host's protein synthesizing machinery, (b) organized from pools of the required materials rather than by binary fission, and (c) located at sites which are non separated from the host cell contents by a lipoprotein, bilayer membrane".

Mycoplasma

Mycoplamas have a prokaryotic cellular organization except the cell wall which is bounded by triple layer unit membrane and lack the ability to synthesize cell wall material. They are highly pleomorphic with small coccoid bodies, ring forms, pear shaped and rarely filamentous (branched or unbreanched). Usually, they are non-motile and filterable through bacterial filters. They are very small and diameter ranges from 0.3-0.9 µm. They produce typical 'fried egg' colonies in cell free solid media and require sterol for growth. They are insensitive to penicillin, but inhibited by tetracycline and some specific antibody.

Viroide

Viroides are smallest known infectious agents. Several plant diseases are known to be caused by viroides. Diener in 1971, observed that potato tuber spindle disease is caused by a small molecule of infectious ribonucleic acid, which he called a viroide.

Symptoms of Plant Diseases

As a result of disease or disorder a visible or detectable abnormality expressed on the plant is called *symptom* and the totality of symptoms are collectively called as *syndrome.* Signs are any physical or direct evidence of a disease-causing agent discernable with the unaided eye. This limits signs chiefly to those agents such as some fungi, nematodes and parasitic plants that produce macroscopic structures. Different types of disease symptoms are cited below:

Rusts

Numerous small pustules growing out through host epidermis, which gives rusty (rust formation on iron) appearance of the affected parts.

Mildew

White, grey or brown coloured superficial growth of the pathogen on the host surface is called mildew.

Necrosis

Plant mortality (death) of an entire plant.

Spots

The cells are killed in a definite limited areas and dead tissue usually turn to different sheds of brown. The shape of the lesion on leaves or fruit of the medicinal and aromatic plants may be round, angular or irregular. The dead areas are often surrounded by red, purple, yellow or at last brown margin.

Rots

The affected tissue decomposes and turns brown. The cell walls are mostly disintegrated. Fleshly tissue of stem, roots, leaves; flower, buds and fruit are affected. Depending of dissolution brought about by the pathogen, the rots may be grouped as soft-rots, wet-rot or dry-rot. According to the organ attacked the rots may be called root rot, stem rot, bud rot or fruit rot.

Lesion

Discrete or localized necrotic area.

Canker

The localized death of woody plant tissue usually resulting in a collapsing of the bark and cambial region. Cankers can be formed from callus tissue, which is often formed by the plant in an attempt to limit spread of the pathogen. In some instances the plant does not have the ability to form callus tissue.

Vascular Discoloration

Browning or darkening seen in the cross-sections of splitting diseased plant part.

Wilts

The plants start wilting and ultimately dry up at seedling and/or at flowering stage. The diseased plants are pulled out easily, due to the loss of rigidity. Sometimes, sudden drooping of leaves and/or only a few branches of a single plant are affected. The disease mostly spreads in patches, but entire field may also be affected in severe conditions. Severe damage has been reported during early pod filling.

Blight

Rapid necrosis of a plant or plant parts. The blackened leaves and twigs developed in a relatively short period of time (several days to a few weeks).

Dieback

Necrosis of terminal portions of the crown and proceeding toward the trunk.

Damping-off

The stem is attacked near the soil surface. The affected portion become weak and is unable to stand in an erect posture. This results in toppling down and death of seedling.

Wilting

Necrosis of plant cells involved in water transport, thus depriving foliage of adequate water and causing loss of turgor pressure.

Chlorosis

Necrosis of plant chloroplasts resulting in a yellow appearance on the foliage of the plants. Chlorosis also can result from the inhibition or retardation of chloroplast function and development.

Atrophy or Hypoplasia

Atrophy or hypoplasia involves the slowing down or retardation of normal plant processes resulting in a stunting of the entire plant or plant parts.

Hypertrophy

Abnormal increase in cell size.

Hyperplasia

Abnormal increase in cell numbers. These conditions frequently occur together and cannot be distinguished without a microscope. Both conditions, separately and together, can cause overgrowths of a plant part or an area on a plant.

Galls

These are malformation of irregular shapes more or less globose due to disorganized growth. Fleshly or woody tissue of stem and roots are affected.

Curl

The leaves twisted, curled and distorted due to growth in tissues in localized areas of the leaf.

Witches-broom

Numerous slender branches arise from particular regions in cluster giving an appearance of broom. The leaves become narrow. The flowers are transformed to small leaves. These symptoms are often referred to little leaf or phyllody.

Abroma augusta Linn.f. (Ulatkambal) (Sterculiaceae)

Found both wild and cultivated throughout hot and moist parts of India. Propagated By seeds, the root and bark of Pivari are uterine tonic and emmenagogue. They contract the uterus and are used for treatment of congestive and nervous dysmenorrhoea, ammenorrhoea, sterility and other menstrual disorders. Powdered root act as an abortifacient and anti-fertility agent. Leaves useful in treating uterine disorders, diabetes, rheumatic paine of joints, and headache with sinusitis. Leaves and stem are demulcent and an infusion of fresh leaves and stems in cold water is very efficacious in gonorrhoea. Ulatkambal is cultivated on commercial scale in all over in several parts of India. The plant is reported to be affected by a number of diseases caused by fungi (Gupta, 2005).

Leaf Spots of Abroma

Cercospora sp.is associated with Ulatkambal (Gupta, 2005). Symptoms are small water soaked, irregular in shape and size sunken, dark brown to black, absence of concentric margin, restricted by vein, isolated (Figure 25.1).

Causal Organism

Cercospora sp. Conidiophores much long, sub-hyaline to light brown, fasiculate, filiform, few, septate, cylindrical, hypophyllous in tufts to rounded, unbranched. 243.00–298.00 × 4.05 µm, av. 265.00 × 4.05 µm. Conidia much long, obclavate cylindric, tapering upward, straight to slightly curved, base truncate dark brown, 12-18 septate, sub-hyaline, broader toward base and acute at tip. 187.92–264.06 × 2.43–5.67 µm, av. 198 × 4.05 µm.

Abutilon indicum (Linn.) Sweet

Leaves are traditionally used to treat bronchitis, gonorrhoea, and as mouthwash in toothache, etc. The results revealed that the leaves contain steroids, sapogenins, carbohydrates and flavonoids. The aqueous extract was most potent in reducing the blood glucose levels. The results revealed that the use of leaves in controlling diabetes mellitus is justifiable.

Leaf Spot of Abutilon

Spots, circular to irregular, grayish white in the centre, surrounded by pinkish to light brown margin with concentric ring. Causal Oraganism: *Cercospora malvacearum* Chiddarwar, Chiddarwar (1959). Conidiophores, pale to light brown, straight, septate, epiphyllous, fasciculate, straight to undulate, irregularly geniculate, rarely blunt apex, 106–146 × 4.05 µm, av. 130.00 × 4.05 µm. Conidia hyaline, acicular, straight to bent base truncate, acute at tip, 2–12 septate, 103.00–131.00 × 4.05–4.86 µm, av. 116.00 × 4.05 µm.

Acorus Calamus (Buch)

Dried Rhizome of the Acorus calamus is nauseant, stomachic, anthelmentic, Stimulants, Emtetic, Expectorant, Carminative, Antispasmodic and nerving sedative. Rhizome is used in the cure of host of diseases such as epilepsy, mental ailments, chronic diarrhoea and dysentery. It is used in incense sticks and dhup and is widely used as insecticide for lice, bedbugs, worms.

Leaf Spot of Acorus

Abundant, purple to dark brown, circular to oval with white purple at centre 3-4 mm dia., surrounded by ill define margin (Figuer 25.2). *Causal Organism: Colletotrichum dematium* (Pers. ex Fr.) Grove Colony on PDA medium blackish with zonate sporulation acervuli, superficial, abundant, aggregated zonate arrangement pinkish mass. Setae dark brown, tapering upward, apex round, septate, sub-hyaline, round at apex, 90.72–158.60 × 6.48 µm, av. 129.60 × 6.48 µm. Conidiophore, sub-hyaline to pale olevaceous 1–2 septate, small tapering upward 24.30–30.78 × 2.43–3.24 µm av 26.00 × 3.00 µm. Conidia hyaline, curved-suckle shaped, acute end, single celled, thin walled 22.68–24.30 × 3.24 µm av. 23.00 × 3.24 µm. Appressoria, poor formation, nipple shape and irregular.

Rust of Acorus

Causal organism *Uromyces acori* Ramakr and Rang. Ramakrishnan and Rangaswamy (1948). Uredosori gregarious, abundant, irregular, vary in shape and size, brown to dark brown, absence of silver papery epidermal wall, mostly like parallel streaks on the host. Sori burst powdery mass of exposing uredospore in the form of brown pastule powder. Uredospore globose, to sub globose flattened at base, narrow truncate at the point of attachment with stalk, orange to light yellow, light brown to brown, patchly minute echinulated, large, except around the germpore and base, two germpore arranged equitorially, stalked, hyaline, thin walled 361.34–392.84 mm in dia. No teleutospores were observed (Figure 25.3).

Atropa belladonna L. and *Atropa acuminata* Royle ex Lendle (Belladonna)

Belladonna is a perrenual herb. Drugs obtain from the flowering tops and roots. It is cultivated a number of countries like Germany, Poland, Russia, USA and India. *Atropa acuminata* occurs naturally in Himanchal Pradesh and North-West Himalayas in Jammu and Kashmir. Leaves and roots contains Alkaloids, such as 1-hyoscymine, atropine and hyoscine. Alkaloids are used in medicine because of there anticholinergic, antispasmodic and mydriate properties. Belladonna and its derivatives are also used as sedative and tranquilizers. The crop is reported to affect by a number of diseases caused by fungi and virus.

Leaf Spot of Atropa

A leaf spot disease of Belladonna was reported from USA (Fenne, 1954). Symptoms are angular to brown spot with chestnut brown colored margins on the leaves. The disease is also reported in India (Singh and Singh, 1984) *Causal organisms*: *Cercospora atropae* Kuashmina. Conidiophores are produced on the spot in tufts olive brown in colored. Conidia are clavate, cylindrical, broadly rounded at the base and tapered towards the apex. The another leaf spot disease are also pathogen is also to be reported in the Belladonna *Ascochyta atropae* Bres. Pycnidia immersed in the necrotic tissue on the leaves oval to globose, light brown, ostiolated, without a beak, 100.3 × 88.6 µm (86.1–115.6 × 68.9–110.7). spores cylindrical to rounded ends, rarely oval, straight, hyaline pinkish in mass, 2- celled, rarely 3- celled, 8.9–3.1 µm (6.8–11.5 × 2.5–3.7).*Control*: on the basis of study that six spray of Blitox-50 might be able to control the disease. The best control was reported by the application of Blitox-50 followed by Bavistin and Dithane M-45 (Singh *et al.*, 1985). *Belladonna Mosaic*: Smith described a new virus disease on an *A. belladonna*. The virus is sap transmitted to tobacco and other solanaceouse crop plants. The virus is inactivated at temperature between 75-80°C. Sap extracted from infected plants remained active for 6-11 days.

Figure 25.1: *Abroma augusta* **Linn.f. (Ulatkambal)**

Figure 25.2: *Acorus Calamus,* **(Buch)**
Causal Organism: Colletotrichum dematium **(Pers. ex Fr.) Grove**

Figure 25.3: ***Acorus Calamus*** **(Buch)**

Figure 25.4: ***Azadirachta indica*** **(Neem)**

Figure 25.5: *Barleria prionitis*

Figure 25.6: *Blumea lacera*

Figure 25.7: *Cassia angustifolia* **(Senna)**

Figure 25.8: *Chlorophytum borivillianum* **Ker Gawl. (Safed Moosli)**

Figure 25.9: *Clitoria ternatea* **L. (Butterfly Pea)**

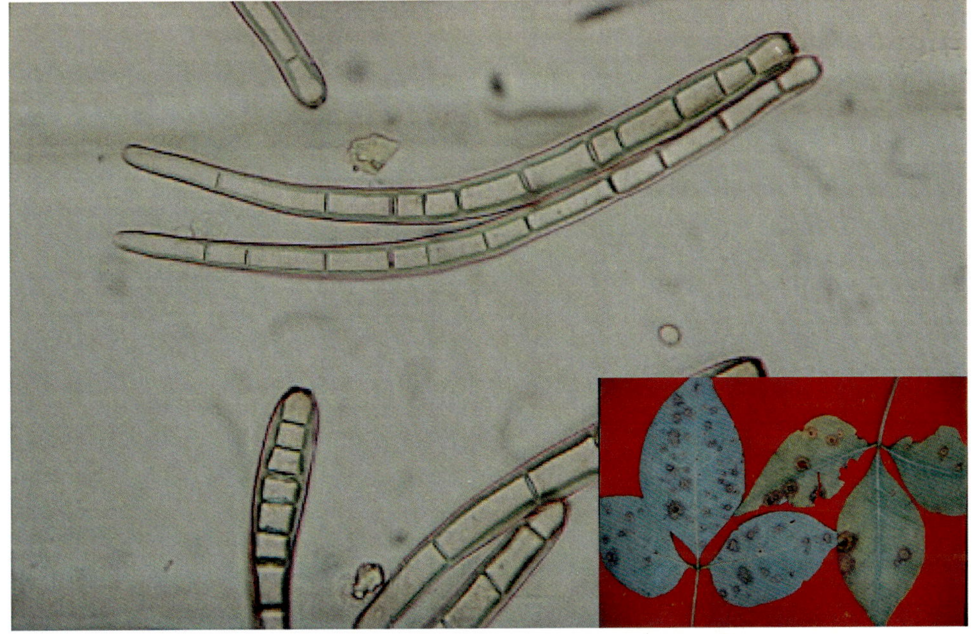

Figure 25.10: *Crataeva nurvala* **Buch. (Varun)**

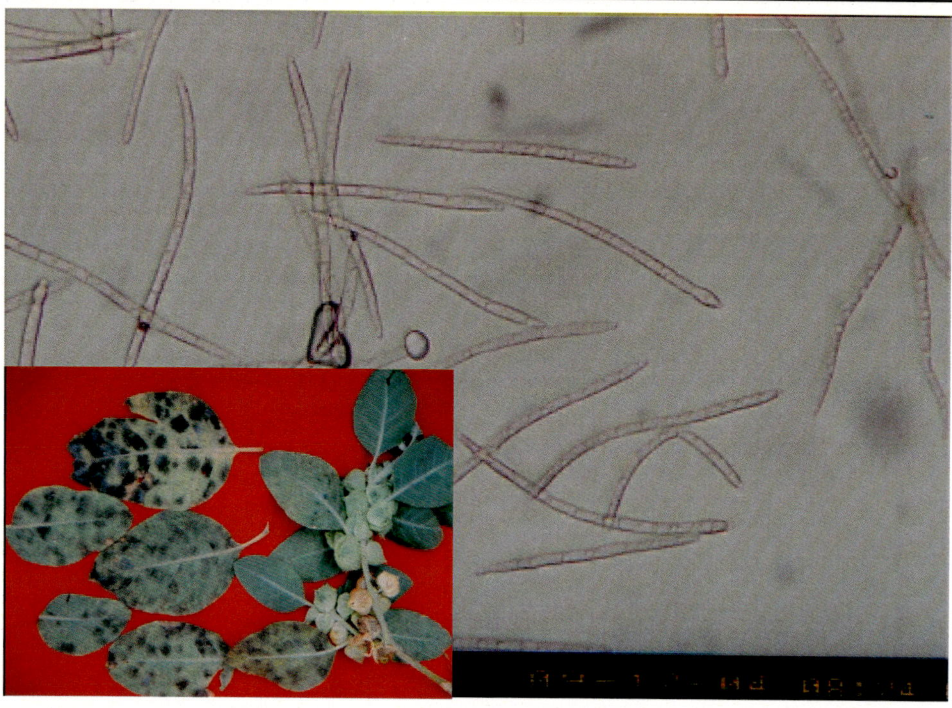

Figure 25.11: *Withania somnifera* **Dunal. (Ashwagandha)** *Cercospora withaniae* **H&P Sydow**

Figure 25.12: *Withania somnifera* **Dunal. (Ashwagandha)** *Alternaria tenuissima* **(Fr.) Keissler**

Figure 25.13: *Tinospora cordifolia* **(Willd.) Hooke & Thoms**

Figure 25.14: *Emblica officinalis* **Gaertn. (Amla)**

Causal Organism

Belladonna mosaic virus. Belladonna plants showing similar symptoms of belladonna mosaic virus is also been reported. (Sol *et al.*, 1961) showed that belladonna mosaic virus and rattle virus are transmitted by nematodes. Belladonna act as a symptoms carried of *Hyoscyamus virus* (Janardhanan, 2002).

Azadirachta indica (Neem)

Vermifuge, Insecticide, Astringent, Tonic and Antispetic. It posseses anti diabitic, anti bacterial and anti viral properties and used successfully in cases of stomach, worms and ulcers. Root bark posseses anstringent, tonic and antiperiodic properties. It is also useful in Malarial fever. The oil is used in making Neem based soaps, shampoos and toothpaste. Uses in pest and disease control both polar extracts and neem oil are of primary interest to research scientists as an insecticide. The active insecticidal ingredient, azadirachtin, is also effective against mites and other arthropods. Many of the tree's secondary metabolites have biological activity, but azadirachtin is considered to be of the most ecological importance.

Leaf Spot of Neem

Spots are light brown, ash gray at centre, circular to irregular, sporulation amphiphyllous on the lower surface, at the centre ash gray mass of conidia present (Figure 25.4). Causal organism *Cercospora subsessilis* H. Syd. Conidiophores arise in cluster, with a dark stroma, hyaline to pale olivaceous, non septate, straight to undulate unbranched, subglobose, fascicles, dense to fairly compact, hyaline at the end point. 21.06–32.40 × 3.24–4.05 µm, av 25.92 × 3.24 µm. Conidia, Hyaline, circular, straight or bent obconically, truncate at base, acute at tip 1-6 septate with a prominent scar. 45.36–66.42 × 3.24–4.05 µm, av. 53.46 × 3.24 µm. Jain *et al.* (1960) and Rao (1962).

Barleria prionitis

Whole-plant extracts of porcupine flower contain iridoid glycosides, barlerin, and verbascoside, which have shown potent activity against respiratory syncytial virus *in vitro* and may account for the plant's use in treating fever and several respiratory diseases in herbal medicine (Balick, *et, al.*, 1998). A mouthwash made from root tissue is used to relieve toothache and treat bleeding gums (Burkill 1985). The whole plant, leaves, and roots are used for a variety of purposes in traditional Indian medicine. For example, the leaves are used to promote healing of wounds and to relieve joint pains and toothache (Parrotta 2001). Because of its antiseptic properties, extracts of the plant are incorporated into herbal cosmetics and hair products to promote skin and scalp health (Prakruti 2002, Probiotics New Zealand 2002, Vaipani 2002).

Leaf Spot of Bacteria

Spots are vary in shape and size, dark brown to blackish with ash colour at centre, surrounded by dark margin at times spots marginal (Figure 25.5). *Causal organism*: *Corynespora cassiicola* (Berk and Curt.) Wei Conidiophores long cylindric, slightly pale olivaceous brown, 1–8 septate, straight, tip round, bearing conidia at apex singly at times in chain. 1–2 proliferations present 152.28–197.64 × 3.24–4.05 µm, Av. 179.82 × 4.05 µm. Conidia, sub-hyaline to light brown, obclavate, cylindric, straight to mildly, curved, 1-11 pseudoseptate, base obconically truncate, tip obtuse, 119.88–142.56 × 6.48–7.10 µm, av. 135.27 × 6.48 µm. Munjal and Gill (1963).

Blumea lacera

Blumea is described in *Ayurveda* as bitter, astringent, acrid, thermogenic, errhine, anti-inflammatory, styptic, opthalmic, digestive, anthelmintic, liver tonic, expectorant, febrifuge, antipyretic, diuretic, deobstruant, and stimulant (Warrier *et al.*, 1996. *Blumea lacera* is considered a valuable homoeopathic drug (Oudhia *et al.*, 1998a) useful in case of enuresis, neuralgia, headache, and cold borne cough. A tincture is useful in case of bleeding piles (Ghosh 1988). Natives of Chhattisgarh use this weed for treating health problems. There is a heavy demand of different parts (fresh and dry both) of this weed in national and international drug markets. Farmers can earn extra income after selling various parts of *Blumea* with the help of co-operatives. Fresh leaves of *Blumea* are the most valuable part.

Leaf Spots of Blumea

Spots are circular to irregular become coalesce blackish with powdery mass of conidia on lower surface of leaves, abundant (Figure 25.6). Causal organism *Cercospora blumeae* de Thumen Conidiophores hyaline to sub-hyaline, pale olivaceous, aseptate, straight, narrowly cylindric, fasciculate, spetate. 25.92–30.78 × 4.05–4.85 µm, av. 28.35 × 4.05 µm. Conidia fusoide to cylindric, curved, 1–7 septate, sub-hyaline to pale olivaceous, sub-acute at both ends. 64.80–74.52 × 3.24–4.85 µm, av. 69.66 × 3.24 µm.

Spots are water soaked, grayish white to ash coloured on lower surface, irregular and abundant. Causal organism *Cercospores blumeicola* Das Conidiophores light brown arise in tufts, geniculate, bent above conidiophores in fascicles few to many, sub-hyaline apically. 87.48–103.68 × 3.24-4.05 µm, av. 93.96 × 4.05 µm. Conidia hyaline obclavate, broadly acicular, base truncate, acute at tip 1–10 septate. 93.96–136.08 × 3.24–4.86 µm, apex 2.43 µm. Das (1958).

Buchanania lanzan (Chironji)

The natives of Chhattisgarh use the seeds in face care. The seed powder is mixed in cow milk and paste in prepared. This paste is applied externally. According to them it improves the complexion and also removes the black spots. The leading beauty-parlours of Chhattisgarh are using the herb for this purpose. In case of Chhappaki (Urticaria), along with other treatments the traditional healers of Bastar region apply the paste prepared by mixing seeds in cow milk, in whole body. In case of dysentery, the Chironji bark is given with milk as promising treatment.

Leaf Spot of *Buchanania*

Sporulation in leaf spot, the *Causal organism: Pestalotiopsis versicolor* (Speg.) acervuli dark brown to subepidermal. discoide to cushion shape, abundant, scattered. Conidiophores hyaline, small, shorter than conidia, subglobose to slender 4.86–6.48 × 3.24-4.86 µm, av. 5.67 × 4.05 µm. Conidia dark, ellipsoid to fusoid 4-5 celled, 2-3 cells dark, 2-3 apical appendases 16.20–2.25 × 5.67–6.48 µm average 17.82 × 5.67 µm. *Buchanania lanzan* Spreng. This species has been report on several hosts from Jabalpur (M.P.). Sahni (1965).

Callicarpa macrophylla

The leaves are heated and applied as a poultice to ease the pain of rheumatic joints. A decoction of the leaves is used in the treatment of diarrhoea and dysentery. A juice made from the leaves mixed with equal portions of Drymaria diandra and Oxalis corniculata is used in the treatment of gastric troubles. The root is chewed to relieve rashes on the tongue. A paste made from the roots is used to treat fevers. The juice of the root is used to treat indigestion. Oil obtained from the roots is aromatic and stomachic. It is assumed that this is an essential oil. The inner bark is pounded and used as a poultice on cuts and wounds. The fruits are chewed to treat boils on the tongue. The juice of ripe fruits is used in the treatment of indigestion and fevers.

Leaf Spot of Callicarpa

Spots are dark brown to black, scattered, irregular, surrounded by light yellow concentric margin, small fruit body, between the infested leaves represent the *Causal organism*: *Septoria* sp. pycnidia. Pycnidia light to dark brown, erumpent, separate, globose, beaked ostiolate, 32.40–42.00 × 25.92–30.78 µm, av. 35.64 × 25.74 µm. Condiophores small, hyaline, tapering upward narrowly flask shaped, 2.43–3.24 × 1.62 µm, av. 2.43 × 1.62 µm. Conidia hyaline, filiform, straight to curved, 1-4 septate usually 3 septate 21.06–35.64 × 3.24 µm, av. 25.92 × 3.24 µm. (Gupta, 2005).

Cassia Angustifolia (Senna)

It is useful in habitual costiveness. It lowers bowels, increases peristaltic movements of the colon by its local action upon the intestinal wall. It is used as expectorant, wound dresser, antidysentric, carminative and laxative.

Leaf Spot of Senna

Leaf spot appear on leaves, petiole, stem, dark brown to black with ash white at centre spot small to large, elongated, circular to irregular, dark brown to brown with concentric margin, severe spot become coalesce and finally dry (Figure 25.7). *Causal organism*: *Alternaria cassiae* Jurair and Khan. Colony on PDA smoky, floccose, dark brown, sporulation abundant. Conidiophore mostly arise singly sometime in group, light brown to pale olivacious, straight, cylindrical, 0–2 septate, dark single scar apically. 35.64–58.32 × 4.86–6.48 µm, av. 53.46 × 6.48 µm scar 3.24 µm. Conidia elongated cylindric to fusiform, born singly on conidiophore, light brown to brown, both transverse and longitudinal septa, present beak long, fragile, sub-hyaline, smooth-walled. Constructed at septum with a prominent scar 6–9 longitudinal to 2–5 transverse septa, 139.32–157.00 × 19.44–21.06 µm, av. 145 × 19.44 µm, beak 64.80–76.14 × 1.62–2.43 µm, av. 71.28 × 2.43 µm. (Gupta, 2005)

Chlorophytum borivillianum Ker Gawl. (Safed Moosli)

Aphrodisiac, tonic, pain reliever and used to cure general debility and impotency. Its powder increases lactation in feeding mothers and lactating cows. It is being increasingly used in Ayurvedic and Pharmaceutical Industries.

Leaf Spot of Moosli

Spots are circular, angular to irregular, reddish brown depressed with light brown to ash colour at centre, surrounded by reddish brown margin distributed around the leaves (Figure 25.8). *Causal organism*: *Colletotrichium chlorophytumie* Chandra and Tandon. Colony on PDA medium, smoke grey, acervuli in culture blackish, superficial. Setae distributed, straight and abundant, dark brown, non septate, tapering upward, acute at tip, distributed around the periphery, 87.48–153.90 × 5.67–9.72 µm. Conidiophores: hyaline, cylindric, tapering upward 5.67–6.24 × 2.43 µm Conidia: hyaline, single celled, curved, acute end, thin walled 19.44–22.68 × 4.05 µm Appressoria: medium, long, germtube, mostly 2 germtube with nipple shape to slightly angular appressoria. Chandra and Tandon (1965).

Clitoria ternatea L. (Butterfly Pea)

Commonly known as Butterfly pea Shankupushpam, is widely used in traditional Indian systems of medicine as a brain tonic and is believed to promote memory and intelligence. The plant is considered as a good brain tonic and is useful for throat and eye infections, skin diseases, urinary troubles even in cattle, ulcer and antidotal properties. Besides its medicinal property butterfly pea is also a good source of phytochemical substances. It contains antifungal proteins and has been shown to be homologous to.

Leaf Spot of Clitoria

Spot on the leaves small, abundant, isolated, confluent, circular to irregular, dark brown to black amphiphyllously, blighted in patch, surrounded by dark yellow margin. *Causal organism*: *Cercospora ternateae* Petch. Conidiophores arise in cluster, pale olivaceous to light brown, 1-3 septate, irregularly bent above, fuscous, and rarely branched. 106.92–152.28 × 4.05–4.86 µm, av. 136.08 × 4.86 µm. Conidia, acicular obclavate, slightly curved, sub-acute hyaline to sub-hyaline, 1–8 septate. 115.02–124.74 × 3.24–4.36 µm, av. 121.50 × 4.05 µm.

Cocculus hirsutus (Jaljamini)

Cocculus is a climbing shrub occurring throughout tropical and sub-tropical tracts of India. In Chhattisgarh, it is a common roadside and wasteland herb. Leaf extract mixed with water, coagulates into a green jelly-like substance in very less time. Due to this unique property it is known as Jaljamini (Jal-water, Jamini-to freeze) in Chhattisgarh. The natives and traditional healers of Chhattisgarh are well aware of its medicinal properties and uses. It is used both internally and externally.

Leaf Spot of Cocculus

Spots on the living leaves, dark brown to blackish, rectangular to irregular on both side, surrounded by grayish white concentric margin, restricted by vein, sporulating mostly hypophyllously (Figure 25.9). Causal organism *Cercospora cocculi* Sydow Conidiophores, sub-hyaline to pale brown, unbranched, non-septate, straight to geniculate, blunt at apex. 38.88–56.70 × 3.24–4.05. Av. 46.98 × 4.05 µm. Conidia cylindric sub-hyaline to pale olivaceous, narrowly obclavate, straight or curved, 1-12 septate, base conic, truncate with a prominent scar apex bluntly round. 69.68–100.44 × 3.24–4.05 µm, av. 84.26 × 3.24 µm. Agarwal and Sharma (1973).

Crataeva nurvala Buch (Varun)

On the basis of principles of Ayurvedic pharmacodynamics and other classical references the clinical trial of Varun (*Crataeva Nurvela*) has been taken up in the treatment of urinary disorders caused by enlarged prostate. Animal experiments have also been carried out to study its anti-inflammatory and anti-lithiatic effect of Varun. The process of stone formation from Ayurvedic as well as modern point of view has been discussed. The anti-inflammatory effect of drug has been confirmed on carrageening induced granuloma pouch, though this effect is less than cortisone.

Leaf Spot of Varun

Spot are scattered, abundant, mostly circular, small to large, yellowish gray, brown at centre with reddish brown concentric margin (Figure 25.10). *Causal organism*: *Corynespora cassiicola* (Berk and Curt.) Wei. Conidiophore brown to light brown, branched 1-6 septate, emerging through leaf epidermis, slightly swollen at the apex, round to cylindric, simple or in tufts, proliferating terminally through scar of previous conidium 155.22–243.00 × 7–10 mm; mid–7.82µm; base–12.96 µm. Conidia 5–18 psuedoseptate, base truncate, top sub-acute to round, tapering upward, pale olivacious to light brown, terminally, single or sometime in short chains, with a thick, colourless exospore, with a dark prominent scar 71.72–220.30 × 7.10 µm. (Gupta, 2005).

Croton tiglium Linn.

Croton oil is a powerful irritant and cathartic. In large doses it is a dangerous poison, occasioning emesis, painful gripings, hypercatharsis, and other serious symptoms. It is principally used as a purgative when the bowels are very torpid. It is admissible in *obstinate constipation* when no inflammation exists, and is the most efficient purgative in *lead colic*. A croton-oil liniment is made by mixing 1 part of croton oil with 4 or 5 parts of olive oil, or 6 parts of turpentine; it is rubbed on the skin several times a day, to cause redness and a pustular eruption; it is very beneficial in *follicular disease of the throat, affections of the larynx, bronchial vessels,* and *lungs, indolent tumors,* and all *painful attacks*. The dose of croton oil is from 1 to 6 drops, which is best given on sugar, or made into a pill with crumb of bread.

Leaf Spot of Croton

Spots on the leaves large marginal, dark brown to grayish in colour, dark reddish brown to brown concentric margin in young infected leaves, black minute dot like structures represent the *Causal organism*: *Macrophoma* sp. pycnidia. Pycnidia innate to partially submerged, globose to subglobose, brown to dark brown, thick-walled, inner wall light brown, ostiolate, 110.50–130.00 µm. in dia., av. 104 µm in dia. Conidiophores small arise from side wall, abundant, tapering upward, hyaline. 6.48–9.72 × 1.62 µm, av. 7.10 × 1.62 µm. Conidia elongated, cylindrical to conic, thin-wall, 1-celled, hyaline, 19-44–22.68 × 5.67–6.48 µm, av. 21.87 × 5.67 µm. (Gupta, 2005).

Withania somnifera Dunal. (Ashwagandha)

Cultivated throughout drier parts of India. It is Tonic, Abortifacient, Astringent, Deobstruent, Nervine, Aphrodisiac and Sedative. It is official Indian Pharmacoepeia. It is popularly known as Indian Ginseng. It gives vitality and vigour and helps in building greater endurance. It has been used in diseases such as rheumatism, leprosy and arthritis. All parts of the plant are used in herbal medicine. In Ayurveda, the fresh roots are sometimes boiled in milk, prior to drying, in order to leach out undesirable constituents. The berries are used as a substitute for rennet, to coagulate milk in cheese making.

Damping-off and Seedling Blight of Ashwagandha

The symptoms are also characterized by plants turn yellowish and finally dried affected plant showed mottling followed by necrotic lesions, curling sand burnt margins. In severe case of infection whole plant was burned and died (Pandey, 1988; Nigam, 1984 and Gupta *et al.*, 1993) *Causal organism*: *Alternaria alternata* (Fr.) Keissler. *Control*. Jain (1985) reported the use of fungicide treatment of seeds for the control of seedling diseases. He found that Delton, Dithane M-45, Brassicol and Bavistin could protect the seedlings. The leaf spot diseases were also reported from Jaipur, India (Mahrishi, 1986). Symptoms were small dirty yellow Sporulation on leaf with black fruit bodies on the lower surface of the leaves. Sporodochia pulvinate, polymorphic, 50–470 µm in dia., av. 280 µm in dia., spore mass becomes hard on drying, shiny black, convex, surrounded by a fringe of entangled white hyphae. Conidiophores in loose fasicles, branched, bearing the phialides, septate, 11–34 × 7143–3.26 µm, av. 21.00 × 3.26. Conidia rod-shaped, ends rounded, one end rarely slightly truncate, smooth walled, hyaline, subhyaline in mass, guttulate, 6.48–9.72 × 2.43 µm, av. 9.0 × 2.43 µm. Root-Knot: A severe root-knot infection was observed on *Withania somnifera*. Plant infected with nematodes showed the symptoms are yellowing and stunting of the whole plants. Roots of infected plants showed formation of galls, swelling or knots (Bhatti *et al.*, 1974; Minz and Harari, 1969).

Alternaria Leaf Spot of Ashwagandha

Leaf spots, vary in shape and size, irregular light brown to brown with black powder at center represent the conidia, with dark concentric margin, spot become coalesce and form a large patches, foliage become dry (FIgure 25.12). *Causal organism: Alternaria tenuissima* (Kunze ex Pers.) Wiltshire. Conidiophores light brown-to-brown, 4-6 transversely septate, small, bearing a chain of conidia, with a prominent scar at the point of attachment. Conidia light brown-to-brown muriform, typically with longitudinal and transverse septa, vary in shape and size, mostly obclavate young conidia ellipsoid with a small to long beak, bearing conidia acropitally. 55.08–63.18 × 16.20–19.44 µm, av. 51.84 × 17.82 µm. Beak, 11.34–16.20 µm. All cells conidia are capable of germinate (Gupta, 2005).

Cercospora Leaf Spots of Ashwagandha

Spots, dark gray, smoke to blackish, irregular, at times vein limited, powdery growth represent the conidia of Cercospora on both surface (Figure 25.11). The *Causal organism*: *Cercospora withaniae* H&P Sydow. Conidiophores pale olivaceous to light brown, unseptate, rarely geniculate, unbranched, straight to bent, bluntly rounded tip, sometimes acute. 21.06–25.92 × 3.24–4.05 µm, av. 21.06 × 3.24 µm, Conidia hyaline to sub-hyaline, 3–7 septate, obclavate, cylindric, straight to mildly curved, base obconically truncate, tip sub obtuse 37.26–72.90 × 2.43–3.24 µm, av 53.46 × 2.43 µm (Gupta, 2005).

Hyoscyamus niger Linn.

The Seeds of Hyocyamus Niger is widely used as a nutritious Bird feed. Gastric, or intestinal cramps, diarrhoea, neuralgia, cough hysteria, manis, skin inflammation and boils. Niger seeds has anodyne, narcotic and mydriatic properties, employed as a sedative in nervous infections. In veterinary practice used as urnary sedative.

Cercospora Leaf Spots of Hyoscyamus

Spots are small, circular to irregular, light brown to brown, partially, sunken, isolated at time coalesce or confluent with light yellow concentric margin. The *Causal organism*: *Cercospora* sp. Conidiophores, arise in fasicles. 1-3 septate, sub-hyaline, to light brown, geniculate, simply or very rarely branched, sub-hyaline at tip with a dark prominent scar at the point of attachment with the conidia, 88.00–115.00 × 4.05–5.67 µm, av. 94.00 × 4.86 µm. Conidia hyaline, linear tapering upward, slightly curved base truncate, acute at tip, 10-12 septate, 120.00–131.00 × 4.05–5.67 µm, av. 123.12 × 4.05 µm, scar 3.24–4.00 (Gupta, 2005).

Tinospora cordifolia (Willd.) Hooke and Thoms

Stem of *Tinospora cordifolia* is an Antiperiodic, Antipyretic, Alterative, Diuretic, Anti-inflammatory. It is a contituent of several compound preparations. It is used in fever, urinary disorders, dyspepsia, general debility and urinary diseases. It is also used in treatment of rheumatism and jaundice.

Cercospora Leaf Spot of Tinospora

Spot are scattered, irregular, yellowish ash white, turn pale brown, surrounded by yellow hallow, centre papery hypophyllously coalescing, form large dry patches (Figure 25.13). The *Causal organism:*

Cercospora tinosporae H&P Sydow. Conidiophores arise in fascicle, hyaline to sub-hyaline, straight, cylindrical, 1–2 septate, stroma 29.06–55.08 µm. Conidia obclavate, apex pointed, sub-hyaline, at time slightly curved, constructed at septum, 1-3 celled, truncate at base and tapering upward, born singly, 19.44–30.78 × 9.72–12.96 µm av. 26 × 12.00 µm, scar of conidia 2.43–3.24 µm (Gupta, 2005).

Emblica officinalis Gaertn. (Aonla)

The fruit of *Emblica ifficinalis* (Aonla) is aperient, carminative, diuretic, aphrodasiac, laxative, astringent and refrigerant. It is the richest known source of vitamin 'C'. It is useful in anaemia, jaundice, dyspepcia, haemorrhage disorders, diabetes, asthama and bronchitis. It cures insomnia and is healthy for hair.

Seed Rot of Aonla

Causal Organism

Phoma putaminum Speg. Pycnidia black dot like sparse to gregarious in sunken, brown spots (Figure 25.14). Pycnidia erumpent partly emerged in host tissue, unilocular, dark brown to black, globose to subglobose, sparse to gregarious or confluent, thick-walled, ostiolate centre not papillae, 145.8 × 152.28 × 152.28–162.0 µm, av. 152.28 × 162.0 µm. Conidia elliptical to pyriform, sometime oval, hyaline, 1-celled, thin-walled, 4.86–6.48 × 3.24 µm, av. 5.67 × 3.24 µm.The species *Phoma putaminum* Speg. was earlier recorded on fruits of *Phyllanthus emblica* by Pandey *et al.* (1980) from Allahabad, U.P. On the basis of morphological characters the fungus associated with *Embelica* has been identified as *P. putaminum,* hence this forms a first record for country (Gupta, 2005).

Terminalia belerica Roxb. (BAEHRA)

The fruit of Terminalia Belerica. Are Astringent, Tonic, Expectorant and Laxative. It is used in coughs and sore throat. Its pulp is used in dropsy, piles and diarrhoea. It is also useful in leprosy, fever and hair care. It is also used in oxalic acid and preparation of ink.

Leaf Spots of Terminalia

Spots are large ash white to light brown, sparse, abundant, circular to irregular with a dark concentric margin, the *Causal organism*: *Phoma* sp. Colony smoke white to light ash colour, hyphae septate, sub-hyaline, branched immersed, small black dots like structures represent pycnidia. Pycnidia partly innate, globose to subglobose, dark brown, aggregated. abundant, sometime sparse to separate, thick walled single ostiolate papillae, 155.52–181.44 × 152.28–181.44 µm, av. 162 × 155.52 µm. Conidiophores small hyaline, tapering upward, basal, 3.24–4.05 × 1.62–3.24 µm, Conidia, hyaline, 1-celled, thin walled often guttulate, oval to or subglobose, 7.10–9.72 × 4.05–4.86 µm, av. 6.48 × 4.86 µm (Gupta, 2005).

Papaver somniferum (Opium Poppy)

Opium poppy is one of the most important medicinal plants used by ancient man the seeds are used as narcotic, sedative, tonic, analgesic and hypnotic. Milky juice from immature fruits yields opium. Ripe and dry fruits contain only small quantity of alkaloids and therefore narcotic properties are mild. Opium latex contains more than dozen of alkaloids. A number of phytopathogenic fungi, bacteria and virus are associated and to infect medicinal as well as commercial value.

Seedling Blight of Opium

Species of Pythium is responsible for the causing seedling blight it first report in Australia to be a serious diseases (Angel, 1950). *Causal organism*: *Pythium ultimum* Trow. *Pythium mamillatum* Meurs. The pathogen cause both pre and post emergence damping-off (Alam *et al.*, 1996) were recently found to be affected in commercial cultivation of opium poppy in Utter Pradesh. The disease occurred during the emerging stage of seedling, mostly 2-4 weeks after sowing. The symptoms are rotting of roots and collar region leading to finally death. *Control*: Seed Treatment with Ridomil MZ-72 WP@0.12g/100g of seeds, or Capton@0.30 per cent or soil drenching with Propamocarb hydrochloride @0.30 per cent also gave satisfactory control of the disease. The damping-off diseases caused by soil born fungi are effective managed by Mancozeb treatment. (Alam *et al.*, 1996). Leaf spot diseases on opium poppy was also observed in Romania and Bulgaria (Savulescu 1932, Kovachevsky 1936, Kochman 1936). Symptoms are cloratic and necrotic area on the foliage followed by curling of the leaves and Sori on the leaves formed round yellow, and finally grey to brownish spots. *Causal organism*: *Entyloma fuscum* Schorter: Uridospore globose to polygonal, hyaline, colourless, then brown, 11–17 µm thick, 2 layered. Conidia on the host on the underside of the leaves, cylindrical, curved, attenuated towards the base or simple septate, hyaline, 10–20 × 3.0 µm. Cercospora leaf spot is also a caused by fungi *Cercospora papaveri. C. purpurea* and *Alternaria brassicae* have also been reported from India, Pakistan and Germany (Pavgi and Upadhyay 1964, Zaman and Khan 1969, Grummer 1953).

References

Agarwal, G.P. and Sharma, N.D., 1973. Fungi causing plant diseases at Jabalpur (M.P.) XIII: Some Cercospora IV. *Indian Phytopath.*, 26: 295–302.

Alam, M., Sattar, A., Chourasia, H.K. and Janardhanan, K.K., 1996. Damping-off: A new disease of opium poppy caused bu *Pythium dissotocum. Indian Phytopath.*, 49: 94–97.

Angell, H.R., 1950. Seedling blight II. Soil in relation of seedling blight of opium poppy and peas. *Aust. J. Agric. Res.*, 1: 132.

Balasubramanian, K.R. and Bhama, V..S., 1977. A new leaf spot disease of *Catharanthus roseus* in India. *Indian Phytopath.*, 3: 556–557.

Balick, M., Blanc, P., Morgan, M., Chen, J.L., Cooper, R., Kernan, M.R., Nanakorn, W., Parkinson, N., Rozhon, E.J., Stoddard, C.A. and Yee, Z.J., 1998. New iridoids from the medicinal plant *Barleria prionitis* with potent activity against respiratory syncytial virus. *Journal of Natural Products*, 61(10): 1295–1297.

Ballarin, C., 1950. Studies of *Helminthosporium papaveries. Phytopath. Z.*, 16: 400.

Bhakuni, D.S. and Jain, S., 1995. Chemistry of cultivated medicinal plants. In: *Advances in Horticulture*, Vol. 11, (Eds.) K.L. Chadha and R. Gupta. Malhotra Publishing House, New Delhi, p. 45–119.

Bhatti, D.S., Gupta D.C., Dahiya, R.S. and Malhan, I., 1974. Additional host of root-knot nematode *Meloidogyne javanica. Curr. Sci.*, 43: 622–623.

Bogarada, A.P., Lyman, V.E. and Taranich, A.P., 1971. On the resistance of poppy to Helminthosporiosis. *Selekts Semenov.*, 36: 78.

Burkill, H.M., 1985. *The Useful Plants of West Tropical Africa*, Vol. 1. Royal Botanic Garden, Kew, U.K., p. 960.

Butler, E.J., 1918. *Fungi and Diseases in Plants*. Thacker Spink. and Co., Calcutta, VI, 547 pp.

Casulli, F. and Ippolito, A., 1995. Observation on liquorice rust [*Uromyces glycyrrhizae* (Rob.) Magn] in south Italy. *Informatore Phytopathologico.*, 45: 27–30.

Chandra, V. and Tandon, R.N., 1965. Two new leaf spot fungi. *Curr. Sci.*, 34: 565–66.

Chandra, V., 1956. Studies on *Rauvolfia.* Part V: *Alternaria* on *Rauvolfia serpentina* Benth. *Ind. J. Pharm.*, 18: 146.

Chastagner, G.A., Ogawa, J.M. and Sammeta, K.P.V., 1979. Cause and control of damping-off in *Plantago ovata. Plant. Dis. Reptr.*, 62: 929–932.

Cheon, S.R. and Kim, H.J., 1999. Comparison of resistance of root-rot caused by *Fusarium solani* in ginseng breeding lines. *Koreon J. Ginseng Sci.*, 14: 50–56.

Chiddarwar, P.P., 1959. Contribution to our knowledge of the *Cercosporae* of Bombay State-II. *Indian Phytopath.*, 12: 110–21.

Chopra, R.N., Nayar, S.L. and Chopra, I.C., 1956. *Glossary of Indian Medicinal Plant*. Council of Scientific and Industrial Research, New Delhi, 330 pp.

Darmono, T.W. and Parke, J.L., 1990. Chlamydospores of *Phytophthora cactorum:* Their production, structure and infectivity. *Can. J. Bot.*, 68: 640–645.

Darpouse, H., 1945. A contribution to the study of the diseases of oleogenous plants in France. *Ann. Epiphyt. N.S.*, 11: 71.

Das, S., 1958. A new species of *Cercospora* on *Blumea lacera* DC. *Curr. Sci*, 27: 260–61.

Diener, T.O., 1971. Potato spindle tuber virus. IV: A replicating, low molecular weight RNA. *Virology*, 45: 411–428.

Farooqi, A.A. and Sreeramu, B.S., 2001. *Cultivation of Medicinal and Aromatic Crops*. Universities Press *Hyderabad*, pp. 1–509.

Fenne, S.B., 1942. *Cercospora* on belladonna in Virginia. *Plant Dis. Repotr*, 26: 280.

Fokin, A.D., 1922. Diseases and injuries of cultivated plants observed during the summer of 1922 in the Govt. of Vyatka. In: *Trans. Fourth All Russsian Entomo. Phyopath Congress,* Moscow. 8–14 Dec. p. 108.

Ganguly, D. and Pandotra, V.R., 1962. *Regional Research Laboratory Jamm. Bull.*, 1: 42–46.

Ganguly, D. and Pandotra, V.R., 1963. Fungi on medicinal and aromatic plants in the north West Himalayas. *J. Mycopath. Mycol. Appl.*, 20: 39–48.

Ghosh, N.C., 1988. *Comparative Materia Medica.* Hannemann Publ. Co. Pvt. Ltd., Calcutta, India, p. 855.

Gill H.S., Riberio, O.K. and Zentmeyer, G.A., 1977. *Phytophthora* blight of periwinkle in Cochella valley of California. *Plant Dis. Reptr.*, 61: 560–561.

Girzitska Mme, Z.K., 1928. Condial stage of *Pleospora papaveracea sacc. Protc. Pan. Soviet Congress Botanists* Leningrad, January, p. 172.

Goto U., 1938. Anthracnose of *Digitalis* caused by *Colletotrichum fuscum* Laubert. *Ann Phytopath. Soc. Japan*, 8: 1–8.

Goyal R. and Pathak, N.C., 1982. A new leaf spot disease of *Catharanthus roseus* (L) G. Don from India. *New Botanist*, 9: 31–32.

Griffin, D., 1993. *Fungal Physiology*, 2nd edn. Wiley-Liss, New York.

Grummer, G., 1953. The influence of *Alternaria* infection of poppy capsules on promptitude of germination by their seeds. *Flora Fauna*, 140: 298.

Gupta, M.L. and Janardhanan, K.K., 1989. A new leaf blight of opium poppy caused by *Alternaria pharmagmospora* van Emden. *Indian J. Plant Path.*, 7: 83–84.

Gupta S., Kumar, A. and Thakur, ,R.N., 1993. Some problems in cultivation of *Withania somnifera* (L) *Dunal* (*Aswagandha*) in Jammu region of India. *J. Res. Edu. Ind. Med.*, 12: 23–27.

Gupta, M.L., Singh, H.B, Kalra, A., Pandey, R. and Singh, S.P., 1997. A New leaf spot of senna caused by *Colletotrichum gloeosporioides*. *Indian J. Plant Pathol.* 15: 95–96.

Gupta, P.K., 2005. Taxonomy of some fungi associated with medicinal plants. *Thesis*, Dept. of Pl. Path., C.O.A., J.N.K.V.V., Jabalpur (M.P).

Gupta, S. and Masood, A., 2003. Leaf spot disease of Indian Aloe (*Aloe vera* Linn. Syn. *Aloe barbadensis* Mill.) caused by *Alternaria alternata*. In: *National Symposium on emerging Trends in Indian Medicinal Plant*, Lucknow, pp. 15(2–0/3), 10–12 Oct.

Halfan-Meir, H., 1973. *Alternaria crassa* a seed born fungus of Datura. *Plant Dis. Reptr.*, 57: 960–963.

Husain, A., 1993. *Medicinal Plants and their cultivation.* Central Institute of Medicinal and Aromatic Plants, Lucknow, pp. 111.

Jain, A.C., Nikam, B.G., Kulkarni, S.N. and Sharma, O.P., 1960. Fungi of Gwalior and Indore region. *The Vikram*, 4: 181–87.

Jain, N.K., 1985. Status report of disease in MP. In: *Proc. V. All India Workshop on MAP*, Banglore, 22–25 December.

Janardhanan, K.K., Ganguly, D. and Husain, A., 1964. Fusarium wilt of *Rouvolfia serpentina. Curr Sci.*, 33: 313.

Janardhanan, K.K., Gupta, M.L. and Husain, A., 1977. *Pythium* die-back, a new disease of *Catharanthus roseus* (L.) G. Don. *Indian Phytopath.*, 39: 427–430.

Janardhanan, K.K., 2002. *Diseases of Major Medicinal Plants*. Daya Publishing House, Delhi, pp. 1–189.

Jin, K.S., Lee, S.W., Kim, J.J. and Ryu, H.Y., 1994. Identification of bacterial isolates obtained from diseased orchid and aloe plants caused by *Erwinia chrysanthemi. RDA Journal of Agricultural Science Crop Protection*, 36: 301–306.

Kapoor, J.N. and Chowdhary, P.N., 1976. Notes on Indian microfungi. *Indian Phytopath.*, 29: 348–352.

Khan, A.S. and Kamul, M., 1982. Note on *Cercospora cavarae* Saccardo. *Curr. Sci.*, 31: 341–342.

Khatua, D.C. and Maiti, S., 1982. Two new leaf blight diseases caused by *Rhizoctonia solani. Indian Phytopath.*, 35: 124–25.

Kneifel, W., Czech, E. and Kopp, B., 2002. Microbial contamination of medicinal plants–A Review. *Planta medica* V. 68(1): 5–15.

Kochman, J., 1936. Grzybe Glowniowe Polski. *Ustilaginales Poloniae Planta Polonica*, 4: 1–161+12 Pls.

Kothari, K.L. and Prasad, N., 1970. Downy mildew of opium poppy in Rajasthan. *Indian Phytopath.*, 23: 674–688.

Kovachevsky, I.C., 1936. Parasitic fungi new for Bulgaria, Fourth Contribution. *Tran. Soc. Bulg. Sci. Nat.*, 17: 13.

Kowalski, J., 1968. Investigations on some biological properties of fungi attacking leaves of *Digitalis purpurea* and *D. lanata. Acta Mycol. Warszawa*, 4: 25–38

Kumar, H.V. and Masood, A., 2003. Leaf spot of *Casearia tomentosa*, a new disease caused by *Alternaria alternata:* First report of occurrence from Uttar Pradesh. In: *National Symposium on Emerging Trends in Indian Medicinal Plants*, Lucknow, 14(2–0/3), 10–12 October.

Kumar, S., Shukla, R.S., Singh, K.P. and Husain, A., 1984. *Alternaria* leaf spot disease of *Hyoscyamus muticus* in India. *Indian J. Mycol. Pl. Path.*, 14: 176.

Kumar, S., Shulka, R.S., Singh, K.P. and Husain, A., 1982. Leaf blight of *Hyoscyamus niger* in India. *Indian Phytopath.*, 35: 335.

Mandal Kunal and Maiti Satyabrata, 2005. Bacterial soft rot of aloe caused by *Pectobacterium chrysanthemi*: A new report from India. In: *New Diseases Reports in British Society for Plant Pathology.*

Laat, P.C.A de, Verhoeven, J.T.W. and Janse, J.D,. 1994. Bacterial leaf rot of *Aloe vera* L., caused by *Erwinia chrysanthemi* biovar 3. *European Journal of Plant Pathology*, 100: 81–84.

Maheshwari, J.K., 2000. *Ethnobotany and Medicinal Plants of Indian Subcontinent*. Scientific Publishers, Jodhpur, 666 pp.

Mahrishi, R.P., 1986. *Withania somnifera:* A new host of *Myrothecium roridum. Ind. J. Mycol. Pl. Pathol.*, 16: 199.

Masurat, G. and Stephan, S., 1961. The occurrence of the most important disases and pests of crop and garden plants in 1960 in German Democratic Republic. *Nachr. Bl. Dtsch. Pfl. Sch. Dienst. Berl.*, NP. 15: 125.

McGovern, R.J., 1994. Target spot of *Catharanthus roseus* caused by *Corynespora cassiicola. Plant Dis.*, 78: 83.

McPartland, J. and Hosoya, T., 1998. Species of *Colletotrichum* on ginseng (*Panax*). *Mycotaxon*, 67: 3–8.

*McRae, W., 1933. *Agric. Res. Inst. Pusa*, 1931–32, pp. 112–140.

Meffert, M.E., 1949. On the history of the helminthosporiosis of the opium poppy. *Nachr. Bl. Dtsch. Pfl. Sch. Dienst. Berl. N.P.*, 3: 104.

Mehrotra, M.D., 1976. Studies on some interesting diseases of *Rauvolfia serpentina* form India I. Contribution to the knowledge of disease development on relation to environment. *Indian Forester*, 102: 119–125.

Miltovic, I., 1952. Contribution to the knowledge of the fungus *Pleospora clavescens* (Fr.) Tulasne on poppy. *Rev. Sci. Agric. Zogreb.*, 13: 121.

Minz, G. and Harari, D.S., 1969. Additional host plant of root-knot nematode, *Meloidogyne* spp recorded in Israel during 1963–1968. *Isr. J. Agri. Res.*, 19: 137–138.

Moesz, G., 1938. Fungi of Hungary H: Archimycetes and Phycomycetes. *Ann. Hist. Nat. Mus. Hung.*, 31: 109.

Mohanthy, N.N. and Addy, S.K., 1957. *Cercospora* leaf spot of *Rauvolfia serpentina* Benth. *Sci. Cult.*, 26: 289–290.

Mohanthy, N.N. and Addy, S.K., 1958. Target spot of *Rauvolfia serpentina. Benth Sci. Cult.*, 23: 608–609.

Mukherji, K.G. and Bhasin, J., 1986. *Plant Disease of India: A Source Book.* Tata McGraw Hill Publishing Comp. Ltd., New Delhi, 468 pp.

Munjal, R.L. and Gill, H.S., 1963. Some Dematiaceous Hypomycetes from India–II. *Indian Phytopath.*, 16: 62–68.

Nair, V.M.G., 1991. Recent advances in ginseng (*Panax* spp) research and the importance of germplasm of wild medicinal plants. In: *Recent Advances in Medicinal, Aromatic and Spices Crops*, Vol. I., (Ed.) S.P. Raychaudheri. Today and Tomorrow's Printers and Publishers, New Delhi, p. 39–51

Nigam, K.B., 1984. Asgandh cultivation. *Indian Horticulture*, (Special issue), pp. 39–41.

Oudhia, P., Joshi, B.S. and Koshta, V.K., 1998a. Chhattisgarh ke kleshkarak kharptwaron se homoeopathic dava nirman kisambhavnayain (The possibilites of preparing homoeopathic drugs from obnoxious weeds of Chhattisgarh. In: *Abstract: Fifty National Science Conference*, Bhartiya Krishi Anusandhan Samittee, JNKVV, Gwalior.

Padwick, G.W. and Merch, J.L., 1943. Notes on Indian fungi I. *Mycological Papers* No. 7, CMI, Kew, U.K.

Pandey, G.N., 1988. Symptomatology and control of Asgandh diseases. *Nat. Sem. Med. Arom. Pl.*, Indore.

Pandey, G.N., 1995. Disease management of medicinal plants. In: *Advances in Horticulture*, Vol II: *Medicinal and Aromatic Plants*, (Eds.) K.L. Chadha and R. Gupta. Melhotra Publishing House, New Delhi, p. 249–269.

Pandey, R.S., Shukla, D.N., Khati, D.V.S. and Bhargava, S.N., 1980. A new fruit rot of *Phyllanthus emblica. Indian Phytopath.*, 33: 491.

Pandotra, V.R. and Husain, A., 1966. Fungi on medicinal and aromatic plants in the north west Himalayas IV. *Mycopath et Mycol. Appl*, 29: 155–159.

Parke, J.L. and Shortwell, K.M., 1989. Diseases of cultivated ginseng. *Univ. Wis. Madison. Coll. Agric. Life. Sci. Res. Bull.*, 3465: 16.

Parrotta, J.A., 2001. *Healing Plants of Peninsular India.* CABI Publishing. Wellington, UK and New York. 917 p.

Patel, D.J. and Patel, S.J., 1984. Senna an additional host of *Macrophomina phaseolina* in Gujrat. *Indian J. Mycol. Plant Pathol.*, 14: 293.

Patel, J.C. and Desai, M.V., 1987. New disease out break (leaf blight of *Plantago ovata. Indian Phytopath.*, 37: 582.

Patel, K.D. and Pillai, S.N., 1979. Effect of leaf spot disease on Sennoside content in seena leaves. *Indian drugs.* 71: 1–2.

Pavgi, M.S. and Upadhyay, 1964. Parasitic fungi from north India II. *Mycopath et Mycol. Appl.*, 24: 347.

Prakruti, 2002. *Suddh Bhangra (maka) oil.* http://www.prakrutiherbals.com/hairoil.htm, 2 p.

Probiotics New Zealand, 2002. Probiotics for life-improving your health and quality of life through good bacteria. http://www.probiotics. co.nz/prdcts_efml.asp?ProductID=5. 3 p.

Ramakrishnan, T.S. and Rangaswamy, G., 1948. *Uromyces acori* Ramakrishnan and Rangaswamy sp. *nov.* on *Acorus calamus. Curr. Sci.*, 17: 240–41.

Rao, P.N., 1962. Some Cercospora species from Hyderabad, India. *Indian Phytopathology*, 15: 112–140.

Reddy, M.A.R., Tewari, R.K., and Bakshi, B.K., 1971. Target leaf-spot disease of *Rauvolfia serpentina* and its chemical control. *Indian For.*, 97: 487–492.

Sahni, V.P., 1965. Deuteromycetes from Jabalpur–I. *Mycopath et Mycol. Appl.*, 27: 342–356.

Saini, S., Sattar, A. and Alam, M., 1996. Occurrence of stem-rot disease caused by *Rhizopus stolonifer* on periwinkle. *Indian J. Plant Pathol.*, 14: 1–4.

Sastry, K.S.M., Thakur, R.N. and Pandotra, V.R., 1977. Diseases of medicinal and aromatic plants and their control. In: *Cultivation and Utilization of Medicinal and Aromatic Plants.* Regional Research Laboratory, Jammu Tawi, pp. 711–733.

Sattar, A., Alam, M. and Janardhanan, K.K., 1986. Isolation of tenuazonic acid, a phytotoxin from *Alternaria crassa* (Sacc.) *Rands* causing leaf blight and fruit rot of *Datura stramonium* Mill. *Curr. Sci.*, 55: 195–196.

Savulescu, T., 1932. Phytosanitary conditions in Romania during the year 1930–31. *Inst. Cerc. Agron. al Romanici Publi.*, 8: 31.

Shakhela, R.R., Jose, V.T. and Tikka, S.B.S., 1985. It is paying to grow Isabgul. *Farmers and Parliament*, 20: 22–32.

Shin, H.S., Lee, H.H. and Lee, M.W., 1986. Studies on the distribution of fungal and *Fusarium* spp/ propagules in ginseng field soil. *Korean J. Mycol.*, 14: 109–119.

Sholberg, P.L., Li, T. and Ginns, J., 1996. First report of powdery mildew on ginseng in North America. *Plant Dis.*, 80: 463.

Shukla, R.S., Kumar, S., Singh, H.N. and Singh, K.P., 1993. First report of aerial blight of *Coleus forskohlii* caused by *Rhizoctonia solani* in India. *Plant Dis.*, 77: 429.

Shukla, R.S., Kumar, S., Singh, K.P. and Husain, A., 1988. Seedling blight of *Hyoscyamus muticus* by *Pythium butleri* in India. *Curr. Sci.*, 57: 40.

Singh, G. and Singh, R.N., 1984. Leaf spot disease of *Atropa belladonna. Indian Phytopath.*, 37: 585.

Singh, G., Singh, R.N. and Bisht, I.S., 1985. Control of leaf spot disease of belladonna caused by *Ascochyta atropae. Indian J. Plant Path.*, 14: 71–71.

Sol, H.H. Van Heuven, J.C. and Seinhorst, J.W., 1961. Transmision of rattle virus and *A.belladonna* mosaic virus by nematodes. *Tijdschr. Plantziekten*, 66: 228–231.

Vaipani, 2002. Vaipani Herbal Ayurvedic Partisthan: A herbal ayurvedic cosmetics company. http:/ /www.vaipani.com/skincare/ skin per cent 20Care–Page3.html. 2 p.

Warrier, P.K., Nambiar, V.R.K. and Ramakutty, C., 1996. *Indian Medicinal Plants.* Vol. 1. Orient Longman, Chennai (India), p. 278–280.

Yossifovitch, M., 1928. Kingdom of the serbs, croates and solvenes: A destructive disease of the poppy in Southern Serbia. *Internal. Bull. Plant Project.* 11: 18.

Zaman, K.B. and Khan, A.S., 1969. Studies on some diseases of *Papaver somniferum* Linn. *Pakistan J. For.,* 19: 185.

*Original not seen

Insect Pest and Disease Management *Pages 392–400*
Editor: **Dr. D. Prasad**
Published by: **DAYA PUBLISHING HOUSE (2016)**

Chapter 26

Ecofriendly Management of Late Blight Disease of Potato in the Plains of West Bengal

Amitava Basu
Department of Plant Pathology, Bidhan Chandra Krishi Viswavidyalaya, Mohanpur – 741 252, Nadia, West Bengal

Introduction

Late blight of potato caused by *Phytophthora infestans* Mont. De Bary came into prominence at the time of Irish Potato Famine in 1844–45. In West Bengal; in the plains, it was first observed in 1898–1900 in Hooghly District (Butler, 1903). It appears nowadays in moderate form in every year, where weather conditions are more favourable, but occasionally assumed in epiphytotic proportions. The disease, however, attains serious proportions at any place, whenever, high humidity (90 per cent R. H.) and cool temperature (15–20°C) conditions prevail for considerable period. The blight pathogen being polycyclic in nature, is capable of multiplying rapidly and may cause a heavy damage to potato crop within a very shortest possible time. Yield losses due to late blight in the plains have been showing steep increase or decline depending upon the varieties grown in the region. Late blight susceptible cultivars were being grown in the plains unto the mid seventies. Consequently, losses were high. Susceptible varieties were gradually replaced by blight resistant varieties like Kufri Jyoti, Kufri Badshah, Kufri Sutlej and Kufri Anand due to which, losses have been minimal. Again the crop losses depend on time of disease appearance, its subsequent build up and spread. In the years, when the disease either does not appear or appears late in the season, the losses are insignificant. Dutt (1979) reported upto 75 per cent crop losses due to late blight of potato in eastern plains of India.

In the Gangetic plains of West Bengal, appearance of late blight generally occurs during 1st to 3rd week of January every year, when weather is congenial. Late blight causes considerable losses in terms of yield by premature defoliation and haulm destruction. During 1994–95 in the plains of West Bengal, highest disease incidence (75 per cent) and disease severity (60 per cent) were recorded (De and Basu, 2002). Chatterjee (1997) reported an epidemic outbreak of late blight of potato in Hooghly, Burdwan and other potato growing areas of West Bengal during 1967 and 1980 due to continuous spell of cloudy and rainy weather for 3–4 days in late January and early February (Basu *et al.*, 2003).

A detailed study was conducted to detect and identify the trends of disease appearance and its further development in the plains of West Bengal taking into account all the intricacies of weather, host and disease. Present investigation reveals that, late blight severity became maximum on 4th week after its onset on susceptible cultivar Kufri Chandramukhi when environmental conditions were favourable for its development 7 days before its onset and may be summarized as follow (*a*) maximum and minimum temperature ranging from 20–21.7°C and 12.5–12.8°C respectively; (*b*) max. and min. relative humidity ranging from 97.0–8.0 per cent and 65.0–67.0 per cent respectively; (*c*) sunshine hrs. 0.5; (*d*) rainfall 5.6–6.3 mm. In susceptible cultivar (Kufri Chandramukhi) the rate of progress of the disease/unit/week steadily increased and reached its maximum during 4th week after its initial appearance. The rate of disease development in resistant cultivar (Kufri Badshah) was remarkably retarded after progressive increase of disease upto 3rd week (Basu and Mukhopadhyay, 2000).

Devastating nature of late blight disease is due to the high vegetative and reproductive ability of the causal fungus *Phytophthora infestans.* A single blight lesion may produce 50–80 thousand zoosporangia and each zoosporangium produces 6–8 zoospores, which are real infecting structures. Thus, a single disease spot can produce 3–5 lakhs infecting spores. If 50 per cent of these spores germinate and developing into disease spot within 4–5 days, from a single spot, 1.5–2.5 lakhs spots will be formed. Each of them capable of producing 50–80 thousand spores. If the cycle is repeated once more within next 4 days, hundreds and thousand crores of blight spots would be formed, which are sufficient to infect almost entire crop of a field (Alexopoulos and Mims, 1983). From this, the extent of crop losses can be predicted in advance.

Besides this, nowdays, introduction of A_2 mating type outside Mexico has increased the possibility of sexual reproduction by the pathogen, which has also great impact and repercussions on disease epidemiology. This will add to one more source of inoculums in the form of thick wall oospores and would result in new virulence A_2 mating type being much more aggressive is fast displacing the old population Le A_1 mating type and results into increase in disease severity. Hence, in these circumstances, effective management of this disease is a vital task for the potato growers.

Disease Management

Late blight of potato is primarily a foliar disease but it readily infects the tubers. Therefore, its effective management would entail reduction of both foliar and tuber infection. This can be achieved through adoption of different strategies *i.e.* by employing chemicals, cultural practices, biological and host resistance. Use of any single method in isolation may prove counter productive. To illustrate, it is possible to manage the disease by applying chemical sprays but the number of sprays required on susceptible variety would be more than double the sprays required on a resistant variety, which is rather uneconomical for most of the farmers. Even if they are able to apply the required number of sprays, it would result in environmental pollution and also the development of fungicide resistant strain of the pathogen. Similarly, resistant varieties may be used for managing the disease but these are a risk of break down of resistance, if the resistance is governed by major genes (R-genes). However, if the resistance is broad based (minor gene resistance) it would still require support of prophylactic chemical sprays.

Chemical Control of Late Blight

Intially, Burgundy mixture ($CuSO_4$ $5H_2O$: Na_2CO_3 : water : : 5 kg : 5 kg : 500 litre) was recommended for late blight management (Vasudeva and Azad, 1952). Subsequently Bordeaux mixture ($CuSO_4$ $5H_2O$: CaO : water : : 5 kg : 5 kg : 500 litre) was found to be more effective than Burgundy mixture (Dutt, 1967). Since then several chemicals were tested against late blight and some of them gave promising results. These are cuprous oxide, mancozeb, captafol, fentin hydroxide, triphenyltin acetate and chlorothalonil, etc. (Dutt *et al.*, 1970; Raj *et al.* 1982). Mancozeb @ 2–3 kg/ha was accepted as most effective one for late blight pathogen control (AICPIP, 1983). The number of sprays varied from 2–3 in the plains. Then Metalaxyl and Oxadixyl had been found highly effective. Their efficacy lasted for 14 days as against 7 to 8 days in cases of mancozeb (Bhattacharya *et al.*, 1990). The fungus had a tendency to adapt more readily to metalaxyl formulations as compared to oxadixyl formulations (Singh *et al.*, 1988). Dipping of infected tubers carrying upto 5 per cent infection in 1500 to 2000 ppm solution of oxadixyl or metalaxyl formulations for 30 minutes could suppress the fungus growth. This treatment was effective only after tubers had overcome the dormancy (Singh and Bhattacharya, 1988). Development of fungicide resistance in *Phytophthora infestans* is a serious matter which need proper attention. Special emphasis is to be given on this type of research work. In the recent times, the systemic fungicides *e.g.* Ridomil MZ 72 WP (8 per cent metalaxyl + 64 per cent mancozeb) have been introduced for late blight management. However, experience world over, wherever systemic fungicides had been used, had shown the fungus developed resistance to those fungicides within a short period of time (Dowley and O'Sullivan, 1981; Schwinn, 1979). Earlier pure systemic fungicides were used. They were later replaced by mixture of systemic and contact fungicides. This was mainly done to reduce the selection pressure on the pathogen. This concept had paid dividends world over (Samoncha and Cohen, 1988). But still the chances of development of fungicide resistance in the fungus have not been ruled out. Alternatively, strategy of alternation of fungicide application was introduced to overcome this problem (Levy *et al.*, 1983). However, experience so far do not support neither of the concept alone for the management of resistance in the fungus. Integration of all the concept if possible is required. For example, instead of one, two or more systemic fungicides should be used in fungicide mixture to further reduce the selection presences. Number of sprays should be restricted (no more than two) in one crop season and a break in the application of systemic fungicide should be given after every 4–5 years. However, these concepts require validation by way of experimentation.

The management of late blight was studied in West Bengal, taking into account large number of chemicals for screening purpose. These are Dithio carbamate, Heterocyclic Nitrogenous compound, Copper oxychloride, Tetrachloro isophthalonitrile, 8 per cent metalaxyl + 64 per cent mancozeb, Zn SO_4 (0.5 per cent) + lime (0.25 per cent), $ZnSO_4$ (0.25 per cent) + lime (0.12 per cent), Oxadixyl + mancozeb, Oxadixyl + Copper oxychloride, Chlorothalonil 80 per cent, Aluminium tris-o-ethyl Phosphonate, Metalaxyl + Ziram. Among 19 chemical tested under All India Coordinated Potato Improvement programme, Diltham M-45 @ 2kg/ha exhibited best performance in terms of high yield and low disease severity in replicated plots as compared to control followed by Dithane Z–78, Ridomil M2-92 WP @ 0.8 kg/ha, Brestan @ 1 kg/ha, Oxadixyl + Mancozeb @ 2 kg/ha, Metalaxyl + Ziram @ 5 litre/ha (Anon, 1972–1998). Spraying of Mancozeb @ 0.2 per cent @ 7–10 days interval or one spray of metalaxyl + mancozeb @ 0.25 per cent followed by Mancozeb at 10 days interval, 2-3 times or more (De and Sengupta, 1988, 1991, 1992, 1993) gave the best performance. De and Mohasin (1993) reported that, the disease can be effectively controlled by Spraying thrice with mancozeb @ 0.2 per cent at an interval of 10 days. Basu and De (1999) found that, suitable fungicide at recommended doze was quite effective for managing late blight of potato. Ridomil MZ 72 WP, had proved to be very appropriate fungicide considering its effectiveness and cost benefit ratio. Spraying twice with Ridomil MZ-72 (8 per cent metalaxyl + 64 per cent mancozb) @ 2.5 kg/ha at 10 days interval managed this disease effective. Otherwise, spraying twice with Antracol (70 per cent propineb) @ 2.5 kg/ha at 10 days interval also managed the late blight. Indofil M-45 (mancozeb) being comparatively cheaper, can also be suggested for control of this disease to get considerable yield and net profit from potato cultivation in the plains of West Bengal.

Cultural Management

This method mainly aims at reducing the initial inoculum load in seed tubers and other sources and arresting the disease spread. Practices need for achieving this objectives are use of disease free seed tubers, burying the infected tubers and vines deep in the soil, removal of ground keepers and volunteer plants, high ridging, restricted irrigation during winter rains and planting of alternate fields, with susceptible and resistant varieties (Bhattacharya *et al.*, 1990).

Planting date can greatly affect the microclimate near the soil level at any given time or reason and there have possibilities for changing the environment so as to nudge the biological balance. Some diseases are very destructive, when susceptible age of the host and optimum soil and atmospheric condition for aggressiveness of the pathogen coincide. Therefore, alteration in date of planting in such a way that the susceptible age of the host does not coincide with favourable environment of the pathogen, helps in reducing losses from such diseases. Keeping this view of above, an investigation on different date of planting was conducted for consecutive 3 years in Hooghly area of West Bengal.

The results revealed that 1st week of December, was not the favourable period of planting. But 4th week of November in every years appeared to be suitable time of sowing because incidence of disease sprouts was low (9.0 per cent) and the tuber rotting was below risk level (around 10.0 per cent) as compared to other dates (2nd and 3rd week of November). Study on soil environment (soil moisture and soil temperature) showed that tuber rotting was higher (50–60 per cent) at higher soil temperature (> 30°C) and soil moisture (>75 per cent) as recorded during early part of November (Basu, 2002).

Biological Management

Several diseases and pests attack potato and large volumes of fungicides/pesticides are used to control them. But injudicious use of synthetic and organic pesticides has resulted in environmental pollution, human health hazards and development of resistance in pests. Therefore, effective safe, eco-friendly and economically viable alternative method is through use of antagonistic microbes against seed and soil borne diseases. Looking into the nature of the late blight fungus, absolute control through bio-agents seems to be remote but there is every likelihood that some bio-agents might provide partial perfection by way of induced resistance. There also exist a possibility of fungicide resistant bioagents which can be mixed with fungicides for providing better disease control. It was with this objective, that a large number of phylloplane mycoflora of potato was screened for its antagonistic effect on P. *infestans* some of them *viz. Penicillium aurantiogriseum* and *Aspergillus terreus* showed antagonistic effect (Roy *et al.*, 1991). Subsequently few more bio-agents *viz. Penicillium viridicatum Myrothecium verrucaria* and *Trichoderma viride* were found promising in laboratory conditions (CPRI 1991–92). Establishment of antagonistic fungi on potato phylloplane is the first pre requisite for providing disease control. It was revealed that *Penicillium* spp. were quick to establish. After 24 hours of their application, they constituted > 96 per cent (CPRI 1995-96) of the total fungus colonies recovered from lower surface of the leaves. Next best result (51 per cent) was obtained from *T. viride.*

Trichoderma spp., *Penicillium viridicatum* and *P. aurantiogresium* were evaluated for their tolerance to fungicides to integrate these with fungicides for control of late blight. *Trichoderma* spp. tolerated commercial formulation *viz.*, mancozeb, ridomil MZ 72 or curzate at 500 ppm without any effect on radial growth or sporulation of the antagonist. Further increase in dosage of mancozeb to 1000 ppm, ridomil MZ 2000 ppm or curzate 2000 ppm delayed or prevented sporulation of the bio-agent without affecting its radial growth. Similarly, *P. viridicatum* and *P. aurantiogriseum* tolerated mancozeb 500 ppm, Ridomol MZ 2000 ppm and curzate 2000 ppm. The results indicated that *Penicillium* spp. can be integrated with systemic fungicides like Ridomil or Curzate but not with mancozeb (CPRI 1997–98).

A spray of *Trichoderma* spp. along with 500 ppm each of mancozeb or Ridomil Mz or curzate increased establishment of the antagonist on foliage and their survival for comparatively longer period. Rodomil MZ, mancozeb and curzate at 500 ppm increased the establishment of *Trichoderma* spp. on upper leaf surface by 41.3, 17.3 and 2.02 per cent, respectively over control on eighth day of spray. Use of mancozeb at 500 ppm alone or in combination with *Trichoderma* also prevented *Aurobasidium pullulans* which otherwise occupied 80 to 100 per cent population of total phylloplane mycoflora and competed with the antagonists for space and nutrients (CPRI 1997–98).

Leaves collected from potato field sprayed with suspension of Trichoderma in mancozeb 500 ppm and inoculated in Laboratory with *P. infestans* resulted in reduction of blight lesions to 0.88 cm^2 from 1.80, 2.75 and 5.27 cm^2 as in case of mancozeb, *Trichoderma* and unsprayed control respectively. Bio-efficacy of three bio-agents *viz. Trichoderma harzianum, T. viride* and *P. aurantiogriseum* was evaluated in continuation with mancozeb under, field conditions. Results revealed that *T. viride* alone or in combination with mancozeb (500 ppm) were either at par or superior to mancozeb (0.2 per cent). Mancozeb (500 ppm) further improved the establishment of the bio-agents (CPRI, 1998–99).

Arora (2000) reported that Interactions between the antagonists and *P. infestans* were studied in laboratory; greenhouse and field. The antagonists either prevented germination of *P. infestans* sporangia or inhibited development of blight lesions or both. Interaction between *P. infestans* and *Penicillium* spp. led to formation of abortive sporangia of the pathogen. *T. viride* prevented differentiation of sporangial cytoplasm in to Zoospores. *Acremonium. strictum* lysed gersn tubers. In greenhouse experiments, a prophylactic spray of *A. Strictum* or *T. viride* at 10^8/ml concentration on two blight susceptible potato varieties 3 to 4 hours before inoculation with the pathogen restricted development of late blight between 1.5 and 14.0 per cent as compared to 61.6 and 88.8 per cent in control, without antagonists.

A different study was conducted at Hooghly area of West Bengal with talc based commercial formulation of four bio-agents namely, *Trichoderma viride, Gliocladium virens, Pseudomonas fluorescens* and *Bacillus subtilis* as seed and soil treatments for managing late blight disease under field condition. The results revealed that seed dressing with *T. viride + P. fluorescens* @ 2.5 g/kg of seed not only reduced the disease intensity, but also increased the yield of tubers. Only soil application of the particular treatment also performed very well under field condition through increasing plant growth (plant weight and tuber number) and yield of tubers (Basu *et al.* 2001). Hence this type of prophylactic management not only increased the productivity, but also proved to be economically profitable to the farmers for producing quality potato without disturbing the soil environment.

Besides these, many soil micro-organisms have ability to stimulate growth of the plant. Plant growth promoting bacteria is due to their ability to fix nitrogen, sequester iron, facilitate phosphorous uptake or produce Phytohormones that triggers responses in a growing plant. These biofertilizers are *Azotobacter* and *Phosphobacteria,* which can reduce the requirement of nitrogen by 17 per cent and phosphorous by 30 per cent respectively (Sunaina *et al.*, 2000). Seed treatment with *Azotobacter* @ 10.0 g/kg, gave significant results in terms of increased seed germination (90.0 per cent), reduction of late blight severity (8.0 per cent), increased tuber yield (260.0 q/ha) and increased plant growth in terms of plant weight (350.0 g), tuber number/plant (15.0) and tuber yield/plant (495.0) (Basu *et al.*, 2003).

Host Resistance for Disease Management

Until early severities, all the parental lines used in late blight resistance breeding programme had *Solaum demissum*, possessing both major and minor gene resistance. However, breeding for exclusive field resistance was initiated by using *Solaum verrucosum* as resistance source (CPRI, 1976). Sources having durable resistance have also been identified (Bhattacharya and Singh, 1986).

Parents possessing similar grades of resistance, when crossed, yielded the progeny having susceptible to highly resistant seedlings. In source of the crosses the progeny seedlings shared even better resistance than their parents (Sharma, 1983) but mere possession of high degree of resistance in parent cultures is not enough. The cultures should be good combiners as well. It was concluded from inheritance studies that varieties seems to differ in their genic constitution for field resistance. It is reflected in great variability in resistant and susceptible seedlings in the progeny, wide range of variability in resistance in progeny shows that resistance is quantitatively inherited. The mean resistance of progeny appears to have direct relationship with degree of resistance of the parents having good combine ability (CPRI, 1961, 62).

Field screening methods have been standardized for grouping potato genotypes into different resistance grades. Mean values and AUDPC were equally effective for this purpose (Singh *et al.*, 1988). Germplasm available at CPRI has a wide range of variability for resistance to late blight. Fourteen genotypes possess very high degree of field resistance, where as 15 are moderately resistant, viz., colonization rate, incubation period, infection efficiency and sporulation were found operative (Singh *et al.*, 1988). Amongst components of field resistance, sporulation was found to have the highest genetic variability, heritability and genetic variance. AUDPC was also found to have high heritability indicating its suitability for estimating field resistance. Path coefficient analysis of resistant components revealed that total sporulation is the most important component for determining field resistance (Birham *et al.*, 1989).

In between 2000 and 2005 thirteen potato genotypes were screened in terms of foliar infection (per cent) tuber infection (per cent) blight lesion area (cm^2) AUDPC and yield (t/ha) against late blight disease of potato in the plains of West Bengal under All India Coordinated Potato Improvement Project programme. The results were very significant. Out of thirteen cultivars Kufri Chandramukhi and Kufri Ashoka were susceptible. But Kufri Jyoti, Kufri Jawhar, Kufri Sutlej, Kufri Punhraj, Kufri Anand, Kufri Giriraj, Kufri Sindhuri and Kufri Lalima were moderately resistant. On the other hand, Kufri Badshah, Kufri Chipsona-1 and Kufri Chipsona-2 were resistant towards late blight disease.

Integrated Management of Late Blight Disease

In West Bengal, intensive agriculture comprising cultivation of crop varieties having a narrow genetic base in monoculture, which are responsive to inputs like irrigation, fertilizers and pesticides has increased the chances of occurrence of plant disease epidemics. IDM is an ecological approach among all beneficial, biological and physical form in the ecosystem to maintain plant health equilibrium. This involves the simultaneous' manipulation of a number of available strategies of reducing the plant diseases with minimum damage to the environment.

Keeping this view of above, a field investigation was done in Hooghly area for assessing five 10M components *i.e.* fungicides, host resistance, bio-control, micronutrients and organic amendments against late blight disease of potato. Investigation revealed that host resistance combined with foliar spray of chemicals *i.e.* Ridomil MZ-72 and Mancozeb 50 WP @ 0.2 per cent each gave best protection against late blight, where disease severity in foliage and tubers was less (7.0 per cent and 5.0 per cent respectively) and yield was comparatively high (260.0 q/ha) as compared to control due to 15 days delayed onset of the disease. On the other hand, host resistance combined with foliar spray of micronutrients *i.e.* $ZnSO_4$ and $CuSO_4$ @ 0.2 per cent had shown 12 days delayed onset of late blight disease, lower disease severity in foliage (10 per cent) and tubers (10.0 per cent) and higher tuber yield potential (235 q/ha) (Basu *et al.*, 2002). Therefore, this type of management has a great significance in terms of sustainable potato cultivation in West Bengal.

References

AICPIP, 1983. *Proceedings of the Tenth Workshop*. All India Coordinated Potato Improvement Project (AICPIP). CPRI, Shimla, 25–28 August, 1981, pp. 3.

Alexopoulos, C.J. and Mims, C.W., 1983. *Sub-division Diplo Mastigomycotina, Class-oomycetis, Introductory Mycology*. Wiley Eastern Limited, Darya Ganj, New Delhi, pp. 145–188.

Anonymous, 1972–2000. *Annual Report*, All India Coordinated Potato Improvement Project, Bidhan Chandra Krishi Viswavidyalaya, Kalyani, Nadia, West Bengal.

Anonymous, 2000–2005. *Annual Report*, All India Coordinated Potato Improvement Project, Bidhan Chandra Krishi Viswavidyalaya, Kalyani, Nadia, West Bengal.

Arora, R.K., 2000. Bio-control of Potato Late blight. In: *Potato: Global Research and Development*, (Eds.) S.M. Paul Khurana, G.S. Shekhawat, B.P Singh and S.K. Pandey. *Proceedings* (*Vol l*) *of Global Conference on Potato*, New Delhi, December 1999, pp. 620–623.

Basu, A., 2002. Influence of date of planting on the development of Late blight of Potato. *Ann. Pl. Prot. Sci.*, 10(1): 164–166.

Basu, A. and De, B.K., 1999. Field evaluation and economic analysis of chemicals against late blight of potato in the plains of West Bengal. *Annals of Pl. Prot. Sci.*, 7(2): 168–171.

Basu, A. and Mukhopadhyay, S.K., 2000. Pattern of late blight development in West Bengal. In: *Proceedings in Global Conference on Potato* (December 11-11-1999), New Delhi, 1: 635–637.

Basu, A., A. Konar, A., Mukhopadhyay, S.K. and Chettri, M., 2001. Biological management of late blight of potato using talc based formulations of antagonists. *J. Indian Potato Assoc.*, 28(1): 80–81.

Basu, A., Hossain, Md. M., Konar, A. and Chettri, M., 2003. Integrated management of late blight disease of potato in West Bengal. *Annals of Plant Prot. Sci.*, 1(1): 64–66.

Bhattacharya, S.K. and Singh, B.P., 1986. Durable resistance against the potato late blight in India. *Indian Phytopathol.*, 39: 615–619.

Bhattacharya, S.K., Shekhawat, G.S. and Singh, B.P., 1990. Potato late blight. Central Potato Research Institute, Shimla, H.P., *Tech. Bull.*, 27: 40.

Birham, R K., Singh, B.P., Bhattacharya, S.K. and Khusu, C.L., 1989. Studies on components of field resistance to late blight in Potato. *J. Indian Potato Assoc.*, 16: 166 (Abstr.).

Butler, E.J., 1903. Potato diseases in India. *Agril. Ledger Crop Dis. Pest. Ser.*, 7: 87–124.

Chatterjee, P.B., 1997. Epidemics of diseases and insect pests in West Bengal with analysis into their causes and impacts (1945–97). *Frontier of Plant Protection,* Department of Plant Protection, Viswa Bharati, 1: 1–12.

CPRI, 1961–62. *Annual Progress Report,* pp. 17–18.

CPRI, 1976. *Annual Progress Report,* pp. 58–61.

CPRI, 1991–92. *Annual Scientific Report,* pp. 9, 77–83.

CPRI, 1995–96. *Annual Scientific Report,* pp. 16, 83–87.

CPRI, 1997–98. *Annual Scientific Report,* pp. 10, 16, 62–65.

CPRI, 1998–99. *Annual Scientific Report,* pp. 2, 16, 70–77.

De, B.K. and Basu, A., 2002. Three decades study on monitoring of late blight disease of potato in the plains of West Bengal. *Environment and Ecology,* 20(1): 216–218.

De, B.K. and Mahasin, Md., 1993. Evaluation of fungicides against late blight disease of potato. *J. Myco. Pathol. Res.,* 31(1): 13–18.

De, B.K. and Sengupta, P.C., 1992. Evaluation of different chemicals for the control of late blight of potato caused by *Phytophthora infestans* in the plains of West Bengal. *Environ. and Ecology,* pp. 179–181.

Dowley, L.J. and O'sullivan, E., 1981. Metalaxyl resistant strains of *Phytophthora infestans* in Ireland. *Potato Research,* 24: 417–421.

Dutt, B.L., 1967. Fungicides and control of potato late blight. In: (*Abstr.*) *Int. Symposium on Plant Pathology,* Indian Phytopathological Soc., ICAR, New Delhi, pp. 63–64.

Dutt, B.L., 1979. *Bacterial and Fungal Diseases of Potato.* ICAR, New Delhi, pp. 1 –17.

Dutt, B.L., Shiv Ram, Mathur, P.N., Sirohi, B.S., Verma, S.S., Sharma, K.P., Azariah, M.D. and Behel, V.K., 1970. Potato late blight (*Phytophthora infestans*). Fungicides and nematicides tests results. *Amer. Phytopathol. Soc.,* 26: 82–83.

Levy, Y., Levi, R. and Cohen, Y., 1983. Build up of pathogen sub population resistant to systemic fungicide under various control strategies: A flexible simulation model. *Phytopathology,* 73: 1475–1480.

Raj, S., Bhattacharya, S.K. and Shiv Ram, 1982. An integrated approach in the control of potato late blight. In: *Potato in Developing Countries,* (Eds.) B.B. Nagaich *et al* Indian Potato Assoc., CPRI, Shimla, pp. 325–332.

Roy, S., Singh, B.P. and Bhattacharya, S.K., 1991. Antagonistic nature of phyllopane mycoflora of potato against *Phytophthora infestans.* In: (*Abstracts*) *41st Symposium on Pathological Problems of Economic Crop Plants and their Management,* CPRI, Shimla, April 29–30.

Samoncha, Y. and Cohen, Y., 1988. Synergistic interactions of cymoxanil mixtures with control of metalaxyl resistant *Phytophthora infestans* of potato. *Phytopathology,* 78: 636–640.

Schwinn, F.J., 1979. Control of Phycomycetes a changing scene. In: *10th Proc. Br. Crop. Prot. Cont.,* 30: 791–802.

Sharma, K.P., 1983. Studies on late blight disease of potato in relation to breeding for resistance. *Ph.D. Thesis,* Bihar University, Muzaffarpur, pp. 171.

Singh, B.P. and Bhattacharya, S.K., 1988. Studies on eradication of late blight tuber infection by systemic fungicides. *J. Indian Potato Assoc.* 15: 190 (Abst.).

Singh, B.P., Birham, R.K. and Bhattacharya, S.K., 1988. Assessment of field resistance to late blight in Potatoes. *J. Indian Potato Assoc.* 15: 190 (Abstr.).

Sunaina, V., Arora, R.K., Kishore, V., Gadewar, A.V., Chandla, V.K., Upadhyay, N.C. and Singh, B.P., 2000. Biocontrol and Biofertilizers. Potato Biotechnology. *Technical Bulletin*, 53: 98–107.

Vasudeva, R.S. and Azad, R.N., 1952. Efficacy of certain fungicides against potato late blight and assessment of loss due to the disease. *Amer. Potato J.*, 29: 60–71.

Insect Pest and Disease Management *Pages* **401–408**
Editor: **Dr. D. Prasad**
Published by: **DAYA PUBLISHING HOUSE (2016)**

Chapter 27

Strategies to Combat Challenges for Management of Red Rot in Sugarcane

Vijai Singh, S.N. Srivastava, B.B. Joshi and S.K. Awasthi
Crop Protection Division, Indian Institute of Sugarcane Research, Lucknow -226 002, U.P.

Introduction

Sugarcane, a long duration clonally propagated crop, is grown in about 3.66 million hectares with an average yield of 64.8 t/ha for the production of crystal sugar, khandsari and gur. The crop is affected by red rot, a most devastating disease (Agnihori, 1996). Several varieties of sugarcane have been phased out from cultivation due to red rot disease. Since, red rot produces complete drying of affected cane, even minor incidence may cause considerable loss in yield. It is often called as cancer of sugarcane. Red rot was first recorded by Went (1893) from Indonesia and Barber (1901) from India. Several epidemics of red rot from U.P., Bihar, Haryana, Punjab, Andhra Pradesh, Kerala, Tamil Nadu, Northern Karnataka and Southern Maharashtra have been reported. Popular varieties *viz.*, Co-213 and Co-312 were badly hit by the red rot in UP and Bihar. Co 1148 in the sub-tropical region has virtually gone out of cultivation due to red rot. CoJ 64, a high sugar and early maturing has also been phased out in sub-tropical India. Several other cultivars *viz.*, Co 205, 213, 281, 290, 312, 313, 419, 421, 453, 527, 658, 975, 997, 1148, 7717, CoC 671, CoH 82, 84, CoJ 64, 82, 84, Colk 8001, 8102, CoS 510, 767, 770, 802, 8436, Bo 3, 10, 11,14, 17, 24, 34, 54 have also gone out of cultivation due to red rot.

Characteristic Symptoms for Diagnosis of the Disease

The disease is recognized by withering of 3rd or 4th leaf of the crown either at the tip or along the margin or both. In severe cases the whole crown dries up. Affected stalks initially exhibit purple discolouration of rind. Pith portion of internodes turns reddish brown and displays white patches.

An affected tissue emits sour smell. Additional symptoms include mortality of buds resulting in to failure of germination; mortality of settling of partially diseased cane; mid rib infection on upper leaves in pre-monsoon season and laminar symptoms as small, pin-head, reddish brown lesions have also been noticed.

Spindle infection appear on lower mid rib of upper leaves even in cases the plants are affected from the bottom. Infected spindles produce millions of spores which can readily infect the cane stalk.

Survival of the Pathogen and Disease Epidemiology

The pathogen penetrates the host surface directly or via wounds. The primary infection of the crop comes through infected setts but rarely through soil. Survival of *C. falcatum* in soil is reported to be 60–90 days. The inoculum in the form of conidia, thick walled mycelia, chlamydospores, etc. is carried in setts both internally and externally. The pathogen penetrates through root primordial leaf scars and bud scales and advanced hyphae induce gum formation in moderately resistant genotypes. The pathogen may attack the cane plant at any stage of crop growth starting from germination. Full expression of the disease depends on the prevailing weather conditions *i.e.* high humidity 90–95 per cent and moderate temperature 28–31°C. Diagnostic symptoms are observed during monsoon or post-monsoon period. The secondary spread of the fungus during monsoon is mediated through conidial dispersal by irrigation, rain water, rain splash at temperature 25–30° C, relative humidity 95 per cent and results in the infection of mid rib, lamina, leaf sheath and stalk, while in winter air currents helps for spread of the pathogen (Satyavir, 2003).

Disease Management

Red rot disease may be managed by quarantine, cultural practices, heat therapy, chemical control, biocontrol and growing resistant varieties. Bio-control agents *viz., Trichoderma, Pseudomonas, Bacillus* and *Acetobacter* have been recently found effective for protecting the disease. The bioagents have been noticed to induce systemic resistance in susceptible varieties and may prolong the break down of resistance.

Quarantine

Quarantine measures are duly adopted during exchange of sugarcane varieties between countries; however, proper care is not been taken while movement of cane seed material between the states. The indiscriminate movement of cane facilitates introduction of dreaded pathogen in to a new area. The recent introduction of the disease in Karnataka and Maharastra was observed because seed material of highly susceptible varieties CoC 92061 and CoC 671 was brought from red rot endemic areas. Thus, strict regulatory measures are required for inter-state and zonal movement of seed material. It is required that seed should be moved with proper phytosanitory certificates (Viswanathan and Samiyappan, 2000).

Healthy Seed and Field Sanitation

Healthy seed, crop rotation, field sanitation and maintenance of efficient drainage have been recommended to reduce inoculums in the field and minimize losses due to the red rot. The seed must be disease free and preferably taken from crop grown after heat treatment in three-tier seed programme (Singh, 1983). One-year crop rotation is effective in eliminating soil borne inoculums of *C. falcatum* in India.

Heat Therapy

Heat therapy is suitable for control of sett-borne infection. Elimination of sett-borne infection of red rot was achieved by hot air treatment at 54°C for 2:30 h (Singh, 1973). Hot water at 50°C for 2 h was also equally effective (Singh and Agnihotri, 1987) but moist hot air treatment (MHAT) was more efficient in deep penetration. It was thoroughly tested in U.P. and Bihar and ascertained as a better practice (Singh, 1983).

Chemical Control

The fungicides have not given satisfactory control because of impervious nature of rind and presence of fibrous nodes of sugarcane. This prevents the low level of absorption of fungicide. Reduction in red rot incidence has been reported when diseased setts were treated with Carbendazim and benomyl for 1–2 h (Satyavir, 2003). Bleaching powder (0.5 per cent) was effective in reducing red rot incidence (Agnihotri *et al.*, 1995). Application of bleaching powder in soil after uprooting the diseased plant reduced the load of inoculums of *C. falcatum*. Secondary spread of red rot by spraying dithiocarbamate and copper fungicides may be reduced (Kirtikar and Verma, 1963). Bavistin (0.2 per cent) reduces sett borne infection of the red rot (Agnihotri *et al.*, 1995). The efficacy of bavistin was improved by adding acetic acid @ 0.2 per cent in solution.

Red Rot Management Through Resistant Varieties

Sources of Resistance

Red rot resistance has been identified in basic *Saccharum* spp. Seven clones of *S. officinarum* were identified as resistant among which 2 were of typical natural hybrid type and 3 were of non flowering type which may not be used in breeding. The *S. robustum*, a wild species is close to *S. officinarum* and has 2 resistant and eight moderately resistant dones. Among *S. sinense* and *S. barberi* only 3 clones were resistant to red rot. *S. barberi* originated in the area of maximum variability of *Colletotrichum falcatum* and must have adequate level of resistance but succumbed to new virulent races of the pathogen. As a result, there were no chance of further genetic recombination and evolution of genes for resistance. The low levels of red rot resistance in cultivars may be due to *S. barberi* initially used in breeding programme. *S. spontaneum* originated from sub-Himalayan region may be utilized for biotic and abiotic stresses. Eastern U.P. and Bihar are among the diversity centers of red rot pathogen. *S. spontaneum* from this region will serve as the natural source of resistance to virulent pathotypes of *C. falcatum* (Sreenivasan, 1995; Viswanathan and Samiyappan, 2000). *S. spontaneum* has comparatively larger number of sources of resistance (109) than 13 in other species out of 1099 screened (Alexander, 1995).

Resistant Varieties Used for Red Rot Breeding Programme

Bo, 12, Bo 91, Bo 102, Bo 104, Co 7314, Co 89003, Co 89015, Co 89016, CoS 91269, CoS 96268, Co Pant 94211 have been utilized for evolving varieties for red rot resistance (Satyavir, 2003).

Vertical and Horizontal Resistance

Red rot resistance is vertical or race specific in sugarcane. Beniwal (1985), studied slow red rot resistance using 22 genotypes. Out of these 4 genotypes *viz.*, Co 1158, CoS 802, S 77/467 and CoJ 64 possessed useful levels of slow red rot. Slow drying of tops, mild spore over the rind and less secondary spread were observed in above varieties. These parameters could be utilized in assessing horizontal resistance of sugarcane. CoJ 64 exhibited useful level of horizontal resistance but it is succumbed to disease in Punjab and Haryana and Western U.P., due to emergence of new virulent pathotypes (Satyavir, 2003). The cultivation of Co J 64 has been partly phased out.

Pathogenic Variability

C. falcatum, a facultative saprophyte, produces new pathotypes in nature due to heterokaryosis, parasexuality and mutation. On the basis of reaction on 13 differentials 10 pathotypes have been identified, so far from India. But few studies were made to identify pathotype with specific host pathogenicity. Work conducted at HAU, Hissar indicated the presence of Cf 01, 02 and 03 (Satyavir *et al.*, 2001). Cf 04,05 and 06 three pathotypes were reported from Anakapalle, A.P (Satyanarayna and Satyanarayana 1976, Satyanayana and Rao, 1995). Differentials *viz.*, Baragua (*S. officinarum*), Khakai (*S. sinense*), SES 594 (*S. spontaneum*), Co S 767, Co 975, BO 91, CoC 671, Co 7717, Co 997,CoJ 64, Co1148, Co 419, and Co 62399 have been utilized for identification of pathotypes (Anon, 1991; Singh and Singh 1993). Recently, Cf 07, Cf 08 and Cf 09 were identified from north Western zone and Cf 10 from tropical and east coastal zone (Singh *et al.*, 1993; Satyavir *et al.*, 2001). Search for better differentials is still required to explain specific host reaction. Identification of filter plants to retain specific pathotypes maybe more useful. Precise molecular identification would be of greater importance.

Inheritance of Resistance

Sugarcane plant is highly heterozygous and complex polyploid crop with meiotic irregularities. Conclusive studies on inheritance of red rot resistance are lacking. Chaudhary *et al.*, 1995 reported that resistance to red rot in sugarcane is governed by the dominant inhibitor genes from *S. officinarum.* Segregation of resistance in progenies was not found to be a Mandelian trait. Indiscriminate distribution of resistance or susceptibility among the progenies was recorded. It is essential to know the combining ability of resistant parents for developing resistant lines. More precise information would be needed to go for directed breeding.

Morphological Features of Host for Red Rot Resistance

Studies on the thickness of epidermis, cuticle, bud scale, rind, relative abundance of vascular bundles under the rind and the presence of septa in vascular bundles have been reported as morphological barriers for red rot entry and movement of *C. falcatum* in sugarcane. But the morphological characters after the origin of new pathotype have not been precisely correlated with the resistance (Duttamajumder, 2002).

Biochemical Characters of Host for Red Rot Resistance

In resistant varieties a gummy substance is formed which seals off further spread of the pathogen but it is not produced in susceptible varieties. Total phenolic contents are increased in resistant but decreased in susceptible genotypes. Though a slight increase at initial stage after infection in susceptible varieties was also noticed. Phenol alone may not be responsible for resistance (Singh *et al.*, 1976). Activities of polyphenol oxidase (PPO) and flavone glycosides were linked with the resistance (Srinivasan, 1995). Activity of phenylanine ammonia lyase (PAL) and tyrosine ammonia lyase (TAL) was higher in resistance but lower in susceptible varieties (Madan *et al.*, 1991). But once resistance of the variety was broken the activities of PAL and TAL had no correlation. Viswanathan *et al.* (1996) indicated the role of the red rot phytoalexin (3-deoxyanthocyanin) in disease resistance. Once the pathogen starts to behave in saprophytic mode indicates the higher susceptibility of the genotype.

Biocontrol Through *T. harzianum*

Different biocontrol agents *viz.*, *Trichoderma*, *Pseudomonas*, *Bacillus* and *Acetobacter* have been found highly antagonistic to *C. falcatum* under *in vitro* conditions. *T. harzianum* applied to the crop through soil and sett treatment was found effective to protect the plants from infection of red rot. *T. harzianum* strains *viz.*, T27, T37 and T38 were more efficient for protecting the infection (Table 27.1). *Trichoderma* may be applied through soil @ 2 kg/ ha mixed with 20 kg sterilized FYM at the time of planting and tiller stage. The sett treatment with spore suspension (10^6 conidia/ml) was equally effective where by red rot was reduced by 40–50 per cent in the susceptible variety. Although *T. viride* was less effective for red rot control but it was at par for promoting the growth of sugarcane (Tables 27.1 and 27.2). *T. viride* metabolites (culture filtrates) 2.5 per cent solutions applied at tiller initiation of ratoon produced significantly higher yield (Srivastava *et al.*, 2006). The formulations of *Trichoderma* biopesticide are non-hazardous, ecofriendly and highly economical (Singh *et al.*, 2006; Srivastava *et al.*, 2006).

Table 27.1: Biocontrol of Red Rot in Co Lk 7701 by *Trichoderma* Species

Strain	*Sett Treatment (Spore Suspension)*			*Soil Treatment (TMC)*		
	Complete Control	*Partial Control*	*Canes Inoculated*	*Complete Control*	*Partial Control*	*Canes Inoculated*
T6	28	72	100	23	77	100
T24	24	76	100	24	76	100
T27	45	55	100	27	73	100
T37	49	51	100	46	54	100
T38	46	54	100	31	69	100
Check	00	3	100	00	13	100

T6: *T. viride*; T24, T27, T37 and T38: *T. harzianum*.

Research Areas for Future

The following areas for future research are important as also envisaged by Yadav (2006) in his vision document:

Table 27.2: Biomanurial Effects of *Trichoderma* Species in CoS 94257

Parameter	*Sett Treatment (Spore Suspension)*			*Soil Treatment (TMC)*			*CD at 5%*
	Check	*Treated*	*Increase (%)*	*Check*	*Treated*	*Increase (%)*	
Germination (%)							
T6	21.3	25.4	4.1	21.3	28.8	7.2	2.20
T24	20.8	27.4	6.5	20.8	28.4	7.5	
Tillers (000)							
T6	83.6	103.7	24.2	83.6	109.2	30.3	3.36
T24	85.8	104.2	21.1	85.8	11.3	29.0	
NMC (000)							
T6	65.5	68.9	5.1	65.5	73.8	13.5	2.18
T24	67.0	71.5	7.6	66.9	77.2	15.3	
Yield (t/ha)							
T6	59.3	62.9	6.0	59.3	68.3	15.1	2.24
T24	59.3	63.7	7.4	59.3	68.3	15.1	

T6: *T. viride*; T24: *T. harzianum*.

1. Prevailing *C. falcatum* pathotypes should be cataloged after molecular characterization of selective pathogenic behaviour.
2. Molecular and biochemical studies on host-parasite interactions are required to understand the operational mechanism of pathogenesis and its utilization for control of red rot.
3. The work on broadening of genetic base/gene pyramiding for red rot resistance from *S. spontaneum* is also important.
4. Disease prediction models with reference to forecast the epidemics need to be worked out for judicious adoption of control measures.
5. Integrated disease management strategy with the utilization of bioagents and/or chemicals for induced systemic resistance would be more useful.

Conclusion

Red rot caused by *Colletotrichum falcatum* Went [Perfect stage *Glomerella tucumanensis* (Speg), Arx and Muller] is a devastating disease of sugarcane in India. Several epidemics of red rot have caused tremendous loss in yield and sugar. Unlike several other crops fungicides were not absorbed to the desired concentration and were degraded quickly in cane plant. Heat treatment failed to protect deep-seated infection and secondary spread of red rot. Host resistance has frequent break down. Biocontrol through *Trichoderma harzianum* and *T. viride* has given encouraging results. The biopesticide applied at the time of planting @ 20 kg FYM based TMC/ha was helpful for preventing the secondary spread of the disease. Metabolites of *T. harzianum* absorbed through roots were effective for longer period (150–200 d). *Trichoderma* biopesticide not only controls secondary spread of red rot but also works as biomanure and improves the growth and yield of sugarcane. The biopesticide may be useful to prolong the life of variety and maintain the fertility of soil.

Red rot can be managed by integrated approach. Diseased cane may not be allowed to move for preventing introduction of the pathogen in new areas. Seed cane should be planted after moist hot air treatment (MHAT) in first year followed by the three-tier seed programme. Seeds may be treated with bavistin + acetic acid (0.2 per cent each). *Trichoderma* biopesticide may be applied at the time of planting. Rouging of diseased plants followed by spot application of bleaching powder (5–10 g/ clump) may be helpful in reducing *C. falcatum* load.

Acknowledgements

Sincere thanks are due to Dr. R.L. Yadav, Director IISR, Lucknow for encouragement and providing facilities for research experiments.

References

Agnihotri, V.P., 1996. Current sugarcane disease scenario and management strategies. *Indian Phytopath.*, 49: 109–126.

Agnihotri, V.P., Singh, N., Lal, R.J. and Singh, V., 1995. Management of red rot disease of sugarcane. In: *Proc. National Seminar on Sugarcane Production Constraints and Strategies for Research and Management of Red Rot*, (Eds.) V.P. Agnihotri *et al.* IISR, Lucknow, pp. 294–312.

Agnihotri, V.P., Lal, R.J. and Singh, V., 1997. Integrated management of red rot disease of sugarcane. *Indian Phytopath.*, MEZ., Ann. Meet. Kanpur, December 18–19, pp. 55.

Agnihotri, V.P., Lal, R.J., Singh, N. and Singh, V., 1997. Effect of bavistin and bleaching powder on conidial germinationof *Colletotrichum falcatum* Went. and development of red rot in sugarcane. *Indian J. Sugarcane Technol.*, 14: 108–111.

Alexander, K.C., 1995. An overview of constraints in sugarcane disease management. In: *Proc. National Seminar on Sugarcane Production Constraints and Strategies for Research and Management of Red Rot*, (Eds.) V.P. Agnihotri *et al.* IISR, Lucknow, pp. 121–129.

Annonymous, 1991. *Proc. AICRP on Sugarcane*, IISR, Lucknow.

Barber, C.A., 1901. Sugarcane diseases in Godaveri and Genjam districts. *Madras, Department. Land Records and Agic. Bull.*, 43: 181–194.

Beniwal, M.S., Satyavir and Virk, K.S., 1989. Pathogenic variability of *Colletotrichum falcatum*, incitant of red rot of sugarcane. *Indian Phytopath.*, 42: 95–98.

Chaudhary, B.S., Mahila, A.S., Virk, K.S. and Rishipal, 1995. Breeding for red rot resistance in sugarcane (*Saccharum* hybrid complex): Review, concept and theory. In: *Proc. National Seminar on Sugarcane Production Constraints and Strategies for Research and Management of Red Rot*, (Eds.) V.P. Agnihotri *et al.* IISR, Lucknow, pp. 221–232.

Duttamjumder, S.K., 2002. A century of red rot disease of sugarcane in India. In: *Sugarcane Crop Management*, (Eds.), S.B. Singh *et al.* SciTech Publishing, LLC, Houston, Texas, USA, pp. 52–108.

Kirtikar and Verma, H.S., 1963. Role of chemicals in sugarcane disease control. *Indian Sugar*, 13: 139–141.

Madan, V.K., Soni, N., Nigam, M., Solomon, S. and Agnihotri, V.P., 1991. Enzyme activity and cane genotype resistant to red rot. *Sugarcane*, 2: 6–8.

Padmanabhan, P., Mohanraj, D., Viswanathan, R, Madhusudanarao, M., Prakasam, N., Joshi, R. and Alexander, K.C., 1996. Differential interaction of sugarcane clones to pathotypes of *Colletotrichum falcatum*. *Sugarcane*, 4: 16–20.

Satyavir, 2003. Red rot of sugarcane: Current scenario. *Indian Phytopath.*, 56: 245–254.

Satyavir, N. Singh, Virk, K.S., Nageswara Rao, G.V., Singh, H. and Mishra, S.R., 2001. Pathogenic variability in sugarcane red rot system. In: *Proc. National Symposium on Role of Resistance in Intensive Agriculture*, (Eds.) S. Nagarajan and A.P. Singh. Kalyani Publishers, Ludhiana, pp. 109–114.

Satyanarayana, Y. and Satyanarayana, B., 1976. Variation in the light race isolates of red rot fungus in Andhra Pradesh. *Coop. Sugar*, 7: 389–392.

Satyanarayana, Y. and Rao, A.M., 1995. Investigations on the race flora of *Colletotrichum falcatum*, incitant of red rot disease of sugarcane. In: *Proc. National Seminar on Sugarcane Production Constraints and Strategies for Research and Management of Red Rot*, (Eds.) V.P. Agnihotri *et al.* IISR, Lucknow, pp. 285–293.

Singh, K., 1973. Hot air therapy against red rot of sugarcane. *Pl. Dis. Reptr.*, 57: 220 –222.

Singh, K., 1983. Seed-piece transmissible diseases of sugarcane and three-tier seed programme. In: *Recent Advances in Plant Pathology*, (Eds.) A. Husain *et al.* Print House (India), Lucknow, pp. 331–343.

Singh, K., Singh, R.P. and Agnihotri. V.P., 1976. Phenolics in relation to sugarcane resistance against red rot disease. *Sugarcane Pathol. News L.*, 15/16: 37–41.

Singh, N. and Singh, V., 1993. *Colletotrichum falcatum* race designation: A methodology. *Curr. Sci.*, 66: 777–779.

Singh, N., Singh, V., Tewari, S.D. and Verma, K.P., 1993. Occurrence and identification of different physiological races of red rot pathogen of sugarcane in Uttar Pradesh. *Indian J. Sugarcane Technol.*, 8: 26–30.

Singh, R.P. and Agnihotri, V.P., 1987. Thermotherapy of sugarcane for disease control. *Tropical Plant Pathology*, 4: 305–329.

Singh, V., Dattamajumder, S.K. and Singh, N., 1993. Possible red rot threat to the cultivation of CoS 767. *Indian J. Sugarcane Technol.*, 8: 81–82.

Singh, V., Duttamajumder, S.K. and Lal, R.J. 2006. Biocontrol of red rot of sugarcane through induced resistance by isolates of *Trichoderma harzianum. Indian J. Agric. Sci.* (under revision).

Sreenivasan, T.V., 1995. Research strategies for breeding for red rot resistance in sugarcane. In: *Proc. National Seminar on Sugarcane Production Constraints and Strategies for Research and Management of Red Rot*, (Eds.) V.P. Agnihotri *et al.* IISR, Lucknow, pp. 233–239.

Srivastava, S.N., Singh, V. and Awasthi, S.K., 2006. *Trichoderma* induced improvement in growth, yield and quality of sugarcane plant and ratoon. *Sugar Tech.*, 8: 166–169.

Virk, K.S., Satyavir and Chaudhary, B.S., 1985. A source of red rot resistance in sugarcane. *Indian Phytopath.*, 38: 551–552.

Viswanathan, R. and Samiyappan, R., 2000. Red rot disease in sugarcane: Challenges and prospects. *Madras J. Agric. Sci.*, 87: 549–559.

Went, F.A.F.C., 1983. Het rood Snot. *Arch. Java Suikeridustrie*, 1: 265–282.

Yadav, R.L., 2006. Research vision to manage red rot disease of sugarcane in India. *Sugar Tech.*, 8: 99–100.

Insect Pest and Disease Management *Pages* ***409–430***
Editor: **Dr. D. Prasad**
Published by: **DAYA PUBLISHING HOUSE (2016)**

Chapter 28

Eco-friendly Management of Insect Pests and Nematodes in Hill Horticultural Crops

R.P. Soundararajan and V. Lakshmanan
Horticultural Research Station, Yercaud – 636 602, Tamil Nadu

Introduction

The horticultural crops includes plantation crops like coffee, tea, spice crops, vegetables particularly cole crops, fruit trees, high value cut flowers like gerbera, carnation, anthurium, orchids, chrysanthemum etc. are suited and grown in high altitude regions. It has high export value and also internal market price. Insect pests are one of the major constraints in the production of these crops and its impact on the quality of produce is much pronounced. The impact of pesticides and hazards caused by the chemicals to health, environment, soil as well as water pollution through different modes of action and application which has forced to search some other alternatives.

Recently several exotic insect pests have entered the country, *viz.*, *Liriomyza trifolii* complex in several hosts, *Hypothenemus hampei* on coffee, *Aleurodicus disperses* on several hosts and *Aceria guerreronis* on coconut. The most recent is the suspected introduction of B-biotype virus caused by *Bemisia argentifolii* in Kolar, Karnataka which could cause tremendous losses to vegetable crops. In fact, exotic pests are recognized today by the environmentalists as a second greatest threat to biodiversity conservations next to habitat destruction. Horticulture industry is not only suffering because of many new exotic pest problems but also due to pesticide-induced resurgent pest problems. The high elevation crops *viz.*, Cocoa, coffee and tea are sprayed regularly resulting of insect pests into major problems (Singh,

2000). Insect pest affecting various crops at high attitude regions of North Western Himalayas was reported by Surjeet Kumar (2004). Several insect pests were recorded on almost all the crops of these areas and most of the species recorded were of common occurrence in plains and low hill areas. On the safe pest management options, cultural operations, mechanical methods and biological suppression is considered as an effective, environmentally non degrading, technically appropriate, economically viable and socially acceptable method of pest management.

In recent years Indian farmers showing considerable interest in developing orchards to overcome adverse environmental conditions like drought, soil erosion, frost etc. in hill regions. Though edible fruits have great demand in abroad, the produces from India were facing problems in the international markets because of pesticides residues. In the past, these orchard eco-systems were very stable but as the time passed by due to misuse of chemical fertilizers and synthetic pesticides aggravated the pest problems. This is also suitable for flower crops especially cut flowers growing under cover and field conditions. It has great demand in European market. To meet the international standards farmers has to adopt and practice organic farming and eco-friendly IPM strategies.

Eco-friendly Pest Management in Plantation Crops

The plantation crops in the hill regions include the beverages like coffee and tea, spice crops such as black pepper and cardamom.

Coffee

Management of Borer Pests

The coffee berry borer, *Hypothenemus hampei* (Scolytidae: Coleoptera) is the most serious pest world over and in India also it is becoming a problem in coffee growing areas. The pest incidence is pronounced at an altitude range of 500–1000 m above sea level. It is a monophagous pest and exclusively feed and breeds on coffee berries. It can cause up to 80 per cent yield reduction. Proper adoption of cultural and phytosanitary measures is foremost in the management of berry borer. Timely harvest and clean harvest is important. No berries should be left over, either on the plant or on the ground. Spreading gunny bags or polythene sheets on the ground while picking will help to minimize gleaning. If gleaning could not be collected, they may be swept along with the mulch and buried below 2.5 feet depth in the soil. Optimum shade and good drainage should be maintained. Biological control of the pest, by using the parasitoids *Cephalonomia stephanoderis*, *Prorops nasuta* (Bethylidae) and *Phymastichus coffeae* (Eulophidae: Hymenoptera) are adopted in many countries. The former two species are introduced in the year 1995 to India and found fairly successful. Application of suspension of the fungus *Beauveria bassiana* also found effective (Anonymous, 1996).

Coffee white stem borer, *Xylotrechus quadripes* (Cerambycidae: Coleoptera) is most serious pest of Arabica coffee in India, particularly in South India it is in alarming situation and every year 10 per cent crops being wiped out by the pest. Though chemical control is the only method for the management of the pest, preventive measures can minimize the damage. Optimum shade should be maintenance in the estates. Removal of loose scaly barks of the main stem and thick primaries (scrubbing) using coir glove or coconut husk to get ride of cracks and crevices to prevent the adult beetles laying eggs on the bark region (Anonymous, 1996).

Sucking pests

Mealy bugs, *Planococcus* sp. and *Pseudococcus* spp. are important sucking pests in coffee. Biological control is effective in the plantation ecosystem as they are perennial crop. Release of *Cryptolaemus montrouzieri* @ 2-10 beetles/infested plant after the blossom is found effective to control mealy bugs (Singh, 2000).

The most successful biocontrol in the coffee ecosystem is the control of scale insect, *Coccus viridis* using the white halo fungus *Verticillium lecanii*. The epizootics of the fungus also observed in the field when climatic conditions are favourable. Inoculative release of the fungus @ 10^4 x 16 spores/ml + quinalphos 0.005 per cent + 0.05 per cent Teepol using knapsack sprayer is effective (Singh, 2000). Use of selective insecticides like neonicotinoids can have less effect on several natural enemies. Spot applications further reduce the effects on ecosystem. Stem application of neonicotinoid, thiamethoxam 0.05 g a.i/plant in the auxillary branches was very effective in suppressing the population of *C. viridus* causing 93.1 per cent reduction (Senthil Kumar and Regupathy, 2006).

Tea

Tea plantation is susceptible to different mite pests. A total of 6 mite species found to attack tea crop during different crop stages. Population of mites is influenced by several physical, biological and cultural factors. Weather plays an important role in their numerical fluctuations and, peaks in populations are always observed during dry weather. Pruning is an important cultural operation which affects the incidence of mites in tea fields. Das (1965) noted that in north-east India pruned tea was considerably less attacked by red spider mite *Polyphagotarsonemus latus*, than the unpruned or stiffed fields. Intensity of shade is yet another fact influencing the abundance of mites. Incidence of red spider mite was always less on tea under heavy shade. Experiments conducted in north-east India suggested that mite was affected by light distribution and leaf temperature which in turn were influenced by the intensity of shade (Banerjee, 1969). Different species of mites exhibit differential feeding preferences for certain jats or cultivars of tea. Generally, the Assam jats are more prone to the attack of pink and purple mites while the China jats and their hybrids are more susceptible to infestation by red spider mite. Rao (1978) reported that the clones B/6/62 (UPASI-10, Pandian) harboured less population density of pink mites, *Acaphylla theae*. Phytophagous mites are preyed upon several predatory mites. The tea mites, *Acaphylla theae* and *Calacarus carinatus* are preyed upon by *Amblyseius berbicolus*, *Agistemus* sp. and *Cunaxa* sp. The former consumes more than 400 pink mites during its life span and eggs of mites are its preferred prey. Formulations containing azadirachtin obtained from neem seed kernel has been found effective against pink mite (*Acaphylla theae* Walt.) in tea (Selvasundaram and Muraleedharan, 1999). Tea mosquito bug, *Helopeltis theivora* Waterhouse is most serious pest on tea plantation in north-East India. A single full grown nymphs or adult can produce 100 spots within 24 hours however, there was varying degrees of associations with different genotypes (Das, 1984). The seasonal incidence studies revealed that August—October months having the average temperature and relative humidity of 24.06–31.6 °C and 70-90 per cent was the most congenial period for development and multiplication of *H. theivora* (Kalita and Karan Singh, 1999). The genotypes TV 6 and TV 18 showed resistance to the bug damage. The resistance might be attributed to repelling and inhibiting actions of total polyphenols in the genotypes (Kalita and Karan Singh, 2004).

Spice crops

Pests and diseases are the major factor responsible for the low productivity of spices in India. Although over 100 plant species are classified as spices, black pepper, cardamom, ginger and turmeric constitute more than 50 per cent of export earnings among different spices. Since spices are of high-value and export oriented in nature, it is also important that the levels of pesticides residues are kept well below the tolerance limits in view of the stringent standards set by the importing countries.

Hence, adoption of eco-friendly integrated management schedules against insect pests of spices is important.

Black Pepper

Black pepper is considered as 'king of spices' in our country which accounts for over 45 per cent export earnings among spices. The pollu beetle, *Longitarsus nigripennis* is the most destructive pest of black pepper, causes 30-40 per cent crop loss in endemic areas. At high altitude above 300 meters and below 750 meters, damage was relatively lower. The experimental results revealed that use of insecticides in soil around the base of vines at the time of spike initiation showed that none of them were effective for more than three months (Nybe, 2001). Shade regulation of standards (support trees) is important since the pest infestation is considerably higher in shaded areas in the plantation. The IPM strategy developed against pollu beetle includes cultural operations and applications of insecticides and neem based products (Devasahayam, 2000). Spraying chemical Endosulfan 0.05 per cent during July followed by 3-4 sprays (at 3 week interval) of neem based products such as Neemgold® 0.6 per cent during August–October or spraying Neemgold 0.6 per cent alone during July–October (4-5 sprays at 3 week interval) along with shade regulation of supporting trees was effective in controlling the pest infestation (Devasahayam and Koya, 1999). Antifeedant activity of seed kernel extract of neem against pollu beetle is also reported (Devashayam and Leela, 1998).

Very few natural enemies have been recorded on pollu beetle in the field (Devasahayam and Koya, 1994). A few black pepper accessions and wild *Piper* species resistant to the pest under field conditions have been identified among the germplasm of black pepper available at Indian Institute of Spices Research, Panniyur (Devasahayam, 2000). Leaf extracts of *Strychnos nux-vomica* L. and *Chromolaena odorata* L. and neem possess appreciable antifeedant activity against the pest indicating their potential in eco -friendly pest management programme (Devasahayam and Leela, 1998).

Scales, *Aspidiotus destructor* and *Lepidosaphes piperis* in pepper can be managed by releasing *Chilocorus nigritus*. The coccinellid predator is easily mass multiplied and produced at Rs. 75 per 100 adults (Singh, 2000). Neem oil 0.3 per cent and neem based insecticide such as neemgold® 0.6 per cent are promising in controlling scale insects in pepper and also found safer to coccinellid predators indicating their potential in the integrated pest management of the pest (Devasahayagam, 1998).

Cardamom

Thrips, *Sciothrips cardamomi* is the most destructive pest of the 'queen of spices' Cardamom and causes up to 90 per cent crop loss in endemic areas. Thrips breed within leaf sheaths especially during summer, pruning of leaf sheaths during February–March before application of insecticides reduces the pest population. Removal of other herbaceous host plants that serve as alternate hosts around the plantation also reduces the pest population (Varadarasan, 1992). Very few natural enemies have been recorded on cardamom thrips. *Chrysopa* sp. and anthocorid bugs which are general predators were observed to predate on adults and larvae in the field (Varadarasan, 1995). In general, the Malabar type of cardamom is more tolerant to thrips than Mysore and Vazukka types. Root grubs, *Basilepta fulvicorne* is a serious pest of cardamom in the nursery and plantation. Collection and destruction of adult beetles using insect net during their peak period of emergence (on receipt of pre-monsoon showers) reduces the population of root grubs to a great extent in the field (Kumaresan *et al.*, 1989). Microbial pathogens like *Metarrhizium anisopliae* and *Beauveria bassiana* and the entomopathogenic nematode, *Heterorhabditis* sp. have been recorded on larva and adult in the field are promising for the management of the pest (Varadarasan, 1995).

Eco-friendly Pest Management in Vegetable Crops

Cruciferous crops

Diamond back moth on Cabbage and Cauliflower

Development of resistance by the devastating Diamondback moth, *Plutella xylostella* to all insecticides is an important constrain in the pest management of cabbage and cauliflower. Moderate infestation of DBM can be controlled by a spray of Cartap hydrochloride (0.05 per cent) a neris toxin extract. Hybrid cabbage KIRCH -5 has tolerance against some insect pests (Barwal, 2001). The activity of natural enemies was higher in the plots treated with *Bacillus thuringiensis* var. *kurstaki* (Adhira *et al.*, 2003). Use of Bt product such as Delfin @ 500g/ha recorded low DBM population and also increase yield of cauliflower (Jayarani and Kennedy, 2004). Promising role of Delfin against DBM was also reported by Gopalakrishnan (2001). *Galleria mellonella* NPV (GmNPV) produced heterologous infection in the lepidopteran pest complex of cruicifers *viz.*, DBM, leaf webber *Crocidolomia binotalis* and the cabbage lopper *Hellula undalis*. The broad spectrum activity of the GmNPV offers wide scope for its use as a noval biopesticide in crucifer crops (Parthasarathy and Rabindra, 2003).

The entomopathogenic fungi, *Paecilomyces farinosus, Zoophthora radicans* was isolated from *P.xylostella*. Application of five rounds of the fungus *P.farinosus* @ 1.7 x 10^8 spores/ml + Triton X-100 (0.01 per cent) at weekly interval significantly brought down the larval population of DBM and increased the marketable yield. *P. farinosus* was also cross infective to *H. armigera, S. litura, C. binotalis, Trichoplusia ni* and *Hellula undalis* (Mani and Krishnamoorthy, 2003a).

As a part of Biointensive Integrated pest management (BIPM) the egg parasitoid genus *Trichogramma* is also utilized for the management of DBM. *Trichogrammatoidea bactrae* Nagaraja has been found to be the most promising egg parasitoid for field use against *P. xylostella* in India (Jalali *et al.*, 2001). *T. bactrae* could be released @ 4.00 lakhs/ha in temperatures from > 20° and < 30° for its effective use in the field as higher temperature drastically reduces its efficacy (Jalali *et al.*, 2004).

Among different plant species tested for antifeedant effect on DBM, the plant species, *Vitex trifolia, Artemisia vulgaris, Ageratum conyzoides* and *Eucalyptus globules* were found to possess strong antifeedant activity on the basis of minimum per cent feeding and maximum protection over control. Further *V. trifolia* recorded lowest PC 95 value of 0.415 per cent (Hemchandra Singh and Singh, 2003). Use of Azadirachtin for the control of DBM was also reported by Raja Reddy and Krishnamohan (2001) on cauliflower. Chandramohan and Nanjan (1992) showed that a Neem oil product 'Biosol' (0.4 per cent active) was 3-4 times more efficacious against *P. xylostella*.

Adopting IPM along with use of bold seeded Indian mustard, *Brassica juncea* as a trap crop is the most successful strategy. Mustard attracts 80-93 per cent of diamondback moth and almost entire population of leaf webber, stem borer and bugs in the cabbage ecosystem. Aphids are also attracted to mustard in large number with the onset of flowering (Srinivasan and Krishnamoorthy, 1991; Srinivasan *et al.*, 1991). Cauliflower when intercropped with coriander harboured significantly less DBM larvae when compared to tomato and mustard as intercrop (Uma Shankar *et al.*, 2005). Talekar *et al.* (1986) observed that the cabbage when planted between tomato, coriander, garlic and marigold had relatively less DBM larvae and pupae, especially when general insect population was high. Their influence in reducing the DBM damage to cabbage appears to be a characteristic repellant odour that each of them posses. Similar results on the useful effects of tomato intercropping in reducing infestation on crucifers were carried out by Kondoria *et al.* (1999). Usually intercropped cabbages remain free from infestation during the early crop stages. But, depending on the severity of DBM, 4 per cent Neem seed kernel extract is applied at primordial or head initiation since it is a critical crop stage. Primordial formation takes place between 17 and 28 days after planting depending on the variety grown (Krishnamoorthy *et al.*, 1998).

Leaf Feeders on Cabbage and Cauliflower

Cabbage butterfly, *Pieris brassicae* is a serious pest in the hill regions of Jammu and Kashmir and Himachal Pradesh. It was found heavily attacked by the larval parasitoids *Cotesia glomerata* and *Hypersota ebenius* causing 60-80 per cent natural parasitism in cabbage and cauliflower (Kakar and Sharma, 1991) and the pest damage was still substantial in these regions, since the parasitoids appeared late. Applications of Bt (Thuricide) at 3×10^8 spores/mg (4 per cent) was more effective in controlling the cabbage butterfly. Avermictin, a microbial toxin @ 1.8 per cent w/v (vertimec) is found to be most toxic compound against 3rd instar larva of *P. brassicae*. Btk products halt (55000 IU/mg), Biolep (32000 IU/mg) and fungal pathogen *B.bassiana* also found effective (Hirak Chatterjee *et al.*, 2003). Neem based formulation Neemarin 0.50 per cent gave 50.00 per cent larval mortality after 72 hours (Singh *et al.*, 2003). Antifeedant activity of petroleum ether extracts of five plant species leaves of big sage brush, *Artemisia brevifolia* Wall. Cypres weed *Eupatorium adenophorum* Spereng, wilt sage *Lantana camara* L., Chinaberry *Melia azadarach* (L.) and roots of broad leaf dock, *Rumex nepalensis* were evaluated against third instar caterpillar of *Pieris brassicae* (L.). Among them, *A. brevifolia* extracts resulted in significantly high antifeedant followed by *R. nepalensis* and *M. azedarach*. All the plant extracts resulted in moderate reduction in larval weight after 24 hours of feeding (Mehta *et al.*, 2001).

Leaf eating caterpillar, *Spodoptera litura* causes severe loss to cabbage and cauliflower. *Chelonus heliopa* @ 75,000 adults/ha when released at 4 times had resulted 15.2 per cent parasitism (Patel *et al.*, 1979). Aphids, *Brevicoryne brassicae* cause severe damage at times on cole crops in different parts of the country. *Diaeretiella rapae* is a major parasitoid on *B. brassicae* infesting cauliflower. Release of *D. rapae* @ 50,000/ha gave significant reduction in the population of aphids infesting cole crops (Desh Raj *et al.*, 2000). The fungus *Erynia neoaphids* was found infecting *B. brassicae* up to 80 per cent during February in Karnataka (Gopalakrishnan and Mohan, 1993).

Other Cruciferous Crops

Brussels' sprout is often attacked by potato aphid *Myzus persicae* and cabbage aphid *Brevicoryne brassicae* during early growth phase. Parasite, *Aphidius* sp. accompanies them in the cold seasons is found to keep them under control. Application of Bt against cabbage butterfly in brussels' sprout provides safety to these parasitoid (Barwal, 2001).

The mustard aphid, *Liaphis erysimi* is an important insect-pest of rapeseed and mustard crops. Traps can be used for the control of the aphids. Two types of traps are used for catching the alate forms, that is water trap and the adhesive trap made by taking yellow colour trays and water is filled in the tray to 2/3 level. The adhesive traps was considered better alternative for catching alate aphids. It is made from glass jars, painted with mustard yellow colour and then smeared with transparent grease on its outer surface to get the flying aphids stuck on the surface of the jar. This can be kept at a height of 1.5m to catch the adults (Prasad and Phakde, 1983). The glass jar used in the trap was replaced by empty tin of 1 kg capacity. The trap was modified telecopic so that the same can be fixed at desired height in the field and it is very cheap (Prasad, 2004). Significant higher aphid incidence was noticed on the crop having no input (Malik *et al.*, 2003). Kalra *et al.* (1983) reported higher aphid incidence with enhancement in nitrogen application, but addition of phosphorus and potash reduced the *L. erysimi* incidence significantly (Singh *et al.*, 1995).

Potato

Potato crop is attacked by several insect pests and tuber yield is much reduced by the insect damage in Mid-hill regions (Mishra, 2002). Aphids are one of the most important pests of this crop and widely distributed all over the world on a large number of host plants. The virus spreading ability of aphids *viz., Myzus persicae* and *Aphis gossypii* play an important role in limited disease-free potato tuber production and cause quick degeneration of seed stocks and yield reduction in potato (Biswal *et al.,* 2004). Among biopesticides *Bacillus subtilis* was most effective against the aphids for a short period (Gujar *et al.,* 2000). The efficacy of *B. subtilis* was due to favourable weather conditions prevailed in this locality for rapid multiplication of the bacterium in winter months. Azadirachtin was also found effective to keep the aphid population below the critical level. To avoid the residual toxicity of chemical pesticide in potato tubers, the biopesticides should be sprayed during later stage of the crop (Amitava Konar and Santanu Paul, 2005). The varieties cv. Kufri Sindhuri and Kufri Lalima did not exhibit any symptoms of hopper burn though it harboured population of aphids (Gulab Ram *et al.,* 1998). Chaudhary *et al.* (1983) have reported that Kufri Sindhuri was a tolerant variety and with stand aphid damage.

Potato crop is also attacked by other sucking pests like leaf hoppers *Amrasca devastans* and mite *Polyphagotarsonemus latus* during rabi season. The integrated management of these pests includes that plant at spacing of 15 cm distance, fertilizer recommendation of 150: 80: 100 (NPK) and foliar treatment with 4 per cent urea after 50 days of planting. Potato tuber moth, *Phthorimaea operculella* is one of the important and serious pests of potato in fields and storage. Chemical control is not the long-term solution. Moreover potato stored can not be used for table purpose. Environmentally safe and low persistent insecticide/biopesticides and plant derivatives can be used. Bio pesticides can be applied to potato tuber either after mixing with dry sand or saw dust or wheat flour at the time of storage of tubers. The finding suggested that cypermethrin with low persistent was at par with other insecticides reducing the potato tuber month incidence in tubers followed by Bt., Ageratum leaves powder application and econeem + cypermethrin (Sharma *et al.,* 2001). Utilization of synthetic sex pheromone for the management of sweet potato weevil, *Cylas formicarius* is of recent technology. Pheromone erected at 10 m distance between the traps recorded higher catches. The septa of the pheromone kept in the traps attracted the male weevils, whereas the sweet potato tuber bits kept in the place septa attracted female weevils (Chiranjeevi *et al.,* 2002). The hybrid JN-2626 and cv. Kufri Sindhuri resistant to cut worm, *Agrotis ipsilon* and red ant, *Dorylus orientalis,* where the both are important pests in north-east region (Gulab Ram *et al.,* 1998). The mean tuber damage due to the greasy cutworm *A.ipsilon* in Uttaranchal hill was to the extent of 27.28 per cent (Mishra and Singh, 1998). Though insecticides are found most effective against the cutworm, application of Bt (Dipel 8 L 1 kg/ha) and NSKE 0.3 per cent (2kg/ha) has recorded less plant damage compared to untreated check (Tripathi *et al.,* 2003).

Among the subterranean pests, whitegrubs have been causing extensive damage right from foot hills to high altitudes. Insecticidal treatments have not come out with sustainable control of these pests. The Entomopathogenic fungi, *Metarhizium anisopliae* and *Beauveria bassiana* were found effective @ 5×10^{13} conidia/ha along with Chlorpyriphos 20 EC at 200 g a.i./ha. It controls the whitegrub population up to 56.5 per cent and increase the tuber yield (Bhagat *et al.,* 2003). Root aphid, *Rhopalosiphum rufiabdominalis* has been reported from various parts of the country. The white skinned varieties Kufri Chandramukhi, Kufri Badshah and Kufri Jyothi were found resistant compared to red skinned varieties (Gulab Ram and Misra, 1998).

Eco-friendly Pest Management in Fruit Crops

Citrus

Sucking Pest on Citrus

Leaf extracts and plant oils were used against mandarin orange aphid *Toxoptera aurantii*, Black flies *Aleurocanthus woglumi* and fruit flies *Bactrocera dorsalis* and *B. cucurbitae*. The results revealed that spraying Jatropha oil @ 3 per cent was as effective as that of Monocrotophos in reducing the population of aphids and black flies. The aphid population at 5 DAT was 6.6 and 6.2 per tree in jatropha oil and neem oil treated trees as against 70 per tree in control. Due to oil emulsion spray, black fly population at 5 DAT was reduced greatly ranging from 3.5–4.1 as against 41.3 in untreated check. Adopting insecticides (Fenthion DDVP @ 1ml/l) as first spray during pea stage and jatropha oil 3 per cent or neem oil 3 per cent for the subsequent sprays at monthly interval is highly useful in reducing the damage by fruit flies (Chandramani and Chezhian, 2003). The best way to avoid fruit flies is trapping the adults. Various trap treatments were tried to monitor and attract adults which includes fermented butter milk, fish meal, jaggery and affected fruits alone. A mean of 356.3 flies were trapped by fish meal as against 40.8 flies in fermented butter milk (Chandramani *et al.*, 2003).

Leaf Feeders

Leaf miner, *Phyllocnistis citrella* is a serious pest in Tamil Nadu and causes 30-40 per cent yield loss. The variety PKM -1 was the least susceptible to leaf miner. The cultivars Pramalini, seedless lime, Sai Sarbati and Vikram were moderately susceptible. Leaf miner was monitored and managed by keeping yellow sticky traps up to some extent (Muthukrishnan *et al.*, 2001). Leaf miner damage is minimized by applying neem derivatives. Two rounds of NSKE 5 per cent reduced the infestation by 50.45 per cent (Philip Sridhar and Umapathy, 2003). Among the natural enemies, the parasitoids *Citrostichus phyllocnistoides* Narayanan and *Cirrospilus quadristriatus* Subba Rao and Rama Mani (Hymenoptera: Eulophidae) and one predatory chrysopid *Mallada boninensis* Okamoto (Neuroptera: Chrysopidae) have been reported as key bioagents (Rao *et al.*, 2003). Release of eulophid parasitoids @ 10 adults followed by the release of predator *M. boninensis* @ 6 larvae per 30 seedlings resulted in 39.7 per cent reduction in leaf-miner incidence (Rao *et al.*, 2005).

Lemon butterfly, *Papilio* sp. is reported through out the year except during winter and the infestation is more pronounced in nurseries and on young trees. Conservation of the indigenous braconid parasitoid *Diatatrix papilionis* through application of less toxic chemicals and botanicals in citrus IPM will ensure natural suppression of citrus butterfly. Use of selective insecticides like Imidacloprid and Dichlorvos have little impact on the parasitoid *D. papilionis* (Jayanthi and Verghese, 1999). Peak parasitisation by the braconids *viz.*, *Apanteles papilionis*, *Apanteles* sp. and *Bracon hebetor* was observed on *P. demoleus* during November to January months. Peak periods of parasitism synchronized with the peak periods of pest activity (Lakshmi Narayanamma *et al.*, 2003). Various citrus species *viz.*, sweet orange, acid lime, sweet lime, kinnow orange are cultivated and yield reduction due to several insect pests observed in citrus orchards. Using chemicals are not succeeding up to expected level. A novel eco-friendly method was tried. Adhesive cage were hanged in the citrus orchards throughout the year. Different insects were found sticking to the adhesive cages. The maximum number of citrus butterfly adhered to the cages in the month of August and May which were followed in the month of September, May and October months in the Kanpur orchard (Sunita Arya and Prasad, 2001).

Pear

Fruit fly, *Bactrocera dorsalis* attained pest status in orchards of pear in various orchards of Punjab. The survey revealed that isolated pear orchards suffered less from the fruit fly incidence than those near the other fruit fly host orchards. Secondly good maintenance of orchards where practices of proper sanitation, fruits were not kept for long on the trees and not allowed to remain on the ground and infested fruits were well disposed off, *i.e.* by deep burying or fed to pigs etc., such adoptions were unable to establish themselves as economic pest (Mann, 2003).

Litchi

Among several insect pests attacking litchi (*Litchi chinensis* Sonn.) the pentatomid bug, *Tessaratoma javanica* Thunberg (Hemiptera: Pentatomidae) is the most devastating one. The stinkbug preferred feeding on flowering and tender branches. Biological control is the only answer for the management this pest. Two parasitoids offer great potential for the control of stick bug during their egg stage are *Anastatus colemani* (Hymenoptera: Eupelmidae) and *Ooencyrtus phongi* (Hymenoptera: Encyrtidae). The mean egg parasitisation by *A.colemani* and *O. phongi* ranged from 12.85 to 52.85 and 20.57 to 55.71 per cent respectively (Vasudha Gautam *et al.*, 2003).

Apple

The productivity of apple in India is considerably low due to various factors among which infestation by insect-pests is a major one. The pest losses can be managed successfully without posing threats to non target organism, if the growers adopt various pest management schedules properly in their orchards right form nursery stage to harvesting. These include grafting of improved and pest free scion varieties on suitable root stock, regular monitoring of orchard to observe appearance of insect-pests and their level of infestation, sanitation of orchard and tree basin (destruction of alternate hosts including weeds and habitats for aestivation or hibernation), tillage and pruning during December-January, conservation of natural enemies, spraying of effective Bt based bio insecticides and neem based formulations, judicious use of selective insecticides and proper fertilizer and water management (Singh *et al.*, 2001). San Jose Scale *Quadraspidiotus perniciosus* is a serious pest of apple. The aphelinid parasitoid *Encarsia perniciosi* and *Aphytis nr. proclia* are found effective against the pest. The pest is found to attack pumpkin and the parasitoid mass reared on pumpkin easily. The cost of production of both parasitoid is Rs.15 per 1000 adult parasitoids which is well suited in the IPM in Apple ecosystem. The predator *Chilocorus infernalis* is also easily mass multiplied and released (Singh, 2000). The European red mite, *Panonychus ulmi* was reported in epidemic form and 72 per cent apple orchards in Himachal Pradesh have been infested (Bhardwaj and Bhardwaj, 2005). The principle environmental factors affecting mite population are temperature and relative humidity (Sangita Sharma and Bhardwaj, 2004). Increase in temperature and relative humidity to a favourable range (18.5–19.8° C and 75-82 per cent respectively) resulted in outbreak of the mite (Sushma Bhardwaj *et al.*, 2006). The shot-hole borer, *Scolytus nitidus* is the main insect pest of apple in the Kashmir region. Nine species of hymenopterous parasitoids, three species of predators and three fungal pathogens are recorded in the region (Abdul Azad Buhroo *et al.*, 2002).

The commodity-based, pathway- initiated, risk assessment identified chances of introduction of a total of 220 pests associated with import of fresh apple fruits into India. Of these, 14 pests have been economic importance and fulfill all the standard conditions for carrying out detailed import risk analysis. Based on the prevailing pest scenario and import risk assessment, it is comparatively safer to import apple fruits from New Zealand than from Australia and China, though it is not absolutely safe (Rajan *et al.*, 2004).

Banana

The banana pseudostem weevil, *Odoiporus longicollis*, is an enigmatic pest beyond any chemical or cultural means of control once the injurious grub are inside the pseudo stems. The parasitic nematode DD-136, *Steinerma feltiae* as a biological control agent against the grubs of pseudostem weevil showed in vitro mortality of the grubs. *Steinerma glaseri* cause mortality within 72 hours but no nematode multiplication was recorded on pseudostem weevil grub in laboratory condition (Gulsar Banu *et al.*, 2001).

Table 28.1: Promising Biological Control Approaches in Fruit Crops (Singh, 2000)

Crop/Pest	*Biocontrol Agent*	*Dosage/ha*	*Frequency and Application Methods*
		Apple	
Eriosoma lanigerum	*Aphelinus mali*	1,000 adults/infested tree	Once as soon as infestation noticed, releasing of adults
Quadraspidiotus perniciosus	*Encarsia perniciosi*	2,000 adults/infested tree	Once releasing adults on infested plants
	Chilocorus infernalis	20 adults or 50 grubs/tree	Releasing adults once during April–May
Cydia pomonella	*Trichogramma embryophagum*	2,000 adult/tree	Releasing adults at weekly interval
		Citrus	
Icerya purchasi	*Rodolia cardinalis*	10 beetles/infested plant	Adults released after the infestation
Planococcus citri	*Cryptolaemus montrouzieri*	10 beetles/infested plant	After the blossom release the adult
	Leptomastix dactylopii	3,000 adults	Need based release of adults
Papilio demoleus	*Bacillus thuringiensis* var. *kurstaki*	0.5 per cent	Single application for each generation using knapsack sprayer

Eco-friendly Pest Management in Flower Crops

Flower crops are important component in hill horticultural systems. Some of the flower crops fetch economically higher value as they are exported. Cut flowers like gerbera, carnation, chrysanthemum and also some variety of roses are raised in poly house condition and some of them also raised in the open field conditions. Insect pests are important constraint to the production of quality cut flowers.

The spiralling Whitefly, *Aleurodicus disperses* Russell poses threat to many horticultural crops in India. The population of spiralling whitefly in flower crops was found to be relatively higher during summer months and the density of the whitefly was positively correlated with maximum temperature and negatively correlated with relative humidity. Chemical control of *A. dispersus* is almost impossible due to its wide host range and heavy wax coating. Application of chemicals like dimethoate, triazophos etc. and mechanical control measures like pruning the heavily infested plants cause only temporary reduction in the population of spiralling whitefly. But the natural enemies chiefly aphelinid parasitoids, *Encarsia qujadeloupae* Viggiani and *Encarsia haitiensis* Dozier proved to be highly useful in suppressing the spiralling whitefly in pacific Islands, African and Asian countries. Initial surveys for natural enemies of this pest in India showed the presence of 42 predators, mainly coccinellids and neuropterans. Amont them, *Axinoscymus puttarudriahi* Kapur and Munshi, *Cryptolaemus montrouzieri* Muls, *Mallada astur* (Banks) and *Cybocephalus* sp. were more frequently observed than the other predators (Mani and Krishnamoorthy, 2003b).

Bud borer, *Helicoverpa armigera* is polyphagous pests affecting rose, carnation, tuberose, china aster, marigold, gerbera, crossandra, dahlia, chrysanthemum, coleus etc. The larva bore into buds and feed on growing petals. In gerbera if prefers centre capitulum causing drying and withering of petals (Soundararajan, *et al.*, 2006). *Trichogramma chilonis*, *T. achaeae* and *Trichogrammatoidea bactrae* and *Telenomus* sp. were reared from *H. armigera* on tuberose (Manjunath, 1972). Mortality of *H. armigera* by the natural enemies chiefly *Campoletis chlorideae* was 34.2 per cent on carnation in Himachal Pradesh. Nuclear polyhedrosis virus @ 250 LE/ha is found to be effective against *H. armigera* particularly under poly house conditions.

Rose

Scale, *Aonidiella aurantii* mainly infest matured shoots and spread to young shoots in case of heavy incidence. Cutting and burning of heavily infested shoots will reduce the infestation. After pruning application of pongamia oil 10 per cent will offer better protection for further infestation of scales (Parvatha Reddy and Verghese, 2002). *Pharoscymnus horni* was found to be an important predator. The chilli thrips *Scirtiothrips dorsalis* has become a major pest of rose flowers (Onkarappa and Mallik, 1998). Application of higher nitrogen level recorded higher population of thrips. And the field without leaf-litter recorded significantly lower population of thrips (Duraimurugan and Jagadish, 2004).

Gerbera

Gerbera crops raised in poly houses caused heavy infestation with the leaf miner, *Liriomyza trifolii*. The damaged leaves turn brown and dry while plants get stunted and produce small flowers in case of severe attack (Jhansi Rani and Jagan Mohan, 1997). Plucking and burning of severely infested older leaves and subsequent spraying with pongamia oil 1 per cent to grown up plants at 15 days interval and setting up of yellow sticky traps control the pest.

Carnation

Carnation is a high value cut flower crop mostly raised in the poly house conditions. Heavy infestation of Red spider mite, *Tetranychus urticae* was observed during summer months. Due to mite feeding, entire plant looks dirty, stunted growth, discolouration of flowers and small flowers produced (Jhansi Rani and Jagan Mohan, 1997). Regular washing of plants with force to dislodge webs will reduce mite population. Spraying of pongamia oil or neem oil 1 per cent at initial stage of infestation and release of predatory mite, *Amblyseius tetranychivorus* @ 20 per plant offered better control. Watering of polyhouse and providing proper ventilation in summer are essential to reduce mite multiplication.

Chrysanthemum

Cut-flower type Chrysanthemum is largely damaged by the pests like aphids, leaf eating caterpillars of moths and butterflies etc. Moths and butterflies lay their eggs on leaves and newly emerging buds cause irreparable damage. The insecticides sprayed on the plant fail to kill them. The growth of plant is also greatly affected by environmental pollutions like toxic gases, smoke, heavy dewdrops, smog, dirt etc. It is not practicable to spray chemicals and insecticides on flower and buds. Use of translucent paper bags or polythene bags to cover the buds and flowers. It protects the buds and flowers from moths and butterflies and does not allow them to lay eggs on them (Pratibha Sharma and Bhattacharjee, 2002). The aphid, *Macrosiphoniella sanborni* is serious pest on cut flower type chrysanthemum during summer months in the Yercaud hill region of Tamil Nadu. *Chrysoperla carnea* can be used to control the pest in glass house. *Endaphis aphidimyza* was found to control this aphid in Maharashtra (Grover *et al.*, 1991). *Cheilomenes sexmaculuta* and the parasitoid *Toxares macrosiphophagum* were recorded on *M. sanborni* in Kashmir (Shujauddin, 1974). The cocinellids, *Coccinella septempunctata* found to be feed on *M. sanborni* (Behera *et al.*, 1999). The feeding potential was high in adult and final instar grubs stage (55.3 and 37.5 aphids/day) Biological control of chrysanthemum aphids has been developed on commercial crops by rearing small parasitic hymenopterans, *Aphidius matricariae*, for release on cuttings before distribution or planting. This technique could be used in gardens but require scientific demonstration (Pratibha Sharma and Bhattacharjee, 2002).

Other Flower Crops

Scale insect, *Coccus herperidum* in tuberose can be suppressed with the encyrtid parasitoid *Metaphycus luteolus* Timb. The other scale insect *Parasaissetia nigra* found infesting Crossandra and ornamental crotons can be managed by *Scutellista caerulea* (Hayat, 1972). Dendrobiums are one of the important and interesting groups of Ochids known for their beautiful flowers and diverse morphological features and essentially epiphytes found growing on trees or rocks. Orchids are often infested with the scale insect, *Diaspis boisduvalli* under poly house conditions and several hymenopteran parasitoids are identified for the suppression of the population in unsprayed condition. Black aphids, *Toxoptera aurantii* is reported to feed on the flowers of *Dendrobium densiflorum* (Nagrare, 2003).

Mealy bugs *Planococcus minor* and *Ferrisia virgata* are found to infest Crotons and Gerbera respectively. *Pseudococcus* sp. cause damage to orchids and the adult and young stage of mealy bugs suck the sap from leaves and shoots. Release of predators *Cryptolaemus montrouzieri* @ 5–10/plant controls the mealy bugs on the flower crops.

Red spider mite, *Tetrananychus urticae* is one of the regular and devastating pest in the poly houses of flower crops. The mites infest lower side of leaves and flowers and covered by dirty white webs. Damage results in discolouration, dusty appearance, drying and dropping of leaves. The cut flowers lost their quality and not useful for export. The pathogenic fungus *Hirsutella thompsoni* was isolated from these mites. Release of indigenous predatory mite, *Amblyseius tetranychivorus* or exotic *Phytoseiulus persimilis* @ 20 adults/plant is recommended to suppress the mites. Effective control was evidenced for *T.urticae* with *Neoseiulus longispinosus* on rose in poly house (Mallick *et al.*, 1998).

Eco-friendly IPM for flower crops under protected cultivation in future should aim for including preventive methods like use of insect proof nets, using double door glass houses/green houses, polyhouses, soil sterilization, monitoring and scouting; soft chemicals like IGR, Avermectins, biological agents like fungus *Verticillium lecanii*, *Beavaria bassiana* and plant products like neem and pongamia oil.

Eco-friendly Management of Nematodes

Generally, nematodes are major problem in horticultural crops in the hilly regions. Particularly banana, citrus, pepper and certain flower crops are succumb to heavy nematode damage. Nematodes are microscopic and are mostly soil inhabitants. Mostly they are root feeders and the symptoms are expressed in the aerial part of the plants.

Nematode Management in Plantation Crops

Nematodes are one of the major pests infecting coffee crop causing considerable yield loss. Kumar (1988) reported that lesion nematode, *Pratylenchus coffeae* is highly destructive to *Coffea arabica* in the coffee tracts of South India while the other commercially cultivated species *C. canephora* exhibited tolerance to *P. coffeae*. Sundaram *et al.* (1990) studies seasonal population fluctuation of nematodes associated with *C. arabica, C. canephora* and *C. congensis* in Tamil Nadu showed that *P. coffeae* population was higher during December. Its greatest relative frequency was observed at 15 cm depth at 800 m above sea level (ASL) and 30 cm depth at 1000 m ASL (Giribabu and Saha, 2003). The depth and season of nematode population and infestation will help to formulate suitable management package for particular regions.

There are several parasitic nematodes associated with Tea and the most important one is root-knot nematode, *Meloidogyne* spp. found to cause damage to nursery and young tea plantations. Mustard oil and neem cake at 10 and 15 g each/seedling were found effective on the growth parameters on the nematode infected tea seedlings (Bora and Neog, 2006).

Fruit Crops

Banana burrowing nematode *Radopholus similis* is an important nematode in banana growing areas. It causes wounding in banana roots usually reddish-brown cortical lesions that are diagnostic symptoms of the disease. Roots and rhizome necrosis is manifested by varying degrees of retarded growth, leaf yellowing and falling of mature plants. Application of neem cake at 800 g per plant (two split applications) or neem coated urea at 110 g per plant reduced the population of burrowing nematode and also increased the yield of banana. Integration of neem cake with *Glomus mosseae* was most effective in reducing the burrowing nematode population both in soil and roots (Parvatha Reddy and Abraham Verghese, 2002). The number of nematodes in soil may be decreased either by following crop rotation or intercropping. The banana bio type of *R.similis* has a narrow host range and limited survival period in soil in the absence of host. Therefore, destruction of the infected area mechanically or by injection of a herbicide into the pseudostem and replanting the area with nematode-free sets seems a practical approach to rehabiliate the infested plantations. After destroying an infested plantation, planting of cover crops like *Panicum maximum* var. *trichoglume* and *Phaseolus atropurpureus* did not support *R. similis* population. Planting sugarcane in soil following the destruction of bananas eliminated *R. similis* after 10 weeks in plain areas.

Citrus nematode *Tylenchulus semipenetrans* Cobb causes slow decline in citrus trees resulted in reduced vigour, chlorosis, falling leaves twigs dieback and reduced fruit production. The nematodes can be controlled by dipping the bare roots of citrus seedlings in hot water at 45° C for 25 min or 46.7° C for 10 min. Neem oil cake integrated with parasitic fungi *Paecilomyces lilacinus, Verticillum chlamydosporium, V. lecanii*, VA mycorrhiza, *Glomus fasciculatum* gave better plant growth and more nematode reduction under greenhouse condition (Parvathy Reddy *et al.*, 1995). The VA mycorrhiza control nematode through improved vigour, physiological alteration of root exudates or through direct role of mycorrhiza in retarding the development and reproduction of nematode within root tissues (Jothi, *et al.*, 2005). Neem cake 1 kg/plant combined with chemical Carbofuran® at 2 kg/ha reduced 33.0 per cent nematode population and increased the yield by 33.8 per cent (Parvatha Reddy and Abraham Verghese, 2002).

Trifoliate orange (*Poncirus trifoliate*) and its hybrids (Citrumello) were highly resistant to the citrus nematode (Parvatha Reddy, 1983). The varieties CRH-3, CRH-5 and CRH-41 were highly resistant to citrus nematode. The lesion nematode, *Pratylenchus coffeae* attack fresh roots of citrus trees. Four selections of a *Microcitrus australis* × *M. australasica* hybrid and Rubidous 70-A5 trifoliate orange show some resistance to lesion nematode (Parvatha Reddy and Abraham Verghese, 2002).

Flower Crops

In commercial flower crop crossandra the root knot nematode *Meloidogyne incognita* cause up to 25.62 and 21.64 per cent loss in number of flower and weight of flowers respectively. Integrated management includes rational combination of a biocontrol agent, *T. harzianum* or *Verticillum chlamydosporium* with a nematicide or neem oil cake. Integration of neem cake, karanj and castor cakes with VA mycorrhiza, *Glomus mosseae* significantly enhanced plant growth parameters and flower yield of crossandra. Combination of *Verticillum lecanii* and *P. lilacinus* with neem leaf extract 5 per cent resulted in significantly higher plant growth parameters and flower yield. Similar combination has been followed for tuberose root knot nematode. Tuberose cv. Shringar (single type) was found resistance, while Suvasini (double type) was tolerant. For the control of *M. incognita* damage in flower crops like gerbera and carnation, application of neem cake at 0.5 kg per plant in combination with *Trichoderma viride* at 100 g per sq.m. was effective (Parvatha Reddy and Abraham Verghese, 2002). Enrichment of FYM with *Trichoderma harzianum* used for large scale field application for root knot nematode in ornamental crops. For this purpose 1 kg of *T. harzianum* with 10 kg of neem cake in 1 tonne of FYM is mixed and cover polythene sheet for 15 days (Pratibha Sharma and Bhattacharjee, 2002). Intercropping of crossandra with marigold significantly reduced the lesion nematode, *Pratylenchus delattrei* population both in soil and roots.

Management of Root-knot Nematode

Root-knot nematode, *Meloidogyne* spp. found to infest citrus, pepper and other vegetable crops. In pepper it was associated with wilt disease. Selection of planting materials is important for the management of root-knot nematode. Avoid growing vegetables as intercrops or as rotational crops in orchards or estates where nematode damage is regular. Crotalaria roots exert a toxic effect on *Meloidogyne* sp. population. A fungal parasite, *P.lilacinus* consistently and effectively controlled the population of *Meloidogyne* spp. (Parvatha Reddy and Abraham Verghese, 2002). In Pepper, *M.incognita, Radopholus similis* and *Helicotylenchus* sp. were found to infest almost all the varieties. Some of the varieties like Ceriyakaniyakadan, Kottaram, Karimunda, Karuvathikotta and Karuvilanchy were found highly resistant to nematodes. Six species of *Bacillus viz., B. subtilis, B. pumilus, B. coaguulans, B. macerans, B. circulans* and *B. licheniformis* were found associated with egg masses, second stage juvenile and adult females of nematodes. The plants received the bacterial load of 1.1–1.2 × 10^8 cells reduced nematode population in soil from third month onwards (Nybe, 2001).

Nematode Management in Medicinal Plants

M. incognita has been also found as great menace in the successful cultivations of different medicinal and aromatic plants including *Ocimum* spp. (Haseeb *et al.*, 1999). *Ocimum canum* has great medicinal value because of its antihelmentic, alexipharmic and antipyretic properties and has a superior position in pharmaceutical industries. No chemicals can be used for the management of nematode pest in the medicinal plant as their active principle has to be extracted for marketing. Botanicals are the only solution for the control of the nematode. Maximum root and shoot growth was observed in the plots applied with neem leaves and mint *Mentha arvensis* leaves (Mittal and Prasad, 2003). Nematode multiplication studies indicated that a significant reduction in root-knot gall system when soil was amended with neem leaves. Application of chopped leaves were highly effective in reducing the plant damage caused by *M. incognita* (Verma and Nehal Khan, 2004). Oils of *Thevetia nerifolia* and *Ricinus communis* (castor) and their esterified products exhibited considerable nematicidal activity to root knot and reniform nematodes. Castor oil has rich source of biologically active molecules. (Prasad *et al.*, 2003). Neem cake @ 2 per cent was effective in reducing the population of *M. incognita* as well as increasing growth parameters of japanest mint, *Mentha arvensis* (Singh and Vinod Kumar, 2000). Jagdale *et al.* (1985) reported considerable increase in number of betelvine leaves with application of karanj cake. Haseeb and Butool (1991) found neem cake to some extent manage root knot nematode in Davana, *Artemisia pallens*.

Conclusion

Regarding commercial availability of bio pesticides, only Bt based bio-insecticides accounts for 90-95 per cent of the world biopesticide market. They are preferred over chemicals owing to their eco-friendly nature and lack of harmful residues on horticultural produce and also due to the slower development of resistance compared to chemicals. The Bt based bioinsecticides meet the needs of the export market on horticultural produce, where importation laws are stringent on chemical residue level. The entomopathogenic fungi have relatively broad host range and are amenable for mass production. The commercial formulations Bovcrol and Boverin from *Beauveria bassiana*, Verlac (for aphids) and Mycotel (for whiteflies) from *Verticillium lecanii*, Metaquino from *Metarhizium anisopliae* and Mycar (for citrus red mite) from *Hirsutella thompsonii* are sold commercially (Asokan *et al.*, 2000).

To achieve expected progress in plant protection and to overcome the on-going ill-effects of plant protection one need to pay sufficient attention in educating the users at all levels. Training/capacity building at every stage starting from school level needs to be strengthened to bring back the healthier situation. Since plant protection was treated as the supporting science for the past several years and neglected, now the community is paying a heavy price for that. Hence it is time to pay appropriate attention to rectify the lacuna (Ranga Rao *et al.*, 2002). The complex systems of biodiversity of host plants, pests their natural enemies and their interaction with environment should be clearly understood to develop sound system of eco-friendly integrated pest management. The growing interest in the market all over the world for organically produced fruits, vegetables, spices and beverages insist the farmers and scientists to take up new eco-friendly pest management techniques and increase the productivity.

References

Abdul Azad Buhroo, Chishti, M.Z. and Masoodi, M. Amin, 2002. Biocontrol agents of shot-hole borer, *Scolytus nitidus* (Coleoptera: Scolytidae) infesting apple orchards. *Indian J. Plant Prot.*, 30: 71–73.

Adhira, D., Kennedy, J.S. and Balasubramanian, G., 2003. Sequential application of Bt and insecticides for the management of the diamondback moth, *Plutella xylostella* L. In: *National Symposium on Bio-Management of Insect Pests*, Annamalai University, Tamil Nadu, March 29–31, pp. 93.

Amitava, Konar and Santanu, Paul, 2005. Comparative field efficacy of synthetic insecticides and bio-pesticides against aphids on potato. *Ann. Pl. Protec. Sci.*, 13: 34–36.

Anonymous, 1996. *Coffee Guide.* Central Coffee Research Institute, Chikmagalur, India, 180p.

Arya, Sunita and Prasad, A., 2001. Studies on the insect adhered to the adhesive cages. In: *National Symposium on Plant Protection Strategies for Sustainable Agri-Horticulture*, Society of Plant Protection Sciences, Public. No. 4, pp. 32.

Asokan, R., Sukhada Mohandas and Lalitha Anand, 2000. Biofertilizers and biopesticides for horticultural crops. *Indian Horticulture*, April–June, p. 44–47.

Banerjee, B., 1969. Shade and manurial experiments at Murmuria: a report on the incidence of mites. *Two and a Bud*, 16: 93–96.

Barwal, R.N., 2001. Managing pests of cole crops scientifically. *Indian Horticulture* April–June, 4–7.

Behera, M.K., Behera, R. and Patro, B., 1999. Biology and feeding potential of *Coccinella septempunctata* (Coleoptera: Coccinellidae) on the chrysanthemum aphid *Macrosiphoniella sanborni*. *J. Biol. Control*, 13: 39–43.

Bhagat, R.M., Gupta, R.B.L. and Yadava, C.P.S., 2003. Field efficacy of two entomopathogenic fungal formulations against whitegrubs in Himachal Pradesh. *Indian J. Ent.*, 65: 76–81.

Bhardwaj, S.P. and Bhardwaj, S., 2005. Insect mite pests. In: *The Apple: Improvement, Production and Post Harvest Management*, (Eds.) K.L. Chadha and R.P. Awasthi. Malhotra Publishing House, New Delhi, pp. 346–396.

Bhardwaj, Sushma, Sharma, Sangita and Bhardwaj, S.P., 2006. Environmental factors affecting the population build–up of *Panonychus ulmi* (Koch) in Apple. *Indian J. Plant Prot.*, 34: 36–39.

Biswas, M.K., De, R.K., Nath, P.S. and Mohasin, Md., 2004. Influence of different weather factors on the population build up of vectors of potato virus. *Ann. Pl. Protec. Sci.*, 12: 352–355.

Bora, B.C. and Neog, P.P., 2006. Effect of oil cakes for management of *Meloidogyne incognita* in tea. *Ann. Pl. Protec. Sci.*, 14: 522–523.

Chandramani, P. and Chezhian, N., 2003. Efficacy of certain botanicals and insecticides against major pests of mandarin orange. In: *National Symposium on Bio-management of Insect Pests*, Annamalai University, Tamil Nadu, March 29–31, pp.84.

Chandramani, P., Chockalingam, Pon and Rajendran, R., 2003. A low cost trap for monitoring fruit flies of orange and guava. In: *National Symposium on Bio-management of Insect Pests*, Annamalai University, Tamil Nadu, March 29–31, pp. 84.

Chandramohan N. and Nanjan, K., 1992. Effect of plant product on the Ovipositional behaviour of Diamondback moth *Plutella xylostella* (L.). *Neem Newsletter*, 9: 8–9.

Chaudhery, R., Singh, M. and Anand, S.K., 1983.Variations in leaf hopper burn incidence in *Solanum tuberosum* and Andigene collection. *Indian J. Ent.*, 45: 493–497.

Chiranjeevi, Ch., Reddy, D.D.R., Gour, T.B., Reddy, Y.N. and Sultana, Abida, 2002. Sex pheromone studies on sweet potato weevil, *Cylas formicarius*. *Indian J. Plant Prot.*, 30: 191–195.

Das, G.M., 1965. *Pests of Tea in North-East India and their Control*. Tocklai Experiment Station Report, Cinnamara, Assam, pp. 115.

Das, S.C., 1984. Resurgence of tea mosquito bug, *Helopeltis theivora* Waterh. a serious pest of tea. *Two and a Bud*, 31: 36–39.

Desh Raj, R.S. Nirmala Devi, Chandel and Kapoor, A.S., 2000. Mass production and utilization of biotic agents of some important pests of cole crops and pea in Himachal Pradesh. Final report submitted to DBT, Ministry of Science and Technology, New Delhi, 87p.

Devasahayam, S., 1998. *Biological Control of Scale insects infesting black pepper*. Final report of ICAR Ad hoc scheme, Indian Institute of Spices Research, Calicut.

Devasahayam, S., 2000. *Insect Pests of Black Pepper (Piper nigrum)*, (Ed.) P.N. Ravindran. Harwood Academic Publishers, Netherlands, pp. 309– 334.

Devasahayam, S. and Koya, K.M.A., 1994. Natural enemies of major insect pests of black pepper (*Piper nigrum* L.) in India. *J. Spices and Aromatic Crops* 3: 50–55.

Devasahayam, S. and Koya, K.M.A., 1999. Integrated management of insesct pests in spices. *Indian J. Arecanut Med. Plants*, 1: 19–23.

Devasahayam, S. and Leela, N.K., 1998. Antifeedant acticity of plant products against pollu beetle (*Longitarsus nigripennis*) In: *Advances in IPM for Horticultural Crops*, (Eds.) P.P. Reddy, N.K.K. Kumar and A. Verghese. Association for Advancement of Pest Management in Horticultural Ecosystems, Bangalore, pp. 172–174.

Duraimurugan, P. and Jagadish, A., 2004. Non-chemical management of *Scirtothrips dorsalis* Hood damaging rose flowers. *J. Appl. Zool. Res.*, 15: 163–165.

Giribabu and Saha, M., 2003. Studies on nematode communities at different depths, altitudes and periods of year in Coffee plantation. *Ann. Pl. Protec. Sci.*, 11: 360–363.

Gopalakrishnan, C., 2001. Field evaluation of commercial formulations of *Bacillus thuringiensis* var. *kurstaki* against *Plutella xylostella* L. on cabbage. *Pestology*, 25: 7–10.

Gopalakrishnan, C. and Mohan, K.S., 1993. Epizootics of *Erynia neoaphidis* (Zygomycetes: Entomophthorales) in field population of *Brevicoryne brassicae* (Homoptera: Aphididae) on cabbage. *Entomon.*, 18: 21–26.

Grover, P., Kshyap, V. and Chandra, R., 1991. Assessment of the efficiency of *Endaphis aphidimyza*, a biocontrol agent against aphids. *Cecidologia International*, 12: 13–29.

Gujar, G.T., Kalia, V. and Kumari, Archana, 2000. Bioactivity of *Bacillus thuringiensis* against the American bollworm, *Helicoverpa armigera*. *Ann. Pl. Protec. Sci.*, 8: 125–131.

Gulab Ram, and Misra, S.S., 1998. Response of different potato hybrids and varieties against root aphid, *Rhopalosiphum rufiabdominalis*. *Indian J. Ent.*, 60: 273–277.

Gulab, Ram, Misra, S.S. and Dhamyanthi, K.P.M., 1998. Reaction of different potato hybrids and varieties against insect pests. *Indian J. Ent.*, 60: 278–285.

Gulsar Banu, J., Subramanian, S. and Rajendran, G., 2001. Laboratory evaluation of entomopathogenic nematodes, *Steinerma* sp. and *Heterorhabditis* sp. against banana pseudostem borer, *Odoiporus longicollis*. In: *National Seminar on Emerging trends in Pest and Diseases and their Management*, Tamil Nadu Agricultural University, Coimbatore, pp.77.

Haseeb, A. and Butool, F., 1991. Evaluation of different pesticide and oil cake on the control of root–knot nematode, *Meloidogyne incognita* infesting Davanna (*Artemisia pallens*). *Current Nematol.*, 2: 210–214.

Haseeb, A., Butool, F. and Shukla, P.K., 1999. Effect of *Meloidogyne incognita* on growth, physiology and oil yield of *Ocimum sanctum*. *Indian J. Nematol.*, 29: 121–125.

Hayat, M., 1972. A new aphelinid genus Eriaphytis (Hymenoptera: Chalcidoidea) reared from *Cerococcus* spp. *Polkil Pismo Ent.*, 1: 151–156.

Hemchandra Singh, O. and Singh, T.K., 2003. Evaluation of antifeedant properties of some plant extracts against diamond back moth, *Plutella xylostella* (Linn.). In: *National Symposium on Bio-Management of Insect Pests*, Annamalai University, Tamil Nadu, March 29–31, pp. 89.

Hirak, Chatterjee and Chowdhury, A.K., 2003. Persistent toxicity of some bio-pesticides against *Pieris brassicae* Linn. infesting cauliflower (*Brassica oleracea* var. *bortrytis* L). *Ann. Pl. Protec. Sci.*, 11: 215–219.

Jagdale, G.B., Pawar, A.B. and Darekar, K.S., 1985. Pathogenicity of *Meloidogyne incognita* on Betelvine. *Indian J. Nematol.*, 15: 210–214.

Jalali, S.K., Singh, S.P. and Venkatesan, T., 2001. Choice of suitable trichogrammatid species for suppression of diamondback moth *Plutella xylostella* (Lepidoptera: Yponomeutidae) on cabbage. *Shaspha*, 8: 161–166.

Jalali, S.K., Venkatesan, T., Murthy, K.S. and Lalitha, Y., 2004. Influence of temperature and dosages on parasitizing ability of *Trichogrammatoidea bactrae* on *Plutella xylostella* eggs. *Ann. Pl. Protec. Sci.*, 12: 425–426.

Jayanthi, P.D.K. and Verghese, A., 1999. Comparative safety of selected insecticides to *Diastrix papilionis* (Vireck) (Braconidae: Hymenoptera), a parasitoid on citrus butterfly. *Pest Manage. Hort. Ecosys.*, 5: 92–98.

Jeyarani, S. and Kennedy, J.S., 2004. Efficacy of certain biopesticides against the Diamondback moth, *Plutella xylostella* (L.) in Cauliflower. *Indian J. Plant Prot.*, 32: 129–130.

Jhansi Rani, B. and Jagan Mohan, N., 1997. Pest management in ornamental crops. In: *Progressive Floriculture*, (Eds.) I.S. Yadav and M.L. Choudhary. The House of Sarpan, Bangalore, pp. 161–181.

Jothi, G., Rajeswari Sundara Babu and Rajendren, G., 2005. Bio-management of nematodes by mycorrhiza. *Agric. Rev.*, 26: 249–260.

Kakar, K.L. and Sharma, J.P., 1991. Role of natural enemies on the mortality of cabbage butterfly, *Pieris brassicae* L. in cauliflower seed crop. *Indian J. Plant Prot.*, 19: 210–212.

Kalita, H. and Singh, Karan, 1999. Seasonal biology of tea mosquito bug, *Helopeltis theivora* on tea. *J. Appl. Zool. Res.*, 10: 29–31.

Kalita, H. and Singh, Karan, 2004. Biochemical basis of resistance of certain genotypes of tea to *Helopeltis theivora*. *Indian J. Ent.*, 66: 176–180.

Kalra, V.K., Gupta, D.S. and Yadava, T.P., 1983. Effect of cultural practices and aphid infestation on seed yield and its component traits in *Brassica juncea* (L.). *Haryana Agricultural University Journal of Research*, 13: 115–120.

Kandoria, J.L., Singh, Gurdeep and Singh, Labh, 1999. Effects of inter cropping cauliflower with tomato on the incidence of diamondback moth. *Insect Environment*, 5: 137–138.

Krishnamoorthy, P.N., Krishna Kumar, N.K., Selvaraj, C. and Daniel, John Sathiasheelan, 1998. Neem seed kernel extract applications for diamondback management, transfer of technology for mechanized farming. *Pest Manage. Hort. Ecosys.*, 4: 128–130.

Kumar, A.C., 1998. Nematode problem of coffee and its management. *Indian Coffee*, 52: 12–19.

Kumar, Surjeet, 2004. Insect-pests of agricultural crops in high altitude dry temperate regions of north Western Himalayas. *Indian J. Ent.*, 66: 138–141.

Kumaresan, D., Gopalkumar, B. and Varadarasan, S., 1989. A noval method for root grub management in cardamom. *Spice India*, 2: 9–11.

Lakshmi Narayanamma, L., Savithri, P. and Ramakrishna Rao, A., 2003. Natural enemies and per cent parasitisation of citrus butterfly, *Papilio demoleus* on sweet orange and acid lime. *Indian J. Ent.*, 65: 461–464.

Malik, Y.P., Kumar, Jitendra, Singh, S.V., Singh, Mahak and Bisen, R.S., 2003. Influence of crop management inputs on aphid infestation in Indian mustard. *Indian J. Ent.*, 65: 439–443.

Mallick, B., Onkarappa, S. and Kumar, M.H., 1998. Management of spider mite, *Tetranychus urticae*, on rose using phytoseiid predator *Amblyseius longispinosus* in polyhouse. *Pest Manage. Hort. Ecosys.*, 4: 46–48.

Mani, M. and Krishnamoorthy, A., 2003a. Biological control of vegetable and ornamental crop pests in India. In: *Biological Control of Insect Pests*. Phoenix Publishing House Pvt. Ltd., New Delhi, p. 98–115.

Mani, M. and Krishnamoorthy, A., 2003b. Biointensive management of the exotic spiralling whitefly, *Aleurodicus disperses* Russell in horticultural crops in India. In: *National Symposium on Bio-Management of Insect Pests*, Annamalai University, Tamil Nadu, March 29–31, pp. 83–84.

Manjunath, T.M., 1972. Biological studies on *Trichogrammatoidea armigera* Nagaraja, a new dimorphic egg parasite of *Heliothis armigera* in India. *Entomophaga*, 17: 131–147.

Mann, G.S., 2003. *Bactrocera dorsalis* (Hendel) management in pear orchards as adopted by the Punjab farmers. In: *National Symposium on Bio-management of Insect Pests*, Annamalai University, Tamil Nadu, March 29–31, pp. 84 –85.

Mehta, P.K., Sood, A.K., Kashyap, N.P.and Parmer, S., 2001. Antifeedant acticvity of some plant of north–western Himalayas against cabbage caterpillar, *Pieris brassicae* (L.). In: *National Symposium on Plant Protection Strategies for Sustainable Agri-Horticulture*, Society of Plant Protection Sciences, Public. No. 4, pp. 64–65.

Mishra, D.N., 2002. Chemical control of cut worm on potato in mid-hill conditions. *Ann. Pl. Protec. Sci.*, 10: 151–153.

Mishra, P.N. and Singh, M.P., 1998. Chemical control of cutworm *Agrotis ipsilon* (Noctuidae: Lepidoptera) damaging potato at Research Station, Kanatal (2300 m ASL) during Kharif, 1998. *Ann. Prog. Report*, AICRIP, pp. 38–40.

Mittal, A. and Prasad, D., 2003. Effect of neem formulations and triazophos on soybean against *Meloidogyne incognita*. *Ann. Pl. Protec. Sci.*, 11: 404– 406.

Muthukrishnan, N., Palanisamy, S., Bhuvaneshwari, R., Logiswaran, R. and Thangaraj, T., 2001. Citrus leaf miner: Survey, bionomics, varietal resistance and IPM. In: *National Seminar on Emerging Trends in Pest and Diseases and their Management*, Tamil Nadu Agricultural University, Coimbatore, pp. 77–78.

Nagrare, V.S., 2003. Occurrence of aphid, *Toxoptera aurantii* on orchid *Dendrobium densiflorum*: A new report. *Ann. Pl. Protec. Sci.*, 11: 379.

Nybe, E.V., 2001. *Three Decades of Spices Research at Kerala Agricultural University*, Directorate of Extension, KAU, Thrissur, India, pp. 23–34.

Onkarappa, S. and Mallik, B., 1998. Distribution and management of *Scirtothrips dorsalis* Hood (Thysanoptera: Thripidae) on rose. *Proc. 1st National Sym. Pest Mgmnt. Hort. Crops*, Bangalore, pp. 165–167.

Parathasarathy, R. and Rabindra, R.J., 2003. Relative susceptibility of three crucifer pests to the nuclear polyhedrosis virus of *Galleria mellonella* (L.), In: *Biological Control of Insect Pests*, Phoenix Publishing House Pvt. Ltd., New Delhi, p. 40–43.

Parvatha Reddy, P., 1983. Reaction of selected citrus rootstocks to the citrus nematode. *Proceedings of International Citrus Symposium*, Bangalore, pp. 379–381.

Parvatha Reddy, P. and Verghese, A., 2002. Integrated pest management in horticultural crops. In: *Resources Management in Plant Protection*, Volume I. Plant Protection Association of India, Hyderabad, pp. 228–254.

Parvatha Reddy, P., Rao, M.S., Mohandas, S. and Nagesh, M., 1995. Integrated management of the citrus nematode, *Tylenchulus semipenetrans* Cobb using VA mycorrhiza, *Glomus aciculatum* (Thaxt.) and oil cakes. *Pest Manage. Hort. Ecosys.*, 1: 37–41.

Patel, R.C., Yadav, D.N. and Saramma, P.U., 1979. Impact of mass release of *Chelonus heliopae* Gupta and *Telenomus remus* Nixon against *Spodoptera litura* (F.). *J. Ent. Res.*, 3: 53–56.

Philip Sridhar, R. and Umapathy, G., 2003. Management of citrus leaf moner, *Phyllocnistis citrella* with botanicals. In: *National Seminar on Bio-management of Insect Pests*, Annamalai University, Tamil Nadu, India, pp.89

Prasad, D., Ray, Deb P. and Kumar, B., 2003. Effect of *Thevetia nerifolia* and *Ricinus communis* seed oil and their esterified products on mortality of *Meloidogyne incognita* and *Rotylenchulus reniformis*. *Ann. Pl. Protec. Sci.*, 11: 341–345.

Prasad, S.K., 2004. Modified teleoscopic adhesive traps for catching alate mustard aphids. *Ann. Pl. Protec. Sci.*, 12: 211–213.

Prasad, S.K. and Phadke, K.G., 1983. Catches of alate *Lipaphis erysimi* Kalt. on adhesive traps and its relationship with aphids on yellow sarson *Brassica campestris* Linn. and weather variables. *J. Ent. Res.*, 7: 25–29.

Raja Reddy, A. and Krishnamohan, K., 2001. Evaluation of Azadirachtin (Fortune Aza) and Monocrotophos in cauliflower. *Pestology* 25: 39–40.

Rajan, Shashi Bhalla and R.K. Khetarpal. 2004. Analysis of pest risk involved in the import of apple fruits into India. *Indian J. Plant Prot.*, 32: 1–10.

Ranga Rao, G.V., Pande, S., Reddy, Y.V.R., Rao, J. Narayana and Rao, V. Rameshwar, 2002. Insect pests and diseases of Agricultural importance and their management in India. In: *Resources Management in Plant Protection*, Volume I. Plant Protection Association of India, Hyderabad, pp. 85–103.

Rao, C.N., Shivankar, V.J. and Singh, Shyam, 2003. Predatory potential and development of green lacewing, *Mallada boninensis* on citrus leaf-miner, *Phyllocnistis citrella* Stainton. *Indian J. Agric. Sci.*, 73: 60–61.

Rao, C.N., Shivankar, V.J. and Singh, Shyam, 2005. Evaluation of bio-agents against citrus leaf miner, *Phyllocnistis citrella* Staint. *J. Appl. Zool. Res.*, 16: 23–25.

Rao, G.N., 1978. *Annual Report* of UPASI Science Department, pp. 82–108.

Selvasundaram, R. and Muraleedharan, N., 1999. Neem formulation for pink mite control in tea. In: *Biopesticides in Insect Pest Management*, (Eds.) S. Ignacimuthu and Alok Sen. Phoenix Publishing Pvt. Ltd., New Delhi, India, pp. 33–37.

Senthil Kumar, C.M. and Regupathy, A., 2006. Stem application of neonicotinoids to suppress *Coccus viridis* (Green) population in Coffee plantation. *Ann. Pl. Protec. Sci.*, 14: 73–75.

Shankar, Uma, Bar, U.K. and Raju, S.V.S., 2005. Impact of intercropping in cauliflower on Diamondback moth, *Plutella xylostella* (L.). *Indian J. Plant Prot.*, 33: 43–47.

Sharma, D.C., Kashyap, N.P. and Chandel, R.S., 2001. Evaluation of Insecticides/Biopesticides and plant products against potato tuber moth, Phthorimaea operculella. In: *National Symposium on Plant Protection Strategies for Sustainable Agri-Horticulture*, Society of Plant Protection Sciences, Public. No. 4, pp. 67.

Sharma, Pratibha and Bhattacharjee, S.K., 2002. Plant protection in ornamental crops. In: *AICRP on Floriculture*, Technical Bulletin No. 19, Indian Council of Agricultural Research, New Delhi, 116 p.

Sharma, Sangita and Bhardwaj, S.P., 2004. Effects of temperature and relative humidity on population build up of European red mite *Panonychus ulmi* on Apple. *Indian J. Ent.*, 66: 215–221.

Shujauddin, 1974. Two new species of genus Toxares Westwood (Aphelinae: Hymenoptera) from India with a note on the genus. *Indian J. Ent.*, 36: 268–274.

Singh, R.P., Yazdani, S.S., Verma, G.D. and Singh, V.N., 1995. Effect of different levels of nitrogen, phosphorus and potash on aphid infestation and yield of mustard. *Indian J. Ent.*, 57: 18–21.

Singh, R.V. and Kumar, Vinod, 2000. Management of *Meloidogyne incognita* infecting *Mentha arvensis* by neem cake and Carbofuran. *Ann. Pl. Protec. Sci.*, 8: 191–193.

Singh, S.P., 2000. Biological control in horticultural crops. *Indian Horticulture*, April–June, p. 54–60.

Singh, S.S., Tiwari, B.C. and Rao, V.K., 2003. Comparative efficacy of some modern insecticides and neem based formulations against cabbage butterfly, *Pieris brassicae*. *Indian J. Ent.*, 65: 264–267.

Singh, S.S., Tiwari, H.C. and Lal, Hira, 2001. Strategies for managing insect–pests of Apple. In: *National Symposium on Plant Protection Strategies for Sustainable Agri-Horticulture*, Society of Plant Protection Science, Public. No. 4, pp. 47.

Soundararajan, R.P., Praneetha, S. and Lakshmanan, V., 2006. Damage of *Helicoverpa armigera* (Noctuidae: Lepidoptera) on *Gerbera jamesonii*. In: *Nat. Sym. Biotec. Insect Pest Manage.*, Entomology Res. Instt., Loyola College, Chennai, pp. 85.

Srinivasan, K. and Krishnamoorthy, P.N., 1991. Indian mustard as a trap crop for management of major lepidopterous pests on cabbage. *Tropical Pest Management*, 37: 26–32.

Srinivasan, K., Murthy, P.N.K. and Krishnamoorthy, P.N., 1991. Mustard plants trap major cabbage pests. *Indian Farming*, 40: 11–12.

Sundaram, R., Sridharan, S. and Thamburaj, S., 1990. Seasonal fluctuation of root-lesion nematode in coffee in Shevaroy hill of Tamil Nadu. *Int. Nematol. Newsl.*, 7: 10–11.

Talekar, N.S., Lee, S.T. and Huang, S.W., 1986. Intercropping and modification of irrigation methods for control of diamondback moth, DBM. In: *Proc. 1st International Workshop*, Asian Vegetable Research and Development Centre, Shauhua, Taiwan, p. 145–151.

Tripathi, D.M., Bisht, R.S. and Mishra, P.N., 2003. Bio-efficacy of some synthetic insecticides and bio-pesticides against black cutworm, *Agrotis ipsilon* infesting potato (*Solanum tuberosum*) in Garhwal Himalaya. *Indian J. Ent.*, 65: 468–473.

Varadarasan, S., 1992. Insect pest management in cardamom. In: *Black Pepper and Cardamom: Problems and Prospects*, (Eds.) Y.R. Sarma, S. Devasahayam and M. Anandaraj. Indian Society for Spices, Calicut, pp. 51–54.

Varadarasan, S., 1995. Biological control of insect pests of cardamom. In: *Biological Control of Social Forests and Plantation Crop Insects*, (Ed.) T.N. Ananthakrishnan. Oxford and IBH Publishing Co. Pvt. Ltd., New Delhi, pp. 109–119.

Vasudha Gautam, Khan, M.A., Kumar, Suneel and Antika, 2003. Ecofriendly approach to control litchi stink bug, *Tessaratoma javanica* Thunberg. In: *National Symposium on Bio-Management of Insect Pests*, Annamalai University, Tamil Nadu, March 29–31, pp. 89.

Verma, A.C. and Khan, M. Nehal, 2004. Potentiality of botanicals for managing *Meloidogyne incognita* in *Ocimum canum*. *Ann. Pl. Protec. Sci.*, 12: 464–465.

Insect Pest and Disease Management
Editor: **Dr. D. Prasad**
Published by: **DAYA PUBLISHING HOUSE (2016)**

Pages ***431–439***

Chapter 29

Entomopathogenic Nematodes: A Potential Biocontrol Agent

D. Prasad
Division of Nematology, I.A.R.I., New Delhi – 110 012

Introduction

Many species of nematodes are endoparasite of insects pest of agricultural importance and broadly termed as entomopathogenic nematodes. Members in families Steinernematidae and Heterorhabditidae have aroused the most interest, and information about them is growing exponentially. They are mutualistically associated with bacteria and are similar in their actions. Interest in entomopathogenic nematodes as biological control agents has heightened in recent years due to the adverse effects of chemicals and poor stability and persistence of other biological control agents. Using entomopathogenic nematodes is an ideal alternative, is economical, has long-term control, without risk to non-target organisms. They potential, agents as they serve as vectors of bacteria, achieve a quick kill of target insect pests, have a broad host range, highly virulent, possess chemoreceptors, can be cultured easily *in vitro,* have a numerical but no functional response, are safe t6 vertebrates, plants and non targets, have been exempt from registration in USA, are easily applied using standard application equipment, are compatible with many chemical pesticides, and are amenable to genetic selection.

Distribution

Since the development of baiting technique in 1975 the EPN have been isolated from all the inhabited continents. Habitat type had little effect on presence of the nematodes In India they were isolated from coastal areas (0 m elevation) to high altitude areas (2000 m). *Steinernema* spp. were isolated from potato cutworms–*Agrotis ipsilon, A. segetum* and *A. biconica* at Shimla (220 m amsl), Jalandhar (240 m amsl) and Rasalpur (1152 m amsl). Annual median temperature and rainfall had little association with the presence of nematodes.

Bacteria

The enterobacteria associated belong to *Xenorhabdus* spp. or *Photorhabdus* spp. They are medium to long motile rods with peritrichous flagellae. They are gram negative facultative anaerobes and form sphaeroplasts. As they are non sporeformers they do not have an environmentally resistant stage and have not been naturally found except in the nematode vectors or insect hosts. Each of the *Steinernema* spp. is associated with a subspecies of *Xenorhabdus* and *Heterorhabditis* spp. with *Photorhabdus luminescens.* The bacteria may be isolated in primary and secondary forms and usually distinguished by the colony morphology and adsorption of dyes. Primary colonies are generally smaller and more convex than secondary, The most reliable indicator is the primary form's adsorption of neutral red from MacConkey agar and bromothymol blue in NBTA. The primary form of *Xenorhabdus* differs biochemically from the secondary, supports greater nematode production for both *in vivo* and *in vitro* cultures and produces antibiotics. Both forms are equally infectious when injected interhaemocoelically into insect hosts. Phase variation is suggested for survival of bacteria as phase I is more susceptible to phage virus. Nematodes do not reproduce in axenic cultures.

Defence Mechanism

Symbiotic bacterial cells found in the gut of infective juveniles are released through anus as soon as juveniles are established in the haemocoel. Entrance of juveniles does not ensure infection as the defence mechanism of host insect comes into operation. Nematodes entering through oral and anal routes may be defaecated. Those reaching the haemocoel are attacked arid melanized at site of penetration. Encapsulation of juveniles does not always provide safety to insect host as the adhering bacteria may establish and kill the insect. A host insect fed on its natural diet is more resistant to nematode attack. Thus the process of infection is the result of interaction between host: vector and the bacterium.

Virulence

Virulence of nematodes is often related to plant variety or plant chemistry of the host fed upon. Nematode progeny production was highest from rootworms (*Diabrotica undecimpunctata*) that had fed on corn, lower for peanut, and lowest for squash rootworms fed on bitter squash than those fed on non bitter squash.

Development

Infective juveniles of *Steinermema* develop into amphimictic females and males and never into hermaphrodites. The *Heterorhabditis* spp. always develop into hermaphrodites in the first generation. Subsequent generations produce amphimictic males and females in both genera. First generation adults are termed as giant adults due to their size. Protandry would be a reproductive strategy for EPN consistent with the cruise foraging strategy, but not with ambushing. Some of the eggs are laid through vulva and later the females become ovoviviparous. They complete 2–3 generations in the insect which depends upon food availability. With the depletion of food in insect cadaver the juveniles develop into the dauer and move out in search of new hosts.

Variations

Different species and strains of pathogens and predators differ in their effectiveness as biological control agents of insects. These differences have been well documented for predatory insects, parasitoids and a number of entomopathogens. The importance of such interspecific and intraspecific variation has been recognized and demonstrated for insect parasitic nematodes. Genotypic characterization of isolates will allow the assessment of genetic diversity within the species and populations and form a basis for the collection, maintenance and selection of insect parasitic nematodes. Diversity within has been analyzed morphologically and with isozyme analysis.

Bioefficiency

Among the factors limiting the efficacy of nematodes is the need for timing of application to coincide with the phenology of susceptible stages.

In India the work on steinernematids started in the sixties. Work on mortality of insects is documented in lab studies. Work on heterorhabditids is of recent origin (Table 29.1).

Table 29.1

Crop	*Pest*	*Nematode Species*	*Reference*
Rice	*Pseudoletia separata*	*Steinernema carpocapsae*	
	Cirphis compta	DD-136	Israel *et al.*, 1969
	Cnaphalocrosis medinalis		
	Chilo suppressalis		
	Scirpophaga (= *Tryporyza*) *incertulas*		
	Orseolia oryzae		Rao *et al.*, 1971
	Cnaphalocrosis medinalis		Srinivas and Prasad, 1991
Maize	*Chilo partellus*	*Neoaplectana* sp.	Mathur *et al.*, 1966
Tobacco	*Spodoptera litura*	DD-136	Gupta *et al.*, 1987
	Spodoptera litura	*Steinernema fettiae*	Narayanan and Gopalakrishnan, 1987
Groundnut	*Amsacta albistriga*	DD-136	
		Heterorhabditis bacteriophora	Bhaskaran *et al.*, 1994
Potato	*Agrotis ipsilon*,		Singh, 1977, 1982,1993
	A. biconica		
	A. segetum		
	Anomala sp.	DD-136	Rajeswari and Sundarababu *et al.*, 1984
	Leucinodes orbonalis		
	Athalia proxima		Singh and Bardhan 1974
	Papilio aristolochiate		
	Ergolis merione		
	Draterius sp.		
	Orthacris simulans		Sivakumar *et al.* 1989

The fitness of a parasite can be adversely affected by increasing population density within the host. Density dependent effects include mortality, reduction in adult size, increased generation time, reduced fecundity, failure to establish, delayed development, reduced oviposition and change in sex ratio. Effects of high density appear to result from competition for limited nutrients within the host.

Nematode infectivity consists of a sequence of events initiated with host location, host attachment, host penetration, bacterial release, bacterial proliferation and host death. Infectivity is influenced by several biotic and abiotic factors like insect, nematode species, production method, bacterial strain, temperature and bioassay method. Filter paper bioassays are commonly used to screen nematodes against target insects. However the ability to kill insects when placed in close contact provides very little information about their field performance. This led to development of a range of sand and soil based assays. Due to the differences in foraging strategies the true infectivity potential of all nematode species cannot be assessed by a single bioassay. Some species such as *S. carpocapsae* and *S. scapterisci,* use an ambush strategy to parasitize hosts, whereas *S. glaseri* and *H. bacteriophora* adopt a cruising approach. Nematode species with apparently similar foraging strategies can also be influenced differently by the same substrate. Differences in infectivity of *in vitro* and *in vivo* produced *S. carpocapsae* have not been recorded, but in vitro produced *S. scapterisci* caused has higher larval mortality than those reard *in vivo.*

Metallic ions Mg and Mn have an effect on the infectivity of the IJs. Under laboratory conditions, the treatment of *S. carpocapsae* and *H. bacteriophora* with Mg or Mn (II) ions produced an increase in their pathogenicity against *Sitona lineatus* weevils, pupae and larvae. This simple chemical method was useful for increasing the virulence of these natural biopesticides. The effect was more consistent for *S. carpocapsae* than for *H. bacteriophora* in which mortality reached 100 per cent. Only pupal stages showed some resistance to the invasion of nematodes pre treated with Mg ions, their mortality on the fifth day of experiment being 91.7 per cent. This was approximate 40 per cent higher than in control experiment.

Host Finding and Dispersal

The host finding ability is an integrated combination of behaviours consisting of active dispersal, nictation and chemotaxis. Substances such as organic solvent washes off the surface of *Galleria mellonella* larvae, nitrogen compounds from faecal pellets, plant root metabolites, bacteria and carbon dioxide. Heterorhabditids have a better host finding ability than the steinernematids. Active dispersal by infective juveniles is minimal. Some of the passive dispersal methods may be water, wind and human activity, although dispersal by these agents especially wind is not adequately documented. Movement of *S. riobrave* in the field (soil type: sandy clay loam) showed that they can move on an average of 4 cm/day.

Effect on Non-target Organisms

They have been found safe against several non target organisms despite their wide host range. Earthworms were safe to nematode application, however injured earthworms are attacked and nematodes reproduced. Several molluscs, arthropods and diplopods are infected and killed. Fishes, birds, reptiles and mammals are not attacked. Frog tadpoles are susceptible. At high dosages, garden millipedes are affected.

Effect on Plant Parasitic Nematodes and Other Arthropods

Plant parasitic nematodes have been found *to* be controlled by application of entomopathogenic nematodes to soil. Infection of *Heterorhabditis bacteriophora, Steinernema rara,* and *S. glaseri,* in head lice, *Pediculus humanus capitis* nymphs and adults have been recorded. Eggs were not affected.

Isolation

The EPN are isolated from soil or directly collected from the infected hosts, either naturally occurring or purposely deployed 'trap' insects. Soil samples can be collected from different habitats, altitudes and rainfall by random sampling technique. The greater wax moth (*Galleria mellonella*) is quite susceptible to steinernematids and heterorhabditids and makes an excellent trap insect for baiting as it is easily reared and widely available.

Infected hosts may often be recognised by their appearance. They are often flaccid and change colour to orange, yellow or brown (steinernematids) or brownish red to brick red that faintly luminesce in the dark (heterorhabditids). Nematodes found in a dead insect may be parasites or saprophytes which can be proved by Koch's postulates.

Extraction of Entomopathogenic Nematodes

It is done by Modified White trap method prepared in petri dish (110 × 25 mm) with an inverted watch glass (52 mm) to serve as a stage. A single filter paper (Whatman # 1, diameter 90 mm) is used to absorb and provide moisture for the nematodes. Infected-larvae are transferred on to the stage of the white trap and incubated at 23°C. When the nematodes are observed in the trap they are extracted in a beaker and cleaned two or three times with distilled water by sedimentation, followed by decantation. The extracted nematodes are stored at 8°C (steinernematids) and 24°C (heterorhabditids) in distilled water for future use.

Maintenance of EPN Culture

Laboratory culture of the isolates of entomopathogenic nematodes can be maintained using final instar larvae of *G. mellonella*. Nematodes are multiplied using the methods of Dutky *et al.* (1964). The nematode suspension is applied to a single layer of filter paper (Whatman # 1) in a petridish. Last instar larvae of *G. mellonella* are conditioned by dropping in hot water (50°C) for 15 seconds to prevent coccoon formation and webbing. The conditioned larvae are introduced into the petridish which is sealed and incubated at 24°C.

Identification

Identification is based on the characters of first and second generation males and females and infective juveniles. Identification upto genus level are done by studying the host colour, nematode behaviour, morphometrics of IJs and adult males while morphometrics of IJs and adult males are taken into consideration for identification upto species level. Cross breeding techniques can be used for species identification. For strain identification within the same species DNA sequence analysis is done (Curran, 1990).

Mounting of Nematodes for Morphometrics

Males and females are obtained by dissecting infected larvae in Ringer's solution 3 or 5 days after infection for the first generation and 6 or 7 days for the second generation. The infective juveniles are collected as it emerged from the cadavers. The different stages of the nematode populations are first fixed in TAF for 24 hrs, and then processed to pure glycerin by slow processing method using Cobb's slow processing method. Glycerin mounts of the processed specimens are prepared and observed under compound microscope.

Screening

Testing of entomopathogenic nematode isolates for bioefficiency is done in the lab by following the filter paper assay, sand or soil column assay and penetration rate assay. Pesticide tolerant strains can be evolved by exposing the IJs of nematodes to different concentrations of pesticides in aqueous solution and soil. Tolerance to desiccation and tolerance to storage conditions are evaluated by subjecting the IJs to drying for certain time interval and storing at different temperatures and substrata respectively.

In vivo Production

Steinernematids and heterorhabditids infect and reproduce on a broad spectrum of insects. They may be reared *in vivo* in the laboratory. *Galleria mellonella* is the most used host allover the world.

Upto 2,00,000 *Steinernema feltiae* and 3,50,000 *Heterorhabditis bacteriophora* infectives have been harvested from a single last instar larva. However, average production ranges 30–50 thousands per host. However other test insects like *Spodoptera littoralis* and *Corcyra cephalonica* are also utilized. Karunakar *et al.* (1992) reared *S. feltiae, S. glaseri* and *H. bacteriophora* on *Chilo sacchariphagus indicus* and *Scirpophaga excerptalis.* Yields *of* 1,27,000 (*S. feltiae*), 3,70,000 (*S. glaseri*) and 2,10,000 (*H. indicus*) IJs from larvae *of Chilo sacchariphagus indicus* were obtained when introduced @ 20 IJs/5th instar larva.

In vitro Production

In 1940–42, Glaser developed an axenic culture process for *Steinernema glaseri* and *S. carpocapsae.* In 1974, beneficial effect of haemoglobin and cholesterol in a yeast and peptone-based medium was reported. Vanfleteren (1976) working from the results of Rothstein, developed an inexpensive medium for *Caenorhabditis elegans* and for the first time, reported high yield of nematodes from 10 lit. fermentors. Aeration was by bubbling and suspension was assisted by slow stirring. The symbiotic bacterium grows on a number of proteins (animal and plant) and the nematode fed on the bacterium along with other nutrients in the media. Production on a commercial scale had been accomplished by using inexpensive protein and sterol sources. Bedding (1984) pioneered the development of an inexpensive monoxenic, 3 dimensional culture matrix using animal byproducts. This method of nematode production is applicable for cottage industries or for scale-up by biotechnology companies. Steinernematids and heterorhabditids also had been produced monoxenically in large, liquid fermentors using a combination of plant and animal proteins subsequently. Synthetic media used *for* mass production are presented in Table 29.2.

Table 29.2: Synthetic Media for Mass Rearing of Entomopathogenic Nematodes

Synthetic Medium	*Nematode*	*Incubation*	*Temp.*	*Nematodes*
Species	Period		harvested	
Wheat bran	*S. feltiae*	3 weeks	25°C	10^4
Wheat bran + salad oil	*S. feltiae*	3 weeks	25°C	10^7
Dog food agar medium	*S. feltiae*	–	–	10^5/g of medium
Beef extract, peptone, corn meal, water on sponge	*S. feltiae*	–	–	–
Nutrient broth, yeast extract, vegetable oil, flour coated on sponge	*H. heliothidis*	4 Weeks	25°C	10×10^6/250 ml flask
Soyapeptone 3% + yeast extract 3% + chick embryo extract 10% medium	*S. glaseri*	–	–	10,051/week for 93 days
Kidney/fat homogenate	*S. feltiae*	2–3 weeks	20–30°C	38×10^6/30 flask
	S. bibionis			29×10^6/73 flask
	S. glaseri			8×10^6/11 flask
	H. bacteriophora			36×10^6/10 flask
	H. heliothidis (NZ)			32×10^6/15 flask
Yeast extract, soyflour, egg and lard	*S. carpocapsae*	20 days	19°C	38×10^6/flask

In the past, steinernematids and heterorhabditids have been cultured on a variety of substances: Potato mash, ground veal pulp, peptone-glucose agar and pork kidney, homogenised animal tissue, dog food and chicken offal homogenate.

Liquid Culture

Formulation of nematode into a stable product has played a significant role in commercialization. Active nematodes are immobilised to prevent depletion of their lipid and glycogen reserves. Immobilization has been accomplished by maintenance in aqueous suspension at low temperature (5–15°C). Steinernematids can be stored at 4–10°C for 6–12 months without much loss of activity whereas heterorhabditids do not store well and 2–4 months of storage at 4–10°C is considered good. A major constraint on commercialisation of these nematodes is their poor storage capacity. They can be stored under refrigeration in oxygenated water for up to 5 years. But this methodology is not commercially feasible. Nematode metabolism is temperature driven and warm temperatures (20–30°C) increase metabolic activities, reducing nematode viability. Subsequently, nematodes formulated in various carriers such as alginate, clay, vermiculite and gel forming polyacrylamides can be stored for atleast 6 months under refrigeration and atleast 3 months at room temperature. Further advancements have been made with the development of a flowable concentrate. This technological advancement, if commercially successful, will revolutionize nematode applications and eliminate many of the constraints associated with current formulation, namely the dissolution of the carrier or separation of the nematodes from the carriers.

Thus, mass production of entomopathogenic nematodes has been evolved from the first large-scale *in vitro* solid media production to 3D solid media *in vitro* process by Bedding and to the *in vitro* liquid fermentation production method by Friedman (1990). Currently, commercial nematodes are produced monoxenically using the solid media process or the liquid-fermentation method. The solid media process has successfully produced pathogenic steinernematids and heterorhabditids, but the high labour costs limit economies of scale. This technology is most adaptable for countries where labour costs are minimal. The liquid fermentation process, highly efficient for several steinernematid species but not for heterorhabditids is economical for industrialized countries. The latter process is' highly efficient for several steinernematid species, and although heterorhabditids can also be produced in liquid fermentors, their quality, thus far, is poor in comparison to steinernematids.

The cost of mass production using these methods is a major constraint for nematode commercialization. Adverse effect of increased medium depth and poorly mixed media indicated that oxygen transfer was a potential limiting factors in large volume cultures.

Formulations and Storage

Formulation of nematodes is important and can affect handling, application, persistence and stability in storage. Most common formulations are made in alginate clays, polyacrylamide gels, clay, vermiculite, activated charcoal *etc.* Formulation stability is also achieved by partially desiccating and immobilizing in specific carriers.

Compared to many biocontrol agents nematodes as biopesticides have a good shelf life. Most can be stored under refrigeration in aqueous suspension under adequate supply of oxygen. Temperature requirement depends on species. General range for steinemematids is 5–10°C and 10–15°C for heterorhabditids. The introduction of commercial nematode formulations has accelerated in recent years, but new formulation technology that produces stable, effective and consistent products would expand markets even further. Spray application using a nematode suspension in water is straight forward, inexpensive and often effective. Novel alginate capsules containing nematodes, wheat bran bait pellets and alfalfa wheat bait pellets have been developed.

Encapsulation of nematodes with seed has potential to protect roots from insect attack Nematodes escaped from the capsule within 7 days, as the seed germinated it have an opening for the nematodes to escape. Pesta, granular product with entrapped nematodes based on a cohesive dough made of wheat flour, fillers and additives. Effect of prior storage conditions of the infectivity was studied and it was found that storing the dciuer larvae in tap water at 9, 20 or 25°C they became progressively more infective. At higher storage temps of 20°C and 25°C infectivity reached a peak after 2–3 weeks. The possibility that the agitation resulting from adequate aeration of large volume cultures would damage the nematodes or prevent them from mating. Nematode reproduction occurred in both rotary shaker and vigorously air sponged systems and the nematodes appeared normal.

Delivery Systems: Product Development, Economics

Biological pesticide must be able to compete in relative terms with chemical pesticide in field efficacy, cost and application methods.

Quality of nematode products is determined by their availability, usability, efficacy and price. The development of cost effective mass production technology has led to the availability of nematode products comparable with standard insecticides in market. Quality control strategy has to ensure preservation of high nematode viability and virulence during large-scale production and formulation. Factors affecting quality include provision of essential nutrients and appropriate bacteria, delivery of oxygen, production and storage temperature and contamination control.

Application of nematodes is mostly targeted, to the soil and against cryptic habitats. Inundative and inoculative approaches have been attempted. In soil application one or two pre and post application irrigations have been found helpful to provide enough moisture.

Timing soil applications with the life cycle of target insect is a key factor. Thus *S. riobrave* was effective in suppressing com earworm under field conditions of high temperature with irrigation. Presently they are being applied through conventional equipments such as high volume sprayers, drip irrigation, food baits or through sound trap. For foliar application of EPN, antidesiccants have been used successfully to retard evaporation of the nematode suspention on foliage and thus to reduce desiccaton of nematodes. Glycerine 10 per cent has been the most effective adjuvant for increasing survival and activity of nematodes on foliage. However with the high cost of glycerine and potential for phytotoxity at high temperatures, search for more suitable adjuvants continued. In laboratory TX 7719, Rodspray oil, and Nufilm P provided the best antidesiccant activity and in greenhouse, TX 7719 and Rodspray oil were more effective. In field trials; the combination of TX 7719 + Blankophor BBH increased nematode persistence on watercess leaves and efficacy against *P. xylostella* significantly. Addition of phagostimulants enhanced the effectiveness of EPN against *Spodoptera litura* feeding on sunflower head in laboratory and field tests.

Genetic Improvement

EPN are extremely promising biological control agents but certain limitations like susceptibility to environment extremes, host-finding behaviour hamper the complete utilisation of EPN potential. One solution to this problem is to develop genetically improved strains of EPN with enhanced traits for biological control. Strategies for genetic manipulation include artificial selection, hybridization and genetic transformation.

Selection and hybridization have been employed in developing genetically improved strains. These approaches may fail if the available nematode population do not possess sufficient genetic diversity for the beneficial traits. A lack of genetic diversity may be overcome by surveying natural population for the desired traits. Selection is employed to find genetically superior nematode species. Steinernematid species with increased host finding ability has been successfully selected. A heat tolerant strain of *Heterorhabditis bacteriophora* designated IS 5 was discovered in Negev desert, Israel in 1996 and was found superior to HP 88, a commercially available strain of *H. bacteriophora.* Stability of the heat tolerance trait and fitness of *H. bacteriophora* IS 5 was tested in the laboratory after multiple passages through the last instar *G. mellonella* larvae.

Hybridization is an alternative approach because of its relative simplicity and its speed in transforming beneficial genes. Hybridization is used for genetic improvement of heat tolerance in *H. bacteriophora*.

To design EPN with specific traits requires the ability to introduce cloned genes of known functions. Genetic transformation methods are required to produce transgenic nematodes carrying the DNA of interest.

Genetic selection can also be used to enhance the resistance of entomopathogenic nematodes to nematicides. Genetic selection of *Heterorhabditis bacteriophora* strain HP 88 to nematicides resulted in an 8–9 fold increase in resistance to Phenamiphos and Avermictin. The enhanced resistance was stable and continued even after the selection was relaxed.

Insect Pest and Disease Management *Pages* ***440–447***
Editor: **Dr. D. Prasad**
Published by: **DAYA PUBLISHING HOUSE (2016)**

Chapter 30

Eco-friendly Management of Plant Parasitic Nematodes in Vegetable Crops

V.K. Singh
Division of Plant Pathology,
Sher-e-Kashmir University of Agricultural Sciences and Technology,
DLRSS, Dhiansar Jammu – 181 133, J&K

Introduction

Vegetable crops are prone to several biotic and abiotic stresses during their growth period. Plant nematodes constitute one of the major causes of crop losses all over the world. Sasser and Freck man (1987) have indicated annual crop losses due to plant parasitic nematodes on world wide basis exceed $100 billion. Root-knot nematodes are major pests of vegetables belonging to leguminous cruciferous, solanaceous and several other root and bulb crops like onion, garlic, carrot, radish etc. The yields of Okra, tomato and brinjal 27.3 per cent losses respectively due to *Meloidogyne incognita* infestation @ 3-4 juveniles/g soil under field conditions (Bhatti and Jain, 1977). Even 50 per cent reduction in potato yields caused by *Globodera rostochiensis* can not be predicted by inspection of the upper plant. Also, a 50 per cent cabbage field reduction caused by sugarbeet cyst nematode, *Heterodera schachtii* in microplots cannot be assessed accurately by inspection alone. A large number of plant parasitic nematodes have been found associated with vegetable in India but pathogenicity has not been proved for all. Most of the commonly occurring vegetables nematodes belong to genera-*Meloidogyne, Globodera, Heterodera, Rotylenchulus, Pratylenchus, Tylenchorhynchus, Ditylenchus, Radopholus, Belonolaimus, Dolichodorus, Hemicydiophora, Pratylenchus, Paratrichodorus, Longidorus, Nacobbus, Helicotylenchus, Rotylenchus, Trichodorus, Xiphinema etc.*

Important Nematodes on Vegetables

Root-knot Nematode (*Meloidogyne* spp.)

Root-knot nematode is most important pathogen of vegetable crops. Almost all the vegetable crops have been found to be susceptible to this nematode to a varying degree. The four major species are *Meloidogyne incognita, M. javanica, M. arenaria and M. hapla*. These four species account for more than 95 per cent of the root-knot nematode population world over (Sasser and Carter, 1985). The above ground symptoms of the infected plants include yellowing of leaves, stunting of shoot and reduced size of various shoot parts symptoms of the disease appear on the underground parts of the plants. Infected roots swell at the point of invasion and develop into the typical root-knot galls that are two to several times as large in diameter as the healthy root. The root system is galled. Several infection take place along the same root, and the developing galls give the root a rough, clubbed appearance. Usually, infected roots remain smaller, show various stages of necrosis, and develop rotting, particularly late in the season. The male and female root-knot nematodes are easily distinguishable morphologically. The males are wormlike and about 1.2 to 1.5 µm long by 30 to 36 µm in diameter. The females are pear shaped and about 0.40 to 1.30 µm long by 0.27 to 0.75 µm wide. A life cycle is completed in 25 days at 27°C, but it takes longer at lower or higher temperatures.

Cyst Nematodes (*Heterodera* spp. and *Globodera* spp.)

Various species of *Heterodera* and *Globodera* are popularly known as the cyst nematodes. The round cyst nematode, known as the golden nematode, is particularly severe on potato. The name golden nematode of potato was proposed in 1944 by chitwood. The above ground symptoms resemble those associated with root damage and include stunting of shoot, yellowing of foliage etc. The severity depends upon density of nematodes attacking roots. The mature female bodies are found attached to roots by their head end embedded almost in the stele. The site of feeding is modified into a syncitium similar to that found in case of *Meloidogyne*. On the surface of infected roots, white to brown bodies of females can be discerned with naked eyes. The intensity of body colour depends upon the maturity stage of the young female or cyst. In case of cyst nematode matrix with eggs may also be found attached to the posterior region of the female body. The male is wormlike, about 1.3 mm long by 30 to 40 µm in diameter. Fully developed females are lemon shaped, 0.6 to 0.8 mm in length and 0.3 to 0.5 mm in diameter. Approximately 21 to 24 days is required for the completion of life cycle of this nematode. The life cycle process of *Globodera* is similar to the other cyst farming species, with one life being completed in 5-7 weeks. However, those which do not emerge can remain in viable condition in side the cysts even for 7-8 years.

Lesion Nematode (*Pratylenchus* spp.)

Lesion nematodes occur in all parts of the world. Amongst vegetables, tomato is an important host for this nematode. The plants show chlorosis, stunting and general lack of vigour resulting into wilt. The plant form patches or zones in the field. The roots show necrosis and lesions which become ideal for infection of other microorganisms. Both male and female of this nematodes are wormlike, 0.4 to 0.7 mm long and 20 to 25 µm in diameter. They are migratory, endo parasitic nematodes. The life cycle of the various species of *Pratylenchus* is completed within 45 to 65 days. Generally, temperature and host are the most important factors which have profound influence on the build up of the nematode populations.

Reniform Nematode (*Rotyenchulus reniformis*)

Reniform nematode is pest of a large number of vegetables. In India, it is considered as pest of great significance after root-knot nematode. The nematode is a semi-endoparasite and remains attached to roots. Infected plants shows stunting in growth with reduced and discoloured root system. Damage during pre and post emergence of seedlings leads to reduction in germination and crop stand. The life cycle is generally completed in about a month. Soil PH is also an important factor affecting reproduction of the nematode. Soil moisture and temperature have profound influence on infectivity, penetration and the biology of the nematode.

Stem and Bulb Nematode (*Ditylenchus* spp.)

This nematode commonly attacks onion, garlic, potato, pea and carrot etc. The symptoms of injury differ from different plants and different parts of the same plant. Infected seedlings become twisted, enlarged and deformed, leading to death of the plant. The plants become stunted. The nematode feeds on stem leaves and bulbs and is rarely found in soil. The nematode is 1.0 to 1.3 mm long and about 30 µm in diameter. The total duration of life cycle ranges from 19-23 days at 15ºC with four moults and four Juveniles stages, the first moult being within the egg (Yuksel 1960).

Cultural Control

The cultural practices include crop rotation, fallowing, flooding, sanitation ploughing during summer season, mulching, inorganic fertilizer, spacing of plants in the field, time of sowing, resistant varieties etc.

In crop rotation for control of nematodes, susceptible (host) crops are rotated with non host crops. Flooding is reported to reduce incidence considerably. The period of flooding appears to vary with several factors such as kind of soil, season etc. Warm conditions are said to reduce the period of time required for control. Juveniles (larvae) are more easily killed by flooding than eggs. The period of flooding needs to be worked out for each condition. Fallow periods in cropping sequences can also reduce nematode populations. The summer ploughing 2-3 times during hottest period of the year help to expose nematodes to the drying action of sun and wind and reduce the population. Haque and Prasad (1982) reported drastic reduction of plant parasitic nematodes and significantly increased the yield with three deep ploughings. Sanitation terms covers a wide range of cultural practices, including weed control, crop residue destruction and disinfestation of farm equipment before moving it from heavily infested fields to uninfested fields. In monocultures, eliminating the weed hosts can be important in reducing the populations of plant parasitic nematodes. Manipulation of sowing dates, although not very effective for nematode control, there are several reports where early planting of potato, sugarbeet and tomato has been found to reduce the infestation of *Globodera rostochiensis, Heterodera schachtii* and *Meloidogyne incognita.* The nematodes can be also controlled by mulching with different available waste organic materials such as straw, grass clipping, weeds, or plastic mulches around the crops. The solar heat can be trapped and conserved using transparent polythene mulch. In summer soils polarization effectively diseases caused by nematodes (Sharma and Nene 1985, Sharma 1985a) control as well as many other soil borne diseases. It induces microbial processes that can contribute to the control of pathogens (Katan, 1981). Prasad and Yaduroju (1993) indicated that 30 days solarization is enough to achieve 95 per cent control of nematodes. Prasad *et al.* (1995) found about 90 per cent kill of root-knot and reinform nematodes when solarised for 30 days during May-June months.

The addition of inorganic fertilizers alone without organic manure usually increase the nematode population and disease intensity. NH_4-N reduces the disease incidence while NO_3-N may increase the same. Particular forms as well as dose and proportion of NPK may also reduce or increase the incidence of disease. Use of tolerant/resistant varieties is most practical approach for the management of nematode diseases. Crop cultivars resistant to phytonematodes can be the most useful and cheapest means of nematode control for the small-scale farmers.

Biological Control

Nematodes have many natural enemies including fungi, bacteria and predacious nematodes. Certain fungi capture and kill nematodes in the soil. *Arthrobotrys* sp., *Dactylaria* sp., *Dactylella* sp., *Catenaria* spp., and *Trichothecium* spp., are the genera most commonly represented. The biological control of root knot on various vegetable crops under green house conditions by using several fungi and bacteria have been reported by Pandey and Trivedi (1992); Zucker man *et al.* (1993). Bandyopadhyay *et al.* (2001), Singh *et al.* (2001) Some fungi capture nematode by adhesion, but many employ specialized devices that include networks of adhesive branches, stalked adhesive knobs, non constricting rings and constricting rings. The surface of the nematode is penetrated and the fungus hyphae grow throughout the nematode body, digesting and absorbing its contents. Under favourable conditions, large numbers of nematodes may be captured, and killed especially by those fungi that form adhesive net works or hyphal loops.

A bacterial parasite of nematodes *Pasteuria penetrans*, has received much attention and research effort in recent years, *P. penetrans* is probably the most specific obligate parasite of nematodes, with a life cycle remarkably well adapted to parasitism of certain phytonematodes. Applications of *P. penetrans* in several green house and microplot tests have consistently suppressed the nematode induced root galling and egg production with consequent enhanced plant growth Walia (1994); Walia and Dalal (1994); Gogoi and Neog, 2005), and Anil Kumar *et al.* (2006).

Nematicidal plants with roots containing nematicidal substances have been investigated. These toxic substances reduce the population level of some nematode species. African marigolds (*Tagetes* spp.), asparagus, crotalaria, mustard and several cruciferous plants have been reported to produce toxic substances.

Soil amendments with green manure, compost, oil cakes of neem, mahua, mustard, groundnut, cotton, linseed, karanj and saw dust etc. have been found to reduce nematode populations. Neem, karanj or groundnut cakes incorporated into soil at the rate of 1,500 kg/ha gives good control of plant parasitic nematodes and could be practised wherever possible. Apart from encouraging the multiplication of natural enemies like nematode trapping fungi, the decomposition products of these organic amendments are toxic to nematodes. Use of organic manures is of great value since the decomposition products and promotion of natural enemies decrease nematode populations (Singh and Sitaramaiah (1973), Singh and Singh (1991), Singh and Singh (1992), Alam (1986), Alam (1990), Goswami and Swarup (1971), Haseeb *et al.* (1984 a), Mankau (1963), Mishra and Prasad (1974), Singh and Singh (2001). The amount of oil cakes or any organic matter to be incorporated depends on various factors like, soil type and texture, crop to be planted, the predominant nematode fauna of the soil and the amount of soil moisture present in the soil. In trap crops when grown in infested soils, the nematodes penetrate into the root system and start multiplying. Before the nematodes complete its life cycle the plants are uprooted and destroyed.

Chemical Control

Nematicides are used to protect highly susceptible varieties of vegetable from nematode attack. The most important consideration in the control of nematode by chemical is the cost/return ratio. Chemicals thus are recommended for highly valuable or cash crops. The older nematicides are mostly fumigants which are applied in soil. These are seldom recommended because of their hazardous nature and high toxicity to non target organisms. Recently, large number of non fumigant and systemic nematicides are available which are safe on plant. DBCP (Nemagon) has been found very effective for standing crops against root-knot nematodes but its use has been suspended due to adverse effect on human beings. Carbofuran 3G (Furadon), Phorate 10G' (Thimet), Fenamiphos (Nemacur), Fensulphothion (Dasanit) have been recommended for a variety of crops. The dose and method of application would vary with crop. Chemicals and Phytoalexins that have implications in the nematode control (Prasad and Swarup 1986), synthetic pheromones to disrupt nematode reproduction Bone (1987), and chemicals that directly interfere with nematode chemoreceptors Jansson (1987) should be exploited for nematode population management. Prasad (1990) studied the control by seed pelleted with carbofuran (40 flowable) and aldicarb sulfone (75 wp) at the @ 3 and 6 per cent W/w. The chemical control of root knot nematode on tomato under green house conditions by using various chemicals has been reported by Singh and Singh (1991), Singh and Singh (1998).

Regulatory Control

Numerous attempts have been made to prevent the introduction of nematodes into countries or provinces by means of quarantine. Quarantines are established by legislative action in parliament, etc., and usually give quarantine authorities power to make and enforce regulations to accomplish the purpose. Such regulations usually prohibit bringing infected plants into protected areas where similar crops might become infected.

Physical Control

Soil treatment with heat for a short period of time kills most of the nematodes. Steam treatment or hot water dips (at 45-55°C for 5-120 min.) and heat treatment of plant materials particularly the bulbs infected with endoparasitic nematodes are suitable methods but are applicable only in capital-intensive crops as in the green houses or for seedlings in the nurseries. Electricity, radiation and X-rays or Y-rays and ultrasonics like heat, these measures can only be recommended as control measure in capital-intensive crops or in certified seed production programmes. Potato tubers could be freed from *G. rostochiensis* by ionizing irradiation. Some nematologists have reported control of nematodes by osmotic pressure. But this is an impractical and uneconomical proposition. These aspects were reviewed by Jenkins (1960), Jonesh (1978), Mass (1987), Southey (1965) and Southey (1978) also reported the role of physical factors in the soil for regulating nematode population.

Integration of Control Measures

Utilization of the best combination of available management strategies for the pest complex at hand (nematodes, insect pests, disease organisms, weeds etc.) constitutes an integrated crop protection system. Resistant cultivars, crop rotation, pesticides, and sanitary and cultural practices can all be employed to the best possible advantage. An integrated management strategy prevents the excessive build up any single nematode, insect, or disease population and minimizes the development of pest resistance to any single tactic. Integrated pest management systems require flexibility and depend upon the specific pest problem and locally available management options. A fixed set of recommendations may keep a pest complex in check for a limited period of time, but as the pest population shifts, recommendations will have to change also. Therefore, system development takes into account many factors including the species and race of pests present, the availability of resistant host plants, the longevity of the pest, and the crops, cropping systems, and climate of the geographical region. The end result is a management strategy tailored to fit the unique circumstances of each pest situation. AICRP (1992) proposed an integrated approach by combining cultural methods with organic amendments, chemicals and inorganic fertilizers to reduce the population of pathogenic nematodes and enhance the yield of vegetable crops. Application of urea @23.8 kg/ha and neem cake @ 200 kg/ ha in combination with carbofuran @ 1 kg a.i/ha in root-knot infested field gave an enhanced yield of tomato (32.2-68 per cent), brinjal (36.7 per cent) and reduced the galls by 0.2 to 0.9 times over untreated crop (AICRP,1992). The control of plant parasitic nematodes by combined application on various vegetable crops have reported by Jain and Bhatti (1985), Singh and Singh (1991), Reddy (1992), Singh (2006) and Singh (2006).

References

AICRP., 1992) Achievements (1977–87). All India coordinated research project on nematode pests and their control. pp 16–26

Alam, M.M., 1986. Possible use of weeds and soil amendment for the management of root-knot and stunt nematodes attacking egg plant. *Agricultural Wastes,* 16: 97–102.

Alam, M.M., 1990. Control of plant parasitic nematodes with organic amendments and nematicides in nurseries of annual plants. *J. Bangladesh Acad. Sci.,* 14: 107–113.

Bandyopadhyay, P., Kumar, D., Singh, V.K. and Singh, K.P., 2001. Eco-friendly management of root-knot nematode of tomato by *Arthrobotrys oligospora* and *Dactylaria brochopaga*. *Indian J. Nematol.,* 31(2): 153–156.

Bhatti, D.S and Jain, R.K., 1977. Estimation of loss in okra, tomato and brinjal yield due to *Meloidogyne incognita*. *Indian J. Nematol.,* 7(1): 37–47.

Chitwood, B.E., 1944. The golden nematode of potatoes. *U.S. Dept. Agr. Circ.,* 875.

Goswami, B.K. and Swarup, G., 1971. Effect of oil cake amended soil on the growth of tomato and root-knot nematode population. Indian Phytopath. 24: 491–494.

Gogoi, B.B and Neog, PP., 2005. Efficacy of *Pasteuria penetrans* against *Meloidogyne incognita* on green gram under field conditions. *Indian J. Nematol.,* 35: 80–81.

Haseeb, A., Alam, M.M. and Khan, A.M., 1984a. control of plant parasitic nematodes with chopped plant leaves. *Indian J. Plant Pathol,* 2: 180–181.

Jain, R.K. and Bhatti, D.S. 1985. In: *Nematol Network Newsl.,* 2: 17–18.

Jenkins, W.R., 1960. Control of nematodes by physical methods. In: (Eds.) *Nematology,* J.N. Sasser and W.R. Jenkins. Chapel Hill, North Carolina.

Jones, F.G.W. and Krempston, R.A., 1978. In: *Plant Nematology,* (Ed.) J.F. Soc. The HMS, London, pp. 440.

Katan, J., 1981. Solar heatings (solarization) of soil for control of soil borne pests. *Annual Review of Phytopathology,* 19: 211–236.

Kumar, Anil, Walia, R.K. and Kapoor, A., 2006. Efficacy of *pasteuria penetrans* as seed treatment in controlling *Meloidogyne Javanica* on three succeeding crops. *Journal of Biological Control* (accepted).

Mankau, R., 1963. Effect of organic soil amendments on nematode population. *Phytopathology*, 53: 881–882.

Mass, P.W., 1987. Physical methods and quarantine. In: *Principles and Practices on Nematode Control of Crops*. Academic press, Australia, p. 265–279.

Mishra, S.D. and Prasad, S.K., 1974. Effect of soil amendments on nematodes on the crop yields. *Indian J. Nematol.*, 4: 1–19.

Pandey, R. and Trivedi, P.C., 1992. *Indian Phytopath.*, 45(1): 134–135.

Prasad, D., 1990. Control of root-knot nematode *Meloidogyne arenaria* on groundnut by chemical seed pelleting. *Current Nematol.*, 1: 31–34.

Prasad, D. and Yaduraju, N.T., 1993. Solarization of field soil for effective management of nematodes and weeds. In: *Nat. Conf. Ecofriendly Approaches Mgmt. Pests, Dis. Ind. Effluents*, December 20–22, 67 (Abstr.)

Rao, M.S. and Reddy, P.P., 1992. Prospects of management of root-knot nematode on tomato through the integration of biocontrol agent and botanicals. In: *Sem. on Current Trends in Diagnosis and Management of Olant Dis.*, IIHR, Bangalore, pp. 10.

Sasser, J.N. and Carter, C.C., 1985. *An Advanced Treatise on Meloidogyne. Vol 1: Biology and Control*. North Carolina State University Graphics, pp. 1–422.

Sasser, J.N. and Freckman, D.W., 1987. A world perspective on nematology: The role of society. *Vistas of Nematology*, p. 7–14.

Sharma, S.B., 1985A. Nematode diseases of chickpea and pigeonpea. In: *Pulse Pathology Progress Report* No. 43 Patancheru, A.P. (India) International Crops Research Institute for the Semi Arid Tropics, 103 pp.

Sharma, S.B. and Nene, Y.L., 1985. Effect of presowing solarization on plant parasitic nematodes in chickpea and pigeonpea fields. *Indian J. Nematol.*, 15: 277–278.

Singh, R.S. and Sitaramaiah, K., 1973. *Res. Bull.* No. 6, G.B. Pant University of Agril. Sci. and Tech., pp. 1–289.

Singh, K.P. and Singh, V.K., 1991. Nematicidal nature of Arjun bark *Terminelia arjuna* on cyst nematode *Heterodera cajani* of pigeon pea. *New Agriculturist*, 2(1): 77–78.

Singh, K.P. and Singh, V.K., 1992. Terminelia arjuna leaf powder reduces population density of *Heterodera cajani*. *International Pigeonpea Newsletter*, 16: 17–18.

Singh, K.P., Bandyopadhyay, P. and Singh, V.K., 2001. Predation and performance of two predacious fungi for control of root-knot nematode of brinjal. *Mycopathological Research*, 39(2): 95–100.

Singh, V.K. and Singh, K.P., 1991. Effect of various nematicides on the control of root-knot nematode, *Meloidogyne javanica* on tomato. *Indian Biologist*, 23(2): 49–50.

Singh, V.K. and Singh, K.P., 1991. Population management of root-knot nematode *Meloidogyne incognita* on cauliflower. *Indian Biologist*, 23(2): 51–52.

Singh, V.K. and Singh, K.P., 1998. Effect of granular nematicides on the biology of *Heterodera cajani* of pigeonpea. *Indian J. Nematol.*, 28(2): 168–173.

Singh, V.K. and Singh, K.P., 2001. Effect of some medicinal plant leaves on the biology of *Heterodera cajani*. *Indian J. Nematol.*, 31(2): 143–147.

Singh, V.K., 2006. Management of root-knot nematode, *Meloidogyne incognita* infecting cauliflower. *Indian J. Nematol.*, 36(1): 126.

Singh, V.K., 2006. Management of root-knot nematode infecting tomato. *Indian J. Nematol.*, 36(1): 127.

Southy, J.F., 1965. Physical method of control. In: *Plant Nematology,* (Ed.) J.F. Southey. GDI Min. Agr. Fish Food, HMSO, London, pp. 323–332.

Southy, J.F., 1978. Regulation controls. In: *Plant Nematology,* (Ed.) J.F. Southey. GDI. Min. Agr. Fish Food, HMSO, London, pp. 326–332.

Walia, R.K., 1994. Assessment of nursery treatment with *pasteuria penetrans* for the control of *Meloidogyne javanica* on tomato in green house. *J. Biological Control,* 8: 68–70.

Walia, R.K. and Dalal, M.R., 1994. Efficacy of bacterical parasite, *Pasteuria penetrans* application as nursery soil treatment in controlling root-knot nematode *Meloidogyne Javanica* on tomato in green house. *Pest Management and Economic Zoology,* 2: 19–21.

Yuksel, H.S., 1960. Observations on the life cycle of *Ditylenchus dipsaci* on onion seedlings. *Nematologica* 5: 289–296.

Insect Pest and Disease Management *Pages* **448–455**
Editor: **Dr. D. Prasad**
Published by: **DAYA PUBLISHING HOUSE (2016)**

Chapter 31

Nematode Egg Parasitic Fungus, *Pochonia chlamydosporia*

I. Cannayane[1]* and E.I. Jonathan[2]
[1]Nematology Section, Department of Agricultural Entomology, Agricultural College and Research Institute, Madurai – 625 104
[1]Department of Nematology, Tamil Nadu Agricultural University, Coimbatore – 641 003

Introduction

Successful cultivation of crops largely depends on their protection, especially in the intensive cultivation system, from insects, diseases and nematodes. The importance of insects and diseases has already been realized and intensive measures are also taken for their management. However, the Nematodes have long been neglected in the IPM system. Nematodes have several forms ranging from free living to microbivorous and plant parasitic to highly specialized animal parasites (Baldwin *et al.*, 2004). Plant parasitic nematodes are of much importance as they cause heavy monitory loss in severe infestation. Recently, this menace has taken as serious threat to crop production system and off-late corrective measures have been initiated. Several methods have been employed for the management of plant parasitic nematodes *viz.*, regulatory measures, physical, cultural, chemical etc. All the measures have their own demerits. For example, The regulatory methods can tackle only specific nematode problems. Physical method of nematode control includes steam sterilization of soil, hot-water treatment of diseased planting materials etc. which are impractical at large scale cultivation of crops. Nematodes can also be controlled to a great extent by cultural practices but some of them cause monetary losses to growers such as fallow, flooding etc. The chemical control leads to environmental hazards and also has profound influence in eliminating natural enemies of crop pests. Moreover, Indian farmers are reluctant to use newer nematicides due to exorbitant cost, health hazards etc. in spite of their immediate effect on nematodes.

In search of safer and target oriented nematode management, biological control proved promising and efficient source of protective biomolecules, since it is not hampering the ecological balance. Large numbers of natural antagonists have been tested against a range of nematodes at various geographical locations. Earlier workers have given many excellent reports about the possibility of biocontrol of nematodes with various antagonists (Duddington, 1951; Mankau, 1980; Jatala, 1986; Morgan-Jones and Rodriguez-Kabana, 1988; Stirling, 1991; Sikora, 1992; Cook, 1993; Kerry, 2000; Araji and Hafez, 2001; Cannayane and Sivakumar, 2001).

Pochonia chlamydosporia Zare *et al.* (2001) (= *Verticillium chlamydosporium*)

Opportunistic nematode egg parasitic soil fungus, *Verticillium chlamydosporium* is also a member of the hyphomycetes having greater scope for use in the integrated management of nematodes (Kerry, 1987; De Leij *et al.*, 1993). This fungus has been recently renamed as *Pochonia chlamydosporia* by Zare *et al.* (2001). For convenience of understanding, *V. chlamydosporium* is used in this review unless and otherwise mentioned in its original literature.

Verticillium chlamydosporium was found to be the principal egg pathogen of *H. schachtii* and *H. avenae* cysts (Kerry, 1975). It can also colonise the rhizosphere of healthy plants, but colonisation of soil organic matter is limited (Kerry, 1995). The fungus proliferates in the rhizosphere and parasitizes the egg masses of root-knot nematodes or female cyst nematodes as they mature and rupture the root cortex and those egg masses exposed on the root surface. The fungus is not producing any specialized structures to capture and infect nematodes and it is only capable of parasitising nematode eggs or the sedentary females of certain nematode groups (Kerry, 1988).

Interaction with Nematodes and Plants in Rhizosphere

The rhizosphere is a nutrient rich region for soil microorganisms, where root exudates provides carbon and nitrogen compounds. In addition, root exudates may contain specific compounds that attract or repel symbiotic or pathogenic microorganisms and microfauna (O'Flaherty *et al.*, 2003). Microbial populations interact in complex ways in the rhizosphere, particularly in the presence of root pests and pathogens (Lynch, 1990; McSpadden Gardener and Weller, 2001).

Studies on the growth of *V. chlamydosporium* isolate in soil have shown that the fungus can proliferate from 10^4 to 5×10^5 cfu g^{-1} soil over a period of 30 days, especially at those sites where gall formation is more (De Leij and Kerry, 1991). The fungus spread approximately to 50 cm depth in soil profile in nine weeks and most inoculum spread to an area of less than 15 cm (De Leij *et al.*, 1993). Conidial spread was considered mainly by the water movement and is limited in undisturbed ecosystem. As described by Kerry and Bourne (1996), soil invertebrates are considered as important agents for the dispersal of microorganisms in soil profile.

Different developmental stages of *V. chlamydosporium* occur in soil and produces chlamydospores that enable the fungus to survive when nematode hosts are rare (Kerry and Bourne, 1996). The reasons for the switch over of the fungus from saprophyte to nematode parasite are not clearly elucidated and may be used to establish the fungus in soil (Kerry and Andersson, 1984).

The nematode suppressive ability of *V. chlamydosporium* is much altered by interaction between the nematode and the host plant. It is essential that the fungus should colonize the rhizosphere for effective nematode control (De Leij and Kerry, 1991). Root ramification of various isolates of *V. chlamydosporium* differs markedly and also plant species differ in their ability to support the growth of fungus in their rhizosphere (Bourne *et al.*, 1994). In general, the fungus caused highest reduction in nematode reproduction on those plant species, which support the extensive colonization of fungus (Kerry and De Leij, 1992). However, on plants, highly susceptible to nematodes, the fungus was less effective in controlling the nematode because of the embedded egg masses in the large galls and these egg masses were not exposed to fungal attack (De Leij *et al.*, 1992). Thus, the host plant has marked influence on the efficacy of *V. chlamydosporium* as a potential biological control agent through its ability to support the proliferation of the fungus in the rhizosphere and its susceptibility to nematode attack, which affects the extent of galling and the reproduction of root-knot nematodes (Bourne *et al.*, 1996).

When the egg masses emerged on the root surface, the fungus colonized the gelatinous matrix and parasitized the eggs. In a pot test, where approximately 10^4 chlamydospores and hyphal fragments were introduced into soil, the number of *M. arenaria* were reduced by around 80 per cent; this significantly decreased the number of galling in the following generations of nematodes (De Leij and Kerry, 1991). Growth chamber and plastic tunnel experiments have shown that profuse root colonization potential of native and introduced isolates of *P. chlamydosporia* and also significantly reduced the *M. javanica* in tomato plants (Sorribas *et al.*, 2003). Cannayane and Rajendran (2001) have observed marked decrease in *M. incognita* population in tomato plants which received *V. chlamydosporium* spores and hyphal mixtures, besides it also enhanced the growth parameters as measured by plant height and weight, root length and weight and fruit yield.

The efficacy of *V. chlamydosporium* as a biological control agent for root-knot nematodes is influenced by three key factors: the inoculum load of fungus in the rhizosphere, the age of eggs in the egg masses and the size of the galls (Bourne, 1995). In large galls female root-knot nematodes may produce egg masses which remain within the gall and are not exposed to parasitism by *V. chlamydosporium*. Hence, *V. chlamydosporium* is less effective in controlling root-knot nematodes in heavily infested soils and also on highly susceptible crops due to larger gall formation leading to deposition of egg masses with in roots and many eggs escape parasitism (Bourne, 1995). Cannayane and Sundararaju (2004) have also observed in the root-knot nematode infested banana roots, where the egg masses remains with in the cortical tissues, thus preventing the eggs from parasitism by the antagonist. Further, most of the eggs in the egg masses are not colonized and the egg masses near epidermis are partially parasitized by the fungus and also only the developing eggs or those in embryonic stages are engulfed by *V. chlamydosporium*.

Compatibility with Other Bioagents

Biocontrol products based upon the use of single strains provide certain practical and economic benefits that promote commercial development but using multiple strains of the same or different species could provide more reproducible control as suggested by Stewart (2001).

The fungi *viz.*, *V. chlamydosporium* and *P. lilacinus* at 100 per cent concentration caused 77.50 and 74.50 per cent mortality of *M. incognita* eggs by the end of 72 h respectively (Wang, 1992). The propagule number of *V. chlamydosporium* and *P. lilacinus* was more in soils with *M. incognita* infected tomato plants than in soil with uninfected tomato plants (Gaspard *et al.*, 1990). *V. chlamydosporium* was found to be superior to *P. lilacinus* in enhancing plant growth and parasitization of eggs and controlled root-knot and *H. cajani* population to an extent of 90 per cent (De Leij *et al.*, 1993; Siddiqui and Mahmood, 1995).

Sasnowska (2000) pointed out the use of *V. chlamydosporium* and *P. lilancinus* and their combination with *P. penetrans* to be highly effective against plant parasitic nematodes. Reduction in number of galls caused by *M. arenaria* has been observed on squash roots in greenhouse tests following addition of oat grains colonized by two isolates of *Gliocladium roseum, G. catenulatum, P. lilacinus, V. chlamydosporium* and *V. lamellicola* (Rodriguez-Kabana *et al.*, 1984).

Integration of *V. chlamydosporium* with *Glomus deserticola* brings down the population of *M. incognita* in tomato nursery with vigorous transplants (Rao *et al.*, 1997). Further, *V. chlamydosporium* in combination with *G. fasciculatum* has resulted in additive effect against *M. incognita* in brinjal. Additionally, *V. chlamydosporium* did not affect the root colonization of *G. fasciculatum* in field conditions, instead it promoted the root colonization (Rao *et al.*, 2003).

Mechanisms of Action on Nematodes

Those fungi that infect nematode eggs produce an appressorium that adhere to the eggshell by mucilaginous material and develop infection peg to penetrate the eggshell (Dunn *et al.*, 1982). The size of an appressorium is dependent possibly on the chemical composition at the point of entry on the surface of the host. The immature and mature appressoria were found both in germ tubes and mycelial hyphae and a mucilaginous material was also seen, where the appressorium contacted the surface of the eggs where the *V. suchlasporium* (=*P. rubescens*) (Lopez-Llorca and Claugher, 1990). However, in a study with the human parasite (*Ascaris lumbricoides*) by Lysek and Kraji (1987), *V. chlamydosporium* did not produced any specialized structures like appresorium to penetrate nematode eggs, instead traces of enzymatic activity was observed.

Verticillium chlamydosporium and *P. lilacinus* secretes basic serine protease (subtilisin family) an extra cellular enzyme (Lopez-Llorca, 1990; Segers *et al.*, 1994). These extra cellular enzymes are induced by the presence of nematode eggs and are considered to be the host range detection and virulence nature of the egg parasitic fungi. Uziel and Sikora (1992) also reported an enzymatic penetration of the egg shell of *Globodera pallida* by *V. lecanii* (= *Lecanicillium lecanii*). They found very clear appressorium formation on the surface of the cyst and eggs. Lopez-Llorca *et al.* (2002) identified the key factors responsible for early infection of nematode eggs by *P. rubescens, P. chlamydosporia* and *Lecanicillium lecanii*. They observed that the appressoria were frequently formed on hydrophobic surfaces and proteolytic activity was also found in conidial and germling extracts of all the egg parasitic fungi, particularly in *P. chlamydosporia* in the absence of *M. javanica* eggs.

Verticillium chlamydosporium penetrated mechanically with the help of appressoria on the surface of cysts and eggs of *G. rostochiensis* and also traces of chemical secretions were observed from the penetrating hyphae which weakened and digested the cyst surface and egg shell (Saifullah and Thomas, 1997). *V. chlamydosporium* and some *Verticillium* spp. are able to degrade chitin layer of the eggs and the former converts the contents of an invaded egg into a mass of hyphae (Jackson, 1965; Baath and Sonderstrom, 1980). A bioactive chemical compound, Phomalactone was isolated and characterized from the metabolites of *V. chlamydosporium* and the compound showed toxicity to *M. incognita* (Khambay *et al.*, 2000).

An Overall interaction between opportunistic soil fungi and their nematode hosts appears to be quite variable. The main type of destruction activity was thought to be direct physical perforation of egg shells and larval cuticles, enzymatic dissolution of nematode structural elements, physiological disturbances brought about by biosynthesis of diffusible toxic metabolites (Stirling and Mankau, 1979; Morgan-Jones *et al.*, 1983; Morgan-Jones and Rodriguez-Kabana, 1981, 1985).

Mass Production

There have been numerous attempts to utilize a range of fungal biocontrol products for the management of nematodes at varied geographical locations. However, widespread use of these fungi in agri-horti production system to combat phytonematodes is negligible in spite of their appreciable performance in laboratory and/or microplot conditions. This is because of the slow progress of suitable formulation technology of these biomolecules. Successful biocontrol agent for soil borne pests or pathogen including nematodes must be able to grow faster, survive in adverse condition and colonise quickly in the rhizosphere and rhizoplane. However, successful establishment of the fungal bioagent has been correlated with the availability of energy source to support the organisms to compete with the saprophytic fungus in soil (Kerry *et al.*, 1993). Selection of virulent biocontrol agents, ingredients in the formulation, method of formulation, time of application etc. plays a vital role in the biomanagement of nematodes with antagonistic products.

Development of inexpensive, easily applied preparations that remain viable under less than optimal conditions are need of time. Large scale mass production of bioagent can be achieved by using readily available agricultural wastes.

Attempts to deliver the nematode biocontrol agents through both organic and inorganic carriers like bark pellets, vermiculite, pyrex and laponite gel with pyrophyllite clay filler have been evaluated (Fravel *et al.*, 1985). Enzymes such as β-1,3 glucanase and chitinase help the fungi in utilization of cellulose and chitin present in organic wastes for its growth and multiplication (Hadar *et al.*, 1979). Connick *et al.* (1991) developed a formulation containing wheat based substrates and clay for delivering fungal propagules. Sankaranarayanan *et al.* (2000) prepared a granular formulation of fungal bioagents *viz., V. chlamydosporium, Arthrobotrys oligospora, P. lilacinus, G. virens* and *Trichoderma harzianum* in a mixture containg wheat bran, wheat germ, vegetable oil and starch. The colony forming units in the formulation were as high as 6.2×10^7 for *V. chlamydosporium*, indicating the supportive nature of substrate for the fungus. Further, this granular formulation effectively controlled the *M. incognita* in tomato under green house conditions. Additionally, the carrier material also enhanced the fungal growth in rhizosphere region, which was evident from the reisolation of antagonist from the experimental area.

Sundararaju and Cannayane (2002) have utilized the locally available plant leaves including banana wastes (leaves, petiole and pseudostem) for the mass production of *P. lilacinus* (also a nematode egg parasitic fungus, closely related to *P. chlamydosporia*). Higher spore load per g of substrate was obtained in a mixture of banana petiole and castor leaves and these mixtures could be effectively utilized for the mass production of other fungal bioagents like *P. chlamydosporia*, as the substrates are abundantly available in banana gardens and ecofriendly as well. Further, profuse growth of *V. chlamydosporium, V. lecanii* and *P. lilacinus* was observed on the locally available substrates such as farm yard manure (FYM), saw dust, vermicompost and groundnut cake (Cannayane and Sundararaju, 2002). On comparing the substrates nature, FYM supported higher growth rate and spore production of all the test fungi. Sasikala (2003) also reported that the wheat bran, cumbu and sorghum grains, FYM, neem cake and banana wastes (petiole and leaves) are the good substrates for *V. chlamydosporium* multiplication and suggested that banana wastes are excellent source for on farm mass production.

Conclusions

Opportunistic colonizers of nematode eggs and cysts are, however, well adapted to their environment, being most successful competitors that are able to survive host free periods. Being an opportunistic fungus, *P. chlamydosporia* is also capable of colonizing plant roots and exerting deleterious effects on nematodes. The extent of rhizosphere and rhizoplane colonization of the fungus depends on the plant type that support for the multiplication of the fungus. Plants that are more susceptible to nematode attack resulting in deposition of egg masses in deeper layer and thus it affect the efficacy of *P. chlamydosporia* on nematodes. Conversely, plant type that allows more colonization of antagonistic fungi leads to the higher parasitization of nematode eggs. Thus, host plant significantly influences the efficacy of *P. chlamydosporia* in its rhizosphere as a potential biocontrol agent on nematodes. Further, *P. chlamydosporia* has shown additional desirable attributes with regard to the compatibility with other bioagents in suppressing nematode population. This could be the additional useful trait in developing rational biomanagement strategies to tackle region specific nematode management. With regard to its mode of action on nematodes, enzymatic degradation of nematode eggs coupled with mechanical force exerted by the fungus is the first step involved in their action.

An efficient biocontrol agent to be successful among the end users, its scale-up of inoculum is necessary, in addition with time of application and method of treatment. It is clearly indicated that *P. chlamydosporia* can be easily mass produced on locally available organic substrates such as banana wastes, cereal grains, organic amendments etc. Such attributes of the fungus make its wider usage with little technical expertise at growers point. However, emphasis has also been placed to generated data on different formulation methods (*viz.*, aqueous, powder, microencapsulation of popagules etc.), viability of propagules on longer storage, identifying rhizosphere-competent strains, influence of abiotic factors (temperature, soil type, moisture etc.) on the efficacy of soil applied inoculum, refining molecular methods for precise identification of elite isolates for location specific nematode problems and effects on non-target organisms.

References

Araji, A.A. and Hafez, S.L., 2001. *Nematropica*, 31: 181–192.

Baath, E. and Sonderstrom, B., 1980. *Can. J. Bot.*, 58: 422–425.

Baldwin, J.G. *et al.*, 2004. *Ann. Rev. Phytopathol.*, 42: 83–105

Bourne J.M. *et al.*, 1996. *Biocontrol Sci. Technol.*, 6: 539–548.

Bourne, J. M., 1995. *Ph.D thesis*, Submitted to Reading University.

Bourne, J.M. *et al.*, 1994. *J. Nematol.*, 26: 587–591.

Bursnall, L. A. and Tribe, H. T., 1974. *Trans. Brit. Mycol. Soc.*, 62: 595–601.

Cannayane, I. and Rajendran, G., 2001. *Curr. Nematol.*, 12: 51–56.

Cannayane, I. and Sivakumar, C.V., 2001. *Indian J. Agril. Res.*

Cannayane, I. and Sivakumar, C.V., 2001. Agric. Res.

Cannayane, I. and Sundararaju, P., 2002. *In: National symposium on Biodiversity and Management of Nematodes in Cropping Systems for Sustainable Agriculture*, 11–13th November 2002, Agricultural Research Station, Jaipur, p. 45.

Connick, W.J. *et al.*, 1991. *Biol. Cont.*, 1: 281–287.

Cook, R.J., 1993. *Annu. Rev. Phytopathol.*, 31: 53–80.

Cook, R.J. and Baker, K.R., 1983. *The Nature and Practice of Biological Control of Plant Pathogens*, APS Press, St Paul, MN.

Dackman, C. and Nordbring–Hertz, B., 1985. *J. Nematol.*, 17: 50–55.

De Leij, F.A.A.M. and Kerry, B.R., 1991. *Rev. Nematol.*, 14: 157–164.

De Leij, F.A.A.M. *et al.*, 1992. *Nematologica*, 38: 65–79.

De Leij, F.A.A.M. *et al.*, 1993. *Nematologica*, 39: 115–126.

Duddington, C.L., 1951. *Trans. Brit. Mycol. Soc.*, 34: 598–603.

Dunn, M.T. *et al.*, 1982. *Scann. Electron Micros.*, 3: 1351–57.

Fravel, D.R. *et al.*, 1985. *Phytopathology*, 75: 774–777

Gaspard, J.T. *et al.*, 1990. *J. Nematol.*, 22: 207–213.

Gintis, S. *et al.*, 1983. *Nematologica*, 13: 181–200.

Godoy, G. *et al.*, 1983. *Nematologica*, 13: 201–213.

Hadar, Y. I. *et al.*, 1979. *Phytopathology*, 69: 64–68.

Hanssler, G., 1990. *J. Pl. Dis. Protec.*, 97: 194–201.

Jackson, R.M., 1965. *NZJ. Agri. Res.*, 8: 878–888.

Jacobs, H. *et al.*, 2003. *Mycol. Res.*, 107: 47–56.

Jatala, P., 1986. *Annu. Rev. Phytopathol.*, 24: 453–489.

Kerry, B.R. (!974. *Nematologica*, 20: 259–260.

Kerry, B.R., 1975. *EPPO Bull.*, 5: 353–361.

Kerry, B.R., 1980. *J. Nematol.*, 12: 253–259.

Kerry, B. R., 1987. *In: Principles and Practices of Nematode Control in Crops* (Brown, R.H. and Kerry, B. R. Eds), Academic Press, pp. 233–263.

Kerry, B.R., 1988. *Agric. Ecosyst. Environ.*, 24: 293–305.

Kerry, B.R., 1995. *Can J. Bot.*, 73: 65–73.

Kerry, B.R., 2000. *Annu. Rev. Phytopathol.*, 38: 423–441.

Kerry, B.R. and Andersson, S., 1984. *Vaxtskyddsnotiser*, 47: 79–80.

Kerry, B.R. and Bourne, J. M., 1996. *Pesticide Sci.*, 47: 69–75.

Kerry, B.R. and De Leij, F.A.A.M., 1992. *In: Biological Control of Plant Diseases.* (Tjamos, E.C. *et al.*, Eds), Plenum Press, pp. 139–144.

Kerry, B.R. *et al.*, 1982. *Ann. Appl. Biol.*, 100: 477–487.

Kerry, B.R. *et al.*, 1993. *Biocontrol Sci. Tech.*, 3: 355–365.

Khambay, B.P.S. *et al.*, 2000. *Pest Mgt. Sci.*, 56: 1098–1099.

Lopez–Llorca, L.V. and Claugher, D., 1990. *Micron. Microsc. Aacta.* 21: 125–130.

Lopez–Llorca, L.V., 1990. *Can. J. Microbiol.*, 36: 530–537.

Lopez–Llorca, L.V. *et al.*, 2002. *Mycol. Res.*, 106: 499–506.

Lynch, J.M., 1990. *The Rhizosphere.* John Wiley, Chickester: 255 pp.

Lysek, H., 1978. *Parasitology*, 77: 139–141.

Lysek, H. and Kraji, D., 1987. *Folia Parasitol.*, 34: 57–60.

Mankau, R., 1980. *Annu. Rev. Phytopathol.*, 18: 415–440.

McSpadden Gardener, B. and Weller, D.M., 2001. *Appl. Environ. Microbiol.*, 67: 4414–4425.

Morgan–Jones, G. and Rodriguez–Kabana, R., 1981. *Nematropica*, 11: 69–74.

Morgan–Jones, G. and Rodriguez–Kabana, R., 1985. *Nematropica*, 15: 107–114.

Morgan–Jones, G. and Rodriguez–Kabana, R., 1988. *In: Diseases of Nematodes II* (Poinar, G.O. and Jansson, H. B. Eds) CRC Press Inc., Florida, pp.39–58.

Morgan–Jones, G. *et al.*, 1981. *Nematropica*, 11: 155–164.

Morgan–Jones, G. *et al.*, 1983. *Nematropica*, 13: 245–260.

O'Flaherty, S.M. *et al.*, 2003. *European J. Soil Sci.*, 54: 759–766.

Rao, M. S. *et al.*, 1997. *J. Plant Dis. Protec.*, 104: 419–422.

Rao, M.S. *et al.*, 2003. *Indian J. Nematol.*, 33: 29–32.

Rodriguez–Kabana, R. *et al.*, 1984. *Nematropica*, 14: 155–162.

Saifullah, and Thomas, B.J., 1997. *Int. J. Nematol.*, 7: 30–34.

Sankaranayanan, C. *et al.*, 2000. *Indian J. Nematol.*, 30: 157–161.

Sasikala, T., 2003. *M.Sc. Dissertation*, Submitted to Bharathidasan University, Tiruchirapalli.

Sasnowska, D., 2000. *Nematol. Abstr.*, 70: 310.

Segers, R. *et al.*, 1994. *Microbiology*, 140: 2715–23.

Siddiqui, Z.A. and Mahmood, I., 1995. *Biores. Tech.*, 57: 227–230.

Sikora, R. A., 1992. *Annu. Rev. Phytopathol.*, 30: 245–270.

Sorribas, F.J. *et al.*, 2003. *Biocontrol Sci. Technol.*, 13: 707–714.

Stewart, A., 2001. *Australasian Plant Pathology*. 30: 127–131.

Stirling, G.R., 1991. *Biological Control of Plant Parasitic Nematodes: Progress, Problems and Prospects*, CABI, 282 pp.

Stirling, G.R. and Kerry, B.R., 1983. *Australian J. Expt. Agri. Ani. Husbandry.*, 23: 318–324.

Stirling, G.R. and Mankau, R., 1979. *J. Nematol.*, 11: 282–288.

Sundararaju, P. and Cannayane, I., 2002. *Indian J. Nematol.*, 32: 188–189.

Sundararaju, P. and Cannayane, I., 2004. In: *International Congress on Musa: Harnessing Research to Improve livelihoods*, 6–9th July, 2004, Penang, Malaysia, p. 158.

Uziel, A. and Sikora, R.A., 1992. *Nematologica*, 38: 127–130.

Wang, M. Z., 1992. *J. Huazhong Agri., Univ.*, 11: 52–56.

Zare, R. *et al.*, 2001. *Nova Hedwigia*, 73: 51–86.

Insect Pest and Disease Management *Pages 456–458*
Editor: **Dr. D. Prasad**
Published by: **DAYA PUBLISHING HOUSE (2016)**

Chapter 32

Anti-nutritional Compounds in Pulses

Amit Kumar Jain[1], Sudhir Kumar[1], Om Prakash Singh[2], and J.D.S. Panwar[3]
[1]Department of Botany, Janta Vedic College, Baraut (Baghpat), U.P.
[2]Division of Plant Pathology, IARI, New Delhi – 110 012
[3]Head, Division of CPBM, IIPR, Kalyanpur, Kanpur – 208 024, U.P.

Introduction

India is the largest producer of pulses in the world. The common pulses grown are chickpea, pigeonpea, mungbean, urdbean, lentil, field pea, moth bean, horse gram and lathyrus. In India, pulses occupy an area of about 22–23 million hectares with annual production of 13-14 million tones. The term pulse" is used for those leguminous plants containing small amount of fat. Pulses offer a relatively cheaper source of protein, which is valuable for developing countries, and they have about the same calorific value per unit weight as cereals and are fairly rich in some vitamins and minerals. Consumption of pulses is highest in India where majority of the population is vegetarian. Pulses contain 18 to 32 per cent protein and 1 to 5 per cent fat. They are also rich in iron, thiamine, riboflavin and niacin and dietary fiber. The crude fiber, protein and lipid components of pulses have shown to have a hypocholestrolemic effect, the ability to reduce cholesterol in body blood.

The nutritional value of pulses may be adversely affected by the presence of anti physiological or toxic substances. These substances are known as anti-nutritional compounds such as trypsin and chymotrypsin inhibitors, phytates, lectins, polyphenols, flatulence causing compounds, cyanogenic compounds, lathyrogens, esterogens, goiterogens, saponins, anti-vitamins and allergens. These substances reduce the nutritive value of foods by inhibiting digestibility and utilization of proteins.

Lectins

Lectins are proteinaceous toxic compounds commonly found in some of the beans. Seeds of some of the edible species of pulses such as lentil and pea also contain phytohernagglutinins. It was observed that extracts from many edible crude bean seeds agglutinated red cells, but no toxic action was detected in these seeds. These are proteins which possess a specific affinity for certain sugar molecules. Most of the lectins contain 4 to 10 per cent carbohydrates. The lectins from mungbean are non toxic, whereas the lectins obtained from immature seeds of pigeonpea and ricebean are harmful.

Polyphenols

Polyphenols (tannins) are known to form complexes with proteins under certain pH conditions. Tannins-protein complexes are reported to be responsible for low protein digestibility, decreased amino acid availability and increased fecal nitrogen. These complexes may not be dissociated and may thus be excreted with the feces. Among pulses, pigeonpea, urdbean and pea have highest tannins in seeds. It appears that tannins content in pulses varies with colour of the seed coat or testa. Lower amount of tannins, in general are observed in light coloured seeds than brown dark coloured pluses. White seeded varieties of *Phaseolus vulgaris* are found to have almost negligible tannins, while coloured varieties contained large quantities of tannin. The tannins content in beans changes during seed maturation. For example, the tannin content of beans decreases as the seed matures.

Tannins form complexes with proteins, carbohydrates and polymers in foods, as well as with certain metal ions such as iron. The greater tendency of tannins to form complexes with proteins rather than carbohydrates and other food polymers is attributed to the strong hydrogen bond affinity of the carboxyl oxygen of the peptide group. Tannins are mainly located in testa or seed coat of pulses, hence physical removal of seed coat by either dehulling or milling and separating hulls decreases the tannin content of pulses and improves their nutritional quality. Dehulling eliminates about 68 to 99 per cent of tannins in seeds. Soaking of seeds before cooking is a common practice and is used to soften the texture and hasten the process of cooking. Leaching of tannins increases with the time of soaking in distilled water. Cooking and discarding the cooking water results in about 37.5 to 77.0 per cent decrease in tannin content of seeds. Overnight soaking in water and subsequent germination for 48 h removes more than 50 per cent of the tannins in pigeonpea, chickpea, mungbean and urdbean.

Protease Inhibitors

Several substances which have ability to inhibit the proteolytic activity of certain enzymes are found in pulses. Trypsin inhibitors belong to a broad class of proteins (protease inhibitor) that inhibit proteolytic enzymes. Trypsin and chymotrypsin inhibitors are of much importance here. Trypsin inhibitor activity increases as seed maturation progresses. Most of the plant protease inhibitors are destroyed by heat resulting in enhancement of nutritive value of protein. Moisture content, time, temperature and pressure are the major factors that influence cooking rates, inactivation of the trypsin inhibitor, and subsequent increase in nutritive value. Moist heat has been shown to be effective in destroying trypsin inhibitor activity in pulses. Germination also influences the trypsin inhibitors activity.

Lathyrogen Toxin

Lathyrogen toxin is one of the natural toxins found in the seeds of lathyrus, which is known to cause lafhyrism, if consumed in excess quantity for long time. It causes paralysis in the legs in susceptible individuals and is believed to be caused by a toxic amino acid known as N-oxalyl amino alanine (BOAA).

Saponins

Saponins are secondary plant metabolites present in pulses. It contains a carbohydrate moiety (mono or oligosacchande) attached to an aglycone, which may be steroidal or triterpenoid in structure. The chickpea, mungbean and pigeonpea contain saponins in the range of 0.05 to 0.23 per cent. The saponins appear to be beneficial as they are responsible for lowering the cholesterol in body. Saponins also seem to be anti-carcinogen; in one study they inhibited colon cancer.

Pulses are generally subjected to various processing before their consumption. These include dehulling, milling, soaking, germination, fermentation andcooking. Such processing has several nutrition advantages. Heat treatment has been well established to destroy pertinacious antinutrients, such as protease inhibitor and lectins, but this treatment may also des troy some of the amino acids and vitamins. For maintaining the nutritional value of food, it is necessary to ensure the heating temperature required to eliminate the effect of inhibitors. Proteins in pulses are known to interact with lipids, tannins, phytates, flavour compounds and pigments. These interactions occur when pulses are processed and converted into various products. This decreases the bioavailability of proteins. Similarly, tannins and phytates interact with minerals arid vitamins resulting in low bioavailability of minerals and vitamins. Thus, bioavailability of nutrients depends not only on content of the nutrients in the seed, but also on the interaction of nutrients under various processing conditions and these processes produce edible product having a reduced level of anti and toxic compounds elimination of anti-nutrients or toxic compounds depends on type of pulses and the processing technique.

Insect Pest and Disease Management *Pages* **459–467**
Editor: **Dr. D. Prasad**
Published by: **DAYA PUBLISHING HOUSE (2016)**

Chapter 33

Root-knot Nematode Problems in Nursery and Young Tea

B.C. Bora and P.P. Neog
Department of Nematology, Assam Agricultural University, Jorhat – 785 013, Assam

Introduction

Tea (*Camellia sinensis* L) is one of the most important plantation crop grown in India and considered to be an important source of revenue and foreign exchange. As well as production of highest quantity and quality tea with generation of highest number of employment (Anon, 1991). Tea is originated in the region around the point of intersection of latitude 29°N and longitude 98°S near the source of river Irrewaddi. In India, it is mainly cultivated in Assam, West Bengal, Kerela, Tamil Nadu and Karnataka.

Tea crop is attacked by a number of pests and diseases which causes considerable loss in tea production. Among them, plant parasitic nematodes are presently considered to be one of the important production constraints of tea (Srivastava and Singh, 1967). Considerable number of plant parasitic nematodes had been recorded to be associated with nursery, young and mature tea grown in different parts of the world. (Kerr and Vythilingam, 1967; Benerjee, 1967; Basu, 1967 and 1968; Whitehead, 1969; Sivapalan, 1972; Basu and Roy, 1976 and 1979; Phukan and Sanwal, 1979).

However, out of different growth phase of tea crop, it is needless to mention that seedlings at nursery stage are particularly vulnerable to attack of plant parasitic nematodes and numbers of soil borne fungal pathogens. By now, large number of plant parasitic nematodes are known to be associated with nursery and young tea (Anon, 1957; Das, 1958; Banerjea, 1966; Basu, 1967, 1968; Sivapalan, 1972; Basu and Roy, 1976; 1979). Out of which root-knot nematode, *Meloidogyne* spp. is considered to be the most frequently encountered plant parasitic nematodes with high density of population in majority of tea growing areas of the world (Das, 1955; Mukherjee, 1963; Thirugnanam and Rangaswami, 1967; Whitehead, 1969; Sivapalan, 1972; Manuelipillai, 1981). Its attack mainly confined to the seedlings at nurseries for about 3–5 months from sowing/planting of seeds/clonal cuttings (Das, 1955; Mukherjee, 1960). Among the different species of root knot nematode, the species like *Meloidogyne incognita, M. javanica, M. hapla* and *M. thamesi* are some of the economically important species causing severe damage to the nursery seedlings of less than one year old except *M. bravicauda.* (Rao, 1970; Sivapalan, 1972). However, *M. incognita* and *M. javanica* were found to be the most harmful nematode species on nursery tea while *M. bravicauda* on mature tea (Whitehead, 1969). The importance of root knot nematode in nursery and young tea have been well documented by Rang *et al.* (1984) who reported the distribution of this nematode in 18 different countries of the world with 100 per cent infection and 50 per cent death of seedlings in a seriously infested area of tea plantation by *M. incognita, M. javanica* and *M. thamesi.* Large scale failure of nursery seedlings raised from both seeds and vegetatively propagated clones due to severe damage caused by root knot nematode also reported by many workers (Banerjea, 1966; Basu, 1967, 1968; Basu and Roy, 1976; 1979) from the North-Eastern region of India showed that. Generally both young and mature tea plant are highly susceptible to *M. javanica* and *Pratylenchus coffeae* (Visser and Vythilingam, 1959).

Distribution

The first report of root-knot nematode in young tea was from South India, where large numbers of tea seedlings were found infected (Barber, 1901). Singh (1989) recorded 18 species of plant parasitic nematodes including *Meloidogyne* spp. as a dominant on for Jorhat district of Assam. Thereafter, occurrence of two important species of root-knot nematode, *viz., M. incognita acrita* and *M. hapla* were reported from Dooars and Aassam Valley (Das, 1955 and Mukherjee, 1960). Subsequently, considerable number of plant parasitic nematodes had also recorded from Assam and adjoining North-East India associated with nursery, young and mature tea (Anon, 1957; Das, 1958; Banerjea, 1966; Basu, 1967, 1968; Basu and Roy, 1976; 1979). Similarly, two important species of root-knot nematode *viz., M. incognita* and *M. hapla* were reported from roots of tea seedlings in the Dooars and Assam Valley (Das, 1955). The occurrence of *M. bravicauda* and *M. javanica* were also made from tea seedlings at Madras state (Seshadri and Kumarswami, 1963; Thirugnanam and Rangaswami, 1967). Occurrence of *M. ornate* and 13 other genera of plant parasitic nematodes in tea plantation including nurseries were also made in West Bengal (Mukherjee and Dasgupta, 1981). The distribution of root knot nematode in tea nursery had been well documented from 18 different countries by Rang *et al.* (1984). Large scale failure of tea nurseries in Sri Lanka due to root knot nematode infestation sometimes destroyed entire nurseries was also reported by many workers (Light, 1928; Gadd and Loos, 1946; Loos, 1951, 1953). Whitehead (1969) reported that *M. incognita, M. javanica* were found to be the most harmful nematode species on nursery tea while *M. bravicauda* on mature tea.

Host Parasite Relationship

In nursery condition seedlings raised from both seeds and vegetatively propagated clones as well as young tea are very much prone to root-knot nematode and severely damaged by them (Visser, 1959). It is observed that its infestation is mostly confined to seedlings in nurseries for about 3–5 months from sowing of seeds, which resulted poor quality and quantity of seedlings (Mukherjea, 1960; Glover, 1961). Seedling plants in which both tap roots and lateral roots are severely attacked and suffer greater damage than do clonal plants of equivalent age, probably because seedling tea plants posses less than half the root bulk of clonal plants of similar age (Kerr and Vythilingam, 1967).

Pathogenecity of root-knot nematode on tea seedlings revealed that one nematode per gram of soil was proved effective to reduced the seedling vigour and death of the seedlings (Dutta, 1995; Kalita, 2000). Higher the nematode inoculum present in soil there is corresponding decrease in seedling growth parameters (Light, 1928; Visser, 1959; Dutta, 1995; Kalita, 2000).

Symptoms

Root-knot nematode attack mainly confined to seedling of 3–5 months old and causes the formation of galls on the roots and decrease the height of tea seedlings (Das,1955; Mukherjee, 1960). Usually, tea seedlings infested with root-knot nematode are not always killed out right and symptoms of attack can mistaken for drought, mineral deficiencies, lack of organic constituent, water logging and so on. In field patches are found of stunted, wilted, chlorotic or generally unhealthy growth of tea seedlings. These are caused by atrophied root symptoms due to rotting, galling or by killing the fine tender roots and lateral rootlets (Mukherjea, 1960). Diseased seedlings look yellowing and unhealthy during the rainy period and suddenly wilt and die when a period of hot, dry weather prevails towards the end of it.

Root-knot nematode, *M. bravicauda* causes severe galling of the roots of mature tea bushes with resultant decline. The above ground symptoms of attack by this nematode are similar to those of the root lesion nematode, *Pratylenchus loosi.* The infected bushes grow more slowly and the leaves tend to become smaller, dull and yellow in appearance. The root show the characteristic swelling and pitting that is typical symptoms of *M. bravicauda* (Sivapalan, 1972).

Interaction with Fungal Pathogen

Of the various soil borne fungi, damping off caused by *Pythium* sp. and *Rhizoctonia* sp. and collar rot caused by *Phomopsis* sp. are some of the important soil borne pathogens associated with nursery and young tea which cause sustainable harm to the tea seedlings (Agnihothrudu, 1995; Venkataramani and Venkata Ram, 1959). These soil borne fungi are very much common in tea nurseries throughout the tea growing areas of North East India, the soils most of which are of a sticky in nature or where a hard crust forms on the surface after heavy rains (Sarmah, 1960; Tunstall and Sarmah, 1947; Agnihothrudu and Baruah, 1960). Infection occurs, while the stem is still soft *i.e.* during the first six months, immediately below the surface crust. (Bordoloi and Baruah, 1970).

Damage of root knot nematode and fungal pathogens are more common in the early stage of growth particularly in nurseries where seedlings are grown in containers which are being used for planting and replanting programme. The reason for such as:

1. Availability of sufficient numbers of succulent roots which provides abundant food for nematodes.
2. Nursery soil often contains soil borne fungi which together with nematodes results in more severe damage than would be caused by other pathogens alone.
3. Irrigation, cultivation and fertility of soils may favour the nematode and fungal activity into several folds.

Although, these two organisms occur independently producing distinct disease symptoms, but most vital and significant events take place when these two groups of pathogens interact together in the seedling rhizosphere and put forwarded a valid complex which have real economic consequences. Hence, several soil borne pathogens causing wilt, collar rot, and damping off on tea seedlings however, cannot be ascribed to the activities of a single pathogen and generally are attributed to the affects of disease complexes. There is an increasing amount of evidences that root-knot nematode facilitate entry and establishment of pathogenic fungi and bacteria (Pitcher, 1965; Powell, 1971). There are extensive amount of literatures available on interactions of root-knot nematode and plant pathogenic fungi and bacteria which have been reviewed by Powell (1963) and Pitcher (1963). The root-knot nematode feeds on tender roots and cause wounds. These wounded roots are predisposed to the infection of several soil borne fungi and thereby increasing the wilt and other disease incidence.

Management

Seedlings rose from both clonal cuttings and seeds are very much prone to the plant parasitic nematodes particularly root-knot nematode including various soil borne pathogens at nursery stage. Management of these pests together at single attempt to get rid of root-knot nematode attack as well as diseases complexes is considered to be most vital at least to produce healthy seedlings by the growers and reduce mortality. Therefore, the seedlings and young plants immediately after planting at nurseries and main field must be protected from the above soil borne pest and pathogens up to a certain periods of its planting.

A number of methods like physical, cultural, biological, legal and development of crop hybrids for disease resistance are being experimented for effective and economic management against this devastating nematode pest of tea seedlings. In tea nursery and young tea plantation different control methods are tested by different workers (Ichinohe, 1965; Basu, 1980; Mukherjea, 1966; Rao, 1970; Dutta, 1995; Neog and Bora, 1999a, 1999b, 1999c).

Use of Chemicals

Earlier DD and EDB was considered as most effective chemical for pre-plant treatment and of DBCP as post plant application against root-knot nematode associated with nursery tea resulting decreases in gall numbers roots (Ichinohe, 1965). Application of methyl bromide fumigation @ 0.5, 1.0 and 2 lb per square feet successfully control the parasitic nematodes at nursery site which increased the higher percent of rooting as well as high growth of seedling (Kerr and Vythilingam, 1966; Sivapalan, 1969). Although, complete eradication of nematode from contaminated nursery plant was not possible by chemical treatment but with chemicals like, nemagon, nematex in combination with basudin and vapam and application of furadon, temik thimet and mocap when tried at different doses against root-knot nematode on nursery and young tea resulted good plant growth with less galling than control. All the chemicals showed no significant difference in respect of suppressing root-knot indices and increased plant stand significantly (Mukherjea, 1966; Rao, 1970; Basu, 1980; Basu and Gope, 1985). Out of all these systemic nematicides tried three per cent granular furadon @ 3–5g per seedling as pre plant treatment in case of nursery seedling and 3–5 kg a.i./ha in planting site of young tea was found very successful in controlling *M. incognita, M. arenaria, M. javanica* and *M. thamse* population and increased the growth of tea seedlings (Basu, 1980; Rang *et al.*, 1984; Neog and Bora, 1999a). Maximum seedling height, length of roots and number of laterals per root system were obtained when soil in sleeves were treated with 2g per seedling. However, same results could be obtained if the dose of carbofuran 3g was reduced to half and combined with 5g of either neem cake or mustard cake per sleeves (Neog and Bora, 1999a; Kalita and Bora, 2006). Phenamiphos 5 per cent granule @ 7g per plant was also proved significant in suppressing the infestation of root-knot nematode up to 60 per cent and stimulated the growth of tea (Gnanapragasam *et al.*, 1987).

Biological Control

Although, application of biological resources are proved economic to the marginal growers, however, not much works have so far been conducted elsewhere on the utilization of antagonistic fungi and other biological organisms in controlling rootknot and other parasitic nematodes at nursery and young tea plantation. The important microbial agents so far have been successfully exploited against tea root pathogen are species of *Trichoderma, Bacillus* and *Pseudomonas flourescens* which proved effective against disease complex on nursery tea (Borthakur and Dutta, 1992; Kumar and Bezbaruah, 1997).

Organic Amendments

Incorporation of organic matters into nematode infested soil is presently considered to an alternative, easy and ecofreindly method of nematode management (Singh and Sitaramaiah, 1970; Sharma *et al.*, 1971). The value of decomposition of organic matter in the reduction of damage caused by plant parasitic nematodes was first demonstrated by Linford *et al.* (1938). Since than the value of organic matter for the management of plant parasitic nematodes, was tested by several workers in different parts of the world (Vandelann, 1956; Mankau, 1963; Singh and Sitaramaiah, 1966; Mobin *et al.*, 1969).

Use of organic amendments and bio agents not only provides unfavourable situation of the pathogens but also enrich the soil by improving the soil organic matter and soil fertility status and thereby enhance seedling vigour which in turn will help the grower and industry.

Visser (1952) emphasized that organic matter and planting of marigold followed by fallowing should be adopted to suppress nematode population at nursery. Application of organic matter and planting of two species of marigold, *viz., Tagetes erecta and T. patula* along with *Crotolaria* spp. reduced the population of root-knot nematode considerably followed by fallowing the soil at nursery (Visser 1952; Visser and Vythilingam, 1959). Significant reduction of gall formation have also been reported on tea seedlings grown on root-knot nematode infested soil by dressing with oil cakes of mustard, sesame and groundnut each at the rate of 0.5, 1.0 and 2.0g of soil and the effects were reported to be more pronounced at comparatively high rates of application (Anon, 1974a). Mustard cake and neem cake and decaffeinated tea waste when applied either @ 3 per cent (W/W) or 10–15g per seedling before planting were proved more effective in respect of suppressing *M. incognita* population and increasing plant growth parameters (Neog and Bora, 1999a; Dutta, 1995; Kalita, 2000).

Certain industrial and agricultural waste like sawdust, oil cakes, wheat bran, press mud, rice bran, vermicompost, decaffeinated tea waste (DCTW) etc. can be effectively utilized as organic amendments against root-knot nematode disease complex (Dutta, 1995; Neog and Bora, 1999a; Kalita and Bora, 2006).

Physical Methods

Soil heating and use of chemicals are the most effective method for the control of *M. incognita* in tea nurseries (Anon, 1980). Heat treatment of root-knot nematode infested soil at 60–67°C for 4–5 minutes before planting of tea seedling was found effective for reduction of root-knot nematode infestation and increasing seedling growth at nursery (Rao, 1976; Basu, 1978; Neog and Bora, 1999c). Steam sterilization of soil at 15 lb pressure per sq. inch for half an hour, heat treatment of soil by hot pan for 3 hours at 60°C and sun drying of soil continuously for 10 days are some of the effective treatment in respect of root-knot nematode control and thereby producing healthy nematode free seedlings of tea. Out of all these, steam sterilization of soil at 15 lb pressure per square inch for half an hour eradicate completely all the nematode population and thereby provides nematode free condition to the seedlings throughout the period (Neog and Bora, 1999c).

Resistant Varieties

Use of resistant varieties/clone is one of the effective and cheapest method of pest management. Limited works have so far been conducted on this aspect in relation to nematode management on nursery tea elsewhere. However, records show that 8 Tocklai released clones *viz.*, TV1, T2, V9, V10, V14, TV16, TV17 and TV18 were tested for their susceptibility to *M. incognita* and found that baring TV9 and TV14, the rest of the clones exhibited varying degree of resistance to *M. incognita* (Anon, 1972). It was also observed that out of 4 Tocklai released clones tested against *M. incognita* TV13 and TV15 were found slightly more susceptible than TV1 and TV18 (Anon, 1974b). Neog and Bora (1999b) found that Tocklai released tea clones TV1 and TV18 were moderately resistant to *M. incognita* than any other clone released by them. Seedlings of six different tea seed stocks when screened for their susceptibility to *M. incognita* and reported that the number of galls per plant was high in stocks 462, 463 and 464 and low in stock 491 and 493 while, stock 461 occupied an intermediate position (Anon, 1981). Gnanapragsam *et al.* (1985) reported six clones free from galling six years after planting in area cleared of old tea seedling heavily infested by *M. bravicauda*. However, after ten years, Clone TRI 214 showed a root-knot index of 3.24, while rest of the clone had indices ofless than 0.6.

It is well known fact that the early aged crops at their initial stage of establishment in nursery and in the planting sites are very much prone to the various soil borne pathogens including plant parasitic nematodes. Therefore, the seedlings and young plants immediately after planting in nurseries and main field must be protected from the above soil borne pest and pathogens up to a certain periods of its planting with a view to reduce mortality percentage and produce healthy and disease free seedlings by the growers. This will in turn increase the production and productivity of tea in long run.

Besides poor growth, transplanting of such type of diseased seedlings from infested nursery to planting sites is considered to be one of the prime causes of dispersal of these economically important disease causing agents. Infect, an infested nursery could serve as a source of primary inoculums to contaminate uninfected main fields. Therefore, adequate control of these soil borne pest and pathogens at this stage are vitally important to provide initial protection to the nursery tea with a view to prevent the dissemination of some of the dangerous species which are not known to occur in planting site earlier. This would definitely help the grower in particular and industry in general.

However, no systematic approach so far made particularly in respect of host parasite relationship and towards exploitation of indigenous materials and potential bio agents in managing the plant parasitic nematodes and soil borne fungal pathogens at nursery tea. It is felt very much essential to strengthening the concept of integrated pest and disease management strategy in nursery tea with a view to increase production and productivity of tea in long-run.

References

Agnihothrudu, V., 1995. Some more important fungal diseases of tea. *Indian Phytopathol.* 48: 229–253.

Agnihothrudu, V. and Baruah, G.C.S., 1960. Notes on fungi from N.E. India VI *Xylobotrymandinum* Pat. on Tea,. 14: 136–137.

Anonymous, 1957. Report, Tocklai Experimental Station, *Tea Res. Assoc.*, pp. 124–130.

Anonymous, 1972. Clonal susceptibility to eelworms. *Annual Scientific Report*, 1971–72, TRA, Tocklai Experimental Station, pp. 53–54.

Anonymous, 1974a. Organic dressing and nematode pathogenecity. *Annual Scientific Report*, 1973–74, TRA, Tocklai Experimental Station pp. 47–48.

Anonymous, 1974b. Clonal Resistance. *Annual Scientific Report*, 1973–74, TRA, Tocklai Experimental Station, 4B.

Anonymous, 1980. Effect of soil heating on *Meloidogyne iuncognita* infestation of tea seedlings and chemical control of *M. incognita.* India, Tea Research Association.

Anonymous, 1981. Susceptibilities of seedlings of different tea seed stocks to *Meloidogyne incognita* (Kofoid and White) Chitwood. *Annual Scientific Report,* 1980–81. TRA, Tocklai Experimental Station, 4B.

Anonymous, 1991. *Annual Scientific Report.* Tea Research Association, Tocklai Experimental Station, Jorhat, Assam, pp. 97.

Banerjea, B., 1966. Report, Tocklai Experimental Station, Tea Research Association, 1966, pp. 75–76.

Banerjea, B., 1967. Eelworm in tea: Some problems. *Two and a Bud,* 14: 82–83.

Barber, C.A., 1901. *Bull. Dept. Land Res. Agric., Madras Branch,* 3: 227–234.

Basu, S.D., 1967. Eelworm in tea: Some important groups. *Two and a Bud,* 14: 84–85.

Basu, S.D., 1968. Eelworms: A progress report on a few more groups found in North East Indian tea soils. *Two and a Bud,* 15: 70–71.

Basu, S.D., 1978. Heat the soil and kill the nematode. *Two and a Bud,* 25: 42.

Basu, S.D., 1980. Control of *Meloidogyne incognita* on tea seedlings with some systemic nematicides. *Two and a Bud,* 49: 25–28.

Basu, S.D. and Roy, S.K., 1976. A preliminary note on nematode survey in Darjeeling. *Two and a Bud,* 23: 58–59.

Basu, S.D. and Roy, S.K., 1979. Sample survey of nematodes in Dooars. *Two and a Bud,* 26: 69–70.

Basu, S.D. and Gope, B., 1985. Control of root-knot nematode, *Meloidogyne incognita* (Kofoid and White) Chitwood on tea seedling with some systemic nematicides. *Two and a Bud,* 32: 32–35.

Bordoloi, D.N. and Baruah, H.K., 1970. Studies on the distribution of certain fungi in tea plantation soil and air. *Univ. Gauhati,* 18–19: 153–170.

Borthakur, B.K. and Dutta, P.K., 1992. Prospect of Biocontrol of tea disease in N.E. India. In: Proc. *Thirty First Tocklai Conference,* pp. 163–167.

Das, S.D., 1955. Effect of infestation of *Meloidogyne incognita* (Kofoid and White) Chitwood on tea plant in North-East India. *Two and a Bud,* 16: 28–29.

Das, G.M., 1958. *Report,* Tocklai Experimental Station, pp. 229.

Dutta, P., 1995. Pathogenicity and management of *Meloidogyne incognita* on tea seedlings. *M.Sc. (Agri) Thesis,* Assam Agril. University, Jorhat, Assam.

Gadd, C.H. and Loos, C.A., 1946. *Tea Quarterly,* 98: 3–11.

Glover, P.M., 1961. Nematodes in North-East India. *Two and a Bud,* 8: 6–7.

Gnanapragasam, N.C., Jabamalai, A.S. and Anpalagan, V.T., 1987. Control of root lesion nematodes *Pratylenchus loosi* in contaminated tea nursery plants. *Sri Lanka J. Tea Sci.,* 56: 95–97.

Gnanapragasarn, N.C., Sivapalan, P., Jabamalai, A.S. and Dharmasena, W.A.M., 1985. Tolerance/ susceptibility of six different clones to *Meloidogyne bravicauda* the root-knot nematode of mature tea. *Int. Nematol. Network Newsl.,* 2: 9–10.

Ichinohe, M., 1965. Nematode menace perennial field crops. *Biokemia,* 9: 10–13.

Kalita, S., 2000. Pathogenicity and management of *Meloidogyne incognita* on tea seedlings. *M.Sc.(Agri) Thesis,* Assam Agril.University, Jorhat, Assam.

Kalita, S. and Bora, B.C., 2006. Effect of nematicides and organic amendments in the management of *Meliodogyne incognita* in nursery tea. *Indian J. Nematol.*, 36: 148–149.

Kerr, A. and Vythilingam, M.K., 1966. Fumigation of nursery soil with methyl bromide. *Tea Quarterly*, 37: 162.

Kerr, A. and Vythilingam, M.K., 1967. The occurrence and control of the root lesion eel worm (*P. loosi*) in nurseries. *Tea Quarterl.*, 38: 22–28.

Ligfht, S., 1928. Tea nematodes. *Tea Quarterly*, 1: 19–22.

Linford, M.B., Ypp, F. and Oliviera, J.M., 1938. Reduction of soil population of the root-knot nematode during decomposition of organic matter. *Soil Sci.*, 45: 127–141.

Loos, C.A., 1951. Eelwoms. *Tea Quarterly*, 22: 27–30.

Loos, C.A., 1953. *Tea Quarterly*, 34: 34–38.

Mankau, R., 1963. Effect of organic soil amendments on nematodes population. *Phytopathol.*, 53: 881–882.

Manuelipillai, N., 1981. Non arthropod pests nematodes and rodents. In: *Pest Management in Tea.* United Planters Association of Southern India, Valparai, India, p. 77–81.

Mobin and Khan, A.M., 1969. Effect of organic amendments on the population of rhizosphere fungi and nematodes around the roots of some fruit plants. *All India Nema. Sympo.* Held at IARI, New Delhi, 21st and 22nd August, p. 68.

Mukherjea, T.D., 1960. Destructive eelworm of tea seedling. *Two and a Bud*, 7(2): 5–6.

Mukherjea, T.D., 1963. Chemical control of root-knot nematodes parasitic on tea seedlings in North-East India. In; *All India Nematology Symposium*, New Delhi, August, 22–23.

Mukherjea, T.D., 1966. Chemical control of root-knot nematodes parasitic on tea seedlings in North-East India. *Trop. Agricultural Trin.*, 43: 335–340.

Mukherjee, B. and Dasgupta, M.K., 1981. New records of plant parasitic nematode infestation and their significance in decline in tea plantations of West Bengal. India. *Current Science*, 50: 824–825.

Neog, P.P. and Bora, B.C., 1999a. Organic amendments for the management of *Meloidogyne incognita* in tea nursery. *JASS*, 12: 203–206.

Neog, P.P. and Bora, B.C., 1999b. Screening of certain tea clones against *Meloidogyne incognita. Indian J. Nematol.*, 29: 92–93.

Neog, P.P. and Bora, B.C., 1999c. Management of root-knot nematode (*Meloidogyne incognita*) in tea nursery with physical treatments. *Indian J. Nematol.*, 29: 101–103.

Phukan, P.N. and Sanwal, K.C., 1979. Taxonomic studies on nematodes from Assam, India. *Indian J. Nematol.*, 9: 20–26.

Pitcher, R.S., 1963. Role of plant parasitic nematodes in bacterial diseases. *Phytopathol.*, 53: 35–39.

Pitcher, R.S., 1965. Inter-relationship of nematodes and other pathogens in plants. *Helminth. Abstr.*, 34: 1–17.

Powell, N.T., 1963. The role of plant parasitic nematodes in fungus diseases. *Phytopathol.*, 53: 28–35.

Powell, N.T., 1971. Interaction of plant parasitic nematodes with other disease causing agents. In: *Plant Parasitic Nematodes*, (Eds.) B.M. Zuckerman, W.F. Mai and R.A. Rhode. Academic Press, New York, 2: 119–136.

Rang, W.Z., Juan, J. and Zhang, K.S., 1984. Study of tea seedling disease caused by root knot nematode (*Meloidogyne* sp.). *Acta Phytopathologica Sinica,* 14: 225–232.

Rao, G.N., 1970. Tea pests in Southern India and their control. *Pest Artic News Summ.,* 16: 667–672.

Rao, G.N., 1976. Control of nematode in nursery soil. *Planters Chronicle,* 71: 257–259.

Seshadri, A.R. and Kumarswami, T., 1963. A preliminary report of the occurrence of root-knot nematode (*Meloidogyne* spp.) in some important crop plants in Madras State. *Madras Agric. J.,* 50: 98–99.

Sharma, K.C., 1960. Disease of tea and associated crops in N.E. India. Memorandum No. 26. Indian Tea Association, Tocklai Experimental Station, Jorhat.

Sharma, A.N., Sharma, V.K. and Singh, R.S., 1971. Oil cake and sawdust control root knot. *Indian Fmg.,* 20: 17–20.

Singh, R., 1989. Plant parasitic nematodes associated with plantation crops of Jorhat districts. *M.Sc. (Agri) Thesis,* AAU, Jorhat.

Sivapalan, P., 1969. Further development in the control of the root lesion nematode (*P. loosi*) in tea nurseries. *Tea Quarterly,* 40: 111–114.

Sivapalan, P., 1972. Nematode pests of Tea. In: *Economic Nematology,* (Ed.) John M. Webster. Academic Press, New York, pp. 285–311.

Singh, S.P., 1998. In: *Biological Suppression of Plant Diseases, Phytoparasitic Nematodes and Weeds.* Project Directorate of Biological control, Bangalore, pp. 1–6.

Singh, R.S. and Sitaramaiah, K., 1966. Incidence of root-knot of okra and tomatoes in oil cake amended soil. *Plant Disease Reporter,* 50: 668–672.

Singh, R.S. and Sitaramaiah, K., 1970. Control of plant parasitic nematodes with organic soil amendments. *Pest Artic. News Summs.,* 16: 287–297.

Srivastava, A.S. and Singh, B., 1967. *Labdev Journal of Science and Technology,* 3: 259–263.

Thirungnanam, M. and Rangaswami, G., 1967. A report on naturally occurrence and control of *Meloidogyne incognita* in tea seedlings *Madras Agril. J.,* 54: 278–282.

Tunstall, A.C. and K.C. Sarmah, 1947. Notes on the stem diseases of tea in N.E., India. Memorandum, Tocklai Experimental Station, Indian Tea Association, 16: 1–17.

Vanderlann, P.A., 1956. The influence of organic manuring on the development of potato eelworm, *Heterodera rostochiensis. Nematologica,* 1: 112–115.

Venkataramani, K.S. and Venkata Ram, C.S., 1959. *Rhizoctonia solani* inciting a collar rot of tea. *Phytopathology,* 49: 527.

Visser, T. and Vythilingam, M.K., 1959. The effect of marigold and some other crops on the *pratilenchus* and *Meloidogyne* population in tea soil. *Tea Quarterly,* 30: 30–38.

Visser, T., 1952. Observation prevalence and control of parasitic eelworm problem in tea. *Tea Quarterly,* 30: 96–107.

Visser, T., 1959. Practical aspects of the eelworm problem in tea. *Tea Quarterly,* 30: 143–149.

Whitehead, A.G., 1969. Nematode attacking coffee, tea and cocoa and their control. Nematodes of Tropical Crops. *Tech Comm. Bur, Helminth,* 40: 238–250.

Insect Pest and Disease Management *Pages* ***468–479***
Editor: **Dr. D. Prasad**
Published by: **DAYA PUBLISHING HOUSE (2016)**

Chapter 34

Ufra: A Nematode Disease in Deep Water Rice and its Management

Debanand Das[1] and Bharot Ch. Bora[2]
[1]Regional Agricultural Research Station,
Assam Agricultural University, North Lakhimpur– 787 001
[1]Department of Nematology, Assam Agricultural University,
Jorhat – 785 013, Assam

Introduction

Deep water rice, commonly known as 'bao rice' or 'floating rice' is considered as one of the important crops in many parts of Asia and West Africa. It is estimated that about 25–30 per cent world's rice land is covered under deep water rice. There are extensive areas in Bangladesh, India, Thailand, Myanmar, Vietnum, the Phillipines, Indonesia, Guinae, Mali, Sierra Leone and Egypt. The largest area of deep water rice is concentrated in Bangladesh with about 2 million ha followed by its neighbouring country India, with about 1.7 million hecteres. Amongst the African countries, Mali is the most prominent covering 70 per cent of rice growing areas under deep water rice. Deep water rice is grown from latitude 27°N in Assam and Uttar Pradesh, India to latitude 3°S in South Kalimanthan, Indonesia. Of the 9–10 million ha of deep water rice in Asia, a region of 63 per cent is occupied by two great rivers, the Gangas and the Brahmaputra in eastern India and Bangladesh, followed by 14 per cent in Mekong flood plains of Kampuchea, Vietnum and 9 per cent in Chao Phraya river basin in Thailand; which annually cause deep and extensive flooding. It is only the deep water rice that provides a means to use this unruly flood water in the service of human beings because of its ability of internodes elongation as flood water deepens. Many deep water rice varieties are able to elongate to 14 ft. Area under deep water rice in India is low as compared to other rice, which is confined within the states of Eastern India like Assam, West Bengal, Bihar, Orissa, and Uttar Pradesh.

'Ufra' in India, *'Dak pora'* in Bangladesh, *'Okhet Pet'* in Myanmar, *'Yad Ngo'* in Thailand and *'Tim Dot San'* in Vietnum are the local name of one dangerous disease of deep water rice, caused by *Ditylenchus angustus* (Butler, 1913) Filipjev, 1936. In 1908, people of rice growing belts of Noakhali and Tippera districts of Bangladesh (then Eastern Bengal) observed a new and virulent type of disease, particularly at the head of Bay of Bengal. *'Ufra'*–the local cultivators named this disease, was then believed to be caused due to atmospheric conditions and the curious natural phenomenon, known as 'Barisal guns'. The term "ufra" also believed to be derived from "upara" meaning "above" (Butler, 1913). Mr. C.W. Mason, Supernumerary Entomologist, Pusa was the first scientific personnel to visit the disease hit areas at Tippera and in 1909 confirmed that the symptoms were not caused by any insect attack. But in 1910, a report threw doubt on the conclusion made by Mr. Mason and in December 1911, a team comprising Mr. E.J. Butler, Mr. B. Fletcher and Mr. A.G. Brit visited the disease hit areas of Chaumuhaf, Bangladesh. After thorough examination of the diseased samples, Butler confirmed the presence of a semi parasitic nematode genus *Tylenchus* and the same was published in the *Bulletin of Bureau of Agricultural Intelligence and Plant Diseases*, 3rd vol. No 7 July, 1912 p. 1661. In 1913, Butler gave precise description of the *Ufra* disease in rice in Bengal and described the causal organism as *Tylenchus angustus*. Later on Filipjev (1936) transferred *T. angustus* to *D. angustus*. This disease has also been recorded in winter rice as well as in transplated rice (Bakr, 1978). Catling *et al.* (1988) recognized *ufra* as one of the eight most important disease of deepwater rice.

Distribution

Occurrence of *ufra* disease is reported from several deep water rice growing countries of the world. However, its occurrence is restricted to certain tropical areas where environmental condition is suitable for its development (Ou, 1985). After Butler (1913) from Bangladesh, occurrence of this disease was reported from Malaysia (Jack, 1923), Irrawady delta of Myanmar (Seth, 1939), the Phillipines (Rayes and Palo, 1956), Egypt (Sasser and Jenkins, 1960), Thailand (Hashioka, 1963), Madagascar (Voung, 1969) and Vietnum (Kinh, 1981).

Singh (1953) reported occurrence of *ufra* disease from Uttar Pradesh, India for the first time. Subsequently, this disease is reported from West Bengal, Orissa, Assam and more recently from Maharastra (Chakraborty *et al.*, 1985; Rao *et al.*, 1986; Rathaiya and Das, 1987; Patil, 1998). In Assam this disease was first reported from Jorhat (Roy,1977) and was known as "Bao paddy malady". Subsequently, the disease was recorded from Lakhimpur, Dhemaji, Sonitpur and Sibsagar districts of the state (Anon, 1991).

Epidemiology

Ditylenchus angustus generally called 'rice stem nematode' or *'ufra nematode'* is an obligate ecto-parasite. This is stubble borne in nature and infested plant residues are believed to be the primary source of infestation. After harvesting of the infested crop, the nematodes survive in the fallow field and wait to the next crop on wild rice, ratoons, weed grass and left over stubbles (Butler, 1919; Miah and Bakr, 1977a). In the infested rice stubbles they remain dormant forming coil, particularly inside peduncles, under upper leaf sheaths and within the glumes of the lower grains of the panicle. The number of nematode in single infested stubbles varies from 1-30,000 (Anon, 1991, Cox and Rahman, 1979a, Catling *et al.*, 1979, Rathiah, 1899) and in seeds varies from 5.3 to 2,400 (Sein, 1977; Pathak, 1992; Ibrahim and Perry, 1993).Though, occurrence of this nematode in soil and seed has been reported (Sein, 1977a; Cuc and Giang, 1982; Cuc, 1982 b), but this nematode is not soil borne (Hashioka, 1964; Pathak, 1992). In seed, they are located at the overlapping region of the rice husk (personal observation). Infestation starts when seedlings are few days old. When atmospheric temperature is 28-30°C and humidity is above 85 per cent (Butler, 1919; Mc Geachie and Rahman, 1983), the active nematodes climb up the surface of the newly germinated rice seedlings and enter into the tender growing tissues. Free water is not absolutely necessary for their movement. Coleoptiles of the germinating seeds are readily infested. They enter into the inner portion of the growing tissue by working their way between folded leaves and sheaths, but they never enter bodily through the tissue. As the seedlings grows the nematode moves up to the new tissue through the space between the leaf sheaths. At four leaf stage of rice plant, nematode usually takes about three days to reach the inner most leaf sheath. They suck sap from the epidermal cells with their stylet from the tender growing tissues and developing inflorescence (Butler, 1919; Padwick, 1950; Ou, 1972; Bridge, 1976) where epidermal cell walls are usually very thin and tender. Compact nature of the inner most leaf fold prevents the growing point from nematode attack and is believed to be the reason that the infested plants are not killed, but grow to flower (Butler, 1913 a). The nematode does not produce any toxic substances while feeding (Butler, 1919), but continues to suck sap through their small stylet.

Reproduction occurs on the host plant and takes place during the month of May-June and November, but number of generations in a single season is not certain and quantitative estimation of eggs laid per female is difficult to ascertain (Padwick, 1950; Perry, 1995). However, three generations in a cropping season and 50–100 eggs per female has been reported by Butler (1913a, b), and this may be confirmed by the observation of three population peaks in a cropping season by Cox and Rahman (1979a). In the artificial condition, the developmental cycle of *D. angustus* takes 15 days from second stage juvenile (J_2) to become adult, 21 days from J_2 to egg and 24 days to complete its life cycle (Cuc, 1989). However, Plowright and Gill (1994) mentioned that period of life cycle of *D. angustus* is 10-20 days at 30°C. When the crops moves towards maturity, nematodes are found in large numbers at the top of the stem, in the panicle and in the seeds, become inactive, each being tightly coiled in a circular fashion with the head at the centre. While place in water, they uncoil and become active and vigorously motile, progressing in a wrinkling or snake like motion. At 31°C they become active much more quickly. They can move on a solid surface at an atmospheric humidity of 85 per cent or more.

Symptoms

The symptoms of *ufra* is significantly characteristic and vary considerably with the degree and time of infestation. In the field, the symptoms develop two months of crop growth but at this stage of crop growth it is difficult to recognize that the crop is infested by *ufra* nematode. In seedling stages, the first and visible symptoms appear are mosaic like discoloration arranged in a splash pattern or chlorosis or white streaks on young leave and sheaths; becoming more evident with time. The entire leaf may become twisted or severely malformed. In some cases, the basal portion of the young leaf becomes wrinkled, following whitish green discoloration. At the advance stage of the disease, the chlorotic portion of the leaf becomes brown to dark brown. In severe case, the margin of the leaf may crinkle and leaf tip is twisted. Many a time, the central leaf may emerge from the side of the stem by forming a loop or the central leaf may be curly, then become dry (Butler, 1913b; Watts Padwick, 1950; Timm and Ameen, 1960; Hashioka, 1963; Ichinohe, 1972; Ou, 1985). Sometimes, several branches are produced from the infested node and make the plant bushy. Of course, some times the symptoms are masked and plant looks healthy. Up to heading stage, little is known about this disease. Most typical and conspicuous symptoms are seen at the heading stage of the crop, which again vary considerably according to degree and time of infestation. (Butler, 1913b) classified the disease into two types: *"thor ufra"* or *swollen ufra* and *"pucca ufra"* or *ripe ufra* are sen with top of the shoot swollen in to a spindle shaped thickening, consists of immature ear enclosed within. The ear remains imprisoned up to harvest and becomes mouldy and rotten. In Bangladesh, the symptom is called *'thor'* (swollen) *ufra*.

'Thor' ufra is observed in the field when infestation occurs from primary inoculums, like left over stubbles of the previous year. Cox (1980) and Cox and Rahman (1980) indicated this symptom as *'ufra I'*. The panicle coils, twisted or distorted and bears sterile or empty spikelets. A strong tendency towards branching at the infested portion and even formation of two to three distorted ear head has been observed (Padwick, 1950; Hashioka, 1963). When the plants are attacked by secondary inoculums, then half of the panicle emerges out of the sheath and other portion remains enclosed within the sheath. The grains are unfilled, sterile and empty. The term *'pucca* (ripe) *ufra'* is used in Bangladesh to indicate these symptoms. The uppermost sheath surrounding the half emerged panicle is withered and bears characteristic brown strains. On removing the sheath, the stalk is found blackened and shrunken. The term *'Ufra II'* is used by Cox (1980) and Cox and Rahman (1980) to indicate this disease symptoms. When rice plants are attacked much late (tertiary inoculums), then panicle completely emerges out of the boot leaf containing healthy grains at the tip of the panicle and false grain at the base. This type of symptoms is termed as *'Ufra III'* (Cox, 1980; Cox and Rahman, 1980).

Ufra symptoms generally appear in patches in the field and enlarged with the advent of crop age. Severely infested plants are noticed at the centre of the patches with *ufra I* symptom than the edges. Field with severe *ufraI and ufra II* symptoms, yield may decline to zero, while only with *ufraIII*, some yields may be expected.

Yield Loss

Although this disease is most devastating to deep water rice, it can also attack the transplanted winter rice and summer rice (Bakr, 1978, Miah and Rahman, 1981; Miah 1984; Chakravarti *et al.*, 1985). Butler (1913 a) observed the devastation of this disease, when he recorded complete grain loss in many fields in Bangladesh and around 200,000 mounds of grain loss in Begumganj Thana in Bangladesh. The disease was severe in some fields that, farmers could not expect any grain to harvest and the crop was cut for cattle feed (Butler, 1913 and Padwick, 1950). Miah and Bakr (1977a) recorded that *ufra* disease affected only 2 per cent deep water rice in national scale in Bangladesh, resulting in average yield loss of 50 per cent in particular fields and even 100 per cent in individual farms. Mondal and Miah (1987) reported that 60-70 per cent of low lying areas in Bangladesh were infested with *D. angustus.* Yield loss of 20-100 per cent in transplanted aman and deep water rice, 10–100 per cent in boro rice and 10-50 per cent in transplanted aus rice had been reported by Rahman *et al.* (1994) in Bangladesh. Loss in grain yield due to *ufra* disease is variable from year to year and country to country depending upon the environmental condition that prevails during period of disease development. In Uttar Pradesh, India and Pathalung, Vietnum the loss in grain yield was estimated to be 20–90 per cent (Singh, 1953; Hashioka, 1963), in West Bengal, India 30 per cent in the surveyed area (Pal, 1970), 10-15 per cent in Assam and West Bengal, India (Rao *et al.*, 1986), 20 per cent in Bangladesh (Catling *et al.*, 1978) and 20–100 per cent in Vietnum (Kinh and Phuang, 1981). Losses ranging from 40–100 per cent have been reported from Vietnum, Bangladesh and India (Cuc and Kinh, 1981; Miah and Bakr, 1977c; Miah, 1984; Chakraborti *et al.*, 1985). If the nematode inoculums present in the field at the time of sowing, the infestation is more (Rahman and Evans, 1987). Field experiment for two consecutive years in Bangladesh (1985 and 1986) revealed that 10 per cent infested seedling at the time of transplanting are sufficient to cause significant yield loss (Mondal *et al.*, 1989). Loss in grain yield depends on the per cent *ufra* infestation also. If there is more than 40 per cent ufra II symptoms in a field, the yield loss may go up to 100 per cent (Cox and Rahman, 1980).

Survival, Perpetuation and Transmission

Ditylenchus angustus is an obligate ecto-parasite. Infested materials are the primary source of inoculums. This parasite can survive in quiescent stage in grains, dry paddy stubbles, in infested field on the surface of soil until the next cropping season comes. They also survive in active stage after harvest of paddy in rations, weeds and wild rice (Sein and Zan, 1977; Cuc, 1982 a; Miah and Bakr, 1977 b; Pathak, 1992). In dry paddy stubbles the nematode can survive desiccation for at least 15 months (Butler, 1913; Miah and Bakr, 1977c; Catling *et al.*, 1979), but after harvest, their population in stubbles reduces drastically to leave no live nematode after 4-5 months of harvest (Cox and Rahman, 1979 b; Kinh, 1981, Annon, 1981). There is no specific 'survival stages' of *Ditylenchus angustus* and in the dry diseased stubbles three stages of *D. angustus*, J_3, J_4 and adult are found, J_4 being the predominant. The preponderance of J_4 in the secescing plants together with superior survival attributes make it possible that J_4 is the principal stage besides J_3 and adult that survive drying and subsequently invades the next season crop (Ibrahim and Perry, 1993, 1994).

Perpetuation of *D. angustus* from one field to another or from one plant to another takes place through plant residues, water (flood or irrigation water), rain splash and stem and leaf contact under high humid condition (Rahman and Evans, 1987; Hashioka, 1963; Sein and Zan, 1977). Though Seshadri and Dasgupta (1975) mentioned that chances of transmission of this nematode through seed is minimal, still there is possibility of dispersal of this nematode through seeds from infested field as considerable number of *D. angustus* is reported in filled grains containing more than 12 per cent moisture (Cuc and Giang, 1982, Ibrahim and Perry, 1993). Mondal and Miah (1987) reported that tidal water plays important role in spread of this nematode.

Influence of Biotic and Abiotic Factors in Disease Development

Humidity, temperature and rainfall plays important role to break the dormant stage of the nematode. *Ditylenchus* angustus can not withstand prolonged desiccation at low humidity as they do not have intrinsic ability to control the rate of water loss from body. Occurrence of spring rainfall which leads to a high relative humidity of more than 85 per cent is essential for nematode revival from quiescent stage, to crawl to the tip of the crop to cause infection (Butler, 1919; Ou, 1985). If spring rain comes late, incidence of *ufra* becomes less (Cox *et al.*, 1980). This disease is severe in wet years. Depth of water is also an important factor in determining the infection and early symptom development. When water level during infection is adjacent to, or just below the collar region at the top of the leaf sheath, then symptom develops more quickly (Plowright and Gill, 1994). Temperature of 35°C or above is always lethal to the nematode. Therefore, winter rice which is sown during March-May and Harvested during November-December is less infected (Ou, 1972). Thi *et al.* (1982) reported that when seed humidity decreases below 12 per cent, nematodes were killed.

Population of *D. angustus* and severity of disease is more in less fertilized or non fertilized soil then fertilized soil (Ahmed, 1989a). Application of potash and sulphur reduces the *ufra* disease by reducing multiplication of the nematode (Annon, 1985). Loss of grain yield due to *ufra* is aggravated when rice is grown in zinc deficient soil (Ahmed, 1989 b; Miah *et al.*, 1984). Zinc deficiency indirectly encourages the accumulation of certain amino acids in plant and helps in disease development.

Host Range

Almost all the cultivated deepwater rice varieties including some of the transplanted rice varieties are good host of *Ditylenchus angusatus.* 13 species of *Oryza viz., O. perennis, O. sativa var. fatua, O. glaberima, O. cubensis, O. officinalis, O. meyaiana, O. latifoila,O. eichingeri, O. alta, O. minuta, O. nivera, O. rufipogon, O. spontanese* (Hashioka, 1963; Miah and Bakr, 1977b; Sein and Zan, 1977) are reported as host crop of *D. angustus.* Apart from *Oryza* spp. *Hygroryza aristata, Leersia hexandra* (Miah and Bakr, 1977a; Voung, 1969; Cuc, 1982 a; Sein and Zan, 1977); *Paspalum scorbiculatum* (Pathak, 1992); *Sacciolepsis interrupta and Echinochola colona* (Cuc.1982a) were recorded as alternate host of *D. angustus.* Five different fungi namely *Curvularia oryzae, Epicoccum purpurascens, Pestalotia oryzae* and *Rhizoctonia solani* can also harbour this nematode.

Management

Field Sanitation

Burning of diseased stubbles, followed by ploughing reduces the incidence of *ufra* disease and is suggested as an effective control measures (Butler, 1919; Padwick, 1950; Hashioka, 1963; Ou, 1972). For effective burning, uprooting and drying of the plants prior to setting fire important. Stubble burning should be carried out in an organized way and in a community basis, because if some infested patches remains unburnt, this will act as source of inoculum for subsequent crop. Ploughing should be carried out soon after harvest of the crop so as to give sufficient time to decompose the stubbles and to expose the nematodes to the soil, where they do not survive for a long period and become unable to transmit the disease. The field should be kept free from any weeds and ratoons. Removal of infected leaves along with upper portion of rice plant also reduces the nematode population in the field.

Crop Rotation

Rotation of deep water rice with non-host crop such as jute (*Oletorious* spp.) and mustard (*Brassica* sp.) reduces the *ufra* disease in the next crop (Butler, 1919). In areas with good irrigation facilities, cultivation of autumn rice (*boro* rice) followed by summer rice (*ahu* rice) will result in control of this disease, as this nematode will find no host during its active period. However, this practice may not be feasible in most of the deepwater rice growing areas, where flood water rises to a high level, resulting cultivation of *ahu* rice is a problem.

Delay in Sowing/Transplanting

The half life of *Ditylenchus angustus* population in infested stubbles is about two weeks and the nematodes just able to manage to survive up to the end of April. Kinh (1981) failed to collect any live nematode by early February from diseased stubbles. The length of the over wintering period is the critical factor in regulating the survival of *D. angustus from one season to the next and* lengthening this period should reduce primary infestation (Cox and Rahman, 1980). If one can prolong the decay phase even by only a few weeks, *ufra* can be controlled. This could be achieved by sowing later, transplanting much later, but before the flood and harvesting earlier (Mc Geachie and Rahman, 1983). Das and Bhagawati (1992) recorded 100 per cent *ufra* incidence in early sowing crop which decreases gradually with delay in sowing time to reach its minimum (37 per cent) on first week of May. Cultivation of 'Padmapani' a popularly growing early maturing variety is found to escape the incidence of this disease.

Application of Fertilizer

Application of balanced fertilizer can reduce the *ufra* infestation. Population of *Ditylenchus angustus* and severity of disease is more in less fertilized or non fertilized soil then fertilized soil (Ahmed, 1989a). Record of zinc deficiency in the endemic areas of *ufra* disease indicated a relationship between zinc deficiency in soil and plant susceptibility to *ufra* disease. Infection was found aggravated when deepwater rice is grown in zinc deficient soil than those grown in zinc added soil (Miah *et al.*, 1984). Application of 50 ppm of zinc and 100 ppm potash and 120 ppm of sulphur to the deficient soil significantly reduces the severity (Mondal and Miah, 1984, 1985, 1986). Application of potash and sulphur reduces the *ufra* disease by reducing multiplication of the nematode (Annon, 1985).

Resistant Variety

According to Butler (1919), every variety of paddy tested were susceptible to *ufra* disease. But, subsequently a good number of rice cultivars are reported to be tolerant/resistant to this disease (Hashioka, 163; Sein, 1977 c; Miah and Bakr, 1977a; Rahman and Mc Geachie, 1982; Rahman, 1987; Pathak, 1992; Das and Sharmah, 1995; Sarmah *et al.*, 1996; Das *et al.*,2000). Some of the varieties resistant to rice stem nematode includes: Bazail-65, Jalamagna, Basudeu, AR-9(C), IR 13437-20-4E-PI, IR 17643-4, Lakhi, Karkati, BR 308-3-3-2, Rayada 16-011, Rayda 16-013, Rayada 16-05, Rayada 16-06, Rayada 16-07, Ba Tuc etc.

Chemical Control

A good number of chemical pesticides have been known to be useful for control of ufra disease. Application of Diazinon at 100 ppm concentration on soil (Srivastava and Saxena, 1956), Hexadrin @ 1 lit/700 lit of water two times at two weeks interval (Pal, 1970), Carbofuran solution @ 30 kg a.i./ ha (Sein, 1977 b) Fensulfothion @ 5 kg a.i./ha on soil and Diazinon @ 15 kg a.i./ha at 15, 45 and 75 days after transplanting (Voung and Rodriguez, 1972), Carbofuran @ 0.67 kg a.i./ha (Miah and Bakr, 1977b), Carbofuran @ 24 kg a.i./ha on soil followed by Benomyl spray @ 2.5 kg a.i./ha (Cox and Rahman, 1979 c) found effective to suppress *ufra* disease in field. Application of Carbofuran either at broadcasted or at transplanted rice significantly decreases *ufra* infestation and increases yield. When Carbofuran is applied at the time of transplanting, maximum yield and minimum *ufra* infestation is obtained (Rahman and Taylor, 1983; Rahman and Miah, 1989; Rahman, 1993).

Seedling root dip treatment either with Carbofuran 3 per cent, Miran 3 per cent or Tecto 40FL @ 2.5–10 per cent also controls *ufra* disease effectively (Mondal and Miah, 1987 b).

Treatment combination of seed dressing with Carbosulfan 25 STD plus soil application of Carbofuran 3G plus foliar spray with Carbosulfan 40 EC at 40, 120 days after sowing the seeds and before panicle initiation stage can effectively reduce *ufra* disease in deep water rice (Das, 2004). Two foliar sprays with Carbosulfan 40 EC @ 0.2 per cent before flowering also found effective in reducing this disease (Das, 1996).

Certain limitations encountered in following the chemical pesticidal schedule in deep water rice. Most of this disease manifest under flooded condition and therefore, any curative measures is not possible. Application of chemical pesticides under such condition will cause serious health and environmental hazards and harm the fish fauna and other aquatic resources. Moreover, as the deep water rice is low yield potential type of rice, use of chemical pesticides will be not economical also.

Botanical Pesticides

Attempts have also been made to control *ufra* through application of bio-pesticides, in order to limit the health and environmental hazards. Chakraborti (2000a) reported that different neem formulations applied in combination, can effectively manage the *ufra* problem. Combination of cultural practices (field sanitation, crop rotation, deep ploughing) and neem based pesticide application can also reduce *ufra* infestation to a greater extent (Chakraborti, 2000 b). Incidence of *ufra* can be substantially reduce by following an integrated approach of burning of diseased stubbles plus application of Carbofuran at soil before sowing of seed along with two spray of neem product (Das and Saikia, 2005).

References

Ahmed, M.U., 1989a. Effect of soil types on the incidence of *ufra* disease of rice. A part of *Ph. D.* research work. Department of Zoology, Dhaka University.

Ahmed, M.U., 1989b. A lecture note prepared for '*International Training Workshop on Nematode Pest of Deep Water Rice*' held on 16-28 September, 1989 at Gazipur, Bangladesh.

Anonymous, 1981. *Annual Report*, BRRI, Jodebpur, Gazipur, 105–106p.

Anonymous, 1985. *Annual Report*, BRRI, Jodebpur, Gazipur.

Anonymous, 1991. *Biennial Report*, AICRP on plant parasitic nematodes with integrated approach for their control. Assam Agricultural Universitt, Jorhat Centre.

Bakr, M.A., 1978. Occurrence of *ufra* disease in transplanted rice. *IRRN*, 3(3): 16.

Butler, E.J., 1913a. Diseases of rice. I. An eelworm disease of rice. *Agr. Res. Inst. Pusa Bull.*, 34B: 1–27.

Butler, E.J., 1913b. Diseases of rice. I. *Agricultural Journal of India*, 8(3): 205–220.

Butler, E.J., 1919. The rice worm (*Tylenchus angustus*) and its control. *Botanical Series*, 10: 1–37.

Catling, H.D., Alam, S. and Miah, S.A., 1978. Assessing losses in deepwater rice due to insects and diseases in Bangladesh. *Experimental Agriculture*, 14: 1–11.

Catling, H.D., Cox, P.G., Islam, Z. and Rahman, L., 1979. Two destructive pests of deep water rice, yellow stem borer and ufra. *The ADAB News*, 6(8): 16–21.

Catling, H.D., Islam, Z. and Rahman, L., 1988. Status of pest and diseases in Asian deepwater rice. In: *Proceedings of the 1987 International Deepwater Rice Workshop*, International Rice Research Institute, P.O. Box 933, 1099, Manila, Philippines, p. 525–537.

Chakraborti, S., 2000a. Evaluation of neem products on management of *ufra* disease of rice. *Indian Journal of Nematology*, 30 (2): 234–236.

Chakraborty, S., 2000 b. Assessment of some management practices on rice stem nematode. *Indian Journal of Nematology*, 30(2): 245–247.

Chakraborty, H.S., Nayak, D.K. and Pal, A., 1985. *Ufra* incidence in summer rice in West Bengal. *IRRN*, 10(1): 15–16.

Cox, P.G., 1980. Symptoms of ufra disease in deep water rice in Bangladesh. *International Rice Research Newsletter*, 5(2): 18.

Cox, P.G. and Rahman, L., 1979a. *Ufra* nematode population in deep water rice in Bangladesh. *International Rice Research Newsletter*, 4(3): 10–11.

Cox, P.G. and Rahman, L., 1979b. The over winter decay of *Ditylenchus angustus*. *International Rice Research Newsletter*, 4(5): 4.

Cox, P.G. and Rahman, L., 1980. Effect of *ufra* disease on yield loss of deep water rice in Bangladesh. *Tropical Pest Management*, 26: 410–415.

Cox, P.G., Rahman, L. and Hannan, M.A., 1980. Set back for *ufra*. *International Rice Research Newsletter*, 5(1): 10–11.

Cuc, N.T.T., 1982a. New weed host of rice stem nematode identified in Vietnam. *International Rice Research Newsletter*, 7 (3): 15.

Cuc, N.T.T., 1982b. Field soil as source of rice stem nematode. *IRRN*, 7(4): 15.

Cuc, N.T.T., 1989. Study on the life cycle of *Ditylenchus angustus*. (Unpublished).

Cuc, N.T.T. and Giang, L.T., 1982. Relative humidity and nematode number of survival in rice seeds. *IRRN*, 7(4): 14.

Cuc, N.T.T. and Dang Ngoe Kinh, 1981. Distribution and damages of rice stem nematode disease in Vietnam. *IRRN*, 6(6): 14–15.

Das, Debanand, 2004. Chemical control of rice stem nematode, *Ditylenchus angustus* in deep ware rice in Assam. *Annals of Biology*, 20(1): 43–45.

Das, Debanand, Sarmah, N.K., Borgohain, R. and Das, P., 2000. Rayada-B_3: A high yielding, *ufra* resistant deep water rice for Assam, India. *International Rice Research Notes*, 25(3): 11.

Das, Debanand and Saikia, P., 2005. An integrated approach for management of *ufra* disease in deep water rice. *Annals of Plant Protection Sciences*, 13(2): 517–518.

Das, P., 1996. An integrated approach for management of rice stem nematode, *Ditylenchus angustus* in deep water rice in Assam. Indian *Journal of Nematology*, 22(1) (Publ. 1997): 222–225.

Das, P. and Bhagawati, B., 2002. Incidence of rice stem nematode, *Ditylenchus angustus* in relation to sowing time of deep water rice in Assam. *Indian Journal of Nematology*, 22(2) (Publ. 1994): 86–88.

Das, P. and Sharmah, N.K., 1995. Sources of resistance to rice stem nematode, *Ditylenchus angustus* in deep water rice genotype. *Annals of Plant Protection Sciences*, 3(2): 135–136.

Hashioka, Y., 1963. The rice stem nematode, *Ditylenchus angustus* in Thailand. *FAO Plant Protection Bulletin*, 11(5): 97–102.

Hashioka, Y., 1964. Nematode diseases of rice in the world. *Riso*, 13: 139–147.

Ibrahim, S.K. and Perry, R.N., 1993. Desiccation survival of rice stem nematode, *Ditylenchus angustus*. *Fundamentals and Applied Nematology*, 16(1): 31–38.

Ibrahim, S.K. and Perry, R.N., 1994. Infectivity and population dynamics of the rice stem nematode, *Ditylenchus angustus*. *Nematologica*, 40(3): 412–422.

Ichinohe, M., 1972. Nematode disease of rice. In: *Economic Nematology*, (Ed.) J. M. Webster. Academic Press, New York, p. 127–143.

Jack, H.W., 1923. Rice in Malaya. *Malaysian Agricultural Journal*, 11: 103–119, 139–169.

Kinh, Dang Ngoe, 1981. Survival of *Ditylenchus angustus* in diseased stubbles. *IRRN*, 6(6): 13.

Kinh, Daog Ngoe and Phuang, C.D., 1981. Reaction of some deep water rice varieties of *Ditylenchus angustus*. *IRRN*, 6(6): 6.

Latif, M.A. and Miah, I. H., 1995. Fungal hosts and gnotobiotic culture of *Ditylenchus angustus*. *Japanese Journal of Nematology*, 25(1): 11–15.

Mc Geachie, I. and Rahman, L., 1983. *Ufra* disease: A review and a new approach to control. *Tropical Pest Management*, 29(4): 325–332.

Miah, S.A., 1984. Disease problems and progress of research on *ufra* disease of rice in Bangladesh. *IRC Newsletter*, 33(2): 35 –38.

Miah, S.A. and Bakr, M.A., 1977a. Sources of resistance to *ufra* disease of rice in Bangladesh. *IRRN*, 2(5): 8.

Miah, S.A. and Bakr, M.A., 1977b. Chemical control of *ufra* disease of rice. *PANS*, 23(4): 412 –413.

Miah, S.A. and Bakr, M.A., 1977c. Nematode diseases of rice. In: *Literature Review of Insect Pests and Diseases of rice. BRRI*, p. 125–129.

Miah, S.A. and Rahman, M.L., 1981. Observation on incidence and chemical control of *ufra* in Boro fields. *International Rice Research Newsletter*, 6: 12.

Miah, S.A., Sahjahan, A.K.M., Sharma, N.R., Shah, A.L. and Bhuyan, N.I., 1984. Interaction of zinc deficiency and *ufra* disease on rice in Bangladesh. *Bangladesh J. Agric.*, 3(2): 61–67.

Mondal, A.H. and Miah, S.A., 1985. Effect of potash on the severity of *ufra* disease. *Bangladesh Journal of Plant Pathology*, 1(1): 67.

Mondal, A.H. and Miah, S.A., 1986. Sulphur reduces stem nematode pathogenecity. In: *Proceeding of 11th Annual Bangladesh Science Conference*, 3(1): 11.

Mondal, A.H. and Miah, S.A., 1987a. *Ufra* problem in low lying areas of Bangladesh. *IRRN*, 12: 29–30.

Mondal, A.H. and Miah, S.A., 1987b. Control of *ufra* of rice by seedling treatment. *Bangladesh Journal of Plant Pathology*, 3(142): 25 –30.

Mondal, A.H. and Miah, S.A., 1984. Effect of zinc on stem nematode infested rice. *International Rice Research Notes*, 9(5): 19–20.

Mondal, A.H., Rahman, L. and Miah, S.A., 1989. Yield loss due to transplanting of *ufra* infected seedlings of rice. *Bangladesh Journal of Botany*, 18(1): 67–72.

Ou, S.H., 1972. Rice stem nematode. In: *Rice Diseases*. CABI Publication, 346–351 pp.

Ou, S.H., 1985. *Rice Diseases*, 2nd edn. Commonwealth Mycological Institute. Kew, Surrey, England, 380 pp.

Padwick, W., 1950. *Manual of Rice Diseases*. Commonwealth Mycological Institute. Kew, Surrey, England. 198 p.

Pal, A.K., 1970. Survey of nematode infection of paddy and successful treatments with Hexadrin. In: *Proceedings of Indian Science Congress Association*, 57(111): 296.

Pathak, D.K., 1992. Some studies on the rice stem nematode, *Ditylenchus angustus* (Butler, 1913) Filipjev, 1936 of deep water rice of Assam. *M. Sc. (Agri) Thesis*, Assam Agricultural University, Jorhat, Assam.

Patil, M.D., 1998. Occurrence of *ufra* disease of rice in Maharastra, India. *Oryza*, 35(1): 86.

Perry, R.N., 1995. Rice stem nematode, *ditylenchus angustus* development and survival. *International Rice Research Notes*, 20(2): 21–22.

Plowright, R.A. and Gill, J.R., 1994. Aspects of resistance in deep water rice to stem nematode, *Ditylenchus angustus*. *Fundamentals and Applied Nematology*, 17(4): 357–367.

Rahman, M.L., 1987. Source of resistance in deepwater rice. *International Rice Research Notes*, 12(1): 8.

Rahman, M.L., 1993. Effect of time of nematicide application to control *ufra* disease. *Bangladesh Journal of Plant Pathology*, 9(1–2): 9–12.

Rahman, M.L. and Evans, A.A.F., 1987. Studies on host parasitic relationship of rice stem nematode, *Ditylenchus angustus* (Nematoda: Tylenchida) on rice (*Oryza sativa* L.). *Nematologica*, 33(2) (Publ. 1988): 451–459.

Rahman, M.L. and Mc Geachie, I., 1982. Screening of resistance of *ufra* disease (*Ditylenchus angustus*) in deepwater rice. In: *Proceeding of the 1981 International Deepwater Rice Workshop*, IRRI, Manila, Philippines, p. 459–466.

Rahman, M.L. and Miah, S.A., 1989. Effect of deepwater rice transplanting and application of urea and carbofuran on *ufra* disease. *Bangladesh Journal of Plant Pathology*, 5(1–2): 59–63.

Rahman, M.L. and Taylor, B., 1983. Nematode pests associated with deepwater rice in Bangladesh. *International Rice Research Notes*, 8(5): 20–21.

Rahman, M.L., Mondal, A.H. and Bakr, M.A., 1994. Widespread *ufra* disease incidence in different rice eco-systems in Bangladesh. *International Rice Research Notes*, 19(3): 27–28.

Rao, Y.S., Prasad, J.S. and Panwar, M.S., 1986. Stem nematode (*Ditylenchus angustus*)a potential pest of rice in Assam and West Bengal, India. *IRRN*, 3(4): 24–26.

Rathiah, Y., 1988. Diseases of deep water rice in Jorhat district, Assam, India. *International Rice Research Newsletter*, 14(2): 32–33.

Rathayah, Y. and Das, G.R., 1987. *Ufra* threatens deep water rice in Majuli, Assam. *IRRN*, 12: 4.

Rayes, G.M. and Palo, A.V. 1956. Nematode disease of rice. *Araneta Journal of Agriculture*, 3: 72–77.

Roy, A.K., 1977. Widespread occurrence of leaf scald and *ufra* diseases of rice in Assam, India. *International Rice Research Newsletter*, 2(6): 8.

Sarmah, N.K., Das, D., Hussain, S.A., Barman, B. and Senadhira, D., 1999. Nematode resistance of IRRI breeding lines in Assam, India. *International Rice Research Notes*, 24(1): 22–23.

Sasser, J.N. and Jenkins, W.R., 1960. *Nematology: Fundamentals and Recent Advances with Emphasis on Plant Parasitic and Soil Forms*. Chapel Hill. University of North Carolina, xv+480 pp.

Sein, T., 1977a. Seed borne infection of *ufra* disease. *IRRN*, 2(2): 6.

Sein, T., 1977b.Testing some pesticides against *ufra* disease. *International Rice Research Newsletter*, 7(7): 2.

Sein, T., 1977c. Varietal resistance to *ufra* disease in Burma. *International Rice Research Notes*, 2(2): 3.

Sein, T. and Zan, K., 1977. *Ufra* disease spread by water flow. *International Rice Research Newsletter*, 2(2): 5.

Seshadri, A.R. and Dasgupta, D.R., 1975. *Ditylenchus angustus. C.I.H. Description of Plant Parasitic Nematodes*. Set 5, No. 64, 4 p.

Seth, L.N., 1939. Report of the Mycologist Burma for the year ending 31st March, 1939, 6 p.

Sing, B., 1953. Some important diseases of paddy. *Agriculture and Animal Husbandry in India*, 3(10–12): 27–30.

Srivastava, A.S. and Saxena, H.P., 1956. Effect of diazinon on paddy nematode. In: *Proceeding 43rd Indian Science Congress*. 286 p.

Thi, N., Cuc, N.T.T. and Giang, L., 1982. Relative humidity and nematode number and survival in rice seed. *International Rice Research Newsletter*, 7(4): 14.

Timm, R.W. and Ameen, M., 1960. Nematode associated with commercial crops in East Pakistan. *Agric. Pak.*, 11(3): 355–368.

Voung, H. Hai., 1969. The occurrence in Madagascar of the rice nematodes, *Aphelenchoides besseyi* and *Ditylenchus angustus*. Technical Communication. Commonwealth Bureau of Helminthology, 40: 274–288.

Voung, H. Hai and Rodriguez, H., 1972. Result of trial for chemical control of nematode parasites of rice in Madagascar: *Aphelenchoides besseyi* Christie, 1942, *ditylenchus angustus* (Butler, 1913) Filipjev, 1936. Institute de Resherches Agronomiquis Madascar, No. 335: 24.

Watts Padwick, G., 1950. *Manual of Rice Diseases*. Commonwealth Mycological Institutem, Kew, pp. 198.

Insect Pest and Disease Management *Pages* **480–500**
Editor: **Dr. D. Prasad**
Published by: **DAYA PUBLISHING HOUSE (2016)**

Chapter 35

Biotechnological Approaches in IPM: Scope and Recent Development

N. Emmanuel and Swaran Dhingra
Division of Entomology, Indian Agricultural Research Institute,
New Delhi - 110 012

Introduction

Biotechnology can be broadly defined as "using living organisms or their products for commercial purposes." As such, biotechnology has been practiced by human society since the beginning of recorded history in such activities as baking bread, brewing alcoholic beverages, or breeding food crops or domestic animals. A narrower and more specific definition of biotechnology is "the commercial application of living organisms or their products, which involves the deliberate manipulation of their DNA molecules".

Combining DNA from different existing organisms (plants, animals, insects, bacteria, etc.) results in modified organisms with a combination of traits from the parents. The sharing of DNA information takes place naturally through sexual reproduction and has been exploited in plant and animal breeding programs for many years. However, sexual reproduction can occur only between individuals of the same species. What's new since 1972 is that scientists have been able to identify the specific DNA genes for many desirable traits and transfer only those genes, usually carried on a plasmid or virus, into another organism. This process is called genetic engineering and the transfer of DNA is accomplished using either direct injection or the *Agrobacterium,* electroporation, or particle gun transformation techniques. It provides a method to transfer DNA between any living cells (plant, animal, insect, bacterial, etc.). Virtually any desirable trait found in nature can, in principle, be transferred into any chosen organism. An organism modified by genetic engineering is called transgenic.

Gene Delivery Techniques

The ability to introduce foreign genes into plants, such that they are stably expressed and transmissible from generation to generation, has revolutionized plant biology. This technology provides a means of investigating the roles of specific genes in pathways of metabolism, growth and development, and opens up new possibilities for using plants as green factories for the production of an increasing range of commercially useful products.

DNA Delivery Systems

Agrobacterium Mediated Transformation

Agrobacterium tumefaciens, a soil plant pathogenic bacterium, has become the most used method for the introduction of foreign genes into plant cells and the subsequent regeneration of transgenic plants. A. *tumefaciens* naturally infects the wound sites in dicotyledonous plant causing the formation of the crown gall tumors. A. *tumefaciens* has the exceptional ability to transfer a particular DNA segment (T-DNA) of the tumor-inducing (Ti) plasmid into the nudeus of infected cells where it is then stably integrated into the host genome and transcribed, causing the crown gall disease. T-DNA contains two types of genes:

1. The oncogenic genes: encoding for enzymes involved in the synthesis of auxins and cytokinins and responsible for tumor formation;
2. The genes encoding for the synthesis of opines.

These compounds, produced by condensation between amino acids and sugars, are synthesised and excreted by the crown gall cells and consumed by A. *tumefaciens* as carbon and, nitrogen sources. Outside the T-DNA are located the genes involved in the process of T-DNA transfer from the bacterium to the plant cell and the genes involved in bacterium-bacterium plasmid conjugative transfer. Virulent strains of A. *tumefaciens* and A. *rhizogenes,* when interacting with susceptible dicotyledonous plant cells, induce diseases known as crown gall and hairy roots, respectively. These strains contain a large megaplasmid (more than 200 kb) which play a key role in tumor induction and for this reason it was named Ti plasmid, or Ri in the case of A. *rhizogenes.* During infection the TDNA, a mobile segment of Ti or Ri plasmid, is transferred to the plant cell nucleus and integrated into the plant chromosome. The T-DNA fragment is flanked by 25-bp direct repeats; which act as a *cis* element signal for the transfer apparatus. The process of T-DNA transfer is mediated by the cooperative action of proteins encoded by genes detennined in the Ti plasmid virulence region (*vir* genes) and in the bacterial chromosome (de la ' Riva *et al.,* 1998).

The initial results of the studies on T-DNA transfer process to plant cells demonstrate three important facts for the practical use of this process in plants transformation.

1. The tumor formation is a transformation process of plant cells resulted from transfer and integration of T-DNA and the subsequent expression of T-DNA genes.
2. The T-DNA genes are transcribed only in plant cells and do not play any role during the transfer process.
3. Any foreign DNA placed between the T-DNA borders can be transferred to plant cells, no matter where it comes from. These well-established facts allowed the construction of the first vector and bacterial strain systems for plant transformation.

Ti Plasmid-based Transformation Vectors

It will be apparent that in order to establish useful transformation systems for genetic engineering purposes based on the *Agrobacteriurn* Ti plasmid, it has been found necessary to overcome a number of inherent limitations of the system while maintaining the extremely advantageous features. Central to the manipulation of this system is the fact that the processes of T-DNA transfer and disease symptom formation can be separated both conceptually and physically. Specifically, the oncogenes and the opine biosynthetic genes can be removed from the T-DNA with no adverse effects on T-DNA excision, translocation and integration, as long, as the T-DNA border repeats (and the right border, in particular) remain intact. By replacing these genes with selectable marker genes that function in plant cells, and by other genes of interest for plant genetic manipulation, an efficient DNA delivery system has been developed.

Binary Vector System

The principal advantage of this development is the small size of the TDNA-containing 'binary' plasmid, which facilitates cloning. The *vir region* I is typically located on a separate Ti plasmid which has had the T-DNA removed. One of the earliest developed such binary vector is Bin19. The important features of binary vectors comprise

1. The presence of a multiple cloning site for the facilitated insertion of transgenes; an antibiotic resistance gene (e.g, *nptll* gene), which confers on transgenic plant cells resistance to the antibiotics (*e.g.* neomycin, kanamycin)
2. The T-DNA left and right border repeats, for efficient T-DNA processing. This vector is commonly used with, for example, the *Agrobacteriurn* strains LBA4404 and C58, which contain *vir* genes in *trans.*

A number of binary vectors based on Binl9 are commercially available, which contain transcription promoters and screenable reporter genes, such as those encoding β-glucuronidase, GUS or firefly luciferase at the multiple cloning site. The presence of screenable markers, "in addition to selectable genes, greatly facilitates the identification of transgenic individuals, particularly important when devising new protocols for plant transformation (Lindsey, 1996).

Vector-free Gene Transfer

Vector-free or direct gene transfer systems have been developed primarily as an answer to the perceived problems of generating stably transformed cereals by the *Agrobacterium* system. The general principle of direct gene transfer is, however, essentially the same as for *Agrobacterium-mediated* transformation namely, to introduce DNA into cell or tissues from which uniformly transformed plants (non-chimeric for the gene) can be induced to regenerate. The main difference, of course, is that the foreign DNA is introduced as a naked molecule without a biological vector. There are broadly three approaches to direct gene transfer, which can be described as chemical, electrical and physical. Each is designed to overcome the barrier to DNA uptake that is presented by the cell wall and the plasma membrane.

Chemical Methods

A commonly used chemical method for direct gene transfer into plant protoplasts involves the use of polyethylene glycol (PEG). Protoplasts are incubated with a solution of DNA and PEG (typically at a concentration of 15-20 per cent w/v), washed and cultured. The objective is to use a PEG concentration that allows high cell survival and DNA uptake, but low fusion frequency.

Electrical Methods

Electroporation: When cells or protoplasts are subjected to short electric current (DC) they may experience an increase in the permeability of the plasma membrane to hydrophilic molecules such as DNA, which are normally unable to enter the cell. Above a threshold field strength (voltage) electrical pulses produce irreversible damage to membrane structure, resulting in cell death. Below this threshold value, however, there may occur a nonlethal permeabilisation of the plasma membrane termed 'reversible dielectric breakdown', which is thought to be due to the formation or enlargement of pores in the lipid bilayer.

Physical Methods

The third approach to overcome the cell wall as a barrier to DNA movement employs what can be described as 'physical' methods, and can be distinguished from chemical and electrical techniques by the use of physical or mechanical means to introduce transgenes directly into the protoplast or cell. The physical methods of direct gene transfer are as follows:

Microprojectile Bombardment

The transformation technique aims to penetrate cells or tissues with accelerated metal microspheres, which are coated with DNA. Such micro spheres, of tungsten or gold (used because of their high densities), and of diameters of 1-4mm, have been shown to penetrate cells when accelerated at high velocities, and the DNA can be released and expressed transiently or can be stably integrated into chromosomes. The DNA-coated microprojectiles are placed on a macroprojectile, which is accelerated towards the tissue following an explosive discharge. The accelerated macro projectile hits a stopping plate and microprojectiles are propelled onwards and penetrate the tissue depositing the DNA as they go. Microprojectile bombardment has been successfully used to transform tissues that show good morphogenetic responses in a wide range of species including tobacco, soybean, rice, barley and wheat (Lindsey, 1996).

Biolistic PDS-1000/He Device

The Biolistic PDS 1000 device is the most widely used particle delivery system for gene transfer. This device is used to transform living organisms with foreign DNA. DNA is delivered in cells in association with microscopic metal; particles called microcarriers that are propelled at high velocity towards target tissues. The micro carriers are accelerated on a plastic cylinder, called macro carrier, which is driven by shock wave of helium gas. A small high-pressure chamber within the gas acceleration tube is led with helium gas to a selected pressure. The chamber has been temporarily sealed by rupture disc. Helium gas is supplied by a high-pressure helium tank. When critical pressure for the rupture disc is reached; it bursts and releases powerful gas shock wave. In the path of the gas shock is the micro-carrier launch assembly. Where wave energy shock is transferred to micro carriers (microscopic gold particles), that have been dried on the face of macro carrier. Upon firing the macro-carrier separates from its The micro-carriers, however, are launched and continue downwards until they impact and penetrate target cells.

Bacillus thuringiensis Berliner and Genetic Engenering

A transgenic plant is simply a normal plant with one or more additional genes from diverse sources. Scientists have bred crops for resistance against insect pests for a long time. These transgenic plants produce insecticidal or antifeedant proteins continuously in the plants under field conditions.

Before taking up any attempt to produce transgenic plants to counter insect attack, the following requirements and priorities need to be identified:

The factors for resistance should be controlled by single genes:

1. Standardization of methods for transfer of such genes can easily be accomplished.
2. Expression of transferred gene should occur in the desired tissues at the appropriate time.
3. The transgenic plant should be safe far consumption.
4. Inheritance of the gene in the successive generations should be very stable.
5. There should be no penalty for yield in terms of other quantitative characters.

Plant Genetic Systems, a Belgian Biotechnology Company, in July 1987 were the first to report development of transgenic plants of tobacco containing delta-endotoxin in enough amount to kill the first instar larvae of *Mdnduca sexta* (Johannsen) and *Heliothis virescens* (Fabricius) within 3 days. Plants infested with tobacco homworm showed very limited feeding damage compared to non-transgenic control plants, which suffered heavy feeding damage and were almost completely defoliated within two weeks.

What is *Bt*

1. A rod-shaped (1.0-1.2 by 3-5 micron), gram-positive, facultative anaerobic, spore forming bacterium.
2. A facultative pathogen of some insect species commonly used as a biopesticide (microbial control agent).
3. More than 150 insects belonging to the orders Lepidoptera, Diptera, and Coleoptera are known to be susceptible to *Bt*. To be effective, *Bt* must be eaten by insects during their feeding stage of development, when they are larvae. *Bt* is ineffective against adult insects.
4. During sporulation, it produces protein inclusions (parasporal inclusions) adjacent to the endospore.
5. The parasporal inclusions consist of one or more insecticidal proteins in the form of a crystal or crystal-complex. These insecticidal proteins are commonly known as Insecticidal Crystal Proteins (ICP) or delta (δ) endotoxin. The primary insecticidal activity of Bt is due to ICP. It is the active ingredient of most (90 per cent) of the microbial insecticides produced in the world. The endospore, highly resistant to environmental stress, provides a mechanism for long-term survival of *Bt*.
6. Besides ICP and endospore, some Bt subspecies produce Beta Exotoxin which is toxic to all forms of life including humans.
7. An insect must have a strongly alkaline mid-gut, specific proteolytic enzymes and suitable tissue receptor sites; to be susceptible.

Classification of *Bt* Subspecies

Based on the serotype (serological analysis of flagella antigens) supplemented by morphological and biochemical criteria. Over 67 subspecies have been identified; subspecies commonly used in agriculture/public health.

1. *Bt* subspecies kurstaki (*Bt* k)–controls various types of lepidopterous insects.
2. *Bt* subspecies israelensis (*Bt* i)–effective against mosquitoes, blackflies' and some midges.

3. *Bt* subspecies tenebrionis (also formerly as subsp. san diego) (*Bt* te)–effective against certain beetle (chrysomelids) species and the boll weevil.
4. *Bt* subspecies Japonensis (*Bt* j)–effective against many species of scarabid beetles.
5. *Bt* subspecies aizawai (*Bt* a)–used against wax moth larvae in honeycombs. Pathogenicity/ toxicity: Toxicity is due to ICP, endospore or Beta Exotoxin.

Insecticidal Crystal Proteins (ICP)

1. Also called as delta-endotoxin. They are protoxin (it must be activated before it has any effect).
2. Commonly designated as "Cry" proteins encoded by "cry" genes.
3. The genes that encode ICP are mostly on plasmids; each ICP is the product of a single gene.
4. A subspecies can synthesize more that one type of ICP.
5. ICPs have various forms (bipyramidal, cuboidal, flate rhomboid, or a composite with two or more crystal types).
6. Size: Protoxin–A large protein of up to 230 kilo Daltons (kDa), Active Toxin–65 kDa. *Extremely potent and toxic to target insect larvae at picomole concentrations.
7. Most susceptible species belong to the orders Lepidoptera, Diptera, and Coleoptera.
8. Most susceptible insects are killed by ingestion of the crystals alone; a mixture of spores and crystals are required for a toxic effect in only a small number of insects.
9. Cry proteins are classified according to their insect specificity (Host spectrum is designated in roman numerals):
   (*a*) CryI Lepidoptera-specific
   (*b*) CryII Lepidoptera- and Diptera-specific
   (*c*) CryIII Coleoptera-specific
   (*d*) CryIV Diptera-specific
   (*e*) Cry V Coleoptera- and Lepidoptera-specific.
10. A mixture of different Cry toxins results in a synergism of insecticidal activity.
11. A mixture of Cry toxins and Bt spores results in a synergism of insecticidal activity in several insect pests including Indian meal moth, Plodia interpunctella (due to either additional toxins in the spore coat or the outgrowth of vegetative cells from ingested spores).
12. Highly insoluble in normal digestive system conditions–safe to humans, higher animals, and most insects.
13. It is soluble in reducing conditions of high pH (above pH 9.5)–a condition commonly found in the mid-gut of lepidopteran larvae. For this reason, Bt is a highly specific insecticidal agent.
14. The midgut of susceptible insects contains the correct pH and enzymes to transform the crystal into an active toxin.
15. Different domains of the ICP are responsible for host susceptibility (receptor recognition) and toxicity (pore formation).

Endospores

1. Endospores are resistant to inactivation by heat and desiccation and persist in the environment under adverse conditions.
2. Provide a mechanism for long-term survival of *Bt*.
3. Are pathogenic for some insects, particularly when combined with ICP.

Beta Exotoxin

1. During vegetative growth, some Bt subspecies (thuringiensis, galleriae, and darmstadiensis) produce Beta Exotoxin, an ATP analogue, water soluble, heat stable secondary metabolite.
2. Beta-exotoxin is an inhibitor of RNA polymerase and acts competitively with ATP in various biological processes.
3. It is toxic to almost all forms of life including humans. It has broad-spectrum insecticidal activity.
4. Products with Beta Exotoxin are prohibited in the USA. Manufacturing process includes monitoring to prevent Beta Exotoxin from appearing in products.

Mode of Action

Bioactivity of *Bt* is dominated by the ICPs and can be summarized in the following stages:

1. Ingestion of spores and ICP by a susceptible insect larva;
2. ICP dissolve in the alkaline midgut;
3. Activation of the ICP by proteases;
4. Irreversible binding of the C-terminal domain of activated ICP to specific receptors in the midgut cell membrane;
5. Insertion of the N-terminal domain of activated toxin in the cell membrane and formation of pores and channels in the gut membrane, followed by destruction of epithelial cells and paralysis of the digestive system;
6. Spore germination and septicemia;
7. Larval death (within an hour to a few days) from starvation or septicemia.

Effects on Insects

1. Immediate cessation of feeding and general paralysis of the larval midgut.
2. Subsequent symptoms include vomiting, diarrhoea, general sluggishness, and general paralysis.
3. Death may follow immediately or in an hour to a few days.
4. Sublethal effects may include: reduced feeding, reduced larval and adult longevity, reduced fecundity, reduced larval and adult body weight.

Using *Bt* in Integrated Pest Management

Transgenic plants carrying *Bt* genes have now been produced in a wide range of crop species including tobacco, tomato, potato, cotton, maize, rice, broccoli, oilseed rape, soybean, walnut, larch, poplar, sugarcane, apple, peanut, chickpea and alfalfa with different crystal protein genes.Many field trials of crops expressing these modified *Bt* genes have been carried out and most of these have produced impressive results, *e.g. CrylA* cotton against *H. zea* and *Pectinophora gossypiella* (Saunders), *Cry3A* potato against *Leptinotarsa decemlineata* (Say) and *CrylA* maize against O. *nubinalis.*

Japanese scientists have introduced a truncated *Bf* delta-endotoxin gene *CrylA(b)* into *alaponica.* rice, which has specific biological activity against two major lepidopteran insect pests, *viz.* striped stem borer, *Chilo suppressalis* and rice leaf *Cnaphalocrocis medinalis.* An aromatic rice variety arom Molaii' has been transformed carrying a synthetic truncated toxin gene based on *Cry(b)* and another variety IR58 has been transformed with *CrylA(b).* Both these transgenic lines showed high levels of tolerance to *Chilo suppressalis* (Walker) and *Scirophaga incertulas* (Walker). IR64 has been transformed using *CrylA* (*c*) gene and transgenic rice plants were highly toxic to S. *incertulas.* In view of the ability of insects to develop resistance to *Bt* toxins, availability of multiple crystal, proteins active against the same insect species, but with different receptor binding characteristics may be important in the long term application of crystal proteins in pest management. This has been made possible by construction of hybrid crystal proteins. A new cry protein with broader activity spectrum has been obtained by in *vivo* recombination of genes of *CrylC* and *CrylE.*A nother drawback in using *Bt* genes has been the low expression level of the toxin (<0.02 per cent of total leaf protein). Through extensive gene synthesis, promoter optimization and protein targeting, researchers have now improved expression of nuclear *cry* genes such that tobacco leaves can accumulate up to 0.8 per cent of their soluble protein as *Bt* toxin in about 10 per cent of the plants while the remaining transformants produced between 0.001-0.06 per cent of their soluble protein as toxin. Recently, it was reported that amplicization of an unmodified *CrylA(c}* coding sequence in chloroplasts up to 10000 copies per cell resulted in the accumulation of an unprecedented 3-5 per cent of the soluble protein in tobacco leaves as protoxin. The technique known as *plasmid transformation* resulted in highest level of toxin accumulation presently reported in plants. The plants were extremely toxic to larvae of *Heliothis virescens* (*Fabricius*), *Helicovepa zea* (Boddie) and *Spodoptera exigua* (Hubner). Also, the isolation of toxin genes from *B. thuringiensis* strains with specificity toward other classes is being pursued.

Agricultural Products–Based on *Bt* Subspecies kurstaki

Biobit® Flowable Concentrate: Bt (15 per cent) and inerts (85 per cent)

Biobit® Wetable Powder: Bt (29 per cent) and inerts (71 per cent)

Dipel® 2X: *Bt* (6.4 per cent) and inerts (93.6 per cent)

Dipel® 4L: *Bt* (1.76 per cent) and inerts (98.24 per cent)

Dipel® ES: *Bt* (3.5 per cent) and inerts (96.5 per cent)

Javelin®, Thuricide® 32B and Thuricide® 32LV: *Bt* (1.6 per cent) and inerts (98.4 per cent) Javelin® WG: *Bt* (6.4 per cent), dispersing agent (15.0 per cent) and fermentation residue (78.6 per cent) Thuricide® 48L V: Bt (2.4 per cent) and inerts (97.6 per cent).

Thuricide® HP: Bt (3.2 per cent) and inerts (96.8 per cent).

Thuricide® HPC: Bt (0.8 per cent), petroleum hydrocarbon solvent (3.0 per cent), and inerts (96.2 per cent).

Bioassay Technique for Estimation of *Bt* Protein Expression in Transgenic Cotton

The Cry proteins of *Bacillus thurlngiensis* have been widely exploited in transgenic approach against Lepidopteran and Coleopteran pests and reviewed by many. An insect-based quantitative bioassay will be more accurate in determining the concentration of active Cry protein in the plant tissue. The assay described here is quite reliable and useful in determining the Cry 1 Ac concentration in various tissues throughout the growth season of transgenic cotton (Greenplate, 1999).

Bioassay Method

1. A semi synthetic diet is prepared, dispensed @200 ml/well and allowed for drying for a few hours.
2. The *CylAc* powder is diluted serially from 5.55 ng/ml to 0.484 *ngl* ml concentration to obtain percentage of thirds ranging from below 90 per cent at 5.55 to about 10 per cent at 0.484 mg/ml concentrations.
3. The lyophilized powdered tissues are suspended in 0.2 per cent agar to the required dilutions, and dispensed @ 50 ml/well and excess moisture is removed by drying.
4. 'Two-day old *E. viffella'eggs* are collected by washing in 0.05 per cent sodium hypochlorite solution, suspended in 0.2 per cent agar and dispensed @25 ml/well maintaining an egg density of 2–4 per well.
5. 'The plates are then heat-sealed with a thin Mylar film and perforations are made with zero number insect pins for aeration. The plates are incubated in the dark at 28°C and 60 per cent humidity. The numbers of insects that are reaching the third instar on the seventh day are counted and the quantity of CrylAc protein is estimated using the standard curve after correcting for the dilution fader.
6. Suitable controls are also run in parallel. For *e.g.* the non *Bt* cotton tissues are also included in the assay wherein the percentage reaching thirds is always high around 90 per cent.

Semi-Synthetic Diet: Fraction A: Chickpea flour- 60g, Sterile water-400 ml

Fraction B: Agar–17g, Sterile water- 400ml. Fraction C: Methylparaben 0.2g, Sorbic acid

0.2g, Streptomycin sulfate-0.5g, Cysteine-0.1g Multivit-2 cups-2.5g, Wesson's salt–2.5g,

Casein-0.1g, Cholesterol-0.5g, Ascorbic acid- 2.0g

Trypsin Modulating Oostatic Factor (TMOF): A New Biochemical Insecticide

Insect neuropeptides are important messenger molecules which influence nearly all physiological processes, including behaviour, even though they are only produced by less than 5 per cent of the cells of the central nervous system. They can act as transmitters, modulators and classical hormones, and often exhibit pleiotropic functions when released into the haemolymph. The large number of neuropeptides that has been identified from some of the model organisms among insects underlines the complexity of the neurosecretory system. Although these signaling peptides are unlikely to be used as insecticides per se, knowledge of their chemistry and biological action can be applied in novel approaches to pest control. Furthermore, the protein nature of neuropeptides makes them amenable to this control by the use of recombinant DNA technology and genetic engineering to have such peptides produced by transgenic crop plants or other suitable organisms (Zhu *et al.*, 2001).

The primary requirements for Neuropeptides as insecticides are:

1. In order to be active, the peptides should not be blocked at either its N- or C-terminus, because plants cannot make these modifications.
2. It should not be quickly degraded by proteolytic enzymes in the gut lumen.

3. The peptide itself should be able to permeate the epithelium of the gut unaltered.
4. If it's partially degraded in the lumen, some of the degraded products should still remain physiological activity upon permeation through the epithelium.
5. Upon arrival in the haemolymph compartment, it should remain active long enough to exert in action.

Dipteran TMOF have been forwarded as potential insecticides since they meet at least some of the criteria mentioned above (De Loof, 1996).

Mode of Action

Trypsin Modulating Oostatic Factor (TMOF) is a new, proprietary approach to controlling pest insect populations without the undesired side effects of current synthetic chemical insecticides or the insect resistance problems anticipated for *Bacillus thurengensis* based insecticides. TMOF is a ten amino acid peptide first isolated from mosquito ovaries. The normal function of TMOF is to signal the epithelial cells lining the midgut, to cease making digestive proteases, when sufficient essential amino acids have been borne to the ovaries to complete egg protein synthesis. When administered to mosquitoes and other insects the resultant lack of trypsin-like digestive enzymes results in no protein digestion, leading to reproductive failure, starvation and death (Figure 35.1). TMOF is active at low concentrations (nanograms per insect).Because of its unique mode of action, development of resistance by the insect is expected to be an extreme low frequency.

Egg development and blood digestion in the mid gut of female mosquito are two inter related processes regulated by complex series of hormonal interactions. Following blood feeding, JH and an

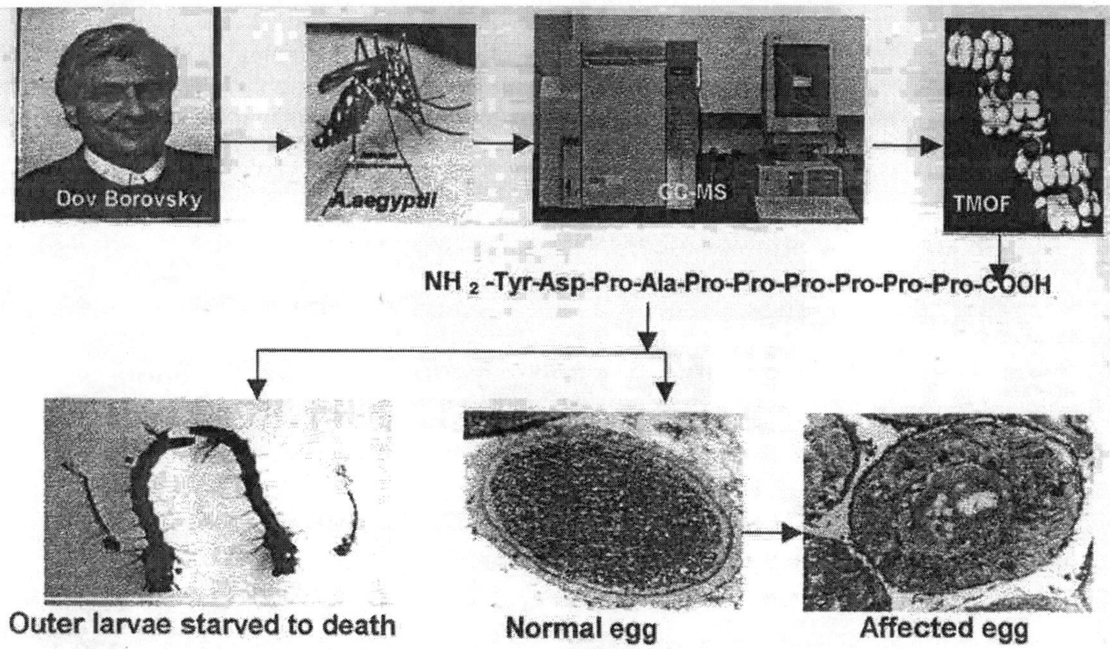

Figure 35.1: Diagrammatic Representation of Mode of Action of TMOF

ovarian factor corpus cardiacum stimulating factor (CCSF) (Borovsky *et al.*, 1982) are released. CCSF releases egg development neurosecretory hormone (EDNH) (Borovsky *et al.*, 1982),which stimulates the ovary to synthesize and release ecdysone (Hagedorn *et al.*, 1979). Ecdysone is converted into 20-OH-ecdysone by the mosquito fat body andin concert with JH, induces the fat body to synthesis vitellogenin (Borovsky *et al.*, 1985). At the same time, the blood meal induces a *de novo* Synthesis of trypsin like enzyme in the mid gut of the female mosquito (Borovsky and Schlien, 1988). Thus, stopping any of these processes may cause arrest of egg development causing sterility (Borovsky, 1988; Borovsky *et al.*, 1990).

Structural Characterization of TMOF

Borovsky-D *et al.*, 1993 reported that Trypsin modulating oostatic factor (TMOF), a decapeptide, directly inhibits the biosynthesis of trypsin- and chymoptrypsin-like enzymes in epithelial cells of mosquito midgut and indirectly inhibits vitellogenesis in a nautogenous females, was sequenced by Fourier transform mass spectrometry analysis (Figure 35.2). The peptide has a primary amino acid sequence ofNH2-Tyr-Asp-Pro-Ala-(Pro)6-COOH and probably exhibits left-handed helical conformation.

Control of Trypsin Biosynthesis in Midgut with TMOF

Synthesis of trypsin-like isozyme was followed in several insect species (*e.g. Culicoides variipennis, Lutzomyia anthophora, Stomoxys calcitrans, Ctenocephalides felis*) using [1,3-3-H]diisopropyl-fluorophosphate (DFP) in the presence of tosylamide-2-phenylethyl chloromethyl ketone. Marked reduction of [1,3-3H]diisopropyl-phosphoryl-trypsin-like derivatives was noted after TMOF treatment, as assessed by polyacrylamide gel electrophoresis.TMOF or TMOF(B) or in combination at 7×10^{-4} M. Inhibited *A.aegyptii* trypin like enz biosynthesis and TMOF(B) at 2.9×10^{-4} gave 61 per cent inhibition in *Culicoides variipennis.* (AEBSF) and N-tosyl-L-phenylalanine chloromethylketone (TPCK) indicate that protease activity is mostly trypsin-like (80 per cent) with a small amount of chymotrypsin-like activity (20 per cent). Injections of late 3rd and 4th instar larval haemolymph into *H. virescens* larvae inhibited trypsin biosynthesis in the larval midgut. Similar results were obtained when highly purified

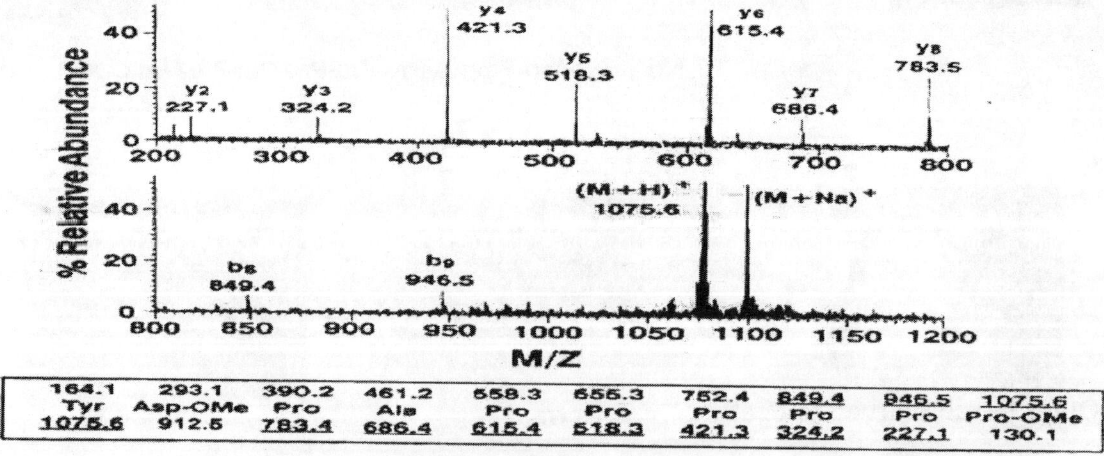

Figure 35.2: Structural Characterization of TMOF by Mass Spectrometry (Borovsky *et al.*, 1993)

4th instar larval haemolymph that crossreacted with Aea- TMOF antiserum using ELISA was injected into the 2nd instar larvae. Injections of Aea- TMOF and its analogues into the 2nd instar, and Aea-TMOF alone into the 4th instar larvae stopped trypsin biosynthesis 24 and 48 h after the injections, respectively. Injections of 4th instar *H. virescens* larval haemolymph into female Aedes aegypti that took a blood meal stopped trypsin biosynthesis and egg development. These results show that biosynthesis of trypsin-like enzymes in the midgut of a lepidoptera is modulated with a haemolymph circulating TMOF-like factor that is closely related to Aea-TMOF.

Aedes aegyptii TMOF modulates ecdysteroid production by prothoracic glands of the gypsy moth. *Lymantria dispar.*

Gelman and Borovsky (2000) reported that the trypsin modulating oostatic factor (TMOF) is a decapeptide that inhibits the biosynthesis of trypsin-like enzymes in the midgut of several insect species and, as such, serves as a dipteran oostatic hormone. In vitro incubation of lepidopteran prothoracic glands with *Aedes aegyptii* TMOF revealed that this decapeptide, in the presence of brain extract, modulates ecdysteroid production. The modulatory effect was highly dependent on both the concentration of TMOF and brain extract. Typically, TMOF was stimulatory in the presence of lower concentrations of Lymantria dispar brain extract (0.01 and 0.025 brain equivalent), and either neutral or inhibitory at higher concentrations (0.25, 0.5, and 1.0 brain equivalent) of extract. In the presence of European corn borer (Ostrinia nubilalis) brain extract, TMOF also exhibited modulatory effects, effects that again were dependent on the concentrations of both brain extract and TMOF present in the incubation medium. At 1.5 brain equivalents, TMOF was inhibitory at all but the highest concentration tested (5×10^{-6} M), at 1.0 brain equivalent, TMOF was stimulatory at 10^{-6} M and at 0.5 brain equivalents, TMOF did not significantly affect PTG synthesis of ecdysteroids. Results suggest the presence of a modulatory peptide(s), which fine-tunes the synthesis and release of ecdysteroids by PTGs in accordance with the insect's developmental physiological requirements.

"Skeetercide" Diet Pill Seen as West Nile Weapons

A new mosquito larvicide developed at the University of Florida's Medical Entomology Laboratory in Vero Beach is expected to be on the market under the brand name "Skeetercide". The "Skeetercide" formulation is based on a naturally occurring mosquito hormone- trypsin modulating oostatic factor– that is produced in mosquito ovaries and acts to control digestion of the blood ingested when a mosquito bites. When ingested by larvae as part of a mosquito control program, TMOF switches off the digestive system, effectively starving the larvae to death in 4-6 days. Dov Borovsky and team have found that yeast serve well for this purpose when bioengineered to include TMOF and bound with cellulose particles to produce granules that are stable in the water for about 20 days. Product developers say it poses no human health concerns and that it is "safe enough to use in drinking water containers." The product is being commercialized by a North Carolina company, Insect Biotechnology Inc., that holds the license to University of Florida patents. The development of TMOF is exciting news because the TMOF at 4.2×10^{-4} M and 8.2×10^{-4} M gave 26 and 46 per cent inhibition–TMOF(B) at 4.2×10^{-4} M and 8.2×10^{-4} M gave 42 and 25 per cent inhibition.–TMOF or TMOF (B) at 4.8×10^{-4} gave 9 and 41 per cent inhibition.

Therefore,TMOF (B) is more effective on *C. felis.*

Specificity

TMOF activity was first isolated from and characterized in adult mosquitoes (*Aedes aegyptii, Culex quinquefasciatus, Anopheles albimanus* and *Anopheles quadrimaculatus*). Additional experiments have extended TMOF activity to a wide range of insect species including *Trichoplusiani* larvae (cabbage looper), *Spodoptra frugiperda* larvae (fall armyworm), *Ctenocephalides felis* (cat fleas), *Musca domestica* (house flies), *Lutzomyia anthophora* (sand flies), *Lutzomyia anthophora* (stable fly), *Neobellieria* (Sarcophaga), *bullata* (flesh fly), *Calliphora* (a blowfly species closely related to *Lucillia cuprnia*, the sheep blowfly) and indications of activity in *Heliothis virescens*, the sweet potato whitefly, Caribbean fruit fly, diamond back moths and cotton boll weevil.

TMOFs and their Relation to Matrix Proteins

DeLoof *et al.*, 1995 determined the sequences of 2 folliculostatic peptides of *N. bullata [Sarcophaga bullataJ*. The 1st peptide (Neb-TMOF: H-NPTNLH-OH), originates from a 75 kDa precursor protein found in vitellogenic oocytes. The hexapeptide directly inhibits the synthesis of trypsin-like enzymes in the gut, and thus lowers the concentration of yolk polypeptides in the haemolymph. It also inhibits the biosynthesis of ecdysone in the larval ring gland. Therefore, it could also be called prothoracicostatic hormone (Neb-PTSH). The 2nd peptide (Nebcolloostatin: H-SIV-PLGLPVPIGPIVVGRP-OH) acts on previtellogenic follicles and is a cleaved product of a collagen-like precursor molecule. The results indicate that peptides which are cleaved from matrix proteins could act as growth-inhibiting factors. Gonadotropin releasing hormone (GnRH)-immuno-like peptides were not identified. Using these results, a novel model of growth control in which matrix proteins play an important role as a potential source of growth regulators.

TMOF-Like Factor Controlling the Biosynthesis of Serine Protease in the Larval Gut of *Heliothis virescens*

Nauen *et al.* (2000) reported that the proteolytic enzyme biosynthesis in the midgut of the 4th instar larva of Heliothis virescens is cyclical. Protease [proteinase] activity increases immediately after the moult from the 3rd to the 4th instar larvae and declines just before the moult into the 5th instar. Characterization of the midgut proteases using soyabean trypsin inhibitor (SBTI), Bowman Birk Inhibitor (BBI), 4-(2-arninoethyl)benzensulfonylfluoride availability of alternatives is an important pest control strategy both for (a) staving off development of insect resistance resulting from exclusive and repeated use of anyone product or method and (b) having products suitable for disparate conditions. For example, *Culex pipiens* and *C. quincquefasciatus* tend to breed in polluted water where the larvicide *Bacillus turingiensis israelensis* (Bti) is less effective.

Future Thrusts

Market Potential

Insect Biotechnology, Inc (University of Florida) holds the exclusive, worldwide rights to all TMOF and related technology. IBI has identified 3 potential markets for its products:

1. Specialty markets–mosquito larvacide–$100 million market in U.S. but rapidly increasing because of vector borne West Nile disease.
2. Professional products markets (sprays/baits)–aphids, wasps, etc., a high mark-up market.
3. Agribusiness products–sprays, baits and possibly (when market climate permits) transgenic plant expressing TMOF-like activity in com borer, cotton borer, caterpillar, etc.- a $1 billion plus market in U.S.

Sequencing and Cloning of the TMOF Receptor

Borovsky has shown there is a TMOF binding site on the hemolymph (insect bloodstream) side of the gut and has identified the TMOF Receptor which binds to it. Sequencing and cloning of the TMOF Receptor is in progress. Once the complete receptor genetic sequence is obtained it will be cloned into a suitable cell line for expression on the cell surface. This will create a highly efficient screening assay for TMOF receptor binding activity for use in screening chemical libraries for compounds with TMOF-like insecticidal activity. Progress in this area is advanced. Intellectual property protection is strong with 8 issued patents and 10 filed and 3 in preparation.

Mammalian and Bird Toxicology has to be standardized by feeding and dermal or ocular studies according to published EPA methods.

Agricultural Pest Toxicology has to be standardized through feeding or injection of the compounds of interest followed by survivorship assays or by monitoring changes in trypsin activity.

Neem (*Azadirachta indica*): A Need for Tissue Culture

The various application of *Azadirachtin,* neem based triterpenoid in the pharmaceutical and agricultural industries needs a large production of high azadirachtin containing seeds, due to numerous applications of neem derived products in various cosmetics, insecticidal, nematicidal products increases the demand of neem seed production. It is therefore necessary to have large scale neem plantation, which can be achieved through tissue culture. Tissue Culture would help in establishing a better productive plantation with uniform, genetically superior, high azadirachtin containing planting material.

Regeneration Protocol Details

1. The shoot tips and nodal segments of the matured tree were collected ftom the healthy trees throughout the year except April-June. Thus, the period of fruiting reduces the number of axillary bud forming nodal explants.
2. The shoot apices and nodal segments were excised ftom the mother plant and were sterilized in a pre-treatment step.
3. A combination of hormones were added to MS media supplemented with sucrose and solidified with agar for bud initiation, multiple shoot induction and further 14 multiplication.
4. Cultures were incubated under light during bud initiation and elongation period.
5. The *in vitro* differentiated, multiplied shoots are further transferred to a ftesh medium and the regenerated shoots after 2-3 weeks of time were transferred to rooting medium. During root initiation and root elongation the light regime was controlled. Hardening:
6. 16-20 weeks juvenile plants are maintained in a polyhouse condition and subsequently in the nursery till they are ready for transplantation.

Transgenic Neem

Please see Table 35.1.

Challenging of Cell Lines: *In vitro* Replication of Insect Viruses

In vitro production of insect virus in cell lines has got several advantages. Cell cultured viruses are free from undesirable viral or microbial contamination, secondary biotypes and insect debris. In insect cell line, virus production can be monitored easily. It is easy to maintain cell lines continuously for virus production. Large scale production is possible using fermentors/bioreactors. Baculovirus replication in insect cell line has been reported for several nuclear polyhedrosis viruses (Lynn *et al.*, 1983).

Table 35.1: Transgenic Neem

Sl.No.	*Research Findings*	*Reference*
1.	Isolation of nimbin from *Azadirachta indica* leaves and its callus cultures	K. Ramesh and Padhya, M.A. (1988)
2.	The production of azadirachtin by *in vitro* tissue cultures of neem, *Azadirachta indica*	Allan, E.J. *et al.* (1994)
3.	Rapid and sensitive analysis of azadirachtin and related terpenoids from Neem (*Azadirachta indica*) by high-performance liquid chromatography atmospheric pressure chemical ionization mass spectrometry.	Schaaf, O. *et al.* (2000)
4.	Somatic embryogenesis and plantlet propagation of *Melia azedarach* L (Ghora Neem) from cotyledonary segments	Deb,C.R. (2001)
5.	Vegetative propagation and genetic improvement of Neem	Palanisamy, K. *et al.* (2001)

Preparation of Virus Inoculum

Fourth instar larvae are inoculated with the homologous NPV by allowing them to feed on diet surface contaminated with 106 polyhedra per bottle. After three days of inoculation, the larvae are washed in sterile distilled water and the haemolymph is drawn by cutting the prolegs and collected in 10ml of TNM-FH medium containing cysteine to prevent melanisation. The medium is then centrifuged at 780 g for 10 min to sediment haemocytes and passed through a 0.45 mm millipore membrane filter under a laminar hood.

Challenging of Cell Lines

At log phase of the cell line, the spent medium is removed and 2.5 ml of medium containing virus inoculum is added to the cells and an adsorption time of two hours is allowed with periodical rocking of flasks at an interval of 10 min. The medium is then removed. The cells washed twice gently with sterile Hanks balanced salt solution (HBSS). Finally, after removing the HBSS completely, five ml of TNM-FH medium containing 10 per cent foetal bovine serum is added and incubated at 27.5°C. The cells were observed daily for virus infection under a microscope. Virus can be harvested after five to six days.

Identification of *In vitro*-cultured Viruses

The cell cultured virus can be subjected to DNA analysis using restriction endonucleases. The methodology is described elsewhere. The DNA profiles of cell cultured virus and *in vivo* produced virus can be compared. Similarity in the DNA profiles of both viruses will indicate that in cell line the virus has not undergone any recombination or mutation.

Biological Activity of *In vitro* Cultured Viruses

The efficacy of cell cultured viruses has to be tested by conducting bioassays. The virus infected cells were harvested 10 days after inoculation and stored at –20°C. Before processing, the cells were thawed out and sonicated in ice bath using a Branson Sonifier 450 to disrupt the cells and release the polyhedra. The cell debris were removed by centrifugation at 200 x g for one min. The supernatant was again centrifuged at 780 x g for five min. to pellet the virus. The polyhedra were suspended in distilled water and the concentration assessed with the help of a haemocytometer under phase contrast microscope.

The virus harvested from the cell line can be bioassayed to compare their biological activity with that of the *in vivo* produced virus. A standard diet surface contamination method is followed starting with a dose of 3.5×10^5 polyhedral occlusion bodies (POB) ml^{-1}. Seven-fold serial dilutions are made to give five doses each. Chickpea seed based-semi-synthetic diet is dispensed into sterile five ml glass vials and allowed to cool down to room temperature under a laminar hood. After 2-3h, an aliquot of 10 ml of the different viral dilutions are applied on the diet and spread uniformly over the entire diet surface using a four mm glass rod with rounded tip. After 15 min., one larva of second instar is introduced into each vial and plugged with sterile cotton. Thirty five to forty larvae are used in each dose. A similar set of larvae without virus treatment is maintained as control. Observations on the larval mortality is recorded daily for ten days and data are subjected to probit analysis. LC_{50} values of the two viruses can be compared.

Applications of PCR Techniques in Entomological Research

The evolution of Polymerase Chain Reaction (PCR), dates back to 1985 just a decade after the invention of gene cloning. Since its conception by Kary B. Mullis, this novel approach has been put into use to study the biological processes in the field of Entomology and allied sciences. In insect Pathology, PCR is used for the molecular characterisation of insect pathogens such as viruses, bacteria, fungi and nematodes.

Principles of PCR Technique

PCR reaction involves the enzymatic synthesis of DNA using two oligonucleotide primers that hybridise to opposite strands and flank the region of interest in the target DNA. DNA, being the source of genetic information when, amplified by PCR decipher the genetic basis of insect pathogens and help in the better understanding and improvement of pathogens. In PCR, as in the normal cells, essentially identical reactions take place but in a test.The simple concept of the PCR relies upon the repeated synthesis of the targeted DNA by DNA polymerase.

Basically, there are three steps needed to achieve this synthesis which are as follows:

Denaturation

High temperature incubation (90–98°C) melts the target-double stranded DNA into separate strands.

Annealing

Low temperature incubation (37–65°C) allows primers to anneal to complementary sites flanking the region of DNA to be amplified.

Extension

Alteration of temperature around 72°C allows abstraction of deoxynucleotides (dNTP) for the 5' and 3' synthesis of DNA using target DNA as template. Incorporation of primers into PCR product occurs at this step.

These three steps are carried out repeatedly as one cycle and the number of cycles required for optimum amplification depend on the amount of starting material and the efficacy of each step.

However, this technique is capable of synthesizing over a million of copies of the specific target DNA in a few hours. This reaction is carried out in thermocycler (automated controlled system). The reaction components include DNA polymerase, dNTPS, buffer reagent, primers, target DNA, $MgCl_2$, non-ionic detergent. The PCR products can be analysed using electrophoresis, hybridisation, nested or labelled primers.

In general, PCR offers unique techniques for the following:

Molecular Systematics

Use of arbitrary primers to produce DNA fingerprints based upon DNA amplification upon digestion provides scope for differentiation and typing of closely related species of insect pathogens.

DNA Sequencing

One highly significant use of PCR is the generation of DNA templates for sequencing. The basis for the utility of PCR in sequencing is the ability to produce large amounts of specific DNA from small amount of genomic template. Single stranded templates developed by asymmetric PCR or SSCP are more amenable to sequencing than double stranded ones.

Gene Cloning

Cloning of PCR products is useful for further characterisation like restriction mapping, sequencing and also for expression and function analysis of insect pathogens.Cloning studies are used in transfer of virulent genes and in baculovirus expression system.

Mutation Studies

Alteration of DNA sequences by election, insertions and point mutations is a powerful approach to the analysis of the structure and function of genes and their products. Mismatches may be included in the primers to give rapid site-directed mutagenesis. The concept of overlap extension, a PCR based technique has been successfully used for engineering of fusion and hybrid genes and for construction of totally synthetic genes from oligonucleotides. In certain cases, mutant strains of insect pathogens are more virulent than parent strains.

Following are some applications of PCR to different groups of insect pathogens:

Insect Viruses

PCR based methods facilitate investigation of phylogeny and grouping of insect viruses by allowing the determination of the nucleotide sequence of a specific gene from any virus isolate. These studies help in the understanding of biological adaptations such as virus host specificity (Bulach *et al.*, 1999). The DNA polymerase and polyhedrin gene are the target sequences for amplification. PCR amplification of specific gene of virus aids in detection of viral DNA in infected insects at early stages of infection, cross infectivity, latent infection and ecological studies. PCR is used for identification of expression of recombinant baculoviruses in insect cell lines.

Bacteria

PCR is widely used in the characterisation of *Bacillus thuringiensis* (*Bt*). Brousseau *et al.* (1993) developed a methodology of arbitrary primers (AP) PCR to distinguish commercial strains and to differentiate *B.t* from *B. cereus.* The purity of commercial *Bt* products can be checked using AP-PCR. A modification of PCR namely Exclusive PCR or E-PCR has been developed to detect novel cry genes by using mixtures of degenerate and specific primers. Electrophoretic patterns of PCR products using specific primers of cry genes predicts the insecticidal activity of *Bt* strains. Besides, PCR techniques are used in the detection of *Bt* strains in the environment and its expression in transgenic plants.

Entomopathogenic Fungi

PCR offers objective and reliable methods for identification of fungal isolates which overcome the limitations of RFLP and RAPD analysis in the classification of entomopathogenic fungi. Combination of PCR with RAPD and SSCP (Single Strain DNA Confirmation Polymorphism) characterise and identify geographic isolate variation of *Beauveria* spp. (Hegedus and Khachatourians, 1996). Geonmic DNA, mitochondrial DNA (mtDNA) and ribosomal DNA (rDNA) are used for synthesis of primer sequence for amplification. Reverse Transcriptase- Differential Display (RT -DD-PCR) technique was used to detect and characterise Prl gene and chit 1 gene of *Metarhizium anisopliae* responsible for the pathogenicity ofthe fungi. Besides, PCR techniques are used for early detection of fungal disease, development and cloning of pathogenesis related gene (Pri gene) from virulent strain in non-virulent strain of fungi.

Basic Protocol

Materials Required

1. Sterile water
2. $MgCl_2$
3. Buffer
4. Deoxynucleotides
5. Primers
6. Template DNA
7. Taq DNA polymerase
8. Ficol1400 (optional)
9. Thin-walled PCR tubes
10. Automated Thermo cycler.

Procedure

1. Produce the reaction master mix
2. Overlay the reaction mixture with 50 to 100ml mineral oil
3. Programme the automated thermal cyclers according to the manufacturer's instruction. Run the programme for the period specified
4. Electrophorese 10 ml from each reaction on an *agarosel* non denaturing polyacrylamide or sieving agaorse gel appropriate for the PCR product size expected. Stain with ethidium bromide.
5. Examine the stained gel to determine which condition resulted in the greatest amount of product.
6. To ensure that the major product is the correct one, digest an aliquot of the reaction with a restriction endonuclease known to cut within the PCR product. Check buffer compatibility for the restriction endonuclease. Electrophorese the digested product on a gel to verify that the resulting fragment has the expected size.

Genetic engineering studies with insect pathogens demand DNA in microgram or still larger quantities. Pathogens, being microorganisms are required in greater number for extraction of sufficient quantity of DNA. PCR which requires only a few nanogram (ng) as a starting material becomes inevitable in insect pathology studies for molecular characterisation of insect pathogens. Theoretically each cycle of PCR can yield 2" copies of DNA where n is the number of cycles.

Using this technique it is possible to elucidate the genomic organisation of insect pathogens and to locate the genes responsible for their pathogenicity. This also facilitates the cloning and recombination studies for the genetic improvement of insect pathogens. Modifications of the traditional PCR techniques such as inverse PCR, asymmetric PCR, exclusive PCR, recombinant PCR have been identified to suit the various needs in the molecular understanding of the pathogens and the basic protocols are available (Coen *et at.*, 1999). It is obvious that use of PCR in insect pathology widens the utility of insect pathogens and makes them powerful tools in insect pest suppression.

Potential in IPM

Biotechnology has provided new avenues for management of insect pests and it holds great potential to be included in IPM system.

1. The low toxicity of proteinase inhibitors and *Bt* a-endotoxin as compared to conventional insecticides would reduce the selection pressure and may slow down the development of resistance.
2. Since all plant parts including growing points would remain covered with toxins, dependence on weather for efficacy of the sprays would be eliminated.
3. Since toxins will always be there, so there will be no need of continuous monitoring of pests.
4. Transgenic plants would also provide protection to those plant parts, which are difficult to be treated with pesticides. Thus, transgenics may prove useful for controlling bollworms and borers, which are difficult to control by means of insecticides.
5. The cost of application in the form of equipment and labour will be nil or negative.
6. The development cost is only a fraction of the cost of development of a conventional pesticide. There would be no problem of contamination in the form of drift and ground water contamination or risk to the field workers.
7. Insecticidal activity would be restricted to those insects which actually attack the plants. Transgenic plants would be safe to non-target species and human beings.
8. Transgenic plants eliminate the problems of shelf life and field stability faced by pesticide formulations as they provide on site biosynthesis of the toxins.
9. Transgenic plants will have inbuilt resistance to various insects replacing some of the current pesticide usage with protection which is intrinsically biodegradable, thus reducing the use of chemical insecticides and minimising the problem of environmental pollution.

In spite of all these advantages, biotechnology programme does face uncertainties.

1. The resistance against a-endotoxin is not ruled out, though it may be slow as compared to insecticides because of continuous exposure of the insect to the toxin.
2. Some weeds may introgress an insect resistance gene from related transgenic plants and make them less susceptible to their usual herbivores, thus exhibiting greater reproductive success and may create a weed problem.

3. If lepidopteran herbivores were removed from plant species, other insects might experience competitive release and become more common.
4. The use of transgenic plants possessing resistance factor in one locality may affect the insect population dynamics in other areas.
5. Although Lepidoptera are herbivores as larvae, many are important pollinators of wild species as adults. Therefore, if larval mortality is high, plant communities might also be disrupted by decreasing availability of pollinators. Further, disruption of pollinators and plant communities could occur if the toxin is expressed in plant nectar and pollen.

Biotechnology has the potential to move farming closer to ecologically sustainable practices, both in developed and developing countries and thus could make a considerable impact on agricultural systems in the future. The strategy to maximise the utility of this approach should involve the use of gene combinations whose products are targetted to different biochemical and physiological processes within the insect. In this way, it is expected to provide a multimechanistic form of resistance which can be tailored to different crops and prevailing insect pests.

References

Allan, E.J., Eeswara, J.P., Johnson, S., Mordue, A.J., Morgan, E.D. and Stuchbury, T., 1994, The production of azadirachtin by *in-vitro* tissue cultures of neem, *Azadirachta indica. Pesticide Science*, 42(3): 147-152; 19 ref.

Borovky, D., Carlson, D., Griffms, P.R., Shabanowitz and Hunt, D.F., 1993. Mass spectroscopy and characterization of *Aedesaegyptii* Trypsinmodulating oostatic factor (TMOF) and its analogs. *Insect. Biochem. Molec. Biol.*, 23(6): 703–712.

Borovsky, D., 1988. Oostatic hormone inhibits biosynthesis of proteolytic enzymes and egg development in mosquitoes. *Archs Insect Biochem. Physiol.*, 7: 187–210.

Borovsky, D. and Schlein, Y., 1988. Quantitative determination of trypsin like and chymotrypsin like'enzymes in insects. *Archs Insect Biochem. Physiol.*, 8: 249–260.

Borovsky, D., Carlson, D. A., Griffm, P.R., Shabanowitz, J. and Hunt, D.F., 1990. Mosquito oostatic factor: A novel decapeptide modulating trypsin-like enzyme biosynthesis in the midgut. *FASEB J.*, 4: 3010–3020.

Borovsky, D., Thomas, B.R., Carlson, D.A., Whisenton, LR. and Fuchs, M.D., 1985. Juvenile hormone and 20-hydroxyecdysone as primary and secondary stimuli vitellogenesis in *Aedes aegypti, Archs Insect Biochem. Physiol.*, 2: 75–90.

Borovsky, D., 1982. Release of egg development neurosecretory hormone in *Aedes aegypti* and *Aedes taeniorhynchus* induced by an ovarian factor. *J. Insect Physiol.*, 28: 311–316.

Brousseau, R.A., Saint-Onge, G., Prefontaire, L. Masson and Cabana, J., 1993. Arbitrary primer PCR, a powerful method to identify *Baccillus thuringiensis* serovars and strains. *Appl. Environ. Microbial.*, 59: 114–119.

Bulach, D.M., Ananthakumari, C., Zaika, A., Liang, H. and Tribe, D.E., 1999. Group II nucleo polyhedrosis virus subgroups revealed by phylogenetic analysis of polyhedrin and DNA polymerase gene sequences. *J. Invertebr. Pathol.*, 73: 59–73.

Coen, D.M., 1999. Short protocols In molecular biology. In: *A compendium of Methods from Current Protocols in Molecular Biology*, (Eds.) F.M. John Ausubel, R. Brent, RE.Kingston, D.O. Moor, J.G. Siedman, J.A. Smith and K. Sturhd. Wiley and Sons, New York, pp.15–31.

Deb, C.R., 2001. Somatic embryogenesis and plantlet propagation of *Melia azedarach* L (Ghora Neem) from cotyledonary segments. *J. Plant Biochem. Biotechnol.*, 10(1): 63–65

De la Riva, G.A., Gonzalez-Cabrera, J., Vazquez-Padron, R. and Ayra-Pardo, C., 1998. Agrobacterium tumefaciens: A natural tool for plant transformation. *Electronic Journal* of *Biotechnology*, 1(3).

Gelman, D.H. and Borovsky, D., 2000. *Aedes agyptii* TMOF maculates ocdysteroid production by prothoracic glands of gypsymoth, *Lymantria dispar. Archs Insect Biochem. Physiol.*, 45(2): 60–68.

Greenplate; J.T., 1999. Quantification of *Bacillus* thuringlensis Insect *ConiF&JI.* Protein *CrylA.c* over time in bongard cotton fruit and terminals. *J. Econ Entomol.*, 92: 1377–1383.

Hagedorn H.H., Shapiro, J.P. and Hanaoka, K., 1979. Ovarian ecdysone secretion is controlled by a brain hormone in an adult mosquito. *Nature* (*London*), 282: 92–94.

Hegedus, D.O. and Khachatourians, 1996. Identification and differentiation of the entomopathogenic fungus *Beauveria bassiana* using PCR and single-stranded conformation polymorphism analysi. *J. Invertebr. Pathol.*, 67: 289–299.

Lindsey, K., 1996. Plant transformation systems. In: *Transgenic Plants: A Production System for Industrial and Pharmaceutical Proteins*, (Eds.) M.RL. Owen and J. Pen. John Wiley and Sons, New York, pp. 5–25

Loof, De.A., Bylemans, D., Schoofs, L., Janssen, I. and Huybrechts, R., 1995. The folliculostatins of two dipteran insect species, their relation to matrix proteins and prospects for practical application. *Entomologia Expermnetalis et Applicata*, 77: 1–9.

Lynn, D.E., Miller, S.G. and Oberlander, H., 1983. Nuclear polyhedrosis virus replication in epithelial cell cultures of lepidoptera. *J. Invertebr. Pathol.*, 42: 424–426.

Nauen, R., Sorge, D., Stemer, A., Borovsky, D., Loof, A., De and Hoffmann, K.H., 2000. TMOF like factor controls the biosynthesis of serin proteases in the larval got of *Heliothis virescens. Archs Insect Biochem. Physiol.*, 47(4): 169–180.

Palanisamy, K., Kumar, Pramod and Kumar, P., 2001. Vegetative propagation and genetic improvement of Neem. *Indian Forester*, 127(3): 347–350.

Ramesh, K. and Padhva, M.A., 1988. Isolation of nimbin from Azadirachta indica leaves and its callus cultures. *Indian Drugs*, 25(12): 526–527.

Schaaf, O., Jarvis, A.P., Esch, S.A., van-der, Giagnacovo, G., Oldham, N.J. and der-Esch, SA van, 2000. Rapid and sensitive analysis of azadirachtin and related triterpenoids from Neem (*Azadirachta indica*) by high performance liquid chromatography atmospheric pressure chemical ionization mass spectrometry. *Journal of Chromatography*, A, 886: 1–2,89–97.

Zhu, W., Vandingenen, A., Huybrechts, R., Baggerrmann, G., Deloof, Poulos, C.P. Velentza and Breurer, M., 2001. *In vitro* degradation of the Neb–TMOF in gut luminal content and hemolymph of the gray *flesh, Neobellieria bullata. Insect Biochem. Molec. Biol.*, 31: 87–95.

Insect Pest and Disease Management *Pages* **501–515**
Editor: **Dr. D. Prasad**
Published by: **DAYA PUBLISHING HOUSE (2016)**

Chapter 36

Management of Rats

S.C. Khanna
Division of Entomology,
Indian Agricultural Research Institute, New Delhi – 110 012

Introduction

Many types of rodents do cause problems in home and may become a nuisance from time to time, but the most troublesome and potentially harmful rodents are rats and mice. Norway rats, roof rats, and house mice are "commensal" rodents–that is, they live in close association with human beings. Rats and mice cause a huge loss of food worldwide. Besides eating our food they spoil it by contamination with their feaces, urine or fur. Rats and mice cause harm by transmitting diseases, destroying property, and biting. There are around 50 diseases which can be transferred to man by rodents, including typhoid, paratyphoid, and scabies. In addition, rodents may be vectors of a large number of diseases affecting domestic animals. They can be found not only in our homes, but supermarkets, restaurants, warehouses, food processing facilities, livestock facilities, and farm fields. They reproduce rapidly with major activity at night. Rats tend to be cautious and mice are more curious. Rats tend to eat most of their food at one time, whereas mice will nibble a little at a time. They also cause damage to our buildings by their burrowing and gnawing activity. They have a pair of incisor teeth in the upper and lower jaws. The incisors are curved inwards and have an extremely hard anterior coating. The softer inside layer is worn down much more rapidly than the hard, outer layer. This means that the teeth are continually kept sharp, enabling them to damage even materials such as masonry and electric cables. The incisors do not stop growing. This means that the animals are forced to gnaw steadily in order to wear them down. They are able to squeeze through very small openings: 1/4 inch for mice and 1/2 inch for rats. A successful rodent control program includes a combination of baiting and trapping for the highest rate of success. The animals that are usually referred to as rat, belong to the family, Muridae, sub-families Murinae, Cricetinae, Microtinae and Gerbillianae.

List of Rats Reported from India

1. Korean field mouse (*Apodemus draco*)
2. Korean field mouse (*Apodemus latronum*)
3. Long-tailed field mouse (*Apodemus orestes*)
4. Wood mouse (*Apodemus sylvaticus*)
5. Indian mole rat (*Bandicota bengalensis*)
6. Bandicoot rat (*Bandicota indica*)
7. Bower's rat (*Berylmys bowersii*)
8. Kenneth's White-toothed Rat (*Berylmys mackenziei*)
9. Manipur White-toothed Rat (*Berylyms manipulus*)
10. Bay bamboo rat (*Cannomys badius*)
11. Pencil-tailed tree mouse (*Chiropodomys gliroides*)
12. Blanford's Rat (*Cremnomys blanfordi*)
13. Cutch rock rat (*Cremnomys cutchicus*)
14. Elvira Rat (*Cremnomys elvira*)
15. Armenian hamster (*Cricetulus migratorius*)
16. Large-toothed giant rat/Millard's rat (*Dacnomys millardi*)
17. Manipur mouse (*Diomys crumpi*)
18. Pere David's vole (*Eothenomys melanogaster*)
19. Indian hairy-footed gerbil (*Gerbillus gleadowi*)
20. Baluchistan gerbil (*Gerbillus nanus*)
21. Indian bush rat (*Golunda ellioti*)
22. Manipur bush rat (*Hadromys humei*)
23. Murree vole/Punjab vole (*Hyperacrius wynnei*)
24. Edward's giant rat (*Leopoldamys edwardsi*)
25. Indian desert jird (*Meriones hurrianae*)
26. Eurasian harvest mouse (*Micromys minutus*)
27. Murray's soft-furred rat (*Millardia gleadowi*)
28. Asian soft-furred rat (*Millardia kondana*)
29. Soft-furred field rat or metad (*Millardia meltada*)
30. Indian field mouse (*Mus booduga*)
31. Fawn-coloured mouse (*Mus cervicolor*)
32. Cook's mouse (*Mus cookii*)
33. House/Rice-field mouse (*Mus dunni*)
34. Servant mouse/Shrew mouse (*Mus famulus*)
35. House mouse (*Mus musculus*)
36. Shrew mouse (*Mus pahari*)

37. Spiny mouse (*Mus phillipsi*)
38. Spiny field mouse (*Mus platythrix*)
39. Grey spiny mouse (*Mus saxicola*)
40. Short-tailed mole (*Nesokia indica*)
41. Brahma White-bellied Rat (*Niviventer brahma*)
42. Smoke-bellied Rat (*Niviventer eha*)
43. Chestnut White-bellied Rat (*Niviventer fulvescens*)
44. Lang Bian White-bellied Rat (*Niviventer langbianis*)
45. White-bellied Rat (*Niviventer niviventer*)
46. Tenasserim white-bellied rat (*Niviventer tenaster*)
47. Malabar spiny dormouse (*Platacanthomys lasiurus*)
48. Polynesian/Pacific rat (*Rattus exulans*)
49. Himalayan rat (*Rattus nitidus*)
50. Brown rat/Norway rat (*Rattus norvegicus*)
51. Palm rat (*Rattus palmarum*)
52. Kerala rat (*Rattus ranjiniae*)
53. Common house rat (*Rattus rattus*)
54. Sikkim rat (*Rattus remotus*)
55. Andaman rat (*Rattus stoicus*)
56. Malaysian wood rat/Malaysian field rat (*Rattus tiomanicus*)
57. Turkestan rat (*Rattus turkestanicus*)
58. Indian gerbille (*Tatera indica*)
59. Indian long-tailed tree mouse (*Vandeleuria nilagirica*)
60. Long-tailed tree mouse (*Vandeleuria oleracea*)
61. White-tailed tree mouse (*Vandeleuria elliotti*)

Types of Rats

On the basis of the relationship of rodents with human being, the rats are grouped into four categories.

Domestic Rodents

Living with man, which nest and live actually in the house. *e.g. Rattus rattus* and *Mus musculus.*

Commensal Rodents

Living on man. Near man but not with man. They prey on man's food and on his waste products and hence enter into the human habitat but do not remain or nest there. *e.g. Rattus frugivorus*

Field Rodents

Living on human but away from man. They live in cultivated fields and consume the produce, never penetrate into human habitat but frequent threshing floors, storage installations, where they come in contact with domestic and commensal rodents. *e.g. Bandicota bangalensis.*

Wild Rodents

Living without man, avoid man and have very little contact with other rodents. Feed on wild plants. *e.g. Meriones hurricanae.*

House Mouse (*Mus musculus*)

Figure 36.1: House Mouse (*Mus musculus*)

The house mouse (*Mus musculus*) is a small, slender rodent having pointed nose, small but somewhat protruding eyes, large sparsely haired ears, hairless tail without obvious scale rings. Tail length equal to head and body length. Generally grayish brown in colour with gray or buff belly. Adult house mouse weighs about 15–30 gm. The house mouse are highly adaptable and live in close association with humans therefore termed as " commensal" rodents. These may be found at temperature as low as –30°C. House mouse lives in and around homes, farms, commercial establishments, in open fields and meadows. They are omnivorous and eat many types of food but prefer grains and seeds. They are very inquisitive, try new foods and nibble many kinds of items. Foods high in fat, protein or sugar are preferred. Adult consumes about 3-4 gm of food per day and intake of water is 3-9 ml/day. It can survive with little or no free water. House mouse is nocturnal in activity but often found active during the day. Territory ranges from 10-30 ft. from the nest. Nests are made from paper, insulation, soft material. House mouse has poor eyesight but highly developed sense of smell, hearing, taste and touch. It breeds round the year. A female is sexually mature in 6-10 weeks of age. A female may have 5-10 litters of 4-8 young ones per year. The gestation period varies from 18-21 days. On average, house mouse lives for one to two years.

Roof Rat (*Rattuus rattus*)/Black Rat/House Rat

The body is slender with large eyes and large ears, pointed muzzle, tail longer than head and body length. It is less adaptable to cooler temperatures. The rat is opportunist, eats small amount of food in many places, carry food, back to nest. It can travel up to 90 meter from the nest in search of food. It is very sensitive, neophobic, good climber, often seen above ground in roof, rubbles on trees in vegetation and sewers/drains. Born in a nest about 21 to 23 days after conception, the young rats are naked and their eyes are closed. There are 5-8 young in a litter, develop rapidly, growing hair within a week. When 9-14 days old, the eyes get opened and begin to explore for food and move about near their nest. In the third week they begin to take solid food. The number of litters depends on the area and varies with nearness to the limit of their climatic range, availability of nutritious food, density of the local rat population and age of the rat. The young may be continued to nurse until 4 or 5 weeks old. Young rats generally cannot be trapped until about 1 month old. At about 3 months of age they are completely independent of the mother and are reproductively mature. Roof rats also have an excellent sense of balance and use their tails for balance and are very agile climbers.

Norway Rat (*Rattus norvegicus*)/Brown Rat/Wharf Rat/Sewer Rat/Water Rat/Grey Rat

Norway Rat is grayish brown in colour, robust body, larger than roof rat and more aggressive. It has blunt nose, small ears, tails scaly, semi naked shorter than head + body. Weighs from 200–500 gm

Figure 36.2: Roof Rat (*Rattuus rattus*)/Black Rat/House Rat

Figure 36.3: Norway Rat (*Rattus norvegicus*)/Brown Rat/Wharf Rat/Sewer Rat/Water Rat/Grey Rat

and is up to 12 inches in length. Lives in sewers and adaptable to cold temperature. It is social in behaviour and lives in colony. Single rat consumes about 25-30 gm food per day. It is omnivorous, prefers carbohydrates, protein rich food. It eats various types of poultry, including ducklings. Food is usually carried in its mouth to a safe place where it is eaten. Large items are dragged to a hiding place. The food the rat leaves uneaten is left behind since, unlike many rodents, the brown rat does not hoard food. The rat lives outdoor and feed at night. It can not survive without water and consumes 15–30 ml of water every day. The rat is not agile but climbs very well. It is neophobic and wary of new food, new objects. A female brown rat is ready to breed when she is 11 weeks old and mates throughout the year. After mating, the male play no part in rearing the young. The female builds a round nest of loose material such as straw. The nest is often located in an underground burrow. After 21-24 days, 6 to 11 blind and hairless young are born. They are totally dependant on their mother, who suckles them for three weeks. At the end of this time, they are ready to leave the nest. A female may sometimes give birth to three to five litters a year. A large colony is often started by a single pregnant female. Lifespan: 1-2 years, but very few live longer than 1 year.

Indian Bandicoot (*Bandicota indica*)

Indian Bandicoot rat is dark grayish brown in colour, measures 30–40 cm in length. The tail is as long as body. It weighs upto 1.4 kg. Territorial very aggressive, grunts and erects hairs on back, fight back when disturbed. It is found, near garbage bins, gardens and poultry farm. The rat makes large burrows.

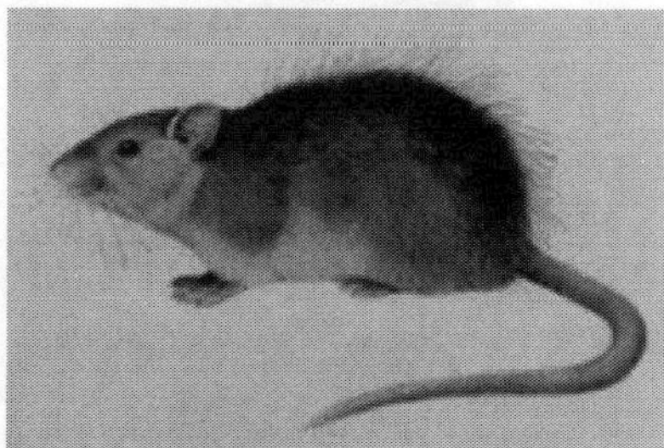

Figure 36.4: Indian Bandicoot (*Bandicota indica*)

Indian Mole Rat (*Bandicota bangalensis*)

Indian mole rat is smaller than Indian bandicoot. It is dark grayish brown in colour. It has rounded head and short broad muzzle, the tail is relatively longer. It is is aggressive and grunts and fights back when disturbed. Found in rural areas, fields in planes. The rat makes extensive burrow pattern with chambers for escape. Indian mole rat stocks grain in burrows. It is known to spread diseases.

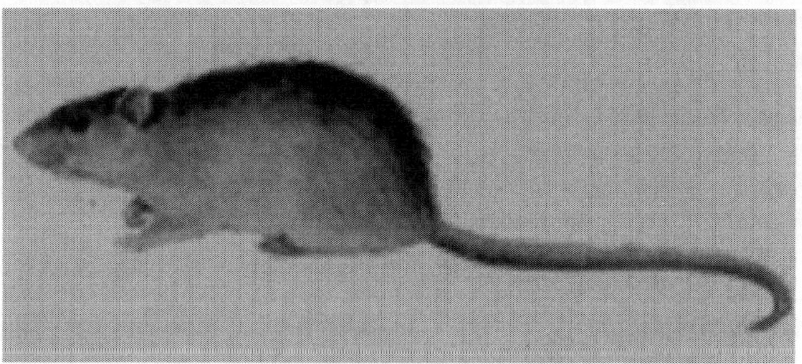

Figure 36.5: Indian Mole Rat (*Bandicota bangalensis*)

Some Interesting Facts About the Rats

1. Excellent swimmers
2. Good climbers
3. Frequency range of 50 kHz versus man's 20 kHz.
4. Poor vision beyond 3–4 feet

5. Incisors teeth grow @ 0.3–0.4 mm per day
6. Can gnaw through lead sheeting, aluminium sliding, glass improperly cured, concrete etc.
7. Can squeeze through small hole of ½ inch for rat and 1/4th inch for mouse.
9. Possess highly developed sense of taste and touch (presence of whiskers- long hair like bristles on the face), odor (urine marks)

Rodent Damage

1. Causes damage to high voltage power cables/plugs
2. Damages fiber optic cables
3. Water pipes
4. High pressure cleaning hose
5. Flexible water hose
6. Lead pipe
7. Burrows in wooden farm houses
8. Library books

Signs of Rodent Infestation

1. Appearance of droppings (faeces)
2. Tracking patches
3. Damage due to gnawing
4. Formation of burrows
5. Presence of runways
6. Marking of grease
7. Presence of urine stains
8. Notice of dead or live rats frequenting in the area
9. Hearing of rat sound
10. Odor emanating from the body of rat (specially mice)

Rodent Control Measures

Sanitation

Sanitation is fundamental to rat control and must be a continuous process. If proper sanitation measures are not adopted and maintained regularly, the advantages of other measures are lost and rats find reentry. Good housekeeping in and around buildings will definitely reduce the shelter and food. Neat and off the ground storage of pipes, firewood, crates, boxes, gardening tools and other household goods would reduce the rat presence.

Prevention of Rat Entry

1. Reduce the food source
2. Use rust resistant garbage cans provided with tight lids
3. Adopt good storage practices

4. Follow regular cleaning and sweeping of the premises
5. Reduce outdoor harbourage, Removal of weeds, cleaning of the perimeter etc.
6. Elimination of water source
7. Do not compost animal products in the area
8. Garbage, trash and debris should be gathered regularly and kept in tight lid containers

Rat Proofing

1. Structural changes as per requirement
2. Tight fitted doors without gaps
3. Wind screens with no gaps and keep them closed
4. Metal plate fixing on the bottom of the wooden doors
5. Concrete perimeter of 45 cm. width and 15 cm. depth all along the external wall
6. Fixing of 1/4th inch wire mesh where possible to prevent squeezing of rats
7. Use mortar to seal any cracks and holes in the basement and house

Trapping

Basic purpose of using traps is to avoid use of hazardous chemicals in the house, to overcome the rat bait shyness and avoid smell that may emanate from the body of dead rats. Trapping is the safest and most effective method for controlling rats in and around homes, garages, and other structures. Because traps can be used over and over again, trapping is less costly than poison baits but more labor intensive. Traps can be set and left indefinitely in areas where rats have been a problem in the past. Generally young rats cannot be trapped until they are about a month old, which is when they leave the nest to venture out for food.

There are many types *viz.* Glue traps, Mechanical traps, Cages are used for the purpose. The best places to set traps are in secluded areas where rats are likely to travel and seek shelter. The number of traps required will depend on the population of the rat in the area. Usually there should be 2 traps for one rat. Place rat traps in the runway without setting the trigger for few days. The rodent will become used to its presence, and trapping will become easier.

Other Control Methods

Rats are wary animals, easily frightened by unfamiliar or strange noises. However, they quickly become accustomed to repeated sounds, making the use of frightening sounds, including high frequency and ultrasonic sounds, ineffective for controlling rats in home and garden situations.

Traps that kill rats by electrocution are available on the market. These traps are considerably more expensive than the common snap trap and can be used safely and effectively only in limited situations. Like the snap traps, these traps catch only one rat at a time and then must be emptied.

Smoke or gas cartridges are sold for the control of burrowing rodents. When placed into the burrows and ignited, these cartridges produce toxic and suffocating smoke/gases. The burrows may extend beneath a residential area and have several open entrances, permitting toxic gases to permeate the dwelling. For this reason and because some fire hazard is associated with their use, smoke and gas cartridges are not recommended for rat control around homes.

The rats may be drowned or flushed from their burrows by flooding them out with water from a garden hose and then closing the holes with dirt.

Predators, especially cats and owls, eat rats and mice. However, much of the time these predators are unable to keep rodent numbers below levels that are acceptable to most people.

Use of Rodenticides

Rodenticides are defined as substance that is used to kill rats, mice and rodent pests. Warfarin, Bromodiolone and Diafenacoum are substances which cause internal hemorrhage and kill by preventing normal blood clotting. Fumigants such as sulfur dioxide, carbon monoxide, hydrogen cyanide and methyl bromide also act as rodenticides. In the past, phosphorus paste, barium carbonate salt and powder such as zinc phosphide, white arsenic, thallium sulphate, strychnine, strychnine sulphate and calcium cyanide have used mixing with bait and placed where rodents will find and eat them. All these compounds are toxic to other animals. Red squill, derived from bulbs of lily like plant is slower in acting and less toxic to animals other than rodents because it is removed from the stomach by vomiting–a reflex absent in rodents.

Table 36.1: Types of Baits Used in Rat Control and Their Active Ingredients

Type of Bait	*Active Ingredients*	*Use Restrictions*
Multiple feeding anticoagulants	Warfarin, Chlorophacinone Diaphacinone	For use in and around buildings and in gardens and landscaped areas
Single feeding anticoagulants	Brodifacoum, Bromodialone Difethialone	For use in and around building only
Non anticoagulants	Bromathalin, Cholecalciferol	For use in and buildings only
Non anticoagulants	Zinc phosphide	For use in around buildings and gardens and landscaped areas

Brodifacoum

Chemical Name

3-[3-(4'-bromobiphenyl-4-yl)-1,2,3,4-tetrahydro-1-naphthyl]-4-hydroxycoumarin

Chemical Formula

$C_{31}H_{23}BrO_3$

Chemical Structure

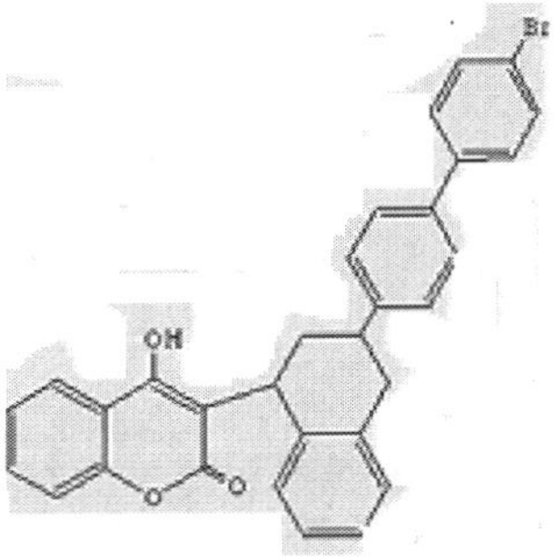

Br

OH

O

O

LD_{50}/LC_{50}

The oral route is the commonest way of entry. LD_{50} oral (female rat) 0.37 to 0.68 mg/kg it is very similar to Difenacoum in its chemical structure and anticoagulant effect (they are both 4-hydroxycoumarin derivatives). Brodifacoum is a weak acid which does not readily form water-soluble salts. It does not lose activity after 30 days in direct sunlight. Commercial formulations are stable at least for two years if protected from extreme temperatures and sunlight. The container should be kept closed to maintain activity of formulation (as a bait). The modern 2nd generation anticoagulant rodenticides such as Brodifacoum have proven severe secondary poisoning properties and have been held responsible for the deaths of many birds of prey, particularly rare Red Kites and Buzzards that habitually feed off dead bodies. This should only be used when there is resistance to other preparations such as Bromodiolone, Warfarin etc.

Bromadiolone

It does not occur naturally. It is used as a rodenticide in urban and farm rodent control and acts by disrupting the normal blood clotting mechanisms causing an increased tendency to bleed. It is a second generation anticoagulant that is effective against rats and mice, including those resistant to first generation anticoagulants. It is used in the form of ready-to-use baits of low concentration containing 0.005 per cent bromadiolone. Bromadiolone is a white to off-white powder. Its solubility in water is very low (less than 20 mg/litre at 20°C). It is slightly soluble in ethanol and ethyl acetate, and soluble in dimethylformamide. The flash-point temperature is 218°C. Bromadiolone is absorbed through the gastrointestinal tract, skin, and respiratory system. The major route of elimination in different species after oral administration is via the faeces. The liver is the main organ of accumulation and storage. Bromadiolone has been found in the liver as the unchanged parent compound. Elimination from the liver is biphasic with an initial rapid 2-8 days and a slower phase with a half-life of 170 days.

Chemical Name

3-[3-(4'-bromobiphenyl-4-yl)-3-hydroxy-1-phenylpropyl]-4- hydroxycoumarin

Chemical Formula

$C_{30}H_{23}BrO_4$

Chemical Structure

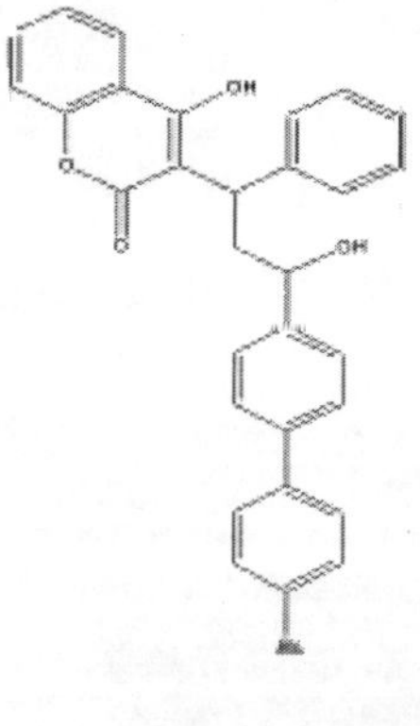

LD_{50}/LC_{50}

Bromadiolone has a high, acute oral toxicity (LD_{50} of 1-3 mg/kg) for various species including rodents and non-rodents. Signs of poisoning are those associated with an increased tendency to bleed. Bromadiolone is non-irritant to the skin. It is a slight irritant for the eye.

Chlorophacinone

Chlorophacinone is an oil concentrate for impregnating the material. Rozol is used also in production of paraffin blocks, water baits and tracking powder. It is an anticoagulant rodenticide and single dose of 50 mg/kg bait. It does not induce bait shyness.

Chemical Name (IUPAC)

2-[2-(4–chlorophenyl)-2-phenylacetyl]indan-1,3-dione

Chemical Formula

$C_{23}H_{15}ClO_3$

Chemical Structure

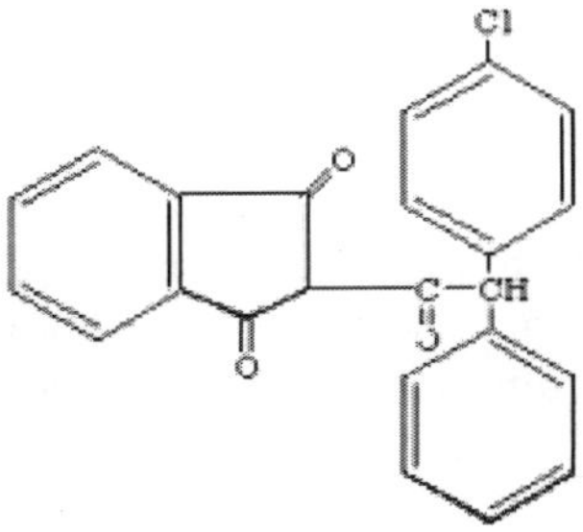

LD_{50}/LC_{50}

100 per cent chlorophacinone (Rat): Oral LD_{50} 3.15 mg/kg (1 dose/21 days). 0.005 per cent chlorophacinone form (Commensal rodents): LD_{50} 63 g/kg. (Field mouse): generally 0.005 per cent - 0.0075 per cent (varies by species/application). Insoluble in water. It is also non-toxic to bees. DERMAL: LD_{50} = 200 mg/kg (albino rabbit). A solution of 5 mg in 2 ml liquid paraffin applied to 100 cm^2 of a rabbit's shaved skin caused only a slight reduction of prothrombin.

Cholecalciferol

Cholecalciferol (known as Vitamin D_3), Rodenticide act by causing vitamin D toxicosis, raising serum calcium levels to dangerously high levels. Clinical signs arise within 18 to 36 hours after ingestion, and include depression, anorexia, increased urination and increased water intake. Muscles are affected by high calcium, and the animal becomes very weak heart conduction becomes slowed, and ventricular fibrillation and cardiac arrest may result. Type of formulations of rodenticide granular (0.075 per cent bait).

Chemical Name

Cholecalciferol; Activated 7-dehydrocholesterol

Other Names

Calciol; (5Z,7E)-(3S)-9,10-Secocholesta-5,7,10(19)-trien-3-ol; 9,10- secocholesta-5,7,10(19)-trien-3beta-ol; oleovitamin d3; CC; duphafral d3 1000; delsterol;Deparal; ebivit; neo dohyfral d3; provitina; ricketon; trivitan; d3-vicotrat; vi-de-3-hydrosol;vigantol; vigorsan; 5,7-cholestadien-3betal-ol; 7-dehydrocholesterol; (3beta, 5Z, 7E)-9,10-secocholesta-5,7,10(19)-trien-3-ol; Dehydrocholesterol; Secocholesta-5,7,10(19)-trien-3-ol, (3beta,5Z,7E)-; Cholealciferol; Colecalciferol; Vitamin D3, crystalline.

Chemical Formula

$C_{27}H_{44}O$

Chemical Structure

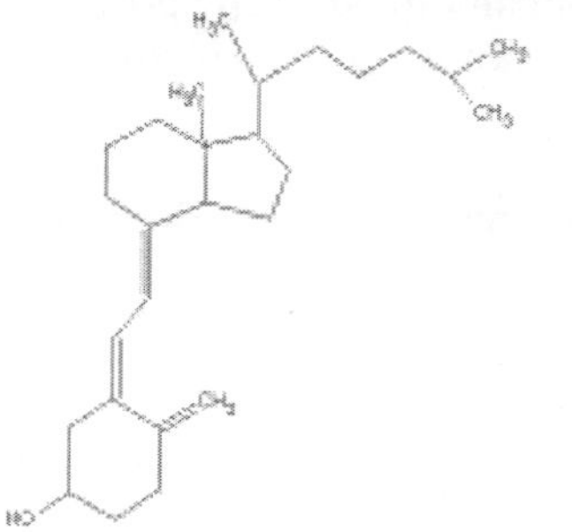

Coumatetralyl

Coumatetralyl is a first generation anticoagulant of the warfarin type. After one exposure the toxicity of coumatetralyl is relatively low, however if over exposure continues over several days the product becomes more toxic. That is, the product must be constantly present in the blood for more than 1 to 2 days in order to be highly toxic. A single exposure, even though relatively large, may not produce toxic sypmtoms as the compound is quite rapidly metabolised. Chronic: animal studies show no evidence of carcinogenic or teratogenic effects.

Chemical Name (IUPAC)

4-hydroxy-3-(1,2,3,4-tetrahydro-1-naphthyl) coumarin

Chemical Formula

$C_{19}H_{16}O_3$

Chemical Structure

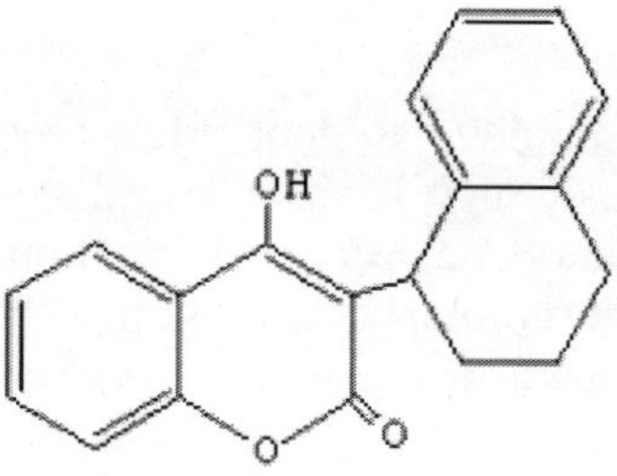

LD_{50} Oral (Ingestion)

Rat 16.5 mg/kg LD_{50} Dermal (skin contact)–rat 25-30 mg/kg. These data refer to the active constituent. The toxicity of the product is greater after repeated exposure than in a single dose. Animal tests have shown coumatetralyl not to have any carcinogenic or teratogenic potential.

Difenacoum

Difenacoum does not occur naturally. It is used as a rodenticide against pest rodents and acts by preventing the production of blood clotting factors. It is an off-white powder. Its solubility in water is very low (less than 10 mg/litre at 20°C and pH 7). It is slightly soluble in benzene, and soluble in acetone and chloroform. It is a weak acid, but its sodium and potassium salts are sparingly soluble in water. Difenacoum does not enter the atmosphere, because of its low volatility. Difenacoum is bound to soil particles and is not taken up by plants. The rate of degradation is relatively slow and depends on soil type. Difenacoum is absorbed through the gastrointestinal tract, skin, and respiratory system. The major route of elimination in different species, after oral administration, is via the faeces. The liver is the main organ of accumulation and storage. Difenacoum has been found in the liver as both the parent compound and metabolites. The elimination from the liver and kidney is biphasic with an initial rapid phase of three days and a slower phase with a half-life of 118-120 days.

Chemical Formula

$C_{31}H_{24}O_3$

Chemical Structure

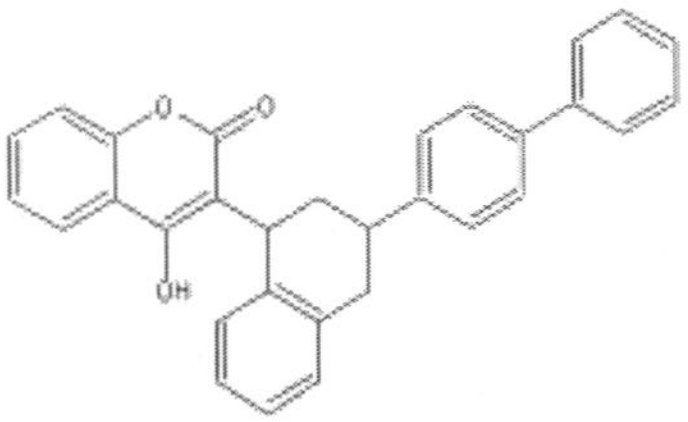

LD_{50}/LC_{50}

The acute oral toxicity of difenacoum is high, *i.e.*, LD_{50} is of 1.8 and 2 mg/kg body weight for rats and rabbits, respectively, and between 50 and 100 mg/kg body weight for various non-rodent mammals. The acute dermal LD_{50} is greater than 50 mg/kg body weight in rats and and rabbits. Signs of poisoning are those associated with an increased tendency to bleed. The main features of difenacoum poisoning in less severe cases are excessive bruising, nose and gum bleeding, and blood in the urine and faeces. Bleeding from several organs within the body, leading to shock and possibly death, occurs in the more severe cases.

Ergosterol

Ergosterol, which is derived from plants, can be converted into ergocalciferol otherwise known as calciferol (vitamin D2) with the aid of ultraviolet irridiation.

Chemical Name (IUPAC)

(5Z,7E,22E)-(3S)-9,10-secoergosta-5,7,10(19),22-tetraen-3-ol

Chemical Formula

$C_{28}H_{44}O$

Chemical Structure

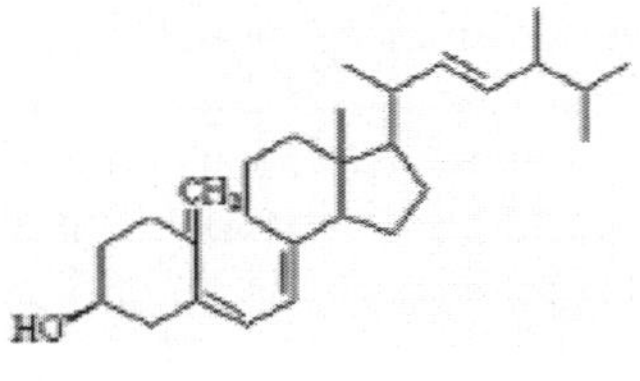

Flocoumafen

Flocoumafen is chemically related to brodifacoum; it is – an off–white powder, almost insoluble in water, slightly soluble in alcohols and soluble in acetone. It is recommended for use at 0.005 per cent in loose grain baits and wax–bound cereal blocks.The modern 2nd generation anticoagulant rodenticides such as Flocoumafen have proven severe secondary poisoning properties and have been held responsible for the deaths of many birds of prey, particularly rare Red Kites and Buzzards that habitually feed off dead bodies. For this reason the use of these pesticides outdoors is prohibited.

Chemical Name

4-hydroxy-3-[1,2,3,4-tetrahydro-3-[4-(4-trifluoro methylbenzyloxy)phenyl]-1-naphthyl]coumarin (mixture of *cis-* to *trans-* isomers in the ratio range 60:40 to 40:60).

Chemical Formula

$C_{33}H_{25}F_3O_4$

Chemical Structure

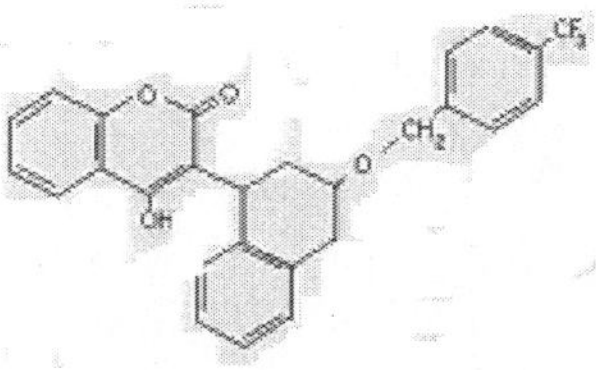

LD_{50}

The acute oral LD_{50} values have been determined to be 0.4 mg/kg for male laboratory *R. norvegicus* and 0.8 mg/kg for male laboratory *M.musculus*. The second generation anticoagulants have proved so lethal to susceptible rats and mice on one feeding.

Yellow Phosphorus

Yellow phosphorus is highly poisonous compared with the relatively non-toxic red phosphorus. When used as a rodenticide, yellow phosphorus is usually mixed with molasses or peanut butter and spread on bread as bait for rodents or roaches. Yellow phosphorous can be released when heating red phosphorous and whilst red phosphorous is considered to be virtually non-toxic, yellow phosphorous is not. If it is ingested symptoms such as nausea, vomiting, abdominal pain or garlic odor on breath will indicate poisoning by the latter. The estimated lethal adult human dose for white phosphorus is 50–100 mg.

Chemical Formula

P (phosphorous is an element and considered a true non metal.

Zinc phosphide is an inorganic chemical that is used to control rats, mice, voles, ground squirrels, prairie dogs, nutria, muskrats, feral rabbits and gophers. It is also uses as a tracking powder for the control of house mice. It is used on crop areas and on non-crop areas including lawns, golf courses, highway medians, and areas adjacent to wetlands. It may be formulated as a grain based bait, as scrap bait or as a paste. Rodenticide baits usually contain 2.0 per cent of zinc phosphide. Upon ingestion, zinc phosphide reacts with dilute acids in the gastrointestinal tract and produces phosphine, which enters the blood stream. Chronic exposure to phosphine may cause nausea, vomiting, diarrhea, tightness of chest, coughing, headaches, and dizziness.

Chemical Name

Zinc Phosphide

Chemical Formula

Zn_3P_2

LD $_{50}$/LC$_{50}$

The LD_{50} for the technical product (80-90 per cent pure) is 45.7 mg/kg. The inhalation of zinc phosphide or its breakdown product phosphine gas may result in acute toxicity.

Strychnine

It is a poisonous alkaloid that is obtained from seeds of the nux vomica tree (*S. nux-vomica*) and related plants of the genus Strychnos. It is practically insoluble in water and is soluble only with difficulty in alcohol and other common organic solvents. It has an exceptionally bitter taste. It is practically insoluble in water and is soluble only with difficulty in alcohol and other common organic solvents. It has an exceptionally bitter taste. Strychnine has been used in rodent poisons and in smaller doses as a stimulant in veterinary practice.

Chemical Name

Strychnine-10-one

Chemical Formula

$C_{21}H_{22}N_2O_2$

Chemical Structure

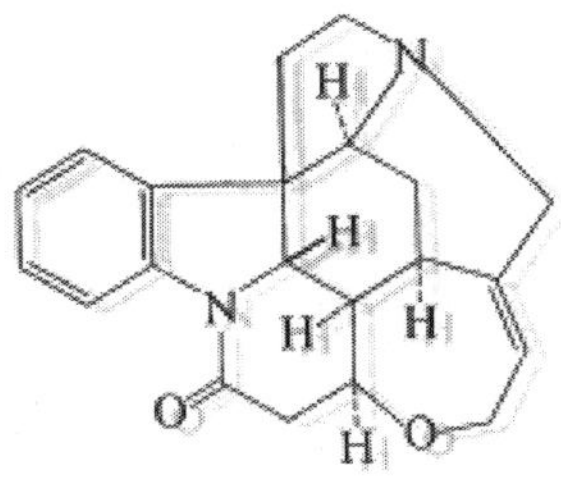

LD_{50}

The acute oral LD_{50} is 2.2 and 6.4 mg/kg for male and female respectively. Strychnine is a convulsive poison. Fatal if swallowed, dusts inhaled or absorbed through the skin.

Insect Pest and Disease Management
Editor: **Dr. D. Prasad**
Published by: **DAYA PUBLISHING HOUSE (2016)**

Pages **516–519**

Chapter 37

Plant Growth Promoting Rhizobacteria in Major Pests and Diseases Control

Amit Kumar Jain[1], Sudhir Kumar[1], Om Prakash Singh[2] and J.D.S. Panwar[2]

[1]Department of Botany, Janta Vedic College, Baraut (Baghpat), U.P.
[2]Division of Plant Pathology, IARI, New Delhi - 110 012

Introduction

Due to frequent use of the chemical pesticides in the recent past for the management of plant diseases and insect pests, several new problems developed mainly due to the development of pesticide resistant insects/strains, environmental pollution and human health hazards. Thus, the use of pesticides for the management of pests will not be a permanent solution. Biological control using microbial inoculants is an alternative method to chemical control and it is safe, cost effect and eco-friendly approach for the management of pests and diseases. Microbial inoculants bacteria, actinomycetes and fungi are part of the natural ecosystem of the healthy plants and they occupy in the habitats of rhizosphere, leaf surface and as endophytes in the plant system. Among the various microbes, some of the fluorescent *Pseudomonas* and the *Bacillus* spp. are commercially being exploited for the management of major soil- borne and foliar diseases of major crop plants in India.

Major mechanisms in PGPR mediated suppression of plant pathogens and insect pests include (*a*) production of antibiotics and lytic enzymes which are related to direct inhibition of the pathogens or pests (*b*) induction of systemic resistance by activating the PGPR treated to express genes encoding pathogenesis related proteins arid genes involved in phenyl propanoid metabolic pathway. Many of the PGPR strains are isolated from disease suppressive soils or rhizosphere of health plants worldwide. Suppressive soils of microbial origins generally contain variety of antagonistic microorganisms responsible for the suppression of soil pathogens (Weller, 1988). Several studies indicate that inoculation of single strain of a biological control agent rarely leads to a level of suppression as observed in naturally disease suppressive soils and positive effect with single inoculants are often inconsistent (Weller, 1988). Thus, application of a mixture of PGPR a likely to more closely mimics the natural situation and represents an effective method of biological control.

Improvement of Efficacy of Formulation by Chitin Amendment

While developing formulations, several molecules have been reported to be added to enhance the survival and efficacy of the PGPR. Chitin, as a carbon source/substrate for the growth of chitinolytic bacteria, increased the chitinase production when bacteria were grown in chitin amended medium (Gooday, 1990). Chitosan, a nontoxic polymer obtained from the chitin of crustacean shell wastes is not only the inhibitor of fungal growth but also activates genes encoding defense-related proteins in plants (Hadwiger *et al.*, 1986). In addition, chitin oligomers which are released during degradation of chitin substrate by chitinolytic bacteria are also found to elicit plant defense reactions (Benhamou and Theriaut, 1992). Tomato plants treated with chitosan showed enhanced protection against crown and root rot caused by Fusarium oxsysporum (Lafontaine and Benhamou, 1996). Recently, Radjacommare *et al.* (2002) demonstrated that chitin amended PGPR formulation significantly controlled the sheath blight disease and leaf folder insect pest in rice besides enhancing the grain yield and attracting more number of natural enemies of the leaf folder insect in rice field.

Combining Microbial Strains for Increasing the Biocontrol Activity

Generally, application or PGPR singly leads to inconsistent performance, because a single PGPR is not likely to be active in all kinds of soil environment and agricultural ecosystems (Raupach and Kloepper, 1998). These genetically modified microbes might evoke an impact on the soil ecosystem and thereby affect soil ecosystem. Moreover, use of genetic engineering techniques for strain improvement is a laborious process. Another approach to improve disease suppression by biocontrol bacteria is the use of combinations of strains which have different modes of actions. It has been well demonstrated in wheat in which application of several fluorescent *Pseudomonas* has greater biocontrol activity against take all disease (Pierson and Weller, 1994). Mixture of microbial inoculants includes mixture of fungi (Paulitz *et al.*, 1990), mixture of bacteria (Raupach and Kloepper, 1998). In addition to disease control, strain mixtures enhanced the plant growth in terms of increased seedling emergence (Dunne *et al*,. 1998), plant height (Raupach and Kloepper, 1998) and yield (Duffy *et al., 1996*). To biocontrol efficacy and growth promotion activity of fluorescent pseudomonas were increased either by mixing two or more strains of Pseudomonas spp. (Raupach and Kloepper, 1998) or mixing with other bacterial or fungal antagonists (Duffy *et al.,*) or mixed with chit or other substances (Benbamou *et al.*, 1998).

Microbial Mixtures Suppression of Major Pests and Diseases

Combination of PGPR antagonistic organisms increases the efficacy and consistency of biocontrol agents (Weller, 1988). *P. fluorescens* is applied as seed application in rice for the control of major diseases such as blast caused by *Pyricularia oryzae*, brown spot caused by *Helminthosporium oryzae*, sheath blight caused by *R. solani* and sheath rot caused by *Sarocladium oryzae* (Samiyappan *et al.*, 1999). The combination of chitinase-producing *Streptomyces* spp. and *B. cereus* with antibiotic- producing *P. fluorescens* and *Burkholderia cepacia* showed a synergistic effect on the suppression of rice sheath blight. The combnation of fungal as *Tñchoderma* spp. and fluorescent *Pseudomonas* are compatible and application of these antagonists together results in greater suppression of the disease incidence. In wheat, take all disease was suppressed greatly due to combined application of koningii and bacterial antagonists (*P. f!uorescens*) than the fungal and bacterial antagonists which were applied separately (Duffy *et al.*, 1996). So, emphasis should be given during development of microbial mixture that has synergistic, broad spectrum effect under different environmental conditions.

Formulation Development

Development of PGPR formulation for biocontrol purpose with suitable carrier is the last step in the commercialization of biocontrol agents, which is practically applicable on a large scale field level. The carrier should be easily available in the particular location, cheap and support the maximum population of microbes during the longer period of storage. Suslow (1980) first developed the formulation of *P. fluorescens* using talc as a carrier. Seed treatment with bacterial cell suspension was found effective in controlling several diseases. However, for commercial and held application, this methodology would be impractical due to difficulties in handling, transport and storage. The formulation also exhibited significant protection against nematode damage and anthracnose in cucumber under field conditions (Kenney *et al.*, 1999). The formulation has also enhanced pine seedling root and shoot growth in the production of containerized seedlings (Enebak and Reddy, 1999). Thus the development of suitable formulation of biocontrol strains mixture as that of chemical pesticide would certainly be beneficial for field application.

Conclusion

In India development of biological control technologies involving PGPR for pest and disease management is a newly emerging field. Most of the research activities involving biocontrol bacteria have been carried out in the laboratory and green house and very rarely field tested for only few crops against only less number of pathogen. Although biocontrol of plant pathogens with PGPR has not dramatically solved the diseases so far, we are at a turning point in this technology to make significant advances. More attention is needed to orient our future research on some of the areas *viz.* PGPR survivability, identification of crop specific strains, mixture formulation, biochemical/molecular markers for identification of effective PGPR and genes activated due to bacterial treatment in order to enhance the performance of biocontrol bacterial inoculants for managing plant pests and diseases in sustainable manner. Further, greater cooperation between industrial and non-industrial scientists is needed for transferring basic scientific information into research and development activities and this will eventually further improve the efficacy and commercial viability of the bio-products.

References

Benhamou, N. and Theriaut, N., 1992. Treatment with chitosan enhances resistance to tomato plants to the crown end root rot pathogen *Fusarium oxsysporum. Physiol. Mol Plant Pathol.*, 41: 33-52.

Duff, B.K., Simon, A. and Weller, D.M., 1996. Combination of *Trichoderma* with fluorescent *Pseudomonas* for control of take-all on wheat. *Phytopathology*, 86: 88-191.

Enebak, S.A. and Reddy, M.S., 1999. Seedling toot and shoot growth of three southern pine species is enhanced with the addition of bacterial amendments to potting media. *Phytopathology*, 89: 5–24.

Gooday, G.W., 1990. Physiology of microbial degradation of chitin arid chitosan. *Biodegrodation*, 1: 177-190

Hadwiger, L.A., Kendra, D.F., Fristensky, B.W. and Wagoner, N., 1986. Chitosan both activates genes in plants and inhibits RNF synthesis in fungi. In: *Chitin in Nature and Technology*. Plenum Press, New York, pp. 209-214.

Kenney, U. S., Reddy, M. S. and K.J.W., 1999. Commercial potential of biological preparations for vegetable transplants. *Phytopathology,* 89: S 39.

Lafontaine, R. and Benhamou, N., 1996. Chitosan Treatment: An emerging strategy for enhancing resistance of greenhouse tomato plants to infection by *Fusarium oxsysporum. Biocontrol Sci. Technol.,* 6: 111–124.

Paulitz, T.C., Ahmad, J.S. and Baker, R., 1990. Integration of Phytium and *Trichoderma harzianum* for the biological control of Phytium damping-off of cucumber. *Plant Soil,* 121: 243–250.

Pierson, E.A. and Weller, D.M., 1994. Use of mixtures of fluorescent *Pseudomonas* to suppress take-all and improve the growth of wheat. *Phytopathology,* 84: 940–917.

Radjacommare, R., Nandakumar, S., Kendan, A., Suresh, S., Sharathi, N.I., Raguchander, T. and Sarn, R., 2002. *Pseudomonas fluorescens* based bio-formulation for the management of sheath blight disease arid leaf folder insect in rice. *Crop Protection,* 21: 671–677.

Raupach, G.S. and Kloepper, J.W., 1998. Mixtures of plant growth promoting rhizobacteria enhance biological control of multiple cucumber pathogens. *Phytopathology,* 88: I 158–1164.

Samiyappa, M.S., Raguchander, T., Jayashree, K. and Nandakumar, R., 1999. Management of major rice diseases with fluorescent *Pseudomonas. Tamilnadu Rice Research Institute,* Aduthurai, pp. 111–121.

Susow, I.V., 1980. Growth and yield enhancement of sugar beet by pelleting with specific *Pseudomonas* spp. *Phytopathol. News,* 12: 40.

Weller, D.M., 1988. Biological control of soil-borne plant/pathogens in the rhizosphere with bacteria. *Ann. Rev. Phytopathol.,* 26: 379–407.

Insect Pest and Disease Management *Pages 520–526*
Editor: **Dr. D. Prasad**
Published by: **DAYA PUBLISHING HOUSE (2016)**

Chapter 38

Present Situation of Crop Losses Caused by Plant Virus

K.K. Biswas
Plant Virology Unit, Division of Plant Pathology,
Indian Agricultural Research Institute, New Delhi - 110 012

Introduction

The science of plant pathology started because of the result devastating crop losses. Estimate of loss is prerequisite to the rational development of any agricultural research program that has plant protection as a component. Crop loss is attributed to reduction in value and/or financial return due to damage; often measured as the difference between actual yield and attainable yield due to the effects of one or more pathogens or pests. Actual yield is the site specific yield obtained when crops are grown using current production practices at the farm level and attainable yield is the site specific yield obtained when crops are grown using all available pest management technologies to minimize the biotic stress, *i.e.*, a measure of the genetic potential of a crop genotype at a specific site. The reliable estimates of losses facilitate the objective of identification of relative importance of biotic pests. Crop loss information provides a means for evaluating the potential benefits of implementing any management decision and to develop decision thresholds.

Degrees of Crop Loss

Viruses are certainly responsible for far greater economic losses than is recognized because the nature of losses caused by viruses is very different than that attributed to bacteria, fungi, nematodes. Looses are often more insidious, frequently less conspicuous and therefore, it go unnoticed. Many viruses are latent in some of there host species even in case of adverse affects. Latent infections usually result in plants that grow more slowly, attain a smaller size, produce less fruit, reduce life span of perennial crops and act in various other ways that go unnoticed. Only when comparative experiments are conducted with virus free and infected plants, we become aware that looses are occurring. It is conceivable that we sustain greater looses due to chronic and widespread latent infections by hundred of viruses in many different crops genera. Therefore, in perennials crops and those that are vegetatively propagated, the term virus "infections" is more appropriate then the term virus diseases in order to consider all losses associated with viruses.

For these reasons mentioned above, there have been very few comprehensive surveys undertaken with loss data with many different viruses in many crops. Data presented in some of the few comprehensive surveys are based on estimations, extrapolations or educated guesses. There are several obvious reasons for not having accurate loss data, among them are variations in losses by a particular virus in a particular crop from year to year, different degrees of losses among regions within the same year, differences in loss assessment methodologies, viral losses are often complicated with other loss factors, variations in definitions of the term 'loss', and some component of loss are virtually unmeasurable. Virus induced symptoms can be confused with disorders due to pesticide toxicities, air pollution, genetic abnormalities, combined infections with the fungi, and mineral excesses or deficiencies.

A few reviews focus on the losses due to plant viruses in general. Reviewers discussed crop loses by plant viruses in different ways: such as types of losses, relationships of losses to mode of virus transmission, economic costs of virus management, experimental methods to determine losses, estimation of potential and actual losses, predication of crop losses and ways of preventing crop losses.

Kinds of Crop Losses

Crop loss is the measurable reduction/damage in quantity and quality of yield provided a thorough description of categories of yield as well as the many forms of direct and indirect losses caused by plant diseases caused by viruses (Table 38.1).

Table 38.1: Some Direct and Indirect Losses Associated with Plant Virus Infections in Crops

Direct Damage	*Indirect Damage*
Reduction in Growth	1. Yield reduction (including symptomless infection)
	2. Crop failure
Reduction in vigor	1. Decreased sensitivity to frost and drought
	2. Increased predisposition of pathogens and pests
Reduction in quality and market value	1. Defects of visual attraction: size, shape, colour
	2. Reduced keeping quality
	3. Reduce consumer appeal like grading, taste, texture and composition
	4. Reduced fitness for propagation
Costs of attempting to maintain crop health	1. Cultural hygiene on farm including vector control
	2. To produce virus free propagation materials
	3. Eradicating programs
	4. Breeding for resistance
	5. Research, extension and education

Plant Vigor Loss

Loss of plant vigor is one of the more common forms in which viruses cause losses. For example, crinkle virus and some mosaic viruses in severe cases reduced plant vigor in Leguminous, Solanaceous and other many crops etc., bushy dwarf virus reduce height and diameter of many perennial fruit crops, necrotic spot viruses caused decline in vigor in papaya and stone fruit trees, latent viruses reduced growth and number of stool bed apple trees and *apple mosaic virus* significantly reduce tree growth and number of laterals produced (Biswas and Gupta, 2007). Many other examples of the detrimental effects of latent virus infection become readily apparent only in controlled experiments.

Productive Life Loss

Virus infection also reduces the productive life of crops, which may be blamed on other causes. This happens in annual crops such as potatoes infected with leafroll but the effect is greater in perennial crops such as orchards where life span is reduced by several years. Leaf curl virus reduces life span of raspberry beds, and latent viruses in citrus and many perennial fruit crops render orchards prematurely unproductive without typical disease symptoms. *Tomato ring spot virus* in peach trees, Cherry leaf roll virus in walnut trees, tristeza virus in citrus, leaf roll viruses in grapevine are examples of viruses that significantly reduce the productive life of perennial crops.

Incompatible Budunion

Grafting and budding failure were often simply explained as incompatibility on the vegetatively propagated of fruits, nut and woody ornamental. It is now known that viruses latent in rootstocks or scion variety were responsible in many cases. Apples, pears, citrus and most of the stone fruits were especially affected by this problem. *Apple mosaic virus* decreased bud take by 2 per cent to 20 per cent among several cultivars tested in trials. Fan leaf virus reduced the ability to produce rooted cuttings and success in grafting grapevines.

Seed Loss

Viruses are responsible for a wide range of economic losses associated with seed production. Hundreds of viruses infect seed and many are transmitted to resulting seedling. Infected seed may appears normal depending on the virus and the host species. Losses also occur when seed is aborted, discolored, cracked or reduced in size, weight, or viability. Neergaard described the many ways in which viruses affect seeds of various crops. For example, germination of wheat seed is reduced by *Barley yellow dwarf virus*, as are apple seed from infected trees, and lettuce seed infected with mosaic virus. Seed is smaller when wheat and barely was infected with BYDV or when soybeans was infected with mosaic virus.

Growth Habit Loss

Viruses also reduce yield in a number of unusual ways by affecting growth habit of their host crops. Viruses in strawberry reduced runner production, pox virus in plums caused 40 per cent to 100 per cent fruits to drop before maturity among several cultivars, as did mosaic virus in apple trees. This virus also was responsible for less branching in young apple trees. Leaf roll virus reduced the number of stems in potato plants while para crinkle virus produced tubers of less uniform size. Therefore infected plants compete more successfully with virus free adjacent plants which results in yield loss.

Loss in Combination with Other Pathogens

Viruses in combination with other kind of pathogens can result in more adverse effect. For example, *Pea enation mosaic virus* reduced yield of broad bean seed by 6 per cent, its aphid vector alone reduced yield by 50 per cent, but both reduced yield by 93 per cent. *Barley yellow dwarf virus* enhanced mildew in oat, ergot disease in wheat, wheat streak mosaic infection was associated with an increase in *Helminthosporium sativum* in wheat. Certain viruses in potato plants cause greater susceptibility to *verticillium dahliae*, late blight. Leaf blight fungus, *Alternaria* was more severe on sugar beets infected with *Beet western yellows virus*. Apple trees with latent viruses were more susceptible to collar rot disease.

Plant Susceptibility

Virus infection can increase plant susceptibility to low temperatures. Leafroll virus in grapevines made them more susceptible to spring frosts and wheat with BYMV did not survive the winter. When growing under heat or drought stress, fanleaf infected vines were killed. Pome and stone fruit crops are more susceptible to frost damage when trees are infected with viruses.

Other Adverse Effects

Viruses are also responsible for some unusual effects on crops such as greater nutrient requirements of fruit trees, and reduced colouration in grapes which adversely fresh market sales and red wine colour. Sugar content can also be reduced by viruses. *Necrotic ring spot virus* in peach trees, and *Beet western yellows virus* each reduced sugar content of the harvest. Leaf roll virus in grapevines reduced sugar content which in turn affected *raisin* production to the extent that some crop was not harvested. Certain viruses in cereal crops reduced milling qualities other examples could be given on some of the unusual effects that viruses have on plants or the quality of their harvest that affect economics of production.

Estimations of Some Natural Crop Loss Caused by Plant Viruses

Virus diseases affect most of the economic crops and are responsible for far greater economic losses of crop productivity and quality not only in India but also worldwide. The extent of losses depends on many factors and may vary widely. Some viruses under appropriate conditions may infect a pant without producing any obvious signs of disease (latent), whereas, other may lead to rapid death of the whole plant (adverse affects). Between these extremes, a wide variety of diseases can be produced. The estimation of crop losses caused by important viral diseases was mentioned in Table 38.2.

Use of Crop Loss Knowledge

Crop loss knowledge is a common approach for managing disease in a given field, season and context, and for continuing the phytopathological research including the research priorities in a particular area. The practical usefulness of crop loss knowledge is for getting real benefit of control against the cost of control. Knowledge about the crop loss is essential to make short-term as well as long-term strategic decisions. The short-term strategic decisions include the choice of crops, cultivars and genotypes, and the choice of tillage etc. They also include other choices, such as, the timing the crop establishment, the type of the seed treatment used, sowing density and spatial pattern of the crop plant etc. The long-term strategic decisions enable decision makers to identify anticipate where IPM is needed for a given disease and when the disease does not represent a threat a crop.

Table 38.2: Important Viral Diseases in India and the Crop Losses by them

Crop	*Virus*	*Yield Reduction*
Apple	Apple mosaic	30-40 per cent
Banana	Banana bunchy top	Up to100 per cent
Bean	Bean golden mosaic	20-40 per cent
Bhindi	Bhindi yellow vein mosaic	50-80 per cent
Blackgram	Mungbean yellow mosaic	50-80 per cent
Cassava	Indian cassava mosaic	18-25 per cent
Chilli	Chilli mosaic	94 per cent
Cotton	Leaf curl	68-71 per cent
Cucurbit	Cucumber mosaic	50-75 per cent
Groundnut	Groundnut bud necrosis	>80 per cent
Maize	Sugarcane mosaic	32 per cent
Melons	Zucchini yellow mosaic	40- 50 per cent
	watermelon mosaic	40- 50 per cent
Mungbean	Mungbean yellow mosaic	21-70 per cent
Orange	Citrus tristeza	10K trees
Papaya	Papaya leaf curl	Upto 100 per cent
Pigeonpea	Pigeonpea sterility mosaic	>80 per cent
Potato	Potato virus X	10-20 per cent
	Potato virus A	10-20 per cent
	Potato virus S	10-20 per cent
	Potato virus M	15-20 per cent
	Potato virus Y	40-80 per cent
	Potato virus X + Potato virus A	40 per cent
	Potato virus X + Potato virus Y	40-100 per cent
	Potato leaf roll	7-16 per cent (Autumn) 39-60 per cent (Spring)
Rice	Rice tungro virus	
Sugarcane	Sugarcane mosaic	30-40 per cent
Subflower	Sunflower necrosis	12-17 per cent
Tomato	Cucumber mosaic	30-60 per cent
	Tomato leaf curl	28-92 per cent

References

Agrios, G.N., 1990. Economic Considerations. In: *Plant Viruses, Vol. II: Pathology*, (Ed.) C.L. Mandahar. CRC Press, pp. 1-22.

Ahlawat, Y.S., 1997. Viruses, greening bacterium and viroids associated with citrus (citrus species) decline in India. *Indian J. Agricultural Science*, 67: 51–57.

Anonymous, 1999-2003. *Annual Report*. Indian Agricultural Research Institute, New Delhi.

Anonymous, 2000. *Diseases and Pest of Potato: A Manual*, (Ed.) S.M. Paul Khurana. Central potato research Institute, Shimla, H.P.

Basu, N. and Giri, B.K., 1993. *The Essentials of Viruses: Vectors and Plant Diseases*. Wiley Eastern limited, 4835/24, Ansari Road, Dayaganj, New Delhi.

Bhargava, K.S., 1975. Sugarcane Mosaic: Retrospect and prospects. *Indian Phytopathol.*, 28: 1–11.

Biswas, K.K. and Chowdhury, A., 2005. Rai mosaic disease in rai (*Brassica juncea*) and Radish (*Raphanus sativus*) in the Himalayan hilly regions of West Bengal. *J. Mycopathol. Res.*, 46: 263–266.

Biswas, K.K. and Gupta, A., 2007. Management of important plant viruses in Indian agriculture. In: *Sustainable Pests Management*, (Ed.) D. Prasad. Daya Publishing House, New Delhi, pp. 208–138.

Biswas, K.K., Das, D.R. and Ahlawat, Y.S., 2005. Viral diseases of fruits, vegetables and spices in North Eastern Himalayan region of India and their management. In: *Crop Protection: Management Strategies*, (Ed.) D. Prasad. Daya Publishing House, New Delhi, pp. 225–246.

Biswas, K.K., Pant, R.P., Das, D.R., Pun, K.B. and Chowdhuri, A. and Ahlawat, Y.S., 2004. Incidence, detection and management of chilli mosaic disease syndrome in capsicum annum cv Kalimpong local in the hills of West Bengal. In: *National Symposium on Molecular Diagnostics for the Management of Viral Diseases*, pp. 89–90 (Abs.)

Biswas, K.K., Pun, K.B., Pant, R.P. and Ahlawat, Y.S., 2005. Mosaic disease in chilli (*Capsicum annuum*) cv Kalimpong Local in Darjeeling Hill of West Bengal and its management. *Indian Phytopath.*, 58: 456–461.

Bos, L., 1982. Crops losses caused by viruses. *Crop Protection*, 1: 263–282.

Bridddon, R.W. and Markham, P.G., 2000. Cotton leaf curl viral disease. *Virus Res.*, 71: 151–159.

Campbell, A.I., 1965. The effect of some apple viruses on clonal root stock production. *Zast. Bilja*, 85–88: 261–265.

Campbell, A.I., 1980. The effects of viruses on the growth, yield and quality of three apple cultivars on healthy and infected clones of four rootstock. *Acta. Hortic.*, 114: 185–191.

Cisar, G., Brown, C.M. and Jedlinski, H., 1982. Effect of fall or spring infection and sources of tolerance of barley yellow dwarf of winter wheat. *Crop Sci.*, 22: 474–478.

Dasgupta, I., Malathi, V.G. and Mukherjee, S.K., 2003. Genetic engineering for virus resistance. *Current Science*, 84: 341–354.

Geske, S. and Riesselman, J., 1988. Barley yellow dwarf and wheat streak mosaic in small grains. Montguide. *MT State Univ. Ext. Service Publ.*, 8802, 3 pages.

Gibbs, A. and Harrison, B., 1976. Ways of preventing crop losses. In: *Plant Virology: The Principals*. Halsted Press, pp. 219–231.

Gill, C.C., 1980. Assessment of losses on spring wheat naturally infected barley yellow dwarf virus. *Plant Dis.*, 64: 197–203.

Hull, R., 2002. *"Matthews' Plant Virology,"* 4th edn. Academic Press, A Harcourt Science and Technology Company, Harcourt Place, 32 Jameston Road, London, NW1 7BY, UK.

Khetrapal, R.K., Maisonneuve, B., Maury, Y., Maury, Y., Chalhoub, B., Dinant, S., Lecoq, H. and Varma, A., 1998. Breeding for resistance to plant viruses. In: *Plant Virus Disese Control*, (Eds.) Hadidi *et al.*, APS Press, pp. 14–32.

Lee, R.F. and Rocha-Pena, 1992. Citrus tristeza virus. In: *Plant Diseases of International Importance*, Vol. III, (Eds.) J. Kumar, H.S. Chube, U.S. Singh, and A.N. Mukhopadhyay. Prentice Hall, N.J., pp. 226–249.

Madden, L.V., 1983. Measuring and modelling crop losses at the field level. The *American Phytopathological Society*, 73: 1591–1596.

Madden, L.V., Pennypacker, S.P., Antle, C.E. and Kingsolver, C.H., 1981. A Loss Model for Crops. *The American Phytopathological Society*, 71: 685–689.

Millikan, D.F., 1955. The influence of infection by ring spot virus upon the growth of one year old Montmorency nursery trees. *Phytopathology*, 45: 565–566.

Mink, G.I., 1993. Pollen and seed transmitted viruses and viroids. *Annu. Rev. Phytopathol.*, 31: 375–402.

Neergaard, P., 1977. Seed borne viruses. In: *Seed Pathology, Vol I*, (Ed.) P. Neergaard. Halsted Press, pp. 71–117.

Nemeth, M., 1986. *Virus, Mycoplasma and Rickettsia Disease of Fruit Trees*. Martinus Nijhoff Publ., 841 Pages.

Nutter, Forrest W., Teng, Paul, S. and Royer Matthew, H., 1993. *Terms and Concept for Yield, Crop Loss, and Disease Thresholds*. The American Phytopathological Society, February, Special Topic, p. 211–215.

Paul Khurana, S.M. and Thind, T.S., 2002. Important biotic diseases of potato and their control. In: *Plant Pathology at Cross Road*, (Eds.) C. Sen and M.K. Dasgupta, S.

Satapathy, M.K. and Anjaneyulu, A., 2001. *Rice Tungro Epidemic Disease: Epidemic Analysis*. Discovery Publishing House, New Delhi.

Singh, B.R., Gupta, S.P., Pandey, R. and Tripathi, D.P., 1990. Response of some chilli lines to mosaic and leaf curl diseases. *Indian J. Virol.*, 6: 113–115.

Singh, S.J., 2003. *Virus and Phytoplasma Diseases of Papaya, Passion Fruit and Pineapple*. Kalyani Publishers, Noida.

Singh, S.J., 1996. Virus and mycoplasma diseases of papaya. In: *Advances in Diseases in Fruit Crops in India*, (Ed.) S.J. Singh. Kalyani Publishers, Ludhiana, pp. 275–296.

Stouffer, R.F. and Lewis, R.H., 1969. The present status of stem pitting in Pennsylvania. *Plant Dis. Rptr.*, 53: 429–434.

Varma, A. and Malathi, V.G., 2003. Emerging geminivirus problems: A serious threat to crop production. *Ann. Appl. Biol.*, 142: 145–164.

Varma, A. and Sinha, S.K., 1992. Sustainable development through long term biotechnological alternatives in agriculture. In: *Proceedings of the International Seminar on Impacts of Biotechnology on Agriculture and Food in Developing Countries*, Committee on Science and Technology in Developing Countries (COSTED), Madras, pp. 44–57.

Verma, H.N. and Prasad, V., 1985/86. Virus diseases in papaw (Papaya). *Rev. Tropical. Plant. Path.*, 2: 311–327.

Walkey, D.G.A. and Payne, C.J., 1990. The reaction of two lettuce cultivers two mixed infection by beet western yellow virus, lettuce mosaic virus and cucumber mosaic virus. *Plant Pathol.*, 39: 156–160.

Insect Pest and Disease Management *Pages 527–528*
Editor: **Dr. D. Prasad**
Published by: **DAYA PUBLISHING HOUSE (2016)**

Chapter 39

Response of *Rhizobium* with Sulphur and Micronutrients on Seed Quality of Black Gram (*Vigna mungo* L. Hepper)

Brijesh Kumar Rathi[1], Amit Kumar Jain[1], Sudhir Kumar[1] and J.D.S. Panwar[2]
[1]Department of Botany, Janta Vedic College, Baraut (Baghpat), U.P.
[2]Division of CPBM, IIPR, Kalyanpur, Kanpur – 208 024, U.P.

Introduction

The *Rhizobium* alone significantly increased the total sugar contents as compared to uninoculated control. However, the combination of *Rhizobium* along with micronutrients and *Rhizobium* along with sulphur and single micronutrient didn't show any impact on the total sugar contents. Plants treated with *Rhizobium*, sulphur and micronutrients gave significant result in relation to total sugar contents. *Rhizobium* along with micronutrients significantly enhanced the N uptake as compared to control. However, the *Rhizobium* along with micronutrient showed a little impact on N uptake by the seeds, but the combination of *Rhizobium*, Zn, Mo and B gave better response in terms of N uptake as compared to *Rhizobium* alone. The combination of *Rhizobium* along with sulphur, Zn, Mo and B gave more significant enhancement in N uptake and total sugars by seeds as compared to other treatments. Similar trends were observed in both the cultivars in both the years, but Cv. PU-19 gave superior response in comparison to Cv. PDU-1.

It is a well known fact that *Rhizobium* application enhanced the nodulation and nitrogen fixation in plants, but the use of micronutrients along with sulphur have been found to be more in enhancing the N and P uptake as well as grain quality. Sulphur is a important constituent of S-containing amino acid (methionine, cystine and cysteine), so its application increased the amount of these amino acids and improved the protein contents as well. Mo replace the nitrogen fertilizers to in legumes and enhance the Rhizobial infection, B helps in the transport of sugars and a close relationship is found between Zn supply and protein contents (Panwar and Thakur, 1995 and Jain, 2004). Thus increased protein content along with total sugars will be helpful in improving the grain quality, which ensures the alleviation of the malnutrition in the balanced diet.

Conclusion

A field experiment was carried out to study the response of sulphur and micronutrients *viz.* Zn, Mo and B with *Rhizobium inoculation* on seed quality of black gram (*Vigna mungo*) Cv. PU-19 and PDU-1 with different combinations at the research farm of Janta Vedic College, Baraut (Baghpat), UP. The nitrogen in the form of urea and potassium as murate of potash were applied as basal dosage. The sulphur as gypsum @ 40 kg/ha, Zn as zinc sulphate @ 4 kg/ha. Mo as sodium molybdate @ 0.6 kg/ha and B as borax @ 0.1 kg/ha were applied at the time of sowing (Ist week of July, 2003). For the comparison, the proper control was maintained. Treatments were comprised with inoculation in different combinations. Each treatment replicated thrice in a randomized in rows 30x10 cm apart. The data was statistically analyzed. The determination of total sugars in the seed was carried out by Nelson (1944) and Somogyi (1952) method and protein content using Kjeldhal method given by Jackson (1958).

References

Jackson, M.L., (1958. *Soil Chemical Analysis*. Prentice Hall Inc., Englewood Cliffs, USA.

Jain, A.K., 2004. Influence of *Rhizobium leguminosarum* inoculation and application of Phosphorus and micronutrients on growth, yield, nodulation and seed quality of green gram (*Vigna radiata* (L.) Wilczek. *Ph.D. Thesis,* CCS University, Meerut.

Nelson, N., 1944. A photometric adaptation of the somogyi method for the determination of glucose. *J. Biol. Chem.,* 153: 375–380.

Panwar, J.D.S and Thakur, A.K., 1995. Physiological and biochemical studies in VAM and *Rhizobium* inoculated mungbean plants under fields conditions. In: *Proc: Mycorrhizae Biofertilizers for the Future.* Tata Energy Res. Institute, pp. 13–15.

Somogyi, M., 1952. Notes on sugar determination. *J. Biol. Chem.,* 195: 19–23.

Insect Pest and Disease Management *Pages 529–544*
Editor: **Dr. D. Prasad**
Published by: **DAYA PUBLISHING HOUSE (2016)**

Chapter 40

Advances in Diagnosis and Management of Banana Bunchy Top Disease

Mohd. Akram and Rajesh Kumar
Department of Plant Pathology,
C.S. Azad University of Agriculture and Technology, Kanpur – 208 002, U.P.

Introduction

The generic name *Musa* is derived from the Arabic word *mouz*. Bananas were known to the early Arabs and appear in the Holy Quran as *"the tree of paradise"*. The banana culture in India is also as old as Indian Civilization. The earliest scientific classification of bananas was made by Linnaeus and suggested the *Musa sapientum* (dessert banana) and *Musa paradisiaca* (plantain group). But now, it is known that these two apparent species are not at all species but closely related to inter specific triploid hybrids of AAB group. All the varieties existing now contain mixed genomes of *Musa acuminata* and *Musa balbisiana* (Robinson, 1999).

Banana has a therapeutic value in many special diets. Ripe mashed banana is an excellent food for babies from the age of three months because of early digestibility and vitamin content. Fruits are low in sodium and protein content, has very little fat and no cholesterol, therefore, is useful in managing patients with high blood pressure, heart diseases and kidney disorders (Srinivasarao, 2003). The capacity of the banana to neutralize free hydrochloric acid suggests its use in peptic ulcer therapy. A fully ripe banana mixed with milk powder is especially recommended for ulcer patients and also in cases of gout and arthritis. The dessert banana has become a very popular fruit in modern westernized diets. It is an important source of vitamins A, B_6 and C and provides instant energy. A medium sized banana gives 280 kilojoules energy, which is more than one citrus fruit. Bananas are high energy fruits, richer than most of other fruits in carbohydrate, phosphorus (P), Iron (Fe), Vitamin A and K, having more than double the concentration in ripe pulp than most other tropical fruits (Morton, 1987).

Bananas and plantains (*Musa* spp.) are the fourth most important global food commodity after rice, wheat and milk in terms of gross value production (INIBAP, 1994). The recent available data in relation to harvested area, yield and production in world and India is presented in Tables 40.1 and 40.2 (FAOSTAT, 2005).

Table 40.1: World Data on Banana Area, Yield and Production During 2000–2005

Year	*2000*	*2001*	*2002*	*2003*	*2004*	*2005*
Area (ha)	4108133	4139407	4303708	4389992	4427813	4439488
Yield (hg/ha) 1 hg = 100g	157192	159802	160501	160710	163569	163227
Production (MT)	64576374	66148554	69075005	70551646	72425334	72464562

Table 40.2: Indian Data on Banana Area, Yield and Production During 2000–2005

Year	*2000*	*2001*	*2002*	*2003*	*2004*	*2005*
Area (ha)	470000	470000	680000	680000	680000	680000
Yield (hg/ha) 1 hg = 100g	300851	302340	247353	247353	247353	247353
Production (MT)	14140000	14210000	16820000	16820000	16820000	16820000

The common diseases occur in banana are panama wilt, sigatoka, fruit rot, bunchy top, banana streak and banana bract mosaic. The viruses known to infect bananas have been designated as *Banana bunchy top virus* (BBTV), *Banana streak virus, Banana bract mosaic virus, Cucumber mosaic virus* (Thomas *et al.*, 1997) and *Banana dieback virus* (Hughes *et al.*, 1998). Though, a wide range of diseases caused by the pathogens including fungi, bacteria, viruses and nematodes limits the production of bananas but Banana Bunchy Top Disease (BBTD) is the most important viral disease among them. The disease becomes very severe in many banana-producing areas, including Asia, Africa and the South Pacific regions (Dale, 1987).

History of BBTD

The documented history of banana bunchy top disease is very recent in comparison to the history of banana cultivation. The first written reports of the disease were from Fiji in 1889 and 1890 although it was probably present there as early as 1879. The disease spread rapidly across Fiji and production declined from 778,000 bunches in 1889 to 114,000 in 1895; however, floods were responsible for at least some of this decline (Magee, 1953).

Banana bunchy top disease was first reported from Taiwan as early as 1900 (Sun, 1961) and Egypt in 1901 (Magee 1953). But the origins of these outbreaks are unknown. The disease was first appeared in Sri Lanka and Australia in 1913 (Magee, 1927). The disease in both the places probably come from the importation of diseased suckers from Fiji. Bunchy top was probably brought to India from Sri Lanka in 1940 (Magee, 1953) and is now present in the states of Andhra Pradesh, Gujrat, Kerala, Maharashtra, Tamil Nadu and Uttar Pradesh (Singh, 1979). The disease was also reported in Gorakhpur in 1971 (Khurana, 1971) and in Kanpur (U.P.) in 2006 (Kumar and Akram, 2006).

Geographical Distribution of BBTD

The "modern" history of BBTD has been reasonably well documented (Dale, 1987). From the first record of BBTD in Fiji in 1889 (Magee, 1953), the disease was subsequently distributed in Taiwan around 1900 (Sun, 1961) and Egypt in 1901 (Magee, 1953). It is believed that BBTD was imported into Australia and Sri Lanka in 1913 in infected suckers from Fiji (Magee, 1927). The disease was probably introduced into India from Sri Lanka in 1940 (Magee, 1953).

BBTD is now present in a number of countries in Africa, Asia and the South-Pacific regions but its geographical distribution is erratic within the regions. The disease continues to move with recent recorded outbreaks in Pakistan in 1988 (Soomro *et al.*, 1992) and in Hawaii in 1989 (Hu *et al.*, 1993). Importantly, the disease has not been recorded in South and Central America and in the Caribbean region and is not universally distributed in Africa or Asia (Dale and Harding, 1998).

Economic Importance of BBTD

There are no accurate estimates of the international economic impact of BBTD despite the disease being widespread in Asia and the South Pacific regions. In most instances, the incidence of BBTD is not well documented. However, there are some examples of outbreaks of BBTD that have been well studied.

BBTD was first recognized in Australia in 1913 in plantations on the border between the states of Queensland and New South Wales. From there plantations, the disease rapidly spread South into New South Wales and North into South-East Queensland. By 1926, 90 per cent of the original banana plantings in New South Wales were out of production, output fell from 460,000 cases of banana in 1922 to 140,000 cases in 1925. BBTD was first recorded in Pakistan in 1988 in Sindh province (Khalid *et al.*, 1993). By the end of 1992, a reduction of about 60 per cent in yield was recorded which has been attributed to BBTD (Soomro and Khalid, 1993). BBTD is wide spreads in Vietnam and the Philippines where the incidence in small holdings varies considerably upto 50 per cent and the disease has been particularly devastating in Tonga.

In India, BBTD was found severe in some areas of the state of Tamil Nadu in 2001 where the disease incidence was found up to 26 per cent (Thiribhuvanamala *et al.*, 2001). According to Kumar and Akram (2006, unpublished data) revealed that during the survey conducted total 1836 plant samples were observed in which 495 samples supposed to be infected/suspected with BBTV. The average disease incidence in all the observed samples was recorded 26.91 per cent. The maximum 41.73 per cent disease incidence was recorded in the CSA University campus and the lowest 7.12 per cent in the farmer's field of village Udanga Khera (Kanpur).

Banana Bunchy Top Virus (BBTV)

Banana bunchy top virus (BBTV) is the cause behind bunchy top disease in banana. It was first classified as a possible luteovirus as the disease had many characteristics of luteovirus infection (Mathews, 1982). Dale *et al.* (1986) extracted dsRNA from BBTD infected plants that was not present in non-BBTD infected plants. Iskra *et al.* (1989) purified 28 nm isometric virus-like particles from BBTD infected plants. In contrast Wu and Su (1990) purified 20-22 nm isometric virus like particles from infected plant and reported that these particles contained ssRNA of about 6.0 kb. Harding *et al.* (1991) and Thomas and Dietzgen (1991) purified 18-20 nm isometric virus like particles from infected plants with ssDNA of about 1 kb. The molecular weight of the coat protein subunit of these particles was approximately 20,000 dalton. BBTV is clearly not a member of any previously described plant virus group. It is most similar to the geminivirus but differs from that group due to the following points:

1. Its virions are isometric and not geminate.
2. Its genome consists of at least six components of circular ssDNA each about 1 kb and not one or two components.

3. Its coat protein has a molecular weight of about 20,000 dalton.
4. It is transmitted by aphids.

Rather, BBTV is probably a member of a new plant virus group (Nanoviruses) which could also contain *Subterranean clover stunt virus* (Boevink *et al.*, 1995), *Faba bean necrotic yellows virus* (Katul *et al.*, 1993), *Coconut foliar decay virus* (Rohde *et al.*, 1990) and *Milk vetch dwarf virus* (Sano *et al.*, 1993). These viruses are isometric, small (about 18-20 nm in diameter), contain ssDNA organized in multiple (at least six) circular ssDNA components and are transmitted by aphids in the persistent manner. To date, these five viruses are considered as possible members of this group (Agrios, 1997). There are strong evidences for the existence of strains of BBTV. Karan *et al.* (1994) has sequenced and compared BBTV DNA component-1 from isolates collected from 10 countries. This comparison clearly indicated that there were two groups of isolates, the Asia group and the South Pacific group. Analysis of DNA components 2 to 6 has confirmed the division of isolates into two groups (Karan, 1995).

Host Range of BBTV

The only confirmed hosts of BBTV are species within the genus *Musa* (*M. acuminata, M. balabisiana* and their hybrids) and *Ensete ventricosum* (Musaceae) (Magee, 1927). Ram and Summanwar (1984) reported that *Colocasia esculenta* could be symptomlessly infected with BBTV and act as a reservoir for BBTV in a symptomless manner. Interestingly, none of the commercial or widely grown cultivars of banana and plantains have been shown to be immune to BBTV, but there does appear to be variation between cultivars in the rate of infection and the severity of symptoms (Jose, 1981; Magee, 1927 and 1948).

There are no reports on the reaction of wild diploid *Musa* sp. to BBTV challenge. If BBTD is a relatively new virus disease of bananas, possibly suggested by Karan *et al.* (1994), there must exist alternate hosts of the disease. However it would appear that if such host exists, they do not play an important role in the epidemiology of BBTD in commercial agriculture. Vakili (1969) reported *Canna* sp. with possible *Banana bunchy top virus* symptoms but Magee (1927) was unable to transmit banana bunchy top to *Canna* or any other genus except *Musa*.

The host range of banana aphid was studied in laboratory on 33 different plant species including shrubs, foliage, flowering and succulent plants and trees. The fresh plant tissues were placed in 250 ml beakers and 5 aphids per plant were released. The aphids were found feeding aggressively on *Alocasia* sp., *Colocasia esculenta* and banana plants only among tested plants (Kumar and Akram, 2006, unpublished data).

A polymerase chain reaction assay was used to study BBTV transmission efficiency, to determine the minimum acquisition-access period, the minimum inoculation-access period, the retention time and to examine the possibility of transovarial transmission in this vector. BBTV was acquired by banana aphids within 4 h and was transmitted within 15 min feeding. On average, more than 65 per cent of single viruliferous adult aphids transmitted BBTV. The aphids retained BBTV for their adulthood of 15-20 days. None of the 131 offspring from adult aphids reared on infected bananas were BBTV positive. Aphid transmission experiments were conducted to determine it taro and gingers are hosts of BBTV. None of the 87 taro and ginger plants exposed to aphid inoculation were infected by BBTV. The BBTV-free status of these plants was verified by PCR assay for 6 months post-inoculation. In addition, none of the taro and ginger samples collected from fields adjacent to BBTV-infected banana plants tested positive for BBTV (Hu *et al.*, 1996).

The Vectors

Banana aphid, *Pentalonia nigronervosa* is the major vector of BBTV. *Pentalonia nigronervosa* occurs on the banana plants usually in small to very large colonies. Both winged and wingless viviparous females and nymphs are present in these colonies, the wingless form being more abundant (Figure 40.1) than the winged forms (Sharma, 1988).

On the plants, aphids occur mainly in sheltered situation, only very rarely the aphids are found exposed. The three important regions on the plant where the colonies concentrate are: (*a*) the base of the pseudostem, (*b*) the top region of the pseudostem and (*c*) the leaf axils (Figure 40.1). Ocfemia *et al.* (1947) found the aphids colonizing on the roots and corms of *Manila hemp* in Philippines during the dry season. Magee (1927) however, could not find the vector on the roots of banana. Magee (1927) and Goddard (1929) based on careful observation made in the field, concluded that the dispersion of *P. nigronervosa* is accomplished by flight. Ocfemia (1930) found that aphid is not capable of prolonged flight and it is spread from plant to plant chiefly by ants. Johnson (1963) found that aphids were never seen flying in the field and winged aphids were more sluggish in their movements than their wingless counterparts.

Males of *P. nigronervosa* have not been observed so far. The female breeds parthenogenetically. Egg stage also is not known. The female gives birth to young ones. These viviparous females may be winged or wingless (Johnson, 1963). The aphid has 4 generation in a year and there are 4 nymphal instars. Its population peaked in April and during September–October (Yang, 1989). A period of 48 hours feeding is required for aphid to become infective after commencing to feed on a newly emerged leaf. In addition to banana, the banana aphid also breeds and infests in nature on several other monocots including *Ravenala* sp., *Strelitiza* sp. (Magee, 1927), wild banana, *Heliconia behai*, *Heliconia* sp. (Wardlaw, 1972), *H. latispatha*, *Xanthosoma roseum* (Evers, 1968), *Tannia* (Wolcott, 1933), *Calla* flowers, *Hedychium coronarium* (Ocfemia and Buhay, 1934), *Languas speciosa* (Sun, 1961), small cardamom (Uppal *et al.*, 1945; Verma and Capoor, 1958); large cardamom (Vasudeva, 1956), *Colocasia* sp. and *Alocasia* sp. (Marathe *et al.*, 1982). None of these hosts is infected by the BBTV except *Colocasia* sp. that has been reported as symptomless carrier of BBTV (Ram and Summanwar, 1984). After 10 days observations in laboratory experiment, it was found that banana aphid population was recorded maximum (134) in *Alocasia* sp. followed by *Colocasia esculenta* (116) and banana (77) from the initial population of 15 aphids released for the study (Kumar and Akram, 2006).

Transmission of BBTV

There is no transmission of BBTV by mechanical means. Magee (1927) showed that the banana aphid, *Pentalonia nigronervosa* Coq. was the major vector. Infective aphids may transmit the virus in 15-20 hours or more of feeding on healthy plant but not by feeding for shorter period. Increase of age of infected plants along with retarded growth rate lead to a fall in the availability of the virus (Stover, 1972). Studies in Taiwan (Sun, 1961) showed that aphids feeding for a period of 2-96 hours on diseased plants increased their transmitting ability from 20-100 per cent. Magee (1940) observed that BBTV was transmitted both by winged and non-winged adults and nymphs were found more efficient than adults. For nymphs, at least 24 hours of feeding was found suitable to acquire the virus for better results. The virus was not transmitted transovarially. Khurana (1971) failed to transmit the virus when 15-25 aphids per banana plant were used for transmission studies. BBTV has also been transmitted by *Myzus persicae*, *Aphis gossypii* and *Aphis spiraecola* to banana and few herbaceous hosts (Rao, 1980; Sastry *et al.*, 1980). Infected suckers mainly transmit the virus. Rao (1980) has also transmitted the virus by leaf patch grafting from banana to *Catharanthus roseus*.

Epidemiology

The basic epidemiology of BBTD appears to be relatively simple. The virus is transmitted in space by its aphid vector in short distance while infected planting material is responsible for both short and long distance movement of the disease. Wild or uncultivated bananas and neglected plantations are also an important source of the disease. There are no confirmed non-*Musa* plants that showed BBTD symptoms (Dale and Harding, 1998). Allen (1978) showed that the mean distance of new infections from their sources of inoculum in on established plantation was 17.2 m, about two-third of new infection was within 20 m and all but about 1 per cent was within 86 m. Allen and Barnier (1977) found that new plantations established from disease free planting material had 88 per cent probability of becoming infected if they were adjacent to an infected plantation whereas this dropped to a 27 per cent probability if the nearest infected plantation was 50 to 100 meters away. The banana aphid, *P. nigronervosa* occurs in all banana-producing countries. Therefore, practically all banana producing countries free of the disease have all the components of a BBTD epidemic except the virus (Dale, 1987).

Symptoms Based Diagnosis

BBTD is identified by its symptoms on the various parts of the plant; *viz.* leaves, pseudostem, flowers and fruits. BBTD is the disease, which produce different types of symptoms. The pathogen causes cytopathological effects in the banana phloem tissues and yellowing symptoms on leaves (Magee, 1927). Infected plants have a bunched appearance at the top of the plant with narrow, upright and erect leaves, which are yellowed at the margins (Figure 40.2). The leaves become increasingly smaller and the plant becomes dwarfed. Due to short and erect leaves, severe resetting occur which leads to bunchy top appearance of the plant (Khalid and Soomro, 1993). These symptoms are usually most visible on young plants and can be subtler on older banana plants (Singh, 1979). Very characteristic symptoms of the disease are the small dark green streaks on the pseudostem, petioles and leaves resulting in a dot-dash or "morse code" pattern (Dale and Harding, 1998). These dark streaks often extend down the midrib and petiole (Thomas *et al.*, 1994).

Green J-hooks symptoms occur where the flat part of the banana leaf meets the midrib. These are the parallel dark green streaks terminate at the leaf midrib with a distinct, J-hooking pattern. Green J-hooks are very reliable indicator of BBTD. Holding a suspected banana leaf between eyes and the sun, best shows these symptoms (Akram and Kumar, 2006). When young plants are infected with BBTV, they usually do not flower. "Dot and dashes" or streaks resemble the "morse code" symptoms appear on the reddish flower bracts of the bellflower (Figure 40.2). Depending on the time of infection, the infected plants do not produce fruits and if produced, the banana fingers and bunches may be stunted, twisted or otherwise deformed and of little use (Dale, 1987; Akram and Kumar, 2006).

Chemical Based Diagnosis

For the control of BBTD, the main problem is how to differentiate between diseased and healthy banana plants or suckers. So a reliable indexing technique is necessary for this purpose. Summanwar and Marathe (1982) developed a diagnostic technique based on the reaction of infected tissues with 2,3,5-triphenyl tetrazolium chloride (TZC) for detecting the latent infection of bunchy top and infectious chlorosis viruses in banana suckers. 1 per cent solution (weight/volume in water) of TZC was used and sections from BBTV-infected plants turn brick red and healthy plants remained colourless at 32-35°C in 1 to 1½ hour (Figure 40.3). It took longer in the case of parenchymatous tissues, than in other tissues for the colour reaction to develop. At lower temperature, the reaction slows down, while it was rather fast at higher temperatures and under shades outside. In direct sunlight, the solutions turn red and the sections become red to black within 30-40 minutes. It was advised to carry out the test in shade. Although tissues from any plant part can be used from distinguish between healthy and diseased, it is preferable to take sections from parenchymatous tissues from midrib of the leaves.

Figure 40.1: Different Stages of Banana Aphid and its Feeding Behaviour

(a) Adult and nymph of banana aphid feeding on banana; (b) Aphid colony feeding on banana plant; (c) Adults and nymph of banana aphid in colony; (d) Aphid population build up on waste/left over suckers of banana

Figure 40.2: Various Symptoms and Suspected Hosts of BBTV

(A) Bunchy top symptoms on young plant; (B) Infected erect leaf; (C): (a) Green J-hooks on leaves, (b) Morse code streaks on pseudostem; (D) Young infected plant emerged from infected mother sucker; (E) Yellowing of leaf due to aphid feeding; (F) Dark streaks on flower; (G): (a) Small and distorted fruits, (b) Mottled flower; (H): (a) *Dieffenbachia* sp., (b) *Colocasia esculenta*, (c) *Alocasia* sp., (d) *Canna indica*

Figure 40.3: Tetrazolium Test Showing Positive Reaction

(A) Midrib pieces of healthy banana with 1% TZC; (B) Midrib pieces of infected banana with 1% TZC; (C) Midrib pieces of healthy banana with distilled water; (D) Midrib pieces of infected banana with distilled water

Ram and Summanwar (1984) used TZC colour diagnostic detection test to confirm the presence of BBTV in *Colocasia esculenta* and result was found positive. However, this technique has been found highly variable and unsuitable for detecting BBTV (Sharma, 1983, 1984, 1985). Sharma (1985, 1987) tested 12 different BBTV free banana cultivars and a number of non-musacious hosts namely, *Canna* sp., *Hippeastrum* sp., *Dracaena* sp., small cardamom, papaya, geranium, fig, tobacco and tomato and observed brick red colour in sections from all these hosts. The vigour of the plant, size of the tissue, concentration of the strain and temperature influenced intensity of the colour and time taken for its development.

Different chemicals ($MgCl_2$, $FeCl_3$, NaCl, NH_4Cl, Copper Oxychloride, TZC, KCl and $HgCl_2$) were tested for the detection of BBTV (Figure 40.4). The study revealed that TZC and $HgCl_2$ were able to develop a distinct colour between infected and healthy samples. The infected plant midrib pieces developed brick red colour at 1 per cent concentration in one hour dipped in Triphenyl tetrazolium chloride (TZC) whereas no colour was observed in healthy samples. In $HgCl_2$, the infected midrib pieces produced light blackish colour at 0.1 per cent concentration in 1.5 hour but the healthy counterparts produced no colour (Kumar and Akram, 2006).

Figure 40.4: $HgCl_2$ Test Showing Positive Reaction

(A) Midrib pieces of healthy banana with 0.1% $HgCl_2$; (B) Midrib pieces of infected banana with 0.1% $HgCl_2$; (C) Midrib pieces of healthy banana with distilled water; (D) Midrib pieces of infected banana with distilled water.

Serological and Molecular Based Diagnosis

Comparatively, research progress on BBTD has been very slow than other serious virus diseases due to the presence of large amount of latex and phenolic compounds in the banana plant, which interfere the virus extraction and purification. Serological, nucleic acid hybridization and polymerase chain reaction assays are now available for the detection of BBTV (Wu and Su, 1990; Thomas and Dietzgen 1991; Harding *et al.*, Dietzgen and Thomas 1991; Harding *et al.*, 1993). Dietzgen and Thomas (1991) developed a triple antibody sand (TAS) ELISA, utilizing polyclonal antibodies fro virus capture and monoclonal antibodies for detection. Greering and Thomas (1996) compared four different serological tests for detection of BBTV in sap of banana and found that ELISA-NPP, ELISA-A and DIBA-ECL had equivalent sensitivity, but DIBA-C was up to 8-fold less sensitive than the fromer 3 assays. ELISA-NPP was adjudged to be the best compromise between sensitivity, cost and completion time.

Xie and Hu (1995) developed a polymerase chain reaction assay for detection of BBTV in banana samples and in single aphid. They revealed that Dot-blot hybridization assays were as sensitive as

ELISA, while PCR was 1,000 times more sensitive than dot-blot and ELISA assays for detection of BBTV in banana. BBTV in banana. Sharman *et al.* (2000), developed a multiplex, immunocapture PCR (M-IC-PCR) for the simultaneous detection of three viruses (*Banana bunchy top virus, Banana bract mosaic virus and Cucumber mosaic virus*) from crude sap extracts of banana and plantain (*Musa* spp.).

A polymerase chain reaction (PCR) assay was used to detect the Indian isolate of Banana bunchy top virus at early stages of infection in banana suckers before symptom expression. The viral DNA was detected from a single viruliferous banana aphid (*Pentalonia nigronervosa*) at 1.0 kb. The six plant species *viz., Zingiber officinale, Colocasia esculenta, Catharanthus roseus, Canna indica, Hedychium coronarium* and *Alphinium* sp. upon inoculation with BBTV showed negative results in PCR assay. Using PCR assay the BBTV could be detected in meristem tip cultured banana plants before symptom expression. In field condition the meristem tip cultured banana plants expressed BBTV symptom 45 days after planting (Manickam et al., 2002)

A comparative cost benefit analysis was conducted for the commercially available and locally produced antiserum and imported antiserum, purchased in 2002, against banana bunchy top virus in Pakistan. The locally produced antiserum is economical, incurring costs of 20000 rupees, which is lower by 2.5-fold than the cost incurred for the imported antiserum kit (45000 rupees). The strain-specific, locally produced antiserum can also be obtained in large quantities (>25 ml), and is sufficient for testing of approximately 65 000 samples, compared to the imported antiserum (1 ml) which is sufficient for testing 1000 samples only. The locally produced antiserum can be made available to provincial agricultural research institutes, extension units and entry ports for checking the health status of the crop Tahira-Yasmin *et al.* (2004).

Indirect ELISA and PCR were used to detect the presence of cucumber mosaic virus (CMV), banana streak virus (BSV) and banana bunchy top virus (BBTV) in cultured banana shoots. Indirect ELISA detected CMV in 2.5 per cent of the shoots, whereas PCR detected BSV in 12.5 per cent. BBTV was not detected by PCR in 25 samples tested (Rao et al., 2005).

Management Practices for BBTV

BBTD has been and continues to be managed by a rang of approaches, however, it remains a major limitation to production in some regions and a continuing threat to others. BBTD therefore has become a constant component of the cost of production either through disease losses or cost of control strategies. The following conventional and non-conventional strategies are now available and can be adopted to manage the BBTD.

Quarantine Regulations and Virus Free Planting Materials

With the availability of a range of sensitive and specific methods for BBTV diagnosis and detection and the advantage of *in-vitro* propagated bananas, now it is possible to provide large quantities of virus tested banana germplasm and planting material. With the adoption of the said material, the risk of introducing BBTV into new areas should greatly be reduced. The International Network for the Improvement of Bananas and Plantains (INIBAP) has established a "transit center" at Leuven in Belgium that maintains an extensive *in vitro* collection of *Musa* germplasm from which virus tested banana and plantain cultivars can be obtained (Dale and Harding, 1998). Various campaigns and regulations were adopted in Australia to managed BBTD (Dale, 1987).

Cultural Control

For BBTD management, it carries both the responsibilities of virus and aphid management. Cultural practices like regular inspection, rouging and digging play an adequate role in minimizing the disease. Early and efficient detection and eradication are the elements of this strategy and require knowledge and ability to diagnose BBTD from early symptoms. Destruction of weeds and alternate hosts, creation of banana free buffer zone around banana farm and adoption of diversified agriculture are other important BBTV reducing factors. Spread of disease could not occur if infected plants could be located and destroyed in the symptomless stage or within a short time of symptoms being displayed. An isolation distance of 100-200 m between any new bananas planting is very beneficial to control BBTD. Removal of diseased banana plants and replant with virus free banana plants leads to an effective management strategy. Cultural practices also play a very important role in aphid management. Destruction of wild or unattended patches of bananas is must to control aphids. A large population of banana aphids can grow on them and the wild patches when infected can serve as a serious and constant source of diseases for farms and residential neighborhoods. Maintenance of good weed control in the banana patches can help in aphid control. Some weeds may host the banana aphid or other banana pests. Aphids prefer to feed on young banana suckers so the banana mats should be pruned of all unwanted suckers.

Eradication

Eradication is being done by the digging and burning of diseased plants and suckers from the past. The herbicide picloram (Keipin formulation) effectively eradicate diseased banana with no detrimental effects to adjacent plants. This technique, coupled with a routine monitoring program, promises to be an effective method for reduction of BBTD. The formulation allows for precise placement of the herbicide with no noticeable effects on adjacent vegetation. It permits selective eradication within a plantation, which can be rapidly followed by plant treatment (Beaver, 1982). Plants and suckers that have previously been effectively sprayed with aphicides when cut down to soil level and sprayed with heavy gal-oil @2 pints/mt also results in complete destruction of stools (Wardlaw, 1972). Use of kerosine and 2, 4-D has also been in practice for eradication of the stools.

Biological Control

Biological control may be use to suppress aphid population in some situation. Ladybird beetle (Coccinella septumpunctata) has predatory potentialities on banana aphid, *Pentalonia nigronervosa*. Introduction of aphid parasitoids as bioagents leads to a good aphid control. Introduction of the polyphagous parasitoid (Aphidius colemani) can be use to control banana aphids. Frequent use of insecticide spray can have a negative, detrimental effect upon aphid predators and parasites. Ants should be controlled as they aggressively protect the aphids from their natural enemies and they feed upon the sweet honeydew that aphids produce. Boric acid baits can be used for ant control.

Resistant Cultivars and Transgenic Resistance

Resistance cultivars to BBTV would be most effective means of disease management. Unfortunately, there are no known commercial cultivars of banana that are immune to BBTV. Transgenic virus resistance, based on virus–derived transgenes, is an effective strategy for the control of plant viruses. These developments should lead to the incorporation of BBTV derived resistance genes into bananas, which hopefully will result in BBTV resistant transgenic bananas. However, researches are in progress at NRC for Banana, ICAR, Truchirapalli and may find these cultivars in future.

Conclusions and Future thrust

Banana bunchy top disease is one of the most important virus diseases of banana and has had a major economic impact on banana production in those countries and regions where it occurs. Diagnosis via symptoms is general. It is must to detect at early ages of infection via molecular and serological methods. Chemodiagnosis by tetrazolium chloride can also play an important role in disease diagnosis. This technique is very simple and less time consuming and very valuable for a certification programmes as well as for farmer's point of view because it requires less expertise.

The availability of specific and sensitive diagnostic methods for BBTV and the development of large scale in vitro micropropagation of bananas should facilitate the international movement of BBTV free banana germplasm and the distribution of BBTV free planting material. This should considerably reduce the risk of introducing BBTV into continents, countries or regions, which are currently free of this virus. The most attractive future strategy for the management of BBTD is to use the resistance cultivars. It is unlikely that such cultivars will be developed through conventional breeding programs at least in the short term as sources of resistance have not been identified because most of the conventional banana breeding programs are in their infancy stage and all of these are located in countries where BBTV does not occur. The components from which to develop transgenic bananas resistant to BBTV are available and this would appear to be the most likely source of BBTV resistant cultivars in the foreseeable future.

References

Agrios, G.N., 1997. *Plant Pathology*, 4th edn. Academic Press, New York, pp. 635.

Akram, M. and Kumar, R., 2006. Banana bunchy top disease: field diagnosis and management. *Indian Farming*, 56: 12–14.

Allen, R.N., 1978. Epidemiological factors influencing the success of roguing for the control of bunchy top disease in New South Wales. *Aust. J. Agri. Res.*, 24: 535–544.

Allen, R.N. and Barnier, N.C., 1977. The spread of banana bunchy top disease between banana plantations in the Tweed River District during 1975–76. *N.S.W. Plant Disease Survey*, 46: 27–28.

Boevink, P., Chu, P.W.G. and Keese, P., 1995. Sequence of *Subterranean clover stunt virus*: affinities with Geminiviruses. *Virology*, 207: 354–361.

Dale, J.L., 1987. Banana bunchy top: an economically important tropical plant virus disease. *Adv. Virus Res.*, 33: 301–325.

Dale, J.L. and Harding, R.M., 1998. Banana bunchy top disease: Current and future strategies for control. In: *Plant Virus Disease Control*, (Eds.) A. Hadidi, R.K. Khetarpal and H. Koganezawa. American Phytopathological Society, pp. 659–669.

Dale, J.L., Philips, D.A. and Parry, J.N., 1986. Double stranded RNA in banana with bunchy top disease. *J. Gen. Virol.*, 67: 371–375.

Evers, C., 1968. Host of sixteen aphids from banana plantations in Honduras. *Fal. Ent.*, 51: 113–118.

FAOSTAT, 2005. FAOSTAT data 2005, *www.fao.org*.

Geering, A.D.W. and Thomas, J.E., 1996. A comparison of four serological tests for the detection of *Banana bunchy top virus* in banana. *Austr. J. Agri. Res.*, 47: 403– 412.

Goddard, E.J., 1929. Bunchy top in bananas. *J. Austr. Council Sci. Indus. Res.*, 2: 21–27.

Harding, R.M., Burns, T.M. and Dale, J.L., 1991. Virus-like particles associated with banana bunchy top disease contain small single stranded DNA. *J. Gen. Virol.*, 72: 225–230.

Hu, J.S., Yu, M.Q., Wu, Z.C. and Wang, M., 1993. Detection of banana bunchy top virus in Hawaii. *Plant Dis.*, 77: 952.

Hughes, J. d. A., Speijer, P. R. and Olatunde, O., 1998. *Banana die-back virus*: A new virus infecting bananas in Nigeria. *Plant Dis.*, 82: 129.

INIBAP, 1994. *Annual Report for 1993*. INIBAP, Montpellier, France, pp.73.

Iskra, M.L., Garier, M. and Bova, J.M., 1989. Purification of *Banana bunchy top virus* (BBTV). *Fruits*, 44: 63–66.

Johnson, J., 1963. Biology of the bunchy top aphid, *Pentalonia nigronervosa* Coq. (Aphididae: Hemiptera). *Agri. Res. J. Kerala*, 2: 45–51.

Jose, P.C., 1981. Reaction of different varieties of banana against bunchy top disease. *Agri. Res. J. Kerala*, 19: 108–110.

Karan, M., 1995. Sequence diversity of DNA components associated with *Banana bunchy top virus*. *Ph.D. Thesis*, Queensland University of Technology.

Karan, M., Harding, R.M. and Dale, J.L., 1994. Evidence for two groups of *Banana bunchy top virus* isolates. *J. Gen. Virol.*, 75: 3541–3546.

Katul, L., Vetten, H.J. Maiss, E. Makkouk, K.M., Lesemann, D.E. and Casper, R., 1993. Characterization and serology of virus–like particles associated with Faba bean necrotic yellows. *Ann. Appl. Biol.*, 123: 629–647.

Khalid, S., Soomro, M.H. and Stover, R.H., 1993. First report of *Banana bunchy top virus* in Pakistan. *Plant Dis.*, 77: 101.

Khurana, S.M.P., 1971. Occurrence of banana aphid and its relation with banana bunchy top disease in Gorakhpur (U.P.). *Indian J. Horti.*, 28: 167–168.

Kumar, R. and Akram, M., 2006. Occurrence and detection of *Banana bunchy top virus* in Kanpur (U.P.), India. In: *Indian Virological Society XVI Annual Convention and International Symposium on Management of Vector-Borne Viruses*, 7–10 Feb., ICRISAT, Hyderabad, pp. 123.

Magee, C.J.P., 1927. Investigation on the bunchy top disease of the banana. *Bull. Comm. Austr. Countries Sci. Industr. Res.*, 30: 64.

Magee, C.J.P.., 1940. Transmission studies on the *Banana bunchy top virus*. *J. Austr. Inst. Agric. Sci.*, 6: 109–110.

Magee, C.J.P., 1948. Transmission of bunchy top to banana cultivars. *J. Aust. Inst. Agri. Sci.*, 14: 18–24

Magee, C.J.P., 1953. Some aspects of the bunchy top disease of banana and other *Musa* spp. *J. Proc. R. Soc. N.S.W.*, 87: 1–8.

Marathe, T.S., Summanwar, A.S. and Ram, R.D., 1982. New additional hosts of *Pentalonia nigronervosa* Coq. *Curr. Sci.*, 51: 435–436.

Mathews, R.E.F., 1982. Classification and nomenclature of virsus. *Intervirology*, 17: 1–200.

Mortan, J.F., 1987. *Fruits of Warm Climates*. Morton, Miami, pp. 505.

Ocfemia, G.O., 1930. Bunchy top of Abaca and Manila Hemp. I. A study of the cause of the disease and its method of transmission. *Amer. J. Bot.*, 17: 1–18.

Ocfemia, G.O. and Buhay, G.G., 1934. Bunchy top of Abaca and Manila Hemp.II. Further studies on the transmission of the disease and a trial planting of abaca seedlings in a bunchy top devastated field. *Philippine Agriculturist*, 22: 567–581.

Ocfemia, G.O., Celini, S. and Garcia, J., 1947. Further studies on transmission of bunchy top and mosaic of Abaca (Manila hemp plant. Separation of the two diseases and mechanics of inoculation by *Pentalonia nigronervosa* Coq. *Philippine Agriculturist*, 31: 87–97.

Ram, R.D. and Summanwar, A.S., 1984. *Colocasia esculenta* (L) Schott. a reservoir of bunchy top disease of banana. *Curr. Sci.*, 53: 145–146.

Rao, D.G., 1980. A short note on bunchy top disease of banana. In: *Proceeding of National Seminar on Banana Production Technology*, T.N.A.U., Coimbatore, pp. 142–143.

Robinson, J.C., 1999. *Bananas and Plantation Crops Production Science in Horticulture*. International Centre for Agriculture and Biosciences, pp.8.

Rohde, W., Randles, J.W., Langridge, P. and Hanold, D., 1990. Nucleotide sequence of a circular single stranded DNA associated with *Coconut foliar decay virus*. *Virology*, 176: 648–651.

Sastry, K.S., Rao, D.G. and Singh, S.J., 1980. Studies on control of bunchy top of banana. In: *Proceeding of National Seminar on Banana Production Technology*, T.N.A.U., Coimbatore, pp. 142–143

Sharma, S.R., 1983. *Annual Report*. Indian Institute of Horticultural Research, Banglore, pp. 114.

Sharma, S.R., 1984. *Annual Report*. Indian Institute of Horticultural Research, Banglore, pp. 89.

Sharma, S.R., 1985. *Annual Report*. Indian Institute of Horticultural Research, Banglore, pp. 60.

Sharma, S.R., 1988. Banana Bunchy Top Virus: A review. *Int. J. Tropical Plant Dis.*, 6: 19–41.

Sharma, S.R., 1987. Efficacy of tetrazolium chloride test for detection of bunchy top and infectious chlorosis viruses in banana. In: *Proc. All India Coordinated Research Project Workshop on Tropical Fruits*, P.K.V., Nagpur, 27–30 July, p. 451–453.

Singh, R.N., 1979. Bunchy top of banana in Uttar Pradesh. *Indian J. Myco. Pl. Patho.*, 9: 253–254.

Singh, S.P., 2004. *Commercial Fruits*, Reprinted first edition. Kalyani Publishers, New Delhi, p. 293.

Sano, Y., Isogai, M., Satoh, S. and Kojima, M., 1993. Small virus-like particles containing single-stranded DNAs associated with milk vetch dwarf disease in Japan. In: *6th International Congress of Plant Pathology*, Montreat, (Abstract).

Soomro, M.H. and Khalid, S., 1993. *Banana bunchy top virus* confirmed in Pakistan. *Infomusa*, 2: 17.

Soomro, M.H., Khalid, S. and Aslam, M., 1992. Out break of *Banana bunchy top virus* in sindh, Pakistan. *FAO Plant Protect. Bull.*, 40: 95–99.

Srinivasarao, A., 2003. Effect of polyamines and alternative oxidase inhibitors in ripening and self-life of banana (*Musa* sp.). *M.Sc. Thesis*, ANGRAU, Rajender Nagar, Hyderabad.

Stover, R.H., 1972. Virus diseases. In: *Banana, Plantain and Abaca Diseases*. Commonwealth Mycological Institute Publications, Kew, p. 316.

Summanwar, A.S. and Marathe, T.S., 1982. Diagnostic technique for the detection of bunchy top and infectious chlorosis in banana suckers. *Curr. Sci.*, 51: 47–49.

Sun, S.K., 1961. Studies on the bunchy top disease of banana. *Spec. Publ. College Agric. Taiwan Univ.*, 10: 82–109.

Thiribhuvanamala, G., Sabitha Doraisamy and Doraisamy, S., 2001. Survey for important virus diseases of banana. *Madras Agricultural Journal*, 88: 148–149

Thomas, J.E. and Dietzen, R.G., 1991. Purification, characterization and serological detection of virus like particles associated with banana bunchy top disease in Australia. *J. Gen. Virol.*, 72: 217–224.

Thomas, J.E., Geering, A.D.W., Gambley, C.F., Kessling, A.F. and White, M., 1997. Purification properties and diagnosis of *Banana bract mosaic potyvirus* and its distinction from *Abaca mosaic potyvirus*. *Phytopathology*, 87: 698–705.

Thomas, J.E., Iskra-Caruana, M.L. and Jones, D.R., 1994. Banana bunchy top disease. *Musa Disease Fact Sheet* No. 2. Montpellier, France: INIBAP.

Uppal, B.N., Verma, P.M. and Capoor, S.P., 1945. A mosaic disease of cardamom. *Curr. Sci.*, 14: 208–209.

Vakili, N.G. (1969. Bunchy top disease of bananas in the Central Highlands of South Vietnam. *Pl. Dis. Reptr.*, 53: 634–638.

Vasudeva, R.S., 1956. Overseas News. *Comm. Phytopathological News*, 2: 25–29.

Verma, P.M. and Capoor, S.P., 1958. Mosaic disease of cardamom and its transmission by the banana aphid, *Pentalonia nigronervosa* Coq. *Indian J. Agri. Sci.*, 28: 97–108.

Wardlaw, C.W., 1972. *Banana Diseases Including Plantain and Abaca*. Longman Group Ltd., p. 878.

Wolcott, G.N., 1933. *An Economic Entomology of the West Indies*, (Eds.) P.R. San Juan. The Entomological Society of Puerto Rico, p. 688.

Wu, R.Y. and Su, H.J., 1990. Purification and characterization of *Banana bunchy top virus*. *J. Phytopatho.*, 128: 203–208.

Yang, L.E., 1989. Bionomics of *Pentalonia nigronervosa* Coq. *Insect Knowledge*, 26: 145–146.

Tahira-Yasmin, Hussain, Shah and Saif, Khalid, 2004. Cost benefit ratio of imported and locally produced antiserum against Pakistani isolate of banana bunchy top virus. *Pakistan Journal of Botany*, 36(3): 621–627.

Rao, Xue Qin, Zhang, Shu Guang and Gao, Qiao Wan, 2005. Detecting viruses in banana differentiated shoots from tissue culture. *Journal of South China Agricultural University*, 26(1): 64–66.

Manickam, K., Sabitha, Doraiswamy, Ganapathy, T., Rabindran, R. and Doraiswamy, S., 2002. Early detection of banana bunchy top virus in India using polymerase chain reaction. *Acta Phytopathologica et Entomologica Hungarica*, 37: 1–3, 9–16.

Kumar, R. and Akram, M., 2006. Studies on host preference and population growth of banana aphids (*Pentalonia nigronervosa* Coq.). *Green Farming* (Communicated).

Kumar, R. and Akram, M., 2006. Occurrence and detection of Banana bunchy top virus. *Annals of Plant Protection Sciences*, 15(1): 263–264.

Insect Pest and Disease Management *Pages* **545–575**
Editor: **Dr. D. Prasad**
Published by: **DAYA PUBLISHING HOUSE (2016)**

Chapter 41

New Paradigms in Weed Management in India

Nisha K. Chopra, Neelam Kumar Chopra, S.N. Sinha and Derhinder Chowdary
Indian Agricultural Research Institute, Regional Research Station, Karnal – 132 001, Haryana

Introduction

Weeds are the integral part of our natural environment. It is difficult to consider them out of environment context. Infect our entire modern crops are derived from wild 'ancestors'. Extensive domestication and breeding program have shaped these wild ancestors in to modern crops suited to our needs with respect to yield, natural values and stress resistance etc.

Economic Impact of Weeds

Weeds are the component of our crop production system. The competitive effects that weeds have on the production of numerous crops are well documented. In general impact of weeds on crop yield varies from crop production systems (high input) to system (low input). It has been estimated that weeds caused 5 per cent loss in agricultural production in most developed countries and 25 per cent loss in least developing countries.

Losses

Of the total annual loss of agricultural produce from various pests in India, weeds account for 45 per cent, Insects 30 per cent, diseases 20 per cent and others 5 per cent. The world food lose due to weeds was estimated to be 287 million tones accounting 11.5 per cent of total food production (Parker and Fryer 1975). Estimates showed that weeds in India cause an annual monetary loss of Rs. 19800 million (Mukhopadhyay 1992). The total loss in pulses under conventional practices has been worked out to be 0.74 million tons worth of Rs. 3251.10 millions. In wheat *Phalaris minor* alone affects yield up to 40 per cent. In rice, when weed pressure is extremely severe at early rice growth stage yield loss can be 100 per cent. Singh *et al.* (1993) reviewed the losses caused by weeds in vegetable crop ranging from 6 to 82 per cent.

Table 41.1: Annual Losses Caused by Different Pests in India (Kulshrestha and Parmar 1992)

Pest	Loss in Value/Year	
	Rs. (in crore)	Loss per cent
Weeds	1980	33
Pathogens (Diseases)	1560	26
Insects	1200	20
Storage Pests	420	7
Rodents	360	6
Other Pests	480	8
Total	6000	100

Weeds are competitive and adaptable to all adverse environments. Weeds cause two types of losses. The most important is the direct yield loss resulting from competition, followed by indirect damage by reducing the crop quality due to competition (De Datta, 1980). All the crops yield less under severe weed stress.

Recent Advances in Weed Sciences

1. Concept has been changed from weed control to weed management.
2. Discovery of safer herbicides.
3. Some of the herbicides are used at extremely low rates of 2-4g/ha as against several kilograms earlier.
4. Management of weeds through bio control agents and bio-herbicides has shown sign of great promise.
5. Impact of biotechnology is also felt in weed science. Development of bio-herbicides, myco-herbicides, herbicide resistant crops etc. are some of the promising areas of application of biotechnology.

Directions in Weed Management

1. Preventive weed management
2. Knowledge of critical period of weed management
3. Chemical approach
4. Weed management for complex weed spectrum
5. New technology for application of herbicides to optimize their performance and minimize their environmental impact
6. Soil solarization a technique for weed management

7. Weed management through agronomic practices
8. Allelopathy
9. Non-living Mulches
10. Organic Weed Management

Preventive Weed Management

Weed Free Seed Crop

Seeds of most crops are contaminated with weed seeds. This happens at the time of harvesting as some weeds have life cycle similar to that of crops and set the seeds at the same time as the crop do so. Seeds of weeds species of same duration resembles in shape and size. The presence of even a small quantity of weed seeds may be enough for serious weed infestation. To obtain weed free crop seeds following cultural and mechanical measures are to be adopted:

1. Cleaning of all farm-produced crop seeds.
2. Separating crop seeds from ad mixture of weed and crop seeds on the basis of physical difference like seed size, weight, shape and colour.
3. Do not feed screenings, grain, or hay containing weed seeds without first destroying their viability by grinding, cooking or ensiling.
4. Do not use manure unless the viability of weed seeds have been destroyed by thorough fermentation, A chicken destroys about 99 per cent of any weed seed in the feed due to grinding in its gizzard. As a result, chicken manure usually has fewer weed seeds than cattle manure.
5. Using well- decomposed weed free farmyard manure eliminates the risk of spreading the weeds seeds.
6. Measures should be adopted tc prevent carrying of weed seeds, rhizomes and other propagating materials to the crop by irrigation water. Preventive measures like cleaning irrigation chanel before irrigation should be adopted.
7. Weed seeds, propagating materials and vegetative propagules adhere to harrows, ploughs, intercultural equipments/hoes during tillage and post-planting operations and get carried to other fields. Hence agricultural equipment should be cleaned before using them.

Source of Weed Infestation

Seed

1. Man often carries weed seeds of his interest from one part of the continent to the other continents. Dispersal of weed seeds through the movement of commercial seed is important example.
2. During combine harvesting weeds seeds are also harvested along with crop. It is particularly true in case of winter grains. Current panic in India about *Parthinium hysterophorus* (Congress grass) and a wide spread occurrence of *Phlaris minor* (Wild canary grass) in winter crop through imported wheat grain in India is the best example of international spread of weed seeds.

Seed Quality

Seed cleanliness, percent germination and vigour are characteristics that can influence the competitive ability of the seedling crop. Planting un-clean (especially farmer-saved) seed which is often of lower quality than certified seed and may contain unknown quantities of weed seed or disease, may result in the introduction of pests not previously observed on the farm. There is also a risk of increasing weed density and moving weeds to parts of fields that were not previously infested. Percent germination and vigour are important to weed management because they collectively affect stand quality and time to canopy closure.

1. In India most of the peasants use their own saved seeds, which are full of weeds, so peasant cannot deny the entry and establishment of new weeds in an area not infected with it yet.
2. Farmers have traditional processing methods, which cannot remove the weed seeds that are equal in size of crop seed.
3. Many crop seeds can pick up weed seeds during uncovered transport.
4. Crop/seed rejects containing weed seeds either reach the soil directly or fed to animals.
5. The weed seeds consumed by cattle while grazing, pass through the digestive canal without loss of their viability. These seeds germinate from the droppings and spread in other areas.
6. Compost and FYM are the notorious source of weed seeds. Most of the weed seeds remain viable in the manure pits and spread like wild fire wherever the manures are applied.

Weed Seed Bank

Weed problem begins with weed seed in the soil. The weed seeds fall to the ground and remain a source of soil seed bank thereby ensuring perpetual establishment of the weed. Hard seeds can also be seen in soil bank, this dormancy can help in spread of seeds over time and space. Some weeds mature their seeds simultaneously or just before them other crop.

1. Weed seeds remain viable for many years in soil.
2. It is common for weed seeds to retain their viability for ten, twenty or even forty years.
3. Dormancy in weeds includes periodic flushes of weeds.
4. By the time seeds from initial weed populations are eliminated from the soil, many more millions are produced and stored in it as subsequent weed flushes. In this context an old saying "one year of seeding is seven years of weeding."

Thus the weed seeds either fall on ground before the crop is harvested or harvested along with the crop.

Knowledge of Critical Period of Weed Competition

To maintain weed free for entire growing season is usually neither economical nor feasible. Therefore, one must distinguish between the period when weed–crop interference is critical and when the weeds must be removed to grow crop without the fear of any appreciable damage. A fundamental principle of plant ecology is that the first plant occupies any area of soil tends to suppress the latter entrants. Thus, the weeds that germinate and grow at the same time, as crop seeds are the most successful competitors usually have negligible competing ability.

Weeds must be controlled with in the critical–period of weed crop competition. Many a times control measures are adopted after the losses have already taken place. The timeliness of the weed control operations is as important as control itself. The period during which weeds can be tolerated by the crop without reduction in the final yield depends not only on the characteristics of weed and crop but also on the period for which weeds are associated with the crop.

The critical period of weed crop competition is between early growth during which weeds can grow without affecting crop yield and the point after which weed growth does not affect yield (Zimdahl, 1980). Establishing the critical period of competition is essential to develop effective and economical weed control measures. Researchers have pin pointed the optimum critical period for different crops (Table 41.2).

Table 41.2: Critical Period of Crop Weed Competition in Seed Crops

Name of Crop	*Critical Period of Weed Competition*	*Number of Days Weeds can be Left in Competition After Crop Emergence*	*Number of Days after which Weeds can be Left to Compete with the Crop*	*Reference*
Rice (Direct seeded)	15-30DAS	15 DAS	30 DAS	Ladu and Singh, 2006
Rice(spring)	40-120DAS	40DAS	120DAS	Pandey *et al.*, 2003
Wheat(irrigated)	30-45DAS	30DAS	45DAS	Chopra *et al.*, 1999
Tomato	15-30DAT	30DAT	15DAT	Prakash and Srivastava, 2006
Onion bulb	20-60DAT	20DAT	60DAT	Chopra and Chopra, 2006
Black gram	10-40DAS	10DAS	40DAS	Kumar and Tiwari, 2004
Cumin	20-40DAS	20DAS	40DAS	Kumar, 2001
Pea	15-60DAS	15DAS	60DAS	Tripathi *et al.*, 2001
Cowpea	15-45DAS	15DAS	45DAS	Tripathi and Singh, 2001
Pigeon pea	30-6-DAS	30DAS	60DAS	Tomer *et al.*, 2003
Sunflower	25-43 DAS	25DAS	43DAS	Wanjari *et al.*, 2000
Chickpea	40-60 DAS	40 DAS	60 DAS	Tiwari *et al.*, 2001
Sorghum	15-45 DAS	15 DAS	45 DAS	Sundari and Kumar, 2002
Maize	15-30 DAS	15 DAS	30 DAS	Kumar and Sundri, 2002

Chemical Approach

Herbicide use to control weeds in crops is an integral part of modern agriculture. Selective herbicides kill weeds but leave the crop unaffected. This crop selectivity is often due to the ability of crop to metabolically detoxify the herbicide, which the weed cannot metabolize. Among various group of pesticides, the growth rate of herbicides was much higher:

Table 41.3: World Pesticide Amount Used by Pesticide Type in 1966 to1987, 2000 and 2001

	*1966 to 1987**	*2000***	*2001***
Herbicides	24.8 per cent	36 per cent	37 per cent
Insecticides	7.04 per cent	25 per cent	24 per cent
Fungicides	3.25 per cent	10 per cent	9 per cent
Others	24.70 per cent	29 per cent	29 per cent

*: (Yadaraju 2002),**: Frederick M. Fishel Web Site at http://edis.ifas.ufl.edu.

Scope of Herbicide Usage

The primary objective of introducing herbicides in our country has been:

1. To replace back breaking manual weeding and to give farmers much needed leisure, of which he has been deprived since ages
2. Spared labour due to herbicidal application can be utilized more efficiently
3. For bringing additional land under cultivation
4. Improving irrigation and fertilizer application
5. Herbicides may be applied to weeds in crop rows when cultivation would be impossible.
6. Herbicides can kill many weeds that survive by mimicry *e.g. Echinochloa* spp. in paddy, *Avena* spp. in oats and wheat, Phalaris in wheat
7. Pre-emergence treatments with herbicides provide early season weed control.
8. Cultivation often injures the crop root system as well as foliage. Selective herbicides reduce the need for tillage.
9. Increased yield is not only reasoned for using herbicides; ease of harvesting, clean seed, and release of land from weed for subsequent crops are also very important. Thus enhancing over all production in the country.
10. Cost of weed management through herbicides is less than the manual weeding.
11. Herbicides are not aimed at substituting for physical, biological or good crop husbandry methods of weed control, but mainly for bridging the gaps in these methods as added production tools in agriculture.

Concerns on Herbicide Usage

Although herbicides are a boon to the agricultural community in substantially increasing crop yields; their use is not without potential problems. Some of the unintended negative impacts of herbicide use are persistence in soil, pollution of ground water, toxic residues in food (contamination), feed and fodder, adverse effect on non-target organisms and development of resistance in weeds.

Following concerns have been shown with the continuous use of herbicides:

1. Shift in weed flora and Herbicide resistance
2. Toxicity/malformation and Residue problems

Shift in Weed Flora and Herbicide Resistance

The greatest area of concern with herbicide weed-management system is the potential for weed shifts and weed resistance. A weed species shift is more common than resistance. It is important to understand the difference between a *weed shifts* and *weed resistance* because the management approach for dealing with each, is different. Weed shifts occur because herbicides are not equally effective against all weed species and differ greatly in the weed spectrum they control. A weed species that is not controlled will increase in density following repeated use of one herbicide. So over time, there is gradual shift or increase in tolerant weed species (Steve Orloff *et al.*, 2006). Therefore, rotating herbicides is recommended. For example, a continuous use of isoproturon in wheat for the control of *Phalaris minor* has resulted in many fold increase in population of broad leaf weeds like *Rumex dentatus,Cornopus didymus, Convolvulus arvensis, Cirsium arvense, Melilotus indica, Anagalis arvensis,* and *Vicia hirsute are the major weeds in* wheat in *rice–wheat rotation.* In India due to continuous use of butachlor in rice, the weed flora is shifting towards sedges like *Cyperus* spp. (Yadaraju 2002) and broad leaf weeds like *Sphenoclea zeylinaca* and *Eclipata alba* (Chopra and Chopra 2003). Similarly application of glyphosate to widely spaced spring crops the weed flora is shifting towards *Parthenium hytophorus.*

In contrast to a weed shift, weed resistance is the inherited ability of a weed to survive an herbicide dose that kills the wild type of that species. In other words, resistance occurs when there is a genetic change so that a weed that is normally controlled by an herbicide is no longer controlled. Rotating herbicides is also an effective strategy for resistance management. Within a weed species there are individuals that have a different genetic makeup that enables them to survive a particular herbicide application. The susceptible weeds in a population are killed, while the resistant ones survive, set seed, and increase over time. Using an effective herbicide with a different site of action controls the resistant biotype, thus preventing reproduction and slowing the spread of the resistant biotype. For the same reasons tank mixing herbicides and herbicides in sequence are also used.

Herbicide Rotations

Herbicide rotation is the practice of following a systematic, rotational sequence of herbicides in the same field (Hauser, 1974). It may involve either- (i) change of herbicide in the same crop, or (ii) change of the crop in the same field so that a different herbicide could be used. The chief objective of herbicide rotation is to curb a wide spectrum of weeds. With repeated use or two kinds of herbicides in the same field, often, some tolerant weed species that were insignificant earlier become prominent. Weed species shifts are also caused by no-tillage planting systems, as well as by periodic changes in rainfall and temperature. Herbicide rotations put a break to undesirable ecological shifts in weed flora in an area. The rotation of herbicides with different modes of action may be important in avoiding the problem of resistance. Following herbicide rotations can help delaying the resistance problem of *Phalaris minor* biotypes:

1. In Punjab and Haryana clodinafop (Topic 15WP) is a boon to farmers in controlling isoproturon resistant biotypes of *P. minor*, but broad leaf weeds were not suppressed by clodinafop (Topic 15WP). *Rumex acetosella* is beyond the control of both of the herbicides, which is creating a major problem in Haryana zero tillage system.
2. Metsulfuron (Algrip) (4ga.i./ha) can give good control *Rumex* spp. in zero tillage system.
3. Fenoxya prop-p-ethyl has also been found to control resistant biotypes of *P.minor*, however, showed its inefficiency towards the broad leaf weeds. Hence Metsulfuron (Algrip) (4 ga.i./ ha) and carfentrazone @20g a.i./ha in mixture can control complex mixed flora of weeds.
4. Sulfosulfuron controls *Phalaris minor* besides broadleaf weeds except *Rumex* species. But shows residual activity and has to be used in areas where Maize and sorghum are not sown as succeeding crops.
5. The best way to delay or prevent weed resistance is through use of tillage, crop rotation and herbicides with different modes of action.
6. Use of labeled rates of herbicides and tank mixes of herbicides with multiple modes of action
7. When weeds are not adequately controlled, they should be removed by hand to prevent seed production.
8. Use alternative methods of control whenever possible such as mechanical cultivation.
9. Rotate crops and their accompanying herbicides' site of action.

10. Limit number of applications of herbicide(s) with same site of action in a given growing.
11. Use mixtures or sequential herbicide treatments having different sites of action that will.

Toxicity/Malformation and Residue Problems

Increasing use of pesticides is causing trepidation as they leave behind their residues in produce and fruits. But herbicides are safer than insecticides as they are either applied before sowing or within one month of sowing. Moreover herbicides are never applied directly on fruits/grains.

Carfentrazone caused white speckling which disappeared 12-18 days without any negative effect on seed yield.Carfentrazone activity is rapid and any crop injury is first noticed within four to seven days of application. The severity of the injury is rate responsive and can be significantly increased by the addition of surfactants, oils and adjuvant such as oil/surfactant blends. Symptoms of Carfentrazone crop injury varies from minor cases of crop injury such as leaf spotting/scalding and plant stunting, to severe cases of total leaf yellowing/whitening and leaf abortion. Recovery from injury is rapid and generally occurs within 14 to 21 days after application.

2,4-Dichlorophenoxy acetic acid an effective and cheap herbicide has been recommended for broad-leaved weed control. However it has been reported to affect the spike morphology of wheat varieties making the roguing and inspection for genetic characters of wheat normally based on these characteristics difficult. Ear deformity due to 2, 4-D spraying is the result of interference from the herbicide in cell division and differentiation of the spike lets in the young ears. Different type of spike abnormalities has been observed like missing spike let, branched spike let, ear inversion, ear breaking and breaking of rachis. The varieties differed in spike malformation and per cent germination which highlights the need for taking into account the germination of progeny seed in addition to the effect of ear deformity and seed yield while determining the sensitivity of wheat cultivars to 2, 4-D (Yaduraju and Ahuja 1993). Herbicides which are highly soluble and less adsorbed to soil such as sulfonylurea (Backie and Mackercher 1990), triazine, alachlor, metalachlor (Hallberg, 1989) contaminate ground water.

Residual Activity of Herbicides on Succeeding Crops

Residual impact of herbicides on the succeeding crops in the rotation is very important for herbicide recommendation for effective weed control. In general crop conditions herbicide molecule should remain biologically active at least up to the critical period of crop weed competition and thereafter it is degraded chemically or microbiologically to form the innocuous products. In the northwestern plains wheat is the main crop which is infested with complex mixed flora (Chopra and Chopra 2005.)

A good number of herbicides with different mode of actions are being used. Since a number of crops followed wheat it becomes necessary to know the residual activity of these herbicides. Yadav *et al.* (2004) reported that recommended doses Diclofop, metsulfuron, clodinafop, fenoxaprop applied in wheat did not exhibit any adverse effect on various kharif crops of moongbean, pigeonpea, maize, cowpea, pearlmillet and cotton. However increasing the dose of metsulfuron to double (8g/ha) reduced the shoot dry weight og pearlmillet and maize crops. Chlorosulfuron at 30g/ha did not leave any residual effect however increasing the dose to 60g showed the residual activity to moongbean crop. Walia *et al.* (2006) in his experiment with sulfosulfuron at 25 and 50 g/ha found that winter maize was more sensitive to vegetable pea and berseem. At 25 g/ha of sulfosulfuron maize showed the reduced height, cob weight whereas both vegetable pea and berseem showed the residual toxicity at 50g./ha dose of sulfosulfuron.

Figure 41.1: Ear Deformity Due to 2, 4-D Spraying in Wheat

Table 41.4: Chemical Control of Major Weeds in Different Crops

Name of Herbicide	*Commercial Name*	*Chemical Family*	*Mode of Action*	*Time of Application*	*Dose/ha (Product)*
			Grass Weed Control in Wheat		
Clodinafop proparygyl	Topic	Cyclohexanediones	Acetyl CoA carboxylase (ACCase) inhibitors	PE-40 DAS	400g/ha
Fenoxaprop-p-ethyl	Puma Super	Cyclohexanediones	Acetyl CoA carboxylase (ACCase) inhibitors	PE-35-40 DAS	1000ml/ha
Sulfosulfuron	Leader	Sulfonylureas	Acetolactate synthase (ALS) inhibitors	PE 35-40 DAS	
Pendimethalin	Stomp	Dinitroanalines	Microtubule assembly inhibitors	Pre-emergence	3.33l/ha
Glyphosate	Round up	–	Prevents synthesis of essential aromatic amino acids needed for protein biosynthesis	Stale Seedbed	1.0 per cent spray
			Broad Leaf Weed Control in Wheat		
Carfentrazone ethyl*	aim	Aryl trizoline	Membrane disruption	PE-35DAS	50g/ha
Metsulfuron methyl	algrip	Sulfonylureas	Stops cell division by inhibiting biosynthesis of the essential amino acids	PE-35-40 DAS	20g/ha
Pendimethalin	stomp	Dinitroanalines	Microtubule assembly inhibitors	Pre-emergence	3.33 l/ha
			Herbicide applied in sequence in wheat		
Clodinafop proparygyl fb Carfentrazone ethyl	Topic fb aim	Cyclohexanediones +Aryl trizoline	Acetyl CoA carboxylase (ACCase) inhibitors and Membrane disruption	PE-35DAS+ 40DAS	400g/ha + 50g/ha
Clodinafop proparygyl fb Metsulfuron methyl	Topic fb algrip	Cyclohexanediones + sulfonylureas	Acetyl CoA carboxylase (ACCase) inhibitors and Stops cell division by inhibiting biosynthesis of the essential amino acids	PE-35DAS+ 40DAS	400g/ha + 20g/ha
Fenoxaprop-p-ethyl fb Carfentrazone ethyl	Puma Super fb aim	Cyclohexanediones + Aryl trizoline	Acetyl CoA carboxylase (ACCase) inhibitors and Membrane disruption	PE-35DAS+ 40DAS	1000ml/ha + 50g/ha
Fenoxaprop-p-ethyl fb Metsulfuron methyl	Puma Super fb algrip	Cyclohexanediones +sulfonylureas	Acetyl CoA carboxylase (ACCase) inhibitors and Stops cell division by inhibiting biosynthesis of the essential amino acids	PE-35DAS+ 40DAS	1000ml/ha + 20g/ha

Carfentrazone ethyl* shows white speckling on wheat crop after 3 to 6 days after spraying. These white speckling disappears after 10 to 15 days.

PE: Post emergence; DAS: Days after sowing

Contd...

Table 41.4–Contd...

Name of Herbicide	Commercial Name	Chemical Family	Mode of Action	Time of Application	Dose/ha (Product)
			Field pea and lentil		
Pendimethalin	Stomp	Dinitroanalines	Microtubule assembly inhibitors	Pre-emergence	3.33 l/ha
			Paddy		
Pyrazosulfuron	Saathi	Sulfonylurea	Inhibits synthesis of the essential amino acids valine and isoleucine, hence stopping cell division	Pre and early post emergence	200-250 g/ha
Fenoxaprop	Puma super	Polycyclic alkanoic acid	Fatty acid synthesis inhibition in grasses	3 leaf stage till early tillering	560-600 g/ha
Butachlor	Machete	Chloroacetanilide	Protein synthesis inhibitor	3DAT	3000 g/ha
Cyhalofop butyl	Clincher	Polycyclic alkanoic acid	Inhibition of acetyl CoA carboxylase	Post-emergence herbicide	700-800 g/ha
Metsulfuron methyl	Algrip	Sulfonylureas	Stops cell division by inhibiting biosynthesis of the essential amino acids	PE-35-40 DAS	20g/ha
Pretilachlor	–	Chloroacetanilide	Cell division inhibitor	Pre-transplanting and any time up to weed emergence	400-500 g/ha
			Cowpea, Pigeonpea, Mungbean, Chickpea		
Pendimethalin	stomp	Dinitroanalines	Microtubule assembly inhibitors	Pre-emergence	3.33 l/ha
			Maize, Bajara and Sorghum		
Atrazine	Atraf	Triazines	Photosystem II inhibitor Trizine	Pre-emergence	1.00kg/ha
			Vegetables		
(A) Rabi Vegetables					
Root crops (Radish, Carrot, Turnip)					
(a) Seed to root	i. Fluchlorolin	Dinitroanalines	Microtubule assembly inhibitors	PPA	1.5
	ii. Pendimethalin	Dinitroanalines	Microtubule assembly inhibitors	Pre-emergence	1.0
	iii. Fluch+Alachlor	Dinitroanalines+ chloroacetamide	Microtubule assembly inhibitors+liquid synthesis inhibitor	PPA	1.0+1.0
(b) Root to seed	i. Fluchlorolin	Dinitroanalines	Microtubule assembly inhibitors	PPA	1.5
	ii. Pendimethalin	Dinitroanalines	Microtubule assembly inhibitors	Pre-emergence	1.0

Contd...

Table 41.4–Contd...

Name of Herbicide	*Commercial Name*	*Chemical Family*	*Mode of Action*	*Time of Application*	*Dose/ha (Product)*
Tomato					
(a) Nursery	i. Fluchlorolin	Dinitroanalines	Microtubule assembly inhibitors	PPA	0.5
	ii. Pendimethalin	Dinitroanalines	Microtubule assembly inhibitors	Pre-emergence	0.5
(b) Seedling to seed	i. Pendimethalin	Dinitroanalines	Microtubule assembly inhibitors	Pre-emergence	1.0
	ii. Fluchlorolin	Dinitroanalines	Microtubule assembly inhibitors	PPA	1.5
	iii. Glyphosate	Isopropylamine salt	Prevents synthesis of essential aromatic amino acids needed for protein biosynthesis	30 DAS Foliar application on weeds after covering seedlings with poly pot	1 per cent
Onion					
(a) Nursery	i. Fluchlorolin	Dinitroanalines	Microtubule assembly inhibitors	PPA	0.5
	ii. Pendimethalin	Dinitroanalines	Microtubule assembly inhibitors	Pre-emergence	0.5
(b) Seedling to bulb	i. Pendimethalin	Dinitroanalines	Microtubule assembly inhibitors	Pre-emergence	1.0
	ii. Fluchlorolin	Dinitroanalines	Microtubule assembly inhibitors	PPA	1.5
(c) Bulb to seed	i. Pendimethalin	Dinitroanalines	Microtubule assembly inhibitors	Pre-emergence	1.0
	ii. Fluchlorolin	Dinitroanalines	Microtubule assembly inhibitors	PPA	1.5
Garden pea	i. Pendimethalin	Dinitroanalines	Microtubule assembly inhibitors	Pre-emergence	1.0
	ii. Fluchlorolin	Dinitroanalines	Microtubule assembly inhibitors	PPA	1.5
(B) Spring Summer Vegetable					
Okra	i. Pendimethalin	Dinitroanalines	Microtubule assembly inhibitors	Pre-emergence	1.0
Ohilli					
(a) Seed to nursery	i. Fluchlorolin	Dinitroanalines	Microtubule assembly inhibitors	PPA	0.5
	ii. Pendimethalin	Dinitroanalines	Microtubule assembly inhibitors	Pre-emergence	0.5
(b) Seedling to seed	i. Pendimethalin	Dinitroanalines	Microtubule assembly inhibitors	Pre-emergence	1.0
	ii. Fluchlorolin	Dinitroanalines	Microtubule assembly inhibitors	PPA	1.5

Contd...

Table 41.4–Contd...

Name of Herbicide	Commercial Name	Chemical Family	Mode of Action	Time of Application	Dose/ha (Product)
	iii. Glyphosate	isopropyl amine salt	Prevents synthesis of essential aromatic amino acids needed for protein biosynthesis	30 DAS Foliar application on weeds after covering seedlings with poly pot	1 per cent
Cucurbits	i. Glyphosate	Isopropyl amine salt	Prevents synthesis of essential aromatic amino acids needed for protein biosynthesis	30 DAS Foliar application on weeds after covering seedlings with poly pot	1 per cent
Cowpea	i. Fluch+ Alachlor	Dinitroanalines+ chloroacetamide	Microtubule assembly inhibitors+ liquid synthesis inhibitor	PPA	1.0+1.0
	ii. Pendi+Alachlor	Dinitroanalines	Microtubule assembly inhibitors	Pre-emergence	0.5+1.0
	iii. Pendimethalin	Dinitroanalines	Microtubule assembly inhibitors	Pre-emergence	1.0
(C) Kharif Vegetables					
Okra	i. Fluch+ Alachlor	Dinitroanalines+ chloroacetamide	Microtubule assembly inhibitors+liquid synthesis inhibitor	PPA	1.0+1.0
	ii. Pendi+Alachlor	Dinitroanalines +chloroacetamide	Microtubule assembly inhibitors+liquid synthesis inhibitor	PE	
	iii. Pendimethalin	Dinitroanalines	Microtubule assembly inhibitors	Pre-emergence	1.0
Chilli					
(a) Seed to nursery	i. Fluchlorolin	Dinitroanalines	Microtubule assembly inhibitors	PPA	0.5
	ii. Pendimethalin	Dinitroanalines	Microtubule assembly inhibitors	Pre-emergence	0.5
(b) Seedling to seed	i. Pendimethalin	Dinitroanalines	Microtubule assembly inhibitors	Pre-emergence	1.0
	ii. Fluchlorolin	Dinitroanalines	Microtubule assembly inhibitors	PPA	1.5
	iii. Glyphosate	Isopropyl amine salt	Prevents synthesis of essential aromatic amino acids needed for protein biosynthesis	30 DAS Foliar application on weeds after* covering seedlings with poly pot	1 per cent
Brinjal					
(a) Seed to nursery	i. Fluchlorolin	Dinitroanalines	Microtubule assembly inhibitors	PPA	0.5
	ii. Pendimethalin	Dinitroanalines	Microtubule assembly inhibitors	Pre-emergence	0.5

Contd...

Table 41.4–Contd...

Name of Herbicide	Commercial Name	Chemical Family	Mode of Action	Time of Application	Dose/ha (Product)
(b) Seedling to seed	i. Pendimethalin	Dinitroanalines	Microtubule assembly inhibitors	Pre-emergence	1.0
	ii. Fluchlorolin	Dinitroanalines	Microtubule assembly inhibitors	PPA	1.5
	iii. Glyphosate	Isopropyl amine salt	Prevents synthesis of essential aromatic amino acids needed for protein biosynthesis	30 DAS Foliar application on weeds after* covering seedlings with poly pot	1 per cent

Table 41.5: Effect of Residual Activity Different Herbicides Applied in Wheat on the Succeeding Crops

Name of Herbicide	Dose g/ha	Succeeding Crop	Effect on Crop	Reference
Metsulfuron	4	Pearl millet and Maize	Shoot dry weight was not affected	Yadav *et al.*, 2004
Metsulfuron	8 and 16	Pearl millet and Maize	Shoot dry weight was significantly reduced	-Do-
Sulfosulfuron	25	Maize	Shoot dry weight was affected	-Do-
Chlorosulfuron	60&120	Maize and Moongbean	Shoot dry weight was affected	Punia *et al.*, 2002 and Yadav *et al.*, 2004
Clodinafop	60	Forage sorghum	Did not affect germination and dry weight	Chopra and Chopra, 2005
Fenoxaprop	100	Moongbean	Did not affect germination and dry weight	-Do-
Metsulfuron	3, 4 and 5	Soybean	Did not affect plant stand and seed yield	Singh and Ali, 2004
Pendimethalin	1500	Maize	Shoot dry weight was affected	Yadav *et al.*, 2004
Sulfosulfuron	25 and 50	Winter Maize	Reduced height, Cob weight and seed yield	Walia *et al.*, 2006
Sulfosulfuron	25	Vegetable pea	Yield attributes were not affected	-Do-
Sulfosulfuron	50	Berseem	Plant population and fodder yield reduced	-Do-

Weed Management for Complex Weed Spectrum

The spectrum of weed flora is changing in such a way that many types of weeds infest the crop. No single herbicide will be able to control all types of weeds without crop injury because of higher dose requirement for increasing the spectrum of weed kill. Such problem of complex weed flora can only be tackled by combining grass herbicides with broad leaf weed killers.

Herbicide Mixtures

Herbicide mixtures are an effective tool to enhance herbicide efficiency, lower optimum doses of herbicides, reduce cost of weed management, arrest weed shifts, prevent herbicide resistance in weeds and facilitate improvement in overall weed management.

Table 41.6: Some of the Herbicide Mixtures Recommended for Different Crops

Crop	*Efficient Herbicide Mixture*	*Dose/ha*	*Reference*
Wheat	Fenoxaprop+carfentrazone	90+20	Singh and Singh, 2005
	Triasulfuron +Clodinafop	20+40	Yadav *et al.*, 2004
	Clodinafop+ Sulfosulfuron	60+20	Punia *et al.*, 2005
Rice	Butachlor+Almix	1000+4	Singh *et al.*, 2005
	Pretilachlor + 2,4-D	300+300	Kathirvelon, 2004
	Anilophos+ Ethoxysulfuron	400+15	Sharma *et al.*, 2003
	Clomazone + 2,4-DEE	175+270	Ghosh and Ghosh, 2005
Maize	Atrazine + Pendimethalin	500+250	Patel *et al.*, 2006
Pea	Alachlor+ Pendimehalin	1000+900	Rana *et al.*, 2004
Potato	Atrazine+Pendimetalin	750+900	Rana *et al.*, 2004
Cotton	Thazopyr+ Paraquat	480+1950	Yadav *et al.*, 2003
Black gram	Pendimethalin + Fluazifop	1250+375	Kumar and Tiwari, 2004
Sugarcane	Sulfentrazone+Atrazine	500+2000	Srivastava, 2003

Mixtures are made up of two or more herbicides with pre emergence activity, post emergence activity or both. These mixtures of herbicides are with different modes of action and varying spectra of an activity. Herbicide mixtures control a broader spectrum of weeds, and they are more consistent l over a wide range of climatic conditions. Herbicide mixtures are applied in lower doses resulting in decreased crop injury and lower costs and lower herbicide residue in crop and soil.

When two herbicides are mixed three types of responses may occur.

1. *Synergetic Type*: When the effect of herbicide mixture is great than when these herbicides were used singly.
2. *Antagonistic*: When there is reduced activity of one herbicide or both the herbicide in mixture than response is called antagonistic.
3. *Additive*: When effect of two herbicides in mixture is equal to the response of herbicide when applied alone.

Managing *Cyperus rotundus* by Application Technique of Glyphosate

Cucurbits are initially slow growing crops and purple nut sedge, which is early emerging weed use the same environmental factors for their growth needs. In these crops the period of weeding is short because of the growth of wines, which interferes with manual weeding, and there is every possibility of damage of wines. Purple Nut Sedge (Motha) (*Cyperus rotundus* L.) is the most troublesome and obnoxious weed of spring summer season. As the temperature increases during first fortnight of March, spring summer crops *viz.*, muskmelon, watermelon and bottle gourd etc. get heavily infested with *Cyperus rotundus* (300 to 500 shoots per m^2). The losses due to this weed range from 10–80 per cent

Figure 41.2: Cucurbit Field Infested with *Cyperus rotundus*

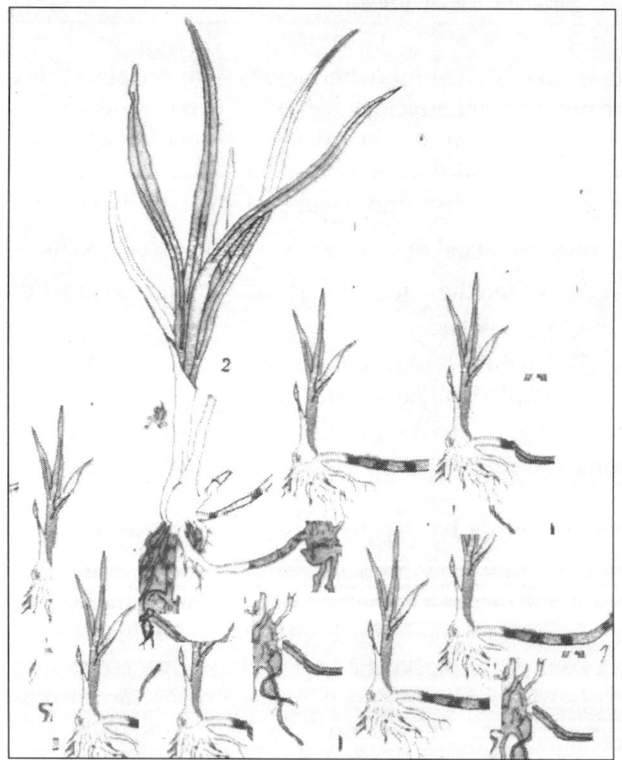

Figure 41.3: Complex Root System

depending upon the weed density and flora. Purple nut sedge rely on vegetative reproduction by rhizome and tubers, and approximately 780 shoots are produced by one tuber (Holm *et al.*, 1977).

The rhizomes produce numerous chains of nuts (tubers). Nuts possess scaly leaves, inconspicuous nodes and internodes and minute buds (which gives rise to new shoots). Tuber produces few rhizomes with basal bulbs. This network consists of 4 to 5 dormant tubers for each 'active' basal bulb. These sequences are repeated, forming a big chain of dozen inter connected tubers. Handweding fails to control *Cyperus rotundus* as the arial shoots sprout again. The use of glyphosate has been reported to be the best means to combat such perennial and persistent weed (Bajwa *et al.*, 1993 and Yadav *et al.*, 1996). Application technique of glyphosate a non selective herbicide, for managing *Cyperus rotundus* has been developed at IARI, Regional Research Station Karnal for long-term control of this obnoxious weed.

Why Glyphosate?

Glyphosate (N–phosphonomethyl glycine–isopropyl amine salt) is a broad-spectrum, nonselective herbicide used for control of annual and perennial plants including grasses, sedges, broad-leaved weeds, and woody plants. It is shoot translocated herbicide that destroys the roots and rhizomes of herbaceous perennial weed like purple nut sedge. Its mode of application is post emergence spray of the foliage. It shows no residual soil activity upon contact with soil, as it is bound by soil particles and is rapidly inactivated (reference). Moreover it does not leach away from the treated area to the adjoining areas. Hence for the control of this obnoxious weed glyphosate a broad spectrum, post emergent, translocated and nonselective herbicide with an application technique is the best choice for the long-term control of *Cyperus rotundus.*

Herbicide Application Technique

In wider spaced crops like cucurbits, (at three to five leaf stage) individual plants are fully covered with PVC pots and than blanket spray of 1 per cent glyphosate (41 per cent SL) @ 600 lit/ha is done on weed foliage after 20 days of sowing. After the spraying, the pots must be allowed to dry to avoid any herbicide droplet adhering to outer surface before shifting to other place. Pots should not be stacked during shifting and must be carried individually. Glyphosate must be applied when the weeds are green and actively growing. Heavy rainfall with in 2 hours of application may require a re-treatment. Since glyphosate is a systemic and translocated herbicide, complete wetting of foliage is not essential (do not spray up to run off point). It did not wash off or leach away from treated area to neighboring areas to damage other vegetation. Unlined steel or galvanized steel drums or containers react with glyphosate and should not be used. Glyphosate has no adverse effect on seed yield and quality.

Mechanization of Spray Technology

Tractor mounted sprayers can be used for the cucurbit crops having wider spacing covering an area more than acre. These sprayers work under a spray pressure of 1.4- 2.8 kg cm^2 and are fitted with multinozzle booms, two rows of the crop outside on either side of herbicide boom are also covered with plastic pots to avoid damage due to any spray drift.

Effect of Glyphosate on weeds

After application on weeds, glyphosate is rapidly absorbed by foliage and is translocated to roots and other underground storage organs in sufficient quantities to kill the entire plant. Visible effects occur with in 3-5 days for most of the annual weeds. *Cyperus rotundus* and other broad-leafed weeds start showing degeneration (Yellowing) with in 6 to 8 days. The early control of weeds resulted in

Figure 41.4: Glyphosate 1 per cent Spray on Targeted Plants

faster growth of wines and higher fruit/seed yield. Least competition from weeds resulted in increased 1000 seed weight and high vigoured seeds (Chopra *et al.,* 2006). The PVC pots did not show any chemical effect of herbicide hence could be reused for years together.

Figure 41.5: Spraying with Track–Mount Sprayer in Larger Fields

Figure 41.6: Comparison of Weedy Check After Glyphosate Spray

Soil Solarization for Nursery Weed Management

Use of herbicides for controlling weeds is very effective and economical but due to associated hazards, development of resistant biotypes and polluting the environment have necessitated development of alternate non hazardous means of weed management. Harvesting of solar energy through soil solarization for controlling of soil borne pests including weeds will be potential step to reduce the dependency on chemicals. Soil solarization (solar heating of soil) as pre planting soil treatment to control soil borne pathogens and weeds, was first described in 1976 by Katan and co-workers in Israel, which involves mulching of the soil with clear plastic films so as to trap the solar heat in the surface soil. The resultant temperature increases would be lethal to soil inhibiting pathogens nematodes and weeds. The common mulch for this purpose is transparent polythene. Soil solarization technique is simple and easy to use by farmers. However its immediate application appears to be more promising in nursery areas and in high value crops.

Principal of Soil Solarization

Soil solarization is a hydrothermal process, which brings about thermal, physical, chemical and biological changes in the moist soil during and even after mulching. Sun is extremely hot object due to which it emits short wave length heat waves. On the other hand environment inside the polythene sheet is very very less hot due to which it emits longer wave length heat waves. A peculiar property of a polythene sheet is that it allows short wave length heat waves (emitted by extremely hot object like sun) to pass through it but does not allow comparatively longer wave length heat waves emitted by less hot object *i.e.* soil to pass through it. A peculiar property of a polythene sheet is that it allows short wave length heat waves (emitted by extremely hot object like sun) to pass through it but does not allow comparatively longer wave length heat waves emitted by less hot object *i.e.* soil to pass through it. Short wave solar radiation of sun and longer wave radiation of earth which are held back are the primary source of heating soil. Short wave solar radiation of sun and longer wave radiation of earth which are held back are the primary source of heating soil. Which result in trapping of heat in the soil (Singh, 2002). Water evaporating from the soil surface will condenses on the mulch and drip back to the soil surface. As a result of formation of water drop lets on the inner side of the mulch will trap the long-wave radiation and thus eliminate radiating cooling of the soil surface.

Plastic mulches have to be intact to reduce the losses through latent heat fluxes. Before covering the seedbeds with polythene sheet mulch of 100 mm, the land has to be irrigated and brought to a good tilth by two disking and planking. Spread Polythene sheet on soil surface with all edges buried in the soil to make it airtight.

Technique is more practical in warmest summer months:

In Northern parts of India: May–June.

In Southern parts of India: April–May

Duration of 4 to 6 weeks is adequate under most of the situation.

Soil solarization effect is restricted to surface only (0-10 cm). Poor effects on weeds at deeper layers.

Size of mulch/Bed's Width

Soil heating at the edges of mulch is lower than in its center. So narrow mulch is less effective in heating the soil than a wide one. Narrow mulch recorded 2 to 4°C lower temperature at the edge of mulch than center at the same soil depth.

Figure 41.7: Size and Direction of Mulch/Bed's Width

Direction of Beds

Beds running N-S direction are preferable to E-W direction to avoid lower temperature on one side of the bed.

Transparent polythene films are more efficient in trapping than black films as they allow maximum transparency of radiation.

Weed Spp. Tolerant to Soil Solarization

1. *Cyperus rotundus* (because tubers of *C. rotundus* are heat resistant).
2. *Melilotus* spp is tolerant to soil heating even 40 days solarization (Yadaraju, 2000).

Weed Spp. Sensitive to Soil Solarization

In North India following weeds were found to be sensitive to soil solarization: *Trianthema monogyna* L., *Digeria arvensis* L., *Echinochloa colona* L., *Elusine indica* L., *Dactyloctenium aegyptium* L., *Phalaris minor* L., *Chenopodium album, Rumex dentatus* L., *Fumeria indica* L., *Avena ludoviciana* L.

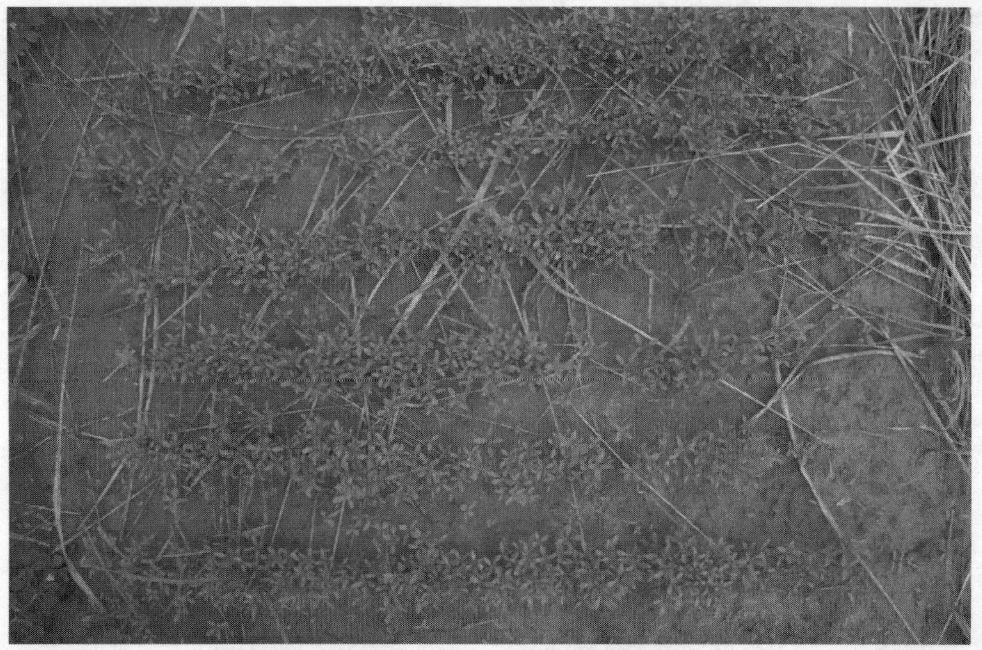

Figure 41.8: Weed Frees Soil Solarized Nursery

Figure 41.9: *Cyperus rotundus* is Tolerant to Soil Solarization

1. Soil solarization is very effective (90 per cent control) against parasitic weeds like Broom rape (*Orobanche* spp.).
2. Weed seeds in the surface 5 cm soil are killed to great extent.
3. Solarization has poor effects on weeds at deeper layers.
4. Soil solarization killed non-dormant seeds but could not eliminate dormant weed seeds from germination zone.

Weed Management Through Agronomic Practices

Crop Rotation

Crop rotation plays an important role in management of weeds in general and also for the problematic weeds which develop resistance to herbicidal control. In monocropping the weed species persists and expands rapidly but crop rotation helps in interrupting the normal life cycle of weeds to become dominant. The incidence of little canary grass is higher in rice wheat rotation than maize-wheat rotation. Changing from rice-wheat and rice-potato to any other rotation not involving rice in kharif tends to reduce population of *Phalaris minor* in wheat. Johnson grass (*Sorghum halipense* L.) becomes predominant weed in continuous maize but can be controlled by rotating with cotton or by inclusion of perennial fodder such as alfalfa (*Medicago sativa* L.) due to combined effect of competition from crop and repeated cuttings. It was observed that with increased use of isoproturon for controlling grass weeds the population of *Convolvulus arvensis* has increased in wheat. The weed twins around stem and completes its life cycle and form seeds; however the population is considerably reduced when wheat was rotated by garden pea. The pea crop grows fast and covers the ground, which does not allow the *Convolvulus arvensis* plant to complete its normal life cycle.

Crop rotations also help increase crop vigour (competitiveness) by decreasing disease and insect pressure. Crop rotations also assist in rotating herbicides. Weed resistances to herbicides is fast becoming a major problem on any farm which has systematically used the same herbicide or family of herbicides year after year. Once the resistant population has become dominant on a farm there are very few good ways of dealing with the problem.

Stale Seedbeds for Weed Free Crop Production

In stale seedbed technique 1-2 flushes of weeds are destroyed before planting of any crop. Most weeds germinate from top 4 to 5 cm of surface layers of soil.

When final preparation of seed bed is with held, a flush of young weed seedlings will appear on it with in a week's time. After 7 to 10 days of good growing conditions there is flush of the weed seedlings which can be killed with herbicides *e.g.* glyphosate or by shallow non-inverting type of tillage implement. Depending upon the time available one or two flushes of weeds can be destroyed. The main advantage of stale seed bed is that the crops germinate in weed free environment. There should be minimal soil disturbance, which reduces subsequent, weed flushes while the crop establishes. Success depends on controlling the first flush of emerged weeds before planting or crop emergence.

Irrigation

Irrigation has direct or indirect effect on weed intensity.

1. Pre-sowing irrigation is always better over dry sowing immediately followed by irrigation because pre-sowing irrigation encourages the weed seeds to emerge, which latter are destroyed by repeated ploughing during field preparations.

Figure 41.10: Blanket Spray of Glyphosate on Stale Seed Bed for Okra Sowing

2. Good crop husbandry and water management favours the crop than weeds. Transplanted rice in puddle conditions and maintaing appropriate plant population gives crop seedlings 14 to 21 days advantage over weeds and covers the ground fast thereby smothering the weed growth. Many weeds cannot germinate or grow under flooded conditions due to their inability to transport oxygen to the roots compared to rice crop. Standing water into which rice is transplanted suppresses weed growth because of the dual effects of the reduction in weed germination and embellishment by water and inherent size advantage of the crop. This is used as weed management methods especially in paddy crop and is becoming popular in perennial weed control
3. In drip irrigation method weed intensity is generally lower than in furrows and check basin method of irrigation; due to wetting of soil near the plant base only, rest of the area remain dry in drip irrigation.
4. Flood irrigation promotes early weed growth and total coverage of space if seed in soil is plenty.

Maximizing Early Crop Vigour

1. Sow seed as shallow as possible. Zero till seedbeds are moist and allow for shallow seeding. Shallow seeded crops emerge quicker, branch or tiller earlier and are less prone to root rots than deeper seeded crops.

2. Use seed treatments to control injury from diseases and insects. Root rots, and damping off can cause major plant stand losses. Beside the seed treatments can help increase root growth, decrease root rot attack and increase crop vigour.
3. Biological seed treatments such as legume inoculants may be beneficial to further enhance the nutrient availability to the crop and increase crop vigour.
4. Leave the stubble standing. Standing stubble will buffer the crop from drying winds thus resulting in less moisture stress and physical damages and greater crop vigour.

Weeds which are left uncontrolled or germinate before the crop exert much more pressure on the crop than weeds which germinate after the crop is established.

Conservation Tillage and Weed Management

In zero tillage (ZT) technology, crop seeds are sown in a single tractor operation using specially designed seed cum fertilizer drill without any field preparation the presence of anchored rice residue at optimum to slightly wetter soil regime.

Major advantages of zero tillage to manage the weeds are as follows:

1. Zero tillage offers a way to manage the problem of resistant weeds.
2. In ZT system wheat can be sown 7-10 days early and is more competitive against *Phalaris minor*.
3. When wheat is sown with ZT, due to less soil disturbance weed seeds from the deeper layers do not come near the surface resulting in to lower weed density.
4. In situations where weed seed bank is enormous, there should be presowing weed germination by irrigating a seedbed that is uncultivated followed by nonselective herbicides to provide stale seed bed.
5. With traditional tillage seed bank of the weeds buried in sub surface soil layers comes to surface. When rice is established as puddle transplanted crop, lighter seed float on water and get deposited on the soil surface. Main reason for change in weed flora seems to be use of herbicides for control of grassy weeds and non-adoption of any measures to control broad leaf weeds in the zero tillage system.

Allelopathy

Allelopathy is an alternative and organic approach to weed control that uses chemicals that are excreted from a plant to cause either direct or indirect harm to weeds by negatively affecting their germination, growth, or development. Nearby weeds can be affected by allelopathic chemicals entering the rhizophere from the roots or the aerial parts of the crop plant. Allelochemicals from the green manure crop in rice inhibits the germination of *Echinochloa* spp. Crop residues from cover crops, such as fall rye, or other organic mulches can also be used to suppress weeds through such allelopathic interactions. Liebl and Worsham (1983) reported that wheat residues suppressed all morning glory (*Ipomea purpurea* L.) and prickly sida (*Sida spinosa* L.) populations in the field. They attributed this effect, in part to release of ferulic acid. Allelopathic crops include rye, oats, sunflowers, barley, wheat, mustard, buckwheat, clovers (red, white, sweet), tall fescue, creeping red fescue, hairy vetch and perennial rye grass. In a season the weeds may be crop specific having greater affinity for a particular crop than others this affinity may have some allelopathic component. In rice wheat rotation *Phalaris minor* is at its maximum in wheat crop, the density of pm is found reduced when wheat is replaced by chickpea, pea or lentil for some time (Swain 2002). Replacement of Soybean with maize crop in Soybean–wheat rotation reduced the incidence of *Euphorbia geniculata*. In rotation, these crops help to suppress weeds in subsequent crops, but they can also suppress subsequent crop growth, therefore caution needs to be employed. The allelopathic effect is complex and requires the correct combination of crop and weather conditions.

Non-living Mulches

Mulching stunts or kills the weeds by cutting light to them. Synthetic mulches, namely black paper or polythene film mulches provide stronger mechanical barriers to all kinds of germinating weeds. Transparent plastic mulches are now preferred on newly prepared fields to suppress weeds. These are made in rolls of different widths can be spread on the seedbeds and then crop seedlings can be transplanted through holes made into them.

Presently mulching with synthetic materials is costly, so it's economic use to certain high value vegetables and ornamentals crops only.

Organic Weed Management

Today, farmers have a renewed interest in organic methods of managing weeds it is important to understand that under an organic system of weed control, weeds will never be eliminated but only managed Melissa Van Tine (2003). The organic farmer has to rely on cultural practices such as crop rotations and mechanical cultivation of the soil for weed control.

1. Maintain proper sanitation on the farm to reduce the introduction and spread of weed seeds.
2. The stale or false seedbed technique of flushing out weed seeds from the soil works by depleting the seed bank. After the soil is cultivated two to three weeks before sowing, emerging weeds are killed by light cultivation. By helping to reduce the seed bank. This technique reduces subsequent emergences of weeds.
3. Managing weeds mechanically is another effective method but it is both time consuming and labour-intensive.The choice of implementation, timing, and frequency will depend on the structure and form of the crop and the type and number of weeds. Cultivation involves killing emerging weeds or burying freshly before they shed weed seeds. In organic rice production use of rotatary tiller is dully benefited:

   (*a*) Break the capillary action

   (*b*) Burry the weeds in the crop

The Rotary Hoe

The rotary hoe is an important tool for the farmer in transition to non-chemical weed control. Under the right conditions, it can give the crop a one to two week head start over weeds. The basis for selective weed control with the rotary hoe is the depth at which weed seeds germinate. As nearly all small-seeded weeds and grasses germinate within the top half inch of soil.

Rotary Hoe Before You See Weeds

When a new flush of weed seed germination in the surface soil, but before the majority of seedlings have emerged, superficial tillage kills the weeds by disrupting water uptake. The hoe works best in somewhat crusted soils that have dried enough to crumble easily. The rotary hoe is also useful to break

Figure 41.11: Transparent Plastic Mulches in Muskmelon

Figure 41.12: Rotary Hoe in Paddy Crop

the soil crust. Once the weed roots penetrate beneath the rotary hoe depth, the weeds are not uprooted and can not be effectively controlled by the rotary hoe. The rotary hoe is least effective on weeds that germinate from deeper soil layers.

Conclusion

Weed management is as old as agriculture itself, only concept has changed from weed control to weed management. It was the bare hands in past, which were used to pull out the weeds. Herbicides since the discovery of 2,4-D has revutionalized agriculture. Discovery of safer herbicides, some of the herbicides are used at extremely low rates of 2-4g/ha as against several kilograms earlier have contributed enormously in increasing food production. Innovative approaches to manage the weeds are in great demand around the world, particularly those which are less expensive and harmful to environment. The search for new control methods, which are effective and economic and have minimal side effect is a continuous process. Preventive weed management. Knowledge of critical period of weed management. new molecules to manage herbicide resistant weeds and complex weed flora, soil solarization for nursery weed management and weed management through agronomic practices *viz.* crop and herbicide rotation alone and integrated with each others are used to manage weeds.

Herbicide use to control weeds in crops is an integral part of modern agriculture. Selective herbicides kill weeds but leave the crop unaffected. This crop selectivity is often due to the ability of crop to metabolically detoxify the herbicide, which the weed cannot metabolize. Modern herbicides are cost effective, efficient, and environmentally benign means for obtaining good weed management. Reliance on herbicides for weed control is expected to continue because there is no attractive superior technology available yet. They are precious resource of great importance to agricultural productivity. They need to be used sparingly.

References

Bajwa, G.A., Bal, J.S., Brar, S.S. and Minhas, P.P. S., 1993. Chemical weed control in ber orchard. In: *Proc. Int. Symp. of Indian Soc. Weed Sci.*, Hissar, November 18–20, 3: 225–227.

Beckie, H.J. and Mackercher, R.B., 1990, *J. Agric Food Chem.*, 38: 310–315.

Carol Mallory–Smith, George, R. Hyslop, 2002. *Herbicide: Resistant Weeds and Their Management*. A Pacific Northwest Extension Publication, University of Idaho.

Chopra, Nisha and Chopra, Neelam Kumar, 2006. Crop weeds competition and determination of critical period in onion (*Alleum cepa*) under north western plains. *Indian Journal of Weed Science,* 38(1&2): 89–91.

Chopra, Nisha, Singh, Harpal and Tripathi, H.P., 1999. Crop weed competition in wheat (*Triticum aestivum*). *Indian Journal of Weed Science*, 31(3&4): 151–154.

David Rourke and Andrew Hargrave www.mandakzerotill.org.

DeDatta, S.K., 1980. ASPAC Food fert. Technol. Center Ext. Bull. Taifei Taiwan 156. *Adv. in Agron.*

Hallberg, G.R., 1989. *Agric. Ecosyst. Environ.*, 26: 299–367.

Holm, L.G., Plunknett, D.L., Pancho, J.V. and Herberger, J.P., 1977. *The World's Worst Weeds.* East-West Center Book, University Press, Hawaii. Honolulu, Hawaii, 609 pp.

Kumar, Suresh S.V. and Sundari, A., 2002. Studies on the effect of major nutrients and crop weed competition period in Maize. *Indian Journal of Weed Science*, 34(3&4): 309–310.

Kumar, Subash, 2001. Critical period of weed competition in Cumin. (*Cuminum cyminum* L.). *Indian Journal of Weed Science*, 33(1&2): 30–33.

Kumar, Arun and Tiwari, A.N., 2004. Crop weed competition studies in summer Black gram (*Vigna Mungo* L.). *Indian Journal of Weed Science*, 36 (1&2): 76–78.

Ladu, M. and Singh, M.K., 2006. Crop weed competition in upland direct seeded rice under foot hills conditions of Nagaland. *Indian Journal of Weed Science*, 38(1&2): 131–132.

Liebl, R.A. and Worsham, A.D., 1983b. Tillage and mulch effects on morning glory and certain weed species. Southern Weed Sci Soc.Proc., 36: 405–414.

Melissa VanTine, 2003. www.wvu.edu/~agexten/ipm/weeds/index.htm.

Mukhopadhyay, S.K., 1992. *Emerging Problems and Advances in Weed Management*. Presidential Address, Agriculture section. Indian Science Congress held at Baroda.

Parker, C. and Fryer, J. D., 1975 Weed control problems causing major reduction in world food supplies. *FAO Plant Protection Bulletin* 23, Food and Agriculture Organization, Rome.

Prakash, Ved and Srivastava, A.K., 2006. Crop weed competition in Tomato (*Lycopersicon esculentum*) under mid hills of North West Himalayas. *Indian Journal of Weed Science*, 38(1&2): 86–88.

Rana, M.C., Kumar, Naveen, Kumar, Akhilesh and Rana, S.S., 2004. Management of complex weed flora in peas with herbicide mixtures under Lahul valley conditions of Himachal Pradesh. *Indian Journal of Weed Science*, 36(1&2): 68–72.

Sharma, H.C., Singh, H.B. and Frisco, G.H., 1977. *Weed Research*, 17: 103–108.

Sharma, S.D., Singh, Samar, Singh, Harpal, Narwal, Sandeep, Malik, R.K. and Punia, S.S., 2003. Evaluation of ethoxysulfuron and its mixture against weeds in transplanted rice. *Indian Journal of Weed Science*, 35(3&4): 201–204.

Singh, K., Pandita, M.L. and Thakral, K.K., 1993 Integrated weed management in vegetable crops. In: *Proceeding of International Symposium on Integrated Weed Management for Sustainable Agriculture*, held at Hissar, during,18–20 November, Vol. 1, pp. 365–368.

Singh, Pratap and Ali, Mashiat, 2004. Efficacy of metsulfuron methyl on weeds in wheat and its residual effect on succeeding soybean crop grown on vertisols of Rajasthan. *Indian Journal of Weed Science*, 36(1&2): 34–37.

Singh, V.P., 2002. Soil solarization new technique for weed management winter school. In: Proceedings of Winter School, on *Recent Advances in Weed Management*, held at NRC Jabalpur from 21 October to 10 November.

Sundari, A. and Kathiresan, R.M., 2002. Crop weeds competition in Sorghum. *Indian Journal of Weed Science*, 34(3&4): 311–312.

Swain, D., 2002. Role of allelopathy in weed management. In: Proceedings of Winter School, on *Recent Advances in Weed Management*, held at NRC Jabalpur from 21 October to 10 November, pp. 29–33.

Tiwari, A.N., Tiwari, S.N., Rathi, J.P.S., Verma, R.N. and Tripathi, A.K., 2001. Crop weed competition studies in chickpea having *Asphdelus tenufolius* dominated weed community under rained conditions. *Indian Journal of Weed Science*, 33(3&4): 198–199.

Tripathi, S.S. and Singh, Govindra, 2001. Critical Period of weed competition in summer cowpea (*Vigna unguiculata* L.Walp). *Indian Journal of Weed Science*, 33(1&2): 67–68.

Tripathi, S.S., Singh, Rohitashav, Singh, Govindra and Singh, Rajesh Kr., 2001. Study on the crop weed competition in tendril pea (*Pisum sativum* L.) under tarai of Uttaranchal. *Indian Journal of Weed Science*, 33(1&2): 46–48.

Vivek, S.S. Tomer, Singh, Gaje and Tripathi, S.S., 2003. Crop weed competition in Arhar (*Cajanus cajan*) under north western plain zones. *Indian Journal of Weed Science*, 35(3&4): 217–220.

Walia, U.S., Singh, Manpret and Sidhu, V.K., 2006. Studies on carry over effect of sulfosulfuron applied to wheat on succeeding crops. *Indian Journal of Weed Science*, 38(1&2): 123–126.

Wanjari, R.H., Yaduraju, N.T. and Ahuja, K.N., 2000. Critical period of crop weed competition in spring sunflower (*Helianthus annuus* L.). *Indian Journal of Weed Science*, 32(1&2): 17–20.

Yadav, Ashok, Balyan, R.A., Malik, R.K., Rathi, S.S., Banga, R.S. and Pahwa, S.K., 1996. Role of Soil solarization and volume of glyphosate spray on the control of *Cyperus rotundus* in ber. *Indian Journal of Weed Science*, 28(1&2): 26–9.

Yadav, Ashok, Malik, R.K., Malik, R.S., Banga, R.S. and Pahwa, S.K., 2003. Evaluation of different herbicides and sowing methods in cotton. *Indian Journal of Weed Science*, 35(3&4): 232–235.

Yadav, Ashok, Malik, R.K., Punia, S.S., Mehta, Ramesh, Dharambir, Amarjit and Balinnder, R.R., 2004. Studies on carry over effect of herbicides applied to wheat on succeeding crops in rotation. *Indian Journal of Weed Science*, 36(1&2): 15–18.

Yaduraju, N.T., 2002. Introduction to weed science. In: Proceedings of Winter School, on *Recent Advances in Weed Management*, held at NRC Jabalpur from 21 October to 10 November.

Zimdahl, R.L., 1980. *Weed Crop Competition: A Review*. Int. Plant Prot. Center Oregon State University, Corvallis, pp. 41–46.

Index

S